THE WORLD OF
PHYSICS

SECOND EDITION

THE WORLD OF
PHYSICS

SECOND EDITION

JOHN AVISON

B.Sc., C.Phys., M.Inst.P.

**Head of Physics
Marton Sixth Form College
Middlesbrough**

Nelson

Thomas Nelson and Sons Ltd
Nelson House Mayfield Road
Walton-on-Thames   Surrey KT12 5PL

51 York Place
Edinburgh EH1 3JD

Thomas Nelson (Hong Kong) Ltd
Toppan Building 10/F
22A Westlands Road
Quarry Bay   Hong Kong

Thomas Nelson Australia
480 La Trobe Street
Melbourne Victoria 3000
Australia

Nelson Canada
1120 Birchmount Road
Scarborough   Ontario
M1K 5G4 Canada
© J.H. Avison 1984, 1989

First published by Thomas Nelson and Sons Ltd 1984
Second edition published 1989
ISBN 0-17-438245-6
NPN 9 8 7 6 5 4 3 2

The inserted photograph on the front cover
shows IRAS, the InfraRed Astronomical
Satellite, which was launched on 26
January, 1983. In a near polar orbit, 900 km
high, it circled the Earth 14 times each day.
It carried a 60 cm infrared telescope which
was supercooled to 2° above absolute zero
(−271°C). Unhampered by the Earth's
atmosphere which absorbs infrared
radiation, the satellite performed an all-sky
survey for sources of infrared emission and
provided a catalogue of infrared sources and
infrared sky maps.

Printed in Hong Kong

To Ben and Janine

Acknowledgements

The Author and Publishers would like to thank the following for their kind permission to reproduce photographs.

Addison-Wesley Publishing Co p 479 (top); Ajax News and Features Services p 103; Allsport pp 91 (top), 112 (top right); John Avison pp 111, 264, 298, 301, 318, 329, 345, 347, 475 (left); Barnaby's Picture Library p 420; BBC Hulton Picture Library p 184 (left); Bett Insulation Ltd p 410; Brewers Association p 102; British Aerospace pp 112 (bottom right), 243; The British Museum p 64; British Nuclear Fuels plc pp 401 (right), 413 (lower right); British Pipeline Agency Ltd p 174 (top right); British Rail pp 108, 174 (bottom left, both), 174 (lower right); British Tourist Authority pp 112 (lower left), 227 (lower); Camera Press p 400; John Carruth p 370 (top right); J Allan Cash pp 121, 141, 412 (right), 420; Central Electricity Generating Board pp 214 (lower), 220, 417 (both), 423, 426; Diane Ceresa pp 95 (top), 98; Chloride Automotive Batteries Ltd p 323; John Cleare/Mountain Camera p 194; Bruce Coleman p 414 (left); Colourspot p 67; Daily Telegraph Colour Library pp 184 (right), 406; CB Daish p 478 (all); James Davis Photography p 90; Earthscan p 96; Education Development Center pp 443, 444, 446; EMI Records (UK) p 352; Energy Technology Support Unit p 427; Eriez Magnetics UK Ltd p 295 (right); Mary Evans Picture Library p 380; Ferranti p 338 (both); Format p 370 (lower right); Jean Fraser pp 129, 172, 231; Barry B Golberg, Thomas Jefferson University Hospital p 456 (centre); Graco UK Ltd p 206; Sally & Richard Greenhill p 469 (top left); Greater London Council p 455; Susan Griggs Agency pp 77 (lower), 91 (lower), 399 (left); Hanford Photography p 77 (top); Barbara Hartley pp 61, 170, 251, 294; Health & Safety Executive p 469 (right); Hewlett-Packard p 370 (left); Honda p 171; Pam Isherwood/Format pp 394, 401 (left); JET Joint Undertaking p 430 (both); Keystone Press Agency pp 72 (left), 400, 413 (top right), 429 (both); Klinger Scientific Apparatus Corp p 473 (left); Frank Lane p 2; Lennox Industries Ltd Corp p 473 (left); Frank Lane p 2; Lennox Industries Ltd p 428; London Electricity Board p 409; London Scientific Photos p 162; London Transport pp 86, 223; Lucas Group Services Ltd p 14; Maclaren Publishers Ltd p 295 (left); Manchester Public Library p 408 (lower); The Mansell Collection pp 100, 387 (left); Moorfields Eye Hospital p 276 (both); Maggie Murray/Format pp 95 (lower), 370 (lower right); Joanne O'Brian/Format p 249; Olympus p 49; Ontario Science Centre p 202; Oxford Scientific Films pp 167, 168, 282; Oxford University Research Laboratory of Archaeology and the History of Art p 398; Dr J Parkyn p 399 (right); Peabody Holmes p 215 (left, both); Phillips Medical Systems p 456 (top); Pittsburg Corning UK Ltd p 412 (left); The Post Office pp 29, 327, 332; Redland Prismo Ltd p 174 (top left); Rists Ltd/Lucas p 322; Rogue Images pp 22, 59, 60 (both), 94, 268; Rolls-Royce Ltd p 147; Alistair Rose-Innes p 215 (right, both); Royal Society for the Protection of Birds p 51; S & G Press Agency p 72 (centre); Science Photo Library pp 4 (top), 4 (lower), 28, 36 (left), 39, 52, 56, 68, 136, 140 (left), 140 (right), 149, 152 (right), 156, 159 (top), 159 (lower), 185, 269, 270, 371 (all), 407 (all), 408 (top), 413 (left), 414 (right), 485 (centre), 485 (bottom), 488; Science Museum, London/British Crown Copyright p 390 (all); Ronald Sheridans Photo Library p 106; Siemans Ltd p 343; Sporting Pictures (UK) Ltd pp 72 (top right), 72 (lower right), 87, 142, 148, 152 (left), 169; Springer Verlag pp 475 (right), 479 (lower); Tektronix p 372; Unilab p 359; United Kingdom Atomic Energy Authority pp 403, 425; University of Dundee Electronics Laboratory p 99; University of Manchester Physics Dept p 381; John Walmsley pp 12, 329; Wembley Stadium Ltd p 383; Val Wilmer/Format pp 10, 36 (right), 299 (top), 434, 450, 469 (lower left); Wormald International Sensory Aids Ltd p 456 (bottom).

All other photographs by Nelson's Visual Resources Unit.

Cover photograph: The Image Bank
Inset: Science Photo Library

Preface to the second edition

Applications of physics can be found in all the machines and inventions of our technological society. The laws and concepts of physics can often help us to understand and describe what we discover about ourselves, the natural world around us and deep space beyond our planet. Whenever we invent and use machines, attempt to explain a natural phenomenon or gaze enquiringly out into space we need to be in touch with physics. In my book I have tried to show how a knowledge of physics is both useful and relevant today and how, through an understanding of physics, we can more fully appreciate our world and more effectively use our discoveries and inventions.

As I believe that everyone should have at least an acquaintance with, if not a sound knowledge of, this important subject, I have tried to write a book which can meet the needs of a wide variety of people. I hope that students preparing for GCSE physics examinations will find that my book provides a detailed and thorough treatment of all the topics required for their courses. Students studying GCSE science syllabuses should find that this book covers not only the specific topics but also the broader background needed for the physics and technological aspects of their studies. I know that my book has also been appreciated and enjoyed by many other readers, including students beginning their A level physics courses, teachers preparing lessons and seeking information and more mature readers discovering the world of physics, perhaps for the first time. I hope that they will all continue to find pleasure and help from this new edition.

The first edition of *The World of Physics* was completed in 1983 and published in 1984. Now, five years later, much has changed in the examination system which students face at 16. Students now need to develop a wider range of skills, be competent at designing experiments and solving practical as well as numerical problems, be able to handle and process data or information in tables, charts and graphs and be more aware of the impact of science on the world around them. However, the basic physics they need has not changed very much and almost all of it was covered in the first edition of *The World of Physics*. As I had been able to anticipate much of the new content and emphasis that is now found in the GCSE physics and science syllabuses, I am glad to say the book was already very suitable for students studying for the new examinations. In particular, the increased emphasis on applications of physics at home, in industry and in medicine was already generously catered for and remains in this edition.

Many of the regular users of *The World of Physics* were consulted about what they would like to see changed in a second edition of the book. Although a few people wanted no change, the majority had very helpful suggestions and almost all preferred the style and structure of the book to be kept essentially the same. Among the selection of topics where revision was called for, I found the greatest number of requests for an expansion of the section on digital electronics. Knowing about logic gates and truth tables is not enough; we must be able to see how they can be applied and combined to do useful jobs. Communication systems also feature in some syllabuses and required some new material. In other parts of the book smaller changes have been made to allow for the inclusion of some work on beams and bridges, on safety in cars and on the economics of different forms of transport.

Almost 100 new GCSE questions have been included to give students a good idea of what is expected of them in the new examinations. In addition the appendix giving answers to the questions has been extended to give more guidance and hints about how to tackle questions and solve problems. I hope this will be helpful for those readers in particular who are studying physics without the regular guidance of a teacher. In response to many requests, four pages of problem-solving questions and tasks have been introduced to give extra practice in the skills required for problem-solving both in written examinations and in practical coursework.

Again I must express my sincerest gratitude to all those people without whose encouragement and support I could never have completed the writing of the first edition of this book. The very kind and positive remarks which they and so many other people have made about my book have led me to spend another summer preparing the manuscript and diagrams for this second edition. Anyone who has ever tried to write a book while doing a full-time job and being a parent of children, will realise just how much I owe to my immediate family for their continued patience over yet more months while writing was in progress. Special thanks continue to go to my wife Carole without whose support neither this edition nor the first edition would have been possible.

My thanks go to everyone who has helped to improve my book by taking the trouble to write to me since 1984 making helpful suggestions. Any errors which remain are, of course, my responsibility.

John H. Avison
Stokesley, September 1988.

To the student reader

Almost all topics are introduced with suggestions for practical work and I hope that if you are at school you will be able to work through the instructions as a part of your study programme. The instructions are all indicated by a black circle at the beginning of the line and are printed in italics. If you are unable to do the experiments it is still worth reading through and thinking about the instructions and associated questions and diagrams. The practical instructions are followed by a description or summary of the important observations, measurements and conclusions from which a full understanding may be gained.

All topics are introduced from a simple and basic starting point and assume that you have very little previous knowledge of physics. As some chapters develop, the treatment of some topics becomes more rigorous and mathematically detailed. These topics are designed to train you to solve numerical problems by giving you an explanation and structure to follow and are illustrated by the worked examples. Make sure you can follow and fully understand these before attempting the questions. At the end of the book you will find answers to all the numerical questions and some other help and hints about these and many of the other questions. Remember that to be completely correct, your numerical answers must also have the correct units. Some topics are essentially descriptive and concerned with applications of physics and its influence on our lives. On these topics you will be expected to have ideas and opinions and to be able to write short statements in support of your views. You will also find sections called 'Assignments' at the end of each major topic. Here are a collection of interesting and useful activities or tasks which round off the work on a topic.

As *The World of Physics* covers a wide range of syllabuses you should not expect to study more than perhaps half or three-quarters of the topics. Your teacher will give you guidance about which topics to study and revise for examinations and the physics syllabus of your Examination Board can also be consulted. I hope, however, that you will enjoy reading about some of the other interesting topics which are not required for your examination.

Key for Examination Boards

The following Examination Boards have kindly given permission to reproduce questions from past or specimen examination papers. The key given for each Board is used in this book to identify the source of each question. Questions without a key have been written by the author.

GCE Boards (Questions from Ordinary Level GCE papers)

The Associated Examining Board	AEB
The Joint Matriculation Board	JMB
The Oxford and Cambridge Schools Examination Board	O&C
The Oxford Delegacy of Local Examinations	O
The Southern Universities' Joint Board	S
The University of Cambridge Local Examinations Syndicate	C
The University of London University Entrance and School Examinations Council	L
Nuffield Physics	Nuffield

To the teacher

As it was my intention that my book would be used as a working companion in practical sessions, experimental instructions are given wherever possible to help both students and teachers carry out successful investigations. If it is likely or desirable that students will carry out the experimental work themselves, the instructions are indicated by a black circle. If however the teacher is expected to demonstrate, perhaps for safety reasons, the circles are deliberately omitted. Advisory warnings are given in those cases where a procedure is known to be hazardous.

The book conforms to the recommendations of the ASE publication: 'SI units, signs, symbols and abbreviations', 1981. I have used quantity algebra for all equations, formulas and calculations. Whenever a letter is written (in italics) to stand for a physical quantity, it also includes the units of the quantity. So when such a letter in an equation or formula is replaced by a numerical value, the units are written down as well. This procedure helps students to appreciate the importance of units, to avoid mixing different units for the same quantity in the same calculation and to see how the units of an 'answer' derive from the units of the data put into the calculation. In many cases the relation between different quantities and their associated units is clarified by this process; for example, a calculation of work done shows clearly that newtons × metres produces joules of work.

The end-of-chapter questions have been extensively revised to include as many suitable GCSE questions as were available after the first GCSE examinations in 1988. The questions are arranged in an order which follows the development of each topic in the text and so the simpler questions are at the beginning and the more comprehensive and demanding questions are towards the end. Questions which involve an element of problem-solving together with a number of practical problem-solving activities have been collected in two new sections on pages 200–1 and 378–9. This type of question is a new feature of both coursework and written papers for the GCSE and I hope that these problems and activities will not only be useful in themselves but will also generate and inspire many new ideas.

GCSE Boards or Groups

(Questions taken from specimen GCSE papers or the first GCSE examinations taken in 1988. Note: the syllabuses marked *are of the 'Nuffield' type.)

London and East Anglian Group, Syllabus A	LEAG
Midland Examining Group	MEG
Nuffield*	MEG[Nuffield]
Northern Examining Association, Syllabus A	NEA[A]
Syllabus B*	NEA[B]
Joint GCE & CSE syllabus	Joint 16+

Associated Lancashire Schools Examining Board
Joint Matriculation Board
North Regional Examinations Board
North West Regional Examinations Board
Yorkshire and Humberside Regional Examinations Board

Northern Ireland Schools Examination Council	NISEC
Southern Examining Group	SEG
Alternative syllabus*	SEG[ALT]
Welsh Joint Education Committee	WJEC

Scotland Standard Grade [General & Credit levels]
(Questions taken from the specimen question papers for examinations to be taken in and after 1990.)

Scottish Examinations Board	SEB

Contents

Light rays and reflection

Early in the day when the sun is low on the horizon and shadows are long, the sunlight streaming through a group of trees is broken into straight beams or rays of light. When sunlight breaks through the clouds after a storm the light rays can be seen clearly against the dark background of the clouds. The beams of light from the headlamps of a car can be seen as they shine out through a heavy rainstorm or fog. Many natural effects of light not only have inspired us by their beauty but also have helped us to understand the basic properties of light. In this chapter we shall discover how many properties and applications of light can be described using the idea of light rays.

1.1
RAYS AND SHADOWS

What is a ray of light?
A ray of light is a narrow beam of parallel light which can be drawn as a single line on a diagram. In diagrams rays are drawn with an arrow on them showing the direction of travel of the light. Rays are produced when light shines through a small hole, which we call a **point source** of light. A beam of light containing many rays is produced by a larger hole or large lamp, which we call an **extended source** of light.

Demonstrating that light travels in straight lines
• *Make a single small hole in each of three screens at exactly the same height above the bench.*
• *Set the three holes exactly in line by threading a length of cotton or string through the holes and pulling it tight.*
• *Carefully remove the thread without disturbing the screens and position a lamp so that it can be seen through the three holes, as shown in fig. 1.1.*
If any one of the screens is moved very slightly then the eye cannot see the lamp. This shows that light can only travel in a straight line. The property of light travelling in straight lines is called **rectilinear propagation** (literally, 'straight-line travel'). Note that the ray of light travels from the lamp to the eye as indicated by the arrows; rays do not come out of the eye.

Shadows
One direct effect of light travelling in straight lines is the casting of shadows by opaque objects. As light cannot pass through an opaque object and, travelling only in straight lines, cannot bend round the object, then the space behind an opaque object must be totally dark.

We can demonstrate how a source of light casts a shadow in the laboratory and use this to explain how eclipses are caused by shadows.

The shadow formed by a point source of light
A point source of light is one which is small enough for all the rays of light to come effectively from a single point. The hole in the screen near the lamp in fig. 1.2a acts as a point source.

Figure 1.1 *Light travels in straight lines*

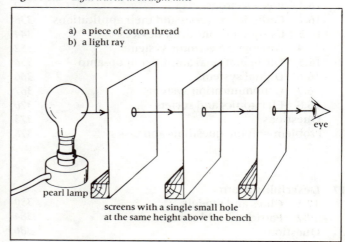

a) a piece of cotton thread
b) a light ray

eye

pearl lamp

screens with a single small hole
at the same height above the bench

• *Hold a circular opaque object like a large coin fairly close to a white screen as in fig. 1.2a and describe the shadow it casts.*

The shadow formed by a point source of light has two important properties:

a) The shadow is uniformly and totally dark all over. It is called the **umbra**, a Latin word meaning shade.

b) The shadow has a sharp edge, supporting the idea that light travels only in straight lines.

The shadow formed by an extended source of light

An extended source of light is large enough for rays to be seen to come from many points. The large pearl lamp shown in fig. 1.2b provides a suitable extended source.

• *Using the arrangement shown in fig. 1.2b, describe the shadow and note the differences between it and the one cast by a point source of light.*

The following points about the shadow formed by an extended source of light should be noted:

a) The centre of the shadow remains uniformly dark as before but is somewhat smaller in size. This part of the shadow, the umbra, still receives no light at all from the source.

b) The edge of the shadow is now blurred and graded, getting gradually lighter further out from the umbra.

c) The region between the totally dark umbra and the fully bright screen is called the **penumbra**, which means partial shade. In this region light from some parts of the extended source reaches the screen, but light from other parts is cut off by the opaque object. Near the umbra very little light reaches the screen and so the penumbra merges into the umbra.

Because the shadows formed by extended light sources are much softer and without sharp edges, we use frosted or pearl light bulbs and lamp shades at home to provide a more pleasant kind of lighting. Fluorescent tubes are usually surrounded by a frosted diffuser to scatter the light and reduce the sharpness of shadows.

Figure 1.2 *Shadows*

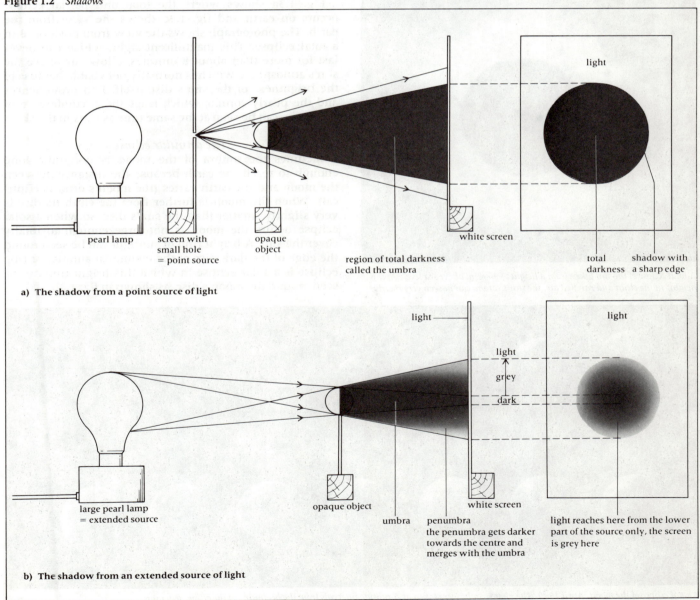

a) **The shadow from a point source of light**

b) **The shadow from an extended source of light**

1.2
ECLIPSES

An eclipse is the total or partial disappearance of the sun or moon as seen from the earth. In this section eclipses are explained in terms of the motions of the earth and moon and the shadow that one casts on the other.

The solar eclipse or eclipse of the sun

Records of solar eclipses have been kept since the time of the ancient Chinese who were afraid of them, thinking that a dragon was trying to devour the sun. The philosophers of ancient Greece understood that the moon was responsible for eclipses of the sun and were even able to predict a solar eclipse. In 1543 Copernicus started a revolution of thought and understanding when he published a book in which he suggested that only the moon went round the earth and that the earth, like all the other

A total eclipse of the sun taken from a height of 9000 m above sea level. At this height, in the clean and rarefied air, the sun's corona can be seen very clearly.

planets, went round the sun. Until this idea of the motion of the earth and moon was accepted it was impossible to explain fully how the different types of eclipses happened.

We now know that the sun is eclipsed when the moon passes between the sun and the earth. When it happens it causes unexpected darkness during the daytime. Solar eclipses are rather rare for two reasons.

a) A solar eclipse can happen only at new moon (when the moon is totally dark). If the orbit of the moon lay in the same plane as that of the earth there would be an eclipse every month. The moon's orbit is, however, inclined at an angle of about 5° to the earth's orbit so that only rarely does the new moon pass exactly through the line joining the earth and the sun, producing a solar eclipse.

b) When a solar eclipse does occur the path of the moon's umbra across the surface of the earth is very narrow, (never wider than 272 km) so that most people on the earth see only a partial eclipse.

Fig. 1.3a shows where the total and partial eclipse occurs on earth and fig. 1.3c shows the view from the earth. The photograph shows the view from position B in a total eclipse. This magnificent sight, which can never last for more than about 8 minutes, allows us to see the sun's atmosphere which is normally not visible because of the brightness of the sun's disc itself. Red prominences and the pearly corona, which rings the circumference of the moon, can be seen at the same time as stars in the sky.

The annular eclipse

Sometimes the umbra of the moon is not quite long enough to reach the earth because the distance between the moon and the earth varies (the moon's orbit is eliptical). When the moon is further from the earth its disc is very slightly smaller than the sun's disc, so when a solar eclipse occurs the moon is not large enough to totally cover the sun. A bright ring of sunlight can be seen round the edge of the dark disc of the moon. An annular or ring eclipse is a solar eclipse in which this bright ring can be seen around the moon's disc, as shown in fig. 1.3b.

A total eclipse of the moon, April 1949. The photographs were taken at 5-minute intervals from the beginning of the eclipse to totality.

Figure 1.3 *Eclipses of the sun (diagrams not to scale)*

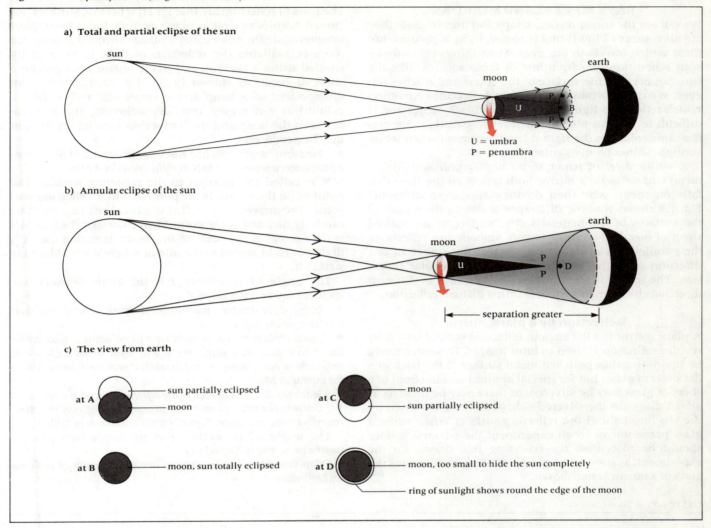

a) **Total and partial eclipse of the sun**

sun

earth

moon

P
A
B
C
U
P

U = umbra
P = penumbra

b) **Annular eclipse of the sun**

sun

earth

moon

P
P
U
D

← separation greater →

c) **The view from earth**

at A — sun partially eclipsed — moon

at C — moon — sun partially eclipsed

at B — moon, sun totally eclipsed

at D — moon, too small to hide the sun completely — ring of sunlight shows round the edge of the moon

The lunar eclipse or eclipse of the moon

The moon does not emit light itself, but only reflects light from the sun; thus when it passes into the earth's shadow its supply of direct sunlight is cut off. A lunar eclipse occurs when the moon passes through the earth's umbra, but it only happens occasionally when the moon is full. Lunar eclipses can last as long as $1\frac{3}{4}$ hours because the moon is much smaller than the earth and takes some time to pass through the earth's umbra. During a total lunar eclipse it is still just possible to see the moon because a small amount of sunlight reaches it by way of the earth's atmosphere. This sunlight, bent or refracted by the earth's atmosphere, reaches the moon turning it a dim coppery colour. Fig. 1.4 explains the lunar eclipse by a ray diagram.

Assignments

Find out

a) When was the last total eclipse of the sun visible in Britain?
b) When is the next total eclipse of the sun expected in Britain?
c) How long does it take the moon to orbit the earth once?
d) Why does a lunar eclipse not occur every month?
e) Why are lunar eclipses more common than solar eclipses?

f) If the ray diagram of fig. 1.3 is drawn to scale and the sun is drawn the same size, how far away should the earth be drawn?

Try questions 1.1 to 1.4

Figure 1.4 *Eclipse of the moon (not to scale)*

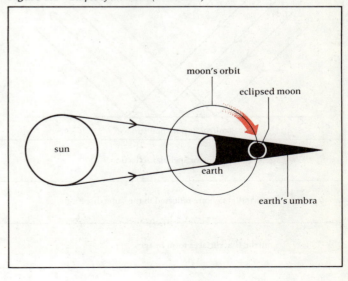

moon's orbit

eclipsed moon

sun

earth

earth's umbra

1.3
RAYS MEET PLANE MIRRORS

We can see the sun and stars, lamps and fires because they are all sources of light; that is to say, light is produced in them which travels to our eyes. Most things can only be seen when the light from one of these sources, like the sun, bounces off the surface of the object and reaches our eyes. We call this bouncing of light **reflection**. An object which reflects no light appears a dull black colour and is difficult to see. An object which reflects all light appears the same colour as the light it is reflecting, so when white sunlight shines on it, its colour is white.

A white sheet of paper and a highly polished silvery metal surface as on a mirror both reflect all the light that falls on them; why then do they appear so different? Fig. 1.5 shows that the difference is due to the nature of the surfaces of the materials. The surface of a polished sheet of metal or a mirror is very smooth and reflects all the parallel rays of light from a particular source in one direction only; this is called **regular** or **specular** reflection. The irregular scattering of the light rays in different directions by a rough surface is called **diffuse** reflection.

Reflection by a plane mirror

A plane mirror is a flat smooth reflecting surface which by regular reflection is used to form images. It is often made by bonding a thin polished metal surface to the back of a flat sheet of glass; but for special applications the front of a sheet of glass may be silvered, or there may be no glass at all. In diagrams the silvered side of a mirror is shown by the shading behind the reflecting surface. When using a glass plane mirror in an experiment the silvered surface should be placed on the reflecting line drawn for the experiment, as shown in fig. 1.6. In many ray diagrams the glass of a mirror is not shown.

Investigating the laws of reflection

The laws of reflection are true for all reflecting surfaces, for curved mirrors as well as plane mirrors, but it is simplest to investigate the laws using a plane mirror. If a darkened room is available the reflection of light is most easily studied using a ray box and plane mirror arranged on a sheet of paper as shown in fig. 1.7. (A ray box is an arrangement of a lamp and a single slit and usually a cylindrical converging lens. By adjusting the distance between the lamp and the lens a thin parallel ray of light may be produced.)

- *First draw a reflecting line XMY on the paper and then, using a protractor, another line MN at right angles to the first.*

MN is called the **normal** to the reflecting surface. The point M on the mirror is the point at which reflection will occur. The arriving ray, called the **incident ray**, must be carefully directed at the point M. We say that the line MN is the normal at the point of incidence, meaning that it is drawn at right angles to the mirror where the incident ray strikes it.

The **angle of incidence** i is the angle between the incident ray and the normal.

- *Using a protractor, measure and mark several angles of incidence on the paper.*
- *Stand a plane mirror upright with its reflecting surface on the line XMY and then shine the ray of light along each of the directions in turn, being careful to see that each time the ray strikes the mirror at M.*
- *Mark the direction of each of the reflected rays with a cross.*
- *Draw in the reflected rays and measure the angles of reflection, recording these in a table. Typical results are shown in table 1.1.*

The **angle of reflection** r is the angle between the normal and the reflected ray.

- *Can you draw any conclusions about the angles of incidence and reflection?*

Figure 1.5 *Types of reflection*

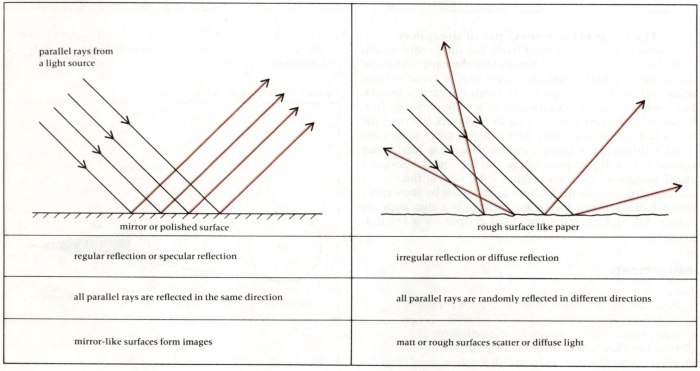

parallel rays from a light source		
mirror or polished surface		rough surface like paper
regular reflection or specular reflection		irregular reflection or diffuse reflection
all parallel rays are reflected in the same direction		all parallel rays are randomly reflected in different directions
mirror-like surfaces form images		matt or rough surfaces scatter or diffuse light

Table 1.1

Angle of incidence i/degree	Angle of reflection r/degree
0	0
15	14
30	32
45	44
60	60
75	72

Figure 1.6 *Reflection by a plane glass mirror (plan view, from above)*

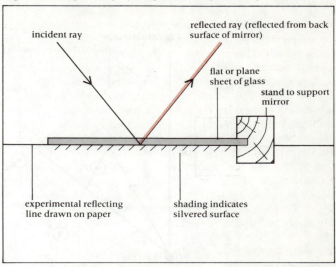

Figure 1.7 *Experiment to measure the angles of incidence and reflection*

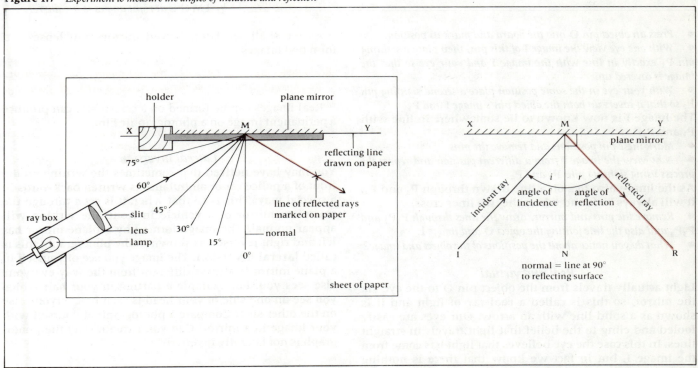

Notice also that an incident ray which travels flat along the surface of the bench will be reflected from the mirror along the bench surface, provided the normal to the mirror is in the same flat surface. If the mirror were leaning backwards, so its normal pointed upwards at an angle to the bench, then the reflected ray would leave the bench surface and not be seen on the paper.

● *Try tilting the mirror and see what happens.*

All these observations lead us to the **laws of reflection:**

The angle of incidence equals the angle of reflection.

The incident ray, the reflected ray and the normal at the point of incidence all lie in the same plane.

(The second law means in effect that they can all be drawn on a flat sheet of paper.)

Images

When we look at the surface of a calm lake, or at ourselves in a mirror, what we see is commonly called a reflection. In physics it is called an **image**.

Finding the image in a plane mirror

A fairly accurate method of finding the position of an image in a plane mirror is to use pins to mark the direction of two or more rays of light from an object which reach the eye after reflection by the mirror. Fig. 1.8 shows the arrangement used to find the image I of the object pin O.

● *Fasten a sheet of paper on a drawing board into which pins can easily be pressed.*

● *Mark the reflecting line on the paper, and stand the reflecting surface of a plane mirror upright on the line.*

Figure 1.8 *Finding the position of the image formed by a plane mirror*

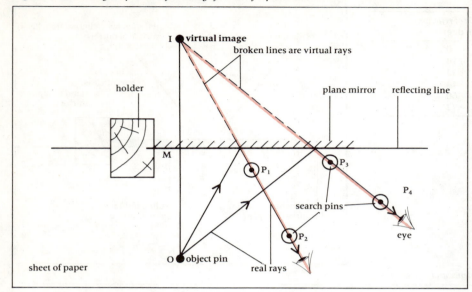

Measurements from this experiment show that:
a) **O M = I M.**
 The object and image are the same distance from the mirror.
b) **I M O** *is at right angles to the mirror.*

- *Press an object pin O into the board and mark its position.*
- *With one eye view the image I of this pin, then place a sighting pin P₁ exactly in line with the image I and your eye so that the image is covered up.*
- *With your eye in the same position place a second sighting pin P₂ so that it covers up both the object pin's image I and P₁.*

The image I is now known to lie somewhere in line with P₁ and P₂.

- *Mark these pin positions and remove the pins.*
- *Now view the image I from a different position and repeat the process using sighting pins P₃ and P₄.*

As the image also lies on a line drawn through P₃ and P₄, it will always be found where the two lines cross.

- *Remove the pins and mirror, draw the lines through P₁P₂ and P₃P₄ and also the line joining the object O and image I.*
- *What do you notice about the positions of the object and image?*

Real and virtual

Light actually travels from the object pin O to the eye via the mirror, so this is called a **real** ray of light and it is shown as a solid line with an arrow. Our eyes are easily fooled and cling to the belief that light travels in straight lines. In this case the eye believes that light has come from the image I, but in fact we know that there is nothing behind the mirror at all and light cannot pass through the reflecting surface of the mirror. The imaginary rays behind the mirror are called **virtual** rays and to distinguish them from real rays we shall always draw them as broken lines.

There are also two kinds of images. An image is formed where the real or virtual rays from an object come together again. The image formed by a plane mirror is called a **virtual image** because it is formed where the virtual rays appear to come from when the real rays are reflected by the mirror. Just as the virtual rays are not there, the virtual image does not exist either; it is an illusion. No light ever reaches a virtual image so it cannot be formed on a screen and it cannot affect photographic film placed at its apparent position.

Virtual images are those which rays of light only appear to come from but no real rays ever reach.

As we shall see later, curved mirrors and lenses can form real images.

Real images are formed when all the rays coming from a point on an object are brought together again at another single point.

Real images can be formed on a screen and can produce a permanent image on a photographic film.

Lateral inversion

You may have noticed that sometimes the writing on the front of a police car or ambulance is written backwards in a special way. This is so that when it is seen through the rear-view mirror of a vehicle in front, the writing will appear normal. The image formed by a plane mirror has left and right reversed as shown in the photograph; this is called **lateral inversion**. The image you see of yourself in a plane mirror is always different from the way everyone else sees you. For example a parting in your hair which you see on one side of your head is seen by everyone else on the other side. Compare a photograph of yourself with your image in a mirror. Can you explain why the photograph is not laterally inverted?

Describing images

When describing an image the following questions should be answered.
1) Is it real or virtual?
2) Is it inverted in any way?
3) How does its size compare with the object?
4) Where is it?

First we can answer these questions for the image formed by a plane mirror. *The image formed by a plane mirror is:*
1) virtual (meaning imaginary),
2) erect (meaning the right way up) but laterally inverted (meaning left and right sides reversed),
3) the same size as the object,
4) as far behind the mirror as the object is in front, and the line joining the object and image is at right angles, or normal, to the mirror.

The periscope

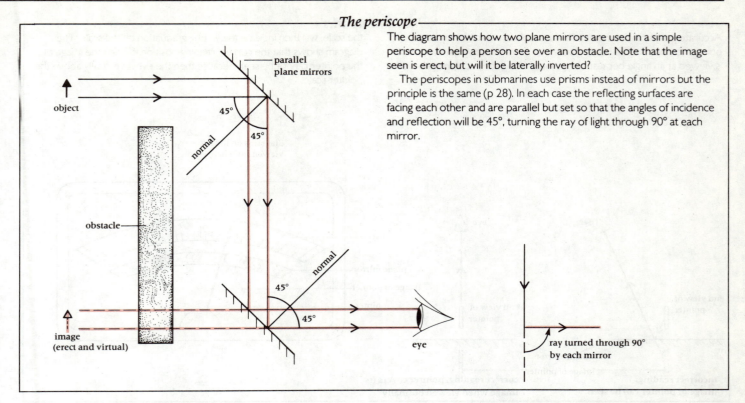

The diagram shows how two plane mirrors are used in a simple periscope to help a person see over an obstacle. Note that the image seen is erect, but will it be laterally inverted?

The periscopes in submarines use prisms instead of mirrors but the principle is the same (p 28). In each case the reflecting surfaces are facing each other and are parallel but set so that the angles of incidence and reflection will be 45°, turning the ray of light through 90° at each mirror.

The kaleidoscope

Two plane mirrors set at any angle will produce, by multiple reflections, more than one image of an object placed in front of them. As the angle between the mirrors is reduced the number of images increases until there are, in theory, an infinite number of images when the mirrors are parallel and facing each other. In the kaleidoscope the two mirrors are usually set at 60° to each other as shown below. At this angle six identical views of an object can be seen, one being the real object and five virtual images. The images are laterally inverted by each reflection forming a symmetrical pattern.

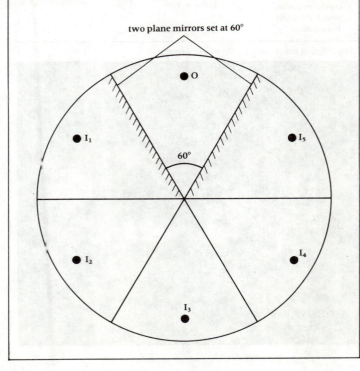

Worked Example
Laws of reflection

The angle between an incident ray and a plane mirror is 25°. Calculate: (a) the angle of incidence, (b) the angle of reflection and (c) the angle turned through by the ray of light.

Referring to fig. 1.9 we can see that:
a) the angle of incidence $i = 90° - 25° = 65°$,
b) the angle of reflection = the angle of incidence = $65°$,
c) the angle the ray is turned through = $180° - (65° + 65°) = 50°$.
Answer: the angles required are: (a) 65°, (b) 65°, (c) 50°.

Figure 1.9

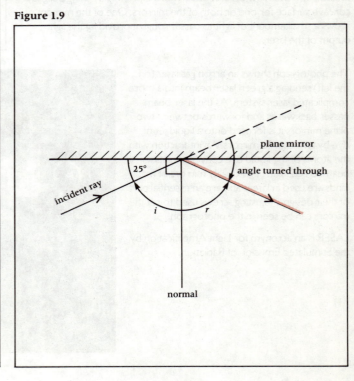

Instrument scales

Accurate pointer instruments often use a plane mirror behind the pointer to improve the reading accuracy. Errors arise when the pointer is viewed at an angle because the pointer, being some distance above the scale, will then indicate the wrong graduation on the scale. The diagram shows that the correct reading position is when the image of the pointer cannot be seen because then the eye is vertically above the pointer scale.

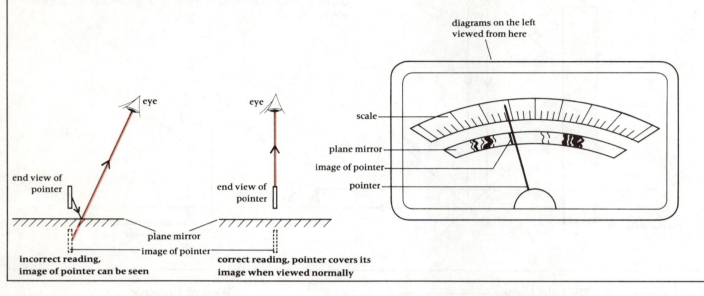

incorrect reading, image of pointer can be seen

correct reading, pointer covers its image when viewed normally

The laser

The laser is a light amplifier. Light is amplified or made brighter each time it is passed through a special crystal or gas which is stimulated into producing more light of the same colour by the light passing through it. To build up the light to high intensities the light must pass through the amplifier many times. This is achieved by placing two exactly parallel mirrors at opposite ends. In the simplest design these would both be plane mirrors, as shown in the diagram. However, the light can be focused into a narrow parallel beam more efficiently by using a slightly concave surface for one or both of the mirrors. One of the mirrors allows a small amount of light to pass through it providing the light output of the laser.

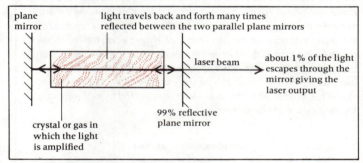

A simple design using two parallel plane mirrors

The photograph shows an argon gas laser (on the left) sending a green laser beam into a more complicated laser system. As the laser beam passes backwards and forwards between two plane mirrors it is focused onto a liquid jet of dye by two curved mirrors. An interaction with the atoms of the liquid dye converts the laser beam to red light. High-quality mirrors of many kinds are used in lasers and are an essential part of their design. Adjusting screws used to set the mirrors can be seen in the photograph.

LASER is an acronym for Light Amplification by the Stimulated Emission of Radiation.

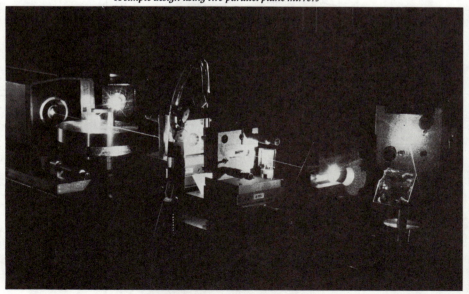

The spot galvanometer

The spot galvanometer uses an optical pointer or ray of light instead of a metal pointer. Apart from the advantages that the ray of light has no mass and can be very long, the angle the ray turns through is always twice the angle the reflecting mirror turns through. This doubles the sensitivity of the instrument (see p 311).

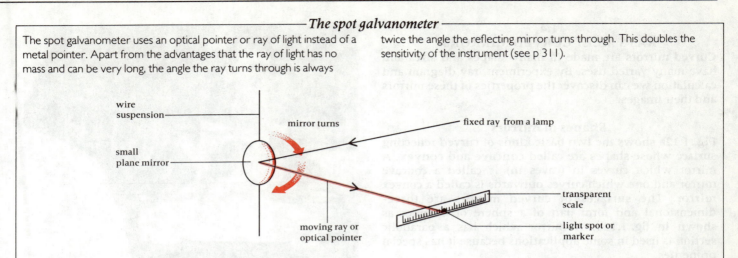

Assignments

Remember
a) The laws of reflection.
b) The description of the image formed by a plane mirror.
c) The meaning of the words real and virtual when used to describe images.

Examine a pointer instrument which uses a plane mirror to help improve reading accuracy.
See p 10 for the arrangement of the mirrors.
d) Notice how you cannot see the image of the pointer when your eye is vertically above the pointer.
e) Draw a ray diagram to show how an image of the pointer can be seen when the pointer is not viewed from vertically above. (It is easiest to draw an end view of the pointer.)

Build yourself a periscope
See p 9 for the arrangement of the mirrors.
f) In the laboratory, mount two plane mirrors on a single stand. Take care to protect the mirrors from breaking by using rubber or cork covered clamps.
g) At home, fix two plane mirrors inside a long cardboard tube with openings cut out where the light enters and leaves the periscope.

Investigate an optical pointer
h) When your teacher has removed the top cover plate from a spot galvanometer, have a look at the lamp, mirror and scale arrangement.

i) Set up a similar arrangement of apparatus as shown in fig. 1.10. Investigate the relation between the angle the mirror is turned through and the angle this causes the reflected ray to turn through. Set the mirror in turn at each of the marked angles with the fixed light ray always directed at the point M on the mirror, then measure the angle the reflected ray turns through from its starting position. Tabulate and compare the two angles.

Investigate the images formed by two plane mirrors
j) Arrange two plane mirrors at the exact angles shown in fig. 1.11. For each angle count how many images you can see. Can you explain how the images are formed when the mirrors are set at 90°?
k) From your observations of two mirrors placed at 60°, draw a diagram of a kaleidoscope as on p 9 and using a letter K as the object in position O, draw the five images of this letter as they would be seen in the kaleidoscope.

Try questions 1.5 to 1.8a

Figure 1.10 *Investigation of the angle turned through by an optical pointer*

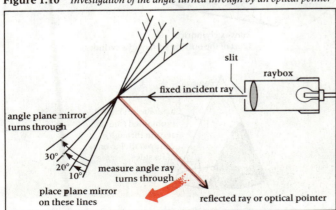

Figure 1.11 *Images in plane mirrors set at various angles*

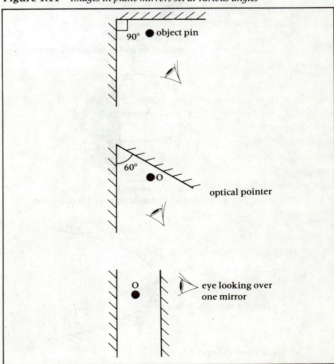

1.4
RAYS MEET CURVED MIRRORS

Curved mirrors are made in many shapes and sizes and have many varied uses. By experiment, ray diagram and calculation we can discover the properties of these mirrors and their images.

Shapes of mirrors

Fig. 1.12a shows the two basic kinds of curved reflecting surface whose shapes are called **concave** and **convex**. A mirror which curves in (caves in) is called a concave mirror and one which curves outwards is called a convex mirror. The surfaces of curved mirrors are three-dimensional and form part of a sphere or cylinder as shown in fig. 1.12b. A mirror which has a parabolic section is used in some applications because it has special properties.

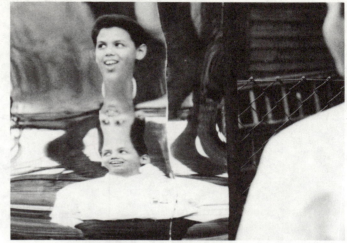

Figure 1.12 *Curved mirrors*

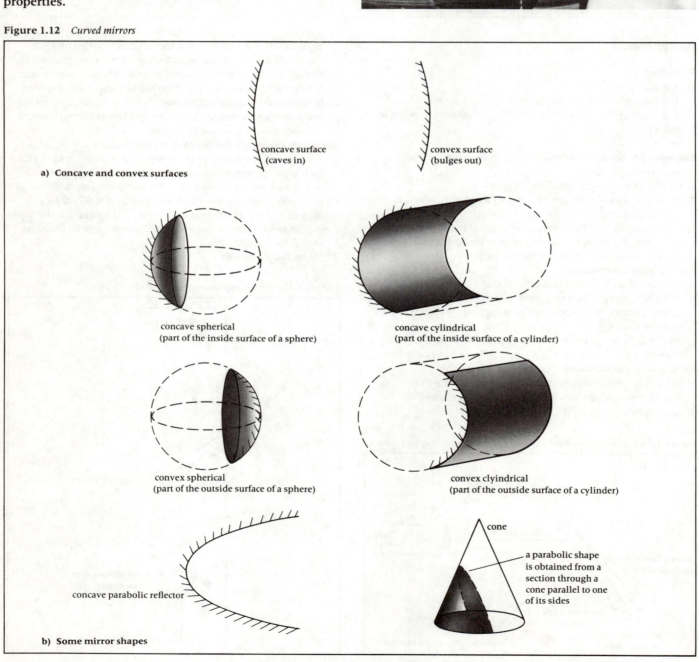

concave surface
(caves in)

convex surface
(bulges out)

a) Concave and convex surfaces

concave spherical
(part of the inside surface of a sphere)

concave cylindrical
(part of the inside surface of a cylinder)

convex spherical
(part of the outside surface of a sphere)

convex clyindrical
(part of the outside surface of a cylinder)

concave parabolic reflector

cone

a parabolic shape
is obtained from a
section through a
cone parallel to one
of its sides

b) Some mirror shapes

Converging and diverging mirrors

When parallel rays of light are reflected by a plane mirror they remain parallel, but curved mirrors reflect each ray in a different direction.

A *concave* mirror **converges** (brings together) parallel rays to a point called a real focus F (fig. 1.13a).

A *convex* mirror **diverges** (spreads out) parallel rays so that they never meet but appear to come from a point called a **virtual focus** F (fig. 1.13b).

The **pole** P of a mirror is the centre of its reflecting surface.

The **centre of curvature** C of a spherical mirror is the centre of the sphere of which the mirror is part.

The **principal axis** of a mirror is the line passing through its pole and centre of curvature.

The **radius of curvature** R of a spherical mirror is the radius of the sphere of which the mirror is part. (PC and MC = R on fig. 1.14.)

The **principal focus** F **of a concave mirror** is the point *through* which all rays close to and parallel to the principal axis *pass* after reflection by the mirror. This is a real focus.

The **principal focus** F **of a convex mirror** is the point *from* which all rays close to and parallel to the principal axis *appear to come* after reflection by the mirror. This is a virtual focus.

The **focal length** f of a mirror is the distance from its pole to its principal focus. (PF = f in fig. 1.14.)

The laws of reflection

The laws of reflection are obeyed by all curved mirrors. In fig. 1.14 the radius CM is also the normal to the surface of the mirror at the point of incidence M. An incident ray LM, parallel to the principal axis is reflected from M to F so that angle *i* = angle *r* (away from F in the case of the convex mirror).

The relation between focal length and radius of curvature for a spherical mirror

If the incident ray LM is close to the principal axis of a spherical mirror it can be shown that, approximately, $f = R/2$.

Referring to fig. 1.14a, in triangle MCF:

      angle *r* = angle *i*

also   angle MCF = angle *i* (LM is parallel to CF)

∴   triangle MCF is isosceles and MF = CF

now if M is close to P: MF = PF (almost)

∴   PF = CF = ½ PC

or
$$f = \frac{R}{2}$$

The focal length is half the radius of curvature.

Figure 1.13 *The converging and diverging action of curved mirrors*

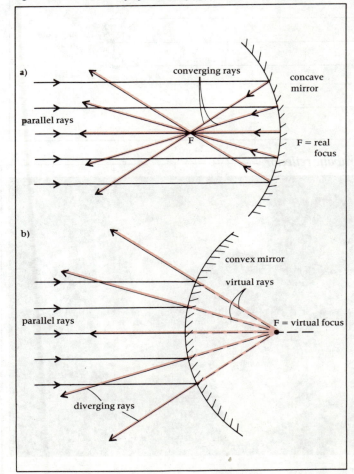

Figure 1.14 *Curved mirror definitions*

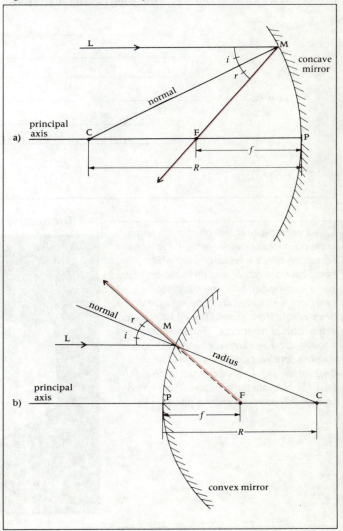

The caustic curve

The light from a single lamp collected inside a cup half full of tea produces a curve of light rather than a focused spot of light on the top of the tea. The inside of a cup, acting as a wide concave cylindrical mirror, forms a curve of light known as a **caustic curve**. Fig. 1.15a shows how rays further out from the principal axis of a concave mirror of circular section pass through points nearer to the mirror and only those rays close to the principal axis pass through a point which is *R*/2 distant from the mirror.

Figure 1.15

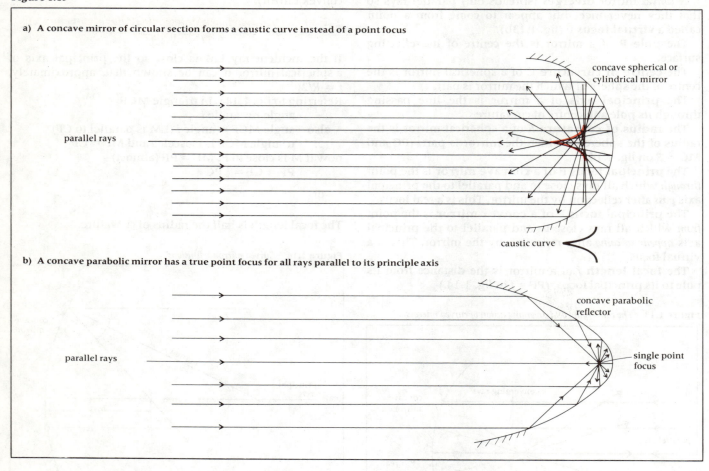

a) **A concave mirror of circular section forms a caustic curve instead of a point focus**

parallel rays

concave spherical or cylindrical mirror

caustic curve

b) **A concave parabolic mirror has a true point focus for all rays parallel to its principle axis**

parallel rays

concave parabolic reflector

single point focus

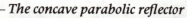

The concave parabolic reflector

Only a parabolic concave reflector has a true single point focus as shown in fig. 1.15b. For this reason the concave reflector used in telescopes has to have a parabolic section to avoid a caustic curve forming. Similarly, only parabolic reflectors are capable of producing a parallel beam of light from a small filament lamp placed at the focus of the mirror. The reflector behind a spotlight, car headlamp or hand torch is usually a concave parabolic mirror. The ray diagram for such a reflector is the same as in fig. 1.15b but with the direction of all the rays reversed. The lamp filament is placed at F.

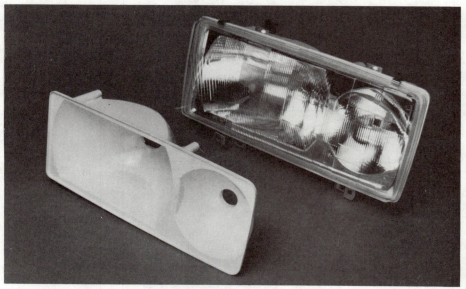

Ray diagrams

Ray diagrams can be used to explain how and where a curved mirror forms an image. They also help us to describe what happens to light rays in experiments. Among the vast number of rays which could be drawn there are three which are particularly helpful in constructing ray diagrams. These special rays, which we shall number ① ② and ③, are shown in fig. 1.16. Two of these rays are needed to find an image.

Figure 1.16 *Three special rays for use in constructing curved mirror ray diagrams*

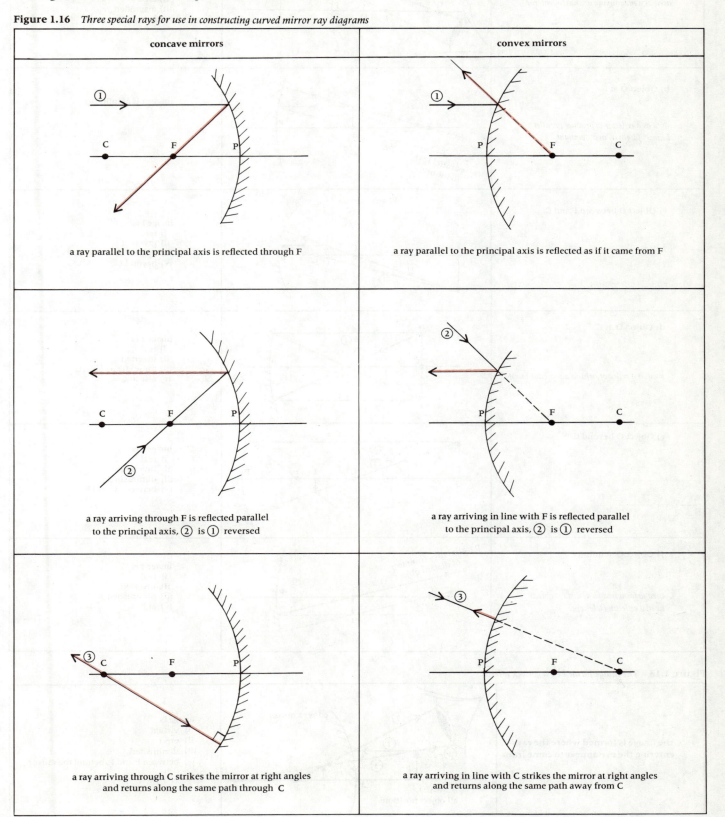

concave mirrors	convex mirrors
a ray parallel to the principal axis is reflected through F	a ray parallel to the principal axis is reflected as if it came from F
a ray arriving through F is reflected parallel to the principal axis, ② is ① reversed	a ray arriving in line with F is reflected parallel to the principal axis, ② is ① reversed
a ray arriving through C strikes the mirror at right angles and returns along the same path through C	a ray arriving in line with C strikes the mirror at right angles and returns along the same path away from C

Figure 1.17 *Images formed by a concave mirror*

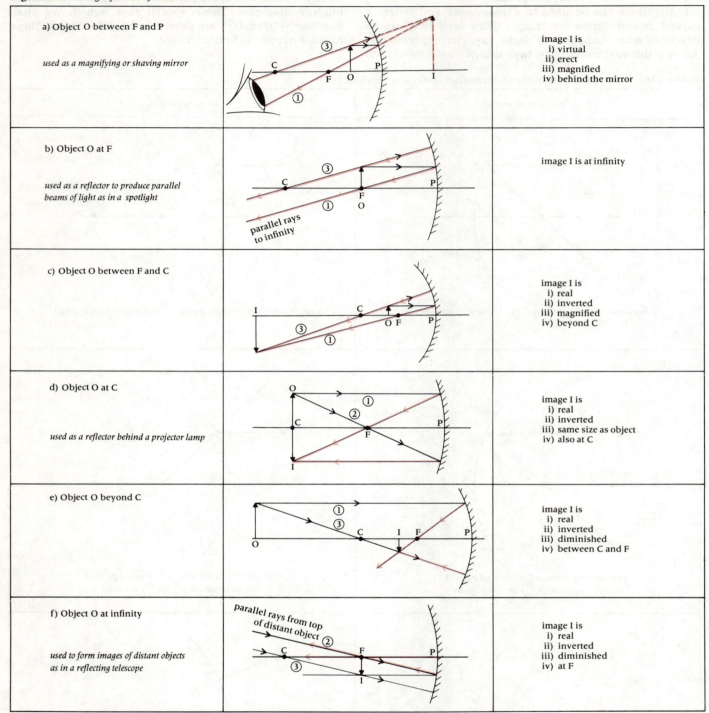

a) Object O between F and P
 	
used as a magnifying or shaving mirror
		image I is
		  i) virtual
		  ii) erect
		  iii) magnified
		  iv) behind the mirror

b) Object O at F *used as a reflector to produce parallel beams of light as in a spotlight*		image I is at infinity

c) Object O between F and C		image I is
		  i) real
		  ii) inverted
		  iii) magnified
		  iv) beyond C

d) Object O at C
 	
used as a reflector behind a projector lamp
		image I is
		  i) real
		  ii) inverted
		  iii) same size as object
		  iv) also at C

e) Object O beyond C		image I is
		  i) real
		  ii) inverted
		  iii) diminished
		  iv) between C and F

f) Object O at infinity
 	
used to form images of distant objects as in a reflecting telescope
		image I is
		  i) real
		  ii) inverted
		  iii) diminished
		  iv) at F

Figure 1.18 *The image formed by a convex mirror*

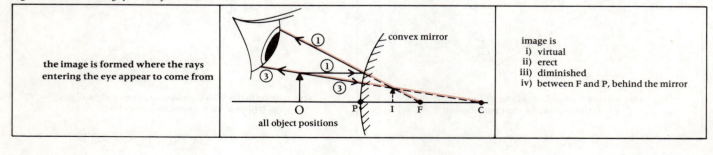

the image is formed where the rays entering the eye appear to come from		image is
		  i) virtual
		  ii) erect
		  iii) diminished
		  iv) between F and P, behind the mirror

Images formed by a concave mirror

The type, size and position of the image formed by a concave mirror depends entirely on how close the object is to the mirror. Fig. 1.17 shows ray diagrams constructed using the three special rays for each of the possible object positions.

For two reasons it is usual to draw an object as an upright arrow ⊥ standing on the principal axis:

a) when the image is formed we can tell which way up it is,

b) we draw rays coming from the tip of the object at a point *off* the axis of the mirror in order to indicate how large the image is compared with the object. Rays from a point *on* the axis return to a point *on* the axis and give no indication of size.

The image is found in each case where the two rays meet again after reflection (or where they appear to come from after reflection when the image is virtual). If the reflected rays are parallel, as in case (b) of fig. 1.17, then they will never meet except at infinity. Thus in case (b) we say that the image is at infinity. In all other cases (c) to (f) the image is real, is formed where rays actually meet, and can be formed on a screen placed at the image position. The image is described by answering the four questions we asked on p 8. Remember that the rays drawn are only two of the very large number which actually leave the object to be reflected by the mirror.

Images formed by a convex mirror

A convex mirror forms virtual images which are always diminished, erect and between the mirror and its principal focus F. The eye shown in fig. 1.18 believes that rays ① and ③ have come from the position of the virtual image I.

A rough method of measuring the focal length of a concave mirror

Rays of light spread out from a point on an object and only a few of them are collected by a mirror or a lens. Fig. 1.19 shows how, as the object is moved further and further away from a mirror or lens, the rays collected from a single point on the object become almost parallel.

In the rough method of measuring the focal length of a concave mirror, we assume that when an object is distant from the mirror, say at the far end of a room, the light rays from a single point on it arrive at the mirror parallel.

● *Hold a concave mirror at one end of a room, facing a distant window.*
● *Hold a white screen in front of and facing the mirror so that it receives rays reflected from it but allows rays to reach the mirror from the window, fig. 1.20.*
● *Move the screen to different distances from the mirror until a sharp image of the window is formed.*
● *Measure the distance from the screen to the mirror with a metre ruler.*

This distance is a rough value of the focal length of the mirror because the roughly parallel light rays from the window will form an image at the principal focus of the mirror. The ray diagram in fig. 1.17f shows how this image is formed.

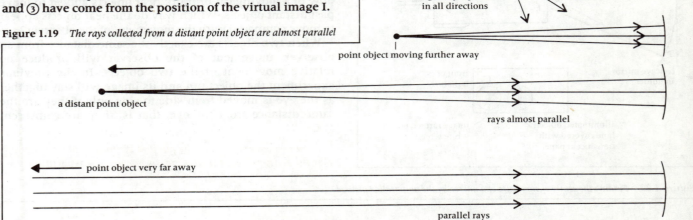

Figure 1.19 *The rays collected from a distant point object are almost parallel*

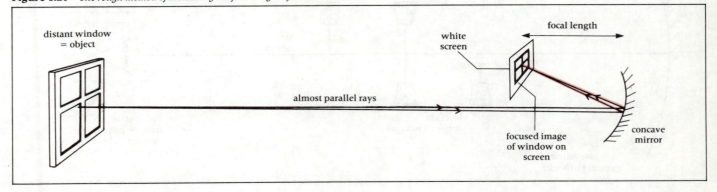

Figure 1.20 *The rough method of measuring the focal length of a concave mirror*

Measuring the radius of curvature of a concave mirror

From fig. 1.17d we can see that if an object is placed at the centre of curvature C of a concave mirror its image will also be formed at C. Thus by finding the position where the object and image coincide we can measure the radius of curvature R of the mirror, half of which is its focal length.

Fig. 1.21 shows a practical arrangement using a light box with an illuminated object. An upright sharp arrow drawn on translucent paper covering a circular hole in front of the light box forms a suitable illuminated object.

● *With the mirror facing the illuminated object, adjust the distance between them until a sharp image is formed on the screen alongside the object. This image will be sharpest when it is exactly the same size as the object.*

● *With a metre ruler measure the distance between the object and back of the mirror.*

● *Make several attempts at finding the position of the sharpest image and repeat the measurement, calculating an average value for the radius of curvature of the mirror.*

● *Divide your result by two for the focal length of the mirror.*

Figure 1.21 *Measurement of R and f of a concave mirror using an illuminated object*

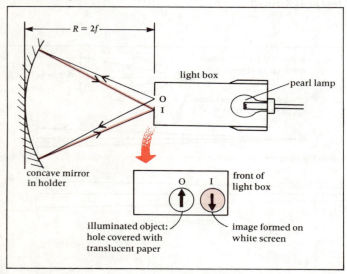

An alternative method using no-parallax

This method also finds the position of the centre of curvature C by arranging for the object and the image to coincide.

● *Support a concave mirror vertically just above bench level so that the centre of the mirror is on the same level as the tip of a pin mounted in a cork, as shown in fig. 1.22.*

● *Move the object pin along the axis of the mirror, viewing it at arm's length along the same line.*

● *When a real inverted image of the pin can also be seen, tilt the mirror carefully so that the tip of the image pin just touches the tip of the object pin.*

● *Now adjust the position of the object pin until it exactly coincides with its image. To do this accurately look for the condition known as no-parallax (see below).*

● *Measure the distance from the object pin to the back of the mirror giving its radius of curvature R.* Again the location of the pin position and the measurement of R should be repeated several times to obtain a more accurate average value.

Parallax and no-parallax

● *Look out of a window and move your head from side to side.*
The near window frame and the distant objects outside appear to move relative to each other although they are actually in fixed positions.

The apparent relative movement of two objects at different distances from an observer caused by movement of the observer is called parallax.

This effect is noticed when looking out of the side window of a moving car or train; near objects appear to move past distant objects. Which way do the near objects appear to move compared with the movement of the observer?

When two objects are exactly the same distance from an observer, movement of the observer will produce no relative movement of the two objects. In the previous experiment the object pin and its image will stay together as the eye is moved from side to side only if they are the same distance from the eye, that is, they are coincident at C.

When two objects (such as the pin and its image) stay together as the eye is moved, the condition is known as no-parallax.

Figure 1.22 *Finding the centre of curvature of a mirror by the no-parallax method*

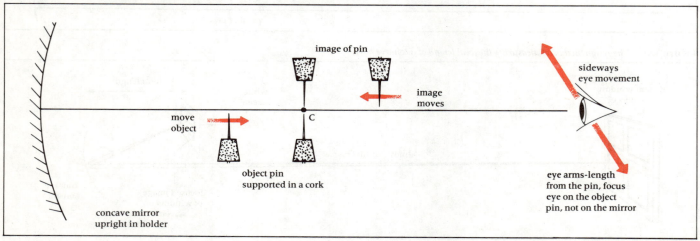

Mirror calculations

All distances are measured from the pole of the mirror:

a) **the focal length** f of a mirror is the distance from its pole to its principal focus,

b) **the object distance** u is the distance from the pole of a mirror to the object, and

c) **the image distance** v is the distance from the pole of a mirror to the image.

The sign convention

The **real-is-positive** sign convention is used in this book:

a) the distances (from a mirror or lens) to *real* objects, images and focuses are *positive*, but

b) the distances (from a mirror or lens) to *virtual* objects, images and focuses are *negative*.

To carry out any mirror or lens calculations successfully we need a system of giving positive or negative values to all distances. There are two systems commonly used but they must be used separately.

The mirror formula

By experiment or by geometrical proof it can be shown that the relation between f, u and v, as defined above for all spherical mirrors, is given by the formulas:

$$\frac{1}{u} + \frac{1}{v} = \frac{1}{f}$$

or

$$f = \frac{uv}{u + v}$$

From measured values of the object and image distances, u and v, the focal length of a mirror can be calculated. The second rearranged formula may be found easier to use when calculating the focal length f. Similarly, knowing the focal length f and the object distance u, we are able to calculate the image distance v. Some worked examples are given below.

Worked Example
Calculating the focal length of a concave mirror

Two sets of results were obtained by different experiments for the same concave mirror. Calculate its focal length from each set of results.

Results a): distance to real object = 30 cm (u)
　　　　　　distance to real image   = 20 cm (v)
Results b): distance to real object = 8 cm (u)
　　　　　　distance to virtual image = − 24 cm (v)

Calculation (a)

Using:
$$\frac{1}{f} = \frac{1}{u} + \frac{1}{v}$$

the results give
$$\frac{1}{f} = \frac{1}{30\,cm} + \frac{1}{20\,cm}$$
$$= \frac{2+3}{60\,cm} = \frac{5}{60\,cm} = \frac{1}{12\,cm}$$
$$\therefore \quad f = 12\,cm$$

Using:
$$f = \frac{uv}{u + v}$$

the results give
$$f = \frac{30\,cm \times 20\,cm}{30\,cm + 20\,cm}$$
$$= \frac{600\,cm^2}{50\,cm}$$
$$\therefore \quad f = 12\,cm$$

Calculation (b)

Using:
$$\frac{1}{f} = \frac{1}{u} + \frac{1}{v}$$

since v is negative, being the distance to a virtual image, we get:
$$\frac{1}{f} = \frac{1}{8\,cm} - \frac{1}{24\,cm}$$
$$= \frac{3-1}{24\,cm} = \frac{2}{24\,cm} = \frac{1}{12\,cm}$$
$$\therefore \quad f = 12\,cm$$

Using:
$$f = \frac{uv}{u + v}$$

since v is negative, $1/v$ is also negative, so we get:
$$f = \frac{8\,cm \times (-24\,cm)}{8\,cm - 24\,cm}$$
$$= \frac{-192\,cm^2}{-16\,cm}$$
$$\therefore \quad f = +12\,cm$$

Answer: The two sets of results agree, the focal length of the mirror is 12 cm and the positive value confirms that it is a concave mirror with a real focus. The two forms of the formula do, of course, give the same answer.

The convex mirror used to give a wide field of view

Convex spherical mirrors are often used for the wing mirror of a car. They give the driver the advantage of a wider field of view than with a plane mirror. This means that the driver can see a greater width of road behind him. On a motorway he has a view of the overtaking lanes as well as the lane behind him. The disadvantage of a convex wing mirror is that it makes the image smaller, giving the impression that the cars behind are further away than they really are. This can be dangerous and a driver has to learn to judge distances differently when using a convex mirror on a car. A quick check in the plane rear-view mirror inside the car is wise before deciding to overtake.

The wide angle of view in a convex mirror means that a large area is included in the image, for which reason they are used in shops so that assistants can have a wide view of a large floor area. They are used also at the top of the stairs in some double-decker buses so that the conductor downstairs can see all of the top deck.

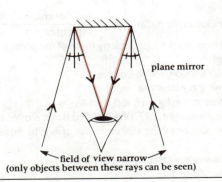
plane mirror
field of view narrow
(only objects between these rays can be seen)

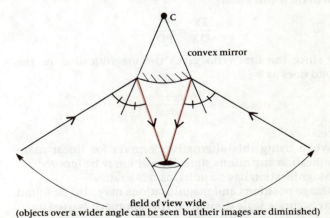
convex mirror
field of view wide
(objects over a wider angle can be seen but their images are diminished)

Magnification

When an image is larger than the object we say it is magnified and when it is smaller we describe it as diminished. The numerical comparison of the image size with object size is always called the magnification. The definition of linear or transverse magnification m (meaning magnification of one dimension: the length or distance across) is given by the formula:

$$\text{magnification } m = \frac{\text{height of image}}{\text{height of object}}$$

Magnification m		Image size
m greater than 1	$m > 1$	magnified: larger than the object
m equals 1	$m = 1$	image same size as the object
m less than 1	$m < 1$	diminished: smaller than the object

Figure 1.23 *Magnification formula $m = v/u$*

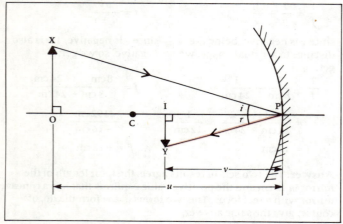

Alternative formula for linear magnification

In the simplified ray diagram of fig. 1.23:

OX = height of object   PO = object distance u
IY = height of image   PI = image distance v

Triangles POX and PIY are similar (angle i = angle r and both contain a right angle), therefore we can say about the ratio of their sides that:

$$\frac{\text{IY}}{\text{OX}} = \frac{\text{PI}}{\text{PO}}$$

and since the first ratio gives the magnification m, the second does as well:

$$\text{magnification} = \frac{\text{image distance}}{\text{object distance}} \quad m = \frac{v}{u}$$

Note:

a) when using this alternative formula for linear magnification any minus signs for u or v may be ignored.
b) Magnification has no units, as it is a ratio.
c) Image positions and magnifications may also be found by accurate scale drawings. This approach is used on p 42 to solve a similar lens problem.

Worked Example
Calculation of image position

An object is placed (a) 15 cm and (b) 4 cm from a concave mirror of focal length 6 cm. Find the position, magnification and nature of the image formed in each case.

a) Using $\frac{1}{u} + \frac{1}{v} = \frac{1}{f}$ and rearranging we have

$$\frac{1}{v} = \frac{1}{f} - \frac{1}{u}$$

Now, for a real object and a concave mirror with a real focus, the distances u and f will be positive so we have $f = +6$ cm and $u = +15$ cm which give

$$\frac{1}{v} = \frac{1}{6\,\text{cm}} - \frac{1}{15\,\text{cm}} = \frac{5-2}{30\,\text{cm}} = \frac{3}{30\,\text{cm}} = \frac{1}{10\,\text{cm}}$$

$$\therefore \quad v = +10\,\text{cm}$$

now using $m = \frac{v}{u}$ to calculate the magnification gives

$$m = \frac{10\,\text{cm}}{15\,\text{cm}} = \frac{2}{3} \quad \text{or} \quad 0.67 \quad \text{(no units)}$$

Answer: The image is real (the value of v is positive), diminished to 2/3 of the object size and is formed 10 cm away from the mirror. Being a real image, it must be in front of the mirror.

b) Using the rearranged formula again, with the values $f = 6$ cm and $u = 4$ cm

from $$\frac{1}{v} = \frac{1}{f} - \frac{1}{u}$$

we get $$\frac{1}{v} = \frac{1}{6\,\text{cm}} - \frac{1}{4\,\text{cm}} = \frac{2-3}{12\,\text{cm}} = \frac{-1}{12\,\text{cm}}$$

$$\therefore \quad v = -12\,\text{cm}$$

now using $$m = \frac{v}{u} \quad \text{(and ignoring the signs)}$$

gives $$m = \frac{12\,\text{cm}}{4\,\text{cm}} = 3 \quad \text{(no units)}$$

Answer: The image is virtual (v is negative), magnified to $3 \times$ the size of the object, and appears to be 12 cm behind the mirror when looking into the mirror.

Assignments

Remember

a) The mirror formula: $\frac{1}{f} = \frac{1}{u} + \frac{1}{v}$

b) The magnification formulas: $m = \frac{\text{height of image}}{\text{height of object}} = \frac{v}{u}$

c) The definitions on p 13, particularly those of the principal focus and focal length.

d) The sign convention 'real-is-positive'.

Practise drawing ray diagrams

e) Try the diagrams in fig. 1.16 and

f) try the diagrams in fig. 1.17. (You will find that it is often possible to find image positions using a different pair of rays to those chosen in fig. 1.17.)

Redraw and complete table 1.2.

Try questions 1.8b to 1.15

Table 1.2

	Plane mirror	Concave mirror	Convex mirror
Shape			bulges out
Effect on parallel rays		converges them	
Type of focus	none		
Type of image		real except when O is nearer than F	
Size of image compared with object			
Field of view compared with plane mirror			
Two applications			

Questions 1

1 Describe the shadow produced by
   a) a small filament lamp,
   b) sunlight on a sunny day,
   c) a fluorescent tube,
   d) a spot lamp,
   e) a candle.
2 Explain why sharp shadows support the theory that light travels in straight lines.
3 Fig. 1.24 shows a very small source of light producing a shadow of the object X on a screen S. What is the size of the shadow, in metres? (Joint 16+)

Figure 1.24

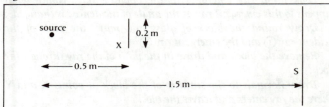

4 a) Draw a diagram to show how a solar eclipse is formed. On your diagram, mark the positions of (i) an observer (P) who sees a partial eclipse and (ii) a second observer (T) who sees a total eclipse.
     Draw two diagrams, one for each kind of eclipse, to show what each observer would see if he looked towards the sun. What property of light have you assumed in drawing these diagrams?
   b) Explain how a single observer may see both a partial eclipse and a total eclipse of the sun while remaining at the same place on the earth.
   c) What change would be necessary for an observer to see an annular eclipse? How does an annular eclipse differ from the eclipses described in (a) and (b)? (JMB)
5 With the aid of diagrams explain the differences between the reflection of light which occurs
   a) from a flat sheet of white cloth
   b) from a flat sheet of shiny aluminium foil.
6 Explain why you might see 𝟛ƆИA⅃UᙠMA painted on the front of an ambulance.
7 Fig. 1.25 shows two plane mirrors set at an angle 100° to each

other. A ray strikes one of the mirrors as shown at an angle of incidence of 45°. Redraw and complete the diagram showing the path of the ray and calculate the angle of reflection at which it leaves the second mirror.

Figure 1.25

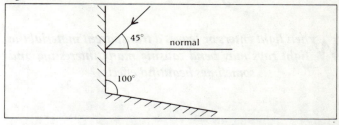

8 a) Fig. 1.26a shows two rays of light leaving an object O and striking a plane mirror.

Figure 1.26

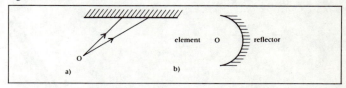

Draw the two reflected rays and use them to find the position of the image.
   b) Fig. 1.26b shows a side view of an electric fire.
      i) What types of waves are given out by the element?
      ii) What name is given to the shape of the reflector?
      iii) The reflector is made of metal. Describe its surface, and explain why metal is used. NEA (A) SPEC
9 The following questions are about curved mirrors.
   a) What is the main advantage of using a parabolic reflector for a headlamp?
   b) Give one disadvantage and one advantage of using a convex mirror as a rear view mirror. (Joint 16+)
10 A dentist has the choice of three small mirrors, convex, concave and plane, to examine the back of your teeth. State which she should use to give the best view. Give your reasons for this choice, rather than the other two.
11 Draw a ray diagram to show how a concave mirror produces a magnified *virtual* image of a suitably placed object, and show where an eye must be positioned in order to see the image.
12 Draw a ray diagram to show how a concave mirror of parabolic section can be used to produce a parallel beam of light from a spotlight.
13 Draw a diagram to show how a hemispherical convex mirror hung from the ceiling of a shop enables customers to be watched over a wide angle of view.
14 A concave mirror has a focal length of 15 cm, calculate
   a) its radius of curvature
      and the position of the image of an object standing
   b) 30 cm from the mirror,
   c) 20 cm from the mirror,
   d) 10 cm from the mirror.
   In each case state the nature of the image formed.
15 The distance between an object and its enlarged *real* image produced by a concave spherical mirror is 200 mm when the object is placed 100 mm from the pole of the mirror.
    Determine the linear magnification of the image and the focal length of the mirror. (L)

Light rays and refraction

When light enters or leaves a transparent material the light rays may bend causing many interesting and sometimes beautiful effects.

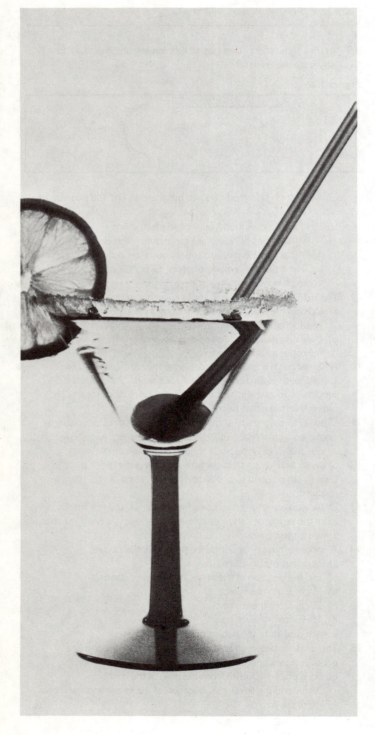

REFRACTION
LIGHT RAYS CHANGE DIRECTION

Investigating light passing through a rectangular block of glass

● *In a darkened room, arrange a ray box with a single slit to send a narrow ray of light into a rectangular block of glass or perspex as shown in fig. 2.1.*

The block should be placed with its largest face on a sheet of white paper so that the ray enters at an angle through a long side, near one end.

Figure 2.1 *A light ray passing through a glass block*

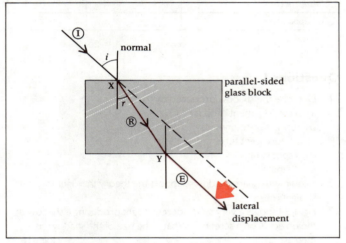

● *Look for a ray emerging at the opposite side. Notice what happens to this emergent ray as the angle of incidence i is changed.*
● *Draw round the block of glass and mark the path of the incident ray ① and the emergent ray ⑥.*
● *Remove the block and draw in the path of the ray through the block.*
● *Draw normals to the surfaces of the block at points X and Y where the ray enters and leaves the block.*
● *Look at the angles between the rays and the normals. As the ray enters the block at X and leaves at Y, which way is it bent, towards or away from the normal? Which angles are equal? What do you notice about the directions of rays ① and ⑥?*

Observations

At X the incident ray ① is bent as it enters the glass. This bending is called refraction and ray ⑧ is called the refracted ray.

Refraction *is the bending of light which occurs when it passes from one transparent material (called a medium) to another.*

The angle of refraction r *is the angle between the refracted ray and the normal.*

At X, as the ray enters the glass, angle r is smaller than angle i and we say that the ray is bent *towards* the normal. This happens when light enters an optically denser medium. The glass is optically denser than air, by which we mean that light travels more slowly in glass than air.

At Y, as the ray leaves the glass, it is bent back to its original direction. The angle of refraction, now in the air, is larger than the incident angle (in the glass).

When a ray of light enters an optically denser medium it is bent towards the normal.

Conversely, when it enters a less dense medium it is bent away from the normal.

If the block of glass has parallel sides, the emergent ray Ⓔ is parallel to the incident ray Ⓘ, but it is laterally displaced. This means the ray is travelling in the same direction but it has been shifted sideways when it emerges. This also happens to light whenever it passes through a plane glass window at an angle to the normal.

Measurements

We can use the same apparatus to measure the angles of incidence i and refraction r, and then investigate the relation between them.

Figure 2.2 *Measuring the angle of refraction r as a ray enters a glass block*

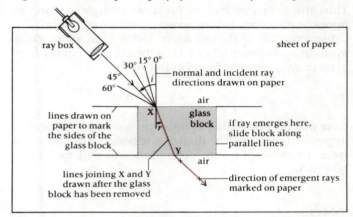

- *Draw accurately two parallel lines on white paper to mark the sides of the glass block.*
- *Using a protractor, draw a normal, in a position as shown in fig. 2.2, and measure from it several angles of incidence i.*
- *Accurately aim the incident ray at X, and for each angle of incidence mark the directions of the emergent ray with two crosses.*
- *Remove the glass block, draw in the emergent and refracted rays and measure the angles of refraction r.*
- *Tabulate the values of the angles i and r and also, using a calculator or tables, the values of sin i and sin r.*

Table 2.1 gives some typical results. It is not easy to see the relation between the angles i and r. Many scientists had failed to find the relation when in 1621 a Dutch mathematics professor called Snell discovered that the ratio of the sines of the angles gave a constant value.

Table 2.1

i/degree	r/degree	sin i	sin r	$\dfrac{\sin i}{\sin r}$
15	10	0.26	0.17	1.53
30	20	0.50	0.34	1.47
45	28	0.71	0.47	1.51
60	35	0.87	0.57	1.53

Calculate the value obtained from your results. The mean value for air to glass is about 1.5 as can be seen from the sample results above. The name given to this constant ratio is the **refractive index**.

The laws of refraction

Snell's discovery is usually called a law of refraction. The other law is based on the observation that, as with reflection, both rays of light and the normal to the surface are all in a flat plane (usually the flat sheet of paper on a bench top in experiments).

1
The incident ray, refracted ray and the normal at the point of incidence all lie in the same plane.

2 *Snell's law*
For light rays passing from one transparent medium to another, the sine of the angle of incidence and the sine of the angle of refraction are in a constant ratio.

Internal reflection

A closer look at the light rays in a glass block reveals reflected rays as well as the refracted rays. In a well darkened room and using the same glass block and ray box as before, it is possible to see two other rays leaving the block.

- *Look at the positions where the rays leave the block and at their directions. Compare the brightness of each of the rays leaving the block. Which angles are equal? Can you explain how each ray is produced?*

Figure 2.3 *Refraction and reflection in a glass block*

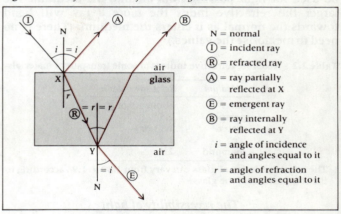

Fig. 2.3 gives an explanation by drawing the path of the rays inside the glass block.

At X the incident ray is divided, part of it being reflected from the surface of the glass and part entering the glass. The reflected ray Ⓐ is fainter than the refracted ray Ⓡ as most of the light enters the glass. The surface of the glass acts like a plane mirror and the reflected ray obeys the laws of reflection.

At Y some of the light is **internally reflected**. The inside surface of the glass also acts as a plane mirror. In this experiment the inside surface, like the outside surface of the glass, is only a partial mirror, that is to say, only a part of the light is reflected from it while the rest passes through the surface and is refracted.

Most of the internally reflected light emerges along ray Ⓑ but some of this light will be internally reflected again. Each time the light is divided up it becomes fainter. In a similar way multiple internal reflections occur in thick glass mirrors and cause multiple images which make them unsuitable for some applications (p 26).

As we shall discover soon, under some conditions the internal reflection of light inside a transparent medium like glass becomes total. When total internal reflection occurs no light emerges through a surface at all.

Refractive index

The value of the ratio $\sin i / \sin r$ indicates how much refraction or bending will occur when a ray passes from one medium to another, hence its name: refractive index. Strictly speaking, the ratio $\sin i / \sin r$ for two media gives their **relative refractive index**. This is because the extent to which a ray is bent depends on both the medium the light is leaving and the one it is entering. For convenience all materials are given a value called their absolute refractive index which indicates the refraction that would occur if a ray of light passed from a vacuum into the medium. In practice, when a ray of light passes into a medium from air the refraction is very nearly the same as it would be from a vacuum. So we often drop the word 'absolute' and assume that light is entering a medium from the air.

$$\text{The absolute refractive index of a medium } n = \frac{\sin i \text{ (in a vacuum)}}{\sin r \text{ (in the medium)}}$$

The (absolute) refractive index, symbol n, has no units. The (absolute) refractive indices of some common transparent materials are given in table 2.2. These values may be used for light passing from air into the medium. The larger the refractive index the more a ray will bend towards the normal as it enters the medium. (There is no need to memorise the values.)

Table 2.2   Absolute refractive indices of some transparent materials

Medium	Refractive index n
glass	about 1.5 or $\frac{3}{2}$*
perspex	1.5
water	1.33 or $\frac{4}{3}$
ice	1.3
diamond	2.4

* The refractive index of glass can vary from 1.48 to 1.96 according to the composition of the glass.

The reversibility of light

The principle of reversibility of light simply states that the paths of light rays are reversible. This means that if a ray of light is sent in the exact opposite direction it will follow the same path.

When a ray of light passes from medium 1 to medium 2 this is indicated by using the symbol $_1n_2$ for the refractive index. Referring to fig. 2.4

$$_1n_2 = \frac{\sin i}{\sin r}$$

where angle i is in medium 1 and angle r is in medium 2.

For a ray travelling in the opposite direction, from medium 2 into medium 1, the symbol used is $_2n_1$. Referring again to fig. 2.4

$$_2n_1 = \frac{\sin r}{\sin i}$$

From these relations we see that:

$$_2n_1 = \frac{1}{_1n_2}$$

Figure 2.4   *The relation between refraction and the speed of light, c*

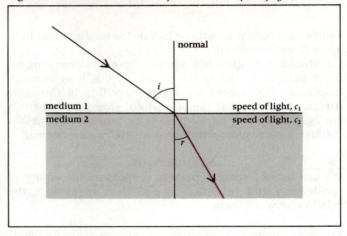

Thus the refractive indices for a ray of light passing in opposite directions between two media are reciprocals.

For example: if the refractive index from air to glass, symbol $_an_g$, is about 1.5 or $\frac{3}{2}$, then the refractive index from glass to air, $_gn_a$ can be found as follows:

$$_gn_a = \frac{1}{_an_g} = \frac{1}{\frac{3}{2}} = \frac{2}{3} \quad \text{or} \quad 0.67$$

Note that rays travelling in opposite directions are bent in opposite directions. The direction of bending can be found from the value of the refractive index as summarised in table 2.3.

Table 2.3

Refractive index		Ray direction	What happens to the ray
n greater than 1	$n > 1$	ray entering an optically denser medium	ray is bent towards the normal
n equals 1	$n = 1$	either direction, no change of optical density	no bending occurs
n less than 1	$n < 1$	ray entering an optically less dense medium	ray is bent away from the normal

Refractive index and the speed of light

The speed of light in a vacuum has been very accurately measured and is about 3.0×10^8 m/s or 300 million metres per second (see appendix C3). Light travels more slowly in transparent materials and it is thought that the bending or refraction of light is due to this change of speed. An explanation of this is given in terms of the wave theory of light (p 444).

For light passing from a vacuum into a medium:

$$\text{the absolute refractive index of a medium } n = \frac{\text{speed of light in vacuum}}{\text{speed of light in medium}}$$

For light passing from medium 1 to medium 2:

$$_1n_2 = \frac{\text{speed of light in medium 1, } c_1}{\text{speed of light in medium 2, } c_2}$$

For example: for light passing from air to glass the speed of light in air is very near to the speed in a vacuum, 3×10^8 m/s, and the refractive index $_an_g = \frac{3}{2}$. This means that light travels $\frac{3}{2}$ times as fast in the air as it does in glass. The above relation shows that the speed in glass is therefore 2×10^8 m/s.

$$_an_g = \frac{3}{2} = \frac{3 \times 10^8 \text{ m/s (speed of light in air)}}{2 \times 10^8 \text{ m/s (speed of light in glass)}}$$

Since we have a relation between the speed of light and refraction, two equal ratios are linked by the refractive index for a pair of media (fig. 2.4).

$$_1n_2 = \frac{\sin i}{\sin r} = \frac{c_1}{c_2}$$

Real and apparent depth

If you look into a clear pool of water it appears to be shallower than it really is. A swimming pool, for example, which is really 4 metres deep will appear to be only about 3 metres deep.

In fig. 2.5 we see how an object O, seen through a transparent medium like water, appears closer than it really is. This effect is caused by refraction at the surface of the water. Rays of light coming from the object O are bent *away from* the normal as they leave the water so that they appear to come from a virtual image I which is above the object O.

Figure 2.5 *Real and apparent depth*

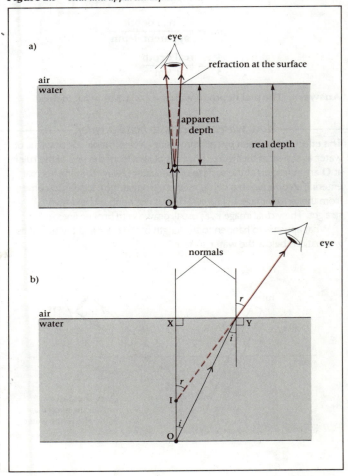

Using fig. 2.5b we can find the relation between the real and apparent depths. Light is passing from water to air and the refractive index is given by:

$$_wn_a = \frac{\sin i}{\sin r}$$

also angle XOY = angle i (alternate angles)
and angle XIY = angle r (corresponding angles)

$$\therefore \quad _wn_a = \frac{\text{sine of angle XOY}}{\text{sine of angle XIY}} = \frac{\text{XY/OY}}{\text{XY/IY}} = \frac{\text{IY}}{\text{OY}}$$

Now when the eye is vertically above the object, Y and X are together, then OY = OX and IY = IX, therefore

$$_wn_a = \frac{\text{IX}}{\text{OX}} = \frac{\text{apparent depth}}{\text{real depth}}$$

Now by reversing the light direction we get the refractive index from air to water. (In the reverse direction the refractive index has the reciprocal value.)

$$_an_w = \frac{\text{real depth}}{\text{apparent depth}}$$

This relation is only accurate when an object is seen normally through a surface, that is when rays of light from the object to the eye pass through the surface at right angles.

We can use this relation in a method of measuring refractive index.

Measuring refractive index by the real and apparent depth method

• *Stand a glass block on end across a straight line drawn on a sheet of paper, as in fig. 2.6.*
• *Support a search pin in a movable holder so that it can be moved up and down and then held steady while measurements are made.*

Looking from above, the line under the glass block will appear to be nearer when seen through the glass than when seen through the air.

Figure 2.6 *Measuring refractive index by the real and apparent depth method*

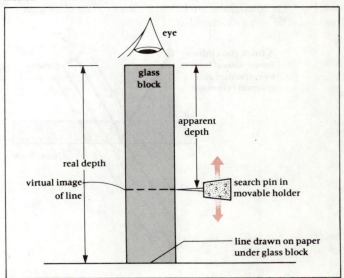

● *To find the apparent position of the line as seen through the glass, move the search pin up and down the side of the block until it is at the same level as the apparent position of the line inside the glass.*

This is done by looking for no-parallax between the line and the search pin. Move your eye across the top of the block at right angles to the line. When no-parallax is found the search pin and line, seen through the glass, should move together. If, however, they separate as your eye moves across then raise or lower the search pin by a small amount and look again for no-parallax.

When the position of no-parallax is found, the search pin indicates the apparent depth of the block.

● *Fix the search pin and measure the apparent depth of the block, measuring down from the top of the block to the pin.*

● *From the measured values of the real and apparent depths of the block, calculate the refractive index of glass:*

$$\text{refractive index of glass} = \frac{\text{real depth}}{\text{apparent depth}}$$

A similar method may be used to find the refractive index of a transparent liquid.

● *Fill a tall beaker or gas jar with the liquid.*

● *Drop a pin to the bottom of the liquid inside its container to act as a marker for the real depth of liquid.*

Seen through the liquid this pin will appear to be nearer than it really is.

● *Locate its apparent position as before with a search pin outside the container.*

● *Measure the distances down from the top of the liquid to the two pins, real depth inside and apparent depth outside.*

How a thick glass mirror forms multiple images

All glass mirrors silvered at the back surface produce faint ghost-like images which blur the main clear image. In the figure, I_2 is the clear image formed by a single reflection of light at the silvered back surface of the mirror. The faint image I_1 is formed by a small amount of light being reflected from the front surface of the glass. Images I_3, I_4, and so on, get fainter as less light remains. Each time the light reaches the inside of the front surface of the glass some of it escapes and is refracted while the rest is internally reflected. Images are seen in line with each refracted ray which leaves the glass, that is to say along the broken lines. To avoid these problems, mirrors used in accurate instruments are made with a polished aluminium film on the front surface of the glass, but this can be very easily damaged.

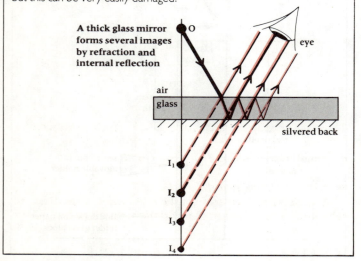

A thick glass mirror forms several images by refraction and internal reflection

Worked Example
Refraction and the speed of light

The speed of light in water is 2.25×10^8 metres per second and in air is 3.00×10^8 metres per second.

a) *Calculate the refractive index from air to water.*

b) *If a ray of light passing from air to water is incident at the surface at an angle of 30°, calculate the angle of refraction in the water.*

a)   The refractive index $\qquad {}_a n_w = \dfrac{\text{speed of light in air}}{\text{speed of light in water}}$
from air to water

$$\therefore \ {}_a n_w = \frac{3.00 \times 10^8\,\text{m/s}}{2.25 \times 10^8\,\text{m/s}} = \frac{3.00}{2.25} = 1.33 \text{ (no units)}$$

b) $\dfrac{\sin i \ (\text{in air})}{\sin r \ (\text{in water})} = \dfrac{\text{speed of light in air}}{\text{speed of light in water}}$

$$\therefore \quad \frac{\sin 30°}{\sin r} = \frac{3.00 \times 10^8\,\text{m/s}}{2.25 \times 10^8\,\text{m/s}}$$

$$\therefore \quad \frac{0.50}{\sin r} = \frac{3.00}{2.25}$$

$$\therefore \quad \sin r = \frac{0.50 \times 2.25}{3.00} = 0.375$$

and so $r = 22°$.

Answer: The angle of refraction $r = 22°$.

Worked Example
Real and apparent depth

A swimming pool appears to be only 1.5 metres deep. If the refractive index of water is 4/3, calculate the real depth of water in the pool.

$$\qquad {}_a n_w = \frac{\text{real depth}}{\text{apparent depth}}$$

$$\therefore \quad \frac{4}{3} = \frac{\text{real depth}}{1.5\,\text{m}}$$

Answer: The real depth of water = $\tfrac{4}{3} \times 1.5\,\text{m} = 2.0$ metres.

How water appears to bend a ruler

This effect can be seen by half immersing a ruler in a sink or bowl full of water as shown in the figure. The rays of light from the end of the ruler at **O** are refracted at the surface of the water (away from the normal, entering an optically less dense medium) so that they appear to come from the virtual image **I**. The real ruler is not bent, and is drawn straight. The virtual image is, as usual, drawn with broken lines.

What appears to happen to the length of the ruler and the size of its graduations below the water surface?

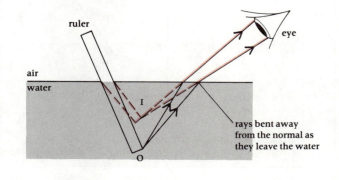

The mirage

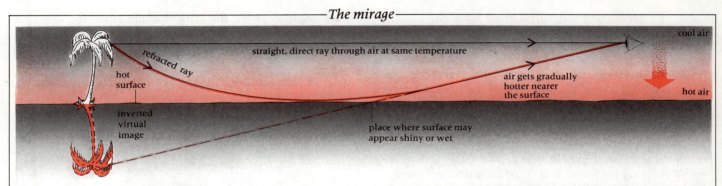

The mirage is produced by refraction in the air as shown. A mirage can happen when the air nearer the surface of the ground is less dense than that above. When the sun has been shining on a desert or a road and has made the surface very hot, the air next to it is heated and expands, becoming less dense. The optical density and refractive index of the air gradually increases with height above the surface as the air gets cooler.

Light from a distant object may reach an observer's eye by the two paths shown in the figure with the result that the object is seen in its true position and also as an inverted image below it. The inverted image is virtual and is called a mirage.

The most likely explanation of the curved ray path is that it is gradually refracted as the light passes through air of gradually changing density. Approaching the surface, as the air gets hotter and less dense, the decreasing refractive index bends the ray away from the normal. After skimming along the surface the ray is gradually bent back towards the normal as it rises through air of increasing density and refractive index.

There is a hazy region on the ground between the object and the observer where the surface seems shiny or reflective and a hot dry road may appear to be wet.

Figure 2.7 *Internal reflection and critical angle*

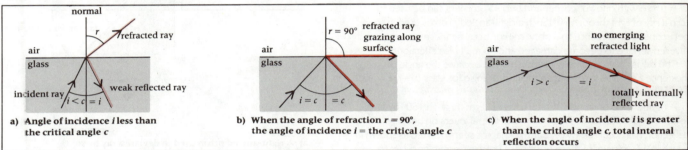

a) **Angle of incidence *i* less than the critical angle *c***

b) **When the angle of refraction *r* = 90°, the angle of incidence *i* = the critical angle *c***

c) **When the angle of incidence *i* is greater than the critical angle *c*, total internal reflection occurs**

Total internal reflection

We have seen that when light passes from air to glass there is always a reflected ray and a refracted ray. Now we shall find out what happens when light tries to pass from glass to air.

In fig. 2.7a a ray meets the surface at a small angle of incidence *i* and a weak internally reflected ray is produced as well as the refracted ray. The angle of refraction *r* is greater than the angle of incidence *i*. It follows that if the angle of incidence is increased it will reach a critical value where the angle of refraction is just 90° and the refracted ray grazes along the surface of the glass, fig. 2.7b. This value of the angle of incidence is called the critical angle:

The critical angle between two media is the angle of incidence in the optically denser medium for which the angle of refraction is 90°.

If the angle of incidence *i* is further increased, becoming greater than the critical angle ($i > c$, fig. 2.7c), it is impossible for the angle of refraction to exceed 90°. Now no light emerges and all the light is totally internally reflected. The inside surface of the glass behaves like a perfect mirror.

Total internal reflection occurs when:

a) a ray of light is inside the optically denser of two media and

b) the angle of incidence at the surface is greater than the critical angle for the pair of media.

Calculation of critical angle

For a ray of light going from glass to air, the angle of incidence *i* equals the critical angle *c* when the angle of refraction *r* is just 90°:

$$i = c \quad \text{when} \quad r = 90°$$

and the refractive index from glass to air ${}_g n_a$ is given by:

$${}_g n_a = \frac{\sin i}{\sin r} = \frac{\sin c}{\sin 90°} = \frac{\sin c}{1}$$

$$\therefore \; \sin c = {}_g n_a = \frac{1}{{}_a n_g}$$

or:

$$\sin c = \frac{1}{\text{refractive index of glass}}$$

Thus if the refractive index of glass ${}_a n_g = 1.5$, we have:

$$\sin c = \frac{1}{1.5} = 0.67$$

therefore the critical angle $c = 42°$.

In general the critical angle *c* between a medium of (absolute) refractive index *n* and the air is given by:

$$\sin c = \frac{1}{n}$$

Measuring the critical angle for glass

A semicircular block of glass or perspex is particularly suitable for this experiment because it allows a ray of light to enter the glass through the curved edge without being refracted. This happens when a ray is directed at the centre of the flat edge along a radius so that it enters the curved surface at 90°, as shown in fig. 2.8.

- *Draw round the semicircular block on a sheet of paper and by measurement draw a normal at the midpoint of the straight side.*
- *Direct a ray of light through the glass to be internally reflected exactly at the midpoint of the straight side.*
- *Move the ray box round until the critical condition is found.*

It should be possible to see a refracted ray grazing just along the surface so that $r = 90°$.

- *Mark the direction of the incident ray and reflected ray with two crosses each.*
- *Remove the block and draw in the rays.*
- *Measure the angle between the incident and reflected rays with a protractor.*

This angle is twice the critical angle since the angle of reflection equals the critical angle of incidence. Thus half this angle gives the critical angle *c*.

Figure 2.8 *Measuring the critical angle*

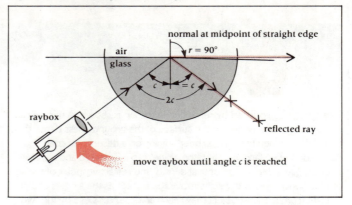

Prisms at work

Prisms are blocks of glass or transparent material with a triangular section. They come in various shapes and sizes with different angles between the three sides.

A prism with one 90° corner and two 45° corners can be used to turn a ray of light through 90°. Such a right-angled prism is used in some periscopes in preference to a plane mirror because there is no exposed silvered surface to become damaged and no multiple reflections.

The ray is totally internally reflected once, because the angle of incidence $i = 45°$ is greater than the critical angle for glass to air, $c = 42°$. The ray is deviated by 90°.

A right-angled prism can also turn rays of light through 180° by two total internal reflections. These eliminate lateral inversion because reflection has occurred twice, but the image is seen inverted as shown in figure (b). Two pairs of these prisms are used in prism binoculars to reduce the length of the instrument and produce an erect final image for the whole instrument (p 51).

A five-sided prism or pentaprism is used in many modern cameras. In a single lens reflex (SLR) camera the pentaprism is used to turn the light rays round inside the camera so that the photographer can see the actual picture he is going to take through the camera lens. By two total internal reflections the pentaprism turns the light through 90° and produces an image which is erect and not laterally inverted. The external surfaces of the prism need to be silvered for total internal reflection to occur because the angles of incidence are less than the critical angle.

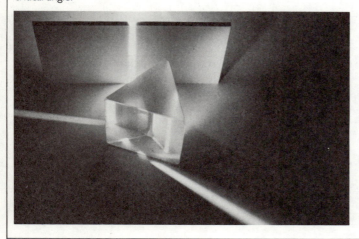

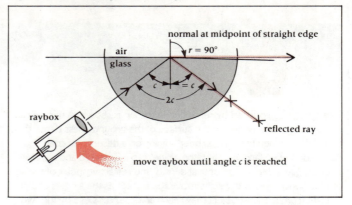

$i = 45°$
$c = 42°$ for glass
∴ $i > c$ and total internal reflection occurs

a) A right-angled prism used to deviate a ray by 90°

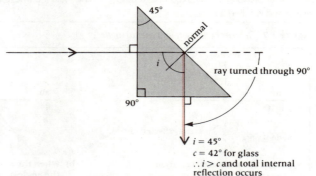

b) A right-angled prism used to turn rays of light through 180°

a section through the pentaprism which forms a characteristic hump on the top of SLR cameras

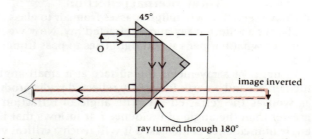

c) The pentaprism used in a single lens reflex camera

An optical fibre or light pipe

An optical fibre is a long glass rod which can be anything from a fraction of a millimetre to 50mm thick. Light can travel along a glass rod even when the rod is bent. When a ray of light strikes the inside of the surface of the rod, if the angle of incidence i inside the glass is greater than the critical angle, then total internal reflection traps the light inside. It works as if the surface of the rod was silvered, but it is not; the solid glass rod has a perfect reflecting surface inside.

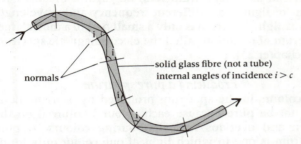

normals

solid glass fibre (not a tube)
internal angles of incidence $i > c$

a) Single optical fibre or light pipe, contains light ray by multiple total internal reflections

light
from
object

flexible bundle of fine
glass fibres

The image is formed from a pattern of light and dark dots, each dot is made by a separate glass fibre:

fibres carrying no light provide dark parts of the image

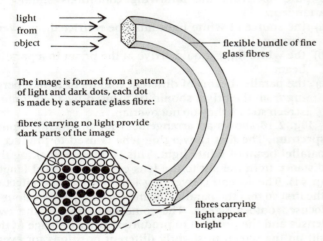

fibres carrying
light appear
bright

b) A bundle of glass fibres used to transmit an image

Optical fibres can be used to carry light round bends and into places where it would be difficult or dangerous to supply electricity. For example an instrument used for viewing inside the body, called an endoscope, passes light along a thin flexible bundle of very fine glass fibres and sends an image of the inside of the body back along another bundle of fibres. This second image-forming bundle may have thousands of individual fibres of diameter about 0.01 mm arranged in fixed

An optical fibre cable has the same capacity for carrying telephone calls as a metal cable ten times as thick.

positions so that the image does not become scrambled. Each fibre forms one dot of the image.

The most important new use for optical fibres is in the communications industry. Optical fibres can carry laser light over great distances. The laser light is made to vary very rapidly to represent information such as telephone conversations, computer data and television pictures. By this method the optical fibre cable can carry more information than a copper wire cable carrying a varying electric current. Optical fibre cables are also much thinner and lighter and, being made of glass, are becoming cheaper while cables made of scarce metals such as copper are becoming more expensive.

Assignments

Remember

a) The laws of reflection.
b) That rays bend towards the normal as they enter an optically denser medium, like glass.
c) The meanings of refractive index, critical angle and total internal reflection.
d) The conditions for total internal reflection to occur.
e) The formulas:

$$\text{refractive index} = \frac{\sin i}{\sin r} = \frac{c \text{ in vacuum}}{c \text{ in medium}} = \frac{\text{real depth}}{\text{apparent depth}}$$

$$\sin c = \frac{1}{n} \quad \text{and} \quad {}_2n_1 = \frac{1}{{}_1n_2}$$

Draw ray diagrams

f) Draw a ray diagram of a periscope using two right-angled prisms. (Replace the plane mirrors in the diagram on p 9.)
g) Have a careful look at the rear reflector on a bicycle or car and explain how it works.
h) Sit at a table and place a coin in the bottom of a cup. Move the cup so that you just cannot see the coin inside. Ask a friend to pour water into the cup without moving the cup. Draw à ray diagram to explain how the coin becomes visible to you. (The ray diagram is similar to fig. 2.5b.)

Try questons 2.1 to 2.7

2.2
COLOUR

The colours of things we see depend on the colours of the light which reaches our eyes from them. Without light nothing has colour. Learning about the physics of colour helps us to understand and appreciate the colourful world we live in.

Newton's experiment

Newton observed that a ray of white sunlight was split into a spectrum of colours as it passed through a glass prism. We can reproduce that original experiment using a ray box as the source of a white light ray, fig. 2.9.

- *Direct the light ray through the prism onto a white screen some distance away in a darkened room.*
- *Where do you find the spectrum?*
- *Which colours can you see in the spectrum?*
- *Which colour has been deviated most and which least?*

Deviation and dispersion

Both these words apply to what happens when white light is split up into a spectrum of colours by a prism, but they have different meanings.

Deviation is the change in direction of a ray of light produced by the prism. Each colour of light has a slightly different deviation which is caused by the refractive index of glass being slightly different for each colour. Violet light has the greatest deviation, and the refractive index for violet light is slightly greater than for the next colour in the spectrum, indigo.

Dispersion is the separation of white light into its constituent colours, that is to say the colours of which it is made. The prism does not add colours to the light, but only separates or disperses all the colours which already exist in white light. White light is not a single colour; it is a combination of all the colours in the spectrum.

The colours of the spectrum of white light

The colours of the spectrum of white light are those seen in a rainbow. They are usually named in order as: red, orange, yellow, green, blue, indigo and violet. These colours gradually change from one to the next and there are no boundaries between the colours. We describe the spectrum of white light as continuous, which means that there is a complete range of colours from the red end to the violet end of the spectrum with no gaps or breaks. Each colour of light has a different frequency and wavelength and the light spectrum is only a small part of a much larger spectrum of radiation called the electromagnetic spectrum (see chapter 23).

Producing a pure spectrum

The colours in the spectrum produced by a prism alone will not be pure because each colour is blurred on the screen and overlaps its neighbouring colours. A pure spectrum is one in which light of one colour only forms each part of the image or spectrum on a screen. To produce a pure spectrum the following conditions should be arranged:

a) the source of white light must be restricted by a very narrow slit,

b) the white light should arrive at the prism in a parallel beam,

c) the parallel beams of different colours of light emerging from the prism should be focused by a lens onto a screen so that they do not overlap.

Fig. 2.10 shows an arrangement for producing a pure spectrum. The first converging lens is used to produce a parallel beam of white light. This is done by making the distance from the slit to the lens equal to its focal length (p 41). The second converging lens reverses the effect of the first so that the parallel beam of each separate colour is focused onto a screen. The combined effect of the two lenses and the prism is to produce a focused image of the slit on the screen in slightly different positions for every colour in the spectrum.

Figure 2.9 *Newton's experiment*

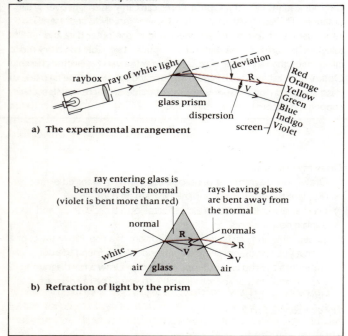

a) **The experimental arrangement**

b) **Refraction of light by the prism**

Figure 2.10 *Producing a pure spectrum of white light*

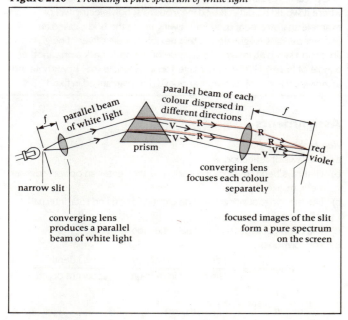

Recombining the spectrum

White light can be separated into a spectrum of different colours of light; it can also be synthesised (meaning put together) from the separate colours. Two methods may be tried. These are shown in fig. 2.11.

Figure 2.11 *Recombining the colours of the spectrum*

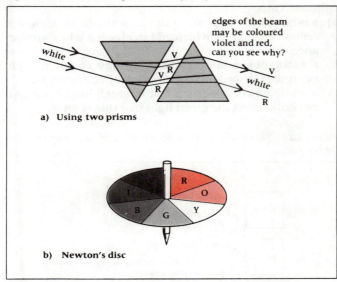

a) Using two prisms

b) Newton's disc

a) The second prism reverses the deviation and dispersion of the first prism so that the colours of the spectrum recombine where the beams overlap. The edges of the emerging beam may be tinged with violet and red.

b) Newton's disc is a card coloured with all the colours of the spectrum in equal areas. If the disc is spun round rapidly the persistence of vision of the human eye remembers all the colours it sees as the colours change places. The colours of the spectrum are added together and appear as nearly white. The off-white colour is due to the imperfect reflection of colours from the disc.

Sums with colours

The human eye has three types of light sensitive cells or colour detectors called cones. Each type of cone detects part of the spectrum of white light as shown in the simplified plan of fig. 2.12.

Figure 2.12 *The three types of cones detect overlapping zones of the spectrum*

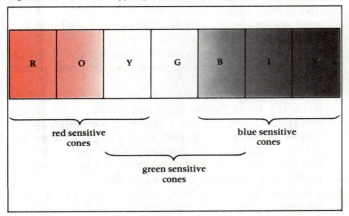

red sensitive cones

green sensitive cones

blue sensitive cones

There is overlapping of the zones of the spectrum detected by the three types of cones. If the eye receives red light, only the red sensitive cones will be stimulated, but yellow light will stimulate both red and green sensitive cones. Although the eye has only three different colour detectors, it can distinguish all the colours of the spectrum from the variety of combinations of responses produced by the overlapping of the zones.

Addition of colours of light

With the eye having three types of cones, or colour detectors, it is not surprising to find that three colours can be used to simulate the other colours in the spectrum. Red (R), green (G) and blue (B) light are called the primary colours of light.

The three **primary colours** of light are those which cannot be made by adding (or mixing) any other colours of light together.

The **secondary colours** of light are made by adding two primary colours together. They are: yellow (Y), magenta (M) and cyan (C) (sometimes also called peacock blue). Table 2.4 shows how the secondary colours are formed.

Table 2.4

Primary colours of light added	Secondary colour made
R + G	= Y
R + B	= M
B + G	= C

Light of the secondary colour yellow (made of red and green) is indistinguishable to the eye from the true yellow light in the spectrum of white light. This is because the red and green sensitive cones in the eye are equally stimulated by a mixture of red and green light or pure yellow light. A prism would distinguish between the two kinds of yellow light by separating the red and green parts of the secondary yellow light. True yellow light is a single colour and cannot be separated.

A combination of all three primary colours will stimulate the three types of cone in the eye in exactly the same way as a continuous spectrum of all the colours in white light. Thus when we see the three primary colours mixed together in equal intensities they give the impression of white light. We can now do some more sums with light colours, as shown in table 2.5. In each case all three primary colours are present, but two may be combined in a secondary colour. The pairs of colours above are called complementary colours.

Table 2.5

Colours of light mixed	Colour made
R + G + B	white
Y + B	white
R + C	white
G + M	white

Complementary colours of light are a primary colour and a secondary colour which when mixed together make white light. There are three complementary pairs and they are given in the table above.

A demonstration of colour addition

Using three projectors or three light boxes we can demonstrate addition of the colours of light on a white screen. The colours to be added can be selected by placing a colour filter in front of each of the projectors or light boxes. It is important to have colours of equal brightness on the screen but colour filters tend to vary in density allowing different amounts of light through them. The brightness of the colours can be adjusted by setting the projectors at different distances from the screen or using rheostats to control the lamp brightness in the light boxes.

A particularly effective way of showing all the possible combinations of primary colours is shown in figure (a). The colour triangle of figure (b) is a way of summarising the results. Any corner and opposite side are a complementary pair making white light.

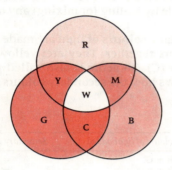

a) **The three primary colours, RBG, on a screen. Two overlapping primary colours make a secondary colour, CMY; all three make white**

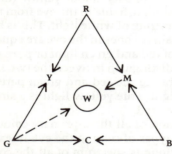

b) **The colour triangle for addition of colours of light**

Subtraction of colours of light

When light arrives at the surface of any object or medium three things happen to it in varying proportions:
a) some light is reflected,
b) some light is transmitted (it passes through the medium),
c) some light is absorbed.
The colour of an object is determined by the colours of light which it *reflects*. The colour of a filter is determined by the colours of light it *transmits* or allows through. The colours of light which are neither reflected nor transmitted must be absorbed into the medium. Light which is absorbed effectively disappears, apart from a slight rise in temperature of the absorbing material. We can think of the absorption of light as the subtraction of certain colours from the spectrum.

The colour of an object

Fig. 2.13 gives examples of how we explain what colour an object will appear in various colours of light. The pigments in dyes and paints are not pure colour reflectors and they usually reflect some of several colours. This complicates the explanations, but the conclusion we reach is that *the apparent colour of an object is the sum of the colours it is reflecting.* Thus:
a) a white object appears white in white light because it reflects all colours which add to give it a white appearance,
b) a black object appears black in any colour of light because it absorbs all colours and reflects none, but
c) a white object appears green in green light because it can only reflect the green light that shines on it.

Figure 2.13   *The colour of an object depends on the colours of light entering the eye*

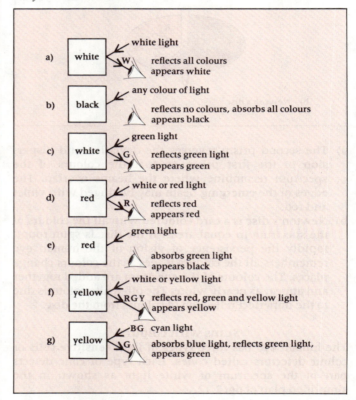

Coloured objects can appear to change colour as in the following examples:
d) a red object appears red in white or red light because it can reflect red light but:
e) a red object illuminated with green light will appear black because it absorbs the green light and reflects none.
f) a yellow object appears yellow in white or yellow light; it reflects yellow light and usually also the two primary colours, red and green, which together appear yellow. If however a yellow object is seen in any colour of light which contains some red, yellow or green light it can reflect those colours and may appear to change its own colour, for example:
g) a yellow object seen in cyan light will appear green because it absorbs the blue light and reflects the green.

Mixing paints

When paints are mixed together their colour absorbing properties are added, that is to say more colours of light are subtracted and less are reflected from the painted surface. The only colours reflected are those not absorbed by either of the paints mixed together. Thus as more colours are mixed more light is absorbed, less is reflected and the colour eventually becomes a dirty grey or black. For example:

yellow paint + blue paint = green paint.

This well-known example appears to disagree with what we have learnt about adding colours of light. The explanation is that in mixing paints we subtract more colours of light rather than add. The blue paint only reflects blue and some green and indigo light. The yellow paint reflects only the green in these colours and absorbs the others; therefore only green is reflected by a mixture of blue and yellow paint.

The colour of a filter

Fig. 2.14 gives some examples of light passing through filters of various colours. In the case of filters, very little reflection occurs and the colours we see are those that are transmitted, the missing colours being absorbed or subtracted by the material of the filter. Filters also allow small amounts of other colours through so that demonstrations of subtraction of colours by filters often do not work very well. In theory any two primary colour filters placed together should transmit no light at all as shown in fig. 2.14b.

Figure 2.14   *Colour filters*

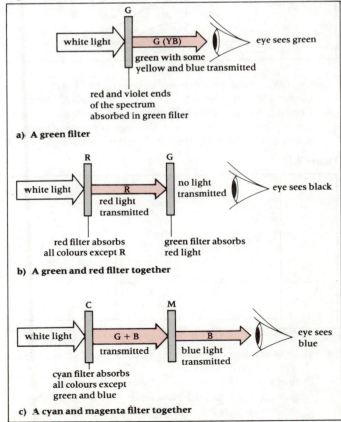

a) **A green filter**

b) **A green and red filter together**

c) **A cyan and magenta filter together**

Colour television

Because all the colours in the spectrum can be synthesised from the three primary colours and our eyes cannot tell the difference, a full colour television picture can be created using dots of only the three primary colours (R, B and G). A close examination of the screen shows it to be made up of thousands of dots of the three colours. A different phosphor is used to produce each colour on the inside of the television screen. The dots of phosphor convert the energy of the electrons striking the screen into each of the three primary colours of light (p 352).

The rainbow

To understand the rainbow we need to know about refraction, dispersion and internal reflection.

White light from the sun, entering raindrops in the sky, is refracted as it passes from air to water and again as it leaves the raindrop going from water to air. Violet light will be bent more than red and so the white sunlight is dispersed into the colours of the spectrum.

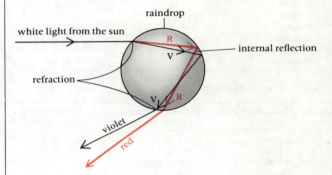

A single raindrop produces a spectrum, by refraction and internal reflection

To reach the eye of an observer the light is internally reflected inside the water drops. The observer will receive light of only one colour from a particular drop because all the other colours from that drop are dispersed in different directions. Other drops in different positions send the other colours to the eye of the observer. Thus each colour comes from drops in a different position in the sky.

The curved, bow shape is the locus (or joining together) of all the raindrops in the sky for which the angle between the rays of light from the sun and the rays entering the observer's eye are the same for each colour.

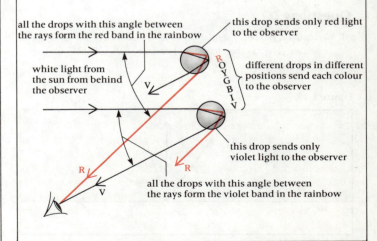

Assignments

Remember

a) The meanings of the words: deviation, dispersion and pure spectrum.

b) The difference between and relations between primary, secondary and complementary colours.

Redraw and complete

c) table 2.6,

d) table 2.7.

Assume that all secondary colours also reflect or transmit the two primary colours which add to make the secondary colour.

Table 2.6   Reflected light

Object colour in white light	Colour of light shining on object	Apparent colour of object
white	red	red
white	blue	
red	blue	
red	magenta	
yellow		red
yellow	cyan	
black	green	
magenta	blue	
magenta	cyan	
magenta		black

Table 2.7   Transmitted light

Colour of light shining on filter	Colour of filter or filters together	Colour of light transmitted
white	red	red
white	green	
white	red + green	
green	red	
green	yellow	
cyan	blue	
cyan	magenta	
cyan	red + blue	
red	red + magenta	
red	magenta + cyan	

Investigate coloured shadows

e) Use two or three light boxes or lamps each with a wide single slit and a colour filter, to produce a broad diverging beam of light of a different colour. Direct the light from each box in a different direction towards an upright opaque object standing on white paper. The shadows cast by the object on the white paper will have interesting colours. Can you explain the colour of each shadow?

Try questions 2.8 to 2.12

Questions 2

1 In fig. 2.15, PO represents a ray of light incident onto an air–glass interface. The ray then follows which path: OA, OB, OC, OD or OE?

Figure 2.15

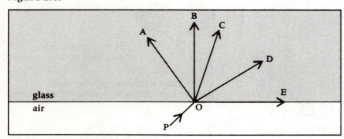

2 a) Fig. 2.16 shows a ray of light passing through a transparent material.

Figure 2.16

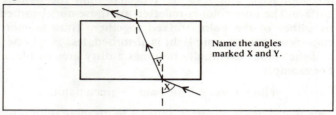

Name the angles marked X and Y.

b) Fig. 2.17a and b each show a ray of light entering and leaving a transparent plastic block.

Figure 2.17

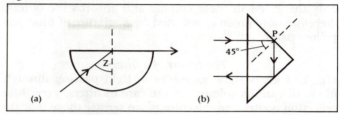

(a)   (b)

i) In fig. 2.17a what name is given to the angle Z?

ii) For fig. 2.17b, A) name the type of reflection at the point P, B) explain why this reflection took place, C) where is this effect used on a bicycle?

c) The type of reflection shown in fig. 2.17b is used to pass a beam of light along a solid glass fibre. This is shown in fig. 2.18.

Figure 2.18

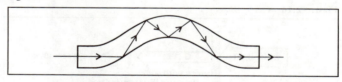

Describe a practical application of a single flexible fibre or a bundle of flexible fibres.   (NEA [A] Q1988)

3 Complete the paths of the two narrow beams of monochromatic light XY shown in fig. 2.19. The critical angle of the glass of both prisms is 42°. Explain the meaning of the term *critical angle*.   (AEB)

Figure 2.19

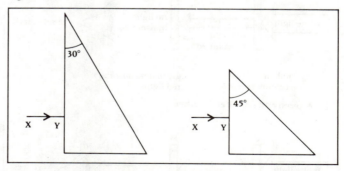

4 a) Draw ray diagrams to illustrate the meaning of (i) critical angle, (ii) total internal reflection.

b) Briefly describe an experiment by which the refractive index of glass could be measured. How would you use the refractive index to calculate the critical angle for a glass/air boundary?

c) Draw a diagram showing how a glass prism can deviate a narrow beam of light through a right angle. Show how this forms the basis for the design of a simple periscope. (JMB)

5 a) Fig. 2.20 shows a torch being held under water in such a way that a ray of light is produced which can strike the surface of the water at different angles *i*. Calculate a value for the critical angle; the refractive index of water is $\frac{4}{3}$.

Figure 2.20

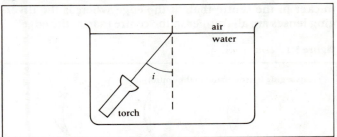

Draw sketches to show what will happen to the ray of light after it has struck the water–air boundary when *i* is about 20° and when *i* is about 60°. Account for the different behaviour of the ray in the two cases.

Why in one case has the resulting ray coloured edges whereas in the other case it has not?

b) In an experiment to determine the refractive index of water a black line is painted on the bottom of a tall glass container which is then partially filled with water. On looking vertically down into the water the black line appears to be closer than it really is. Explain, with the help of a ray diagram, why this is so.

The following results were obtained in such an experiment.

Real depth/cm	8.1	12.0	16.0	20.0
Apparent depth/cm	5.9	9.1	12.0	15.1

Plot a graph of real depth (*y*-axis) against apparent depth (*x*-axis) and hence determine a value for the refractive index of water. (L)

6 Fig. 2.21a represents a ray of light falling normally on the curved face of a semicircular plastic block at X, meeting the opposite face at an angle of incidence of 30° at O, and emerging into the air at an angle of 40°.

a) Explain what happens to the ray at X, between X and O, and at O. Why has the emerging ray been refracted?

b) Calculate the refractive index of the plastic.

c) Describe how the apparatus could be used to find the critical angle experimentally. Calculate its value for this plastic.

d) Glass prisms are sometimes used to turn light through 90° as shown in fig. 2.21b. Would prisms of this plastic be able to do the same? Explain. (O)

Figure 2.21

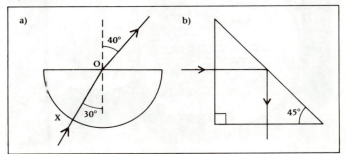

7 a) Draw a labelled diagram to show the subsequent path of a narrow beam of light which is incident in air at about 45° on one face of a rectangular glass block. Use your diagram to explain the meaning of the term 'refractive index'.

If the refractive index for glass is 1.50, calculate the speed of light in glass. (Speed of light in air $= 3.0 \times 10^8$ m s^{-1}.) Calculate also the critical angle for a glass/air interface and draw a diagram to show the meaning of this angle.

b) Show how a suitable glass prism may be used to deviate a ray of light through 90°. Why is this method preferable to using a plane glass mirror for the same purpose? (S)

8 a) Sketch a ray diagram illustrating how a single ray of white light is refracted and dispersed as it passes through a 60° triangular glass prism.

b) Sketch a ray diagram showing how you would produce a pure spectrum of white light on a screen. Explain the purpose of any additional optical apparatus besides the prism. Explain why the spectrum is described as *pure*. (JMB)

9 Draw ray diagrams to show how (i) a narrow beam of red light, (ii) a narrow beam of white light, is refracted through a 60° glass prism. What can you deduce from the diagrams about the refractive index of glass? (JMB part)

10 a) State what is meant by *the refractive index of a solid*. Explain why white light is dispersed (split into its constituent colours) when it passes through a glass prism.

Draw a labelled diagram showing how you would produce a *pure* spectrum of white light on a screen. Your diagram should show the paths of rays of light of at least two different colours.

b) Explain the following.

i) When a blue book is viewed in pure yellow light it appears to be black.

ii) When a pure blue filter is placed behind a pure yellow filter and both are put into the path of a beam of white light, no light passes through.

iii) When blue and yellow paints are mixed, the resulting colour is green. (L)

11 a) Describe an experiment which checks that light passing through a red filter is a primary colour, but that light passing through a magenta filter is a secondary colour.

b) Explain the effect you would expect to see if the red and magenta filters were used simultaneously, completely overlapping, in the experiment you have described. (JMB)

12 Fig. 2.22 shows a piece of white card with red and black discs A and B fixed on it. By completing the table below, state how the discs and card appear when viewed through the filters stated. (AEB)

Figure 2.22

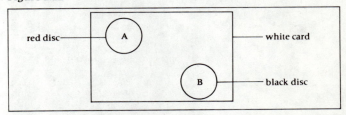

Filter	Appearance of		
	disc A	disc B	white card
transmits red light only			
transmits green light only			

Lenses and optical instruments

In this chapter we shall study the properties of lenses and some of their applications in optical instruments. The action of lenses, like prisms, depends on the refraction of light.

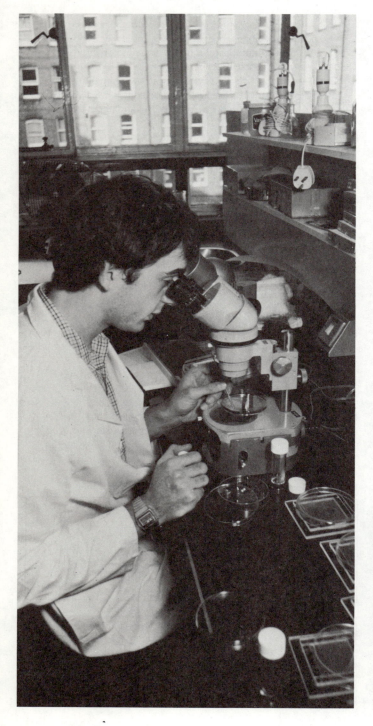

3.1
LENSES

Lenses are made with a great variety of shapes and of different kinds of glass but they all belong to either the **converging** group or the **diverging** group of lenses. The converging type of lens converges (brings together) rays of light and the diverging type diverges (spreads out) rays of light.

As can be seen in fig. 3.1, the converging lenses are all thicker in the centre than at the edge whereas the diverging lenses are all thinner in the centre than at the edge.

Figure 3.1 *Lens shapes*

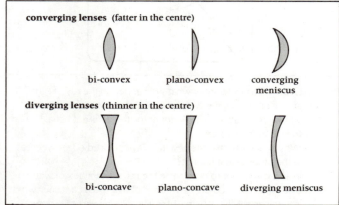

A meniscus lens has both a concave and a convex surface and so, to avoid confusion, we shall refer to lenses by their converging or diverging property rather than their shape. Converging lenses which are bi-convex (i.e. have both surfaces convex) are often called convex lenses. But notice also that confusion can arise because a concave mirror converges rays of light but a bi-concave lens diverges rays of light.

Investigating the properties of lenses

First we shall compare the magnifying properties of some lenses with their shapes.

- *Hold a lens of each shape in your hand over a page of print and observe what happens.*
- *Borrow some spectacles and hold them over the print and decide what type of lens is in the spectacles.*
- *Draw a summary table which gives the name of each lens examined, draw its shape and state its magnifying properties.*
- *Which type of lens is usually called a magnifying glass?*

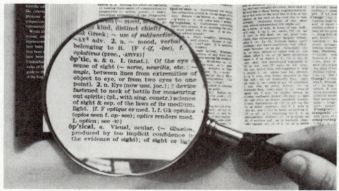

The magnifying glass is a reading aid for people with poor eyesight.

Now compare the converging and diverging properties of different lenses.

● *Set up an experiment in a darkened room as shown in fig. 3.2a.*

● *Use a light box or lamp in a holder to produce multiple rays by passing light through a multiple slit.*

Cylindrical lenses, as shown in fig. 3.2b, are the most suitable type to use because they have the same cross-sectional shape down to paper level on the bench, and they will stand on their flat edge.

● *Mark on the sheet of paper the paths of the rays before and after passing through each lens.*

● *Try some combinations of two lenses in close contact.*

Figure 3.2 *The action of a lens*

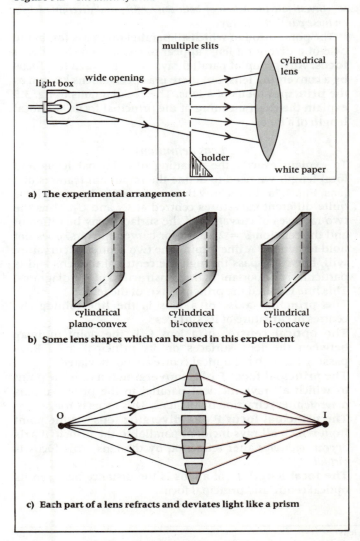

a) **The experimental arrangement**

cylindrical
plano-convex   cylindrical
bi-convex   cylindrical
bi-concave

b) **Some lens shapes which can be used in this experiment**

c) **Each part of a lens refracts and deviates light like a prism**

The lens as a group of prisms

● *Compare the ray paths you have obtained for each type of lens.*

Can you explain the action of the lenses in terms of refraction of light rays at the surfaces of the lens? It helps to think of each lens as being made from several prisms, fig. 3.2c. Each prism produces a different amount of deviation. This diagram gives an explanation of the converging action of a bi-convex lens by considering it as made up of several prisms.

● *Try drawing a similar diagram to explain the action of a bi-concave lens based on the results obtained in your experiments.*

As a ray of light enters each prism it is bent towards the normal and as it leaves, away from the normal. The larger angle prisms at the edges of a lens deviate the light rays most, while the flat piece of glass in the centre of the lens produces no deviation at all.

How does a lens form an image?

We do not normally see light rays unless there is something in the air to make them visible. We can see the path of a light ray in fog or smoke; particles of dust in the air can show the rays of sunlight streaming into a room through the window. The light rays become visible when small amounts of light are reflected or scattered from the small particles of dust or water as they drift through the path of the light. Thus what we actually see is the small particles brightly illuminated by the beam of light. Light rays travel through air whether we can see them or not.

We can use a large box filled with smoke particles to make the paths of light rays visible, to help explain how a lens forms a real image, fig. 3.3 The box has a transparent front and sides (made of glass or perspex) and a means of blowing smoke into it.

Figure 3.3 *The smoke box*

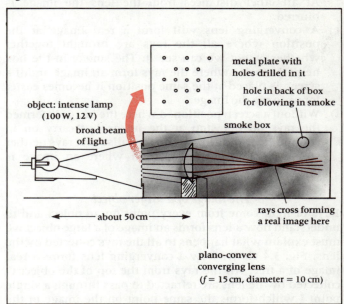

metal plate with holes drilled in it

hole in back of box for blowing in smoke

object: intense lamp (100 W, 12 V)

broad beam of light

smoke box

plano-convex converging lens
(f = 15 cm, diameter = 10 cm)

rays cross forming a real image here

about 50 cm

An intense light source, which sends out light rays in a broad beam through a circular hole in its housing, is used as an object. A few rays can be selected by placing a metal sheet with holes drilled in it at one end of the smoke box. Rays can only enter the box through these holes. A number of rays reach a large converging lens mounted inside the box.

The action of the lens on these rays can now be investigated. The rays are all brought together and pass through a single point in the air on the other side of the lens. There is only one position where this happens and this is where the real image is formed by the lens. An image is a copy of an object which is formed by a lens or mirror as a result of its action on rays of light coming from the object.

- *Try moving the converging lens to different positions inside the smoke box.*
- *Bring the intense light source (the object), right up to the end of the smoke box and then move the converging lens gradually nearer to the object. What happens to the image?*
- *With the object back in its original position, remove the metal plate with holes in it so that all light rays enter the smoke box as a broad beam. What do you see?*
- *Insert a white screen in the box at the position where the rays cross. What do you see?*
- *Reassemble the arrangement of object, lens and screen outside the smoke box and investigate where the image is formed by moving the screen to various distances from the lens.*

Some observations

a) The distance from the object to the lens affects the position of the image. We can see that the rays cross at a different place in the smoke box.

b) When the object gets very close to the lens the rays cannot be brought together by the lens and no real image can be formed.

c) Removing the metal plate shows that all the rays in the beam which pass through the lens cross at the same place.

d) A sharp image of the object can only be formed on a screen at one position, which is where the rays cross. At all other distances from the lens the image is blurred.

e) A converging lens will form a real image at the position where all the rays are brought together whether or not we can see it. The smoke in the box helps us to see where the rays form an image and if a screen is placed at the same position it becomes easier for us to see the image.

f) Without a screen positioned where the image is formed the rays do not stop at the image but carry on in straight lines. It is important to realise that rays neither stop nor bend at the position where an image is formed.

The image of a larger object

Rays of light come from every point on an object and to understand how a lens forms an image of a large object we must explain what happens to all the rays collected by the lens. Fig. 3.4 shows how a converging lens forms a real image of a tree. All the rays from the top of the object O collected by the lens are refracted to pass through a single point I which forms the same point on the image of the tree. Similarly, all the rays from the bottom of the tree at P are brought together again at J.

This diagram only shows the rays from the top and bottom of the tree, but the same happens to all the rays collected by the lens from every point on the tree. Thus for every different point on an object there is a corresponding different point on the image where the lens brings all the rays together again.

The action of a lens on parallel light rays

Arrange the smoke box apparatus shown in fig. 3.2 to produce parallel light rays as follows.

- *Place an additional converging cylindrical lens in front of the opening of the light box.*
- *Adjust the distance between the lamp in the light box and this additional lens until the broad beam of light emerges parallel and the rays produced by the multiple slits are also parallel.*
- *Investigate the action of both converging and diverging lenses on these parallel light rays.*

The point through which all parallel rays pass (or, in the case of a diverging lens, appear to come from) is called a **focus**. Any group of parallel rays are converged to a focus by a converging lens, but there is a special focus known as the **principal focus** of a lens. The ray diagrams of fig. 3.5 explain the exact meaning of the principal focus and focal length of a lens.

Lens definitions

To explain exactly the meaning of principal focus and focal length we need to define the important features of a lens. Fig. 3.5a shows how the two faces of a lens may have quite different curvatures centred at C_1 and C_2. Thus the two radiuses of curvature of the surfaces may be different and the relation $R = 2f$, used for curved mirrors, does not hold for lenses. A line joining the two centres of curvature will, however, pass through the centre of the lens and is particularly important in the construction of ray diagrams. This line is called the principal axis of the lens.

The principal axis of a lens is the line joining the centres of curvature of its surfaces.

The optical centre of a lens L is the point midway between the lens surfaces on its principal axis. Rays passing through the optical centre are not deviated.

The principal focus F of a converging lens is the point to which all rays incident parallel to the principal axis *converge* after refraction by the lens. This focus is *real*.

The principal focus F of a diverging lens is the point from which all rays incident parallel to the principal axis *appear to diverge* after refraction by the lens. This focus is *virtual*.

The focal length *f* of a lens is the distance between its optical centre and principal focus.

Figure 3.4 *How a lens forms an image of a large object*

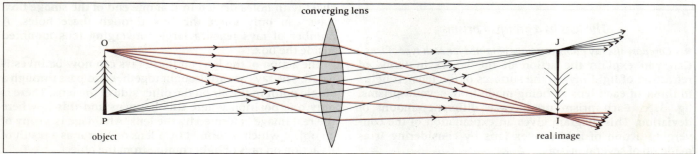

converging lens

object

real image

Figure 3.5 *Properties of lenses*

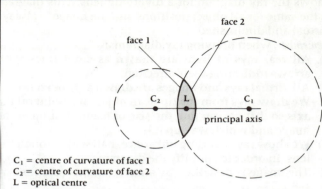

C_1 = centre of curvature of face 1
C_2 = centre of curvature of face 2
L = optical centre

a) The principal axis and optical centre, L

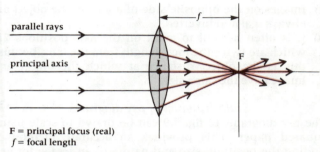

F = principal focus (real)
f = focal length

b) The principal focus F, and focal length f of a converging lens

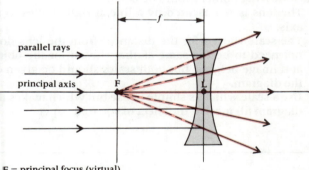

F = principal focus (virtual)
f = focal length

c) The principal focus F, and focal length f of a diverging lens

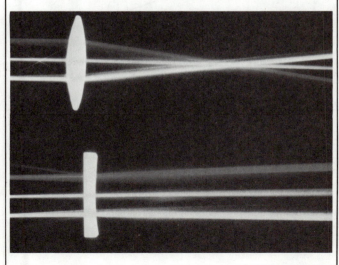

Two principal focuses

A lens has two principal focuses, one on each side of the lens. Thin lenses and those with the same curvature for both faces have equal focal lengths for light passing in opposite directions. When drawing ray diagrams the principal focus F is marked on both sides of the principal axis.

Measuring the focal length of a converging lens

A rough method

We sometimes need a quick method of finding the focal length of a lens. For example we may need to sort out lenses of different focal lengths which have become mixed up. As we saw in fig. 1.19, as an object is moved away from a lens or mirror the light rays collected from the object become almost parallel to each other. So the image of a distant object formed by a converging lens will be roughly at the principal focus of the lens.

• *Using the window frame at the far end of a room (or better still something like a building visible through the window) as a distant object, obtain a sharp image on the wall, as shown in fig. 3.6.*

Figure 3.6 *A rough method of measuring the focal length of a converging lens*

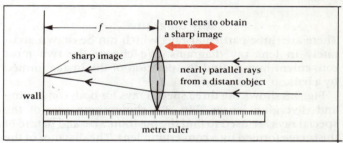

• *Measure the distance from the lens to the image to obtain a rough value of the focal length of the lens.*
Note that the image is real, inverted and diminished.

An accurate method

• *Draw an object (cross-wires) on a piece of translucent paper and fit it over a circular hole in a white screen.*
When placed in front of a light box this will form an illuminated object as shown in fig. 3.7.

Figure 3.7 *Measuring the focal length of a converging lens using a plane mirror and an illuminated object*

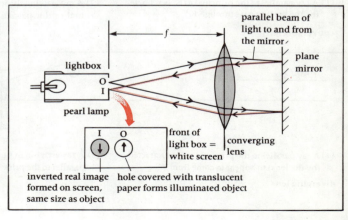

● *Mount a converging lens and plane mirror in holders so that light passing through the lens will be sent back by the mirror along the same path. Arrange the lens and mirror facing the illuminated object.*

● *Move the lens until a sharp image of the object is formed alongside the illuminated object. The image will be sharpest when it is exactly the same size as the object.*

● *Measure the distance between the optical centre of the lens and the front of the light box.*

Rays of light from the object are converged by the lens into a parallel beam of light. The plane mirror returns the parallel beam, passing it back through the lens again where it is converged to form the image. Since parallel rays converge to the principal focus, the distance between the optical centre of the lens and the front of the light box is the focal length of the lens.

An alternative to the light box and illuminated object is to use a pin in a holder as the object. For the same reasons as above, the pin will coincide with an image of itself only when placed at the principal focus of the lens. The pin and its image are made to coincide by adjusting the pin position until no-parallax occurs between pin and image. This method was described in detail for a curved mirror on p 18.

Ray diagrams

There are three particular rays which can be drawn accurately in lens ray diagrams. We choose the two most convenient rays to find the position of the image formed by a lens.

Fig. 3.8 shows the three special rays for both converging and diverging lenses. Fig. 3.9 shows how two of the special rays are used to find the position, size and nature of the image formed by a converging lens. The distance of the object from the lens is critical in deciding the nature, size and position of the image and for this reason it is vital to mark clearly the position of the principal focus of a lens. When the object is nearer to a converging lens than its

principal focus it can only form a virtual image. Fig. 3.10 shows the ray diagram for a diverging lens. This diagram is the same for all object positions and the image is always virtual and diminished.

Remember, when drawing ray diagrams:

a) All real rays of light are drawn as solid lines with arrows indicating their direction.

b) All virtual rays and images are drawn as broken lines.

c) We draw rays from the tip of an object at a point off the axis so that we can find the size or magnification of the image and which way up it is.

d) We show rays bending only once, half way through the lens, in order to simplify the diagram.

e) The image is described by answering the questions on p 8.

f) Images on the same side of a lens as the object are always virtual and erect.

g) Images on the opposite side of a lens to the object are always real and inverted.

h) It is often helpful to mark on the axis positions 2F, which are twice the focal length from the lens, because this is the object distance at which the object and image have the same size.

Scale drawings of ray diagrams

The ray diagrams of fig. 3.9 can be drawn to scale using squared paper. This provides an accurate method of finding the position, size and nature of an image formed by a lens and may be used to answer some lens questions. The following points should be noted:

a) The lens is represented by a line at right angles to its axis.

b) The scale chosen for the distances from the lens does not need to be the same as that chosen for the object and image heights, but both scales should be given on the diagram.

c) Use a scale which is simple to use but which makes the diagram fill most of the sheet of squared paper.

Figure 3.8 *Three special rays used in lens ray diagrams*

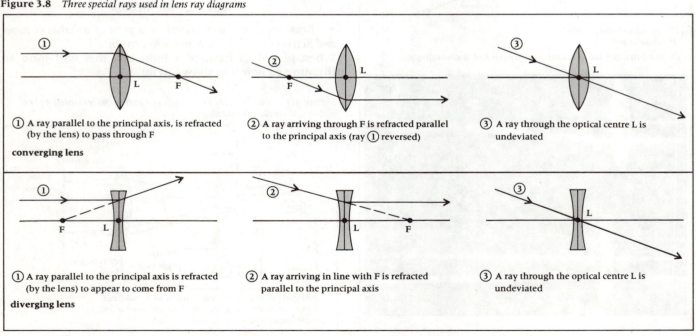

① A ray parallel to the principal axis, is refracted (by the lens) to pass through F

② A ray arriving through F is refracted parallel to the principal axis (ray ① reversed)

③ A ray through the optical centre L is undeviated

converging lens

① A ray parallel to the principal axis is refracted (by the lens) to appear to come from F

② A ray arriving in line with F is refracted parallel to the principal axis

③ A ray through the optical centre L is undeviated

diverging lens

Figure 3.9 *Images formed by converging lenses*

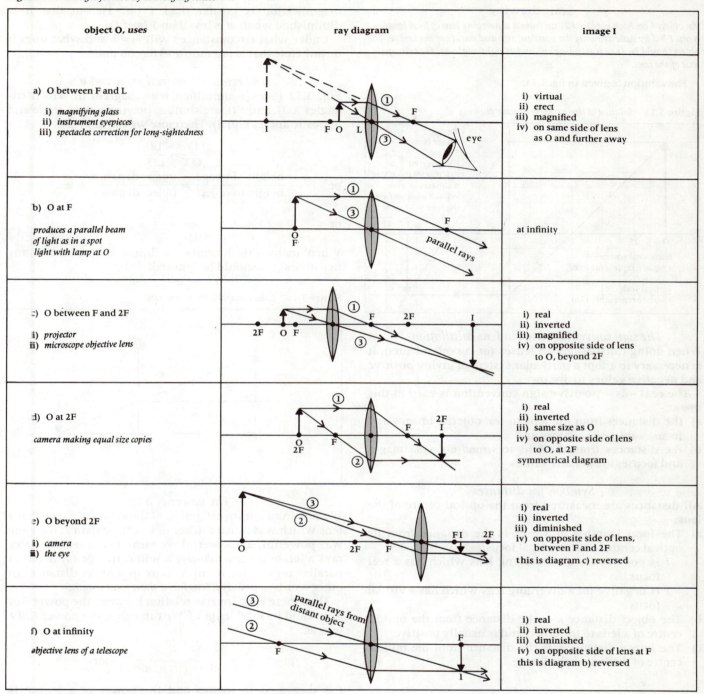

object O, *uses*	ray diagram	image I
a) O between F and L i) *magnifying glass* ii) *instrument eyepieces* iii) *spectacles correction for long-sightedness*		i) virtual ii) erect iii) magnified iv) on same side of lens as O and further away
b) O at F *produces a parallel beam of light as in a spot light with lamp at O*		at infinity
c) O between F and 2F i) *projector* ii) *microscope objective lens*		i) real ii) inverted iii) magnified iv) on opposite side of lens to O, beyond 2F
d) O at 2F *camera making equal size copies*		i) real ii) inverted iii) same size as O iv) on opposite side of lens to O, at 2F symmetrical diagram
e) O beyond 2F i) *camera* ii) *the eye*		i) real ii) inverted iii) diminished iv) on opposite side of lens, between F and 2F this is diagram c) reversed
f) O at infinity *objective lens of a telescope*		i) real ii) inverted iii) diminished iv) on opposite side of lens at F this is diagram b) reversed

Figure 3.10 *Image formed by a diverging lens*

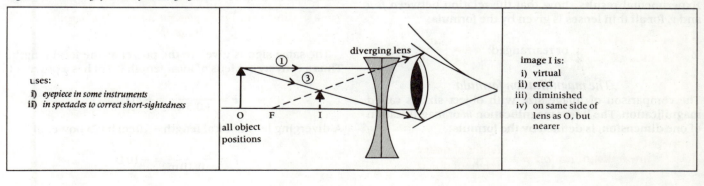

uses: i) *eyepiece in some instruments* ii) *in spectacles to correct short-sightedness*		image I is: i) virtual ii) erect iii) diminished iv) on same side of lens as O, but nearer

Worked Example
Scale drawing

An object 6 cm high is placed 20 cm from a converging lens of focal length 8 cm. Find by scale drawing the position, size and nature of the image. The object should be drawn standing on, and at right angles to, the principal axis of the lens.

The solution is given in fig. 3.11.

Figure 3.11 *Solution of lens problem by scale drawing*

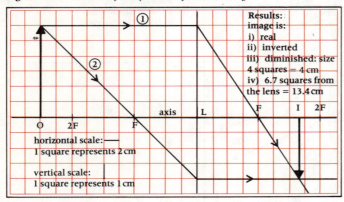

The sign convention used in lens calculations

When doing calculations on lenses (or curved mirrors), it is necessary to adopt a particular system of giving positive and negative values to distances.

The **real – is – positive** sign convention is used in this book:

a) the distances from the lens to *real* objects, images and focuses are *positive* values and

b) the distances from the lens to *virtual* objects, images and focuses are *negative* values.

Symbols for distances

All distances are measured from the optical centre of the lens:

a) The focal length *f* of a lens is the distance from its optical centre to its principal focus:

   f is positive for a converging lens which has a real focus,

   f is negative for a diverging lens which has a virtual focus.

b) The object distance *u* is the distance from the optical centre of a lens to the object and is usually positive.

c) The image distance *v* is the distance from the optical centre of the lens to the image.

The lens formula

Experimental results show that the relation between *f*, *u* and *v*, for all thin lenses is given by the formula:

$$\frac{1}{u} + \frac{1}{v} = \frac{1}{f} \quad \text{or rearranged:} \quad f = \frac{uv}{u + v}$$

The magnification formula

The comparison of image size with object size is called magnification. The linear magnification *m* or magnification of one dimension, is defined by the formula:

$$\text{linear magnification } m = \frac{\text{height of image}}{\text{height of object}}$$

We say that an image is magnified when the linear magnification is greater than 1 ($m > 1$), but that it is diminished when *m* is less than 1 ($m < 1$).

Under what circumstances will $m = 1$ and what does it mean? Note that *m* is a ratio and has no units.

The relation between *m*, *u* and *v*

Fig. 3.12 gives a simplified lens diagram in which triangles XOL and YIL are similar, (having a right angle and opposite angles equal). Therefore we can say that:

$$\frac{IY}{OX} = \frac{LI}{LO}$$

or

$$\frac{\text{height of image}}{\text{height of object}} = \frac{\text{image distance}}{\text{object distance}} = m$$

or

$$m = \frac{v}{u}$$

When using this formula for linear magnification any negative signs should be ignored.

Figure 3.12 *Linear magnification m = v/u*

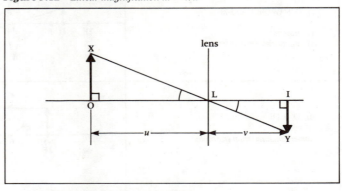

The power of a lens

We saw from our investigation of the action of a lens that a lens which was much thicker in its centre than at the edge was powerful. By powerful we mean that a lens deviates rays a lot, or more precisely, it will converge (or diverge) parallel rays to (or from) a focus in a short distance. In other words: *a powerful lens has a short focal length.*

Thus there is an inverse relation between the power *F* of a lens and its focal length *f*. We can define the power *F* of a lens as:

$$\text{power} = \frac{1}{\text{focal length in metres}} \quad \text{or} \quad F = \frac{1}{f}$$

f is measured in metres and the power of a lens *F* is measured in dioptres, symbol D.

A dioptre (D) *is the power of a lens of focal length one metre*

The same sign is given to the power as the focal length. Thus a converging lens of focal length 0.5 m has a power of

$$F = \frac{1}{+0.5\,\text{m}} = +2\,\text{D}$$

A diverging lens of focal length −10 cm has a power of

$$F = \frac{1}{-0.10\,\text{m}} = -10\,\text{D}$$

Lenses in contact, measuring the focal length of a diverging lens

It can be shown experimentally that when thin lenses are placed together in contact their powers can be added. Thus a converging lens of power +10D in contact with a diverging lens of power −7D will act as a single lens of power +3D. This simple relation provides a method of measuring the focal length of diverging lenses. A converging lens of known greater power is placed in contact with the diverging lens of unknown power. The focal length of the converging combination of lenses is then found by any of the methods we have described. This gives the power of the combination and hence the power and focal length of the diverging lens.

Measuring the focal length of a converging lens using the lens formula

This experiment is sometimes used to establish the lens formula. However, using the lens formula we can calculate the focal length of a lens from measurements of the object and image distances, u and v. The most understandable form of the experiment uses an illuminated object in a darkened room to form a real image on a screen as shown in fig. 3.13. The arrangement is the same as in a projector. The illuminated object takes the place of the film or slide in the projector. The image on the screen is real and inverted but its size depends on the distances u and v as we have learnt from the magnification formula. It is probably simplest to obtain a set of results for u and v as follows.

Figure 3.13 *Measuring the focal length of a converging lens using the lens formula*

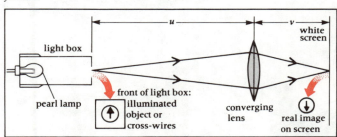

• *Keep the illuminated object in a fixed position at one end of a bench or table and then for various positions of the screen move the lens until the image is sharp on the screen.*

• *Decide on the best position for the lens by finding the point midway between the positions where you can tell that the image is just going out of focus.*

There are two lens positions for each screen position. Do you notice anything about the values of u and v for these two positions?

• *Tabulate your results as shown in table 3.1.*

• *Find values of $1/u$, $1/v$ and $1/u + 1/v$ to two significant figures using a calculator or tables of reciprocals. Average the results for $1/u + 1/v$ to give a mean value of $1/f$.*

• *Now find the focal length f from the reciprocal of this mean value.*

Table 3.1

u/cm	v/cm	$\dfrac{1}{u}\Big/\dfrac{1}{\text{cm}}$	$\dfrac{1}{v}\Big/\dfrac{1}{\text{cm}}$	$\dfrac{1}{u} + \dfrac{1}{v} = \dfrac{1}{f}\Big/\dfrac{1}{\text{cm}}$
60	40	0.017	0.025	0.042

Worked Example
The lens formula and magnification

An object is placed in front of a converging lens of focal length 12 cm. Find the nature, position and magnification of the image when the object distance is (a) 16 cm and (b) 8 cm.

a)
$$u = +16\,\text{cm (real object)}$$
$$f = +12\,\text{cm (converging lens)}$$

using: $\dfrac{1}{u} + \dfrac{1}{v} = \dfrac{1}{f}$ and rearranging, we have:

$$\frac{1}{v} = \frac{1}{f} - \frac{1}{u} = \frac{1}{12\,\text{cm}} - \frac{1}{16\,\text{cm}} = \frac{4-3}{48\,\text{cm}} = \frac{-1}{48\,\text{cm}}$$

therefore $\qquad v = +48\,\text{cm}$

now using: $\qquad m = \dfrac{v}{u}$

we have: $\qquad m = \dfrac{48\,\text{cm}}{16\,\text{cm}} = 3$

Answer: The image is real (v is positive) and inverted, 48 cm from the lens on the opposite side to the object and magnified 3×.

b)
$$u = +8\,\text{cm (real object)}$$
$$f = +12\,\text{cm (converging lens)}$$

using: $\dfrac{1}{v} = \dfrac{1}{f} - \dfrac{1}{u}$ as above, we have:

$$\frac{1}{v} = \frac{1}{12\,\text{cm}} - \frac{1}{8\,\text{cm}} = \frac{2-3}{24\,\text{cm}} = \frac{-1}{24\,\text{cm}}$$

therefore $\qquad v = -24\,\text{cm}$

now using: $\qquad m = \dfrac{v}{u}$

we have: $\qquad m = \dfrac{24\,\text{cm}}{8\,\text{cm}} = 3$

Answer: This image is also magnified 3× but it is virtual (v is negative), erect and on the same side of the lens as the object, 24 cm from the lens.

Worked Example
The power of a diverging lens

A diverging lens of power −2.0 dioptres is used in a pair of spectacles to correct the sight of a short-sighted person. Calculate (a) its focal length and (b) the position and nature of the image it forms of an object 2.0 metres away from it.

a) $F = -2.0\,\text{D}$. Since $F = 1/f$ then $f = 1/F$, therefore

$$f = \frac{1}{F} = \frac{1}{-2.0\,\text{D}} = -0.5\,\text{m}$$

Answer: The focal length of the lens is −0.5 m or −50 cm and it is diverging.

b) $f = -0.5\,\text{m}$ (diverging lens), $u = +2.0\,\text{m}$ (real object).

Using $\dfrac{1}{v} = \dfrac{1}{f} - \dfrac{1}{u}$ we have:

$$\frac{1}{v} = \frac{1}{-0.5\,\text{m}} - \frac{1}{2.0\,\text{m}} = \frac{-4-1}{2.0\,\text{m}} = \frac{-5}{2.0\,\text{m}}$$

$$\text{therefore } v = -\frac{2.0\,\text{m}}{5} = -0.4\,\text{m}$$

Answer: The negative value of v indicates that a virtual image is formed 0.4 m from the lens on the same side of the lens as the object. Thus the eye sees the virtual image by looking through the lens.

Assignments

Remember

a) The formulas:

$$\frac{1}{u} + \frac{1}{v} = \frac{1}{f}, \quad f = \frac{uv}{u+v}, \quad F = \frac{1}{f}$$

b) The magnification formulas:

$$m = \frac{\text{height of image}}{\text{height of object}} = \frac{v}{u}$$

c) The definitions on p 38.

d) The sign convention: real – is – positive.

Draw

e) Practise drawing the ray diagrams of figure 3.9.

f) Practise drawing the ray diagram of figure 3.10.

g) If you have used a ray box or similar arrangement containing a lens and a slit to produce a single ray of light for several of the ray experiments, draw a ray diagram showing how the ray box is adjusted to produce a narrow parallel ray of light. Name and label the type of lens and any important distances.

Try questions 3.1 to 3.9

3.2
RAYS ENTER THE EYE

When we look at an object our eyes form an image of it at the back of the eye like a camera does, but our eyes are much more than an optical instrument. They are our window on the world, and what we see and do not see through them affects our whole view of life. With our eyes we send and receive messages; in them we see love and hatred, hope and fear. By learning about the eye we realise even more what an amazing instrument it is.

The structure and action of the eye

The human eye is spherical in shape except for a slight bulge at the front. A horizontal section is shown in fig. 3.14a with the optic nerve leading to the brain on the nose side of the eye.

The outside of the eye is a white, tough and fibrous layer called the **sclera** with a transparent part at the front called the **cornea**. Inside the sclera is a layer of tissue called the **choroid**, which supplies blood to the eye and also contains black colouring which reduces reflection of light within the eye.

Figure 3.14 *The human eye*

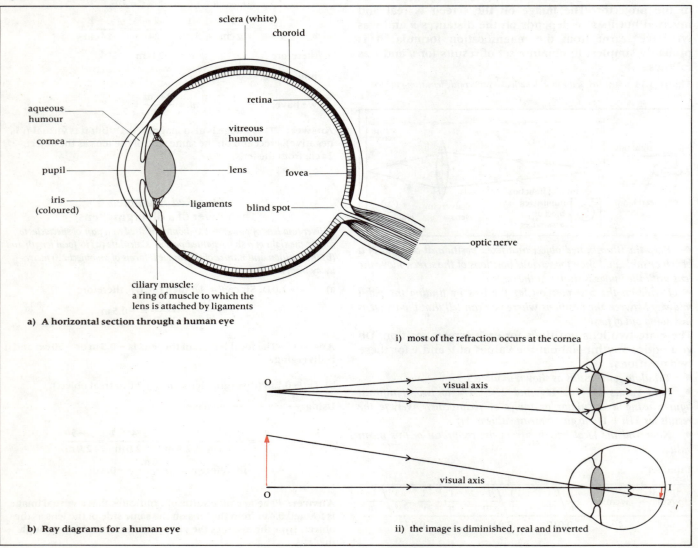

a) **A horizontal section through a human eye**

b) **Ray diagrams for a human eye**

i) most of the refraction occurs at the cornea

ii) the image is diminished, real and inverted

The retina

The retina is the inner layer of the eye which contains light sensitive cells and nerve fibres. Light falling on the retina produces chemical changes in the cells which then send electrical signals along the nerve fibres via the optic nerve to the brain.

The retina contains two types of light sensitive cells which, due to their shapes, are called **rods** and **cones**. Over the whole of the retina the majority of the cells are rods. These are sensitive to a low level of light but do not give much detail or sharpness to the image. At the centre of the retina, at a place called the **fovea**, cone shaped cells are packed closely together. Around the fovea our eyes have the best detail and colour vision. The point where the millions of nerve fibres leave the retina is a **blind spot** because it contains no light sensitive cells.

The lens and focusing

The lens is a bi-convex converging lens of a jelly-like, flexible and transparent material. The lens has a higher refractive index ($n = 1.44$) than the surrounding transparent medium ($n = 1.33$), which maintains the eye's spherical shape. The lens, together with refraction which occurs at the cornea (the front of the eye), forms a real, inverted and diminished image on the retina, fig. 3.14b.

The lens is suspended inside the eye by a circular band of ligaments. The ligaments are attached to a circular ring of muscle called the **ciliary muscle** which controls the shape of the lens. When the ciliary muscle is relaxed the lens has its longest focal length and focuses rays from distant objects onto the retina. Contraction of the ciliary muscle reduces tension in the lens making it more curved and more powerful. The shorter focal length lens now focuses images of near objects on the retina.

Accommodation is the name given to the ability of the lens of the eye to change its focal length and produce focused images of both distant and near objects on the retina.

The eye lens changes its focusing distance by changing its shape. The change in the shape of the lens inside the eye only slightly adjusts the focal length of the whole eye. As the greatest difference in refractive index occurs between the air and the cornea, it is the front of the eye which causes the majority of the bending of the light rays as they enter the eye, fig. 3.14b.

The iris and light control

The iris is the coloured ring, which has a circular hole in its centre called the **pupil**. By adjusting the size of the pupil, the iris can control the amount of light reaching the retina. When a bright object is viewed, the iris closes the pupil to reduce the amount of light entering the eye.

The closing of the iris can, at best, reduce the amount of light entering the eye by a factor of about 30. However, bright sunlight may be over 100 000 times brighter than moonlight, so the change in the size of the pupil alone is clearly not enough to cope with such extremes. The eye also changes its sensitivity to light by varying the concentrations of chemicals in the rods and cones. This change, however, usually requires several minutes to be completed. In some cases a longer time, extending to a few days, may be required for full adjustment. This might happen after an eye operation in which the eye has been covered for a long time, or when someone moves quickly from a dim wintery region to a bright sunny region, like an African desert, or a snowy mountain peak. So if you go into a dark room straight from bright sunlight, it will be some time before your eyes adjust to the very different light level. We can compare the change in sensitivity of the retina with a change of film 'speed' or sensitivity in a camera.

Binocular vision

The advantage of having two eyes spaced about 7 cm apart is that they give our brain two slightly different images of the world. Thus we are able to judge the position and speed of objects quite accurately up to a distance of about 60 m away. Without two eyes our judgment of distance has to be based on the apparent size of objects and the effect called parallax (p 18). Try judging distances with one eye closed, but not when riding a bicycle or driving a car!

Persistence of vision

An image formed on the retina leaves an impression which lingers for about $\frac{1}{10}$ second. This persistence of vision makes television and cinema pictures appear to change smoothly from one image to the next. During the gap between pictures the eye 'remembers' the previous picture. Find out how many times a television picture is renewed every second.

Demonstrations with a model eye

Fig. 3.15 shows a model eye which can be used to demonstrate what happens to rays of light as they travel through the eye both when it is working correctly and also when it is faulty. Three different meniscus lenses acting as the cornea are attached to the side of the flask to represent a normal, a short-sighted and a long-sighted eye (see appendix D1).

Figure 3.15   *A model eye*

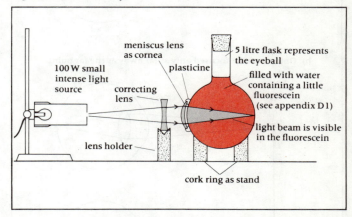

Adjust the distance between the light source and the 'eye' so that, using the 'normal cornea', the 'eye' converges the light beam to a spot at the back of the flask. This represents an image of the light source being focused on the retina.

Now turn the flask so that light enters the 'eye' through each of the other two cornea lenses in turn. These lenses have different curvatures and hence different powers to the normal cornea lens.

What happens to the beam of light inside the eye? The more powerful lens converges the light beam to a focus inside the flask (in front of the retina). This represents what is called a short-sighted eye. The less powerful lens converges the light beam to a point beyond the back of the flask (behind the retina). This represents what is called a long-sighted eye.

In each case investigate how a correcting lens can be used to refocus the beam on the retina. Which type of correcting lens is needed for each type of defect of the eye?

Despite the ability of the eye to adjust its focal length by changing the lens shape, some eyes cannot produce sharp images over the full normal distance range. This type of defect may be due in part to the eyeball being slightly too long or too short or alternatively, as in the example of the demonstration eye, due to the curvature of the cornea being wrong.

Normal vision range

A normal human eye can accommodate the range of distances from about 25 cm (the near point) to infinity (the far point). The near point starts nearer than 25 cm from the eye in young people and moves further away as we get older. Our accommodation becomes more and more limited as we get older.

The **near point** (NP) of the eye is the position closest to the unaided eye to which an object can be brought, and be seen clearly, without noticeable strain of the eye.

The **far point** (FP) of the eye is the position farthest from the unaided eye at which an object can be seen clearly.

Myopia or short-sightedness

A myopic or short-sighted eye can only see nearby objects clearly. Images of distant objects are formed in front of the retina, fig. 3.16a. The eyeball may be too long or the cornea lens too powerful. The far point (FP) of the myopic eye is nearer than for a normal eye, fig. 3.16b.

The defect is corrected using a *diverging lens* (fig. 3.16c), which diverges the rays from a distant object so that they appear to come from a virtual image at the far point. The eye can focus on this virtual image. A meniscus-shaped lens is usually used to match the curvature of the eye; in the case of contact lenses which float on the natural lubrication of the eyeball, the lens shape must fit the curvature of the cornea accurately.

Hypermetropia or long-sightedness

A long-sighted eye can only see distant objects clearly. Images of nearby objects are formed behind the retina, fig. 3.17a. The near point of the hypermetropic eye is farther away than for a normal eye, fig. 3.17b.

This defect is corrected using a *converging meniscus lens* which converges the rays from a near object so that they appear to come from a virtual image at the near point. The eye can focus on this image because it is further away than the real object, fig. 3.17c.

Figure 3.16 *Myopia or short-sightedness*

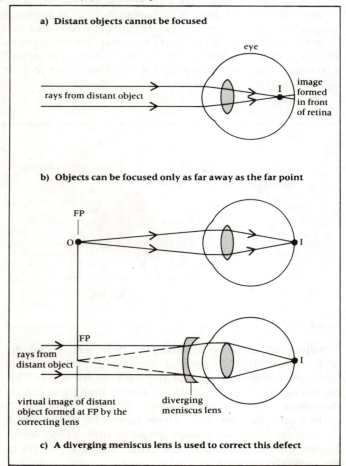

Figure 3.17 *Hypermetropia or long-sightedness*

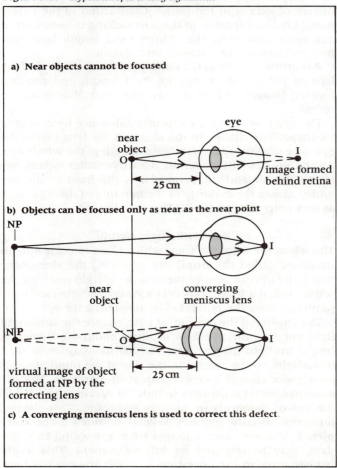

Some other defects

Lack of accommodation or presbyopia affects us all as we get older. The ciliary muscle gradually becomes less able to change the shape of the lens enough to focus objects which are either far away or very close. This inability to adjust the focusing of the eye, called lack of accommodation, can only be corrected by using two pairs of spectacles or bi-focals (two-in-one), containing a converging lens for seeing close up and a diverging lens for seeing far away.

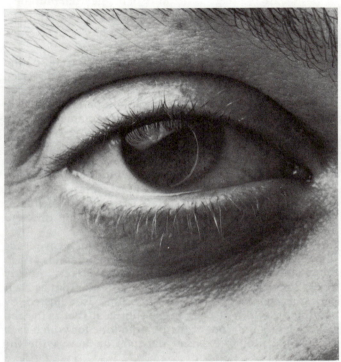

This person is wearing a contact lens. A glass or plastic lens, shaped to fit the curvature of the cornea and floating on tears, gives clear vision in all directions.

Astigmatism is a defect in which the eye has two different focal lengths in two different planes. This is caused by the cornea being slightly barrel shaped rather than true spherical. The defect is corrected using a cylindrical lens of the opposite power to the defect in the cornea. The effect of astigmatism is to make the eye less able to see clearly in one particular plane or direction.

Colour blindness. Total colour blindness is very rare, but there are a few people who see no colours at all and their world is monochromatic, like a black and white television picture. Many more people cannot tell the difference between two or more colours or find some difficulty in doing so. One common inconvenient form of partial colour blindness makes it difficult to distinguish between red and green. The defect may be caused by a deficiency of one of the types of cone cells in the retina.

Detached retina. Sometimes the light sensitive layer at the back of the eye, called the retina, becomes detached from the choroid behind. This defect can now sometimes be treated using a laser to 'spot-weld' the retina back. A brief intense pulse of laser light burns the cells in the retina in very small patches which attach them to the cells behind. The operation leaves small blind spots at the welding points.

Assignments

a) *Find your own blind spots.* Look at the cross and dot in fig. 3.18a with your left eye and close your right eye. Starting with the book about 30 cm from your eye, concentrate on the dot and slowly bring the book towards you. When the image of the cross falls on your blind spot what happens?

b) *Look at* the diagrams in fig. 3.18b. These pictures show how illusions can be created which mislead the eye and the brain.

Figure 3.18

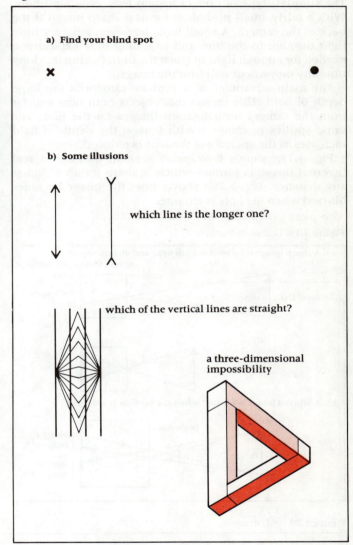

c) *Observe* the iris and pupil of someone near you as they look at a lamp behind you. What happens to the pupils of their eyes when the lamp is switched on and off?

d) *Explain* why it is necessary to wear sunglasses with dark lenses when you first arrive at a place where the sun is very bright. For example, if you go on a skiing holiday, or visit a very hot country near the equator, you will need sunglasses, but someone living there all the time does not.

e) *Estimate* your own near point distance. Look at some print on a page and bring it slowly closer to your eye. The eye should not be under obvious strain to focus. If you are short-sighted, find your far point as well.

Try questions 3.10 to 3.13

3.3
OPTICAL INSTRUMENTS

Optical instruments extend the capabilities of our eyes. A camera makes a permanent record of what we see and a projector reproduces it on a screen. Microscopes and telescopes expand our knowledge and understanding of the universe by enabling us to see both the very small and the very distant.

The pinhole camera

The simplest type of camera has no lens, only a pinhole. With a fairly small pinhole it forms a sharp image at the back of the camera. A small hole, however, lets very little light through to the film and so a long time exposure is needed for enough light to reach the film. During this long time any movement will blur the image.

The main advantage of a pinhole camera is the large depth of field. This means that objects both near and far from the camera form focused images on the film. The same applies to cameras with lenses; the depth of field increases as the aperture of the lens is reduced.

Fig. 3.19a shows how, with a small pinhole, a real inverted image is formed which is sharp for an object at any distance. Fig. 3.19b shows how the image becomes blurred when the hole is enlarged.

Figure 3.19   *The pinhole camera*

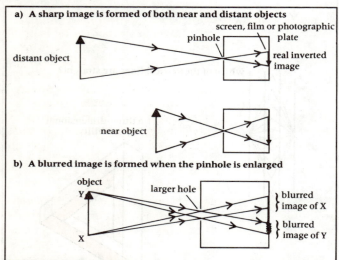

a) A sharp image is formed of both near and distant objects

b) A blurred image is formed when the pinhole is enlarged

Figure 3.20   *A Lens camera*

The sharpness of the image formed by a small pinhole depends on the fact that light rays travel in straight lines and can only reach one point of the image from one point of the object when the hole is small. If the pinhole is made extremely small, diffraction of light makes the image less sharp (p 479).

From fig. 3.19 we can also see that the image size increases as the object is brought nearer. By similar triangles we may prove that the magnification is given by:

$$m = \frac{\text{height of image}}{\text{height of object}} = \frac{\text{distance of image from pinhole}}{\text{distance of object from pinhole}}$$

The lens camera

This camera uses a converging lens (usually a combination of several lenses or elements) to produce a real inverted (and usually diminished) image on a light sensitive film at the back of the camera. Some important parts of a camera are shown in fig. 3.20. For a simple ray diagram of how the image is formed in a camera see fig. 3.9e.

The film, usually in a long thin strip, is wound from a cassette by a sprocket wheel onto a take-up spool and is held flat by a pressure plate from behind. Some cameras use a light sensitive plate which is replaced for each picture. The light sensitive surface of the film, which faces forwards, can have different sensitivities for different applications.

The shutter is a spring-loaded blind which covers the film, except during exposure when a photograph is taken. Some shutters are made of cloth and travel very quickly across the back of the camera in front of the film as their spring is released. The exposure time is adjusted by varying the width of a gap in the shutter so that as the gap travels across the front of the film the time for which each part of the film is exposed depends on how wide the gap is.

Shutter speed, or exposure time, which may vary from $\frac{1}{1000}$ second to several seconds, has two main effects.

a) The time of exposure is one factor determining how much light reaches the film.

b) When an object is moving the time of exposure determines how far the image will move on the film, and hence the amount of blurring.

The iris diaphragm is usually a multi-leaved metal structure which adjusts the aperture (size of hole) of the lens. When the aperture ring is turned the overlapping

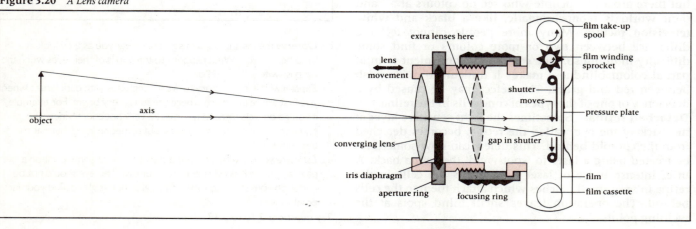

leaves of the diaphragm change the size of the hole in their centre and thus control the amount of light entering the camera.

The different settings of the aperture are referred to as 'stops' and are labelled with values known as f-numbers. The range of f-numbers is from near 1 for a very wide diameter lens to 16, 22 or even 32 for a very small diameter (nearly as small as a pinhole). Two factors determine the effective aperture of a lens, its diameter and its focal length. The f-number is the ratio of the focal length of the lens to its diameter. Thus a stop labelled f4 means that the focal length of the lens is 4 times its diameter. A lens of focal length 50 mm and maximum aperture f-number 2 will have a maximum diameter of $\frac{50}{2} = 25$ mm.

There are two important effects of the aperture setting.
a) The aperture controls the amount of light entering the camera.
b) Small apertures, like the pinhole camera, have a large depth of field (objects over a large distance range form focused images on the film), but wide apertures produce sharp images only for objects at a particular distance from the camera.

Focusing is done by turning the focusing ring which moves the lens nearer to or further from the film. A distant object requires the lens-to-film distance to be the focal length of the lens, but a nearer object requires an image distance greater than f and so the lens is moved outwards.

Table 3.2 compares the eye, the pinhole camera and the lens camera.

This cut-away view of a camera shows how the lens is built from six elements, or different pieces of glass. You can also see the pentaprism and mirror.

Table 3.2

	Eye	Pinhole camera	Lens camera
Type of lens	converging	none	converging
Method of focusing	change of lens shape: thicker for near objects	all distances focused if pinhole is small	lens moves away from the film for near objects
Light control	a) iris b) sensitivity of retina	a) hole size b) exposure time c) sensitivity of the film	a) iris diaphragm b) exposure time or shutter speed c) sensitivity of the film

The projector or enlarger

A projector uses several lenses and a concave mirror (fig. 3.21). The projection lens is a converging lens which produces on a screen a real, magnified and inverted image of the slide or film. (For the image to be the right way up the object, that is the slide or film, must be upside down.) Focusing the image on the screen is achieved by screwing the projection lens in or out, thereby adjusting the distance of the lens from the slide or film. The slide, being the object, is positioned between f and $2f$ from the lens, as in fig. 3.9c.

The rest of the projector is designed to give the maximum possible even illumination of the slide. The lamp is positioned at the centre of curvature of a concave mirror. This reflects light travelling away from the slide through the centre of curvature to the condenser lens. The condenser lens is usually made of two plano-convex converging lenses, separated by a heat shield which protects the film or slide from the heat of the lamp. The condenser lens converges the light evenly onto the whole area of the slide which now forms the illuminated object for the projection lens.

The magnifying glass

The simplest of optical instruments is a hand-held converging lens used to produce a magnified erect and virtual image. This common application of a converging lens, known as a magnifying glass or simple microscope, is explained by the ray diagram of fig. 3.9a. The object must be placed between F and the lens and the virtual image is seen by looking through the lens.

Figure 3.21 *The projector*

concave mirror | condenser lens | projection lens

lamp | film or slide | real inverted magnified image

real object

screen

R

The compound microscope

To increase the magnification of a microscope two lenses are used, fig. 3.22. The first lens, nearest the object O is called the objective lens.

The objective lens is a powerful converging lens of short focal length (principal focus at F_o). It is used to produce a real, inverted and magnified image I_1. Thus the object must be placed between F_o and $2F_o$ of the objective lens and the arrangement is similar to that of the projector.

The eyepiece lens is the lens nearest the eye. It has a short focal length (principal focus at F_e), and is used to magnify the first image I_1, its action being that of a magnifying glass. Thus the first image I_1 is the object for the eyepiece and must be positioned between the lens and F_e. The final image, formed by the eyepiece, I_2, is a virtual, erect and magnified image of I_1.

The combination of the two lenses produces a final image I_2 which is inverted compared with the object O and (in normal adjustment) will be at the near point of the eye.

Notes about drawing the ray diagram

a) Trace the path of two rays completely through the instrument from object to eye. These rays should bend only twice, as they pass through the two lenses, but they should not bend where I_1 is formed.

b) In order to find the path of the rays between the eyepiece and the eye it is necessary to draw two construction lines as shown in fig. 3.22. These lines obey the rules of rays coming from I_1 but should not be shown as rays (no arrows on them).

c) The positions of O and I_1 are critical and the positions of F_o and F_e on both sides of each lens must be shown on the diagram first.

d) The virtual image I_2 and the virtual rays from it should be shown as broken lines.

The astronomical telescope

This is a refracting telescope which uses two converging lenses to produce an inverted virtual image at infinity (fig. 3.23).

Figure 3.22 *The compound microscope*

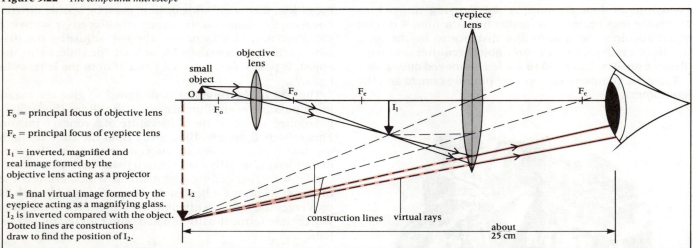

F_o = principal focus of objective lens

F_e = principal focus of eyepiece lens

I_1 = inverted, magnified and real image formed by the objective lens acting as a projector

I_2 = final virtual image formed by the eyepiece acting as a magnifying glass. I_2 is inverted compared with the object. Dotted lines are constructions draw to find the position of I_2.

Figure 3.23 *The astronomical telescope*

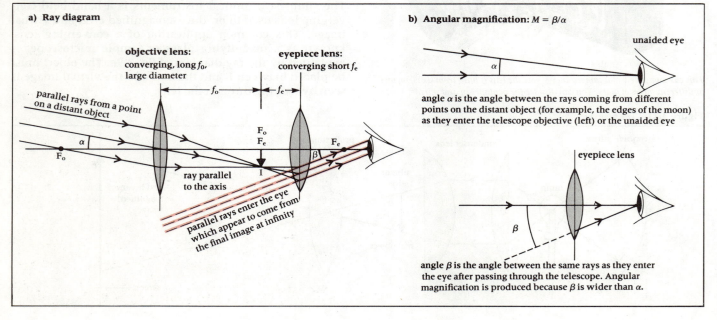

a) **Ray diagram**

objective lens: converging, long f_o, large diameter

eyepiece lens: converging short f_e

parallel rays from a point on a distant object

ray parallel to the axis

parallel rays enter the eye which appear to come from the final image at infinity

b) **Angular magnification:** $M = \beta/\alpha$

unaided eye

angle α is the angle between the rays coming from different points on the distant object (for example, the edges of the moon) as they enter the telescope objective (left) or the unaided eye

eyepiece lens

angle β is the angle between the same rays as they enter the eye after passing through the telescope. Angular magnification is produced because β is wider than α.

The objective lens is a wide aperture converging lens of long focal length. The large aperture enables the telescope to collect as much light as possible from weak distant sources and also improves the resolving power of the instrument. (Resolving power can be thought of as the ability of the instrument to form separate images of two distant objects like two stars which are very close together.) Acting like a camera taking a picture of a distant object, the objective forms a real inverted image I_1 at its principal focus F_o. This image is diminished.

The eyepiece lens is a short focal length converging lens which is used, as in the compound microscope, as a magnifying glass, but with one difference. It is usual to adjust the telescope so that the final image, like the object, is at infinity. This is achieved by positioning the first, real image I_1 so that it forms a real object at the principal focus F_e of the eyepiece. The rays reaching the eye are parallel, appearing to come from a virtual inverted final image at infinity.

Magnification

As we cannot compare directly the size of the object and the final image I_2 since both are at infinity, how does the telescope magnify? Fig. 3.23b shows that the telescope magnifies by refracting rays of light so that they enter the eye at a wider angle than they would without the telescope. The wider angle between the rays coming from the edges of the image make it appear larger. We define the angular magnification M of the telescope as $M = \beta/\alpha$, where the angles are those shown in the figure. The long focal length of the objective f and the short focal length f_e of the eyepiece make β much larger than α so that the telescope magnifies.

Prism binoculars

The length of an astronomical telescope and the fact that its final image is inverted make it unsuitable for some applications. Two prisms are used in binoculars to overcome these problems, as shown in fig. 3.24. Placed between the objective and eyepiece lenses, they both shorten the length of the instrument by passing the light along the tube three times and also produce an erect final image. The total internal reflection used in the prisms was described on p 28.

Figure 3.24 *Prism binoculars*

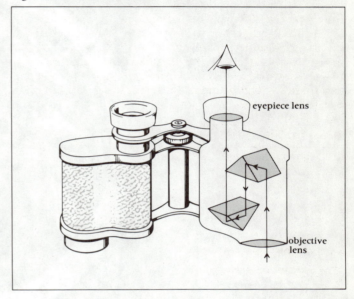

The reflecting or Newtonian telescope

Improving the performance of a telescope needs a larger diameter objective lens. This allows the telescope to collect more light and improves its resolving power. The largest objective of a telescope is about 1 metre in diameter since a larger lens would distort under its own weight and, besides, would be extremely difficult to make.

For telescopes with larger apertures we use a concave parabolic mirror in place of the objective lens (fig. 3.25). The problem with this arrangement is getting the final image in a position where it does not obstruct the incoming light rays. Newton solved this problem by placing a small plane mirror at an angle in front of the concave reflector, just short of its focus.

The eyepiece works the same way as in a refracting telescope, forming a final virtual image at infinity.

The Hale telescope at Mount Palomar, California uses a concave parabolic mirror of diameter 200 inches, or about 5 metres. This large mirror collects enough light to make very weak stars visible. The observer is sitting at the focus of the mirror. Light enters the telescope from behind the observer to reach the mirror, which can be seen at the far end of the structure.

Figure 3.25 *The reflecting (Newtonian) telescope*

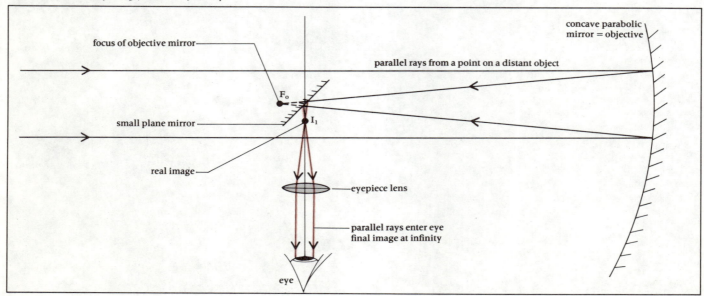

Assignments

Investigate a pinhole camera

A cardboard box of dimensions $15 \times 10 \times 10$ cm is suitable (see fig. 3.19). Cut out a square hole about 7 cm by 7 cm at one end and form a screen by gluing a sheet of translucent paper over the hole. At the other end cut out a circular hole about the same diameter as the lenses which are available. Over this hole glue black or opaque paper. Prick a single small pinhole in the centre of the black paper.

Hold the camera at arm's length and pointing the pinhole towards a bright lamp, view the image on the screen. Describe the image fully. (A suitable bright lamp is a 200 W carbon filament lamp mounted in a stand on a bench.) Things to try:

Turn the camera upside down.

Change your distance from the lamp.

Make the pinhole larger.

Add extra pinholes in the front of the camera.

Now place a converging lens in front of the camera and observe the image-forming property of the lens.

Remove the black paper over the front hole of the camera and again place the lens in front of it.

Move the camera nearer to the lamp. What must you do with the lens to refocus the image on the screen?

Build a compound microscope

Attach the lenses to the ruler as shown in fig. 3.26 using holders or plasticine. Move the object along the ruler until its image can be seen clearly. Now you can make two adjustments.

Figure 3.26 *Constructing a compound microscope*

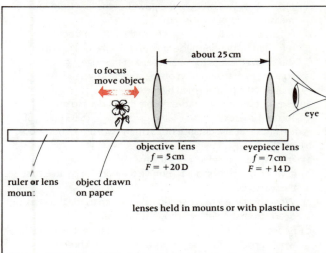

a) Move the object slightly until the final image coincides with the object. Do this by looking for no-parallax. In this position the final image will be roughly at your near point which is considered normal adjustment for a microscope.

b) Try to find the best position for your eye by moving it to different distances from the eyepiece. The best position is found when you have the greatest field of view. Describe the final image fully.

Build an astronomical telescope

The main difference from the compound microscope is that here the objective lens has a long focal length and the object should be a long way off. The distance between the lenses should be set at the sum of their focal lengths. Fine adjustment of this distance should bring the virtual image into focus. Normal adjustment for this instrument is with the final image at infinity.

Try questions 3.14 to 3.21

Questions 3

1 Fig. 3.27 shows two rays from the top of an object **OA** which pass through the lens **L** to the image **IB**. On a sheet of graph paper redraw the ray diagram and draw two rays from the top of the object O_1A_1 which pass through the lens; hence find the image of O_1A_1 and label it I_1B_1.

State the difference between the image **IB** and the image I_1B_1.

Figure 3.27

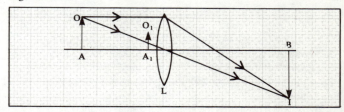

2 A projector is used to produce a magnified image of an object which is 1 cm tall on a colour slide. If the focal length of the projector lens is 20 cm and the colour slide is placed 25 cm from the lens, find by scale drawing the position and size of the image formed on a screen. (Use a scale of 1 cm representing 10 cm for the distances *u, v* and *f.*)

3 An object 2 cm tall is placed on the axis of a converging lens of focal length 5 cm at a distance of 3 cm from the optical centre of the lens. Draw a ray diagram to show how the image is formed and observed. Find also the position and height of the image.

Without drawing further ray diagrams, state the nature and position of the image, saying also whether it is large or small, when the object is at a distance from the lens of (a) 5.5 cm, (b) 10 m. (NB Exact numerical answers are not required.)

State a practical application of the use of such a lens with the object in each of positions (a) and (b). (S)

Figure 3.28

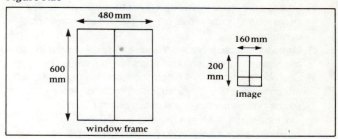

4 Fig. 3.28 shows scale drawings of a window frame and the image of the frame produced on a screen by a converging (convex) lens.

a) Calculate the linear magnification of the image.

b) The image of the frame was produced 500 mm from the lens. Either by calculation or by scale drawing (using 1 cm to represent 200 mm), determine
  i) the distance of the actual frame from the lens
  ii) the focal length of the lens. (AEB)

5 A camera is used to take a close-up picture of an object 3 cm tall. If the object is positioned 24 cm in front of the lens and a focused image is formed on the film 12 cm behind the lens, calculate the focal length of the camera lens and the height of the image formed on the film.

6 A magnifying glass of focal length 5 cm is used to magnify a small object held 4 cm from the optical centre of the lens. Calculate the position and magnification of the virtual image seen.

Figure 3.29

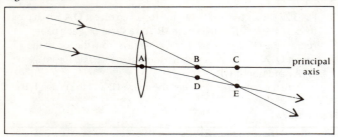

7   Fig. 3.29 shows parallel rays of light from a small distant object arriving at a converging lens.
    a) Which of the labelled positions represents the principle focus of the lens? Give your reasons for rejecting each of the other positions.
    b) At which position is the image of the small distant object formed? Describe the image.
    c) Which distance is the focal length of the lens?

8   a) Copy and complete the diagram below by drawing two suitable rays to show how the converging (convex) lens forms an image of the object. Clearly mark the position of the image. (F is the principal focus.)

Figure 3.30

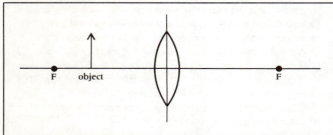

    b) Describe the image by selecting the appropriate words. The image formed is *real/virtual, upright/inverted* and *magnified/diminished*.
    c) Name the optical device which uses a single converging lens to produce an image of this description.   (NISEC SPEC part)

9   a) Describe how you would determine accurately the focal length of a converging lens using a laboratory method. (Assume that you cannot see the sun or any distant object.)
    b) The focal length of a lens is found to be 10.0 cm. Find how far you would place it from an illuminated slide to obtain an image on a screen which was magnified five times.
       Name two characteristics of the image (other than it is magnified).
       Suppose you drop the lens and it breaks so that only half of it is intact. You replace it in the same position to throw an image on the screen. State what effect, if any, this would have on the size and brightness of the image. (Give reasons for your answers.)   (L)

10 A person viewing a distant object switches his attention to look at an object near to him.
    State the change which occurs in his eye to enable him to see clearly the nearby object. Indicate how this change is produced.
    What additional change will occur in his eye if the nearby object becomes much brighter?   (C)

11 This question is about lenses.
    a) i) Fig. 3.31 shows rays of light approaching three different lenses. Redraw the diagram to show how these rays might

Figure 3.31

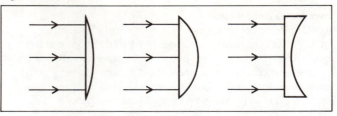

       emerge from the lenses.
    ii) Lenses are used in a number of optical instruments. Name one (other than spectacles) for which a large focal length is required.
    b) A long-sighted student needs to wear spectacle lenses with a power of +2 dioptres in order to be able to focus on close objects like books.
       i) How would you check the **focal length** of one of the lenses experimentally?

Figure 3.32

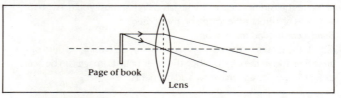

    ii) How would you calculate the **power** of the lens from this measurement of focal length?
    iii) Fig. 3.32 shows how the light from the page of a book is refracted by one spectacle lens. Redraw the diagram to show how this lens forms an image of the page. Mark on your diagram the position of: the student's eye and the image of the page formed by the lens.
    iv) Explain how the spectacles enable the student to see focused images.   (SEG [ALT] 88)

12 A man finds that at a distance of 25 cm the words in a book look blurred.
    a) From what eye defect does the man suffer?
    b) In which direction should he move the book to be able to see the words clearly?
    c) What spectacle lens is needed to allow him to see the words clearly from the distance shown in fig. 3.33?
    d) Redraw fig. 3.33 and sketch how the rays of light from the book would produce the final image, using the lens chosen in (c).   (NEA [B] P88)

Figure 3.33

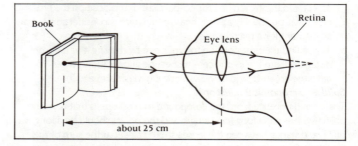

13 A converging lens of power +3.0 dioptres is used in a pair of spectacles. Calculate (a) its focal length and (b) the position of the image it forms of an object placed 25 cm from its optical centre.

55
Questions

14 A pinhole camera is used to look at a distant filament lamp. Which of the following procedures on their own would reduce the size of the image?
a) decreasing the size of the pinhole,
b) shortening the length of the camera,
c) decreasing the distance between the camera and the lamp.
(Nuffield)

15 A man 1.75 m tall stands at a distance of 7.0 m from the pinhole of a pinhole camera. The distance of the film from the pinhole is 0.20 m. Find the length of the image of the man which is formed on the film.
State two ways in which the appearance of the image will change if the size of the pinhole is increased. (C)

16 A person 2 m tall is standing in front of a pinhole camera. The camera is 0.2 m long. If the person is 5 m away from the pinhole of the camera, what will be the size of the image? (Joint 16+ part)

17 a) Draw a labelled ray diagram to show how a converging lens produces a real image of a *distant* object.
b) Describe and explain, with the aid of diagrams, how adjustment is made to bring near and distant objects into focus in a camera and by the human eye (details of the anatomical structure are not required).
c) Describe and explain two ways in which camera settings can be adjusted to take account of an increase in the brightness of the scene to be photographed.
How does the eye adjust to take account of an increase in brightness?
d) The lens of a camera is replaced by a piece of metal with a pinhole in it. Is it now necessary to make any adjustment to obtain a clear picture on the film to take account of the distance of the object from the camera? (O & C)

18 This question compares a pinhole camera with a lens camera. Each camera is to be used to photograph a tree. Copy the figures and:

Figure 3.34

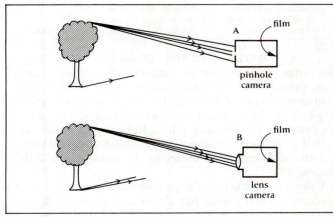

a) Complete the paths of any rays which go through the pinhole in A. Mark on the film the 'image' produced by the pinhole camera.
b) Name the type of lens used in the lens camera.
c) Complete the paths of any rays hitting the lens in B. Mark on the film the image produced by the lens camera.
d) State two ways in which the 'images' produced on the film in these cameras are different from the tree itself.
e) Although 'good' pictures can be obtained using a pinhole camera, most modern cameras are lens cameras. Give some advantages of the lens camera over the pinhole camera. (The figures may give you some clues.) (LEAG SPEC B part)

19 Explain, with the aid of ray diagrams, how a lens, of focal length 100 mm, can be used
a) in a simple camera to photograph a distant object,
b) in a slide projector to project the image of a transparency ('slide') on to a screen,
c) as a magnifying glass to examine the transparency.
In each case show clearly where the photographic film or transparency should be placed in relation to the focal plane of the lens, and state the type of image being formed by the lens. The details of the other parts of the instruments are not required.
When the lens is used as a magnifying glass, the final image produced has a linear magnification of 2.5. Determine, by using a ray diagram drawn to scale, or otherwise, the distance from the lens to the object. (O & C)

20 a) What is meant by the terms principal focus and focal length when applied to (i) a convex (converging) lens, (ii) a concave (converging) mirror?
Draw two ray diagrams, one for each of (i) and (ii).
b) i) If the components in (a)(i) and (a)(ii) were to be used in turn as the main component in each of two different types of telescope, what additional components would be needed in each case?
ii) Draw a ray diagram of one of the completed telescopes. Show how and where the images are formed, labelling the points used in your construction.
iii) For the telescope illustrated in (ii), what factor has the greatest effect on the overall magnification and how could the telescope be designed so that the magnification was large? (JMB)

21 The table below provides information about three telescopes used to examine different sources of radiation from space.

Location	Objective	Objective diameter (D)	Source of radiation	Wavelength detected (λ)
Yerkes Observatory U.S.A.	convex lens	1.0 m	Sirius	500 nm
Mount Palomar U.S.A.	parabolic mirror	5.0 m	Polaris	600 nm
Jodrell Bank U.K.	parabolic dish reflector	75 m	Inter-stellar hydrogen	0.21 m

a) Explain why different types of telescope are needed to detect signals from space.
b) Which of the above telescopes: i) is a refracting telescope? ii) could detect microwaves?
c) The smaller the ratio of the wavelength (λ) to the objective diameter (D), the more effective is the telescope in clearly separating the images of distant sources. This ratio λ/D is called the resolving power of the telescope.
Using the data from the table, show by calculation which telescope has the best resolving power.
d) Apart from improving the resolving power, give another reason why any telescope should have a large value of D.
e) The higher the frequency of radiation emitted by a source, the higher the source temperature.
Which radiation source given in the table has the highest temperature? (SEB SPEC 1990)

4
Measuring

*H*uman beings have always been extremely curious about their surroundings. We ask many varied questions about the world and the things we see and experience around us. Most of the answers which physicists can give boil down to information about three basic ideas which we call length, mass and time.

4.1
LENGTH, AREA AND VOLUME

The length, area and volume of an object all tell us about its size. In the past we have usually compared the size of an object with something familiar like the length of a man's foot but today, in science, we need very precise ways of both making and stating these measurements.

Stating lengths in SI units

The SI (see appendix) unit of length is the metre (symbol m). **One metre** is the length of the path travelled by light in a vacuum during a time interval of 1/299 792 458 of a second. This very technical definition (which need not be memorised!) is used for very precise measurements of length because it is believed to be absolutely constant.

Previously, 1 metre was defined as the distance between two marks on a standard platinum–iridium bar which was kept at 0°C.

Multiples and submultiples of the metre

When we are measuring lengths which are much greater or much less than a metre we often use multiples and submultiples of the metre as a more easily recognisable size. For example, we understand how thick a $\frac{1}{2}$p coin is more easily when we state that it is 1 millimetre (1 mm) than if we state that it is 0.001 metres or $\frac{1}{1000}$ of a metre.

The most common multiple and submultiple units of the metre are given in table 4.1. A fuller list is given in appendix B.

Table 4.1

Unit	Equivalent in metres
1 kilometre (km)	= 1000 m (or 10^3 m)
1 centimetre (cm)	= 0.01 m or $\frac{1}{100}$ m (or 10^{-2} m)
1 millimetre (mm)	= 0.001 m or $\frac{1}{1000}$ m (or 10^{-3} m)
1 micrometre (μm)	= 0.000 001 m or $\frac{1}{1000000}$ m (or 10^{-6} m)

Powers of ten and standard form

The range of lengths in the universe is extremely large, from the diameter of an atomic nucleus (p 382) to the estimated size of the universe, fig. 4.1. Writing down these very large and very small lengths in metres is very clumsy. For example, the distance which light travels in one year (called 1 light year) is about 10 000 000 000 000 000 metres and the diameter of a molecule is about 0.000 000 001 metres.

A short way of writing large and small numbers is used in science so that we can handle these measurements more conveniently. The scale in fig. 4.1 and the figures in brackets in table 4.1 both use the shorthand method. The smaller figures printed above the 10 are called **powers of ten**. A *positive* power of ten is the number of times the number has to be *multiplied* by 10. A *negative* power of ten is the number of times it has to be *divided* by ten.

For example, the speed of light mentioned in chapter 2 was given as 3×10^8 m/s. This means multiply 3 by 10 eight times, which gives 300 000 000 m/s. Similarly, 3×10^{-2} m means $\frac{3}{100}$ or 0.03 metres and 10^{-2} m means 1×10^{-2} m which is $\frac{1}{100}$ or 0.01 metres. Note that 10^0 means 1 not multiplied by ten, so $10^0 = 1$.

This shorthand form of writing numbers is called **standard form**. See appendix C3.

Figure 4.1 *Lengths in the universe*

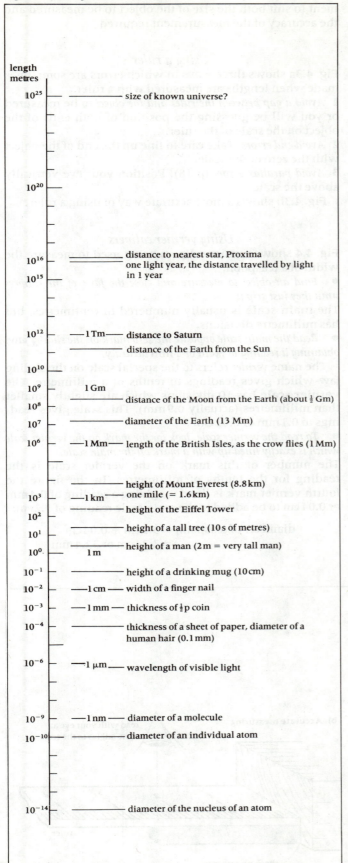

Units of area and volume

Area

When we need to calculate the area of a surface we can use standard formulas for particular shapes. (For example, rectangle $A = lb$, fig. 4.2a, disc $A = \pi r^2$.) Whatever the shape, the SI unit of area is the **square metre**, written m². Sometimes, in questions on pressure for example, we use square centimetres, written cm², and it is often necessary to convert them to square metres.

As there are 100 cm in 1 m, there are 100 cm × 100 cm = 10 000 cm² in 1 m², fig. 4.2b. So

$$1\,\text{m}^2 = 10^4\,\text{cm}^2$$

Figure 4.2

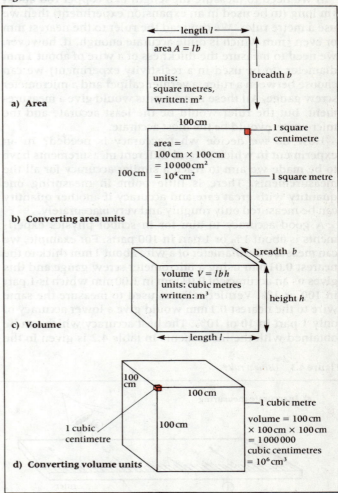

a) **Area**

b) **Converting area units**

c) **Volume**

d) **Converting volume units**

Volume

For regularly shaped solids such as rectangular blocks we can also calculate their volumes using standard formulas, fig. 4.2c. The SI unit of volume is the **cubic metre**, written m³. One cubic metre is a rather large volume and gives high values for densities (see next section), so we often use cubic centimetres, written cm³, for practical measuring.

As there are 100 cm in 1 m, there are 100 cm × 100 cm × 100 cm = 1 000 000 cm³ in 1 m³, fig. 4.2d. So

$$1\,\text{m}^3 = 10^6\,\text{cm}^3$$

Measuring length

When we need to measure a length we must choose a measuring instrument which is suitable for the length to be measured and will give the required accuracy. Table 4.2 gives some instruments and the lengths which they are suitable for measuring.

Table 4.2

Length to be measured	Measuring instrument	Best accuracy
several metres	steel tape measure	1.0 mm
about 1 cm to 1 m	ruler	0.5 mm
about 1 mm to 10 cm	vernier calipers	0.1 mm
about 0.1 mm to 2 or 3 cm	micrometer screw gauge	0.01 mm

If we need to measure the length of a copper rod about $\frac{1}{2}$ m long (to be used in an expansion experiment) then we use a metre ruler. We can read the ruler to the nearest mm or even $\frac{1}{2}$ mm which is quite accurate enough. If, however, we need to measure the thickness of a wire of about 1 mm diameter (to be used in a resistivity experiment) we can choose between a ruler, a vernier caliper and a micrometer screw gauge. All these instruments would give a measurement, but the ruler would be the least accurate and the micrometer would be the most accurate.

How do we decide what accuracy is needed? In an experiment in which several different measurements have to be made we aim to reach a similar accuracy for all the measurements. There is little point in measuring one quantity with great care and accuracy if another quantity can be measured only roughly and very *inaccurately*.

A good accuracy to aim for in school physics experiments is about 1% or 1 part in 100 parts. For example, we can measure the diameter of a wire about 1 mm thick to the nearest 0.01 mm using a micrometer screw gauge and this gives us an accuracy of 0.01 mm in 1.00 mm which is 1 part in 100 or 1%. Vernier calipers used to measure the same wire to the nearest 0.1 mm would give a lower accuracy of only 1 part in 10 or 10%. The best accuracy which can be obtained with the instruments in table 4.2 is given in the last column. So you should choose your measuring instrument to suit both the size of the object to be measured and the accuracy of the measurement required.

Using a ruler

Fig. 4.3a shows three ways in which errors are sometimes made when lengths are measured with a ruler.

1 *Avoid a gap between the ruler and the object* to be measured or you will be guessing the position of both ends of the object on the scale of the ruler.

2 *Avoid end errors.* Take care to line up the end of the object with the zero of the scale.

3 *Avoid parallax errors* (p 18). Position your eye vertically above the scale.

Fig. 4.3b shows a more accurate way of using a ruler.

Using vernier calipers

Fig. 4.4 shows vernier calipers being used to measure the width of a teaspoon.

● *Find an object to measure and close the jaws of the calipers until they just grip it.*

The **main scale** is usually numbered in centimetres, but has millimetre divisions.

● *Read the main scale opposite the zero mark on the sliding jaw, obtaining a reading to $\frac{1}{10}$ cm or 1 mm accuracy.*

The name *vernier* refers to the special scale on the sliding jaw which gives readings to tenths of a millimetre. The **vernier scale** has 10 divisions which are slightly smaller than millimetres (actually 0.9 mm). This scale gives readings to 0.1 mm or 0.01 cm.

● *To read the vernier scale, look for the mark on the vernier scale which is exactly lined up with a mark on the main scale.*

The number of this mark on the vernier scale is the reading for the tenths of millimetres. In the figure the fourth vernier mark is lined up giving a reading of 0.4 mm or 0.04 cm to be added to the main scale reading of 3.2 cm:

$$\text{diameter of teaspoon} = 3.2\,\text{cm} + 0.04\,\text{cm}$$
$$= 3.24\,\text{cm or } 32.4\,\text{mm}$$

Figure 4.3 *Using a ruler*

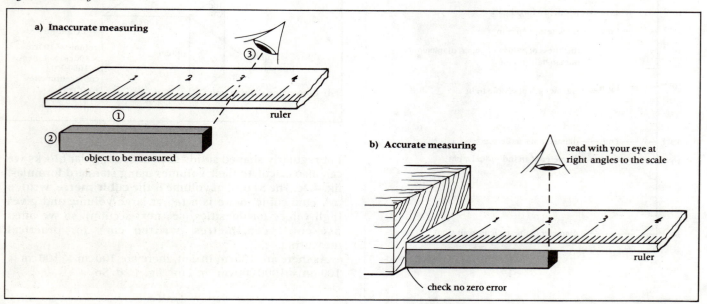

a) **Inaccurate measuring**

ruler

object to be measured

b) **Accurate measuring**

read with your eye at right angles to the scale

ruler

check no zero error

Figure 4.4 *Vernier calipers*

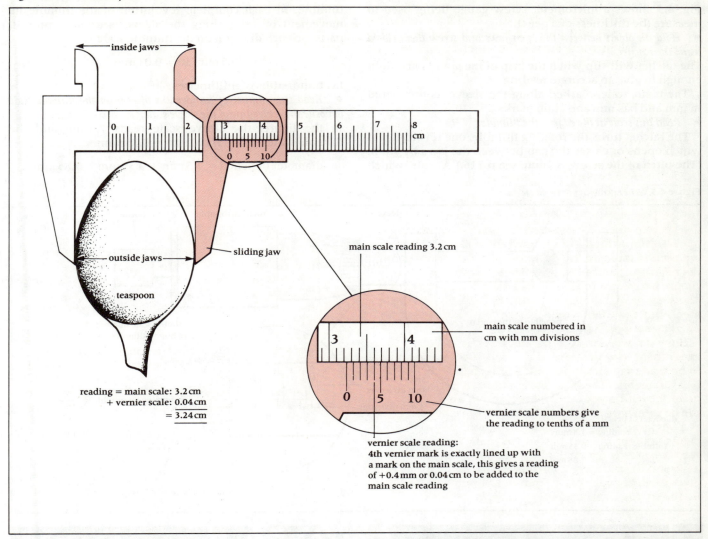

inside jaws

0 1 2 3 4 5 6 7 8 cm

0 5 10

outside jaws

sliding jaw

teaspoon

reading = main scale: 3.2 cm
+ vernier scale: 0.04 cm
= 3.24 cm

main scale reading 3.2 cm

3 4

0 5 10

main scale numbered in
cm with mm divisions

vernier scale numbers give
the reading to tenths of a mm

vernier scale reading:
4th vernier mark is exactly lined up with
a mark on the main scale, this gives a reading
of +0.4 mm or 0.04 cm to be added to the
main scale reading

Measuring long distances

Tape measures can be used to measure distances up to several hundred metres with good accuracy. For example they are used for measuring the distance an athlete throws a discus or javelin. Beyond this, measurements are made indirectly, that is to say, they are calculated from other measurements. For example, when radar is used to find the distance of an aircraft the quantity measured is the time taken for a pulse of radio waves (which travel at the speed of light) to go out to the aircraft and be reflected back again. The distance is then calculated from the measured time and the known speed of the radio wave pulses (distance = speed × ½ time).

Surveyors use a method known as **triangulation** to calculate long distances between three points. From measurements of the angles and one side of a triangle with the points at its corners, the lengths of the other two sides can be calculated.

This architect is using a steel tape graduated in centimetres and millimetres.

Using a micrometer screw gauge

Fig. 4.5 shows a micrometer screw gauge being used to measure the thickness of a pencil.

- *Hold an object between the open jaws and screw them closed using the ratchet.*

The ratchet will slip when the grip of the jaws is just tight enough to give an accurate reading.

The main scale, marked along the sleeve, is numbered in mm and has mm and $\frac{1}{2}$ mm marks.

- *Read this scale at the edge of the thimble.*

The ratchet turns the rotating thimble, one revolution of which opens or closes the gap between the jaws by $\frac{1}{2}$ mm. (The pitch of the screw is $\frac{1}{2}$ mm, see p 116.) A scale, which has 50 divisions, is marked around the edge of the thimble. In each complete revolution of the thimble its movement of $\frac{1}{2}$ mm along the sleeve is divided into 50 parts. So each division on the thimble scale is

$$0.5\,\text{mm}/50 = 0.01\,\text{mm}$$

i.e. hundredths of millimetres.

- *Read the number of hundredths of mm on the thimble scale opposite the centre line of the sleeve scale.*

In the figure, the sleeve scale reads 7.5 mm and the thimble scale reads $35/100$ mm = 0.35 mm:

$$\text{diameter of pencil} = 7.5\,\text{mm} + 0.35\,\text{mm} = 7.85\,\text{mm}.$$

Figure 4.5 *The micrometer screw gauge*

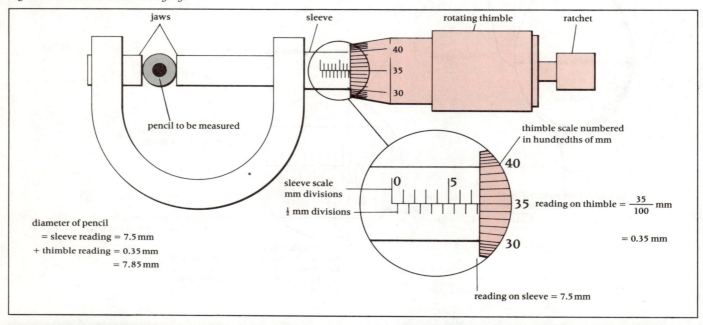

diameter of pencil
= sleeve reading = 7.5 mm
+ thimble reading = 0.35 mm
= 7.85 mm

reading on thimble = $\frac{35}{100}$ mm

= 0.35 mm

reading on sleeve = 7.5 mm

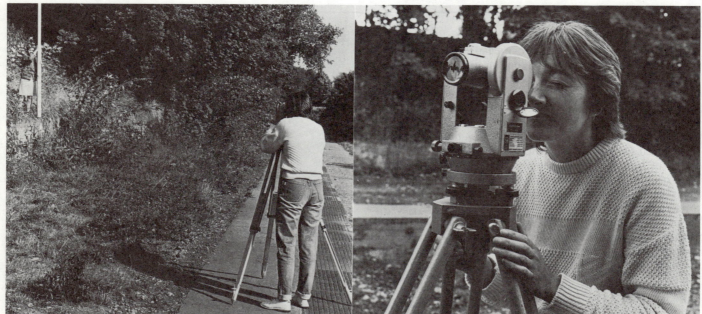

Surveyors use this instrument, called a theodolite, to measure horizontal and vertical angles. A telescope is mounted on a tripod so that it can be turned and tilted. When a site is surveyed, vertical scales are positioned at strategic points and are viewed through the telescope. Angle readings obtained on the theodolite allow heights and distances between these points to be calculated.

Measuring volume

Solids and liquids have almost constant volumes but a great variety of shapes. This makes measuring volumes quite difficult except for regular shaped solids such as rectangular blocks, cylinders and spheres. Table 4.3 summarises the basic methods used for measuring the volumes of solids and liquids.

Table 4.3

Volume to be measured	Instrument
regular shaped solid object	ruler, calipers or micrometer, using a formula
irregular shaped solid object	displacement can and measuring cylinder
liquids (large volumes)	measuring cylinder
liquids (small or accurate volumes)	calibrated burette, pipette or flask

The SI unit of volume is the cubic metre (or $1\,m^3$). The volume of a cube of sides 1 metre is very large and we often use the cubic centimetre (cm^3) as a more convenient unit. One cubic centimetre is equivalent to 1 millilitre (or $1\,ml$). $1\,cm^3 = 10^{-2}\,m \times 10^{-2}\,m \times 10^{-2}\,m = 10^{-6}\,m^3$ and $1\,m^3 = 10^6\,cm^3 = 1\,000\,000\,cm^3$.

Calculating the volume of a regular shaped solid

Volume of a rectangular block = length × breadth × height

$$V = lbh$$

Volume of a cylinder of radius r = height × area of end

$$V = h(\pi r^2)$$

Volume of a sphere of radius r:

$$V = \tfrac{4}{3}\pi r^3$$

Measuring the volume of a liquid

Reading a meniscus

Looked at from a side view, the surface of a liquid inside a tube is not a straight line. A liquid surface curves at the edges where the liquid 'wets' the glass. This curved surface is called the **meniscus**.
Correct readings are taken
a) when the instrument is vertical,
b) when the reading is taken at the **bottom** of the meniscus as shown in the diagram below and
c) when your eye is level with the meniscus.

Reading a meniscus

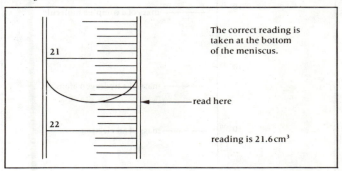

The correct reading is taken at the bottom of the meniscus.

read here

reading is $21.6\,cm^3$

Using a measuring cylinder

Large or small volumes of liquid can be measured using a measuring cylinder but this instrument is less accurate than either a burette or a pipette. Although the scale reads upwards, remember to read the liquid level at the bottom of the meniscus. Note whether the graduations are marked in steps of $1\,cm^3$, $5\,cm^3$ or larger.

Measuring the volume of an irregular shaped solid

A simple method of measuring the volume of a small solid object is shown in fig. 4.6.
● *Partly fill a measuring cylinder with water and take an initial reading: level 1 in the figure.*
● *Now drop the object into the water so that it is completely covered and read water level 2.*
The volume of the object is the difference between the two water level readings.

Solid objects which float

A metal object such as a brass weight can be used as a 'sinker'. By attaching a sinker, the floating object can be pulled down below the surface of the water in a measuring cylinder or displacement can.
● *Find the volume of the sinker alone.*
● *Find the volume of the object plus sinker.*
● *Subtract readings to find the volume of the object alone.*

Figure 4.6 *Measuring volume*

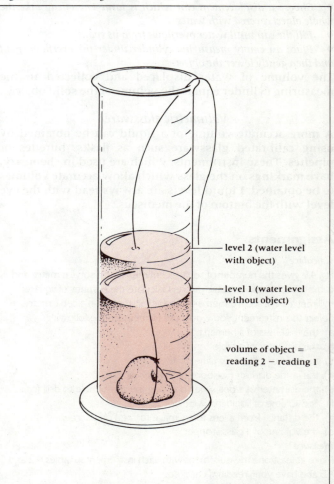

level 2 (water level with object)

level 1 (water level without object)

volume of object = reading 2 − reading 1

Using a displacement can

Another simple method of measuring the volume of any shape or size of solid object is shown in fig. 4.7.

Figure 4.7 *Measuring the volume of a small irregular shaped object*

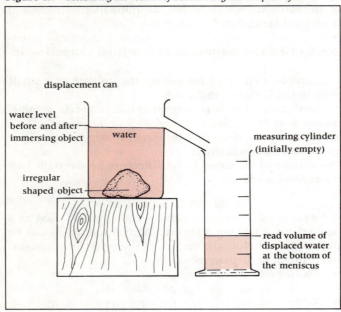

- *Choose a displacement can which is large enough to take the whole object covered with water.*
- *Fill the can until water overflows from its spout.*
- *Place an empty measuring cylinder under the overflow spout and then gently lower the object.*

The volume of water displaced and collected in the measuring cylinder equals the volume of the solid object.

Volumetric glassware

A more accurate volume of a liquid can be obtained by using calibrated glassware such as flasks, burettes or pipettes. These instruments, which are used in chemistry, have markings on the glass which allow accurate volumes to be obtained. Liquid levels are always read with the eye level with the bottom of the meniscus.

Assignments

Calculate

Fig. 4.2 gives the number of square centimetres in a square metre and cubic centimetres in a cubic metre. Calculate the number of square millimetres in a square metre and cubic millimetres in a cubic metre.

Select the instrument you would use to measure the following:
a) the thickness of a human hair,
b) the diameter of a table-tennis ball,
c) the area of a window pane,
d) the inside diameter or bore of a water pipe,
e) the diameter of a bolt when deciding the size of hole to drill for it,
f) the volume of liquid in a wine bottle,
g) the distance from a lens to an image formed on a screen,
h) the volume of a glass stopper.

Measure

Make at least one measurement with each instrument in tables 4.2 and 4.3 and have your readings checked.

Try questions 4.1 to 4.3

Figure 4.8 *Masses in the universe*

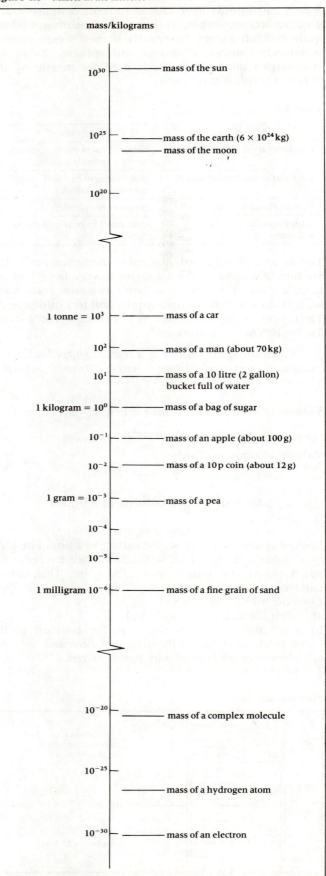

4.2
MASS, INERTIA AND DENSITY

The amount of matter in an object affects certain properties of that object wherever it may be – on the earth, on the moon or in deep space. Two such properties are the inertia mass of an object and its density.

Mass and inertia

Both these words refer to the matter in an object but, although they have slightly different uses, they amount to the same thing.

The mass of an object is a measure of the matter in it and depends on the number of atoms it contains and the size of those atoms.

The inertia of matter is the 'laziness' of matter, or the tendency of a mass to resist changes in its motion. Inertia makes an object difficult to start or stop moving, difficult to change its direction of motion or difficult to accelerate.

The mass or inertia of an object, being dependent only on the amount of matter it contains, is the same wherever it is. For example, an astronaut has the same mass (or inertia) on earth, on the surface of the moon or inside a space ship. That is to say, he needs the same size push to get him moving wherever he is.

Units of mass

The SI unit of mass is the **kilogram**. Note that the symbol is kg with a small 'k' for kilo (= 1000). There is a standard 1 kg mass made of platinum–iridium alloy kept at Sèvres near Paris at the International Office of Weights and Measures. All other masses are, in theory, measured by comparison with this standard mass. Table 4.4 gives the more commonly used multiples and submultiples of the kilogram.

Table 4.4

Unit	Equivalent in:	
	kilograms (kg)	*grams* (g)
1 tonne (t)	$1000\,kg = 10^3\,kg$	$10^6\,g$
1 kilogram (kg)	SI unit of mass	$1000\,g = 10^3\,g$
1 gram (g)	$0.001\,kg = 10^{-3}\,kg$	$1\,g = 10^0\,g$
1 milligram (mg)	$10^{-6}\,kg$	$0.001\,g = 10^{-3}\,g$
1 microgram (µg)	$10^{-9}\,kg$	$10^{-6}\,g$

The enormous range of masses in the universe is shown in fig. 4.8. It is useful in physics to have some idea of the size of an object and its mass. Quoting a measurement or size to the nearest power of ten is called its **order of magnitude**. So, for example, a car of mass about $10^3\,kg$ is one order of magnitude greater than a man of mass about $10^2\,kg$. An increase of one order of magnitude means a ten-fold increase.

Mass and weight

Unfortunately in everyday language we often use the word 'weight' when strictly we should use 'mass' and we talk about 'weighing' an object when we are finding its mass.

The pull of the earth on an object is called its weight.

This pull or weight is caused by what we call 'gravity'.

The weight of an object varies a little from place to place on the surface of the earth; on the moon the weight is only $\frac{1}{6}$ of its value on the earth. In deep space, away from all gravity, an object has no weight at all. We shall learn more about this in chapter 5.

The link between mass and weight is that *at a particular place* the weights of objects are proportional to their masses. Using this proportional relation it is easy to find masses by comparing weights.

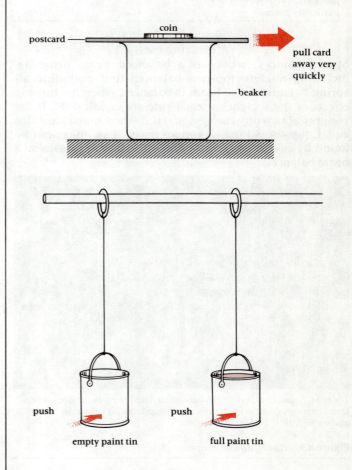

―――――*Some effects of the inertia of matter*―――――

- *Rest a coin on a postcard over a beaker.*
- *Pull the card away very slowly. What happens?*
- *Pull the card away very quickly. What happens?*

The inertia of the coin limits its motion when the frictional force between it and the fast moving card acts for only a very short time.

- *Hang up two similar paint tins, one full and one empty. Try pushing them and then stopping them and see how different they feel.*

coin
postcard
pull card away very quickly
beaker

push
empty paint tin

push
full paint tin

The two tins look alike but they feel different to push or to stop. As you are not lifting the tins against gravity (at least, not much if the strings are long), the difference in their 'feels' is not due to a difference in weights. If they were hovering in front of you in space with no weight you would get the same different 'feels' when you pushed them. The full tin of paint feels more reluctant to move and more difficult to stop moving than the empty one. The full tin has a greater inertia or mass than the empty one.

Table 4.5 compares mass and weight. (Some of the ideas in this table refer to things we shall discuss later, but they are given here for reference.)

Table 4.5

Mass	Weight
is measured in kilograms	is measured in newtons (p 73)
is a measure of the amount of matter in an object	is the pull of the earth on the object
is a scalar quantity	is a vector quantity (a pull is a force and has a direction, p 75)
is constant everywhere	changes slightly when an object moves to different places on earth, is reduced to zero in deep space
can be measured by comparison with a standard mass or by measuring weight and assuming that mass is proportional to weight at a particular place	can be measured by the extension of a spring balance or by comparison with another weight on a beam balance

Measuring mass

Most 'balances' work on a balanced beam principle, including modern top-pan balances (but excluding all spring balances). The beam is balanced when the turning effects of the weights at each side are equal, p 80. If the weights of two objects are equal then their masses are also equal, fig. 4.9. So the unknown mass of an object can be found by comparing it with known, standard masses on a beam balance.

The beam balance has been used since ancient times. Here, in the Book of the Dead *papyrus of the Egyptian scribe Ani, the god Anubis weighs Ani's heart against the feather of Maat (truth).*

Figure 4.9 *Finding an unknown mass*

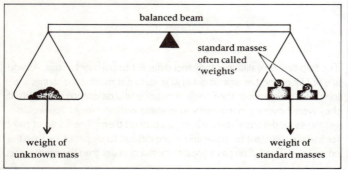

Notes:
1 Standard masses are often called 'weights'.
2 A beam balance would work on the moon and give correct values for unknown masses, but it would not work in deep space where objects have no weight.
3 A spring balance, used to measure the weight of an object on the moon, would give a correct reading of its weight (about $\frac{1}{6}$ of the object's weight on earth) but could not be used to measure its mass directly. (See p 74 for spring balances.)
4 When finding the mass of some liquid, weigh the container when empty and dry and subtract the mass of the container from the mass of liquid + container.

Density

Fig. 4.10 shows some samples of various common substances.

Figure 4.10 *Comparing masses of a cubic centimetre of various substances*

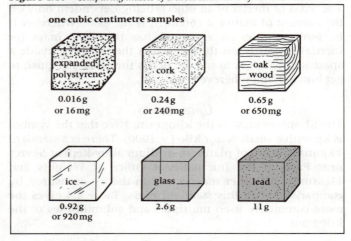

● *Compare the samples and make a list of their similarities and differences.*

Have you said that they are all the same size? Have you said that the lead is heavier than the glass and the cork is lighter than the wood?

In everyday language it is common to use words like 'size' and 'heavier' or 'lighter' where in science we must be more precise. The samples have equal **volumes** of 1 cubic centimetre each. But when we compare the masses of equal volumes of substances we are comparing their **densities.**

The density of a substance tells us how much matter is contained in a certain volume of it (usually $1\,cm^3$ or $1\,m^3$). Compare this idea with the mass of an object which tells us how much matter is contained in the whole object. In dense substances matter is compressed or closely packed, but in a substance of low density matter is loosely packed or expanded (as in expanded polystyrene).

The density of a substance is defined as its mass per unit volume.

(Per means 'of each' and 'per unit' means 'divided by' in a formula.)

$$\text{density} = \frac{\text{mass}}{\text{volume}} \qquad \rho = \frac{m}{V}$$

The Greek letter ρ (rho) is the symbol used for density.

The units of density in SI are: $\dfrac{\text{kilogram}}{\text{cubic metre}}$ or kg/m^3.

But it is often convenient to use:

$$\dfrac{\text{gram}}{\text{cubic centimetre}} \quad \text{or} \quad \text{g/cm}^3.$$

Measuring density

All that is needed to find the density of a particular substance is the measurement of the mass and volume of an object made of that substance, or of a sample of the substance. We have seen earlier in this chapter how to measure volumes and masses.

If you are describing how to find the density of a substance, explain how you would measure both the volume and the mass of a suitable sample and why you would use the instruments you suggest.

Table 4.6 gives some useful values of densities. Note how to convert between the different units:

$$\text{density of water} = 1.0 \times 10^3 \, \dfrac{\text{kg}}{\text{m}^3} = 1.0 \times 10^3 \times \dfrac{10^3 \, \text{g}}{10^6 \, \text{cm}^3}$$

$$\therefore \text{ density of water} = 1.0 \, \dfrac{\text{g}}{\text{cm}^3} \text{ or } 1.0 \text{ g/cm}^3.$$

Table 4.6 Densities of some common substances

Substance	Density g/cm³	Density kg/m³
platinum	21	21×10^3
gold	19	19×10^3
mercury	14	14×10^3
lead	11	11×10^3
steel	7.9	7.9×10^3
average density of the earth	5.5	5.5×10^3
glass, brick, stone and concrete	approx. 2.6	2.6×10^3
water	1.0	1.0×10^3
ice	0.92	920
alcohol, petrol and paraffin oil	approx. 0.8	800
oak wood	0.65	650
cork	0.24	240
expanded polystyrene	approx. 16×10^{-3}	16
air (at sea level)	1.3×10^{-3}	1.3

Worked Example
Calculating density

A glass stopper has a volume of 16 cm^3 *and a mass of* 40 g. *Calculate the density of glass in* g/cm^3 *and* kg/m^3.

Using $\rho = \dfrac{m}{V}$

we have $\rho = \dfrac{40 \text{ g}}{16 \text{ cm}^3} = 2.5 \text{ g/cm}^3$

and converting the units:

$$\rho = \dfrac{2.5 \text{ g}}{1.0 \text{ cm}^3} = \dfrac{2.5 \times 10^{-3} \text{ kg}}{1.0 \times 10^{-6} \text{ m}^3} = 2.5 \times 10^3 \text{ kg/m}^3$$

Answer: The density of the glass is 2.5 g/cm^3 or $2\,500 \text{ kg/m}^3$.

Worked Example
Calculating mass

Calculate the mass of a gold coin of volume 2.1 cm^3. *The density of gold is* 19 g/cm^3.

Rearranging the density formula we have: $m = \rho V$, so

$$m = 19 \, \dfrac{\text{g}}{\text{cm}^3} \times 2.1 \text{ cm}^3 = 40 \text{ g}$$

Answer: The mass of the gold coin is 40 grams.

Worked example
Calculating volume

Calculate the volume of a block of expanded polystyrene of mass 400 g *if its density is* 16 kg/m^3.

The mass (400 g) must be converted to kilograms to match the density units: $m = 400 \text{ g} = 0.40 \text{ kg}$. The rearranged density formula gives:

$$V = \dfrac{m}{\rho}$$

so $V = \dfrac{0.40 \text{ kg}}{16 \text{ kg/m}^3} = 0.025 \text{ m}^3$

Answer: The volume of the polystyrene is 0.025 m^3 or $25\,000 \text{ cm}^3$.

Assignments

Remembering and rearranging a formula
Can you remember and rearrange the formula

$$\rho = \dfrac{m}{V}$$

To help rearrange formulas like this one which have three related quantities, you can use the following triangle symbol:

When you have remembered that mass m is on the top of the density formula you will also know that m goes in the top of the triangle symbol. To find the formula for mass, cover up m in the triangle. You are left with $\rho \times V$, i.e. mass = density $\times$ volume. Similarly to find the formula for volume, cover up V. Now you are left with m/ρ, which gives you the formula:

$$\text{volume} = \dfrac{\text{mass}}{\text{density}}.$$

But this is only an aid to help you rearrange a formula; it is not physics and you still have to know that density is mass divided by volume.
Explain
a) why cork, wool and expanded polystyrene have very low densities.
b) why the mass of an object is the same on the moon as on earth, but its weight is different.
Build a stack of wooden blocks about five high. Try pulling out the bottom one without touching the rest of the stack and without it falling over. Can you replace the bottom block with another block of the same size, again without touching or knocking over the blocks above? (A hammer might be helpful). Explain what you do and what happens.
Try questions 4.4 to 4.9

4.3
TIME

Methods of telling the time, or measuring time, all depend on some regular event, either natural or devised. The sun has provided us with a natural clock which counts in years, as the earth travels in orbit round it, and in days, as the earth rotates on its own axis. The sundial was devised to divide up the day using the sun-cast shadow of a rod as a slowly moving pointer across a dial. But we needed clocks that ran in the dark and clocks that were accurate. As technology developed we needed clocks of ever greater accuracy until now our computers need clocks which measure time in fractions of a millionth of a second.

Measuring time

The second

The standard unit of time used in science is the **second**. Previously we defined the second as a precise fraction of the solar year 1900. But we cannot go back to the year 1900 to check its length and so this length of time is not a reproducible one. We now use an atomic clock to give us an unvarying time standard against which other clocks are checked. The atomic clock chosen is the caesium-133 atom which emits electromagnetic radiation of a precise and unvarying frequency. *The second is defined as a precise number of time periods of the atomic oscillator emitting this radiation.* The time period or time interval measured on the caesium atomic clock is about 10^{-10} seconds. (You do not need to memorise this definition but you should understand that seconds are now timed by an atomic clock.)

We need many kinds of clocks

Different kinds of clocks are needed because of the wide range of times which we try to measure, as shown in fig. 4.11. Some particularly useful clocks are the following:
a) the oscillations of a crystal such as the quartz crystal used in watches,
b) the oscillations of electrons in an electric circuit such as the 50 hertz frequency of the mains electricity which is used to drive clocks and record players at a constant speed (p 313)
c) the mechanical oscillations of a pendulum or a balance wheel in a clock or watch,
d) the time for some grains of sand to fall through a narrow neck in a tube as used in an egg timer,
e) the rotation of the earth on its axis,
f) radioactive decay clocks: the half-life of a slowly decaying radioactive isotope such as carbon-14 can be used to measure the age of remains from many thousands of years ago (p 398).

Measuring time in the laboratory

On many occasions in physics experiments we shall need to measure time intervals rather than the time of day or 'clock time'. A *time interval* is the length of time between the beginning and end of some event. Now we are briefly going to look ahead to some of the timing methods we shall use. At this stage do not worry about the details of the experiments or how the clocks work, but take note of what regular time interval each kind of clock uses, how long it is and how it is read or recorded.

If we write t as the symbol for the time of day or 'clock time' then we need a different symbol for a time interval. The symbol Δt is used for a time interval or change of time. The Δ symbol is a Greek capital D which we use to stand for *difference*. So Δt is the *time difference* between the beginning and end of an event.

Figure 4.11 *Some times and some 'clocks'*

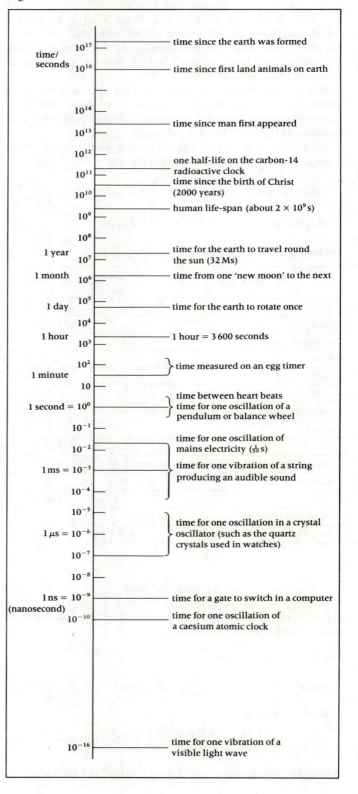

Stopwatches and timers

Mechanical, 'clockwork' stopwatches and stopclocks are usually graduated in $\frac{1}{5}$ or $\frac{1}{10}$ second time intervals. Electronic stopwatches and centisecond timers can measure time in intervals of $\frac{1}{100}$ second. Millisecond timers are also used which measure $\frac{1}{1000}$ second intervals of time.

Foster (Great Britain), Vasala (Finland) and Keino (Kenya) in the 1500 metres race in the 1972 Olympics. The timing is accurate to 0.01 s.

Although a clock may give readings for a time interval in hundredths or thousandths of a second, the measurement is not necessarily accurate to such a small fraction of a second. The accuracy of measurement of a time interval depends on both the clock and the method of starting and stopping the clock. For example, if the clock is operated manually by pressing a button then there will always be an error called the **human reaction time**, which can be a quite large fraction of a second. If an electronic clock is operated electrically then there may be some delay in the working of the switches. Even the fastest light-operated solid-state switches take a few nanoseconds to switch on or off, but these very fast switches are effectively instantaneous when used with a laboratory centi- or millisecond timer. An electric timing method is used in chapter 8 to time a falling object.

The timing accuracy using a manually operated stopwatch can be greatly improved when a regular event is timed. To understand how, try the following experiment.

Timing a pendulum

● *Set up a pendulum as shown in fig. 4.12a with a length l of about* 50 cm. *A small metal object or a glass stopper can be used as a bob.*
● *Fix a pointer opposite the position of the bob when it hangs at rest.*
● *Set the pendulum swinging and check with a protractor that the angle of swing on each side is not more than about 10°. (For a pendulum 50 cm long, the bob should be displaced no more than 10 cm to one side.)*
● *Sit in front of the pendulum so that your eye is level with the bob and at right angles to its swing.*
● *As the bob passes the pointer, start the stopwatch.*
● *When it next passes the pointer going in the same direction, stop the stopwatch and read the time interval, fig. 4.12b.*
For a 50 cm long pendulum the correct time for one oscillation, which you have just measured, is 1.4 seconds. How accurate was your timing?
● *Repeat the timing several times.*
● *By how much do your measurements vary?*

Figure 4.12 *Timing a pendulum*

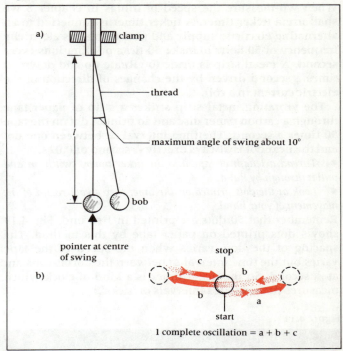

● *How could you reduce the effect of the human reaction time and so obtain a more accurate result?*
● *Try timing ten oscillations instead of just one.*
When you divide your measurement by 10 you will also reduce the error due to your reaction time by 10. If you repeat the timing of 10 oscillations several times and each time divide by 10, the values you get for the time of 1 oscillation will vary far less than before. In other words, these values are more reliable and more accurate.

We are able to improve the accuracy of this time interval measurement because it repeats regularly. In fact a pendulum is used in some clocks to regulate the timing of the clock because a pendulum of constant length always takes exactly the same length of time for each oscillation or swing, even if the swing becomes smaller in amplitude.
● *Change the mass on the end of your pendulum and, keeping the length l constant, check whether the time for one oscillation has changed.*

Notes:

1 The time interval equal to the time for one complete oscillation of a pendulum is called its **period** T, see p 435.
2 The pendulum is timed at the centre of its swing with a reference pointer because this gives the most accurate timing. [The pendulum passes through the centre of its swing in less time than it spends near the end of its swing.]
3 You can also vary the length of a pendulum and time its swings, or oscillations, to find out how its period T depends on its length l. We find that T is proportional to $\sqrt{l}$, i.e., if we make a pendulum 4 times longer it takes twice as long to make each oscillation.
4 A similar method of improving the accuracy of a manual stopwatch timing method is described on p 452 when the speed of sound is measured using multiple echoes.

Using a ticker timer

When we measure the speed of things in chapter 8 we shall use a ticker timer. A ticker timer is connected to an alternating electicity supply and uses the mains electricity frequency of 50 hertz to make 50 ticks or vibrations every second. A metal strip is made to vibrate up and down 50 times a second driven by the changes of direction of the electric current in a coil.

The vibrating metal strip strikes a strip of paper tape through a carbon paper disc and so prints a dot on the tape 50 times a second. The time interval Δt between one dot and the next dot is always exactly $\frac{1}{50}$ second or 0.02 s.

- *Thread a length of tape into the ticker timer, switch on and pull it through by hand.*
- *Look at the dots printed on the tape; they give a record of the movement of your hand.*

Remember that 50 dots are printed in 1 second. Fig. 4.13 shows dots printed on paper tape by this method. The spacing of the dots varies when the speed of the tape varies but the time interval Δt between the dots is constant at $\frac{1}{50}$ second. So the ticker timer is a kind of clock which measures time in 0.02 s intervals or 'ticks'.

Figure 4.13 *A ticker timer*

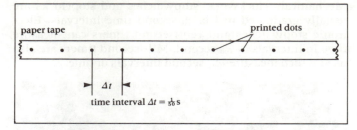

Using multiflash photographs

The time interval between flashes of light can be used to measure motion in a similar way to the ticker timer. Fig. 4.14 shows a multiflash photograph of a girl skipping. The camera shutter is kept open while a series of flashes of light form multiple images on the film. The time intervals between one flash and the next are exactly constant so that the images on the photograph can be used as timing marks on a clock. If the number of flashes in one second is known, called the flash frequency f, then the time between images is $1/f$ seconds.

Figure 4.14

Using a cathode-ray tube

The electron beam in a cathode-ray tube (TV tube) can be made to travel across the screen horizontally in a certain time. The horizontal axis across the screen then becomes a time scale or an electronic clock (p 350).

Figure 4.15 *Timing using a cathode ray tube (CRT)*

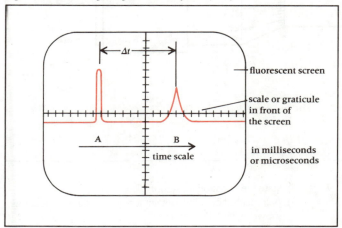

Fig. 4.15 shows two spikes or pulses on the screen produced by a radar system. A radar pulse (a kind of very brief radio signal) sent out from a transmitter produces the first spike A on the screen at the instant it sets off. When some of the radio signal is reflected back by an aircraft the second spike B is shown on the screen at the instant the reflected signal or echo is received back at the transmitter. The time interval Δt between the two spikes A and B is a measure of the time taken by radio waves to travel out to the reflecting aircraft and to come back again. From this time interval measurement we can calculate the distance away of the aircraft. Time intervals shorter than 1 microsecond can be measured by this method.

Assignments

Find out

a) What kind of a clock a clepsydra was.

b) How long the time interval between seeing and acting is, when a car driver has to stop a car in an emergency.

c) How to make a candle clock and how accurate it is.

d) How many different kinds of clocks are in use today.

Try question 4.10.

The shadow of a pointer moving round a sundial provides a simple kind of clock. Through what angle does the shadow move in 1 hour? What else happens to the shadow during the day?

Questions 4

1 Express the quantities in column 1 of the table in the units given in column 2. Use standard form for your answers. (See also appendix C3.)

1	2
0.3 mm	in metres
2.3 hours	in seconds
10 g	in kg
220 mg	in kg
0.03 kg	in grams
0.03 m	in mm
25 cm	in metres
40 cm²	in m²
9.2 mm²	in m²
0.01 m²	in mm²

2 Fig. 4.16 shows vernier calipers being used to determine the diameter of a cylindrical rod. What is the reading shown by the calipers?   (O & C)

Figure 4.16

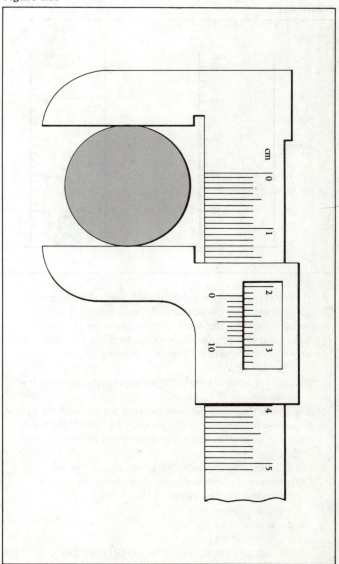

Figure 4.17

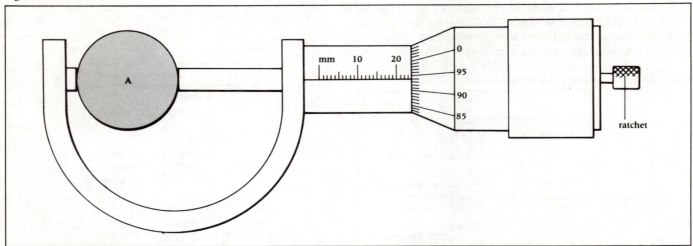

3   Fig. 4.17 shows an enlargement of a micrometer screw gauge set to measure the diameter of the ball bearing A. Write down the reading of the gauge. What is the purpose of the ratchet?   (C)

Figure 4.18

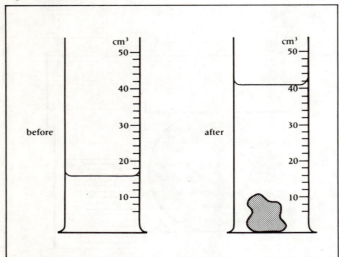

4   A lump of plasticine was weighed and found to have a mass of 85.0 g. It was then lowered carefully into a measuring cylinder containing some water. Fig. 4.18 shows the levels of the water before and after the plasticine was lowered into the cylinder.
   a) Write down the reading on the measuring cylinder
      i) before ii) after.
   b) Calculate i) the volume of the plasticine, ii) the density of the plasticine.   (W 88)

5   Explain the difference between the mass and the weight of a body. State how these quantities change when the same object has its mass and weight measured, first on the earth and then on the moon.   (JMB part)

6   A pile of 500 sheets of paper has a mass of 2 kg. The pile is 300 mm long, 200 mm wide, and 50 mm thick. Calculate:
   a) the thickness of one sheet,
   b) the mass of one sheet,
   c) the volume of the pile,
   d) the density of the paper,
   e) the mass of a one square metre sheet of this paper.   (O)

7   a) i) State the equation for *density*.
      ii) A graduated cylinder contains a certain volume of liquid, as shown in fig. 4.19.

Figure 4.19

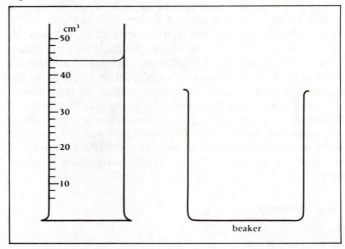

This volume of liquid is poured into a beaker. Using the additional information given below find the mass and the density of the liquid, showing clearly how you get your answer.
      mass of empty beaker   = 205.1 g
      mass of beaker plus liquid = 235.9 g
   b) A graph showing how the mass and volume of aluminium objects are related is shown in fig. 4.20.
      i) What is the mass in g of a block of aluminium of volume 20 cm³? Show how you get your answer.
      ii) Find the density of aluminium. Show how you get your answer.
      iii) In what way would the graph differ for a material of greater density?   (NISEC 88 part)

8   A tin containing 5 000 cm³ of paint has a mass of 7.0 kg.
   i) If the mass of the empty tin, including the lid, is 0.5 kg calculate the density of the paint.
   ii) If the tin is made of a metal which has a density of 7 800 kg/m³ calculate the volume of metal used to make the tin and the lid.
   (JMB)

Figure 4.20

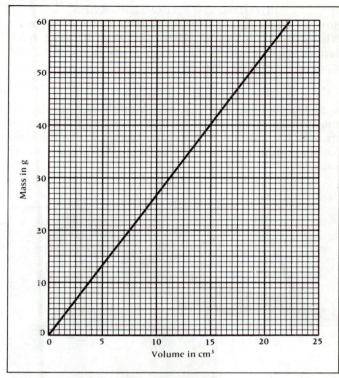

9   Describe how you would find the density of a dense solid of irregular shape; make clear the measurements to be made and the method of calculating the result.

An astronaut on the moon is measuring the density of an irregularly shaped rock. He finds that the volume of the rock is 80 cm³ (8.0×10^{-5} m³) and, by comparison with standard masses on a lever balance, that the mass of the rock is 0.44 kg. What is the density of the rock?

When the experiment is repeated on earth with the same piece of rock, the values obtained for the mass and for the volume are the same as those found on the moon. Explain this.   (C part)

10   A pendulum is a simple oscillating system.
a) The **period** of oscillating is the time for one complete oscillation. How would you measure the period of oscillation of the pendulum?

Figure 4.21

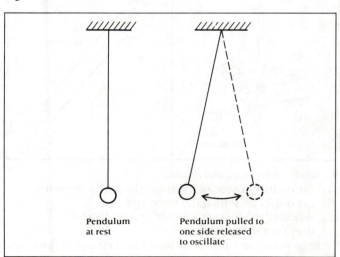

b) Here are some results which were obtained using pendulums of different lengths.

Length of pendulum, *l*, in m	0.30	0.40	0.50	0.60	0.70	0.80
Period of oscillation, *T*, in s	1.09	1.26	1.40	1.54	1.66	1.78

On the graph paper similar to that shown in fig. 4.22.
i) label the vertical axis correctly,
ii) plot the points (the first two have been plotted for you),
iii) draw a smooth curve through the points.

Figure 4.22

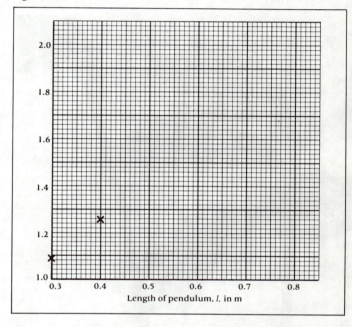

c) A straight-line graph can be obtained if (period of oscillation)² is plotted against the length of pendulum, fig. 4.23.
i) Use the triangle to find the gradient (slope) of the line.
ii) Calculate the acceleration due to gravity, *g*, using the formula: $g = 40/\text{gradient}$.   (SEG 88)

Figure 4.23

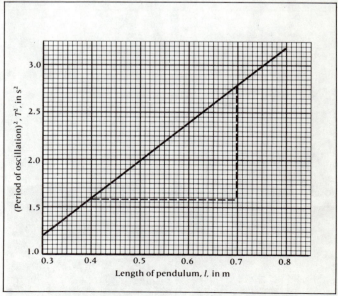

5
Forces

We use and feel forces all the time but we are not always consciously aware of them. Forces which push and pull and get things moving or slow things down are easy to spot, but forces which hold things still and keep things balanced are less obvious. Whatever its effect, a force is the action of one object on another object.

5.1
FINDING OUT ABOUT FORCES

How do we recognise a force and how do we describe it? What causes a force and what effect does it have? What happens when several forces act at the same time and place on a single object?

What is a force?

A force is a *push* or a *pull* which one object applies to another object. In fig. 5.1 several forces are indicated by arrows. The tail-end of the arrow begins where the force pushes or pulls and the arrow points in the direction of the force.

Figure 5.1 *Forces*

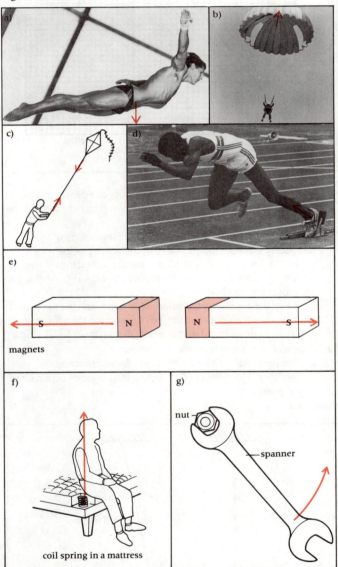

- *Study each picture and decide:*
  - i) the name or cause of the force shown by the arrow,
  - ii) the object on which the force acts,
  - iii) whether the force is a 'push' or a 'pull' and
  - iv) the effect of the force.

To help you, some names of forces and effects of forces are given below.

Naming forces

You will meet all these kinds of forces in this book.

Weight The weight of an object is the pull of the earth (or another large object such as the moon) acting on it. Any two masses are attracted to each other by what is known as gravitational attraction and the pull of the earth on an object is sometimes called the force of gravity.

Tension A stretched rope or spring pulls at both of its ends as it tries to reduce its length back to normal. The pull of a string or spring is called its tension.

Contact force A push produced when two objects are pressed together and their surface atoms try to keep them apart is called a contact force.

Expansion force A push found in a compressed spring or a squashed material and the push produced when a heated object tries to expand may be called an expansion force.

Upthrust force The buoyancy which all objects find when immersed in a liquid or gas is caused by an upwards push called an upthrust force (see Archimedes' principle, p 100).

Resistance Forces which oppose or prevent motion, such as air resistance or drag, are called resistance forces. The drag of the air resistance against a car increases with the speed of the car.

Friction Frictional forces are also forces which resist motion. The forces between the surfaces of two objects which act parallel to the surfaces and prevent or resist them sliding or slipping are called friction.

Magnetic forces Magnetic forces act on magnetic materials and on electric currents as in the electric motor (see chapter 15).

Electric forces Forces between electric charges are called electric forces. These forces make your hair cling to a plastic comb or your clothes stick together (see chapter 11).

What do forces do?

Here are some of the effects which forces have on the objects which they push or pull.

Forces get objects moving and cause acceleration.
Forces stop moving objects and cause deceleration.
Forces change the direction in which objects are moving.
Forces stretch objects.
Forces squash or compress objects.
Forces bend or distort objects.
Forces turn or twist objects.
Forces prevent movement of objects by balancing other forces acting on the same object.

● *When you have studied the pictures in fig. 5.1 check your answers against these:*

a) The arrow shows the diver's *weight*.
   This force is the pull of the earth on the diver.
   Gravity always pulls.
   This force is getting the diver moving and gives a downwards acceleration.

b) The arrow shows the air *resistance* or *drag*.
   The drag force of the air acts on the moving parachute.
   The force pushes on the inside of the parachute.
   The drag opposes the motion of the parachute and the weight of the parachutist and prevents acceleration.

c) The arrows both show the *tension* in the string.
   At one end the string tension pulls the child and at the other end it pulls the kite.
   Strings can only pull, they never push!
   The tension force on the kite balances the pull of the kite due to the wind.

d) The arrow shows the *contact force* of the starting block.
   The contact force of the block pushes the sprinter.
   This is a forwards push.
   The push gets the sprinter moving and gives a forwards acceleration.

e) The arrows show the *magnetic force*.
   Magnet X pushes magnet Y and magnet Y pushes magnet X.
   Like poles repel, i.e. push each other (p 287).
   The forces push or hold the magnets apart.

f) The arrow shows the *expansion force* of a compressed spring.
   The spring pushes the sitter.
   This is an upwards push.
   The upwards force balances the weight of the sitter.

g) The arrow shows the pull of the hand.
   The pull acts on the spanner.
   This pull is known as a **turning force** or a **moment**.
   The turning force turns the spanner and the nut and so it overcomes the frictional grip between the nut and the bolt.

Measuring forces

Some forces are weak and others are strong. This is a way of saying that forces have sizes or **magnitudes**.

The unit of force, called a **newton** (symbol N), is based on the acceleration of an object caused by an unbalanced force and will be explained and defined in chapter 8. For now it is important to get some idea of the magnitude of different forces expressed in newtons.

Fig. 5.2 shows a range of forces from about one hundredth of a newton (0.01 N) to one hundred newtons (100 N).

● *Lift each object and feel its weight.*

Figure 5.2 *Forces measured in newtons*

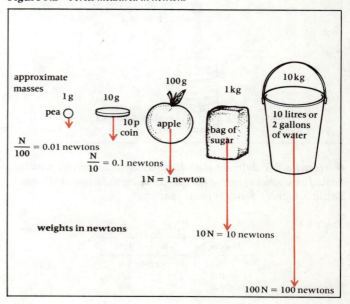

approximate masses

pea 1g
10p coin 10g
apple 100g
bag of sugar 1kg
10kg 10 litres or 2 gallons of water

weights in newtons

$\frac{N}{100} = 0.01$ newtons
$\frac{N}{10} = 0.1$ newtons
$1 N = 1$ newton
$10 N = 10$ newtons
$100 N = 100$ newtons

The force you use to hold the object and the force you feel pressing against your hand are equal to the weight of the object. The force of gravity pulling the apple towards the earth is about one newton. In other words, the weight of the apple is about 1 N.

Note the clear distinction between the *mass* of the apple which is about 100 *grams* and its *weight* which is about 1 *newton*. Notice also that there is a proportional relation between the weights of the objects and their masses. On the earth there is a force of about 10 newtons acting on each kilogram of mass.

Using a spring balance to measure forces

Spring balances are calibrated using known forces to stretch or compress their springs. Since spring balances measure forces, *they should be calibrated in newtons* but they often have scales in grams or kilograms. (This is so that they can also be used to find masses on earth. The readings in grams or kilograms are correct only on the surface of the earth.)

We can test that weight is proportional to mass by weighing some known masses on a newton spring balance, fig. 5.3.

Figure 5.3 *Measuring weight on a spring balance*

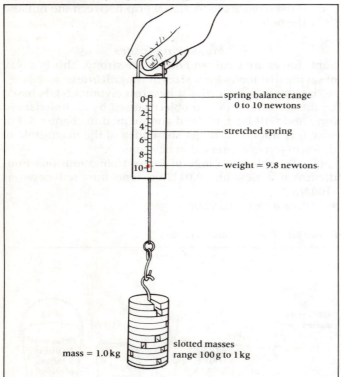

spring balance range
0 to 10 newtons

stretched spring

weight = 9.8 newtons

mass = 1.0 kg

slotted masses
range 100 g to 1 kg

● *Hang a few different slotted masses on their hanger which is hooked on a newton spring balance and note the weight readings.* Table 5.1 gives some typical readings.

Table 5.1

Mass m/kg	Weight W/N
0.1	nearly 1.0
0.5	4.9
1.0	9.8

We do find that the weight of an object is proportional to its mass; in other words, if we double the mass we double the weight and so on. And we find that (on the surface of the earth) the weight of an object of mass 1 kg is 9.8 N.

This means that the earth's gravitational pull on an object of mass 1 kg is 9.8 N at the earth's surface. We can express this fact another way and say that:

The strength of the earth's gravitational field (symbol g) at the surface of the earth is 9.8 newtons per kilogram.

The relation between mass and weight is given by:

$$W = mg$$

and at the surface of the earth:

$$W = m \times 9.8 \, \text{N/kg}$$

or roughly

$$W = m \times 10 \, \text{N/kg}$$

For example: the weight of a girl of mass 35 kg is given by

$$W = 35 \, \text{kg} \times 10 \, \text{N/kg} = 350 \, \text{N}.$$

(Simply multiply a mass in kg by 10 to find its weight in newtons.)

Weighing in different places

If we could weigh the same apple on the earth, on the moon and in deep space away from all detectable gravity forces we would find the results shown below. We see that the spring balance does not measure the *mass* of the apple but just its *weight*. The spring balance gives the correct reading of zero weight for the apple in space. But an astronaut holding the top of the balance with one hand and its hook with the other hand could still stretch the spring and measure the pull of his hands against each other. In this case there is no weight involved.

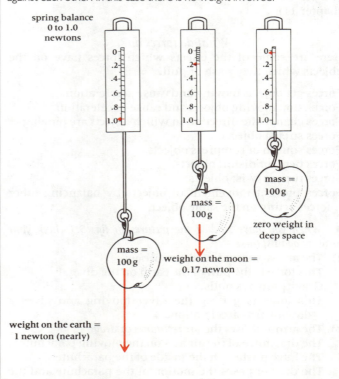

spring balance
0 to 1.0
newtons

mass =
100 g

mass =
100 g

mass =
100 g

zero weight in
deep space

mass =
100 g

weight on the moon =
0.17 newton

weight on the earth =
1 newton (nearly)

Adding forces

We have measured forces in newtons and know that they have a size or magnitude. We have also noticed that all forces act in a particular direction. For example, weight always pulls towards the earth and acts downwards. So how do we add two forces which act on the same object taking into account their directions, which may be different?

Vectors and scalars

A physical quantity which has no direction is called a scalar quantity.

Time, temperature, mass, volume and density are examples of quantities without any direction. They are all scalars. *Scalar quantities are added by the normal rules of arithmetic.* For example, a tank holds 5 cubic metres of water. If 1.5 cubic metres of water are run out of it, there will be 3.5 cubic metres remaining. The rules of arithmetic allow only one calculation and only one answer:

$$5\,\mathrm{m}^3 - 1.5\,\mathrm{m}^3 = 3.5\,\mathrm{m}^3$$

A vector quantity is one which has both magnitude and direction.

So force is a vector quantity. Other quantities we shall meet which are vectors include velocity, acceleration and momentum.

All vector quantities obey a special rule for addition and subtraction which takes account of direction as well as magnitude. A vector can be represented by a straight line with an arrow on one end. The length of the line represents the magnitude of the vector quantity (sometimes drawn to scale), and the direction of the line gives the direction and line of action of the vector. The special rule is known as the parallelogram law.

The parallelogram law for adding forces

Two forces acting on the same object can be represented by a single equivalent force. In this experiment two forces F_1 and F_2 produce a certain extension of a spring S. They are then replaced by a single force F_3 which produces the same extension of spring S. So F_3 is equivalent to F_1 and F_2 added together.

• *Pin a large sheet of paper to a drawing board and fix one end of the spring S by a nail or clamp near one end of the board, fig. 5.4a.*
• *Attach a small metal ring to the other end of spring S and to the hooks of two spring balances using thread.*
• *Attach another length of thread to the loops on the top ends of each of the spring balances and pull these threads in different directions.*
• *When the spring S is stretched and the balances give readings which are about half the full scale, clamp the threads at the edges of the drawing board and mark the direction of each force F_1 and F_2 with a cross.*
• *Draw round the outside or inside of the metal ring to mark its position precisely.*
• *Record the readings on the spring balances for F_1 and F_2.*
• *Now replace the two forces with a single force F_3, fig. 5.4b. Pull the thread attached to the balance until the metal ring is centred on the circle previously drawn round it.*

The spring has exactly the same extension as before. If the metal ring is pulled to exactly the same position, the single force F_3 is equivalent in both magnitude and direction to the combined pulls of F_1 and F_2. In other words, F_3 is the sum of F_1 and F_2.

• *Mark the direction of F_3 and record the reading on the balance. Remove the apparatus from the sheet of paper.*

Figure 5.4

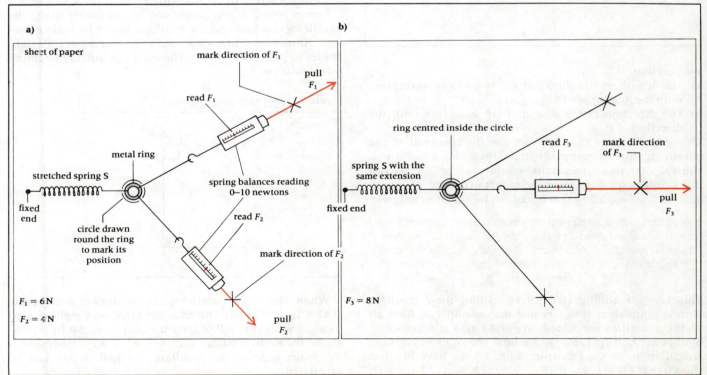

Constructing the force parallelogram

- *Draw lines from the markers for F_1 and F_2 through the centre O of the circle.*
- *Using a suitable scale (such as 1 cm representing 1 N), draw arrows OA and OB to represent F_1 and F_2.*
- *Complete the parallelogram OAPB as follows. With a pair of compasses, draw an arc, centred on A, of radius equal to OB (4 cm in fig. 5.5a). Draw another arc, centred on B, of radius equal to OA (6 cm in the figure). The two arcs cross at point P on the parallelogram. Draw the sides AP and BP to complete the parallelogram and then draw its diagonal OP extending the line beyond P, fig. 5.5b.*
- *Measure the length of the diagonal OP and note its direction.*

Figure 5.5

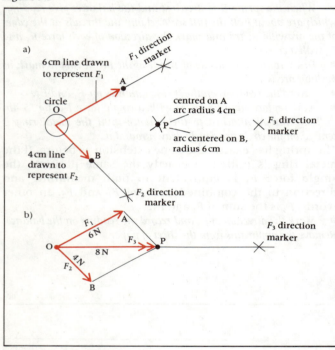

We find that:

a) The length of the diagonal OP is in close agreement with the magnitude of F_3.

b) The direction of the diagonal OP is in line with the direction of F_3.

This demonstrates that we can use the diagonal of the parallelogram to represent the single force which is equivalent to two forces acting on the same object. We call this single equivalent force the **resultant force**. *The parallelogram law* for adding two forces can be stated as follows:

If two forces, acting at one point on the same object, are represented in magnitude and direction by the sides of a parallelogram drawn from the point, their resultant is represented in both magnitude and direction by the diagonal of the parallelogram drawn from the point.

This law for adding two forces to find their resultant (single equivalent force) is just one example of how all vector quantities are added. We shall also add velocities by this method (p 126). Notice how the rules of addition are different: in the example of fig. 5.5 we have in effect shown that 4 N + 6 N = 8 N!

Special cases: parallel forces and forces at right angles

Parallel forces which act in the same line on the same object can be added arithmetically taking account of their directions. The parallelogram is completely flat so that the diagonal equals the length of its two sides added together. Fig. 5.6 gives examples of addition of parallel forces. Notice how forces in opposite directions have opposite signs so that when they are 'added' their magnitudes subtract (examples (b) and (c)).

Figure 5.6 *Adding parallel forces*

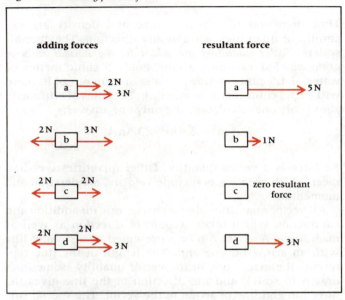

To specify the resultant force you must give both its magnitude and direction. For example, the resultant in example (b) is a force of 1 N acting in the same direction as the 3 N force.

Forces acting at any angle can be combined using the parallelogram law and the resultant found by scale drawing. However, when two forces F_1 and F_2 act at right angles to each other their resultant R can quickly be found by calculation, fig. 5.7.

Figure 5.7 *Adding forces at right angles*

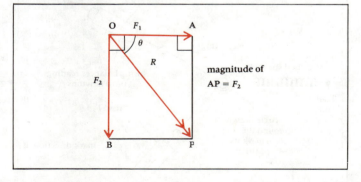

When the force parallelogram is drawn a rectangle OAPB is produced. The triangle OAP is a right-angled triangle with side AP of length equal to F_2. So by Pythagoras' theorem, we have $OP^2 = OA^2 + AP^2$, from which the magnitude of the resultant R equal to OP can be calculated.

When F_1 and F_2 are perpendicular forces acting on the same object we can write for the magnitude of the resultant R:

$$R^2 = F_1^2 + F_2^2$$

The direction of the resultant R can be specified by the angle θ, that is, the angle between the resultant force and one of the original forces. Angle θ is found by trigonometry as follows:

$$\tan \theta = \frac{AP}{OA}$$

or
$$\tan \theta = \frac{F_2}{F_1}$$

So the angle θ is the angle whose tangent is equal to F_2/F_1 (This angle is found using the inverse or arctan key on a calculator.)

Worked Example
Adding forces at right angles

Two forces of 3 N and 4 N act at right angles at the same point on an object, fig 5.8a. Find by calculation the resultant force which is equal in magnitude and direction to the combined effect of the two forces.

Figure 5.8 *Adding forces at right angles*

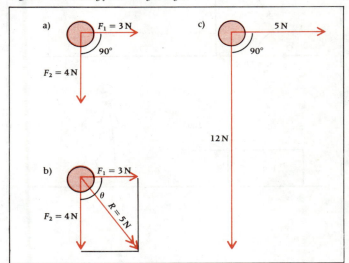

The forces are at right angles so we may use Pythagoras' theorem:

$$R^2 = F_1^2 + F_2^2$$
$$\therefore R^2 = (3\,N)^2 + (4\,N)^2 = 25\,N^2$$
$$\therefore R = 5\,N$$

Using the tangent relation we have (fig. 5.8b):

$$\tan \theta = \frac{F_2}{F_1} = \frac{4\,N}{3\,N} = 1.33$$
$$\therefore \text{angle } \theta = 53°.$$

Answer: The resultant force has a magnitude of 5 newtons and acts in a direction between the two forces at an angle of 53° from the original 3 newton force.

Note you can use this calculation method to find the resultant of the two forces shown in fig. 5.8c.

Practical examples of adding forces

The photograph of two tugs pulling a ship shows how two towrope forces acting at an angle to each other combine to give a resultant force which pulls the ship along a line between the two towropes.

At a point in a steel structure where several girders are joined together there will be different pulls and pushes in each of the girders. To calculate the strength required for each girder, the engineer must calculate the resultant force on each girder by adding the forces in the other parts of the structure which are joined to it. These very complicated vector additions are done at the design stage by computers. The computer programs are based on the parallelogram law for addition of forces.

Resolving a force

The parallelogram law is used to combine two forces to find an equivalent single force. When we try to analyse the effect of a *single* force acting on an object it is often helpful to use a process which is the reverse of finding a resultant force.

The reverse process, called **resolving a force**, divides a single force into two parts which act at right angles to each other. The two parts of a resolved force, called its **components**, have exactly the same effect on an object as the single force. The following example shows how we can resolve the weight of an object resting on a slope into two components and why this is a useful process.

Fig. 5.9 shows an object O resting on a slope inclined at an angle α to the horizontal. The weight W of the object acts downwards at an angle θ to the slope. The weight of the object has two effects. Part of the weight pushes against the surface of the slope and causes a contact force C_1 normal to the slope and another part of the weight tries to pull the object down the slope with a force C_2 which is parallel to the slope. The two components C_1 and C_2 are both provided by the weight W and so we can consider them as being equivalent to W, fig. 5.9b.

Figure 5.9 *Resolving the weight of an object into components*

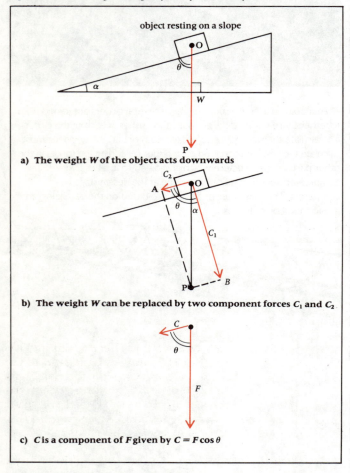

a) **The weight W of the object acts downwards**

b) **The weight W can be replaced by two component forces C_1 and C_2**

c) **C is a component of F given by $C = F\cos\theta$**

Calculating components

From the angles of the triangle in fig. 5.9a we can see that $\alpha + \theta = 90°$. Since the components C_1 and C_2 are perpendicular we can see that angle BOP = α.

In $\triangle$ BOP, $\cos\alpha = \dfrac{BO}{PO} = \dfrac{C_1}{W}$

$\therefore$ component $C_1 = W\cos\alpha$.

In $\triangle$ AOP, $\cos\theta = \dfrac{AO}{PO} = \dfrac{C_2}{W}$

$\therefore$ component $C_2 = W\cos\theta$.

We can summarise these results as follows: To find the magnitude of a component C, multiply the force F by the cosine of the angle θ between F and C, fig. 5.9c. In general the component force C is given by:

$$C = F\cos\theta$$

Worked Example
Finding a component of a force

A man attempts to pull a box along the ground. He can pull a rope attached to the box with a force of 100 N. A force of 70 N is needed to overcome frictional forces along the ground and just get the box moving. If he uses a short rope to pull the box it makes an angle of 60° with the horizontal, fig. 5.10a, but if he uses a longer rope it makes an angle of 20° with the horizontal, fig. 5.10b. Explain why he fails to move the box in case (a) and succeeds in case (b), even though in both cases he pulls the rope with a force of 100 N.

Figure 5.10 *Pulling a box along the ground*

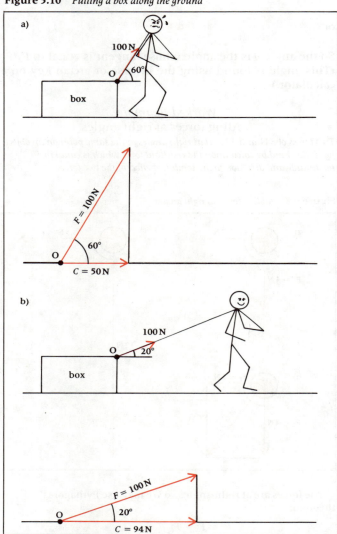

The force in the rope can be considered to have two components, one which pulls the box along the ground and acts parallel to the ground and the other which pulls against the weight of the box and acts upwards.

The component C which pulls the box along the ground is given by $C = T\cos\theta$.

In case (a) $C = 100\,N\cos 60° = 100\,N \times 0.5 = 50\,N$
In case (b) $C = 100\,N\cos 20° = 100\,N \times 0.94 = 94\,N$

So in case (a) the component force parallel to the ground is less than the 70 N needed to move the box. In case (b) the lower angle of pull by the longer rope gives a much larger component force in the horizontal direction. 94 newtons can get the box moving.

Describing a force

We now know enough about forces to be able to describe a force more precisely. The following information can be given about a force:

a) its size or magnitude measured in newtons,
b) its direction,
c) wherever there is a force, two objects are involved and it is always possible to say that one object pushes or pulls the other object,
d) the point on the object at which the force acts.

Assignments

Explain the meanings of the words:
a) vector, scalar, resultant and component.
Remember
b) the relation between weight and mass: $W = mg$
c) how to add two forces together which are (i) parallel, (ii) at right angles and (iii) at any angle.
d) the formulas for calculating the magnitude and direction of a resultant when the forces are at right angles:

$$R^2 = F_1^2 + F_2^2 \quad \text{and} \quad \tan\theta = \frac{F_2}{F_1}$$

e) a component force is given by: $C = F \cos\theta$.
Try questions 5.1 to 5.5

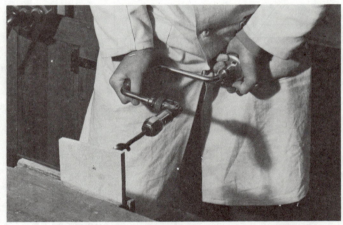

The turning effect produced by the person's hand is increased by the crank of a brace. The crank (the bent part of the tool) increases the distance of the force from the axle of the tool and so increases its moment.

Figure 5.12   *Defining the moment of a force*

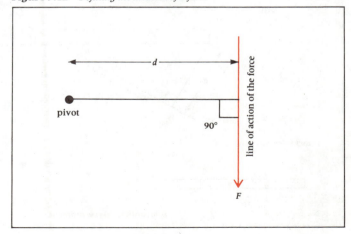

5.2
TURNING FORCES

Why is it easier to loosen a tight nut using a long spanner? How can a seesaw balance when people of different weights sit on opposite sides? What makes a racing car less likely to turn over than an ordinary car?

These are the kinds of questions which we can answer when we understand the effects of turning forces.

The moment of a force

We call the *turning effect* of a force the **moment of the force**. To find out what factors are involved in the moment of a force try the simple experiment shown in fig. 5.11.

Figure 5.11   *Feeling the moment of a force*

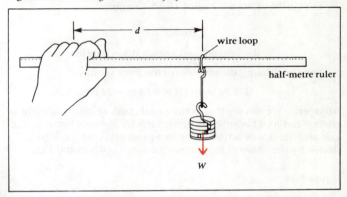

●   *Hang a slotted mass holder from a wire loop attached to a half-metre ruler.*
●   *Hold one end of the ruler in your hand so that you keep the ruler level.*
●   *Try changing (a) the weight W of the slotted masses and (b) the distance d of the hanger from the middle of your hand.*
●   *What can you feel?*

The twisting or turning effect that your hand feels and has to resist is the moment of the force. The moment of the force depends on both the magnitude of the force and how far away it is from the turning point. The turning point has several names and can be several things. Names which are used include pivot, axis and fulcrum. Knife-edges, axles, hinges and the edges or corners of objects can act as turning points.

We measure the moment of a force by multiplying the magnitude of the force by its distance from the pivot. To be more precise we define the moment of force M as follows:

The moment of a force is the product of the magnitude of the force F and the perpendicular distance d from the pivot to the line of action of the force (fig. 5.12).

$$M = F d$$

where F and d are perpendicular.

The unit of the moment of a force is the **newton metre** (N m). (Note: although newton × metre = joule (p 107), which is the unit of work and energy, the concept of the moment of a force is quite different from energy and so we write its units as N m and do not use the joule.)

You can now explain why a longer spanner will make it easier to loosen a tight nut using the same force: when you apply that force at a greater distance from the nut, the moment of the force will be greater.

Worked Example
Calculating the moment of a force

The crank of a bicycle pedal is 16 cm long and the downwards push of a leg is 400 N. Calculate the moment of the force when (a) the crank is horizontal and (b) the crank has turned to an angle of 60° below the horizontal, fig. 5.13.

a) When the crank is horizontal its length l is perpendicular to the downwards foot push F. So the perpendicular distance d from the pivot to the line of action of the force is $l = 0.16$ m.

Using $\qquad M = Fd$
we have $\qquad M = 400\,\text{N} \times 0.16\,\text{m} = 64\,\text{N m}.$

b) When the crank has turned to 60° below the horizontal the perpendicular distance from the pivot to the line of action of the foot push F is d.

The cosine of the 60° angle in fig 5.13b gives:

$$\cos 60° = \frac{d}{l}$$

$$\therefore\ d = l\cos 60° = 0.16\,\text{m} \times 0.5 = 0.08\,\text{m}$$

Now calculating the moment of the force we have:

$$M = Fd = 400\,\text{N} \times 0.08\,\text{m} = 32\,\text{N m}.$$

Answer: We can see that as the pedal crank turns, its turning effect is reduced because the perpendicular distance between the axle and the line of action of the foot push gets smaller. This is shown by the answers for the two moments: 64 N m and 32 N m.

Figure 5.13

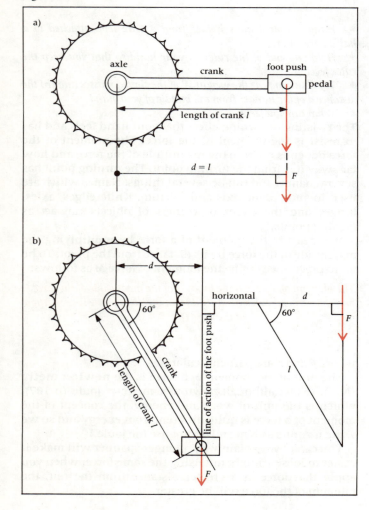

Investigating the law of moments

The law of moments is also known as the **law of the lever** and as the **principle of moments**. This law is about objects balancing when the moments of all the forces acting are balanced. To investigate the law, set up a metre ruler (or half-metre ruler) as shown in fig. 5.14 and try the following:

• *Hang two slotted mass hangers from small wire loops fitted on the ruler, as in fig. 5.14a.*
• *Set mass m_1 at a fixed distance d_1 from the pivot and slide mass m_2 along the ruler until it balances.*
• *Record the values of m_1, m_2, d_1 and d_2 in a table.*
• *Repeat the procedure with different values of each of the four quantities. Complete the table as shown in table 5.2.*

Figure 5.14 *Experiments on the laws of moments*

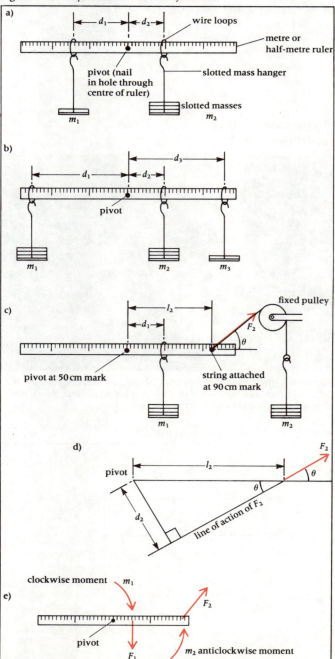

Table 5.2

Mass m_1/g	Force F_1/N	Perpendicular distance d_1/m	Moment of force M_1/Nm	Mass m_2/g	Force F_2/N	Perpendicular distance d_2/m	Moment of force M_2/Nm
200	2.0	0.30	0.60	400	4.0	0.15	0.60

To calculate the forces use $F = mg$,

example: $F_1 = 0.2\,\text{kg} \times 10\,\text{N/kg} = 2.0\,\text{N}$.

To calculate the moment of the forces use $M = Fd$,

example: $M_1 = 2.0\,\text{N} \times 0.3\,\text{m} = 0.6\,\text{Nm}$.

The results for the moments of the forces on each side of the ruler show that when the ruler is balanced the moments are equal: $M_1 = M_2$.
- *Repeat the experiment using two forces on one side of the ruler as shown in fig. 5.14b.*
- *Complete another table of results calculating the moments of forces F_2 and F_3 separately:*

example: $M_2 = F_2 \times d_2$ and $M_3 = F_3 \times d_3$.

The results this time show that $M_1 = M_2 + M_3$. (Note that it would be incorrect to add the two forces F_2 and F_3 together because they act at different distances from the pivot and therefore have different turning effects.)
   The third experiment changes the direction of one of the forces so that it is not perpendicular to the ruler.
- *Set up an arrangement like the one shown in fig. 5.14c using a pulley to turn the weight of the mass m_2 into a tension force F_2 acting at an angle θ to the ruler.*
   If the string is attached to the ruler at the 90 cm mark then the distance from the pivot l_2 is 40 cm, but this is not the perpendicular distance d_2. Fig. 5.14d shows how the perpendicular distance d_2 from the pivot to the line of action of F_2 is found.
- *Measure the angle θ between the string and the ruler.*
In the triangle in fig. 5.14d we can see that $\sin \theta = d_2/l_2$

$$d_2 = l_2 \sin \theta$$

- *Complete another table of results showing values for l_2 and d_2.*
The moments of the two forces are again equal, but as they act on the same side of the ruler we describe M_1 as a **clockwise moment** and M_2 as an **anticlockwise moment**, these being the directions in which the forces try to turn the ruler, fig. 5.14e.
   Summarising the main points we note that:
a) all the perpendicular distances are measured from the pivot,
b) when two forces produce a moment in the same direction we add the moments of the forces and not the forces themselves,
c) whenever the ruler is balanced the moments turning it in the clockwise direction are equal to the moments turning it in the anticlockwise direction.
   We describe a balanced object which is neither moving nor turning as being in **equilibrium** and the *law of moments* describes such an object:

When an object is in equilibrium, the sum of the clockwise moments about any point (acting as a pivot) equals the sum of the anticlockwise moments about the same point.

Worked Example
Using the law of moments
A boy of weight 500 N sits on the left side of a seesaw a distance of 2.4 metres from its pivot. If a girl can balance the seesaw by sitting 3.0 metres from the pivot on the right side, what is her weight?

Figure 5.15

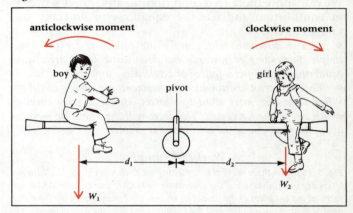

Referring to fig. 5.15 and using the law of moments when the seesaw is balanced, we can write:

anticlockwise moment of the boy's weight = clockwise moment of the girl's weight

and
$$W_1 d_1 = W_2 d_2$$
$$500\,\text{N} \times 2.4\,\text{m} = W_2 \times 3.0\,\text{m}$$
$$W_2 = \frac{1200\,\text{Nm}}{3.0\,\text{m}} = 400\,\text{N}$$

Answer: The girl's weight is 400 newtons.

Worked Example
Testing the law of moments
Some results obtained from an experiment set up as shown in fig. 5.14c were as follows:

$d_1 = 0.20\,\text{m}$, $l_2 = 0.40\,\text{m}$, $m_1 = 600\,\text{g}$, $m_2 = 420\,\text{g}$, $\theta = 45°$.

Do these results agree with the law of moments?

$$F_1 = m_1 g = 0.60\,\text{kg} \times 10\,\text{N/kg} = 6.0\,\text{N}$$
$$F_2 = m_2 g = 0.42\,\text{kg} \times 10\,\text{N/kg} = 4.2\,\text{N}$$
$$d_2 = l_2 \sin \theta = 0.40\,\text{m} \times 0.71 = 0.28\,\text{m}$$

clockwise moment of $F_1 = F_1 d_1 = 6.0\,\text{N} \times 0.20\,\text{m} = 1.2\,\text{Nm}$
anticlockwise moment of $F_2 = F_2 d_2$
$= 4.2\,\text{N} \times 0.28\,\text{m} = 1.176\,\text{Nm} = 1.2\,\text{Nm}$.

Answer: The moments are equal to two significant figures.

Parallel forces
When two or more parallel forces act on an object several different things may happen. First we decide whether the object is in equilibrium. If it neither moving nor turning it is definitely in equilibrium. (The forces acting on an object can also be in equilibrium if the object has motion which is not changing, but you will not meet questions of that difficulty.)

Parallel forces in equilibrium

When an object is in equilibrium and two or more parallel forces are acting on it we can say that:

1

The sum of the forces acting on it in one direction must equal the sum of the forces acting on it in the opposite direction.

2

The sum of the clockwise moments about any point on the body must equal the sum of the anticlockwise moments.

We can apply these two conditions to any object which is in equilibrium and use the equations to find any unknown forces.
• *Choose directions which make the equations for the forces simple. For example, upwards and downwards forces are always equal unless an object is falling or accelerating upwards.*
• *Choose a pivot about which the moments are easy to calculate. When there are more than two forces, choose a pivot through which one of the forces acts. This force will then have no moment about the pivot and so the calculation is simpler.*

Worked Example

Fig. 5.16 shows three vertical forces acting on a wheelbarrow. These three forces are in equilibrium if the wheelbarrow neither moves up or down nor turns about the pivot at the axle of its wheel. Find the value of the force F needed to hold the handles and the reaction force R at the axle of the wheel.

Figure 5.16

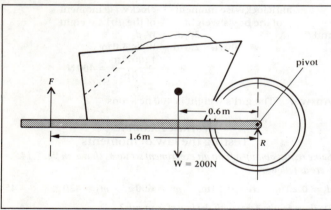

Using condition 2, we can apply the law of moments about the pivot at the wheel.

$$\frac{\text{clockwise moment}}{\text{of lifting force } F} = \frac{\text{anticlockwise moment of weight } W}{\text{of wheelbarrow and load}}$$

$F \times$ distance from pivot $= W \times$ distance from pivot
$$F \times 1.6\,\text{m} = 200\,\text{N} \times 0.6\,\text{m}$$
$$\therefore F = \frac{200\,\text{N} \times 0.6\,\text{m}}{1.6\,\text{m}} = 75\,\text{N}$$

Using condition 1, we can now find the reaction force R at the wheel.
Sum of the upward forces = sum of the downward forces
$$F + R = W$$
$$75\,\text{N} + R = 200\,\text{N}$$
$$\therefore R = 200\,\text{N} - 75\,\text{N} = 125\,\text{N}$$

Couples

When you use two hands to turn the handlebars of your bicycle or the steering wheel of your car you are applying two parallel forces to the same object. These two forces are not in equilibrium because they do not act in the same straight line. They are called a couple and have a turning effect or moment.
A couple is a pair of forces acting on an object which:
• *are equal in magnitude and opposite in direction,*
• *do not act along the same straight line,*
• *apply a moment to the object and so tend to turn it,*
• *do not produce a single resultant force and so do not tend to move it from one position to another.*
In fig. 5.17a the total moment of the two forces about the pivot P is given by:

$$\text{moment of the couple} = Fy + Fx = F(y + x) = Fd$$

This means that the combined turning effect of the two forces (the moment of the couple) is given by $F \times d$.

The moment of a couple is the product of the magnitude of one of the forces F and the perpendicular distance d between the two forces.

Figure 5.17   *The moment of a couple*

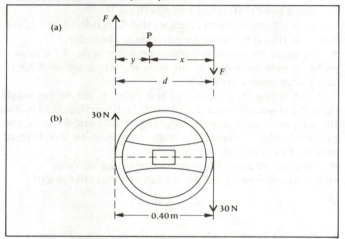

Worked Example
The moment of a couple

The hands of a motorist apply equal and opposite forces of 30 N to the steering wheel of a car, fig. 5.17b. If the diameter of the steering wheel is 0.40 m find the moment of the couple applied to the steering wheel.

Moment of couple = force × perpendicular distance
between the forces
$$= 30\,\text{N} \times 0.40\,\text{m} = 12\,\text{Nm}$$

Assignments

Explain
a) why it is easier to loosen a tight nut using a longer spanner.
b) why it is easier to push a door open when you push at a point near the edge of the door which is farthest from the hinge.
Find
d) two examples of single forces which are used to turn things.
e) two examples of couples which are used to turn things.
Estimate the moment of the force which your foot applies to your bicycle pedal.

Try questions 5.6 to 5.8

5.3
BEAMS AND BRIDGES

Beams

Beams are the most common part of large structures such as bridges and buildings. All beams supporting a load tend to bend a little. The forces which cause a beam to bend set up stresses in the edges of the beam. Typically, one edge becomes compressed and the other stretched. Fig. 5.18a shows where the stretching and compression forces occur in a loaded beam. A beam which is supported at one end only or sticks out beyond its vertical support is called a **cantilever** (5.18b). Examples of cantilevers include the support beams for overhanging buildings, the wings of an aircraft and the horizontal jib of a crane used on building sites.

Figure 5.18 *Loaded beams*

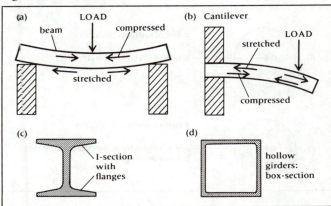

The shapes of beams

As the central part of a bent beam remains almost unstressed, the cross-sectional shape of a beam can be designed with much of the central part removed. This keeps the strength in the edges of a beam where it is needed while reducing its weight. Examples of common cross-sections are shown in fig. 5.18. I-shaped steel girders (5.18c) are an ideal shape because the top and bottom flanges have the strength needed to withstand the compression and stretching forces while the central part improves the 'stiffness' of the beam. Box girders (5.18d) are often used for building bridges.

The choice of materials

Materials such as stone, brick and concrete are strong when they are compressed by a load but relatively weak if stretched. Such brittle materials tend to crack easily when under tension and are weakened by surface scratches and flaws. These materials are chosen for use in structures where they will be compressed by the loads they have to support. Some concrete beams are strengthened by steel rods set inside the concrete. These rods are stretched while the concrete is set around them and then, when released, compress it and so increase its strength. Steel is particularly valuable as a material for building structures because it is strongest when stretched. Steel girders and steel cables are used in many kinds of buildings and bridges. Steel is an elastic material and returns to its original length and shape after being stretched (providing it's not stretched too much).

Bridges

The arches in the simple bridge designs shown below are compressed by the loads above. Stone, brick or concrete are suitable materials for constructing the arches.

Bridges with arches

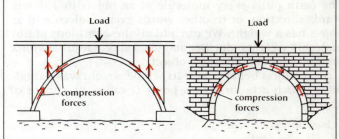

In a suspension bridge the beam carrying the vehicles is suspended by steel cables which are strong under *tension*. The concrete towers which support the cables are strong under *compression*. Here both kinds of materials are used in the way they are best suited for greatest strength.

Suspension bridges

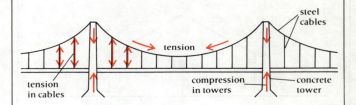

Structures built from girders such as the bridge design below, the jibs of cranes and electricity pilons (photographs, pages 77 and 220) are strengthened and stiffened by extra diagonal girders. These frameworks of girders give extra strength for minimum weight and allow the structure a certain amount of flexibility.

Girder structures

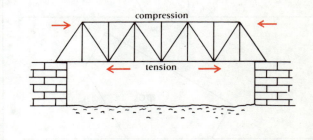

Assignments

Carry out a survey of your local bridges. For each bridge:
a) describe the kind of structure,
b) find out what materials it is made of,
c) for each main part of the structure, decide whether the material is under tension or compression,
d) comment on the suitability of the materials used.

Try questions 5.9 to 5.10

5.4
CENTRE OF GRAVITY

When gravity pulls an object towards the earth it always appears to pull at the same point on the object.

Where is the centre of gravity of an object?

The earth pulls every molecule of an object in a downwards direction, or in other words every molecule in an object has a weight. We can add all these millions of tiny molecule weights together and get a single resultant force for the weight of the whole object.

So an object behaves as if its whole weight was a single force which acts through a point G called its **centre of gravity.**

We define the centre of gravity of an object as the point through which its whole weight acts for any orientation of the object.

Fig. 5.19 shows an object orientated in three different ways. The centre of gravity G stays in the same position on the object (although it may change its height above the ground) and the weight of the object always acts through it.

Figure 5.19 *The weight of an object acts through its centre of gravity for all orientations of the object*

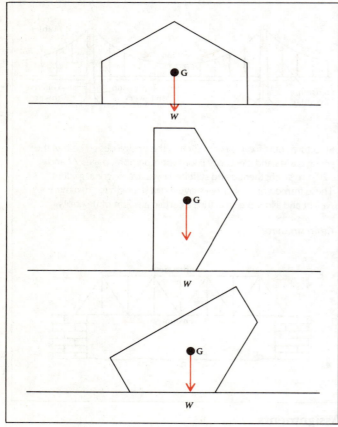

The centres of gravity of regular shaped objects

An object of uniform thickness and density has its mass evenly spread throughout and its centre of gravity is at its geometrical centre. Some examples of objects with regular shapes and uniform densities are shown in fig. 5.20. It is interesting to note that the centre of gravity of an object is not necessarily inside that object, fig. 5.20e.

Figure 5.20 *Centres of gravity of some objects with regular shapes and uniform densities*

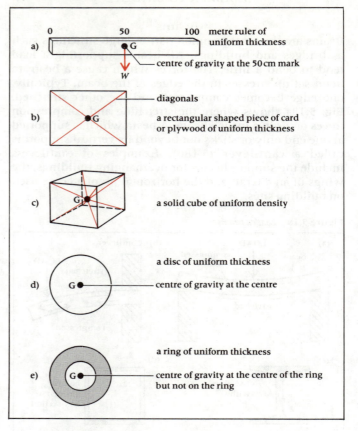

Finding the centre of gravity of an object with an irregular shape or non-uniform thickness or density.

- A suitable object is an irregular shaped sheet of card or wood. Make three small holes in it near its edges.
- Put a strong pin through one of the holes and fix it in a clamp so that the object can swing freely.
- Attach a plumbline (a length of thread with a heavy mass on the end) to the pin as shown in fig. 5.21.
- When the object and the plumbline have both stopped swinging, mark a cross on the object exactly behind the plumbline and near the opposite edge to the hole.
- Repeat this procedure in all three holes.
- Remove the object from the pin and draw straight lines with a ruler from each hole to the opposite cross.

The point where the three lines cross is the centre of gravity of the object.

Figure 5.21 *Finding the centre of gravity of an object using a plumbline*

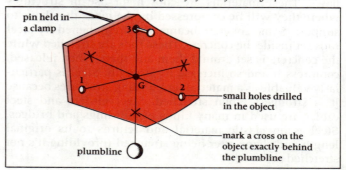

As the weight of the object always acts through G and acts downwards, G will always be pulled to its lowest possible position directly below the pivot or supporting pin. The plumbline also hangs straight down towards the earth so that G will always lie somewhere along the line of the plumbline. If G is somewhere on each of the three lines drawn in the experiment, it can only be at the one point where those lines cross.

Investigating stability

● *Try balancing a solid wooden cone in the three ways shown in fig. 5.22.*

If a suitable cone is not available, you can use a Bunsen burner instead. As you make your tests think about the following:

i) What forces are acting on the cone and which way and where do they act?
ii) When you tilt or displace the cone what happens to its centre of gravity G?
iii) When you then let go of the cone what happens to it and to its centre of gravity?

In all three cases there are two vertical forces acting on the cone (ignoring any frictional forces). They are the weight *W* of the cone, which always acts downwards through G, and the contact force *C* from the table's surface which pushes upwards against the cone's weight. The cone can be in equilibrium and not fall over only if it is possible for these two equal and opposite forces to act in the same line and stay there.

In case (a) balance is effectively impossible. As soon as the cone has the slightest tilt its weight has a moment about the point of the cone which makes it fall over. Note that

i) as it tilts, G goes lower and continues to get lower as it falls over,
ii) the line of action of the weight *W* passes outside the (very small) area of contact with the table's surface,
iii) this orientation of the cone is described as **unstable equilibrium**.

In case (b) the cone rests easily in what is called **stable equilibrium**. The two equal forces *W* and *C* act in opposite directions in the same line. Note that when you slightly tilt the cone:

i) its centre of gravity is *raised* and the contact force moves to the edge of its base,
ii) the moment of the weight provides a turning effect which tries to lower the centre of gravity and makes the cone fall back to its stable position and
iii) the line of action of the weight *W* passes inside the base area of the cone.

In case (c) it is possible to roll the cone to many new positions and let it rest there. It will neither roll back to where it came from nor roll on any further. The centre of gravity neither rises nor falls and so cannot gain any greater stability by being lowered. The two forces remain equal, opposite in direction and act along the same line in all positions of the cone. In no position can the weight provide a moment which will turn the cone to a new position. We describe this condition as **neutral equilibrium**.

Figure 5.22 *Kinds of equilibrium*

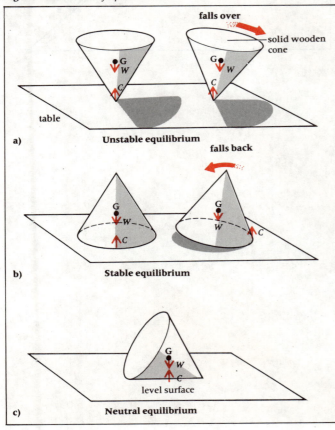

This desk lamp is made very stable by its wide and heavy base. Where do you think its centre of gravity is?

Designing stable objects

Why are some objects more stable than others and how can the design of an object improve its stability?

Fig. 5.23 shows an object of rectangular shape in one unstable position and two more stable positions.

In case (a) the rectangle topples over because the line of action of its weight *W* passes outside the corner of its base B. *W* has a clockwise moment *M* about B which topples the rectangle and lowers its centre of gravity.

In case (b) there is a low centre of gravity G. Now the line of action of *W* passes inside the base and provides an anticlockwise moment *M* about B which prevents it from toppling over. We can see that *lowering the centre of gravity of an object makes it more stable.*

This idea is used in the design of a double-decker bus. The bus is made more stable (i.e. less likely to turn over on its side when going round a corner), by keeping its centre of gravity very low down. Light-weight materials are used for the construction of its upper deck and it has a heavy chassis and engine mounted as low as possible. When the bus is fully loaded with sitting passengers on the top deck with only the driver and conductor downstairs, it should not topple over when the chassis tilted up to 28° from the horizontal.

In case (c) the object rests on a wider base which also causes *W* to provide an anticlockwise moment about B and prevents it from toppling over. We can see that *a wider base makes an object more stable.*

Figure 5.23 *Improving the stability of an object*

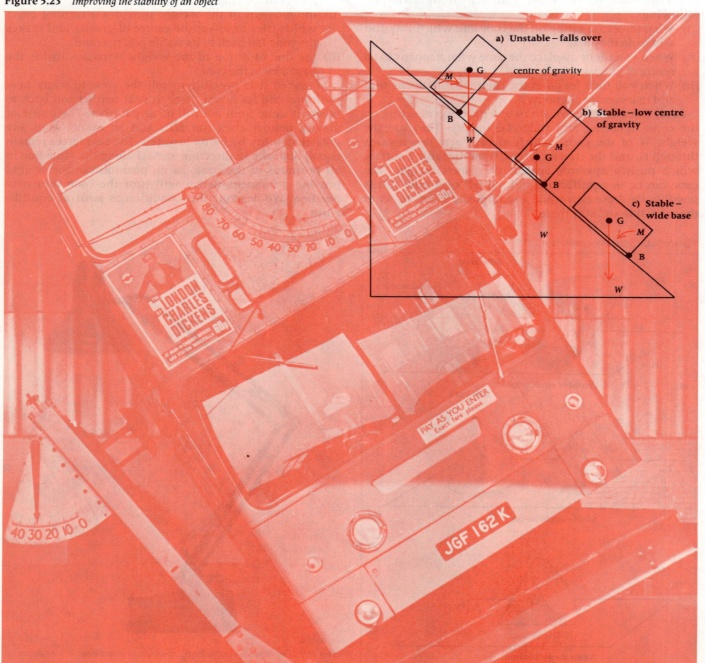

a) **Unstable – falls over**

centre of gravity

b) **Stable – low centre of gravity**

c) **Stable – wide base**

Designing stable objects

A racing car has both a low centre of gravity and a wide wheel base
(long axle length) to improve its stability.

Hanging objects are stable because their centres of gravity hang
below the pivot or point of suspension (a). Any displacement
of a hanging object causes its centre of gravity to rise and provides a
moment which pulls it back down again (b). Many balancing tricks and
toys have a centre of gravity lower than the balancing point or pivot and
so the object is really hanging.

In (c) the centre of gravity of the pencil and penknife is below the
pivot or balancing point. Similarly the beam and scale pans of a beam
balance must have a centre of gravity lower than the knife-edge pivot
on which the beam balances.

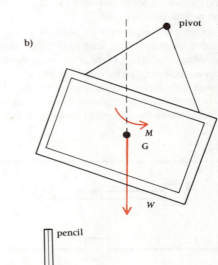

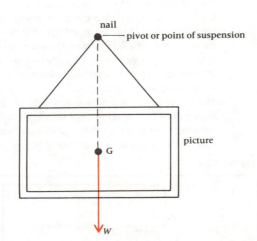

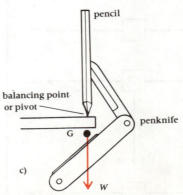

Assignments

Draw a balancing toy and explain how it balances.

Explain why it would be unwise to allow standing passengers on the top deck of a double-decker bus.

Try questions 5.11 and 5.12

Questions 5

Assume g = 10 N/kg throughout.

1 Calculate the weight (in newtons) of
 a) a girl of mass 40 kg,
 b) a car of mass 1 tonne,
 c) a pin of mass 300 mg.
2 Find the resultant of a force of 5 N and a force of 12 N acting at the same point on an object if
 a) the forces act in the same direction in the same straight line,
 b) the forces act in opposite directions but in the same straight line and
 c) the forces act at right angles to each other. Remember to give both the magnitude and the direction of the resultant force.
3 Why is *force* referred to as a vector quantity?
 Two forces acting at a point have magnitudes 5 N and 8 N. Explain why their resultant may have any magnitude between 3 N and 13 N.
 Forces 7.0 N and 11.0 N act at a point so that the angle between their lines of action is 35°. By means of a scale diagram, determine the magnitude of the resultant of these two forces. (C part)
4 When you measure a force you should measure its strength and its direction. Some other quantities in physics also have a direction. These quantities are called **vectors**.
 a) Which of the following quantities are vectors: mass, weight, power, time, density, velocity?
 b) Forces are vector quantities. If we draw an arrow to represent a force we can draw its length to a scale which tells us the size of the force. What other aspect of the arrow must we get right?
 c) Fig. 5.24 shows a box being pulled by two or three forces. The arrows represent the forces. Redraw each of the diagrams showing only the resultant force (a single arrow) acting on the box in each case. Write the strength of the force along side the arrow.

Figure 5.24

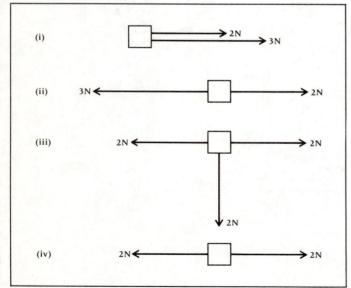

5 a) State, with a brief explanation, whether it is possible, given forces of 3 N and 8 N, to produce a resultant force of
 i) 5 N,
 ii) 15 N,
 iii) 8 N.
 b) An elephant is dragging a tree trunk along a horizontal surface by means of an attached rope which makes an angle of 30° with the horizontal. The tension in the rope is 4000 N.
 i) By scale drawing, or otherwise, determine the horizontal force exerted by the rope on the tree trunk.
 ii) Determine also the vertical component of the force exerted by the rope on the tree trunk. Explain why it serves a useful purpose.
 iii) Name the other forces which act on the tree trunk and show clearly the directions in which they act. (L)
6 A girl uses a spanner of length 20 cm to tighten a nut. If she pulls at right angles to the end of the spanner with a force of 50 N, calculate the moment of her pull. What difference would it make to the moment if the angle between her arm and the spanner was increased from 90° to 120°?
7 A boy of mass 40 kg and a girl of mass 30 kg play on a seesaw of negligible weight. If the boy sits 270 cm from the pivot of the seesaw, where must the girl sit to make it balance?

Figure 5.25

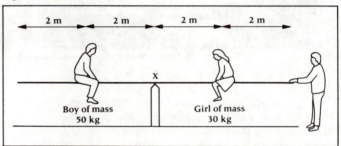

8 a) Fig. 5.25 shows a boy and a girl on a playground see-saw. The see-saw has a mass of 30 kg and is pivoted at its centre. Their mother has to hold the girl's end in order to keep the see-saw level.
 The boy's mass is 50 kg and the girl's mass is 30 kg.
 All the distances are shown on the diagram.
 The strength of the Earth's gravitational field is 10 N/kg.
 Calculate i) the boy's weight; ii) the moment (turning effect) of the boy's weight about the point X; iii) the girl's weight; iv) the moment (turning effect) of the girl's weight about the point X; v) the force their mother must apply on the end of the see-saw in order to keep it level; vi) the total downward force on the central support of the see-saw.
 b) Many road accidents are caused by high-sided vehicles like that shown in fig. 5.26 which are not sufficiently stable. Although manufacturers of these vehicles always pay particular attention to their stability, windy conditions or incorrect loading can

Figure 5.26

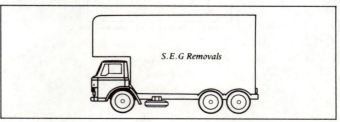

reduce the stability to dangerous levels.
i) What problem might occur if the stability of such a vehicle is low? ii) Why is the stability of the vehicle particularly important in 'windy conditions'? iii) How can 'incorrect loading' reduce the stability of the vehicle?

One way of testing the stability of a vehicle would be to load up a high-sided vehicle in such a way that its stability is just above the minimum standard. 'Suspect' vehicles could then be compared with this 'standard vehicle'.

  iv) How would you compare the stability of the two vehicles? (You should state what equipment you would use, how you would use it, and what measurements you would make.)
  v) It would be very expensive to have a 'standard vehicle' at every testing station. Suggest how this problem might be overcome. (SEG 88)

Figure 5.27

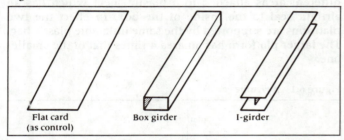

Flat card (as control)   Box girder   I-girder

9 A student was asked to investigate the strength of two girder shapes. Using a piece of card and glue the student made the above shapes. (Fig. 5.27).
  a) Draw a diagram to show a suitable arrangement of apparatus to carry out the experiment.
  b) Describe how the student could compare strength of each girder with the flat card. (SEG 88)

10 Fig. 5.28 shows a cantilever bridge in which the centre span of the bridge is supported by a cantilever arm at each side of the bridge. The cantilever arm is the shorter half of a concrete beam B which is shaded in the diagram. The weight W of this beam is 10MN. W acts through the centre of gravity of the beam at a point 4 m from the support P. Half of the weight of the centre span acts as a load L on each cantilever arm.
  a) If we assume that the weight of the beam B must balance the load L without any additional force being applied to the anchor arm, find the weight of the centre span which the two cantilever arms can just support. Take the support P as a pivot.
  b) When the centre span is carrying vehicles an extra force will be needed at the end of the anchor arm. In which direction will this force act?
  c) If the beam across the centre span is made of concrete, where will it experience compression forces and where tension forces? Explain what gives the concrete beam extra strength where it is under tension.

Figure 5.28

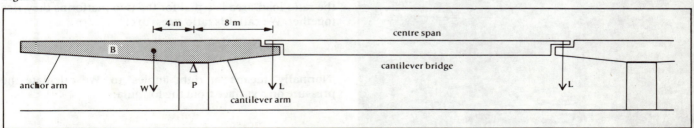

11 What is meant by the *centre of gravity* of an object?
   Describe how you would find by experiment the centre of gravity of a thin, irregularly shaped sheet of metal.
   Explain why a minibus is more likely to topple over when the roof rack is heavily loaded than when the roof rack is empty.
   A metre rule is supported on a knife edge placed at the 40 cm graduation. It is found that the metre rule balances horizontally when a mass which has a weight of 0.45 N is suspended at the 15 cm graduation, as shown in fig. 5.29.

Figure 5.29

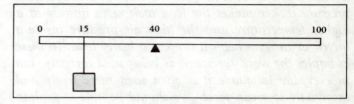

Calculate the *moment*, about the knife edge in this balanced condition, of the force due to the mass of the rule.
   If the weight of the rule is 0.90 N, calculate the position of its centre of gravity. (C)

12 A table lamp (see fig. 5.30) has a circular base of diameter 120 mm and a height of 300 mm. It stands on a rough horizontal surface. The centre of mass of the table lamp is 90 mm above the base.

Figure 5.30

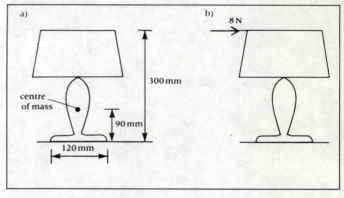

a) With the help of a diagram, explain why the table lamp topples when a certain angle of tilt is exceeded. Find the value of this angle.
b) Explain why it is possible for the centre of mass to be only 90 mm above the base, and describe a simple experiment to check the accuracy of this position.
c) When a horizontal force of 8 N is applied at the top of the table lamp as shown in fig. 5.30b, the table lamp just begins to pivot about its base. Calculate the mass of the table lamp.
d) Give **two** reasons why fitting a thin but heavy metal disc of diameter 160 mm to the base would improve the stability. (O)

6
Pressure

The pressure under your feet can compress soft ground or snow so that your feet sink in. The pressure of the air can push mercury up an evacuated barometer tube or crush an evacuated can. The pressure of water can throw a fountain high in the air. The increased pressure inside a pressure cooker makes the food cook more quickly at a higher temperature and the high air pressure inside a bicycle or car tyre helps it support a heavy load. In these examples the word 'pressure' is being used correctly, but in everyday language it is often used more freely and incorrectly to mean force. We should be able to explain what pressure means in science and to distinguish between pressure and force.

6.1
FORCE AND PRESSURE

If you walk on soft snow in ordinary shoes you sink in, but if you wear skis or snow shoes you are less likely to sink in. If you walk on a wooden or plastic floor surface in ordinary shoes you leave no impression on the surface, but if you wear shoes with very narrow and pointed stiletto heels then you may damage the floor surface and leave a permanent impression or dint. In each of these examples your weight does not change but the *pressure* under your shoes does. This is because the pressure under your shoes depends on the *area* of shoe in contact with the ground as well as your weight.

Pressure depends on force and area

The pressure box shown in fig. 6.1 has two platforms of different areas attached to a plastic sheet which has an airtight seal to the inside of the box. In effect the two platforms are supported by the same inflatable plastic bag. The larger platform has an area 4 times that of the smaller one.

Figure 6.1   *A pressure box*

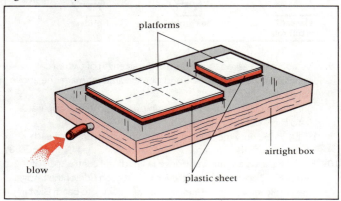

- *Place a $\frac{1}{2}$ kg load on each platform and blow into the tube, increasing the air pressure inside the box. Which platform rises first?*
- *Increase the load on the larger platform in $\frac{1}{2}$ kg steps up to six $\frac{1}{2}$ kg masses (total 3 kg), but leave the load on the smaller platform constant at $\frac{1}{2}$ kg. After each extra load is added blow into the box and note which platform rises first.*

When the load on the larger platform is less than 2 kg it rises before the smaller platform. When the load on the larger platform is 2 kg both platforms rise together. When the load is increased to $2\frac{1}{2}$ kg or 3 kg the smaller platform rises first.

The same pressure below the two platforms lifts them together when there is a 4 times greater load on the platform with a 4 times greater area. We can see that when the ratio load/area is equal for the two platforms they rise together. We call this ratio **pressure**.

Pressure is defined as the force acting normally per unit surface area.

'Normally' means 'at right angles' to. We calculate the pressure on a surface from the formula:

$$\text{pressure} = \frac{\text{normal force}}{\text{area}} \qquad p = \frac{F}{A}$$

Racing bicycles need very high air pressure inside the tyres, because the narrow tyres have a very small contact area with the road. The hard road surface can support the high pressure under the wheels. However, heavy vehicles which travel over soft ground need very wide tyres. This provides a large contact area and a lower pressure against the ground and prevents them sinking in.

The units of pressure are:

$$\frac{\text{newton}}{\text{square metre}}$$

or newton per square metre (N/m²). In SI the pressure unit has a special name: 1 newton per square metre is called 1 pascal.

1 pascal (Pa) *is a pressure of 1 newton per square metre.*

$$1\,\text{Pa} = 1\,\frac{\text{N}}{\text{m}^2}$$

If we rearrange the formula for pressure we get a useful formula for calculating force:

force = pressure × area $F = pA$

Note that there are many other units of pressure in everyday use, for example:
a) atmospheres of pressure = number of times a pressure is greater than atmospheric pressure,
b) mm of mercury = gas pressure read on a mercury barometer (p 97),
c) pounds per square inch is an old unit still used for stating pressure for car tyres etc.

Calculating the pressure under your own feet

● *Place one shoe on a sheet of (A4) graph paper (or two sheets if you have big feet) and draw round it.*
● *Count the number of square centimetres in your footprint (estimate half squares round the edge).*
● *Calculate the area of your shoe in square metres.*
(1 cm² = 10^{-4} m², or in other words there are 10 000 square cm in 1 square m.)
● *Weigh yourself and calculate your weight in newtons.*
($W = mg$, where m is your mass in kilograms and $g = 10\,\text{N/kg}$.)
● *Now calculate the pressure under one foot when it supports all your weight using $p = F/A$.*

Worked Example
Calculating pressure
Calculate the pressure under a girl's foot in pascals if her mass is 33.6 kg and the area of her shoe is 168 cm².

Area = 168 cm² = 168×10^{-4} m²
Weight = mg = 33.6 kg × 10 N/kg = 336 N
Now using $p = F/A$ we have:

$$p = \frac{336\,\text{N}}{168 \times 10^{-4}\,\text{m}^2} = 2.0 \times 10^4\,\text{N/m}^2 = 20\,\text{kPa}$$

Answer: The pressure under the girl's shoe is 2.0×10^4 pascals or 20 kilopascals.

Worked Example
Calculating force from fluid pressure
The pressure of the hydraulic fluid in a car braking system rises to 1000 kPa when a man puts his foot on the brake pedal. If the area of the piston in the brake slave cylinder (which presses the brake pads against the rotating disc brake) is 8 cm², calculate the force with which the brake pad is pressed against the disc.

Area of piston on which the fluid presses = 8 cm² = 8×10^{-4} m²
Now using the rearranged formula for force, we have:

$$F = pA = 1000\,\text{kPa} \times (8 \times 10^{-4}\,\text{m}^2)$$
$$\therefore\ F = (10^6\,\text{N/m}^2) \times (8 \times 10^{-4}\,\text{m}^2) = 8 \times 10^2\,\text{N}$$

Answer: The force applied to the brake pad is 800 newtons.

Assignments

Explain
a) the difference between force and pressure,
b) why skis should have a large surface area,
c) why stiletto heels on shoes can damage floors.
Estimate
d) the pressure under an elephant's foot,
e) the pressure under a chair leg when you are sitting on the chair.
Remember
f) the pressure formula

$$p = \frac{F}{A}$$

[You can use the triangle symbol to help you rearrange this formula, p 65.]
g) pressure is measured in pascals and 1 Pa = 1 N/m².
Try questions 6.1 to 6.5

6.2
PRESSURE IN LIQUIDS AND GASES

You can feel the pressure of water as it squirts out of the end of a hose pipe or a tap turned fully on. You can also feel the pressure of air as it rushes out of a balloon or a tyre valve. What do these pressures depend on and how are they measured?

Investigating the properties of liquid pressure

Pressure and depth

Fig. 6.2 shows a simple way of demonstrating that the pressure in a liquid increases with depth. The water comes out of the lowest tube in the tank fastest due to the greatest pressure. The pressure is caused by the weight of liquid above the level of the tube, and the weight of liquid is proportional to the height of liquid above that level. The distances reached from the base of the tank by the jets of water are roughly proportional to the heights of water h_1, h_2 and h_3.

Figure 6.2 *Pressure in a liquid increases with depth*

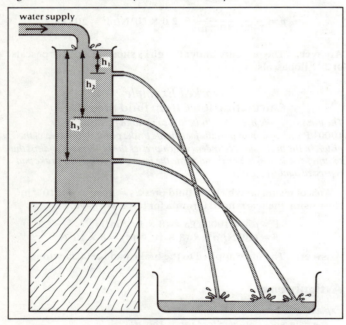

The pressure in a liquid at a certain level is proportional to the height of liquid above that level.

Or more briefly, the pressure in a liquid increases with the depth below its surface.

The U-tube shown in fig. 6.3, containing coloured water, can be used to measure pressure. Air pressure acts downwards on the water surface in both sides of the U-tube. If the pressure at the two sides is equal then the water levels are equal. If, however, P_1 is greater than P_2 the water levels change until the pressure of the extra water on the right-hand side balances the pressure difference between P_1 and P_2. As we have seen, the pressure below the extra height of water is proportional to the height h, and so h can be used to measure the pressure difference between P_1 and P_2. This U-tube pressure gauge is called a **manometer**.

Figure 6.3 *Pressure acts equally in all directions at the same depth in a liquid*

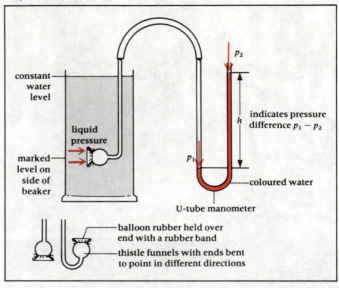

Pressure and direction

- *Make some pressure sensors by fitting a sheet of balloon rubber over the ends of some thistle funnels using rubber bands, fig. 6.3.* The thistle funnel tubes are bent at various angles so that the liquid pressure acting against the rubber sheet can be measured in different directions. The liquid pressure causes a normal force against the surface of the rubber sheet.
- *Mark a level low down on the side of a tall beaker and measure the pressure at this level in different directions. The pressure can be measured by connecting the thistle funnels to a U-tube manometer and measuring the height difference h as shown.*

We find that the pressure in a liquid is equal in all directions at the same depth.

Pressure and liquid density

If the pressure is measured at the same depth below the surface of different liquids we find that:

The pressure is proportional to the density of the liquid.

A similar experiment to the one shown in fig. 6.3 could be done to test this, but large volumes of most liquids may not be available.

Liquid levels

When a liquid is poured into a set of connected tubes of various shapes the liquid flows round the tubes until all the liquid surfaces are at the same level, fig. 6.4. We say that *a liquid finds its own level*.

Figure 6.4 *A liquid finds its own level*

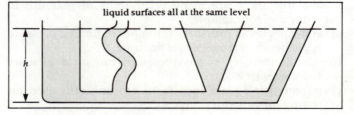

The pressures at the open tops of the tubes are all the same, being the air pressure. The liquid pressures at the bottom of each tube must also be equal otherwise the liquid would flow to equalise the pressures.

Even though the tubes have different shapes and different cross-sectional areas, for the pressures at the bottom of the tubes to be equal, they require only the same vertical height *h* of the same liquid.

The liquid pressure depends only on the height of the particular liquid and not on the shape or width of the tube.

Calculating pressure in a liquid

Fig. 6.5 shows a column of liquid of height *h* and base area *A*. The volume *V* of the liquid in the column is given by:

$$\text{volume} = \text{base area} \times \text{height}$$
or
$$V = A\,h$$

Figure 6.5 *A formula for liquid pressure*

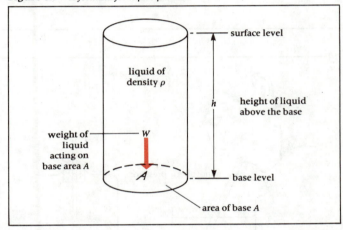

The mass *m* of the liquid of density *ρ* is given by:

$$\text{mass} = \text{density} \times \text{volume}$$
or
$$m = \rho V$$

The weight *W* of the liquid is given by: $W = mg$. So we have:

$$W = mg = (\rho V)g = \rho(A\,h)\,g.$$

The pressure at the base of the liquid column is found from:

$$\text{pressure} = \frac{\text{normal force}}{\text{area}} = \frac{W}{A}$$

$$p = \frac{A\,h\,\rho g}{A} = h\,\rho g$$

$$p = h\,\rho g$$

The pressure below a liquid surface is proportional to the height h of liquid above and the density ρ of the liquid.

The pressure is independent of the area.

Note that this formula does not include any air pressure acting on the surface of a liquid. It gives the extra pressure in the liquid at a depth *h* below its surface. Some applications of liquid pressure are described in chapter 7, which is about machines.

Worked Example
Pressure in a liquid

The density of liquid mercury is 13.5×10^3 kg/m³. Calculate the liquid pressure at a point 0.76 m below the surface of mercury. (g = 10 N/kg.)

Using the liquid pressure formula $p = h\,\rho g$, we have:

$$p = 0.76\,\text{m} \times (13.5 \times 10^3\,\text{kg/m}^3) \times 10\,\text{N/kg} = 1.0 \times 10^5\,\text{N/m}^2$$

Answer: The liquid pressure is 10^5 pascals or 100 kPa.

Notes

a) the units of liquid pressure work out to be N/m² which is the same as for $p = F/A$.

b) the pressure under a column of mercury 0.76 m high is equal to normal atmospheric pressure.

A mercury barometer records atmospheric pressure simply as the height of a column of mercury. This calculation shows how to convert the barometer reading from mm of mercury into pascals. See p 97.

Demonstrating some effects of air pressure

The milk bottle experiment

- Fill a milk bottle full of water by immersing it in a bowl of water.
- Keeping the top of the bottle below the water surface, lift the rest of the bottle out of the water.

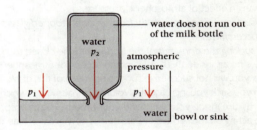

The water does not run out of the bottle. Why?

The atmospheric pressure P_1 (or air pressure) on the surface of the water balances the pressure of the water P_2 inside the bottle. If the water began to run out of the bottle then, without any air in the bottle, P_2 would become less than P_1. The atmospheric pressure will not allow this to happen.

The can-crushing experiment

- Put a small volume of water in a metal can and boil the water for several minutes to drive out most of the air (a).
- Stop heating and immediately seal the can with a well-fitting rubber stopper.

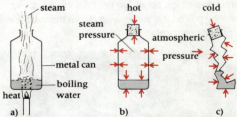

At the moment you close the can the steam pressure inside exactly balances the atmospheric pressure outside (b).

As heat is lost from the can the steam inside condenses and the inside pressure falls. The atmospheric pressure is now much greater than the pressure inside the can, so it crushes the can and makes its volume very small (c).

The Magdeburg hemispheres

Two metal hemispheres, named after the Mayor of Magdeburg who invented the vacuum pump, can be used to demonstrate the enormous strength of the atmospheric pressure. With air inside them, the two hemispheres can easily be pulled apart.

- *Seal the rims of the two hemispheres together with grease and connect them to the vacuum pump.*
- *Open the tap and pump out the air.*
- *Close the tap, remove the pressure tubing and screw on the removable handle.*
- *Try pulling the two hemispheres apart.*

When Otto von Guericke, Mayor of Magdeburg, first demonstrated this experiment, two teams of eight horses failed to pull the hemispheres apart. Although your laboratory hemispheres are smaller than the original ones, if the seal between their rims is good and the pressure inside is very low, you will not be able to separate them without allowing air back inside through the tap.

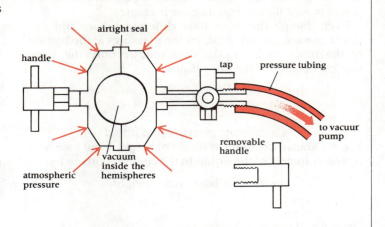

Sucking

We think of sucking a drink up a straw as being a result of our action rather than an effect of atmospheric pressure.

- *Try sucking a drink up a straw from an open-topped glass and you will be successful (a).*
- *Try sucking the drink out of the bottle with the closed top (b).*

As there is no air inside this bottle and no access for atmospheric pressure, you will not succeed in sucking up much of this drink.

When you suck you increase the volume of your lungs, which reduces the air pressure inside your lungs and your mouth. The atmospheric pressure acting on the surface of the liquid is now greater than the reduced air pressure inside your mouth, so drink is *pushed* up the straw by the pressure excess of the atmosphere over your mouth pressure. The absence of atmospheric pressure on the surface of the liquid in the closed bottle means that there is no excess pressure to push the liquid up the straw.

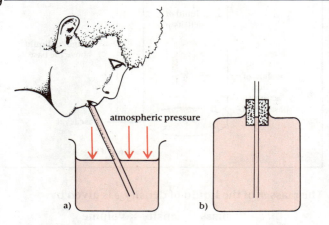

Rubber suckers

Rubber suckers can be used for lifting heavy objects with flat smooth surfaces and for hanging things on walls and windows. The sucker is pressed against the surface to squeeze out the air from behind it. The atmospheric pressure on the outside of the sucker holds its rim firmly against the smooth surface. As the smooth rim forms an airtight seal with the smooth surface, no air can return and a vacuum exists behind the sucker. Any pull on the sucker away from the surface is opposed by the atmospheric pressure.

These rubber suckers can lift up to 3 tons, but the sheet of glass must be kept vertical to prevent it breaking under its own weight.

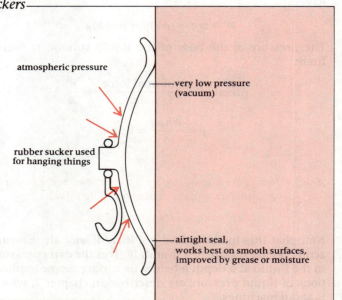

The syringe

A syringe has a piston which slides smoothly inside a cylinder making an airtight seal.

● *To fill a syringe, start with the piston at the bottom of the cylinder. Place the nozzle below the liquid surface and pull the piston upwards.* This produces a low pressure in the cylinder below the piston.

The greater atmospheric pressure on the surface of the liquid pushes it up the nozzle into the cylinder. When the syringe is removed from the liquid, as air is unable to get back into the cylinder below the piston, the atmospheric pressure at the opening of the nozzle helps to keep the liquid inside.

When the syringe is used the piston is pushed down the cylinder applying increased pressure to the liquid and forcing out of the nozzle against the atmospheric pressure.

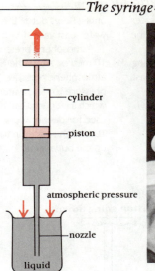

A syringe being used to give a cat medicine.

The force pump

The force pump is like a syringe with a separate outlet for the water and two valves added.

Piston going up: As the piston rises it reduces the pressure in the cylinder below the piston and atmospheric pressure on the water in the well below pushes water up past valve 1 into the cylinder.

Piston going down: As the piston is pushed down valve 1 falls closed and the increased pressure of the water in the cylinder opens valve 2 by pushing water past it. In this way the valves prevent water from going back down to the well and allow it to be raised to a greater height.

Above valve 2 there is a volume of air which becomes compressed while the piston is going down. This air acts as a shock-absorbing spring as valve 2 opens and closes.

Piston going up again: The reduced pressure of the water below valve 2 allows it to fall closed and stop the water flowing back into the cylinder. The compressed air maintains a flow of water up the outlet pipe. Atmospheric pressure on the water in the well pushes more water up into the cylinder.

This force pump can raise water from great depths because after atmospheric pressure has raised water into the cylinder, the piston is used to force water further up the outlet pipe. The pressure under the water in the outlet pipe increases as the water goes higher in the pipe. There is a limit to the pressure which the valves and piston can withstand, so there is also a limit to the height the outlet pipe can be.

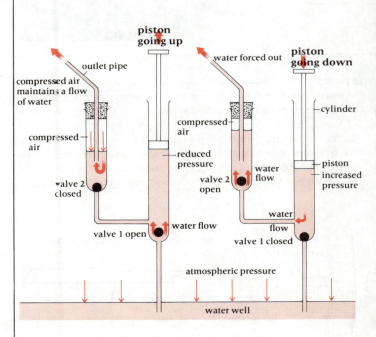

The lift pump

It is usually necessary to 'prime' a lift pump. To do this, pour some water into the top of the cylinder so that a good airtight seal is made round the piston and in valve 2.

Piston going up: The reduced pressure in the cylinder below the rising piston allows atmospheric pressure to push water up from the well past valve 1 into the cylinder.

Piston going down: Valve 1 falls closed and as the piston pushes down into the water in the cylinder, the water opens valve 2 and flows through the piston into the cylinder above.

Piston going up again: The lift pump now *lifts* the water above its piston. As valve 2 falls closed, the rising piston lifts the water above it

until it flows out of the spout. Atmospheric pressure is again pushing water past valve 1 into the cylinder below the piston.

Atmospheric pressure can raise water only to a height of about 10 metres. At this height the pressure under the water equals the atmospheric pressure. A lift pump relies entirely on the atmospheric pressure to raise water into its cylinder so the the maximum depth from which a lift pump can raise water is about 10m. A force pump must be used for deeper wells. The cylinder of a force pump must also be within 10m of the water in the well, but extra height is gained by forcing water up the outlet pipe.

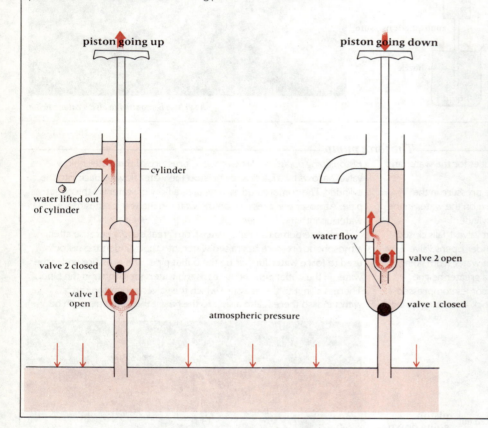

The bicycle pump

Two valves are used when a bicycle tyre is blown up. One, the tyre valve, prevents air escaping from the inflated tyre and the other, a flexible greasy leather washer, forms both a valve and a piston inside the pump barrel.

When the pump handle is pushed in, the air in the pump barrel is compressed (a). The high pressure of the air in the barrel presses the leather washer against the inside of the barrel, closing the pump valve.

When the pressure of the compressed air becomes greater than that of the air already in the tyre, air is forced past the tyre valve into the tyre.

When the pump handle is pulled out, the pressure of the air in the barrel is reduced (b). The higher pressure of the air in the tyre closes the tyre valve preventing air escaping. The atmospheric pressure, being greater than the reduced pressure in the barrel, forces air past the leather washer, opening the pump valve and refilling the barrel with air.

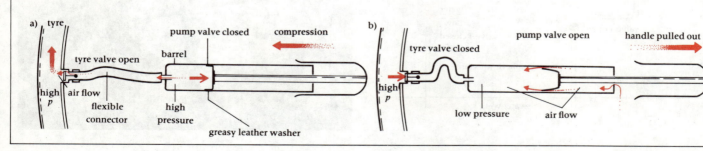

Measuring pressure

U-tube manometers

We used a U-tube manometer as a pressure gauge in an earlier experiment, see fig. 6.3. Now that we know that the pressure in a liquid is given by $p = h\rho g$, we can calculate a value for pressure from the height reading obtained on a manometer.

Figure 6.6 *Measuring pressure with a U-tube manometer*

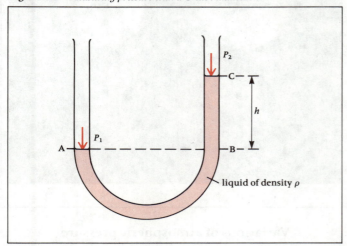

Fig. 6.6 shows a U-tube containing a liquid of density ρ. The pressure difference between P_1 and P_2 is indicated by the difference between the liquid levels h. The pressures at A and B are equal because they are at the same level in the liquid and any pressure difference at the same level causes a flow of liquid to equalise that difference:

the pressure at A = P_1.
the pressure at B = surface pressure + liquid pressure
$$= P_2 + h\rho g$$

$$P_1 = P_2 + h\rho g$$

When P_2 is atmospheric pressure, $h\rho g$ gives the pressure excess of P_1 over atmospheric pressure.

The mercury barometer

We can measure atmospheric pressure by using the principle of the U-tube manometer.

If we reduce the pressure at point C in fig. 6.6 to zero, i.e. if we have a vacuum above C and $P_2 = 0$, then $P_1 = h\rho g$. So the atmospheric pressure can be calculated from $h\rho g$. Atmospheric pressure is so large that if we used water in the manometer it would need to be over 10m high. Instead we use mercury which, being 14 times more dense than water, has the same pressure for a height 14 times shorter.

Demonstrating a simple mercury barometer
Warning *This experiment should only be demonstrated using a fume cupboard to remove the mercury vapour.*

To produce a vacuum above C we use a strong glass tube closed at one end and about 80 to 100 cm long, fig. 6.7.

Connect a glass funnel by rubber tubing to the open end of the tube and slowly pour in mercury. When it is almost full, stop and remove air bubbles from the tube. With your thumb firmly over the end of the tube allow the bubble of air at the end to float up and down the tube until it has gathered up all the small bubbles clinging to the sides of the tube. Now completely fill the tube with mercury. With your thumb over the end again, invert the tube and put the end well under the mercury in the dish before removing your thumb.

The mercury level inside the tube drops until the pressure at B is equal to atmospheric pressure P_1 at A. The labelling in fig. 6.7 corresponds to those on the U-tube manometer of fig. 6.6.

The atmospheric pressure measured on this barometer is quoted as the height of the mercury column h, normally about 76 cm. We mean by this that the atmospheric pressure is equal to the pressure under a column of mercury 76 cm high.

To convert the height reading on the mercury barometer we use the liquid pressure formula $h\rho g$. This calculation was done, in the worked example on (p 93), giving a value for the pressure below 0.76 m of mercury or 100 kPa.

So atmospheric pressure, measured on a mercury barometer, is normally about 76 cm or 0.76 m of mercury, which is equal to 100 kPa.

Tilting the barometer tube: fig. 6.7 also shows the effect of tilting the tube. In (a), when the tube is vertical, the mercury height is 76 cm above the mercury level in the dish. In (b) the mercury level in the tube remains at 76 cm above the dish level showing that the pressure in a liquid depends only on the vertical height of liquid. In (c) there is no vacuum and mercury fills the tube because its vertical height is less than 76 cm.

An accurate mercury barometer used in laboratories, known as the Fortin barometer, includes a vernier scale (p 58) for accurate measurement of the mercury height.

Figure 6.7 *The mercury barometer*

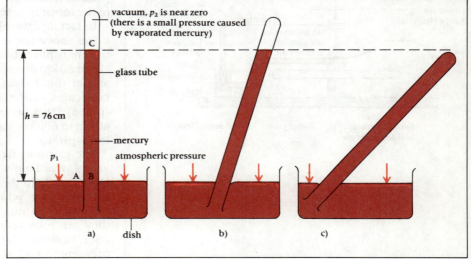

The Bourdon gauge

When the gas pressure inside the curved metal tube increases, the tube tries to uncurl, or straighten out. The end of the tube is linked to a pointer which reads pressure on a circular scale. Gas pressure gauges are usually of this type. They are fitted to gas cylinders and the Boyle's law apparatus used in chapter 10.

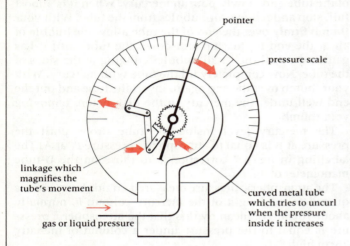

pointer

pressure scale

linkage which magnifies the tube's movement

curved metal tube which tries to uncurl when the pressure inside it increases

gas or liquid pressure

The aneroid barometer

An aneroid barometer literally means one without any liquid in it. Some aneroid barometers work on the principle shown below.

A flat, partially evacuated, cylindrical metal box is sensitive to changes in atmospheric pressure. Any increase in pressure squashes the box very slightly. The small movement of the top surface of the box must be magnified to be seen easily. The system of levers, spring and chain that is usually used to produce this magnification is represented by the simple lever pointer in the diagram.

Aneroid barometers are suitable for use as altimeters in aircraft. Their scales can be calibrated in height above sea level based on the reduction in atmospheric pressure with increasing height.

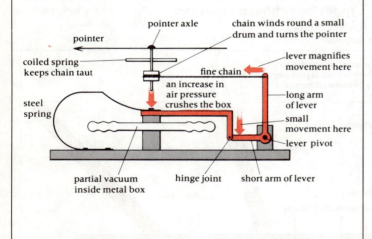

pointer axle

chain winds round a small drum and turns the pointer

pointer

coiled spring keeps chain taut

lever magnifies movement here

fine chain

an increase in air pressure crushes the box

long arm of lever

steel spring

small movement here

lever pivot

partial vacuum inside metal box

hinge joint

short arm of lever

Variations of atmospheric pressure

Height above sea level

If the air in the atmosphere was uniformly dense then the pressure of the air would be directly proportional to the height of atmosphere above, in the same way that the pressure in a liquid is proportional to the height of liquid above.

The atmospheric pressure at sea level is about $100\,\text{kPa}$ or $10^5\,\text{N/m}^2$ and the density of air at sea level is about $1.3\,\text{kg/m}^3$. If all the air in the atmosphere had this density we could calculate the height of the atmosphere. Re-arranging the pressure formula we have:

$$h = \frac{p}{\rho g} = \frac{10^5\,\text{N/m}^2}{(1.3\,\text{kg/m}^3) \times (10\,\text{N/kg})} = 8 \times 10^3\,\text{m}$$

So the height of the atmosphere would be about $8\,\text{km}$ if the air had the same density all the way up to the top. The peak of Mount Everest, about $9\,\text{km}$ high, would stick out of the top of the atmosphere!

In fact *the atmosphere gets gradually less dense with increasing height* above sea level. There is no definite upper limit to the atmosphere because it gradually merges into space. At about $80\,\text{km}$ above sea level, the pressure has fallen to $1\,\text{Pa}$ and at this height radiation from space has ionised the molecules of the air and made them into charged ions (p 268). This layer of atmosphere, extending from about $60\,\text{km}$ to $600\,\text{km}$ above the earth, is therefore known as the **ionosphere**.

The graph in fig. 6.8 shows how atmospheric pressure varies with height above sea level. The greater density at lower levels in the atmosphere is caused by the weight of the air above, which compresses the lower layers of air. This effect is not usually noticeable in liquids because they are relatively incompressible. Because of the variation in density of the atmosphere the formula for liquid pressure ($p = h\rho g$) cannot be used.

Changes in the weather

Barometers kept in the same place at the same height above sea level show some variation in atmospheric pressure from day to day. These pressure variations are shown on weather maps.

All places on the weather map which have the same atmospheric pressure are joined together by lines called isobars. Pressures are quoted on weather maps in pressure units called **millibars**,

1 bar = 100 kPa = normal atmospheric pressure,

1 bar = 1000 millibars.

So the atmospheric pressure on an average day near sea level may be quoted as either 100 kPa, 1 bar, 1000 millibars, or 76 cm of mercury!

Regions where atmospheric pressure is lower than average are called **cyclones** or **depressions** and winds blow spirally inwards towards the low pressure centre. Regions where atmospheric pressure is higher than average are called **anticyclones**. In these anticyclones, winds circulate round the high pressure centre, spiralling outwards.

Different weather conditions occur in low- and high-pressure areas and changes in atmospheric pressure are used as a guide for predicting future weather. Aneroid barometers which some people have at home often indicate how the weather may change with different pressures.

The range of atmospheric pressures over the British Isles is usually well within the range 960 millibars to 1040 millibars (or 96 kPa to 104 kPa).

The satellite photograph and weather map shown were obtained on 27 June 1979 at about 1500 GMT. They show a low pressure area (990 millibars) between Greenland and Iceland. The winds, blowing anticlockwise round the centre of this depression and from the west over the British Isles, are bringing rain clouds to the west coasts. A few hours later the clouds completely covered the British Isles.

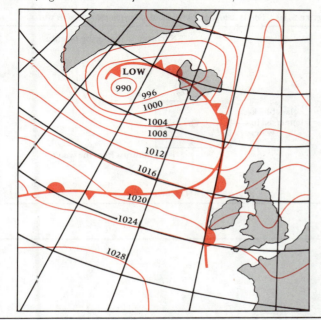

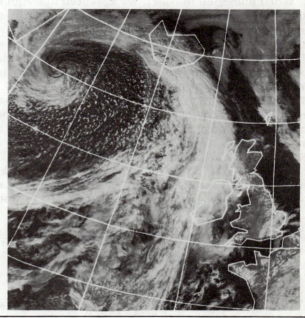

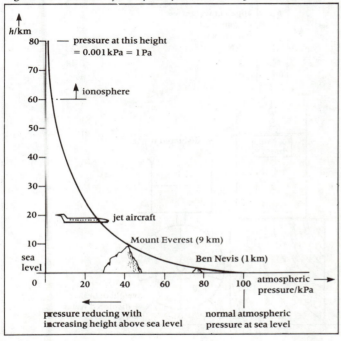

Figure 6.8 *Variation of atmospheric pressure with height above sea level*

Assignments

Explain

a) why it is difficult to remove the lid from a preserving jar which was closed when the space above the food was full of steam.

b) why evaporated milk flows out of a can more easily if two holes are made at opposite sides of the can top.

c) why diving bells for exploring deep water are built in a spherical shape.

d) why the water storage tank is in the roof of your house.

e) why dams which hold water in reservoirs must be much thicker at the base of the dam than at the top.

f) why high-flying aircraft need to be airtight and have pressurised cabins for the people.

g) why a force pump must be used instead of a lift pump to raise water from a deep well.

Remember the liquid pressure formula $p = h\rho g$. Do you know the units of pressure?

Try questions 6.6 to 6.12

6.3
ARCHIMEDES AND THE UPTHRUST FORCE

Heavy steel ships can float on water, but as steel is more dense than water we might expect them to sink. If you try to lift a heavy object which is under water you find it surprisingly light and much easier to lift than when it is out of water.

The Greek scientist Archimedes was the first person to realise that there is an upwards force on an object placed in a liquid which comes from the liquid itself and makes the object appear to lose weight.

Investigating Archimedes' principle

We need an arrangement for weighing an object in both air and water to find out how much weight it appears to lose when immersed in water. We also need to weigh the water displaced (that is, pushed out of the way) by the immersed object. The apparatus shown in fig. 6.9 allows both these measurements to be made.

The readings on scale A give the weight of the glass block in air (8 N) and then its apparent weight when immersed in a can of water (5 N).

The readings on scale B give first the weight of the empty beaker (2 N) and then the weight of the beaker plus the water displaced from the displacement can by the glass block (5 N).

Scale A shows an apparent loss of weight by the glass block of 8 N − 5 N = 3 N. Since the mass of the block is constant, its real weight is still 8 N. Fig. 6.10 shows the forces acting on the block when it is immersed in water.

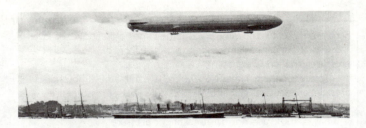

There is equilibrium between the three vertical forces acting on the block. The downwards weight of 8 N is equal and opposite to the sum of the two upward forces. These are a 5 N tension force measured on the spring balance and a 3 N **upthrust force** provided by the water and causing the apparent loss of weight of the block.

Figure 6.10 *Forces acting on the glass block when immersed in water*

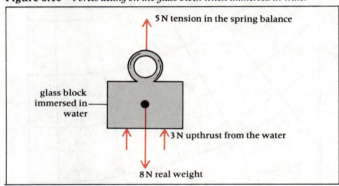

Figure 6.9 *Investigating Archimedes' principle*

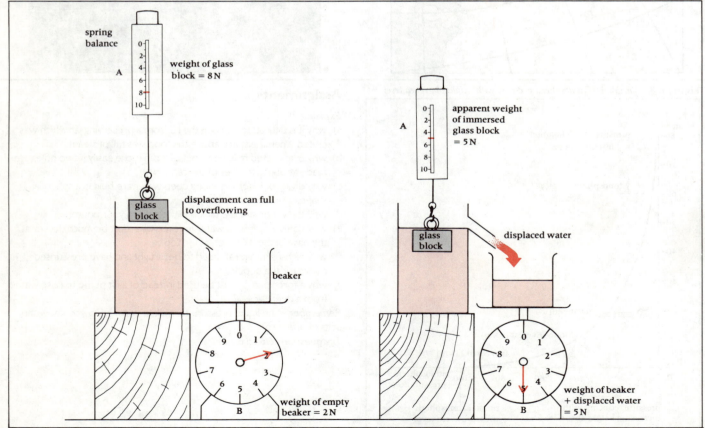

Scale B shows an increase in weight of $5\,\text{N} - 2\,\text{N} = 3\,\text{N}$, which is the weight of water displaced by the block.

We can see that the upthrust force (of 3 N) provided by the liquid is equal to the weight of liquid displaced (3 N) by the glass block. This discovery can be demonstrated for all fluids, i.e. for all gases as well as liquids. The same result is produced whether the object is wholly or partially immersed in a fluid and whether it sinks or floats. This is the principle of Archimedes which can be stated as follows:

The upthrust force on an object wholly or partially immersed in a fluid is equal and opposite to the weight of the fluid displaced by the object.

The cause of the upthrust

Fig. 6.11a shows how the increase in pressure with depth affects an object immersed in a fluid. The pressures on the sides of the object will balance each other, but there will always be a greater pressure p_2 below the object than above it p_1. This pressure difference is the cause of the upthrust force.

Figure 6.11   *Pressure and upthrust*

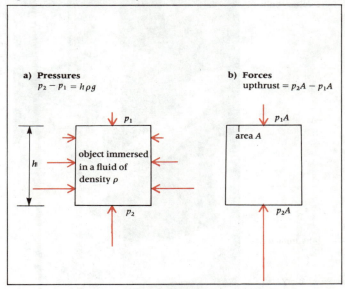

If the difference in height between the top of the object and its bottom is h, then the pressure difference in the fluid is given by:

$$p_2 - p_1 = h\rho g$$

The upthrust force can be calculated from the pressure difference. If the object has a top and bottom area of A, the forces on those areas are given by $p_1 A$ and $p_2 A$, fig. 6.11b (force = pressure × area). The difference between these two forces: $p_2 A - p_1 A$ is the upthrust force:

$$\text{upthrust force} = p_2 A - p_1 A = A(p_2 - p_1) = A(h\rho g)$$

Now Ah is the volume of the object and the volume of the fluid displaced. Since mass = density × volume, $\rho \times Ah$ is the *mass* of the displaced fluid and $Ah\rho g$ is the *weight* of the fluid displaced.

In other words: *the upthrust force equals the weight of fluid displaced*, which is Archimedes' principle.

Sinking and floating

The vertical forces acting on an object immersed in a fluid can be represented by the two forces shown in fig. 6.12. W is the weight of the object and U is the upthrust of the fluid. The vertical forces acting on a floating object such as a ship (fig. 6.12a) are in equilibrium as the ship neither rises up out of the water nor sinks down into it.

Figure 6.12   *Sinking and floating*

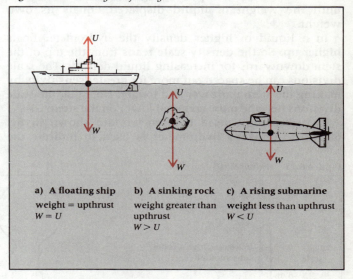

a) A floating ship
weight = upthrust
$W = U$

b) A sinking rock
weight greater than upthrust
$W > U$

c) A rising submarine
weight less than upthrust
$W < U$

So for a floating object we can say:

the upthrust = the weight of the object

or $\qquad\qquad U = W$

Since the upthrust also equals the weight of fluid displaced (Archimedes' principle), we can also say:

The weight of fluid displaced = the weight of the object.

This is usually stated as the **law of flotation**:

A floating object displaces its own weight of the fluid in which it floats.

This provides the explanation of how a steel ship floats. The hollow steel hull of the ship sinks down into the water and displaces water until the weight of water displaced is as great as the weight of the ship. Then the upthrust equals the ship's weight and it floats.

A rock also experiences an upthrust in water and appears to weigh much less in water than in the air. But the weight W of a rock is greater than the upthrust U and so it sinks, fig. 6.12b.

An object made of a material less dense than water, such as wood or cork, or a submarine which has filled its flotation tanks with air, will experience an upthrust U which is greater than its weight W and so will rise to the surface of the water, fig. 6.12c.

A balloon filled with hot (expanded) air or a gas of low density (hydrogen or helium) will rise in the atmosphere for the same reason as the submarine rises. The weight of the balloon filled with a low density gas is less than the upthrust on it caused by the displaced denser air. A balloon will continue to rise in the atmosphere until the reduced density of the air has decreased the upthrust to a value equal to the weight of the balloon.

Hydrometers

Hydrometers are floating instruments used to measure the density of liquids. A hydrometer has a long neck or stem with a density scale reading in grams per cubic centimetre (g/cm³). A large bulb filled with air displaces the liquid and provides an upthrust to make the hydrometer float. Lead shot glued in the bottom keeps it upright.

In a liquid of low density the hydrometer sinks further down in the liquid, displacing a greater volume of liquid until the weight of liquid displaced equals its own weight.

In a liquid of higher density the hydrometer floats higher up. So the density scale reads from the top of the stem downwards for increasing liquid density. The scale divisions can be spaced out more for greater sensitivity by making the stem narrower and longer. Note that the scale divisions become more cramped lower on the stem.

Two commonly used hydrometers are shown in fig. 6.13. The brewer's hydrometer is used to indicate the sugar and alcohol content in wine or beer. Sugar dissolved in water increases its density. As the sugar is converted into alcohol by fermentation, the density of the liquid falls because alcohol is less dense than water.

The sulphuric acid concentration in a car battery is a good guide to the state of charge or discharge of the battery. Concentrated acid is denser than dilute acid. When the acid density is about 1.30 g/cm³ the battery is fully charged, but when it falls to about 1.18 to 1.15 g/cm³ the battery needs recharging. As sulphuric acid is corrosive and dangerous, a special pipette arrangement is used which has a small hydrometer inside a large bulb.

To use this hydrometer, first squeeze the rubber bulb at the top of the pipette then lower the rubber tube into the battery acid. Release the bulb slowly and allow the atmospheric pressure to push acid up into the pipette until the hydrometer floats. Read the acid density on the hydrometer scale before carefully returning the acid to the battery by squeezing the rubber bulb again.

Figure 6.13 *Hydrometers*

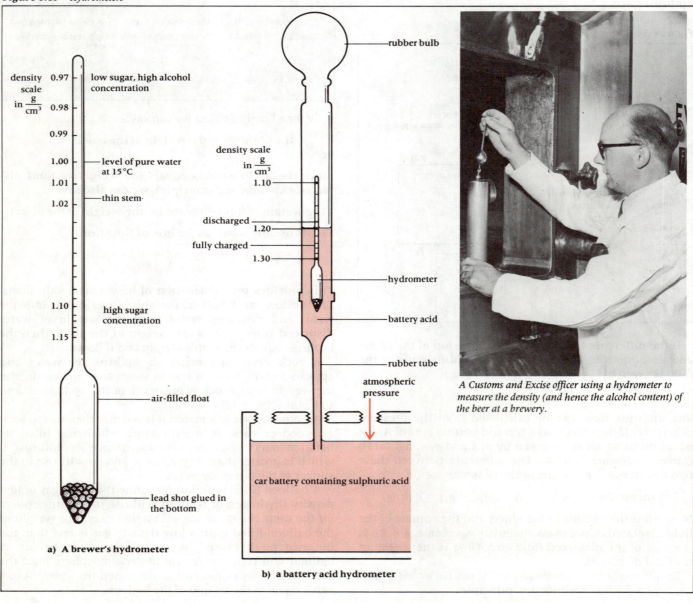

A Customs and Excise officer using a hydrometer to measure the density (and hence the alcohol content) of the beer at a brewery.

a) A brewer's hydrometer

b) a battery acid hydrometer

Worked Example
The volume of water displaced by a boat

A boat of mass 400 tonnes floats in sea water of density 1030 kg/m³.
Calculate the volume of sea water displaced.

The boat will displace a weight of sea water equal to its own weight. But weight is proportional to mass so we can say that the mass of water displaced equals the mass of the boat $= 400 \times 10^3$ kg. Now for sea water:

$$\text{volume} = \frac{\text{mass}}{\text{density}} = \frac{400 \times 10^3 \,\text{kg}}{1030 \,\text{kg/m}^3} = 388 \,\text{m}^3$$

Answer: The volume of sea water displaced = 388 cubic metres.

Worked Example
Apparent weight in water

A ship's anchor made of steel has a weight of 5000 N. If steel has a density of 8.0 × 10³ kg/m³ and the ship is floating in fresh water of density 1.0 × 10³ kg/m³, calculate the apparent weight of the anchor when it is below the water.

Note: to find the weight of water displaced we need to know the volume of water displaced. The volume of a liquid displaced is always equal to the volume of the object which displaces it. So first we calculate the volume of the anchor.

The weight W of the anchor = 5000 N, so the mass of the anchor is given by

$$m = \frac{W}{g}$$

where g = 10 N/kg.

$$\therefore \text{ mass } m = \frac{5000 \,\text{N}}{10 \,\text{N/kg}} = 500 \,\text{kg}$$

Now the volume V of the steel anchor = m/ρ

$$\therefore V = \frac{500 \,\text{kg}}{8.0 \times 10^3 \,\text{kg/m}^3} = 6.25 \times 10^{-2} \,\text{m}^3$$

This is the volume of water displaced by the anchor. Now the mass of water displaced = density of water × V

$$\therefore \text{ mass of water displaced} = (1.0 \times 10^3 \,\text{kg/m}^3) \times 6.25 \times 10^{-2} \,\text{m}^3$$
$$= 62.5 \,\text{kg}$$

The weight of water displaced = mg = 62.5 kg × 10 N/kg
$$= 625 \,\text{N}$$

By Archimedes' principle the upthrust is also 625 N, and the anchor will appear to weigh 5000 N − 625 N = 4375 N.

Answer: The apparent weight of the anchor when below water is 4375 newtons.

Plimsoll lines on the hull of a coaster trading between the UK and northern continental ports.

Assignments

Remember
a) the principle of Archimedes,
b) the law of flotation.
Explain
c) why a steel ship floats.
d) why a loaded ship floats lower in the water than an empty ship.
e) why it is easier for a swimmer to float in sea water than in fresh water.
f) why a balloon blown up with cold air falls to the ground but one blown up with helium gas floats upwards.
Find out
g) about the Plimsoll line on a ship. How is it used?
Try questions 6.13 to 6.16

Questions 6

Assume g = 10 N/kg.

1 Calculate the pressure on a surface when a force of 48 N acts on an area of (a) 12 m², (b) 96 m², (c) 4.0 cm², (d) 10 mm².
2 If the pressure at the base of a water tank is 5 kPa (5000 Pa), calculate the weight of the water acting on (a) 1.0 cm² of the tank bottom, (b) the whole of the tank bottom if it measures 80 cm by 40 cm.
3 A stone pillar has a mass of 3.0 tonnes. If the area of its base is 0.3 m², calculate the pressure under the pillar. Give your answer in pascals.
4 A block of metal of density 3000 kg/m³ is 2 m high and stands on a square base of side 0.5 m, as shown in fig.6.14.
   a) What is the base area of the block?
   b) What is the volume of the block?
   c) What is the mass of the block?
   d) What is the weight of the block?
   e) What is the pressure exerted by the weight of the block on the surface on which it stands? (O)

Figure 6.14

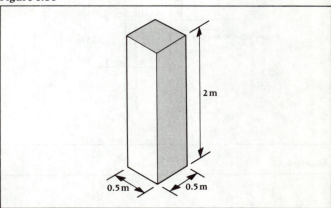

5 A rectangular block measures 8 cm by 5 cm by 4 cm, and has a mass of 1.25 kg.
   a) i) If the gravitational field strength is 10 N/kg, what is the weight of the block? ii) What is the area of the smallest face of the block? iii) What pressure (in N/cm²) will the block exert when it is resting on a table on its smallest face? iv) What is the least pressure the block could exert on the table?
   b) i) What is the volume of the block? ii) Calculate the density of the material from which the block is made. (NEA [A] SPEC)

6 a) Explain clearly what is meant by pressure.
Describe a simple experiment to show how the pressure exerted by a liquid varies with the depth below the liquid surface. State the result you would expect to obtain.

b) A rectangular block 0.01 m by 0.02 m by 0.04 m has a mass of 0.064 kg. Calculate (i) the density of the material of the block, (ii) the weight of the block, (iii) the pressure the block would exert when resting on its smallest side.

[Assume that the acceleration of free fall (due to gravity) is 10 m/s².]

(JMB)

7 If the density of sea water is 1150 kg/m³, calculate
a) the pressure below 40 m of sea water due to the water alone,
b) the pressure at the same depth when the atmospheric pressure of 100 kPa on the surface of the sea is included,
c) the depth at which the pressure due to water alone is 92 kPa,
d) the depth at which the total pressure is double the surface atmospheric pressure of 100 kPa.

8 a) What is meant by *pressure*? Explain simply how the pressure exerted on a surface by a rectangular block depends on which face of the block rests on the surface.

b) Describe **three** experiments, **one** for each case to show that the pressure exerted by a liquid
   i) varies with the depth of the liquid;
   ii) varies with the density of the liquid,
   iii) is the same in all directions at a given depth.

c) A tank with a base area of 4 m² is connected at the bottom to a vertical tube of cross-sectional area 0.01 m² by a horizontal tube. A liquid density 1000 kg/m³ is poured into the tank until the depth of liquid in the tank is 0.5 m.
Sketch the arrangement of the tank and tubes (not to scale) showing clearly the depth of liquid in the tank and in the vertical tube. Calculate
   i) the pressure due to the liquid on the base of the tank,

ii) the pressure due to the liquid at the base of the vertical tube.
If the atmospheric pressure at the time were 120000 Pa (N/m²), what would be the total pressure on the base of the tank?
[Assume that the acceleration of free fall is 10 m/s² (N/kg).]

(JMB)

9 Fig. 6.15 shows a manometer with limbs of area of cross section 0.012 m², and contains liquid which weighs 8000 N/m³. The manometer is connected to the laboratory gas supply and the tap turned on. As shown in the diagram there is a difference in the liquid levels in the two limbs of 0.25 m. Calculate
a) the volume of liquid between the levels AB and CD in the right-hand tube, neglecting the meniscus,
b) the weight of this liquid,
c) the excess pressure, in N/m², of the gas supply above atmospheric.

(C)

10 Fig. 6.16a shows a simple mercury barometer.
a) What is in the space labelled A?
b) Fig. 6.16b shows the same barometer. The tube of the barometer has been tilted from the vertical. Mark on fig. 6.16b the level of mercury in the tube.
c) The barometer height, *h*, is 0.75 m. What is the value of atmospheric pressure in Pa? The density of mercury is 13.6 × 10³ kg/m³.

(O & C)

11 In fig. 6.17, the drawing shows a foot-pump which may be used to pump air into car tyres and the diagram shows the internal details of the pump. As the pedal is pressed down the cylinder is pushed and moves down the outside of the piston. (This is equivalent to pushing the piston into the cylinder.) This causes air to be forced along the connecting tube to the tyre.

a) If the area of the piston is 25 cm² and the force applied to the cylinder is 800 N, show that the pressure exerted on the air by the piston is 320 kPa (that is, 3.2 × 10⁵ Pa).

b) With the type of pump shown the force exerted by the foot is not the same as the force exerted on the cylinder. Why is this?

c) It is important that the area of the piston (and cylinder) is not too large or too small. i) What problem would arise if the area of the piston was too large? ii) What problem would arise if the area of the piston was too small?

d) A car rests on four wheels, and each tyre is in contact with the ground over an area of 0.0075 m². If the pressure in each tyre is 200 kPa, calculate the mass of the car. (*g* = 10 N/kg)

e) The manufacturer's handbook for this car says that the tyre pressures should be increased when it is heavily loaded. The owner thinks that this is unnecessary because the extra weight will automatically increase the pressure. What is wrong with his reasoning?

(SEG 88)

Figure 6.15

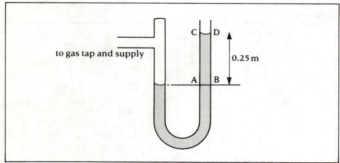

Figure 6.16

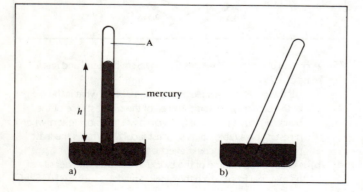

Figure 6.17

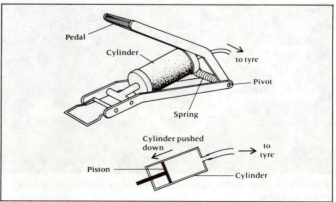

12 a) Complete the following equation which relates the mass, volume and density of a solid object.
The density =
b) The level of water in a measuring cylinder is at the 50 cm³ mark. A small solid object is put into the cylinder so that it is submerged. The water level is now at the 90 cm³ mark. Take $g = 10\,N/kg$, i.e. $10\,m/s^2$.
   i) What is the volume of the object?
   ii) If the density of the object is 2.5 g/cm³ what is (A) the mass of the object and (B) the weight of the object?
c) A hydrometer is placed in water and the stem is half submerged. Will the length of the stem below the liquid level increase, decrease or remain constant if the hydrometer is placed in a liquid which
   i) is slightly more dense than water,
   ii) is slightly less dense than water?
d) What is always true about the hydrometer of part (c) and the liquid it displaces if it floats?   (Joint 16+)

13 Fig. 6.18 shows a block of length 20 cm, uniform cross-sectional area 4 cm², and density 1.25 g/cm³. The block is suspended from a spring balance and fully immersed in a liquid of density 0.8 g/cm³.
a) Calculate the mass of the block, in g.
b) Calculate the weight of the block in N.
c) What would be the reading, in N, on the spring balance if the block were half immersed in the liquid?   (AEB)

14 Fig. 6.19 shows a flat-bottomed test tube containing lead shot floating upright in a liquid. Draw a diagram showing the forces per unit area acting on the test tube caused by the liquid at the points indicated by A_1, A_2, B_1, B_2 and C. (The relative sizes of the forces should be indicated.)

The following readings were obtained for the total mass, M, of the test tube and lead shot, and the depth, h, of the test tube immersed as lead shot was added to the tube.

M/g	48	55	60	65	73	77	84
h/cm	8	9	10	11	12	13	14

Plot a graph of these readings. (You are advised to start your M-axis at 40 g and your h-axis at 8 cm.)
From your graph find the depth immersed when M is 90 g. Use this result to find the area of the base of the test tube.
(Density of the liquid = 1.2 g/cm³, or 1 200 kg/m³.)   (L)

15 The table below gives the densities of some solids and liquids in kg/m³.

Solids		Liquids	
Concrete	2400	Carbon tetrachloride	1630
Cork	240	Paraffin oil	800
Perspex	1190	Turpentine	870
Ice	920	Water	1000

a) Ice floats in water. Is the density of ice smaller, the same, or greater than the density of water?
b) From the list of solids above which would:
   i) float in water;
   ii) float in paraffin oil;
   iii) sink in carbon tetrachloride;
   iv) sink in turpentine?

16 Fig. 6.20 shows an inverted test tube, containing some air, inside a large glass jar full of water.
If the bung is pressed down, the test tube moves to the bottom of the jar. Give an explanation of this.   (AEB part)

Figure 6.20

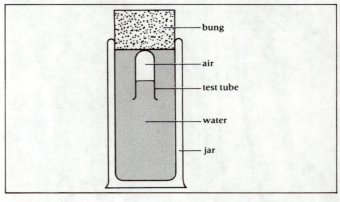

Figure 6.18

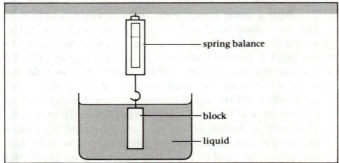

Figure 6.19

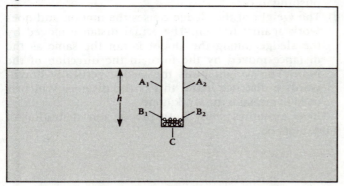

Machines

*T*he earliest stone tools ever discovered, thought to have been used by our ape-like ancestor Homo erectus, are over two million years old. Since that time long ago, our development and progress have been closely linked with making and using tools and machines.

WORK, ENERGY AND POWER

All machines, whether simple or complex, allow a force applied at one place to overcome another force at a different place. Overcoming a force involves doing work. Machines in action do work by taking in energy at one end and feeding it out at the other end, perhaps in a different form. So to understand machines we must look carefully at what we mean by work, energy and the power of a machine.

Work

In science we use the word **work** in a precise way. For example, if you lift a brick from the ground and put it on a wall or if you climb up the stairs you are 'working' in the scientific sense of the word. Similarly if you push a pram or a bicycle and it moves you are also working, but if you push a wall and it remains standing, although you may get tired, you are *not* working. For work to be done a force must produce motion.

We say that *work is done when a force moves its point of application*. This involves movement of the object on which the force acts and also movement of the force itself. When an object is stationary none of the forces acting on it are doing work. Some of the forces acting on a moving object also do no work. This happens when a force acts at right angles to the motion.

Fig. 7.1 shows several forces doing work. Look for other forces acting on the objects which do not do work.

a) Force F is a pushing force which moves the pram. Notice that:

the pram moves in the same direction as the force, a distance s,

the pushing force moves with the pram,

the weight of the pram (acting downwards) and the normal contact force of the ground (acting upwards) act on the pram but do no work because they act at right angles to the motion.

b) The lifting force F raises the brick a vertical height h. The force does work while lifting the brick vertically, in the direction of the force, but does no work while the brick is moved horizontally, at right angles to F.

c) The normal contact force of the stairs does not do the work of raising the man because it does not move upwards with the man. Internal forces, mainly within the muscles of the man's legs, push his body upwards. The internal forces, acting upwards through his centre of gravity, are equal and opposite to his weight and overcome the pull of the earth (his weight). Note that the work done depends on the vertical height h climbed or the distance moved upwards by the pushing force.

d) The weight of the sledge causes the motion and does work against friction. The actual distance moved by the sledge (along the slope) is not the same as the distance moved by the force in the direction of the force. The weight pulls the sledge vertically downwards a distance h and this is the distance which is used to measure the work done.

These examples show how precise our definition of work must be

When a force moves its point of application, the work done by the force is given by the magnitude of the force multiplied by the distance moved in the direction of the force.

Figure 7.1 *Measuring work done*

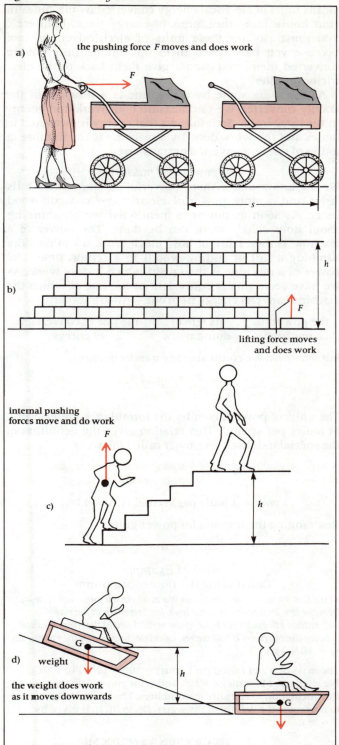

a) the pushing force *F* moves and does work

b) lifting force moves and does work

c) internal pushing forces move and do work

d) weight

the weight does work as it moves downwards

A briefer statement for use when calculating work done is:

work done = force × distance moved in the direction of the force

$$W = F s$$

We use *s* to mean displacement, or the distance in the direction of the force.

Since *W* is used as a symbol for both work and weight, when there is any risk of confusion write the words and avoid the symbol.

The unit of work, given by force × distance, is the newton × metre. This relation provides the definition of the special unit used for work, the joule. The SI unit of work is called the **joule (J)**.

A joule is the work done when the point of application of a force of 1 newton moves through a distance of 1 metre in the direction of the force.

The relation between the units is therefore:

$$1 \text{ joule} = 1 \text{ newton} \times 1 \text{ metre}$$
$$1 \text{ J} = 1 \text{ Nm}$$

Worked Example
Work done, motion in the same direction as the force

If a force of 50 N is used to pull a box along the ground a distance of 8 m and the box moves in the same direction as the force, calculate the work done by the force.

Using $W = Fs$ we have
$$W = 50 \text{ N} \times 8 \text{ m} = 400 \text{ Nm} = 400 \text{ J}$$

Answer: The work done by the force is 400 joules.

Worked Example
Work done, motion and force at an angle

A man of mass 60 kg walks up a track inclined at an angle of 30° to the horizontal, fig. 7.2. If he walks 400 m along the track, how much work does he do? Take g = 10 N/kg.

Figure 7.2

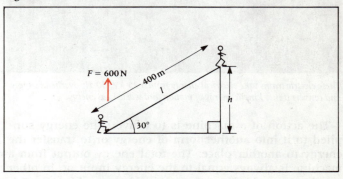

The man must use an upwards force equal and opposite to his weight to climb up the track.
The man's weight = mg = 60 kg × 10 N/kg = 600 N,
∴ the upwards force used = 600 N.
The distance moved in the direction of this force is the vertical height *h* climbed up the track (not the length of the track). In the triangle shown in fig. 7.2,

$$\sin 30° = \frac{h}{l}$$
$$\therefore \ h = l \sin 30°$$
$$\therefore \ h = 400 \text{ m} \times 0.5 = 200 \text{ m}.$$

Now using: work = vertical force × vertical distance,
we have: work = 600 N × 200 m = 120 000 Nm
                    = 120 kJ.

Answer: The man does 120 kilojoules of work.

Energy and power

Machines and energy

Neither people nor machines can do work without a supply of energy. We get our energy supply from the food we eat and machines are 'fed' with energy in many forms. For example some machines are fed with fuels such as coal, oil and gas. Since the energy stored in these fuels is released by chemical reactions such as burning, we describe them as chemical forms of energy.

Some machines take their energy supply in the form of electricity from power stations. But electrical energy is a form of energy which has first to be produced from another form of energy, such as chemical energy or nuclear energy. A few machines take their energy directly from the sun or the wind. Can you think of examples of these? All these sources of energy are discussed in more detail in chapter 19.

These electric trains take electrical energy from a 25 kV, 50 Hz overhead supply and convert it into kinetic energy of motion and waste heat energy.

The action of a machine is to convert the energy supplied to it into another form of energy or to transfer the energy to another place. The total energy output from a machine is always equal to the energy input or, in other words, energy is always **conserved** by a machine, fig. 7.3. In this sense machines do not 'use up' or 'consume' energy even though they need energy to be able to do work. It is more accurate to think of the work done by a machine as a measure of the energy it has **transferred** or **converted** rather than consumed.

Figure 7.3 *Energy is conserved in a machine*

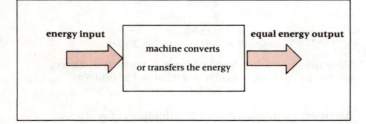

When an electricity bill refers to 'units used' it really means units of electrical energy converted by machines in your house into other forms of energy (heat, light etc.). You must pay for these units of electrical energy not because you have 'used them up' but because, having converted them, you cannot give them back to the electricity supplier!

As the work done by a machine is a measure of the energy converted we can measure both work and energy in the same units. So it follows that energy is measured in joules and the work done by a machine is the number of joules of energy converted or transferred.

The power of a machine

We all know that a more powerful car can climb hills faster and a more powerful electric saw can cut wood faster. As soon as power is mentioned we are thinking about how *quickly* work can be done. The power of a machine is a measure of how much work it can do (like climbing a hill or cutting wood) in a certain time. The power of a machine is the rate at which it does work. As we have seen, this is equivalent to the rate at which the machine converts energy from one form to another:

$$\text{the power of a machine} = \text{its rate of doing work} = \text{the rate of conversion of energy}$$

For *conversion* we could also say *transfer* or *change*.

$$\text{power} = \frac{\text{work}}{\text{time}} = \frac{\text{energy converted}}{\text{time}} \qquad P = \frac{W}{t} = \frac{E}{t}$$

The units of power given by the formula are joule/second or joules per second. This relation gives the definition of the special unit used for power called the **watt**.

A watt is the rate of working or energy conversion of 1 joule per second.

1 watt = 1 joule per second (1 W = 1 J/s).

Rearranging the formula for power gives:

$$\text{work} = \text{power} \times \text{time} \qquad W = Pt$$

Worked Example
Calculating the power of a pump

At the Dinorwig reservoirs in North Wales an electric pump can raise water from the low-level reservoir to the high-level reservoir at the rate of 200 million kilograms per hour. If the vertical height the water is raised between the reservoirs is 530 metres, calculate the power of the pump. (g = 10 N/kg.)

The mass of water raised (in 1 hour) = 200×10^6 kg or 200 Gg We shall use M for mega = 10^6, and G for giga = 10^9, throughout the calculation to simplify the numbers. The upwards force used to overcome the weight of the water, i.e. to lift it, is given by $F = mg$,

$$\therefore F = 200 \,\text{Gg} \times 10 \,\text{N/kg} = 2\,000 \,\text{MN}$$

The work done (in 1 hour) raising this water

$$= F \times \text{vertical distance.}$$

$\therefore$ Work done = $2\,000 \,\text{MN} \times 530 \,\text{m} = 1\,060\,000 \,\text{MJ}$, and the time taken to do this work = 1 hour = $3\,600$ s, so

$$\text{the power of the pump} = \frac{\text{work done}}{\text{time taken}} = \frac{1\,060\,000 \,\text{MJ}}{3\,600 \,\text{s}} = 294 \,\text{MW}.$$

Answer: The power of the pump is 294 megawatts.

Worked Example
Work done = power × time

Calculate the work done in one hour by an electric motor in a washing machine which has an output power rated at 1.5 kW.

The time of working = 1 hour = 3 600 s.
the rate of working = 1.5 kW = 1 500 W = power,
work done = power × time.
∴ work done = 1 500 W × 3 600 s = 5 400 000 J or 5.4 MJ.

Answer: The work done by the motor in 1 hour is 5.4 megajoules.

Measuring your own power output

Your rate of working is probably at its highest when you are running up a hill or up stairs lifting your own weight.

● *Get another person to time how long it takes you to run up a long flight of steps.*

● *Measure the height of one step and count the number of steps you climbed. You also need to know your own weight.*

Here are some results for a fit girl of mass 44 kg.

time to climb the steps = 11 s
height of 1 step = 20 cm
number of steps climbed = 60

The vertical height climbed = 60 × 0.20 m = 12 m,

the girl's weight = mg = 44 kg × 10 $\frac{N}{kg}$ = 440 N.

$$\begin{matrix}\text{The work} \\ \text{done by} \\ \text{the girl}\end{matrix} = \begin{matrix}\text{force used to} \\ \text{overcome her} \\ \text{weight}\end{matrix} × \begin{matrix}\text{vertical} \\ \text{distance} \\ \text{climbed}\end{matrix}$$

So work done = 440 N × 12 m = 5 280 N m = 5 280 J,

and the power of the girl is:

$$\frac{\text{work done}}{\text{time taken}} = \frac{5\,280\,J}{11\,s} = 480\,W$$

The girl's power was 480 watts.

Assignments

Measure your own power for some of the following tasks:
a) running up stairs as described above,
b) climbing a local hill,
c) cycling up a hill (include the weight of the bicycle),
d) lifting a heavy object up and down a measured height for a large number of times.

Remember the definitions and formulas for calculating work and power:
e) work W = force F × distance s moved in the direction of the force.

f) the formula for power:

$$\text{power } P = \frac{\text{work } W}{\text{time } t}$$

or $$W = Pt$$

Try questions 7.1 to 7.4

FORCE MULTIPLIERS AND DISTANCE MULTIPLIERS

Machines may be designed either to increase the size of a force or to increase the distance or speed with which something moves. In a mechanical machine the energy input is supplied by a force called the effort and the energy output is obtained as the machine is used to do work in moving the load.

Force multipliers

Machines which allow a small effort to move a larger load are called **force multipliers**. Some examples of force multipliers include: a crowbar, wheelbarrow, nutcracker and bottle opener. The number of times a machine multiplies the effort is called its **mechanical advantage**.

The mechanical advantage of a machine is the number of times the load moved is greater than the effort used.

$$\text{mechanical advantage (MA)} = \frac{\text{load}}{\text{effort}}$$

● *Mechanical advantage is a ratio and so has no units.*
● *Force multipliers have a mechanical advantage greater than 1 (MA > 1).*

Figure 7.4 *The lever, a simple machine*

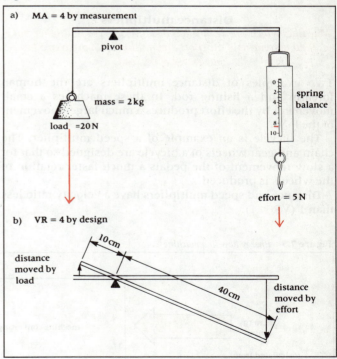

The mechanical advantage of a machine can only be found by measurement. Fig. 7.4a shows a simple lever with unequal arms being used as a force multiplier. The effort needed is being measured by a spring balance. The load, a 2 kg mass, has a weight of 20 N. So the lever has a mechanical advantage given by:

$$\text{mechanical advantage} = \frac{\text{load}}{\text{effort}} = \frac{20\,N}{5\,N} = 4$$

Velocity or distance ratio

When a machine is used to multiply the effort, energy conservation is maintained by making the effort move a greater distance than the load. This is necessary because the energy output from the machine (load × distance load is moved) cannot exceed the energy input (effort × distance effort moves).

The number of times further the effort moves than the load is called the **distance ratio**. It is also called the **velocity ratio (VR)** because an effort which moves further than the load in the same time also moves faster.

$$\text{distance or velocity ratio} \qquad VR = \frac{\text{distance moved by the effort}}{\text{distance the load is moved}}$$

- *The distance or velocity ratio of a machine has no units.*
- *The distance or velocity ratio of a force multiplier is always greater than 1.*
- *The distance or velocity ratio of a machine is always greater than its mechanical advantage.*

Finding a distance or velocity ratio

The distance or velocity ratio of a machine can usually be calculated exactly from its design or geometry. For example, the distances moved by the effort and load for the lever shown in fig. 7.4b are in the same ratio as the lengths of the arms, i.e. 40 cm to 10 cm or 4:1. This gives the lever a velocity ratio of exactly 4.

Distance multipliers

Machines which are designed as distance or speed multipliers take a small movement of the effort and multiply it to produce a larger movement of the load.

Two examples of distance multipliers are the human forearm and a fishing rod. In these machines a small movement by the effort produces a much larger movement of the load.

The bicycle is an example of a speed multiplier. The chain and gear wheels of a bicycle are designed so that for a slow movement of the pedals a much faster rotation of the wheels is produced.

Distance and speed multipliers have a velocity ratio less than 1 (VR < 1).

The efficiency of a machine

Energy conservation demands that the total energy output of a machine must equal its energy input. However, when we measure the energy output as work done on the load by a machine, we find it is less than the energy input.

The work done by a machine against its load (moving, lifting, cutting it etc.) is called its **useful work** or **useful energy output**. In a simple mechanical machine we can measure this useful energy output as the load × the distance the load is moved by the machine.

The machine also does work against frictional forces and sometimes does work in moving itself. For example, in a pulley machine work is done in lifting the moveable pulley block and hook to which the load is attached.

The work done against friction converts input energy into wasted heat energy and a little noise energy (which also eventually becomes heat energy). The energy equation now looks like this:

$$\text{energy input} = \frac{\text{useful energy}}{\text{output}} + \frac{\text{wasted energy}}{\text{output}}$$

As a machine wastes some of its input energy, fig. 7.5, it is not completely efficient at converting the input energy into the desired output form. We measure the efficiency of a machine, usually stated as a percentage, by the ratio:

$$\text{efficiency} = \frac{\text{useful energy output}}{\text{energy input}} \times 100\%$$

The efficiency of a mechanical machine can be calculated from the input and output work done:
useful output work = load × distance load is moved,
the input work = effort × distance effort moves,

so that efficiency
$$= \left[\frac{\text{load}}{\text{effort}}\right] \times \left[\frac{\text{distance load is moved}}{\text{distance effort moves}}\right] \times 100\%$$

Looking at this formula we can see that it contains (in the brackets) the relations for MA and 1/VR. It is often useful to be able to calculate the efficiency of a machine from values of its MA and VR. So we have the alternative formula for the efficiency of a machine:

$$\text{efficiency} = \frac{\text{MA}}{\text{VR}} \times 100\%$$

Figure 7.5 *Energy flow in a machine*

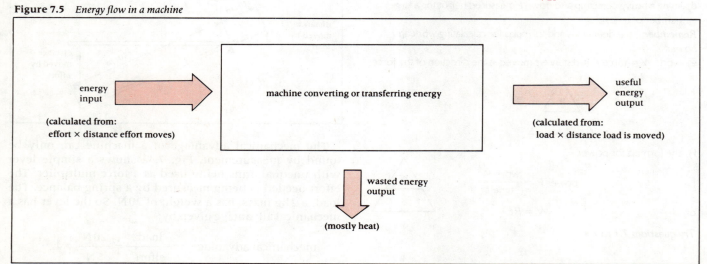

Worked Example
Calculating the efficiency of a machine

A mechanic uses a pulley machine with a velocity ratio of 6 to raise an engine out of a vehicle. The engine, which has a weight of 2800 N, is raised a vertical distance of 1.5 m by the machine. If the mechanic pulls with an effort of 500 N, calculate:

a) *the work done by the mechanic,*
b) *the useful work done by the pulley machine,*
c) *the mechanical advantage of the machine,*
d) *the efficiency of the machine.*

a) Using the formula for VR rearranged we have:
distance effort moves = VR × distance load is raised.
(The VR of the pulley machine tells us that the effort will move 6 times further than the load is raised.)
∴ distance effort moves = 6 × 1.5 m = 9.0 m
Now the work done by the mechanic, i.e. by the effort, is given by the effort × distance effort moves:
∴ work done by effort = 500 N × 9.0 m = 4500 J.

b) The useful work done by the pulley machine = force to overcome the load × distance load is raised
∴ useful work done = 2800 N × 1.5 m = 4200 J

c)
$$MA = \frac{load}{effort} = \frac{2800\,N}{500\,N} = 5.6$$

d) The efficiency of the machine is given by:

Method 1
$$efficiency = \frac{useful\ output\ work}{input\ work} \times 100\%$$

$$\therefore\ efficiency = \frac{4200\,J}{4500\,J} \times 100\% = 93\%$$

Method 2
$$efficiency = \frac{MA}{VR} \times 100\%$$

$$\therefore\ efficiency = \frac{5.6}{6} \times 100\% = 93\%$$

Answers:
a) The work done by the mechanic, which is the energy input to the machine, is 4500 joules.
b) The useful work done by the pulley machine i.e. the useful energy output is 4200 joules.
c) The MA of the machine allowed the mechanic to lift an engine 5.6 times heavier than the effort force the mechanic used.
d) The efficiency of the machine is 93%.

Efficiency and friction

Friction between moving parts in a mechanical machine is the main cause of its wasted work which produces unwanted heat energy. Reducing friction in a machine improves its efficiency and saves energy.

Friction between two surfaces may be caused by roughness of the surfaces in which points or projections from one surface bump into similar projections from the other surface. These projections may be so small that they can be seen only with a microscope. Obviously if the surfaces can be made smoother then they will be able to slide over each other more easily.

However some surfaces which are very flat and smooth, such as the metal surfaces of a bearing in a car engine, will 'stick' together even though they are as smooth as a polished mirror. The frictional force between these surfaces arises because the smooth surfaces can come very close together and the molecules in one surface are strongly attracted to molecules in the other surface. So strong is this mutual attraction between the two surfaces that at the points of closest contact the surfaces are effectively welded together. When they are pulled apart molecules are 'torn' from each surface producing visible damage or scratches. Damage to surfaces in contact as they move against each other is called **wear** and is said to be caused by friction.

The surface of this bearing should be smooth and polished. Instead it is rough and worn where the surface molecules have been torn away.

Assignments

Remember the definitions or formulas of MA, VR and efficiency.
Find more examples of methods of reducing friction in machines.
Explain why the MA of a machine cannot be greater than its VR.
Make yourself a simple hovercraft as shown in fig. 7.6.

Figure 7.6 *A homemade hovercraft*

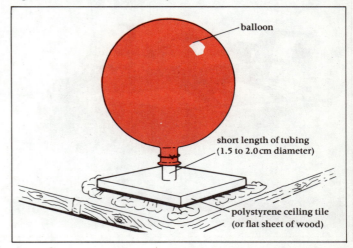

balloon

short length of tubing
(1.5 to 2.0 cm diameter)

polystyrene ceiling tile
(or flat sheet of wood)

Reducing friction

Most problems with friction in machines are reduced or overcome in one way or another by keeping the moving surfaces apart.

The following list gives most of the methods used to reduce friction in machines.

Moving surfaces are made as smooth as possible.

Lubricants such as oil and silicone are used to separate surfaces.

Machines are moved on rollers and wheels to reduce friction with the ground; roller and ball bearings are used to separate rotating axles from their mountings.

Machines are held above the ground by cushions of air as in the hovercraft and some bearings use compressed air as an elastic lubricant. Air has the advantage over oil that, being a gas, it is compressible and therefore has extra useful elastic properties which cushion vibrations in a rotating axle.

Machines which move through fluids such as water and air are made streamlined in shape to reduce the frictional drag.

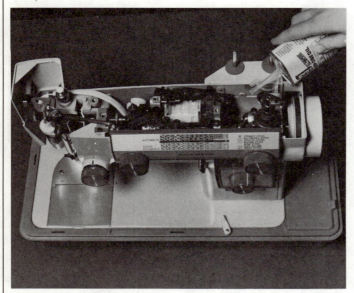

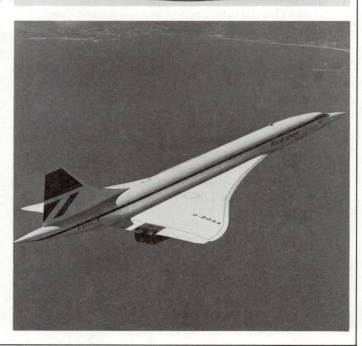

7.3
THE PRINCIPLES OF SOME MACHINES

Machines convert energy from one form to another and allow a supply of energy at one point to do work at another point. Some machines, with a mechanical advantage greater than 1, multiply the strength of the effort-maker so allowing him or her to move larger loads. Other machines, with a velocity ratio less than 1, allow a small movement to be magnified into a larger one. The machines we shall look at in this section show how these principles are applied.

Levers

Levers are simple machines which use a **pivot** or **fulcrum** to transfer the work done by the effort at one point to a load at another point.

We have already used the crowbar type of lever as an example of a simple machine, fig. 7.4. Other types of levers are shown in fig. 7.7. In each case note the following:

a) where the pivot is,
b) whether the effort or the load is nearer to the pivot; and so decide:
c) whether its velocity ratio is greater or less than 1,

d) whether you expect its mechanical advantage to be greater or less than 1.

The first type of lever has the pivot in between the load and the effort. Both the scissors and the screwdriver used to remove a tin lid have a velocity ratio greater than 1 and, using the principle of the crowbar, *magnify the effort.*

● *Try to cut a piece of string near the pointed end of the scissors. What happens and why?*

The second type of lever has the load between the effort and the pivot. The wheel barrow and the bottle opener show how this arrangement also gives a velocity ratio greater than 1 and the lever *magnifies the effort.*

The third type of lever has the effort between the load and the pivot. Fishing rods are designed like this so that a small movement of the effort produces a *magnified movement of the load.* In the case of the tweezers the force applied to the load is much smaller than the effort and so allows for a fragile object to be held very gently. The biceps muscle which raises the human forearm with a load in its hand must apply a much greater effort force than the load it lifts. The mechanical advantage of the human forearm is about $\frac{1}{7}$.

In which category do you place the spanner?

Figure 7.7 *Examples of lever machines*

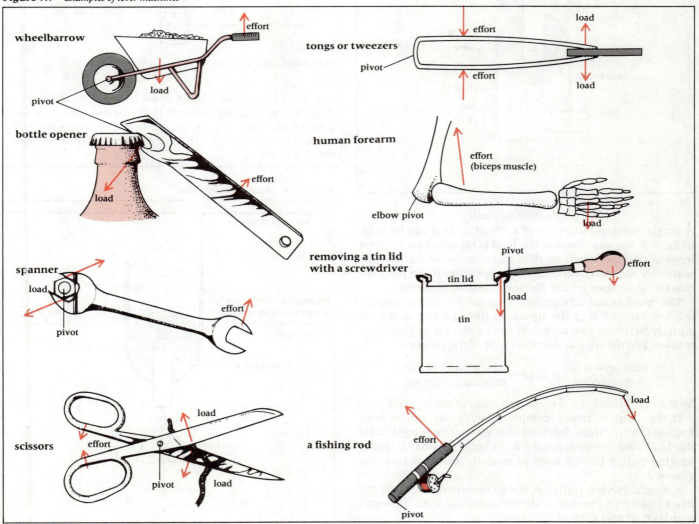

Pulleys

Pulleys are used to change the direction of a force and to gain a mechanical advantage greater than 1.

A single fixed pulley

A fixed pulley is one with a fixed support which does not move with either the effort or the load. The pulley itself should turn on its axle as freely as possible for maximum efficiency. Fig. 7.8 shows a single fixed pulley which is used to change the direction of the effort force *E* from a downwards pull to an upwards lift. The tension *T* in the string or rope applies the upwards force to the load *L*. It is often easier to pull a rope downwards than to lift a load upwards.

The VR of a single fixed pulley must be exactly 1 as the load will rise by the same distance as the effort moves.

The MA will be almost 1, there being only a small amount of work wasted against friction on the pulley bearing and in lifting the weight of the rope.

Figure 7.8 *A single fixed pulley*

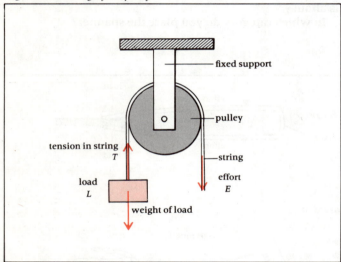

A single moving pulley

A single moving pulley gives a VR of 2. This can be seen in fig. 7.9. For any distance the load is raised, there are two lengths of rope equal to that distance to be pulled upwards by the effort. So both the ropes supporting the load must be shortened by the distance the load is raised.

The mechanical advantage is found by measurement, but we can see that the upwards lifting force is shared equally between two upwards forces, the effort *E* and the tension *T* in the rope at the other side of the pulley.

$$\text{total upwards force needed} = \text{load} + \text{the weight of the moving pulley etc.}$$

So the effort needed = $\frac{1}{2}$ (load + weight of pulley etc.)

If the load is heavy compared with the pulley and frictional forces then the effort needed will be roughly half the load and the MA nearly 2. In other words, a single moving pulley can be used to magnify an effort force by almost 2.

A single moving pulley is often combined with a single fixed pulley to provide a simple machine with a downwards effort and a VR of 2.

A block and tackle

Pulleys are usually used in sets of two or more to gain a higher VR and MA. Two sets of pulleys are used, one fixed and one moving. The pulleys are mounted, usually side-by-side, in a **block** or frame and the apparatus of pulleys and ropes is generally called the **tackle**. The whole system or machine is known as a block and tackle. Many large machines use 'block and tackle' pulley systems, for example cranes and lift mechanisms.

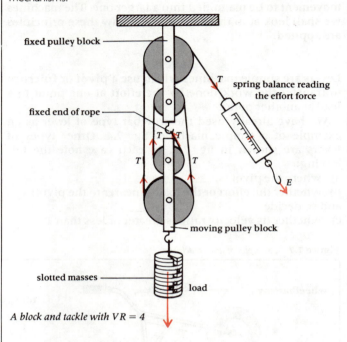

A block and tackle with VR = 4

So that we can see more clearly where the ropes go, we usually draw the pulleys below each other in each block. The lower, moving pulley block is supported by 4 ropes. To raise the load by 1 metre will require 4 metres of rope to be pulled out of the machine by the effort and so the VR is exactly 4.

Figure 7.9 *A single moving pulley*

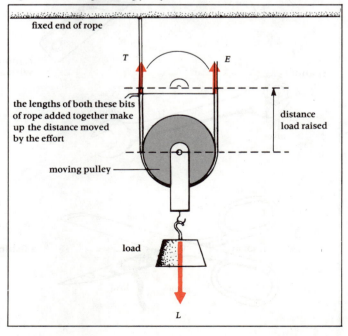

Measuring the MA and efficiency of a block and tackle pulley system

In a perfect machine the four tension forces T which support the load would take exactly a quarter of the load each. The effort force providing the tension would also be $\frac{1}{4}$ of the load and so the mechanical advantage would be exactly 4. But we can only find the MA of a particular machine by measuring the effort needed to overcome or lift a particular load.

Using a block and tackle with a VR of 4 and slotted masses from 1 to 10 kg, a suitable spring balance would measure over the range 0 to 30 N.

• *Record readings of the spring balance for the effort needed to just raise the load for loads from $\frac{1}{2}$ kg to 10 kg.*
• *Make a table of results as in table 7.1.*

Table 7.1

Mass of load/kg	Load/N	Effort/N	$MA = \dfrac{L}{E}$	Efficiency = $\dfrac{MA}{VR} \times 100\%$
0.5	5	2.5	2.0	50%
1	10	3.5	2.9	72%

• *Flot a graph of MA against load and of efficiency against load.* Some typical results, shown in fig. 7.10, indicate an increase in MA and efficiency as the load increases. This is to be expected because at high loads the weight of the moving pulley block and frictional forces are very small compared with the load. Note that the VR, which is constant at 4 in this example, limits the MA to a maximum which is always less than the VR.

Figure 7.10 *The MA and efficiency of a pulley system of VR = 4*

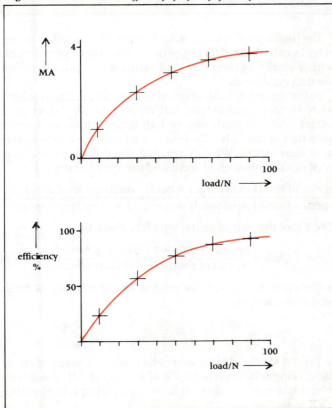

Inclined planes

As we have already seen (p 107) the work done in lifting a load depends on the *vertical distance* moved.

An inclined plane is a slope or ramp which allows a load to be raised more gradually and by using a smaller effort than if it were lifted vertically upwards. Bolts and screws are based on the principle of an inclined plane. Inclined planes, bolts and screws are machines which magnify the effort, i.e. they have a MA greater than 1.

In fig. 7.11 the effort E, pushing up the inclined plane or slope, moves a distance d along the slope. The load L is raised a vertical height h against gravity.

$$VR = \frac{\text{distance effort moves along the slope}}{\text{vertical height load is raised}} = \frac{d}{h}$$

Figure 7.11 *An inclined plane*

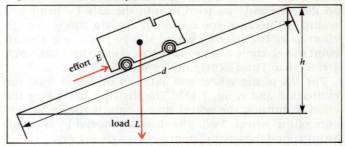

It follows that the longer and more gradual the slope, the greater its velocity ratio will be.

Measuring the MA of an inclined plane

• *Find the weight of a small trolley by hanging it on a 0 to 10 N spring balance.*
• *Now pull the same trolley up a sloping ramp or runway using the spring balance. Read on the spring balance the effort force which is just large enough to keep the trolley moving up the slope at a constant speed.*
• *Change the gradient of the slope and investigate how the effort, MA and efficiency are affected.*
• *Do the calculations in the same way as for the block and tackle pulley system.*

Bolts and screws

• *To discover how a bolt or screw is based on an inclined plane cut a sheet of paper into a long 'ramp' and wrap it round a short length of dowel rod or a pencil, as shown in fig. 7.12.*
• *Compare the paper ramp as it winds round the rod with the thread on a bolt as it winds up the bolt.*

Figure 7.12 *A bolt is an inclined plane*

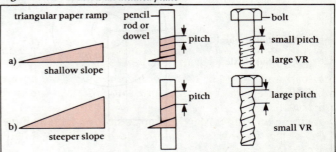

The distance between adjacent threads (ones next to each other) is called the **pitch** of the thread. When the bolt is turned through one full turn it moves up or down a distance equal to the pitch of its thread.

The threaded bolt can be pulling or lifting a load as it turns. The distance the load is moved for each turn of the bolt equals the pitch of the thread. The distance round one turn of the thread, along the ramp, is the corresponding distance the effort moves.

Fig. 7.12b shows a bolt with a larger pitch thread. This bolt will have a smaller VR: the slope of the ramp is steeper. Wood screws use the same principle but the thread is tapered to make it easier to enter the wood.

The wheel and axle

A car steering wheel provides a good example of the wheel and axle principle, fig. 7.13a. If you had no wheel on the axle and you tried to turn the axle by hand you would fail because you could not apply a strong enough effort to it. The steering wheel allows you to use a small effort to overcome a large load. A steering wheel has a MA greater than 1 and so magnifies your effort.

The VR of the wheel and axle can be found from the radiuses R and r, fig. 7.13b. The distance moved by the effort in turning the wheel round once is the circumference of the wheel, $2\pi R$. The distance moved by the load acting round the circumference of the axle is $2\pi r$.

$$VR = \frac{\text{distance moved by effort}}{\text{distance moved by load}} = \frac{2\pi R}{2\pi r} = \frac{R}{r}$$

Figure 7.13

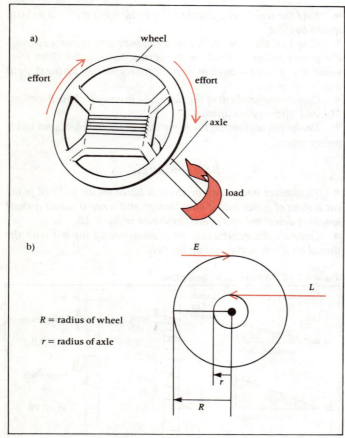

a)

wheel

effort

effort

axle

load

b)

E

L

R = radius of wheel

r = radius of axle

r

R

Gears

Unlike pulleys which turn freely on their axles, most gears are rigidly fixed to and turn with their axles. Gears are designed with VR both greater and less than 1. When the VR is greater than 1 the effect is to slow down the speed of rotation and to magnify the effort force, fig. 7.14a. When the VR is less than 1 the effect is to speed up the rotation and magnify the distance moved, fig. 7.14b.

Figure 7.14   *Gears*

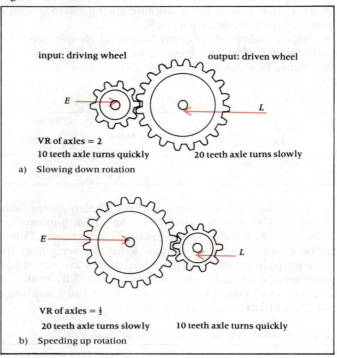

input: driving wheel   output: driven wheel

E   L

VR of axles = 2

10 teeth axle turns quickly   20 teeth axle turns slowly

a)   Slowing down rotation

E   L

VR of axles = $\frac{1}{2}$

20 teeth axle turns slowly   10 teeth axle turns quickly

b)   Speeding up rotation

The *input* gear wheel, which receives the effort at its axle, is called the **driving wheel**. The *output* gear wheel, which works against the load acting at its axle, is called the **driven wheel**.

In the figure, in both cases, when the smaller gear wheel with 10 teeth makes one full revolution the larger gear wheel with 20 teeth makes half a revolution and turns only half as quickly. *The larger wheel with more teeth always turns more slowly*. We can see that there is an *inverse ratio* between the number of teeth and the speed of rotation:

$$\frac{\text{speed of rotation of larger wheel}}{\text{speed of rotation of small wheel}} = \frac{\text{teeth on smaller wheel}}{\text{teeth on larger wheel}}$$

The VR of the pair of gears, which is given by

$$VR = \frac{\text{speed of rotation of driving wheel}}{\text{speed of rotation of driven wheel}}$$

is therefore also inversely related to the number of teeth on the two wheels:

$$VR = \frac{\text{number of teeth on the driven wheel}}{\text{number of teeth on the driving wheel}}$$

In fig. 7.14a, the driven wheel has twice as many teeth as the driving wheel giving a VR of 2. In fig. 7.14b the driven wheel has half as many teeth as the driving wheel giving a VR of $\frac{1}{2}$.

To gain a MA greater than 1 a small gear wheel is used to turn a larger one. For example, when you drive a car up a steep hill you change into a low gear. This means that the output gear wheel will have a low speed and the car will move more slowly. But the effort from the car engine will be turning the input gear wheels quickly. This is arrangement (a) in fig. 7.14. A VR greater than 1 slows down the rotation of the driven wheel and gains a MA greater than 1 so magnifying the effort of the engine which is raising the car up the hill.

Hydraulic machines

Hydraulic machines are able to use liquid pressure to transfer energy from one place to another because of the following liquid properties:
a) liquids are incompressible,
b) liquid pressure acts equally in all directions (at the same level) and
c) changes in liquid pressure are transmitted instantaneously to all parts of a liquid.

The advantages of using hydraulic equipment include:
a) the ability to magnify a force simply by using a piston of larger area, see fig. 7.15,
b) the ability to apply a force in any direction at any point (by using a flexible pipe) and
c) the ability to apply forces to several points simultaneously (e.g. brakes to all four wheels of a car).

The principle of a hydraulic machine is shown in fig. 7.15. The effort E applied to a piston of small area A_1 produces a pressure p in the liquid given by:

$$\text{pressure} = \frac{\text{force}}{\text{area}} \text{ or } p = \frac{E}{A_1}$$

This pressure acts throughout the liquid and against the piston applied to the load. The magnitude of the force applied to overcome the load L is given by:

$$\text{force} = L = \text{pressure} \times \text{area} = p A_2 = \left(\frac{E}{A_1}\right) A_2$$

$$\therefore \text{ MA} = \frac{L}{E} = \frac{A_2}{A_1}$$

The MA gained by a hydraulic machine is the ratio of the areas of the pistons.

The piston to which the effort is applied is called the **pump** piston and the one which is applied to the load is called the **ram** piston. So the MA can be stated as:

$$\text{MA} = \frac{\text{area of ram piston}}{\text{area of pump piston}}$$

Figure 7.15 *Magnifying a force using a hydraulic machine*

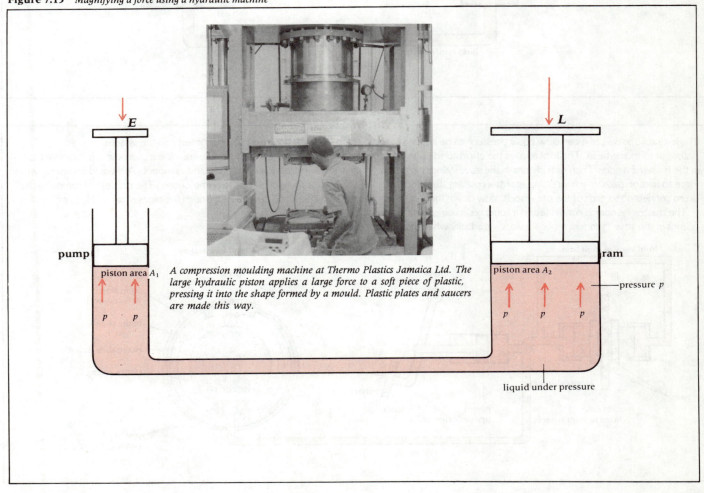

A compression moulding machine at Thermo Plastics Jamaica Ltd. The large hydraulic piston applies a large force to a soft piece of plastic, pressing it into the shape formed by a mould. Plastic plates and saucers are made this way.

The hydraulic jack

A reservoir of liquid is needed if a pump piston in a small cylinder is to raise a ram piston in a large cylinder because of the volume of liquid required.

The pump piston acts like a force pump, p 95. The *down-stroke* of the pump piston applies the effort which raises the load. The liquid pressure, being greater than atmospheric pressure keeps valve A closed and the flow of liquid under pressure opens valve B and raises the ram piston. On the *up-stroke* of the pump piston, the pressure below the piston is reduced and atmospheric pressure opens valve A and keeps

the cylinder filled with liquid. The high pressure of the liquid below the ram piston keeps valve B closed and prevents the ram falling while the pump piston rises.

To lower the ram piston and its load, the pressure of the liquid below it is released slowly through a release valve which allows liquid to return to the reservoir. The release valve must be closed again before pumping can begin.

A hydraulic press uses exactly the same arrangement with a strong enclosed chamber in which pressing occurs above the ram piston.

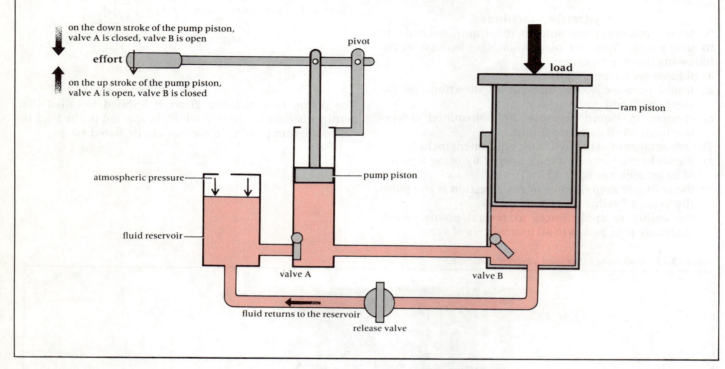

Hydraulic brakes

The hydraulic brakes of a car allow equal pressure to be applied to the pistons at all four wheels. The foot applies the effort to the pump piston in the master cylinder. Then ram pistons in the slave cylinders apply a force to friction pads which, pressing against a rotating steel disc or drum connected to each of the car wheels, slow down the car.

The master cylinder is connected to a liquid reservoir so that, as the pistons in the slave cylinders move forwards gradually when the friction

pads wear down, more liquid is fed into the system.

Special non-corrosive brake liquids are used and it is important to remove all air from the pipes and cylinders. (Air would compress when pressure is applied by the pump piston.) The process of removing the air by pumping new liquid through the pipes is called 'bleeding' the system.

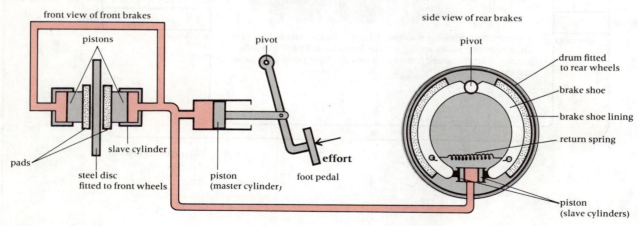

Questions 7

Assume $g = 10\,\text{N/kg}$.

1   Calculate the work done when a force of 20 N moves an object a distance of
    a) 5 m in the direction of the force,
    b) 2 km in the direction of the force,
    c) 20 m in a direction which makes an angle of 60° with the direction of the force.

2   A boy of mass 50 kg runs up a hill of vertical height 300 m in 20 minutes. Calculate
    a) the average vertical force he uses to lift himself up the hillside,
    b) the work he does climbing the hill,
    c) his average power.
    (Use $g = 10\,\text{N/kg}$.)

3   Calculate the power of the following machines:
    a) an electric fire which converts 1.2 MJ of electrical energy into heat in 20 minutes,
    b) a motor which raises a lift cage of mass 1 000 kg a vertical height of 40 m in 40 s.

4   The electric motor, in fig. 7.16, completely lifts the box in 5 s, at a steady speed. If the box weighs 100 N, calculate
    i) the work done by the motor,
    ii) the power of the motor.   (W 88)

Figure 7.16

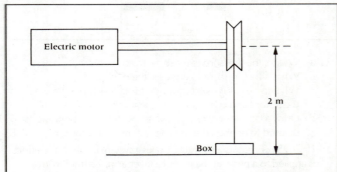

5   Fig. 7.17 shows a wire of British Rail's electrification system, being held taut by a load L and a pulley system P.
    a) By what factor is the force multiplied?
    b) What is the purpose of pulley 1?
    c) Why are pulleys used at all?
    d) If the load L is 2000 N, what is the tension in the wire W?

Figure 7.17

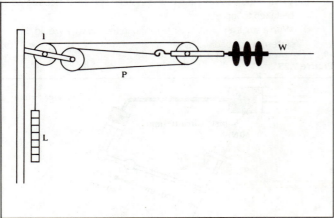

6   Fig. 7.18 shows an object of weight 1 000 N being pushed up a ramp of length 15 m.

Figure 7.18

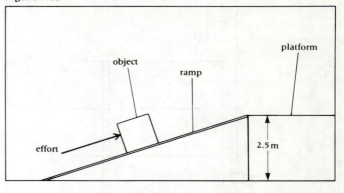

a) Calculate the useful work done, in J, in taking the object from ground level to the platform.
b) If 500 J of energy is wasted in this operation, calculate the effort required.
c) Give a reason for this wasted energy.   (AEB)

7   A block and tackle pulley system with a velocity ratio of 5 and 60% efficiency is used to lift a load of mass 60 kg through a vertical height of 2 metres.
    i)   What effort must be exerted?
    ii) How much work is done lifting the load?
    iii) How much work is done by the effort?
    (Assume that the acceleration of free fall (due to gravity) is $10\,\text{m/s}^2$.)   (JMB part)

8   A crate, of mass 70 kg, is pulled a distance of 12 m up an inclined plane and in the process its centre of gravity is raised 2.0 m, as shown in fig 7.19. In order to do this a force of 150 N is applied to the crate in a direction parallel to the inclined plane.

Figure 7.19

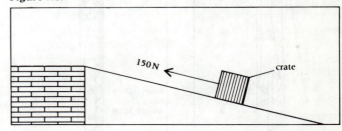

a) What is the increase in the potential energy of the crate?
b) What is the work done by the force?
c) Why do your answers to (a) and (b) differ?   (O & C)

9   It may be said that machines make work easier but do not make it any less. If you have, for example, a lifting job to do you may use a pulley system to do it, as in fig. 7.20.
    a) There are at least two good reasons for using a simple pulley to do this job. Let us see what they are.
        i)   Calculate the work done in raising the load directly if it weighs 300 N and has to be raised through a height of 20 m.
        ii) Calculate the work you would do in raising yourself through the same height if you weighed 600 N.
        iii) Can you see one good reason for using the pulley? Explain your answer.
        iv) To raise the load yourself, you have to climb the scaffolding ladders. What is the other good reason for using the pulley?

Figure 7.20

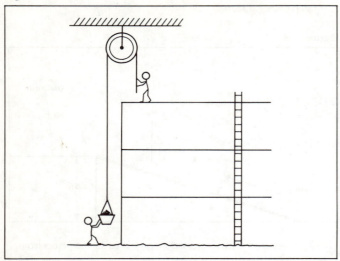

b) If you used a pulley and a force of 350 N to raise the load at a steady speed, at what power would you work to raise it through 20 m in 35 s? (Use the equation: Power = energy transferred/time taken.)

c) In reality, a block and tackle or compound pulley system would be used. The system shown in fig. 7.21 has **four** "lifting" strings. Calculate i) the force in **one** lifting string if the total weight of the load and the lower pulleys is 400 N, ii) the fraction of the total weight represented by the 300 N load. (NEA [B] 88)

Figure 7.21

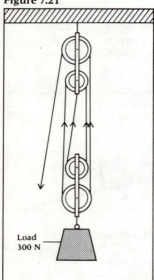

Load
300 N

Figure 7.22

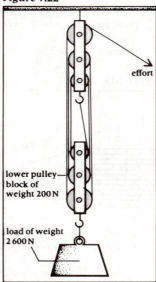

effort

lower pulley block of weight 200 N

load of weight 2 600 N

10 a) Fig. 7.22 shows a pulley system. What is the velocity ratio of this system?

b) The minimum effort required to lift the load is 500 N. When the load is raised through a vertical distance of 1.5 m, by how much does the work done by the effort exceed the total work done in raising the lower pulley block and the load?

c) Suggest a reason for this difference. (AEB)

11 Fig. 7.23 shows a set of gears used to lift a load attached to the axle P by applying an effort to the axle Q.

a) In order to lift the load through a distance of 2 m, the axle P must rotate 5 times. How many times must axle Q be rotated?

b) Through what distance must the effort be applied, if the axles P and Q have the same diameter?

c) Assuming that the set of gears is 100% efficient, calculate the effort required to lift the load. (AEB)

Figure 7.23

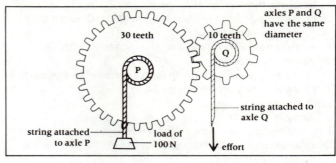

axles P and Q have the same diameter

30 teeth

10 teeth

Q

P

string attached to axle P

load of 100 N

string attached to axle Q

effort

12 Fig. 7.24 represents a hydraulic system in which two cylinders with closely fitting pistons A and B are linked by a tube filled with oil. The area of A is $5\,cm^2$ and the area of B is $50\,cm^2$. A force of 10 newtons is exerted downwards on A.

Figure 7.24

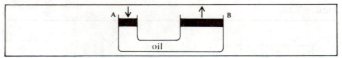

A   B

oil

a) What is the pressure exerted on the oil by A?

b) What is the pressure exerted by the oil on B?

c) What force will B exert upwards if frictional forces between the pistons and cylinders can be neglected?

d) Draw a labelled diagram showing how the same principle can be used to operate the braking system in a motor car. (JMB)

13 Fig. 7.25 represents a hydraulic braking system in which the effort is applied to a pedal at the end of a lever arm, 200 mm from a pivot. On the other side of the pivot, 40 mm away, the lever connects to a piston of area $50\,mm^2$. The piston transmits pressure through oil to another piston, of area $100\,mm^2$, connected to the brake.

a) When the effort applied is 60 N, what is:
   i) the force applied to the first piston;
   ii) the pressure on the oil?

b) What force is applied to the brake at the end of the system if frictional forces can be ignored?

c) What is the velocity ratio of this system, and what steps could be taken to increase it?

d) With a small amount of air trapped in the oil, why would the system work badly, if at all? (O)

Figure 7.25

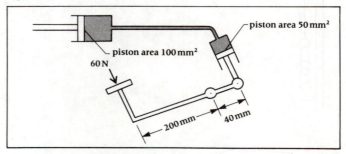

piston area $50\,mm^2$

piston area $100\,mm^2$

60 N

200 mm

40 mm

8
Motion

The motion of a satellite in orbit round the earth and that of a spaceship travelling through deep space are quite different but neither of them needs its engines switched on to keep moving at a constant speed. However, any cyclist knows that he must keep pushing on the pedals to keep his bicycle moving at a steady speed even on a flat road. If a man jumps out of an aeroplane at a great height above the ground his speed increases as he falls so that when he reaches the ground he will be moving so fast that he is likely to be killed. But if he uses a parachute his speed does not increase and he should land safely. A swimmer swimming across a river finds that he lands some distance downstream on the opposite bank even though he swims in a direction perpendicular to the river banks. Each of these examples of motion is different in some way and in this chapter we learn how to describe and explain each kind of motion.

8.1
SPEED AND VELOCITY

When we describe how fast an object is moving we are describing its speed. When we say which direction it is moving in as well we are stating its velocity.

Average speed

As a runner completes a lap of a running track we can calculate his average speed from the distance round the track and the time taken to run round. During the lap his speed will vary and our calculation only gives the *average* value of his speed. A cyclist who travels 200 km in 4 hours will, on average, travel 50 km in each hour. So '50 km in each hour' is called his average speed but we know that for all sorts of reasons he will not travel at a steady speed; for example, as he becomes tired he will probably cover less kilometres in the last hour than in the first hour.

The speed of a car calculated from the time taken to travel a full journey or a particular distance is its average speed. However, the speed of a car as indicated by its speedometer at any instant in time is called its instantaneous speed.

An average speed might be written as 50 km/hour where the slanting stroke is short for 'per'. So we can read 50 km/hour as '50 kilometres per hour' or 50 kilometres in each hour. When we are describing motion, *per* means *in each* (unit of time). When we calculate an average speed we are finding the distance gone in each second or hour (i.e. unit of time). So to find the average speed we must divide the total journey travelled or distance moved by the number of seconds or hours taken.

$$\text{average speed} = \frac{\text{distance moved}}{\text{time taken}}$$

Using v as the symbol for speed (or velocity) and s as the symbol for the distance moved and t as the symbol for the time taken to move that distance:

$$\text{average } v = \frac{s}{t}$$

Rearranged, the formula gives

$$\text{distance moved} = \text{average speed} \times \text{time taken}$$

$$s = \text{average } v \times t$$

Rearranging again we have:

$$\text{time taken} = \frac{\text{distance moved}}{\text{average speed}}$$

$$t = \frac{s}{\text{average } v}$$

The SI units of distance and time are the metre and the second, but many other units are in everyday use. So speed is measured in metre/second, or m/s in SI, but some useful equivalents are:

$$1 \text{ mile/hour} = 0.45 \text{ m/s}$$

$$1 \text{ km/hour} = \frac{1000 \text{ m}}{60 \times 60 \text{ s}} = 0.28 \text{ m/s}.$$

Worked Example
Calculating average speed

A racing car completes a 12 km lap of a course in 4 minutes exactly. Calculate its average speed in km/s, m/s and km/hour.

We use

$$\text{average speed} = \frac{\text{distance moved}}{\text{time taken}}$$

First working in seconds, the time taken = $4 \times 60\,\text{s} = 240\,\text{s}$

$$\therefore \text{ average speed} = \frac{12\,\text{km}}{240\,\text{s}} = 0.05\,\text{km/s}$$

$$\text{and } 0.05\,\frac{\text{km}}{\text{s}} = 0.05 \times 1000\,\frac{\text{m}}{\text{s}} = 50\,\text{m/s}$$

Now working in hours, the time taken = $\frac{4}{60}$ hour = $\frac{1}{15}$ hour.

$$\therefore \text{ average speed} = \frac{12\,\text{km}}{\frac{1}{15}\,\text{hour}} = 12 \times 15\,\frac{\text{km}}{\text{hour}} = 180\,\text{km/hour}.$$

Answer:
The average speed = 0.05 km/s = 50 m/s = 180 km/hour.

Worked Example
Calculating distance travelled

If the average speed of a cyclist is 50 km/hour, how far will he travel in (a) 6 hours and (b) 360 seconds?

Using $s = \text{average } v \times t$ we have:

for (a) $s = 50\,\dfrac{\text{km}}{\text{hour}} \times 6\,\text{hour} = 300\,\text{km}$

for (b) $s = 50\,\dfrac{\text{km}}{\text{hour}} \times \dfrac{360}{60 \times 60}\,\text{hour} = 5.0\,\text{km}$

Answer: The distances travelled are (a) 300 km in 6 hours and (b) 5.0 km in 0.1 hours.
Note: the units of time used for speed and time must be the same in the calculation.

Measuring speed using a stop watch

●   *Measure the time taken to move a certain distance; running, swimming or cycling (measure the distance).*
●   *Calculate the average speed using the formula:*
$$\text{average } v = s/t.$$
●   *List the problems you found in measuring the time and the distance.*

Note the following points:
  The distance moved must be measured accurately.
  The clock must be started and stopped at the precise moments that motion along the measured distance begins and ends.
  The speed of the motion measured is changing in each case and the result obtained is the average speed for the whole motion measured.
  More accurate timing methods are needed for high speeds and short times of motion.

Measuring speed using a ticker timer

A ticker timer which prints dots on paper tape was described on p 68 (fig. 4.13). Synchronised to the mains electricity frequency of 50 Hz, the ticker timer prints 50 dots every second giving a small time interval Δt of $\frac{1}{50}$ second (0.02 s) between one dot and the next.

The small distance between one dot and the next is the change of distance or difference of distance Δs which the object (pulling the tape) has moved in the small time interval Δt. By measuring the small distance moved Δs in a small time interval Δt we can find the actual speed or instantaneous speed at that particular moment of time. This is because during a very small time interval there is little likelihood that the speed will vary much.

Figure 8.1   *Measuring speed with a timer*

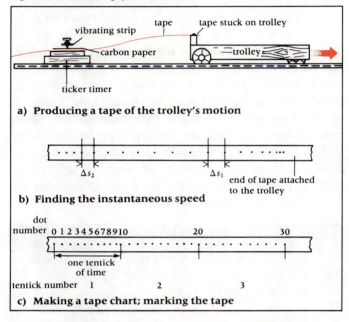

a) **Producing a tape of the trolley's motion**

b) **Finding the instantaneous speed**

c) **Making a tape chart; marking the tape**

The **instantaneous speed** v of an object is given by:

$$v = \frac{\Delta s}{\Delta t}$$

which can be stated as the rate of change of distance moved with time.
●   *Attach about 2 m of paper tape to a trolley and thread the tape through a timer, fig. 8.1a.*
●   *Switch on the ticker timer and give the trolley a push to send it along the bench top.*
●   *Examine the paper tape. What do you notice about the spacing of the dots? Can you explain what you see?*
●   *Find the average speed of the trolley by measuring the total distance s covered by dots.*
The total time t is given by the number of spaces between dots, each worth $\frac{1}{50}$s or 0.02 s. For example, if the total distance moved $s = 142.5\,\text{cm}$ and there are 95 spaces between dots, the time taken $t = 95 \times 0.02\,\text{s} = 1.9\,\text{s}$, and the average speed is

$$\frac{s}{t} = \frac{142.5\,\text{cm}}{1.9\,\text{s}} = 75\,\text{cm/s}$$

●   *Find the instantaneous speed at two times, one near the beginning of the motion and one near the end.*
●   *Measure Δs_1, the distance moved in 0.02 s early in the motion, and Δs_2, the distance moved in 0.02 s late in the motion, fig. 8.1b.*
●   *Calculate the instantaneous speeds using*

$$v = \frac{\Delta s}{\Delta t}$$

For example, if $\Delta s_1 = 1.8\,\mathrm{cm}$ and $\Delta s_2 = 1.2\,\mathrm{cm}$, the instantaneous speed

$$v_1 = \frac{\Delta s_1}{\Delta t} = \frac{1.8\,\mathrm{cm}}{0.02\,\mathrm{s}} = 90\,\mathrm{cm/s}$$

$$v_2 = \frac{\Delta s_2}{\Delta t} = \frac{1.2\,\mathrm{cm}}{0.02\,\mathrm{s}} = 60\,\mathrm{cm/s}$$

We can see that the trolley slows down (due to friction) and the average speed is somewhere in between the early and late instantaneous speeds.

Making a tape chart

The vibrations of the ticker timer are similar to the ticking of a clock, but the timer makes very rapid ticks: fifty every second. We can call the time from one dot to the next a 'tick' of time. So 1 tick of time $= \frac{1}{50}\mathrm{s} = 0.02\,\mathrm{s}$.

Usually the distance from one dot to the next dot on the tape is quite short, often less than 1 cm. So we find that a useful length of time is ten spaces between dots which we shall call a 'tentick' of time. One tentick of time is the time interval Δt from dot number 0 to dot number 10 on the tape:

1 tentick of time $= 10 \times 0.02\,\mathrm{s} = 0.2\,\mathrm{s}$ or $\frac{1}{5}\mathrm{s}$.

• *Mark out all the tenticks of time along your tape as shown in fig. 8.1c Number the tentick strips so that you know their order in time.*

The length of each tentick strip is the distance moved in one tentick of time, that is 0.2 s. So the length of a tentick is a measure of instantaneous speed v:

$$\text{instantaneous speed } v = \frac{\text{length of a tentick strip}}{0.2\,\mathrm{s}}$$

• *Cut up the tentick strips and stick them side by side on a sheet of paper to make a tape chart as shown in fig. 8.2.*

The lengths of the tentick strips may change as shown in fig. 8.2. This shows you how the speed of the trolley changed. The strips show you the distances moved in equal times of one tentick.

Figure 8.2 *Making a tape chart*

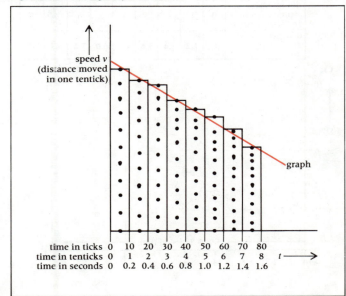

Speed–time graphs

The tops of the strips form a graph showing how the speed of the trolley changes with time. Such a graph is known as a **speed–time** (or velocity–time) graph. Fig. 8.3 shows some basic speed–time graphs which you might obtain from a tape chart of a moving object.

Figure 8.3 *Speed–time graphs*

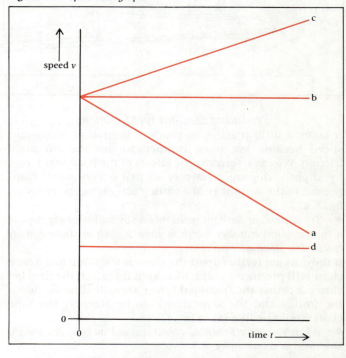

Graph (a) is the same as the tape chart shown in fig. 8.2. It shows a decreasing speed. This happens on a flat surface because friction slows down the motion.

Graph (b) with a horizontal line would be obtained if all the tentick strips had the same length.

If a moving trolley produces equal length tentick strips then it moves equal distances in each tentick of time and has a *constant speed*. We can call tenticks taken in order 'equal successive time intervals' and then we have a more formal definition of motion at constant speed:

An object has a constant speed if it moves equal distances in equal successive time intervals.

Constant speed appears as a horizontal line on a speed–time graph.

Graph (c) shows an increasing speed as the distances moved in successive tenticks of time get longer. This is an example of *accelerated motion*.

Graph (d) compared with graph (b) shows an object also moving at a constant speed but the speed of (d) is smaller than (b).

The slope or gradient of a speed–time graph shows how the speed of an object changes.

A horizontal line (b) and (d), of zero gradient (no slope), shows no change of speed, or no acceleration. An upwards sloping line (c) shows an increasing speed, or acceleration. A downwards sloping line (a) shows a decreasing speed, or negative acceleration (also called deceleration).

Figure 8.4 *Compensating for friction to produce constant speed*

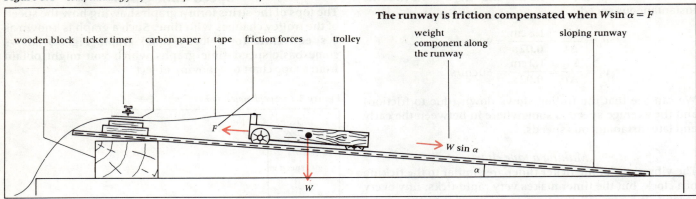

The runway is friction compensated when $W\sin\alpha = F$

Producing constant speed motion

It takes a little practice to produce motion at a constant speed because we must compensate for the effects of friction. We can overcome the effects of the frictional forces by slightly sloping a runway so that a very small component of the weight W of a trolley acts along the runway, fig. 8.4.

● *Try to produce constant speed motion by adjusting the slope of a runway until a trolley which is given a push produces equally spaced dots along a length of tape.*

If the dots get further apart the slope is too steep and a tape chart will produce a graph like (c) in fig. 8.3. If the dots get closer together the frictional forces are still slowing down the trolley and the slope needs to be steeper. The tape chart would be like (a) in fig. 8.3.

● *When you have produced constant speed motion, construct a tape chart as shown in fig. 8.5.*

The area under a speed–time graph shows the total distance moved.

We can see from the tape chart shown in fig. 8.5 that the area covered by paper tape depends on the total length of tape used. In other words, the total distance moved is cut up into tentick lengths and covers the area of the graph. The graph area is given by the area of the rectangle, length × breadth.

$$\frac{\text{area of}}{\text{rectangle}} = \frac{\text{constant}}{\text{speed}}\, v \times \text{time } t = \frac{\text{total}}{\text{distance}}\, s$$

Figure 8.5 *The area under a speed–time graph shows the total distance moved*

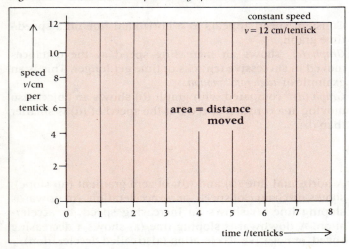

In fig. 8.5, $v = 12$ cm/tentick and $t = 8$ tenticks, so the total distance moved, $s = vt$, is given by:

$$s = 12\,\frac{\text{cm}}{\text{tentick}} \times 8 \text{ tenticks} = 96\,\text{cm}$$

96 cm is the total length of tape on the chart and the total distance moved.

Distance–time graphs

Another kind of graph can be constructed by making distance measurements along a length of tape.

● *Produce a length of tape for a trolley moving with constant speed, fig. 8.6a.*

● *Number the dots and measure the distance to each dot from the start of the motion at dot 0.*

Figure 8.6 *Distance–time graphs*

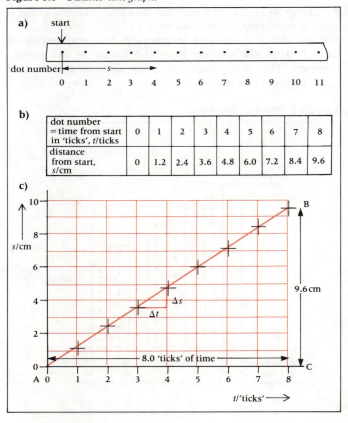

dot number = time from start in 'ticks', t/ticks	0	1	2	3	4	5	6	7	8
distance from start, s/cm	0	1.2	2.4	3.6	4.8	6.0	7.2	8.4	9.6

● *Complete a table showing distances moved s from the start and the time taken t from the start. The dot number gives the time in 'ticks' of time (i.e. intervals of 0.02 s).*

● *Draw a graph of distance moved s on the vertical axis against time taken t on the horizontal axis, fig. 8.6c.*

For motion at constant speed the distance–time graph is a straight line of constant gradient. The gradient of the full line is given by the large triangle ABC:

$$\text{gradient} = \frac{BC}{AC} = \frac{9.6\,\text{cm}}{8.0\,\text{ticks}} = 1.2\,\text{cm/tick}$$

This is the average speed of the trolley.

The gradient for a time interval Δt is given by the small triangle:

$$\text{gradient} = \frac{\Delta s}{\Delta t} = \frac{4.8\,\text{cm} - 3.6\,\text{cm}}{4\,\text{ticks} - 3\,\text{ticks}} = \frac{1.2\,\text{cm}}{1\,\text{tick}} \text{ or } 1.2\,\text{cm/tick}$$

This is the instantaneous speed of the trolley. The instantaneous speed

$$v = \frac{\Delta s}{\Delta t}$$

is also the rate of change of distance with time.

In this case the average speed and the instantaneous speed are the same because the motion is at constant speed. An object moving with constant speed has a constant rate of change of distance with time, that is, the gradient of its distance–time graph, $\Delta s/\Delta t$, is constant.

Fig. 8.7 compares the speed–time graphs with the distance–time graphs for constant low-speed and constant high-speed motions. Notice that the gradient of a distance–time graph indicates speed, and the gradient of a speed–time graph indicates the rate of change of speed.

Figure 8.7 *Comparing speed–time and distance–time graphs for constant speed motion*

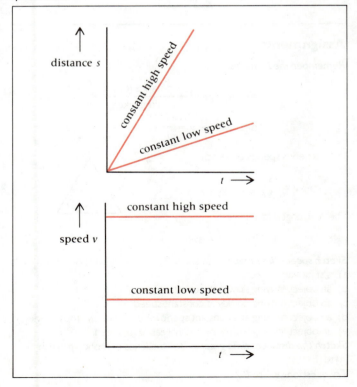

Speed and velocity

In chapter 5 we heard about the difference between scalar and vector quantities. Now that we are studying motion we should expect to meet more vector quantities because objects have motions in particular directions. We use *speed* as a scalar, which has only magnitude, and *velocity* as a vector, which has both magnitude and direction. So velocity is both the speed and direction of the motion of an object.

For example, the speed of sound is 330 m/s but it has no particular direction. On the other hand, when we say the velocity of the north wind is 90 km/hour we are giving a magnitude of 90 km/hour and also a direction from north to south.

We define the velocity of an object as:

Velocity is the rate of change with time of distance moved in a particular direction.

Distance and displacement

We also use two words to distinguish between scalar and vector distances. We define the displacement of an object as:

Displacement is the distance moved in a particular direction.

For example we may be able to travel a distance of 10 km using 1 litre of petrol in a car (and direction is not mentioned). The distance of 10 km is a scalar quantity. However, the information that the City of York is 110 km north of Nottingham gives both the distance and the direction. The displacement of 110 km north is a vector quantity with a specified direction.

Now we can define velocity more briefly as:

Velocity is the rate of change of displacement with time.

When we quote a vector quantity we must give both its magnitude and its direction. Answers to questions about the velocity of an object should always include its direction as well as its speed. When we add or combine two velocities we must take into account the directions of the two velocities.

Adding velocities in a straight line

Fig. 8.8 shows a train moving at a velocity of 80 km/hour in a direction along the railway track (to the right in the figure). This is the velocity of the train compared with the stationary track or ground beneath it and it is the velocity a stationary observer on the ground would see as the train went past.

A man inside the train can walk along the train either to the right (a), or to the left (b) at a velocity of 5 km/hour. How fast is the man on the train moving relative to the railway track?

Clearly the direction in which he walks makes a difference because in one case his velocity adds to the velocity of the train and in the other it subtracts. We usually take velocities to the right as positive and velocities to the left as negative. This gives the train a velocity of +80 km/hour in both cases, and the man a velocity of +5 km/hour in case (a) and −5 km/hour in case (b).

Figure 8.8 *Adding velocities*

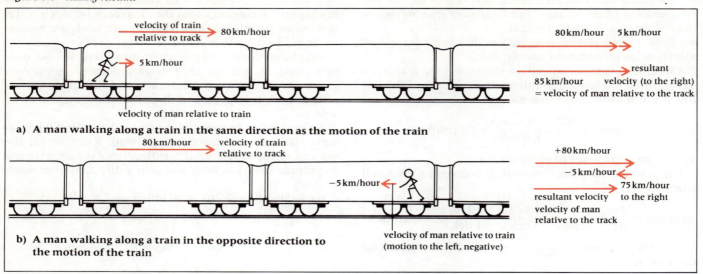

a) **A man walking along a train in the same direction as the motion of the train**

b) **A man walking along a train in the opposite direction to the motion of the train**

So we can add the velocities of the train and the man in both cases to find their **resultant**, or combined, velocity relative to the track.

a) resultant v = 80 km/hour + 5 km/hour = 85 km/hour to the right.
b) resultant v = 80 km/hour − 5 km/hour = 75 km/hour to the right.

These additions of vectors are also represented by arrows in fig. 8.8.

Adding velocities by the parallelogram law
Two velocities which are not in the same straight line can be added by using the same parallelogram law that we used for adding forces, p 76. We can learn how to use this law by studying the worked example.

Worked Example
Adding velocities using the parallelogram law
A swimmer swims across a river at a velocity of 0.8 m/s in a direction perpendicular to the river banks. The water flows down the river at 0.6 m/s. Find the resultant velocity of the swimmer.

Fig. 8.9 shows how the resultant velocity v_r is found using the parallelogram law. The magnitude of the resultant velocity v_r is given by Pythagoras':

$$v_r^2 = v_1^2 + v_2^2$$

$$v_r^2 = (0.8\,\text{m/s})^2 + (0.6\,\text{m/s})^2$$

$$v_r^2 = 0.64(\text{m/s})^2 + 0.36(\text{m/s})^2 = 1.00(\text{m/s})^2$$

So the magnitude of the resultant velocity v_r is 1.0 m/s. The direction of v_r is given by

$$\tan \alpha = \frac{v_1}{v_2} = \frac{0.8\,\text{m/s}}{0.6\,\text{m/s}} = 1.33$$

and using 'inverse tan' on your calculator, this gives

$$\alpha = 53°.$$

Answer: The resultant velocity of the swimmer is 1.0 m/s pointing downstream at an angle of 53° to the river bank.

Figure 8.9

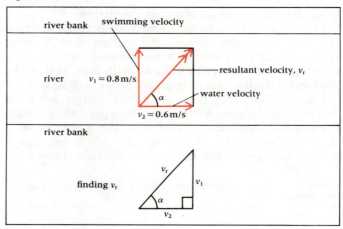

Assignments

Remember the formulas:

$$\text{average speed} = \frac{\text{distance moved}}{\text{time taken}}$$

$$\text{or} \quad \text{average } v = \frac{s}{t}$$

Instantaneous speeds or velocity:

$$v = \frac{\Delta s}{\Delta t}$$

The rearranged formula:

$$s = \text{average } v \times t.$$

Sketch speed–time graphs to show the following kinds of motion:
a) a stationary object,
b) an object moving slowly at a constant speed,
c) an object moving quickly at constant speed,
d) an object moving at a constant speed which suddenly stops moving,
e) an object moving from rest and increasing its speed.
Sketch the distance–time graphs for the kinds of motion given in (a) to (e).
Try questions 8.1 to 8.7

8.2
ACCELERATION AND THE EQUATIONS OF MOTION

Whenever the velocity of an object is changing it has an acceleration. The object could be speeding up, slowing down or changing direction for in each of these cases its velocity is changing. The equations of motion we shall meet in this chapter describe the motion of objects which have a constant acceleration in a straight line.

Acceleration

We have seen trolleys increasing their speed and slowing down on a runway. We know that a change of speed is called acceleration. When we want to measure acceleration we need to know how much time it takes for a certain change of speed or velocity and we define acceleration as follows:

Acceleration is the rate of change of velocity with time.

So we calculate acceleration from the change of velocity which occurs in unit time:

$$\text{acceleration} = \frac{\text{change of velocity}}{\text{time taken for the change}} \qquad a = \frac{\Delta v}{\Delta t}$$

Measuring the acceleration of a trolley

- *Set up a runway with a slope which makes a trolley accelerate, i.e. move with increasing speed down the runway.*
- *Attach a length of tape to the trolley and thread the tape through a ticker timer, and obtain separate tapes for each of the following motions.*
- *Let the trolley run from rest down the slope from the top.*
- *Change the slope of the runway and let the trolley run down it again.*
- *Give the trolley a push up the slope from the bottom, letting go quickly.*
- *Choose for your number 0 dot one which is clear and separate from the muddle of dots at the very beginning of the motion.*
- *Mark out and number tentick strips along the tape, fig. 8.10a.*
- *Construct a tape chart for each motion using 6 tentick strips of tape.*

An example of a tape chart obtained from an accelerating trolley is shown in fig. 8.10b. We can make the following observations about this tape chart:

a) The tentick strips of tape which show the distance moved in one tentick of time (0.2 s) are increasing in length so the velocity of the trolley down the runway is increasing i.e. the trolley is accelerating.

b) The steps up from the top of one tentick strip to the top of the next are equal. These steps show the increase in velocity from one tentick to the next and so the equal steps show equal increases in the velocity of the trolley. We call this **constant acceleration**, or **uniform acceleration**.

Uniform, or constant, acceleration

An object has uniform, or constant, acceleration if its velocity changes by equal amounts in equal successive time intervals. The equal changes in velocity are the equal steps up from the top of one tape strip to the top of the next strip. The equal successive time intervals are the tentick time intervals of the tape strips placed in order along the

bottom of the tape chart. The uniform acceleration of the trolley is also shown by the upwards sloping straight line which can be drawn through the tops of the tape strips.

The change of velocity Δv

The change in velocity is given by the difference between the final velocity v and the initial velocity u. (We usually write u for initial velocity and v for final velocity. Remember that u comes before v in the alphabet and the initial velocity comes before the final velocity.) Difference of velocity or change of velocity:

$$\Delta v = v - u$$

From fig. 8.10c we can see that the change of velocity is given by:

$$\Delta v = 14 \frac{\text{cm}}{\text{tentick}} - 4 \frac{\text{cm}}{\text{tentick}} = 10 \frac{\text{cm}}{\text{tentick}}$$

or 10 cm per tentick. Now as one tentick of time $= \frac{1}{5}$ s or 0.2 s ($10 \times \frac{1}{50}$ s), we can convert the units of Δv from cm per tentick to cm per second:

$$\Delta v = 10 \text{ cm per tentick} = 10 \text{ cm in each } \tfrac{1}{5}\text{ s}$$
$$\therefore \ \Delta v = 5 \times 10 \text{ cm in each full second}$$
$$\therefore \ \Delta v = 50 \text{ cm/s}.$$

Figure 8.10 *Measuring acceleration from a tape chart*

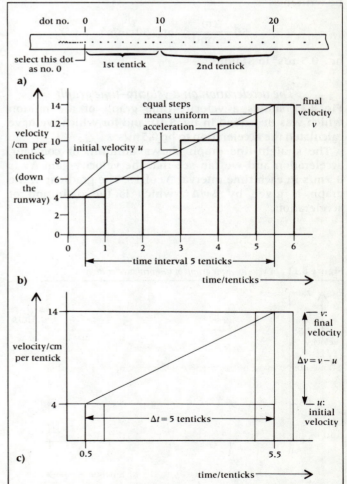

The time taken for the change Δt

The length of each tentick tape strip gives the average speed during the tentick of time. So we take the length of a tape strip to represent the speed of the trolley at the moment half way through the tentick of time. The middle of the sixth tentick is 5 tenticks later than the middle of the first tentick, figs. 8.10b and 8.10c.

The time interval $\Delta t = 5$ tenticks $= 5 \times \frac{1}{5}$s $= 1$s

Calculating the acceleration a

In fig. 8.10 the acceleration shown is given by:

$$a = \frac{\Delta v}{\Delta t} = \frac{50\,\text{cm per second}}{1\,\text{second}}$$
$$= 50\,\text{cm per second per second}$$

The units of acceleration

At first sight the units of acceleration may seem strange. We read our result as '50 cm per second per second'. This means that there is a change of velocity or a gain of velocity of 50 cm per second *in each second*.

The SI unit of acceleration is the 'metre per second squared' written as

$$\frac{\text{metre}}{\text{second}^2} \quad \text{or} \quad \text{m/s}^2.$$

So the acceleration of 50 cm per second per second can be written as

$$50\,\frac{\text{cm}}{\text{second}^2} \quad \text{or} \quad 50\,\text{cm/s}^2$$

or 0.5 m/s² in SI units.

The acceleration on a velocity–time graph

Fig. 8.11 shows a velocity–time graph of the motion which was described in fig. 8.10, and for which we have calculated the acceleration to be 0.5 m/s².

The straight-line graph shows uniform, or constant, acceleration and we can see that the velocity gain Δv is 0.5 m/s in each time interval Δt of 1 s. The gradient of the graph is given by $\Delta v/\Delta t$, which is the definition of acceleration:

$$\text{gradient} = \frac{\Delta v}{\Delta t} = \text{acceleration}$$

Figure 8.11 *A velocity–time graph of uniform acceleration*

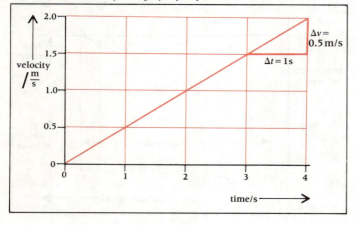

The graphs shown in figs. 8.10 and 8.11 have positive gradients, i.e. they show a gain of velocity in a certain time. A *positive* gradient indicates a positive acceleration and means that the velocity is *increasing*. If you have done the experiment in which a trolley is pushed up a sloping runway you should obtain a graph with a negative or downwards slope, fig. 8.12. A *negative* gradient indicates a negative acceleration and means that the velocity is *decreasing*. Negative acceleration is sometimes also called **deceleration** or **retardation**.

Figure 8.12 *A velocity–time graph for uniform deceleration*

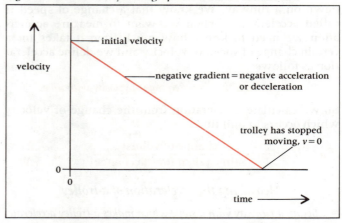

The equations of uniformly accelerated motion

The useful formulas we are now going to obtain describe the motion of an object in a particular direction or in a straight line. Because we are describing motion in a particular direction we use the *vector* quantity velocity rather than the scalar quantity speed. Note that these formulas or equations can only be used to calculate the motion of objects which have a constant or uniform acceleration.

Velocity and acceleration

Acceleration is defined by the formula $a = \Delta v/\Delta t$. If we use the symbols:

u = initial velocity,
v = final velocity and
t = time taken for the change from u to v,

then the change of velocity $\Delta v = v - u$, and the acceleration is given by

$$a = \frac{v - u}{t} \qquad \text{①}$$

If you want to calculate a final velocity v, when you know the initial velocity u and the acceleration a, you can rearrange this formula to give another formula for v. Multiply both sides of the equation by t, and then cancel t:

$$a \times t = \frac{(v - u)}{t} \times t = v - u$$

Now add u to both sides of the equation:

$$u + at = v - u + u = v$$

which gives equation ②

$$v = u + at \qquad \text{②}$$

Worked Example
Using equation ①

A cyclist starting from rest with uniform acceleration can reach a velocity of 20 m/s *in 25 seconds. Calculate her acceleration.*

Initial velocity $u = 0$ m/s,
final velocity $v = 20$ m/s,
time taken $t = 25$ s.

using equation ①

$$a = \frac{v - u}{t}$$

we have $a = \dfrac{20\,\text{m/s} - 0\,\text{m/s}}{25\,\text{s}} = \dfrac{20\,\text{m/s}}{25\,\text{s}} = 0.8\,\dfrac{\text{m}}{\text{s}^2}$

Answer: Her acceleration is 0.8 metres per second per second or 0.8 m/s².

Worked Example
Using equation ②

If a car can accelerate uniformly at 2.5 m/s² *and starts from a velocity of* 36 km/hour, *find its velocity after* 8 *seconds.*
Notice that the units are mixed, so we convert to SI first.

Initial velocity $u = 36\,\dfrac{\text{km}}{\text{hour}} = \dfrac{36 \times 1000\,\text{m}}{60 \times 60\,\text{s}} = 10\,\dfrac{\text{m}}{\text{s}}$

acceleration $a = 2.5$ m/s²,
time taken $t = 8$ s.
using equation ②, $v = u + at$,

we have $v = 10\,\dfrac{\text{m}}{\text{s}} + \left(2.5\,\dfrac{\text{m}}{\text{s}^2} \times 8\,\text{s}\right) = 10\,\dfrac{\text{m}}{\text{s}} + 20\,\dfrac{\text{m}}{\text{s}} = 30\,\dfrac{\text{m}}{\text{s}}.$

Answer: The final velocity of the car is 30 m/s.

Formulas for distance moved

We have already seen that the area under a speed–time or velocity–time graph represents the distance moved. Fig. 8.13 shows the motion of an object with uniform acceleration a, which starts at an initial velocity u and accelerates to a final velocity v in a time interval t.

The area under the graph is made up of a rectangular patch and a triangular patch shaded differently in the figure.

The area of the rectangle $= ut$
and the area of the triangle $= \frac{1}{2}(v - u)\,t$.
So the total area = distance moved s is given by

$$s = ut + \tfrac{1}{2}(v - u)\,t$$

Figure 8.13 *The area under a velocity–time graph represents distance moved*

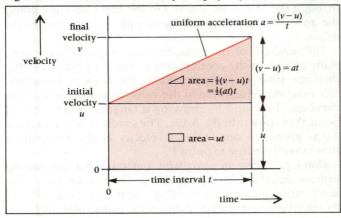

Two formulas can be obtained from this equation.
First we can multiply out the bracket as follows:

$$s = [ut - \tfrac{1}{2}ut] + \tfrac{1}{2}vt = \tfrac{1}{2}ut + \tfrac{1}{2}vt$$

which gives the equation

$$s = \left(\frac{u + v}{2}\right) t \qquad ③$$

where $\left(\dfrac{u + v}{2}\right)$ is the *average velocity.*

So this is the basic formula for distance moved which we have already met (p 121): $s = $ average $v \times t$. Use formula ③ to calculate the distance moved s when you know u, v and t.

Second we can replace the change of velocity $(v - u)$ as follows: multiply both sides of equation ① by t, this gives

$$\frac{(v - u)}{t} \times t = a \times t \quad \text{and} \quad (v - u) = at$$

Now replacing $(v - u)$ in $s = ut + \frac{1}{2}(v - u)\,t$

we obtain $\qquad s = ut + \tfrac{1}{2}(at)\,t$

which gives equation ④

$$s = ut + \tfrac{1}{2}at^2 \qquad ④$$

Use formula ④ to calculate the distance moved s when you know u, a and t, but not v.

Try to sketch a velocity–time graph of the motion of this child accelerating from rest down a slide.

Worked Example
Using equations ① to ④

A car accelerates uniformly from rest for 20 s with an acceleration of 1.5 m/s². It then travels at a constant speed for 2 minutes before slowing down with a uniform deceleration to come to rest in a further 10 s. Sketch a velocity–time graph of the motion and find
a) *the maximum speed,*
b) *the total distance travelled and*
c) *the acceleration while slowing down.*

A sketch of the motion is shown in fig. 8.14. While the car is accelerating we have:
$u = 0$ (starts from rest),
$a = 1.5\,\text{m/s}^2$
$t = 20\,\text{s}$
Using equation ② we can find the maximum or final velocity v

$$v = u + at = 0 + \left(1.5\,\frac{\text{m}}{\text{s}^2} \times 20\,\text{s}\right) = 30\,\frac{\text{m}}{\text{s}}$$

Answer (a): The maximum speed reached after 20 s is 30 m/s.

The distance travelled while accelerating can be calculated using equations ③ or ④. We have $u = 0$, $v = 30\,\text{m/s}$, $a = 1.5\,\text{m/s}^2$ and $t = 20\,\text{s}$. Equation ③ gives:

$$s = \left(\frac{u+v}{2}\right)t = \left(\frac{0 + 30\,\text{m/s}}{2}\right)20\,\text{s} = 15\,\frac{\text{m}}{\text{s}} \times 20\,\text{s} = 300\,\text{m}$$

Alternatively equation ④ gives:

$$s = ut + \tfrac{1}{2}at^2 = 0 + \tfrac{1}{2}\left(1.5\,\frac{\text{m}}{\text{s}^2}\right)(20\,\text{s})^2$$

$$= 0.75\,\frac{\text{m}}{\text{s}^2} \times 400\,\text{s}^2 = 300\,\text{m}$$

The distance travelled at a constant speed of 30 m/s for 2 minutes is given by:

$$s = \text{constant speed} \times t = 30\,\frac{\text{m}}{\text{s}} \times 120\,\text{s} = 3\,600\,\text{m}.$$

The distance travelled while decelerating can be calculated from the following data using equation ③:
$u = 30\,\text{m/s}$
$v = 0$ (comes to rest)
$t = 10\,\text{s}$ (time taken to come to rest)

$$s = \left(\frac{u+v}{2}\right)t = \left(\frac{30\,\text{m/s} + 0}{2}\right)10\,\text{s} = 15\,\frac{\text{m}}{\text{s}} \times 10\,\text{s} = 150\,\text{m}.$$

Answer (b): The total distance travelled during the motion is $300\,\text{m} + 3\,600\,\text{m} + 150\,\text{m} = 4050\,\text{m}$.

The acceleration while slowing down can be calculated from $u = 30\,\text{m/s}$, $v = 0$ and $t = 10\,\text{s}$. Using equation ①:

$$a = \frac{v-u}{t} = \frac{0 - 30\,\text{m/s}}{10\,\text{s}} = -3.0\,\frac{\text{m}}{\text{s}^2}$$

Answer (c): There is a negative acceleration of $-3\,\text{m/s}^2$, or a deceleration while the car slows down of 3 m/s in each second.

Figure 8.14

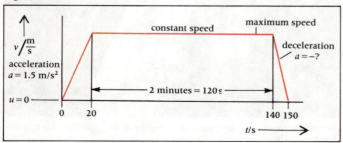

Another useful formula

We shall use this formula later for calculating the energy of a moving object.

Equation ① is $\quad a = \dfrac{v-u}{t}$

and equation ③ is $\quad s = \left(\dfrac{u+v}{2}\right)t$

By multiplying these equations together we can eliminate t:

$$a \times s = \frac{(v-u)}{t} \times \frac{(u+v)}{2} \times t = \frac{(v^2 - u^2)}{2}$$

Multiplying by 2 we get $2as = v^2 - u^2$ and rearranging we have equation ⑤:

$$v^2 = u^2 + 2as \qquad\qquad ⑤$$

Worked Example
Using equation ⑤

If a train accelerates uniformly from rest at 0.2 m/s² over a distance of 1 km, calculate the velocity it reaches.

We have $u = 0$ (starts from rest), $a = 0.2\,\text{m/s}^2$ and $s = 1\,\text{km} = 1\,000\,\text{m}$, and using equation ⑤

$$v^2 = u^2 + 2as = 0 + 2\,(0.2\,\text{m/s}^2) \times 1\,000\,\text{m} = 400\,\text{m}^2/\text{s}^2$$
$$\therefore\ v = 20\,\text{m/s}$$

Answer: The train reaches a maximum velocity of 20 m/s.

Interpreting graphs of motion

Graphs of motion can show very clearly how an object moves but we need to understand how to read and interpret them. We must take care to notice whether velocity or distance is plotted on the vertical axis.

Several different kinds of motion are shown as graphs in fig. 8.15.

Graph (a) shows a stationary object for which neither v nor s changes. On the velocity–time graph zero velocity is a horizontal line along the time axis. On the distance–time graph the object is shown stationary at a distance s above the time axis, i.e. where observations or measurements are made.

Graph (b) shows an object with constant, or uniform, velocity. The gradient of the distance–time graph gives the velocity.

Graph (c) shows an object with uniform acceleration starting from rest. On the velocity–time graph the line passes through the origin for $u = 0$. On the distance–time graph the upwards curving graph starts with zero gradient at the origin.

The gradient of the velocity–time graph $\Delta v/\Delta t$ is constant and gives the acceleration a; but the gradient of the distance–time graph increases as the velocity increases. The instantaneous value of the velocity at a point P on the curve can be found by drawing a tangent to the curve at point P as shown in fig. 8.15c. The gradient of the tangent $\Delta s/\Delta t$ gives the *instantaneous* velocity at the moment in time corresponding to point P.

Graph (d) shows an object with uniform acceleration starting with an initial velocity u. On the distance–time graph the acceleration again leads to an upwards curving graph and its initial positive gradient shows the initial velocity u.

Graph (e) shows an object with a negative uniform acceleration, or deceleration. The curved distance–time graph suddenly becomes a horizontal line (*v* = 0) when the motion stops.

Graph (f) shows an object with a non-uniform, or irregular, acceleration, which is typical of a real car or train accelerating. The acceleration gets smaller at higher speeds as it approaches the maximum speed. It is difficult to distinguish between uniform and non-uniform acceleration on a distance–time graph because both produce curved graphs.

Table 8.1 compares the important features of these two kinds of graphs of motion.

Table 8.1

	Velocity–time or speed–time graphs	Displacement–time or distance–time graphs
Gradient	$\dfrac{\Delta v}{\Delta t} = a$ (fig. 8.11)	$\dfrac{\Delta s}{\Delta t} = v$ (figs. 8.6 and 8.15c)
Intercept on vertical axis	initial velocity *u*	initial distance from observer *s*
Area under graph	distance moved	—

Figure 8.15 *Graphs of motion*

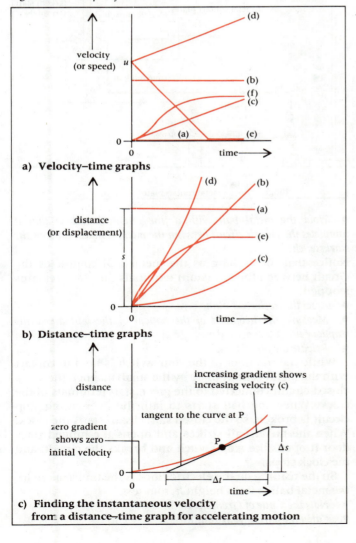

a) **Velocity–time graphs**

b) **Distance–time graphs**

c) **Finding the instantaneous velocity from a distance–time graph for accelerating motion**

Measuring the acceleration of a falling object

- *Support a ticker timer some 2 or more metres above the ground so that there is a clear drop below it as shown in fig. 8.16.*
- *Thread* 2 m *of tape into the ticker timer and attach a* 200 g *mass to the downwards leading end.*
- *Arrange the rest of the tape so that it can flow into the ticker timer smoothly and avoid getting any folds or twists in it.*
- *Switch on the timer and allow the* 200 g *mass to drop to the ground.*
- *Construct a tape chart using* twotick *strips of tape.*

For a large acceleration ten ticks produce strips which are too long; twotick strips should fit on a page of your book. The time interval of a twotick is

$$2 \times \tfrac{1}{50}\text{s} = 0.04\,\text{s}, \quad \text{or} \quad \tfrac{1}{25}\text{s}.$$

Figure 8.16 *Measuring the acceleration of a falling object*

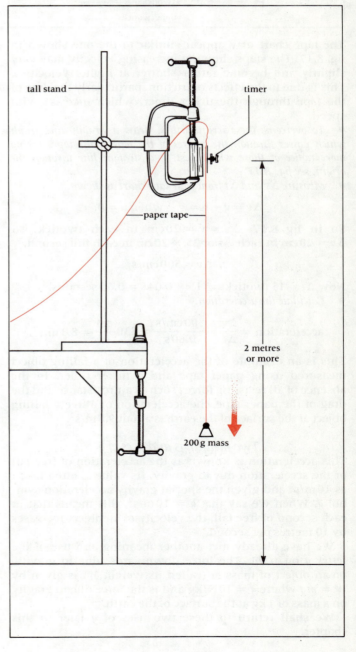

tall stand

timer

paper tape

2 metres or more

200 g mass

Figure 8.17 *A tape chart for a falling object*

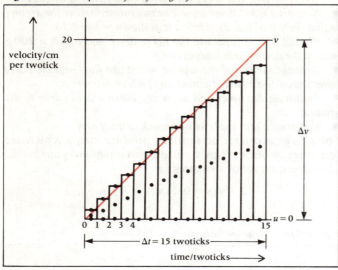

The tape chart may appear similar to the one shown in fig. 8.17. The steps showing increasing velocity may vary slightly and become rather shorter at higher velocities. This is due to the effects of friction, particularly the drag of the tape through the ticker timer, which increase with speed.

● *To overcome these irregularities, draw a straight-line graph which passes through the centres of the tops of the tapes taking more notice of those which best fit a straight line through the origin, see fig. 8.17.*

● *Measure Δv and Δt from your tape chart as shown.*

$$\Delta v = v - u = v, \quad \text{when } u = 0.$$

So in fig. 8.17, $\Delta v = v = 20$ cm in each twotick, so $\Delta v = 20$ cm in each $\frac{1}{25}$ s or 25×20 cm in each full second.

$$\therefore \quad \Delta v = 500 \text{ cm/s.}$$

Now $\Delta t = 15$ twoticks $= 15 \times 0.04$ s $= 0.60$ seconds.

● *Calculate the acceleration.*

$$\text{acceleration} = \frac{\Delta v}{\Delta t} = \frac{500 \text{ cm/s}}{0.60 \text{ s}} = 830 \frac{\text{cm}}{\text{s}^2} = 8.3 \text{ m/s}^2.$$

This is an example of the acceleration of a falling object measured using paper tape and a ticker timer. In the absence of all resisting forces such as air resistance and the drag of the paper tape, the acceleration of a freely falling object at the surface of the earth is about 9.8 m/s².

Two meanings and uses of g

This acceleration is known as the acceleration of free fall or the acceleration due to gravity. Its value is often taken as 10 m/s² and given the special gravity acceleration symbol g. When we say that $g = 10$ m/s², this means that in each second of free fall, the velocity of an object increases by 10 metres per second.

We have already met another meaning and use of the letter g on p 74. The force downwards, due to gravity, on an object of mass m (called its weight W) is given by $W = mg$ where $g = 10$ N/kg and is the force due to gravity on a mass of 1 kg at the surface of the earth.

We shall return to these two uses of g later in this chapter.

Measuring g using an electric stopclock

The value of g obtained using paper tape attached to a falling object is very inaccurate due to the effects of friction and drag on the tape. The effect of air resistance alone on a freely falling metal object such as a steel ball or pendulum bob is very slight and with accurate timing a good value of g can be obtained.

The arrangement shown in fig. 8.18 can give very accurate results. An electric stopclock such as a **centisecond timer** (reading to 0.01 s) or a **scaler timer** (reading to 0.001 s) is started and stopped by two separate switches. The pair of red terminals are connected to a *start switch* where the ball is released, and the pair of green terminals are connected to a *stop switch* where the ball strikes and opens a metal trapdoor.

Figure 8.18 *Measuring g using an electric stopclock*

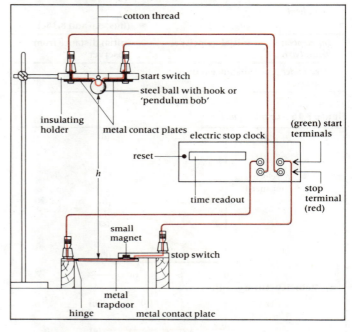

● *Hold the metal ball with a fine cotton thread so that it completes the electric circuit between the metal contact plates on the start switch.*

Notice that for as long as the metal ball completes the circuit between the red (stop) terminals the clock remains stopped.

● *Reset the clock read-out to zero.*

● *Measure the height h of the bottom of the ball above the trapdoor using a metre ruler.*

● *Release the metal ball.*

While the trapdoor of the stop switch is held in contact with the metal contact plate by the small magnet, there is a closed circuit connected to the green start terminals of the clock. When the ball starts to fall, the over-riding stop circuit is broken, so the closed start circuit starts the clock. When the metal ball strikes and opens the hinged trapdoor it opens the stop switch and breaks the circuit, and the clock stops.

So the reading on the electric clock is the time t taken for the metal ball to fall a height h from rest.

● *Note the value of t from the clock.*

● *Calculate g from your result.*

Using the following specimen results we can see how to calculate *g*. If *h* = 925 mm and *t* = 435 ms, we have:

$u = 0$ (falls from rest)
$s = 925\,\text{mm} = 0.925\,\text{m}$
$t = 435\,\text{ms} = 0.435\,\text{s}$

Using equation ④ and writing $a = g$ gives:

$$s = ut + \tfrac{1}{2}at^2 = \tfrac{1}{2}gt^2 \quad (\text{since } u = 0)$$

Rearranging this gives a formula for *g*:

$$g = \frac{2s}{t^2}$$

Substituting for the specimen results:

$$g = \frac{2 \times 0.925\,\text{m}}{(0.435\,\text{s})^2} = 9.78\,\frac{\text{m}}{\text{s}^2}$$

The result should be correct to the first decimal place, giving the acceleration due to gravity $g = 9.8\,\text{m/s}^2$.

• *Repeat the experiment for different values of h and calculate g from each result. Find an average value for g.*

Using multiflash or stroboscopic photography to study motion

A xenon stroboscope produces flashes of light at regular time intervals and so can be used to time the motion of a moving object. Each flash lasts for a very short time, perhaps 0.001 s or less. Each flash produces a sharp image of the moving object because the flash is so brief.

The independence of horizontal and vertical motion

The multiflash photographs show the motion of two balls which were both released at the same time. One ball was dropped from rest and the other was thrown sideways.

The horizontal lines on the upper photograph, show that both balls fall with the same acceleration, and that the horizontal motion of a ball does not affect its vertical motion.

The equally spaced vertical lines on the lower photograph, show that although a ball is falling freely and is accelerating vertically it continues to move at a constant horizontal velocity.

We can see from these photographs that the vertical motion of a ball does not affect its horizontal motion; and likewise its horizontal motion does not affect its vertical motion.

Horizontal and vertical motions are independent and do not disturb each other.

Measuring the acceleration of a falling golf ball

• *Position the stroboscope so that its light does not shine directly into the camera lens and so that it illuminates the full height of the motion.*

A good position is vertically above, as in fig. 8.19, or below the motion. However, if the stroboscope is mounted on the floor below the falling ball, use a perspex screen to protect the glass lens of the stroboscope.

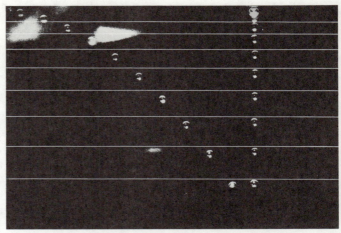
An object moving horizontally while it falls has the same downwards acceleration as an object which falls straight down.

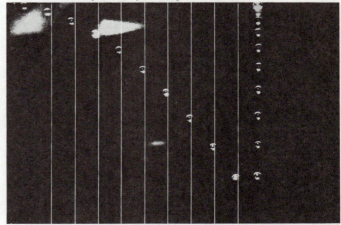
An object moves at a constant horizontal speed while it falls and accelerates downwards.

Figure 8.19 *Using multiflash photography*

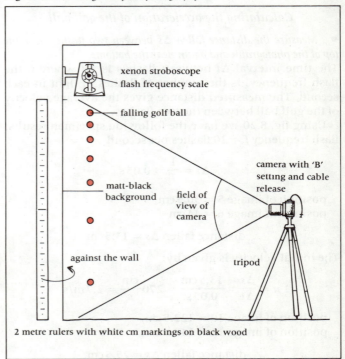

xenon stroboscope
flash frequency scale
falling golf ball
matt-black background
field of view of camera
camera with 'B' setting and cable release
against the wall
tripod
2 metre rulers with white cm markings on black wood

- *To decide what flash frequency to use, estimate the time of falling the full height and decide the number of images required.* This is done as follows. Rearranging

$$s = \tfrac{1}{2}gt^2$$

gives $t^2 = \dfrac{2s}{g}$ and $t = \sqrt{\dfrac{2s}{g}}$

For example, if $h = 2$ metres, and using $g = 10\,\text{m/s}^2$, we have

$$t = \sqrt{\frac{2 \times 2}{10}} = 0.6\,\text{s}$$

If we decide that 10 images of the falling ball would be suitable, then 10 flashes in 0.6 seconds requires a flash frequency given by

$$f = \frac{\text{number of flashes}}{\text{time taken}} = \frac{10}{0.6\,\text{s}} = 17\ \text{flashes/second.}$$

- *Set the calculated flash frequency on the stroboscope scale.*
- *Set up and focus the camera and select 'B' on its shutter control. (For details of camera and film see Appendix D3.)*
- *Use a 'count-down' procedure for operating the camera:* On the count of 1 open the shutter. On the count of 0 drop the ball.
- *Close the shutter when you hear the ball strike the ground.* This procedure keeps the camera shutter open for the minimum length of time which produces the best contrast between the white golf ball and the dark background.
- *Repeat the whole procedure several times and vary the stroboscope flash frequency to give a choice of photographs. Display the flash frequency on a card in each picture.* When the film has been processed, take readings from a print or from a white screen on which a negative is projected. The details of a typical photograph are shown in fig. 8.20.

Calculating the acceleration of the golf ball

- *Measure the distance fallen Δs between two flashes near the top of the photograph and again near the bottom.* The time interval Δt between flashes is $1/f$ where f, the flash frequency, is the number of flashes of light in each second. The measured distance gives the average velocity of the golf ball between the two flashes.

Using fig. 8.20 we have the following specimen results: flash frequency $f = 20$ flashes per second

$$\therefore\ \Delta t = \frac{1}{f} = 0.05\,\text{s},$$

position of image 5 = 30.5 cm, position of image 6 = 44 cm

$$\therefore\ \text{distance fallen } \Delta s = 13.5\,\text{cm.}$$

The initial velocity is given by:

$$u = \frac{\Delta s}{\Delta t} = \frac{13.5\,\text{cm}}{0.05\,\text{s}} = 270\,\frac{\text{cm}}{\text{s}} = 2.7\,\text{m/s.}$$

position of image 10 = 122.5 cm, position of image 11 = 148 cm,

$$\therefore\ \text{distance fallen } \Delta s = 25.5\,\text{cm.}$$

Figure 8.20 *Stroboscopic images of a falling golf ball*

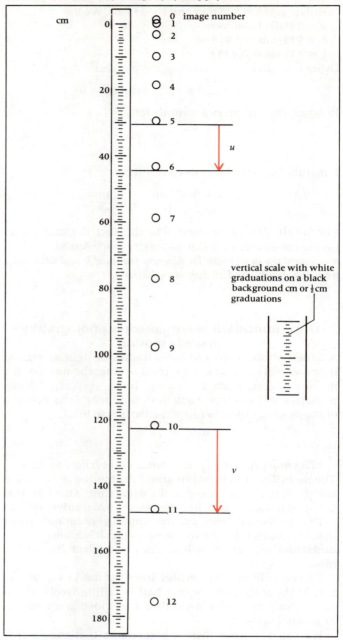

The final velocity is given by:

$$v = \frac{\Delta s}{\Delta t} = \frac{25.5\,\text{cm}}{0.05\,\text{s}} = 510\,\frac{\text{cm}}{\text{s}} = 5.1\,\text{m/s.}$$

The time taken for this increase in velocity is the time interval t from the 5th flash to the 10th flash, which is

$$5 \times \Delta t \quad \text{or} \quad 5 \times 0.05\,\text{s} = 0.25\,\text{s.}$$

Now acceleration is given by equation ①:

$$a = \frac{v - u}{t}$$

$$\therefore\ a = \frac{5.1\,\text{m/s} - 2.7\,\text{m/s}}{0.25\,\text{s}} = \frac{2.4\,\text{m/s}}{0.25\,\text{s}} = 9.6\,\frac{\text{m}}{\text{s}^2}$$

The acceleration of the golf ball is $9.6\,\text{m/s}^2$.

Plotting a graph

This is an alternative method of measuring *g*.

Measure all the distances between images and record the results in a table as shown in table 8.2. (The data shown in table 8.2 is obtained from fig. 8.20.)

Table 8.2

Image number	Position of image on scale/cm	Δs (distance fallen between images)/cm	$v = \dfrac{\Delta s}{\Delta t}$ $\Big/ \dfrac{m}{s}$	t/s* from image no. 0
0	0	—	($\Delta t = 0.05$ s)	
1	1	1	0.2	0.025
2	5	4	0.8	0.075
3	11	6	1.2	0.125
4	19.5	8.5	1.7	0.175
5	30.5	11	2.2	0.225

* Note that *t* is the time from image number 0 to halfway through each time interval. Δt, i.e. 0.025 s after each flash. This follows from the fact that *v* is the average velocity halfway through each time interval.

- Plot a graph of *v* against *t*, fig. 8.21.
- Find the acceleration due to gravity from the gradient of the graph.

In fig. 8.21, $\Delta v = 4.9$ m/s and $\Delta t = 0.5$ s

$$\therefore a = \frac{\Delta v}{\Delta t} = \frac{4.9 \, m/s}{0.5 \, s} = 9.8 \, \frac{m}{s^2}$$

The gradient of the graph gives $g = 9.8 \, m/s^2$.

Figure 8.21   *Velocity–time graph from stroboscopic photography*

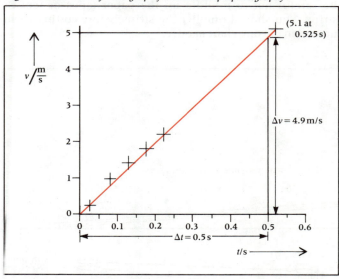

Uniform acceleration

The images of the falling golf ball obtained by stroboscopic photography show that the acceleration due to gravity is uniform. We can see this by looking at the increases in the velocity of the golf ball between each of the time intervals. The increase in velocity is always near to +0.5 m/s in each 0.05 s time interval, see table 8.2. Similarly the points plotted on the velocity–time graph, fig. 8.21, closely follow a straight line which shows constant, or uniform, acceleration.

Worked Example
Calculating distance fallen using equation ④

If a stone is dropped from rest down a well and a splash is heard after 2.5 seconds, how deep is the well? (Assume that sound travels very quickly and g = 10 m/s².)

We have: $u = 0$ (drops from rest),

$\qquad\quad t = 2.5$ s,

$\qquad\quad a = g = 10 \, m/s^2$.

Using equation ④, $s = ut + \frac{1}{2}at^2$, we have:

$$s = 0 + \tfrac{1}{2}(10 \, m/s^2)(2.5 \, s)^2 = 31.25 \, m.$$

Answer: The depth of the well is 31.25 metres.

Worked Example
Calculating velocity and time for vertical motion

A ball is thrown vertically upwards and reaches a height of 28.8 m. Ignoring the effects of air resistance and taking g = 10 m/s², find (a) the initial upwards velocity and (b) the time taken to return to the hands of the thrower.

Since the ball comes to rest momentarily at the maximum height, we have: $v = 0$

$\qquad\quad a = -g = -10 \, m/s^2$ (deceleration is negative),

$\qquad\quad s = 28.8$ m

a) Using equation ⑤, $v^2 = u^2 + 2as$, we have:

$$0 = u^2 + 2\,(-10 \, m/s^2)\,28.8 \, m$$
$$\text{or } u^2 = +576 \, m^2/s^2$$
$$\therefore \; u = 24 \, m/s$$

Answer: The initial upwards velocity of the ball is 24 m/s.

b) Using equation ②, $v = u + at$, for the upwards motion, we have:

$$0 = (24 \, m/s) + (-10 \, m/s^2)t$$
$$\text{or } + (10 \, m/s^2)\,t = 24 \, m/s$$
$$\therefore \; t = \frac{24 \, m/s}{10 \, m/s^2} = 2.4 \, s$$

Answer: The downwards motion can be shown (using the same formula) to take exactly the same time as the upwards motion. So the total time taken is 4.8 seconds.

Assignments

Remember

a) Can you remember the equations of motion, numbered ① to ⑤? In some examinations you are given these equations when you need them, but in others you are expected to know them. In either case, knowing these formulas helps you to tackle questions quickly and confidently. They are used frequently in many parts of physics.

$$a = \frac{v - u}{t} \quad ① \qquad s = \left(\frac{u + v}{2}\right)t \quad ③$$
$$v = u + at \quad ② \qquad s = ut + \tfrac{1}{2}at^2 \quad ④$$
$$v^2 = u^2 + 2as \quad ⑤$$

b) Equation ③ gives a formula for distance as

$$s = \text{average velocity} \times t$$

You cannot use $s = vt$ if the velocity is changing.

c) Acceleration is a *vector* quantity in the same way as velocity and displacement. The importance of this was shown in the last worked example where the acceleration was given a negative value. You can explain this by saying that the upwards direction is taken as positive and so the downwards acceleration is negative.

d) The information about graphs of motion given in table 8.1.

Try questions 8.8 to 8.13

8.3
NEWTON'S LAWS OF MOTION

In physics we study the motion of many things to help us understand and explain their behaviour. Three hundred years ago, Sir Isaac Newton formulated his three famous laws of motion which describe the nature of forces and their effects on the motion of all kinds of objects both large and small. His laws can be applied to the molecules in gases, to people and their machines, and to satellites, moons and planets in their orbits.

Newton's laws of motion help scientists to understand the motion of the 17 moons and the millions of particles in the rings of the planet Saturn. This picture was taken in 1979 by the Pioneer spacecraft, 2 500 000 km from Saturn.

Investigating how the motion of a trolley depends on the force applied to it

● *Set up a trolley and ticker timer on a runway and adjust the slope of the runway to compensate for friction, so that the trolley runs down the runway with a constant velocity when given a gentle push.*

● *Fit a short dowel rod in each of the holes at the ends of the trolley and hook an eyelet of an elastic cord over the anchor dowel rod as shown in fig. 8.22.*

An elastic cord is somewhat shorter than a trolley when unstretched. By stretching an elastic cord by a constant extension we can pull a trolley with a constant force. (The tension in the elastic cord is proportional to the extension, which is Hooke's law, p 164.)

With practice you will be able to walk along the side of a runway pulling the trolley with an elastic cord stretched by a fairly constant amount. It helps to use a marker dowel rod as a guide to keep the elastic cord stretched to a fairly constant length (equal to the length of the trolley). You must not touch the trolley while it is being pulled by the cord and obviously it must not bump into the sides of the runway.

● *Tie a string across the end of the runway to stop the trolley running off the end and being damaged.*

● *Two or three elastic cords may be used in parallel to double or triple the pulling force, but remember that they must always have the same constant extension.*

● *When you have practised pulling the trolley with a constant force, attach paper tape to it and thread the tape through the ticker timer.*

● *Make a tape chart for the motion of the trolley pulled by 1, 2 and 3 stretched elastic cords.*

Be careful not to muddle up the paper tapes or tentick strips. You should number the strips before cutting them up and make one chart at once.

Figure 8.22 *Pulling a trolley with elastic cords down a runway*

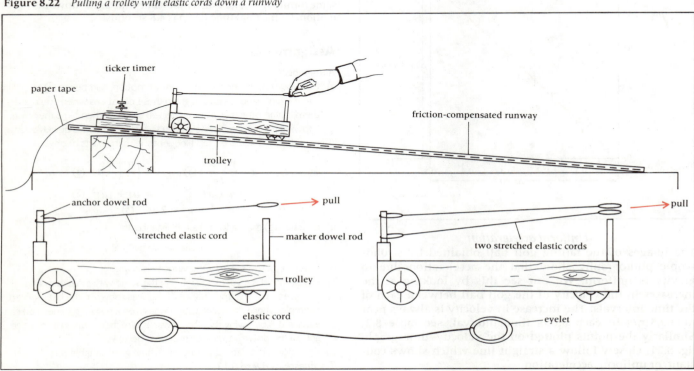

Figure 8.23   *Tape charts for an accelerated trolley*

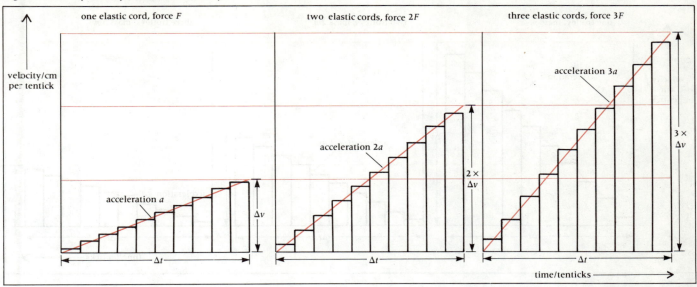

Some typical tape charts are shown in fig. 8.23. The same number of tentick strips has been used on each tape chart so that Δt is the same for all three charts.

The gain of velocity measured on the first chart is Δv. The gain of velocity measured on the second and third charts is found to be 2 and 3 times greater, i.e. $2\Delta v$ and $3\Delta v$. The acceleration produced by:

a single force (F) $\qquad \dfrac{\Delta v}{\Delta t} = a$

two forces ($2F$) $\qquad \dfrac{2\Delta v}{\Delta t} = 2a$

three forces ($3F$) $\qquad \dfrac{3\Delta v}{\Delta t} = 3a$

These results show that the acceleration of the trolley is directly proportional to the force applied to it. Or, using the proportional symbol $\propto$, we have (for constant mass)

acceleration ∝ force $a \propto F$

Unbalanced force

It is important to remember that we have used a friction compensated runway so that all of the force provided by the elastic cords was able to produce acceleration. We can call this force an **unbalanced force**, by which we mean that there was no force acting on the trolley opposing this force or cancelling it out. So it would be more accurate to write that:

The acceleration of an object is directly proportional to the unbalanced force acting on it.

Investigating how acceleration depends on mass

In the last experiment the mass of the trolley was constant. In this experiment we shall keep the force constant and vary the mass. The simplest way of changing the mass is to stack trolleys on top of each other, fig. 8.24. The mass of each trolley is used as a unit of mass; two similar trolleys will have a mass of two of these units and so on. There are holes in the tops and bottoms of trolleys to allow them to be stacked with short lengths of dowel rod.

- *Repeat the last experiment using a constant force of 2 stretched elastic cords for one, two and three trolleys stacked.*
- *Construct tape charts and measure the acceleration in each case.*

Some typical results are shown in fig. 8.25. The gain of velocity measured on the first chart for a single trolley of mass m is Δv. The gain of velocity can be seen to decrease to a half and a third of Δv when the mass is increased to $2m$ and $3m$, respectively. The acceleration of:

one trolley of mass m $\qquad \dfrac{\Delta v}{\Delta t} = a,$

two trolleys of mass $2m$ $\qquad \dfrac{\frac{1}{2}\Delta v}{\Delta t} = \dfrac{a}{2}$

three trolleys of mass $3m$ $\qquad \dfrac{\frac{1}{3}\Delta v}{\Delta t} = \dfrac{a}{3}$

These results show that the acceleration of the trolley is inversely proportional to its mass when the same force is applied.

acceleration ∝ $\dfrac{1}{\text{mass}}$ or $a \propto \dfrac{1}{m}$

Figure 8.24   *Stacking trolleys to change the mass*

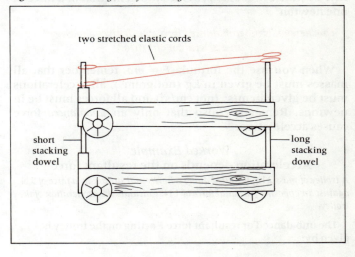

two stretched elastic cords

short stacking dowel

long stacking dowel

Figure 8.25 *Tape charts for stacked trolleys pulled by a constant force*

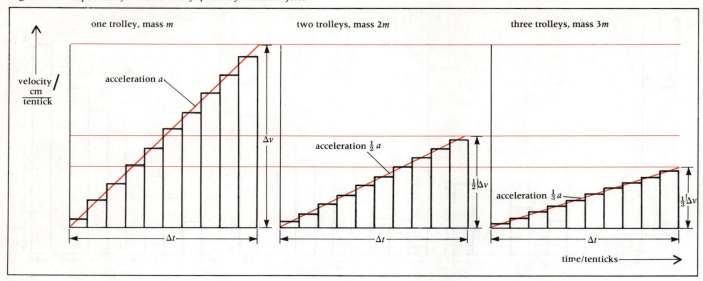

Newton's second law of motion and the newton

Combining the results of the last two experiments we have:

$$\text{acceleration} \propto \frac{\text{unbalanced force}}{\text{mass}} \quad \text{or} \quad a \propto \frac{F}{m}$$

and rearranging this gives the relation $F \propto ma$. Now we can write an equation $F = Kma$ where K is a constant number which depends only on the units we use to measure F, m and a. We can make this formula simpler by choosing our units so that $K = 1$, always. In SI units, mass m is measured in kilograms and acceleration a is measured in metres per second per second so we invent a new unit of force called the newton which makes $K = 1$.

We have already met the newton many times but now we can see how it is defined. When $K = 1$, we have:

$$F = ma$$

So if $m = 1$ kg and $a = 1$ m/s^2, then

$$F = 1 \text{ kg} \times 1 \frac{\text{m}}{\text{s}^2} = 1 \text{ newton.}$$

The formula is one way of stating **Newton's second law of motion**. It also provides the definition of a force of one newton:

One newton is the unbalanced force which gives a mass of 1 kg an acceleration of 1 m/s^2.

When you use the formula $F = ma$, remember that all masses must be given in kg (not grams), all accelerations must be given in m/s^2 (not cm/s^2) and all forces must be in newtons. Remember also that only an *unbalanced* force causes acceleration.

Worked Example
Acceleration depends on the resultant force

A trolley of mass 2 kg is pulled along a horizontal surface by a force of 5 N against an opposing frictional force of 1 N. Calculate the acceleration of the trolley.

The unbalanced or resultant force F acting on the trolley is given by:

resultant force F = applied force − frictional force
$$\therefore \ F = 5 \text{ N} - 1 \text{ N} = 4 \text{ N}$$

Using $F = ma$ rearranged, we have:

$$a = \frac{F}{m} = \frac{4 \text{ N}}{2 \text{ kg}} = 2 \text{ m/s}^2$$

Answer: The acceleration of the trolley is 2 m/s^2.

Worked Example
Calculating deceleration and the braking force

A car of mass 900 kg travelling at 30 m/s must stop
a) in a time of 6 seconds,
b) in a distance of 50 metres.
Calculate the average deceleration of the car in each case and the average force applied by the brakes in each case.

a) we have: $m = 900$ kg, $u = 30$ m/s, $t = 6$ s and $v = 0$ (comes to rest). Using equation of motion ①, we have

$$a = \frac{v - u}{t} = \frac{0 - 30 \text{ m/s}}{6 \text{ s}} = -5 \frac{\text{m}}{\text{s}^2}$$

Now the average braking force is given by $F = ma$,

$$F = 900 \text{ kg} \times (-5 \text{ m/s}^2) = -4500 \text{ N.}$$

Answer: The car has an acceleration of -5 m/s^2 (i.e. a deceleration) produced by an average braking force of -4500 newtons.

b) we have: $m = 900$ kg, $u = 30$ m/s, $v = 0$ (comes to rest) and $s = 50$ m. Using equation ⑤:

$$v^2 = u^2 + 2as$$

with $v = 0$ this gives $2as = -u^2$. Then

$$a = -\frac{u^2}{2s} = -\frac{(30 \text{ m/s})^2}{2 \times 50 \text{ m}} = -\frac{900}{100} \frac{\text{m}}{\text{s}^2} = -9 \frac{\text{m}}{\text{s}^2}$$

The average braking force:

$$F = ma = 900 \text{ kg} \times (-9 \text{ m/s}^2) = -8100 \text{ N}$$

Answer: The car has an acceleration of -9 m/s^2 produced by an average braking force of -8100 newtons.

Two meanings for *g*

1   *g is the strength of the earth's gravitational field*

In chapter 5 we found that the weight *W* of an object is proportional to its mass *m*; also the relation between *W* and *m* is given by $W = mg$, where *g* has a constant value in a particular place. At the surface of the earth, if *m* is given in kg then the value of *g* is 9.8 newton per kilogram, giving *W* in newtons.

In other words, we find that there is a *force* due to gravity of 9.8 newtons acting on each kilogram of mass in an object. Fig. 8.26a shows a mass *m* hanging on a spring balance and its weight is given by $W = mg$. The meaning of this use of *g* is not acceleration. The object is not accelerating because there is zero resultant force acting on it. Its downwards weight is balanced by the tension force in the spring balance.

So what does this use of *g* mean? One of the basic forces in the universe is the attraction between two masses or lumps of matter. For example, the earth attracts the moon and keeps it in orbit round the earth and the moon attracts the earth in return, producing the visible effect of the tides. A small object such as a stone of mass *m* near the earth's surface is attracted to the earth and the pull of the earth which we can feel is called its weight *W*. This small stone will also pull the earth but, being so massive, the earth will not be affected noticeably.

When we pick up a stone and feel its weight we might wonder about how it gets its weight. A simple answer is that it is attracted by the earth and that the earth pulls it with the force of gravity. But we notice that the action of the earth on the stone involves no visible link, no string or connecting rods, and works over (almost) any distance.

We imagine that the earth produces all round itself an invisible effect which is 'waiting to pull' on matter and to describe this 'action at a distance' we invent the idea of a **gravitational field**. Near the surface of the earth the earth's gravitational field pulls with a force of 9.8 newtons on every kilogram of matter. This force acting on each and every kilogram is called the **gravitational field strength**

and is given by the relation:

$$\text{gravitational field strength} = \frac{\text{force (weight)}}{\text{mass}} \qquad g = \frac{W}{m}$$

We can see that *g* has units of newtons per kilogram and that the earth's gravitational field strength (near its surface) *g* is 9.8 N/kg. By rearranging the formula for *g* we again find that the weight of an object of mass *m* is given by $W = mg$.

Fig. 8.26b shows the earth's gravitational field pointing inwards (called a radial field). The field direction is the direction of the pull of the earth, which is towards the centre of the earth.

2   *g is the acceleration caused by gravity*

When the object of mass *m* is no longer supported by the spring balance it falls freely in the earth's gravitational field, fig. 8.26c. The unbalanced weight force causes an acceleration *a*, given by the Newton formula $F = ma$. In the earth's gravitational field we can write:

$$\text{unbalanced force } F = \text{weight } W$$

and the acceleration *a* caused by this force:

$$a = g$$

where *g* is the acceleration due to gravity.

So again we have $W = mg$. Although this is the same formula as we had for the first meaning of *g*, this time *g is an acceleration*. The value of *g* as the acceleration due to gravity at the surface of the earth has been found by experiment to be $9.8 \, \text{m/s}^2$. Although this value of *g* is numerically the same, it has a different meaning and a different use indicated by its different units.

earth's gravitational field strength, 9.8 N/kg is used for calculating weights	$= g =$	the acceleration due to earth's gravity, $9.8 \, \text{m/s}^2$ is used for calculating velocities and distances fallen

Figure 8.26

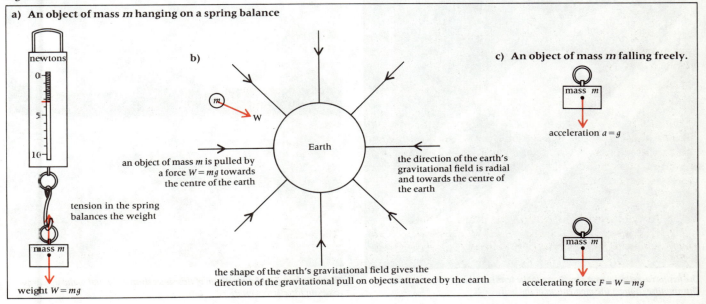

a) **An object of mass *m* hanging on a spring balance**

newtons

tension in the spring balances the weight

mass *m*

weight $W = mg$

b)

m

W

an object of mass *m* is pulled by a force $W = mg$ towards the centre of the earth

Earth

the direction of the earth's gravitational field is radial and towards the centre of the earth

the shape of the earth's gravitational field gives the direction of the gravitational pull on objects attracted by the earth

c) **An object of mass *m* falling freely.**

mass *m*

acceleration $a = g$

mass *m*

accelerating force $F = W = mg$

Momentum and impulse

Now that we have met Newton's second law of motion in the form of the equation $F = ma$, we can use it to study another useful idea in physics called **momentum**. Using equation ①:

$$a = \frac{v - u}{t},$$

we can write

$$F = ma = \frac{m(v - u)}{t}$$

and multiplying both sides by t gives

$$Ft = mv - mu$$

Impulse is force × time. For example, when you kick a football your boot is in contact with the ball for a time t, during which the kick force F acts on the ball. The value of $F \times t$ is called the **impulse** applied to the ball. (F is taken as the average force because the actual force would vary during the time of contact.)

Impulse has the units of force × time, i.e., newton × second or N s, but no special name.

Momentum is mass × velocity. Momentum is a very useful concept in physics because it helps us to understand and calculate what happens in collisions and explosions. For example, we can calculate the motion of a rocket from the mass and velocity of the exhaust gases ejected from its engines, or the pressure of a gas from the collisions of molecules with the walls of the containing vessel. The special thing about momentum is that it is conserved in collisions. We shall see soon how important this is.

The units of momentum are those of mass × velocity, i.e., kg × (m/s), or kg m/s. Again there is no special name for the units of momentum but the equation above shows that the units of momentum, kg m/s, are equal to the units of impulse, N s.

Momentum and Newton's second law of motion

For an object of mass m, if u is its initial velocity and v is its final velocity then

  mu = its initial momentum,
  mv = its final momentum

and we can say that its change of momentum $\Delta(mv)$ is:

$$\Delta(mv) = mv - mu$$

If this change of momentum is produced by a force F, which acts on the object for a time interval Δt, then we can rewrite

$$Ft = mv - mu$$
$$\text{as} \quad F\Delta t = \Delta(mv)$$

In words this says:

Impulse = change of momentum

Rearranging the equation we have

$$F = \frac{\Delta(mv)}{\Delta t}$$

This relation is an alternative form of **Newton's second law of motion**, which states in words that:

the resultant force acting on an object = its rate of change of momentum

Note that the change of momentum occurs in the same direction as the resultant or unbalanced force acts. Both force and momentum are vector quantities.

The rate of change of momentum of this space shuttle is equal to the upwards resultant force acting on it.

The tennis racket delivers a large force to the tennis ball in a very short time. This impulse changes the momentum of the ball.

Practical effects explained by Newton's second law

In each of the following examples a force causes a change of momentum. The size of the force depends on the length of time for which the force acts.

Hammering a nail

When a hammer hits a nail, the change of momentum of the hammer occurs in the very short time during which the nail is driven into the wood. So the hammer has a very rapid loss of momentum caused by the force of the nail acting on the hammer. This force, which is equal and opposite to the force of the hammer acting on the nail, must be very large to cause the rapid change of momentum of the hammer (by Newton's second law.) So the impulse which drives the nail into wood is a *large* force acting for a *short* time.

Packaging eggs in boxes

Eggs are packed in soft, shock-absorbing boxes so that when they suddenly stop or start moving they do not get cracked. How does this work? A moving egg with a certain amount of momentum requires a force to stop it moving. If the force acts for a short time (in a hard box) then the force on the egg will be large and will crack it. If, however, the force is spread out by the cushioning effect of a soft box and so lasts a longer time, it will be weaker and will not break the egg. The impulse needed to stop an egg moving should be a *small* force acting for a *long* time, rather than a large force acting for a short time. Both cause the same change of momentum, but a slow rate of change of momentum does not break the eggs.

Crumple-zones and seat belts in cars

It is more difficult to package human beings inside cars so that they will not be seriously damaged in a crash. The aim must be to make the forces which act on a person in a crash as small as possible. This means that the forces must be spread out over a longer time during the impact. The same change of momentum must occur but the deceleration needs to be more gradual.

The first safety features of a car are the crumple-zones at the front and rear. The metal bodywork of a car is designed to crumple or crush gradually during a collision. This action is the same as the egg box idea in which the force acting on the people inside the car is spread out over a longer time and so is made smaller. The car decelerates

The result of a head-on collision.

more gradually as the crumpling bodywork absorbs the kinetic energy of the car.

The second safety device is the seat belt. This is designed to spread the force which slows down a passenger over a longer time so reducing it to a safe level. The seat belt also spreads the force over a larger area of the person's body covering a broad band across the chest and over the hip bones. This reduces the pressure applied to the person and so further reduces the risk of injury. When not wearing a seat belt a passenger strikes the very hard windscreen. Apart from the risk of being cut by broken glass, the windscreen applies a very large force (for a very short time — the hammer and nail example) and so causes serious injuries, often to the head. Protruding objects such as the steering wheel can apply large forces to small areas of the body also causing serious injuries. The seat belt should be worn as tightly as is comfortable, leaving as large a space as possible between the passenger and the windscreen.

Calculations on Newton's second law

We now have two equations which give the resultant force acting on an object when its motion changes and both represent Newton's second law of motion:

$$\text{Resultant force } F = ma \text{ and } F = \frac{\Delta(mv)}{\Delta t}$$

Your choice of which formula to use will depend on the information you are given in a question.

Worked Example
Force changes momentum

A boy catches a cricket ball of mass 0.14 kg which has a velocity of 20 m/s. Calculate (a) the momentum of the ball, (b) the average force used by the boy's hands to stop the ball in (i) 0.5 seconds and (ii) 0.01 seconds. Can you explain why stopping the ball in 0.01 seconds hurts the boy but stopping it in 0.5 seconds does not?

a) Momentum $= mv = 0.14 \, \text{kg} \times 20 \, \text{m/s} = 2.8 \, \text{kg m/s}$.

b) When the ball is brought to rest the change of momentum $\Delta(mv) = -2.8 \, \text{kg m/s}$, because its momentum is reduced to zero.

Using Newton's second law of motion:

$$F = \frac{\Delta(mv)}{\Delta t}$$

In case (i) $\Delta t = 0.5 \, \text{s}$, so

$$F = \frac{-2.8 \, \text{kg m/s}}{0.5 \, \text{s}} = -5.6 \, \text{kg} \frac{\text{m}}{\text{s}^2} = -5.6 \, \text{N}$$

(The minus sign means a force causing a loss of momentum.) In case (ii) $\Delta t = 0.01 \, \text{s}$, so

$$F = \frac{-2.8 \, \text{kg m/s}}{0.01 \, \text{s}} = -280 \, \text{N}$$

When the boy applies a force of 280 N to the ball, the ball applies an equal and opposite force to his hands. A force of 280 N acting on the boy's hands hurts him. When he stops the ball more

Newton's first law of motion

In chapter 4 we saw some effects of the inertia of matter. The inertia of matter is its 'laziness' or reluctance to be moved. Because of inertia, *stationary things do not move on their own and they stay where they are unless a resultant, or unbalanced, force is applied to them.*

Similarly the inertia of a moving object makes it difficult to change its motion in either speed or direction. Everyone knows that you must keep pushing your bicycle to keep it moving along a level road and that a horse must keep pulling a cart otherwise it will stop moving. So it seems that a force is needed to keep things moving. But this is a misleading idea. These forces are actually needed to overcome or balance the frictional forces which oppose the motion.

There are many examples of motion which show that a force is *not* needed to keep the motion going and that *moving things naturally go on moving if they are left alone, or if no resultant forces act on them.*

For example, a spaceship in deep space does not use its engines to keep moving. Away from the influence of gravity, a spaceship moves at a constant speed in a straight line unless it does use its engines. When it does, the ejected exhaust gases apply a force to change the speed or the direction.

Similarly an ice skater can glide over the surface of ice at an almost constant speed in a straight line without any effort because there is very little friction to change the motion.

Newton expressed this understanding of motion in his **first law of motion** which can be stated simply as follows:

Stationary objects do not move on their own and moving objects keep on moving at a constant speed in a straight line if you leave them alone.

Or more formally as:

An object at rest will stay at rest and a moving object will continue to move with uniform velocity unless an external resultant force acts on it.

(Uniform velocity = constant speed in a straight line.)

Measuring the speed of a glider on an air track

We can investigate low friction motion in the laboratory using a hovercraft principle. Fig. 28 shows a glider floating almost free of friction on a cushion of air. The air is blown out of small holes along an air track.

A. Using multiflash (stroboscopic) photography.
- *Take multiflash photographs of a pointer on the glider.*
- *Measure the distances between the images of the pointer.*
- *Calculate the speed of the glider by dividing the distance between images by the time between flashes.*

B. Using an electronic timing method, e.g. light-operated switches connected to a Vela or electronic clock.
- *Arrange for the pointer on a glider to break a beam of light which operates an electronic switch and timer.*
- *To find the speed of the glider, divide the width of the pointer by the time for which the beam of light was broken.*

On a level track the glider will have an almost constant speed confirming Newton's first law.

A skater naturally goes on gliding over the ice at a nearly constant speed and does not need a force to keep him moving.

Figure 8.27   *A glider on an air track*

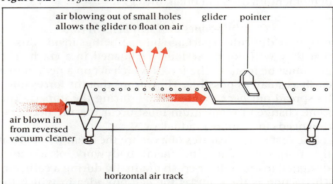

Frictional forces acting against gravity

Fig. 8.28 shows two examples of frictional forces acting against gravity as an object falls through a fluid. In fig. 8.28a the air resistance is the frictional force F opposing the weight W of the man and in (b) there is a frictional force F due to the viscosity or 'stiffness' of the oil which opposes the weight W of the marble.

In all cases the frictional forces oppose the motion. The magnitude of the frictional force increases from zero with the speed of the falling object, fig. 8.28c.

When an object is released from rest and $u = 0$, there is no frictional force and the resultant force is W. So at the moment of release the acceleration of the object is given by:

$$a = \frac{\text{resultant force}}{\text{mass}} = \frac{W}{m} = \frac{mg}{m} = g$$

i.e., the full acceleration due to gravity of $g = 9.8\,\text{m/s}^2$.

Figure 8.28 *Frictional forces acting against gravity*

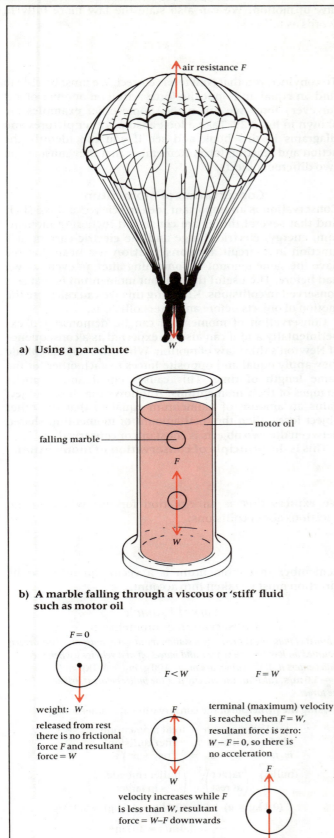

a) **Using a parachute**

b) **A marble falling through a viscous or 'stiff' fluid such as motor oil**

$F = 0$

weight: W

released from rest
there is no frictional
force F and resultant
force $= W$

$F < W$

velocity increases while F
is less than W, resultant
force $= W-F$ downwards

$F = W$

terminal (maximum) velocity
is reached when $F = W$,
resultant force is zero:
$W - F = 0$, so there is
no acceleration

c)

When *the object has gained velocity* a frictional force F opposes its weight W. Now, the downwards resultant force acting on the object is $W - F$ and its acceleration is given by:

$$a = \frac{\text{resultant force}}{\text{mass}} = \frac{W - F}{m}$$

So the acceleration of the falling object is now reduced to less than $9.8\,\text{m/s}^2$.

As the velocity of the falling object increases so the magnitude of the frictional force increases until it reaches a value equal and opposite to W; so

$$\text{the resultant force} = W - F = 0$$

and the acceleration is also zero.

So the object continues to fall at a constant speed, known as the **terminal velocity**, which is the maximum downwards speed possible for a particular object falling through a particular fluid. This illustrates Newton's first law of motion. For example, a falling parachutist continues to fall at a constant speed in a downwards direction because there is no resultant force acting.

(Note that in all these examples there is also a small upthrust force equal to the weight of the fluid displaced, p 101.)

Worked Example
Calculating frictional drag

A man of mass 70 kg jumps out of an aeroplane and before he opens his parachute he is falling with a reduced acceleration of 6.0 m/s². Calculate the frictional drag of the air which is reducing his acceleration if g = 10 N/kg.

The resultant force acting on the man is $m\,a$,

$$\therefore \text{ resultant force} = 70\,\text{kg} \times 6.0\,\frac{\text{m}}{\text{s}^2} = 420\,\text{N}$$

$$\frac{\text{The resultant force}}{\text{acting on the man}} = \frac{\text{the man's}}{\text{weight}} - \frac{\text{the frictional drag}}{\text{of the air}}$$

The man's weight $= mg = 70\,\text{kg} \times 10\,\text{N/kg} = 700\,\text{N}$

so we can write: $420\,\text{N} = 700\,\text{N} - \text{frictional drag}$.
So the frictional drag force $= 700\,\text{N} - 420\,\text{N} = 280\,\text{N}$.

Answer: The drag of the air causes an upwards force of 280 N.

Gliders floating on a cushion of air allow us to study motion under very low friction conditions.

Newton's third law of motion

We have already seen that when a force acts there are always two objects involved. A force can best be described as the *action of object* A *on object* B. Newton was the first person to realise that forces always come in pairs and that a single force is an impossibility.

A simple example is shown in fig. 8.29a. When you lean against a tree you are pushing the tree and the tree pushes back at you. The push of the tree is provided by a contact force at its surface. Your push is called the **action** force and the contact force of the tree is called the **reaction** force.

Figure 8.29 *Equal and opposite pairs of forces*

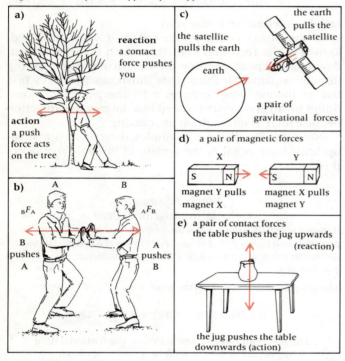

The action force acts on the tree and the reaction force acts on you. These two forces are equal and act in opposite directions. This illustrates the basic law about forces which **Newton stated in his third law:**

For every action force acting on one object there is an equal and opposite reaction force acting on another object.

It is important to understand why these two equal and opposite forces do not produce a zero resultant force and so always prevent acceleration happening. The reason is that the action and reaction forces always act on *two different objects*, not on the same one.

Fig. 8.29b shows a person A pushing another person B. The two pushes, although equal and opposite, do not cancel each other out, because they act on different objects. You can test this by pushing someone else when you are both standing on roller skates. The pushes cause both of you to move away in opposite directions.

If we write $_AF_B$ for the force an object A exerts on another object B and $_BF_A$ for the reaction force that object B must exert on object A, then since these two forces are equal and opposite:

$$_AF_B = -_BF_A$$

This relation is a more precise statement of Newton's third law of motion. We can also state the law more fully in words as follows:

If object A exerts a force F on object B, then object B exerts a force −F (of equal size but in the opposite direction) on object A.

To convince you that Newton is correct, we must be able to find an equal and opposite force acting on another object for every force we can name. Some other examples are shown in fig. 8.29. Take a look at some other pictures and diagrams in this book and see if you can identify the action and reaction force. Remember that they must act on two different objects.

Conservation of momentum

Conservation is an important idea in science and we shall find that several things are conserved including momentum, energy, electric charge and the electric current at a junction in a circuit. By **conservation** we mean that we have the same amount of something after an event as we had before. The useful thing about momentum is that it is conserved in collisions. Knowing this we can calculate the motion of objects before and after collisions.

Conservation of momentum can be demonstrated experimentally and it can also be expected as a consequence of Newton's third law of motion. When two objects collide they apply equal and opposite forces to each other for the same length of time. This causes equal and opposite changes of their momentums. It follows that if one object gains an amount of momentum equal to that the other object loses, then the total amount of momentum shared between the two objects is constant, or conserved.

This is the **principle of conservation of momentum:**

When two or more objects interact, their total momentum remains constant, providing no external resultant force is acting on them.

We express this as an equation for use when we do questions about collisions:

$$\frac{\text{total momentum}}{\text{before collision}} = \frac{\text{total momentum}}{\text{after collision}}$$

Remember that momentum is a vector quantity, so its direction must be taken into account.

Worked Example
Conservation of momentum

A bullet of mass m_1 is fired into a stationary target of mass m_2. The target is mounted on low-friction wheels and moves off at a velocity v when the bullet enters it. If the values are $m_1 = 100$ g, $m_2 = 4.0$ kg and v = 5.0 m/s, calculate the velocity u of the bullet before it strikes the target.

Using the principle of the conservation of momentum,

$$\frac{\text{total momentum}}{\text{before collision}} = \frac{\text{total momentum}}{\text{after collision}}$$

$$m_1 u + m_2 \times 0 = (m_1 + m_2) v$$

$$\text{(bullet)} \begin{pmatrix} \text{target} \\ \text{at rest} \end{pmatrix} \begin{pmatrix} \text{bullet embedded} \\ \text{in target} \end{pmatrix}$$

$$(0.1 \text{ kg} \times u) + 0 = (0.1 \text{ kg} + 4.0 \text{ kg}) \times 5.0 \text{ m/s}$$

$$u = \frac{4.1}{0.1} \times 5.0 \text{ m/s} = 205 \text{ m/s}.$$

Answer: The velocity of the bullet was 205 m/s before the collision with the target.

Demonstrating conservation of momentum for an elastic collision

Elastic collisions are ones in which two objects collide and then move apart again having lost little or none of their total motion energy. We can make two trolleys collide elastically by using the spring-loaded piston of one trolley as a 'springy buffer' in the collision. The idea is to make one trolley 'bounce' or spring off the other one. Fig. 8.30 shows a suitable arrangement.

- *Set up the runway and compensate for friction.*
- *Thread two tapes through one ticker timer using two carbons.*
- *Place one trolley A halfway down the runway. It should just stay at rest.*
- *Give the second trolley B a quick push aiming its piston towards the end of trolley A. Allow both trolleys to continue moving down the runway after the collision.*

From the tapes, you can measure the following constant velocities:

velocity of trolley B before the collision = u_B,
velocity of trolley B after the collision = v_B,
velocity of trolley A after the collision = v_A.

The velocity of trolley A before the collision is $u_A = 0$.

Using equal trolleys we can use the mass of a trolley as the unit of mass.

- *Measure the velocities in cm/tentick and make a table of results like table 8.3.*
- *As suggested in the table of results you can repeat the experiment using stacked trolleys to double or triple the mass of trolley A and trolley B.*

These results are rather rough, but what do you notice about the total momentum before the collision and the total momentum after the collision? They should of course be equal.

Demonstrating conservation of momentum for an inelastic collision

A good example of an inelastic collision is one in which two trolleys stick together during the collision. This can be arranged by mounting a pin on one trolley and a cork on the other so that on collision the pin sticks into the cork and holds the trolleys together, fig. 8.31. A strip of Velcro tape glued on the ends of two trolleys can also be used to hold them together after a collision.

- *Using a friction-compensated runway, set trolley A halfway down it, initially at rest.*
- *Attach a single length of paper tape to trolley B and, aiming its pin towards the cork in the end of the stationary trolley, give it a quick push.*

From the tape, you can measure the following constant velocities in cm/tentick:

velocity of trolley B before collision = u_B,
velocity of trolley B + A (joined after collision) = v

- *Again you can vary the masses by stacking trolleys.*
- *Measure the velocities and compile another table of results to compare the total momentum before and after the collision.*

Momentum is conserved in all kinds of collisions if no external force acts on the colliding objects.

Table 8.3

	Before collision							After collision						
	Trolley A			Trolley B			Total momentum		Trolley A			Trolley B		Total momentum
m	u_A	mu_A	m	u_B	mu_B	$mu_A + mu_B$	m	v_A	mv_A	m	v_B	mv_B	$mv_A + mv_B$
1	0	0	1	10	10	10	1	9	9	1	1	1	10
2	0	0	1				2			1			
1	0	0	2				1			2			

Figure 8.30 *An elastic collision of trolleys*

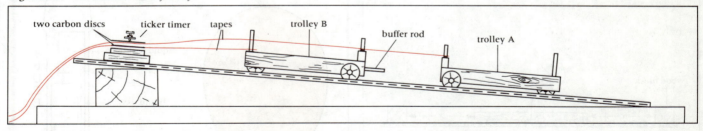

Figure 8.31 *An inelastic collision of trolleys*

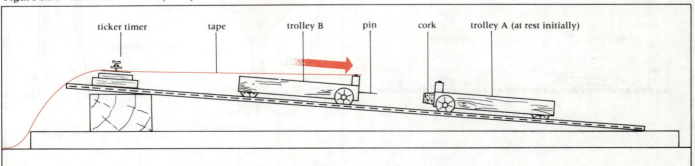

Momentum is conserved in head-on collisions and explosions

If two cars of equal mass collide head-on at equal speeds, they will both stop dead. At first sight it appears that momentum is destroyed in this type of collision because both cars stop and thus lose momentum.

To explain this collision in terms of the conservation of the total momentum of the two cars we must remember that momentum is a *vector* quantity.

If the momentum of the car moving to the right is $+mv$, then the momentum of the car moving to the left is $-mv$. It follows that their total momentum before the collision is $mv + (-mv) = 0$. After the collision both cars are at rest so that their total momentum is again zero and the total momentum is conserved, fig. 8.32.

Figure 8.32 *Momentum is conserved in a head-on collision*

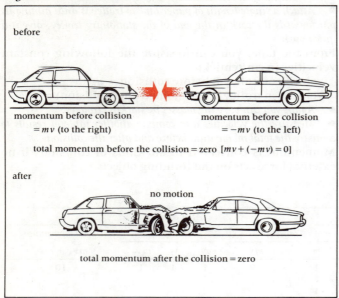

before

momentum before collision = mv (to the right)

momentum before collision = $-mv$ (to the left)

total momentum before the collision = zero $[mv + (-mv) = 0]$

after

no motion

total momentum after the collision = zero

Figure 8.33 *An explosion between two trolleys*

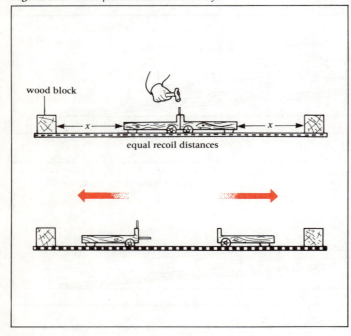

wood block

x x

equal recoil distances

- Test this by sending two trolleys fitted with a pin and a cork (as in fig. 8.31) towards each other with equal speeds along a horizontal runway.
- Now set up two trolleys facing each other as shown in fig. 8.33. One of them has its spring-loaded piston under compression.
- Release the piston by tapping the release dowel rod in the hole above it.

The trolleys fly apart with equal and opposite momentums so that their combined momentum remains zero after the explosion as it was before.

- You can test whether the trolleys have equal speeds by arranging blocks of wood at equal distances from the two trolleys. Do the trolleys reach the wooden blocks simultaneously?

The same principle applies to all explosive 'collisions', such as a gun firing a bullet or a jet or rocket engine firing hot gases at high speed behind the engine.

The rocket engine

We can demonstrate the principle of a rocket engine by allowing an inflated balloon to escape from our hands. The *action* of the air escaping causes the equal and opposite *reaction* of the movement of the balloon. The momentum of the balloon and air is conserved.

A rocket engine carries its fuel with it and can work in space as well as in the atmosphere. Two fuels, such as liquid hydrogen and liquid oxygen, burn together explosively and force the gases produced out of the rocket nozzle at high speed. The gain of momentum of the rocket is equal and opposite to the momentum of the ejected fuel. Clearly, when the fuel is ejected at high velocity after burning explosively it has a large momentum. Again the total momentum of the rocket and ejected fuel is conserved.

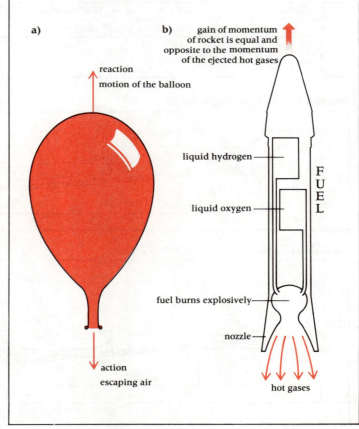

a)

reaction

motion of the balloon

action

escaping air

b) gain of momentum of rocket is equal and opposite to the momentum of the ejected hot gases

liquid hydrogen

liquid oxygen

FUEL

fuel burns explosively

nozzle

hot gases

The jet engine

A jet engine uses the same principle of momentum conservation as the rocket engine except that it takes in air to burn the fuel. So a jet engine will only work in the atmosphere. The figure shows a simplified diagram of a jet engine. The sequence of events inside the jet engine is as follows:

1) Air is drawn in through the front of the engine.
2) This air is compressed by the compressor blades.
3) Fuel is injected and burnt with the compressed air.
4) The exploding hot gases are forced through the engine turning the turbine, which turns the compressor.
5) High-speed gases are ejected from the back of the engine with high momentum, producing an equal and opposite increase in the forward momentum of the engine.

A Rolls Royce RB211–524 jet engine during final assembly at Derby.

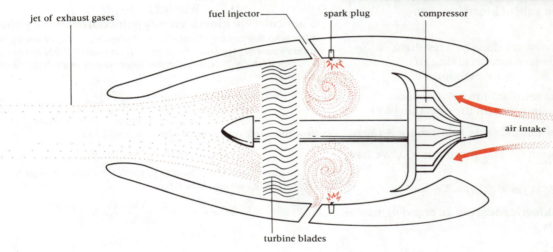

jet of exhaust gases   fuel injector   spark plug   compressor

air intake

turbine blades

Worked Example
A spaceship changes velocity

To make a course correction, a spaceship fires 5 kg of fuel out of its rocket engines as hot gases moving at 10 000 m/s. If the mass of the spaceship is 20 000 kg, calculate its change of velocity. Assume that the mass of the spaceship remains constant.

The total momentum of the spaceship and hot gases is conserved and therefore does not change. It follows that the change in momentum of the spaceship alone is equal and opposite to the change in momentum of the hot gases.

$$\text{change of momentum of spaceship} + \text{change of momentum of hot gases} = 0$$

$$\text{or} \quad \text{change of momentum of spaceship} = - \text{change of momentum of hot gases}$$

$$(m \Delta v)_{\text{spaceship}} = -\Delta (m v)_{\text{hot gases}}$$

$$20\,000\,\text{kg} \times \Delta v_{\text{spaceship}} = -5\,\text{kg} \times 10\,000\,\text{m/s}$$

$$\therefore \Delta v_{\text{spaceship}} = -\frac{5}{20\,000} \times 10\,000\,\text{m/s} = -2.5\,\text{m/s}$$

Answer: The change in velocity of the spaceship is 2.5 m/s in the direction opposite to the velocity of the hot gases fired out of its rocket engines.

Assignments

Make a simple model hovercraft and test Newton's first law of motion. Fig. 7.6 shows how you make a hovercraft using a balloon and a flat ceiling tile.

Describe another example of motion which illustrates Newton's third law or the conservation of momentum.

Explain why a parachutist allows his legs to bend and rolls over onto the ground when he lands.

Explain why the front of a motor car is designed to crumple up gradually when it hits something, rather than staying stiff and rigid.

Remember the formulas which we can use when a resultant or unbalanced force acts on an object (Newton's second law of motion.)

$$F = ma \quad \text{and} \quad F = \frac{\Delta (m v)}{\Delta t} \quad \text{or} \quad F = \frac{mv - mu}{t}$$

where F = the resultant or unbalanced force, a is the acceleration and mv = the momentum.
Try questions 8.13 to 8.21

8.4
KINETIC ENERGY AND POTENTIAL ENERGY
Kinetic energy and potential energy are two different forms of mechanical energy. Kinetic energy is the energy possessed by an object because it is moving and potential energy is the energy possessed by an object because of its position or condition.

Kinetic energy E_k
We saw in chapter 7 that when work is done energy is converted from one form to another. For example, when work is done pushing an object which accelerates this gains motion energy, which we call **kinetic energy**. *A resultant, or unbalanced, force applied to an object makes it accelerate and increases its kinetic energy.*

However, a force which is used to overcome friction does work against friction, but the energy is converted into heat energy by the friction process. Where there is no acceleration there is no gain of kinetic energy.

So we can write:

kinetic energy E_k gained by an object of mass m	=	work done by a resultant force F, giving an acceleration a

$$\text{gain in } E_k = \text{resultant force } F \times \text{distance moved } s \text{ (in the direction of } F)$$

$$\text{gain in } E_k = (ma) \times s = m \times (as) = \tfrac{1}{2}m \times (2as)$$

Using equation of motion ⑤, and putting $u = 0$, we have $2as = v^2$,

$$\therefore \text{ gain in } E_k = \tfrac{1}{2}m \times (2as) = \tfrac{1}{2}m \times v^2$$

So the formula for the kinetic energy of an object of mass m moving at a velocity v is

$$E_k = \tfrac{1}{2}mv^2$$

When the mass is given in kg and the velocity in m/s, the kinetic energy will always be in joules.

Two important equations can now be compared:

$$\text{impulse} = Ft$$
$$\text{work} = Fs$$

We have found that impulse causes a change of momentum, and that the work done by a resultant force causes a change of kinetic energy. Writing these two ideas as equations we have:

$$Ft = mv - mu \quad \text{(change of momentum)}$$
$$Fs = \tfrac{1}{2}mv^2 - \tfrac{1}{2}mu^2 \text{(change of kinetic energy)}$$

Worked Example
Calculating kinetic energy
Calculate the kinetic energy of a sprinter of mass 60 kg running at 10 m/s.

Using $E_k = \tfrac{1}{2}mv^2$ we have

$$E_k = \tfrac{1}{2} \times 60\,\text{kg} \times (10\,\text{m/s})^2 = 3\,000\,\text{J}$$

Answer: The sprinter has a kinetic energy of 3 000 joules.

Worked Example
Calculating kinetic energy gained from work done
A free-wheeling motor cyclist of mass (including her machine) 100 kg is pushed from rest over a distance of 10 m. If the push of 250 N acts against a frictional force of 70 N, calculate her kinetic energy and velocity when the push ends.

Resultant force causing acceleration	= push − frictional force

$$F = 250\,\text{N} - 70\,\text{N} = 180\,\text{N}$$

Now the work done which causes acceleration is

$$F \times s = 180\,\text{N} \times 10\,\text{m} = 1\,800\,\text{J}$$

So the kinetic energy gained E_k is 1 800 J. By rearranging the kinetic energy formula we have:

$$v^2 = \frac{2E_k}{m} = \frac{2 \times 1\,800\,\text{J}}{100\,\text{kg}} = 36\left(\frac{\text{m}}{\text{s}}\right)^2$$
$$\therefore \quad v = 6\,\text{m/s}$$

Answer: The motorcyclist and her machine reach a kinetic energy of 1 800 joules and a velocity of 6 m/s in the direction of the push.

Potential energy E_p

Potential energy is stored energy which an object has because of its position or condition. An object can have stored energy if it is held in a position from which it can fall under the influence of gravity. This potential energy, called **gravitational potential energy**, is possessed by all objects which are raised above the earth's surface.

We can calculate the gravitational potential energy possessed by an object from the work done in lifting it to a position above the earth's surface, fig. 8.34. A force equal and opposite to the weight of an object is needed to lift it. So if the object has a mass m, the force needed is mg:

$$\text{work done lifting an object of mass } m = \text{vertical force} \times \text{vertical distance}$$

$$\therefore \text{ work done} = mg \times h$$

The gravitational potential energy E_p gained by the object is equal to the work done in lifting it and is given by:

$$E_p = mgh$$

When m is in kg, h is in metres and g is in N/kg, then energy is in Nm or joules.

A grandfather clock is 'wound up' by raising a large mass to the top of the mechanism. As the mass slowly falls, its gravitational potential energy is converted into motion energy of the clock mechanism and works against frictional forces.

Objects can also possess potential energy called **elastic potential energy** or **strain energy** because of their condition. For example they can be stretched, compressed, twisted or distorted. A stretched catapult, a bent bow and a wound up spring all possess elastic potential energy.

Worked Example
Calculating gravitational potential energy

A grandfather clock uses a mass of 5 kg to drive its mechanism. Calculate the gravitational potential energy stored when the mass is raised to its maximum height of 0.8 m ($g = 10$ N/kg).

Using $E_p = mgh$ we have

$$E_p = 5\,\text{kg} \times 10\,\frac{\text{N}}{\text{kg}} \times 0.8\,\text{m} = 40\,\text{Nm} \quad (\text{or } 40\,\text{J})$$

Answer: The gravitational potential energy stored in the clock is 40 joules.

Conservation of mechanical energy

In many machines there is a constant interchange between kinetic energy and potential energy. In a frictionless machine the total of the kinetic energy + potential energy would remain constant. This is an example of **energy conservation**. In real machines frictional forces are always converting some mechanical energy into heat energy. However, we often use the conservation of mechanical energy to do calculations in mechanics.

If the total mechanical energy is conserved we can write:

$$\text{change of kinetic energy } E_k = -\text{ change of potential energy } E_p$$

gain of E_k = loss of E_p
loss of E_k = gain of E_p

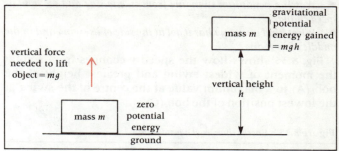

The potential energy stored in the bent bow is converted into kinetic energy of the arrow when the string is released.

Conservation of energy in collisions

The total momentum and total energy (of all kinds) is conserved in all collisions in which no external force acts on the colliding objects, but mechanical energy is conserved only in special cases. These special cases are called **perfectly elastic collisions**. If two objects collide and bounce off each other so that the total kinetic energy after the collision equals the total kinetic energy before it and no energy is converted to any other form, the collision is said to be perfectly elastic.

The collisions between gas molecules are normally perfectly elastic, but most collisions in everyday events are far from being perfectly elastic. The mechanical energy lost in collisions is converted into other forms of energy, which all eventually become heat energy. Energy conservation will be studied more fully in chapter 19.

Worked Example
Conservation of mechanical energy

Assuming conservation of mechanical energy, find the velocity with which a stone will strike the ground when it is dropped from a height of 80 m ($g = 10$ m/s²).

Using gain of E_k = loss of E_p

$$\tfrac{1}{2}mv^2 = mgh$$

and rearranging we have

$$v^2 = \frac{2mgh}{m} = 2gh$$

$$\therefore \quad v^2 = 2 \times 10\,\frac{\text{m}}{\text{s}^2} \times 80\,\text{m} = 1\,600\,\frac{\text{m}^2}{\text{s}^2}$$

$$\therefore \quad v = 40\,\text{m/s}$$

Answer: The stone will strike the ground with a vertical velocity of 40 m/s.

Energy changes in a swinging pendulum

● *Watch a pendulum swinging from side to side and see how its speed changes.*

● *What kind of energy has it got at the side of its swing and in the middle of its swing?*

Fig. 8.35 shows how the speed v changes from zero at the moment of widest swing and greatest height of the bob (A) to a maximum value at the centre of the swing at the lowest position of the bob (C).

Figure 8.35 *Energy changes of a pendulum*

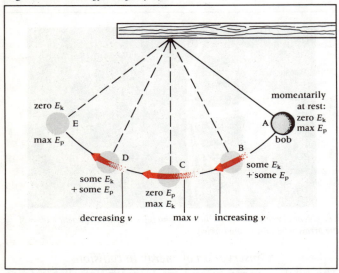

The kinetic energy E_k is greatest when v is greatest, at C. The potential energy E_p is greatest when the bob is at its greatest height h, in positions A and E. At intermediate positions, such as B and D, the energy is changing between E_k and E_p, depending on whether v is increasing or decreasing.

Stopping a car

What decides how quickly a car can stop and, perhaps more importantly, the shortest distance it can stop in? A moving car has kinetic energy all of which must be converted into another form of energy (usually heat) for the car to stop. The work done by the brakes of a car in slowing it down must equal the amount of kinetic energy lost. The brakes must turn the car's kinetic energy rapidly into heat using the frictional forces between the brake linings and the steel brake discs or drums. The whole brake assembly can get very hot when the brakes are used a lot.

Figure 8.36

The distance travelled by a car during emergency braking has two parts: the thinking distance and the braking distance.

● *The thinking distance is the distance travelled at a constant speed while the driver is reacting but before he applies the brakes.*

Even a driver who is alert takes at least 0.6 seconds to 'think' and press on the brake pedal. A driver who is tired or under the influence of alcohol will take much longer to react.

For example, at 15 m/s (30 m.p.h.) the distance the alert driver will travel in 0.6 seconds is given by:

$$s = vt = 15\,\text{m/s} \times 0.6\,\text{s} = 9\,\text{m}$$

but if he takes 2 seconds to 'think', he will travel 30 m.

● *The braking distance is the distance travelled while the brakes are doing work turning the kinetic energy of the car into heat.*

If the braking force applied to the car is F, the work W done by the brakes is given by:

$$W = Fs = \Delta E_k = \Delta(\tfrac{1}{2}mv^2)$$

Referring to fig. 8.36, compare the stopping distances of the two cars X and Y. They are travelling at 15 m/s and 30 m/s, respectively. The cars each have a mass of 960 kg. Their brakes can apply a braking force of 8 kN or 8000 N. For car X,

thinking distance = 9 m (as above)
braking distance, using: $Fs = \Delta(\tfrac{1}{2}mv^2)$
$8000\,s = \tfrac{1}{2} \times 960 \times 225$
∴ braking distance $s = 13.5$ m
∴ overall stopping distance = 9 m + 13.5 m = 22.5 m

For car Y,

thinking distance = $vt = 30\,\text{m/s} \times 0.6\,\text{s} = 18\,\text{m}$
braking distance, using $Fs = \Delta(\tfrac{1}{2}mv^2)$
$8000\,s = \tfrac{1}{2} \times 960 \times 900$
∴ braking distance $s = 54$ m
∴ overall stopping distance = 18 m + 54 m = 72 m

In the graphs of fig. 8.37 we can see how the thinking, braking and overall stopping distances depend on speed. Graph (a) is a straight line since the distance travelled at constant speed during the thinking time is just directly proportional to the speed.

Graph (b) curves upwards because the braking distance depends on the kinetic energy of the car which is proportional to the square of the speed: $E_k = \tfrac{1}{2}mv^2$.

For example, although car Y is travelling *twice* as fast as car X, it has *four times* as much kinetic energy and therefore travels *four times* further during braking.

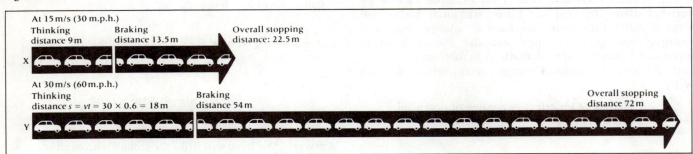

At 15 m/s (30 m.p.h.)
Thinking distance 9 m   Braking distance 13.5 m   Overall stopping distance: 22.5 m

X

At 30 m/s (60 m.p.h.)
Thinking distance $s = vt = 30 \times 0.6 = 18$ m   Braking distance 54 m   Overall stopping distance 72 m

Y

Figure 8.37 *Stopping distances*

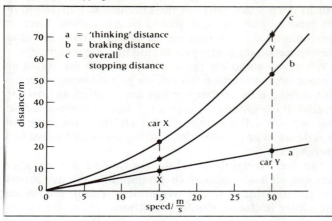

Assignments

Remember the formulas for mechanical energy:

a) ● kinetic energy $E_k = \frac{1}{2}mv^2$

b) ● potential energy $E_p = mgh$

Investigate

c) The distances given in fig. 8.36 are the '*shortest* stopping distances' for an average family car. Find out two conditions of the car and road which would increase these distances and suggest what action should be taken in each case.

d) The changes of energy that occur as:
   i) a lift goes up a lift shaft carrying people;
   ii) two children play on a see-saw.

Calculate

e) For the car described in fig. 8.36, calculate the shortest stopping distance if the car is travelling at 45 m/s (about 90 m.p.h.). How much further would the car travel if the driver had a thinking time of 2 seconds?

Try questions 8.22 to 8.24

8.5
MOTION IN A CIRCLE

Many objects move in circular paths or orbits. There are examples from the largest objects such as planets and moons to the smallest objects such as electrons and fragments of atomic nuclei moving in a magnetic field; examples from everyday life include clothes in a rotating spin-dryer, a car turning a corner or a person having a ride on a roundabout at a fun fair.

In every case there must be a resultant or unbalanced force producing the motion in a circle because otherwise the objects would all travel in a straight line as described by Newton's first law of motion.

An orbit needs an inward or centripetal force

● *Attach a ball to the end of a piece of string and whirl it round and round in a circle above your head.*

● *Feel the force acting in the string. Which way does it pull your hand and which way does it pull the ball?*

● *How is the force affected by a change of the speed of the ball?*

● *What happens to the ball when you let go of the string?*

Fig. 8.38a shows the orbit of the ball seen from above. The string is always in tension and can exert only pulling forces on the objects at its ends. A string cannot push! The tension force F which pulls the ball points inwards towards the centre of the orbit.

The other end of the string pulls with an equal and opposite force $-F$ on your hand. This is an example of Newton's third law of motion. Note how the two equal and opposite forces act on different objects.

There is only one force F acting on the ball. So this force is an unbalanced or resultant force which will change the motion of the ball.

The resultant force which acts on an object and is directed towards the centre of its orbit is called a centripetal force.

The centripetal force changes the motion of the ball by changing its direction. In fact, the ball has an acceleration towards the centre of its orbit in agreement with the relation: resultant force $F = ma$.

Figure 8.38 *An orbit needs an inwards or centripetal force*

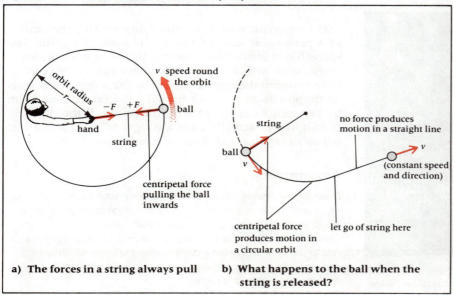

a) The forces in a string always pull

b) What happens to the ball when the string is released?

The framework of the big wheel provides an inwards or centripetal force on the people in the chairs as they fly round in a circle.

The acceleration can be shown to be equal to v^2/r, where v is the speed around the orbit and r is the orbit radius. So the centripetal force F given by $F = ma$ can be found using

$$F = m\frac{v^2}{r}$$

When the speed v of the ball is increased you can feel that a stronger force F is needed to keep it in its orbit.

The acceleration of a satellite in orbit round the earth is the acceleration caused by the pull of gravity which always acts in a direction towards the centre of the earth. The orbiting satellite is continually falling towards the earth with an acceleration g, but this falling and accelerating is just enough to keep the satellite in its circular orbit. (The value of g is smaller than $9.8\,\text{m/s}^2$ for an orbiting satellite because the pull of gravity gets weaker further away from the earth.)

Fig. 8.38b shows what happens when you let go of the string. The ball flies off in a straight line at a constant speed v because no resultant force is acting on it. But the direction of the line is at a tangent to the circular orbit. There is no outwards force acting on the ball and in the absence of the pull of the string the ball simply carries on moving in a straight line in the direction in which it was moving when you let go of the string.

If there is no centripetal force then there can be no orbit. So for every example of circular motion we should be able to identify the source and nature of the inwards acting centripetal force. The following are some examples.

The moon and satellites are held in orbit around the earth by the invisible but real force called gravitational attraction. An electron is held in orbit around an atomic nucleus by the attraction between unlike electric charges (p 203). The washing in a spin-dryer is held in orbit by the inwards push of the rotating drum wall. (The water flies off at a tangent to the orbit when it escapes through holes in the drum wall.) When you are whirled round on a roundabout there is a push from your seat which pushes you from behind towards the centre of the orbit. Electrons travelling down a television tube are pulled into curved paths by a magnetic field (p 346).

The hammer thrower has to provide a large inwards or centripetal force to keep the hammer travelling round in a circle above his head.

Weightlessness

Out in deep space far from the gravitational attraction of any stars or planets there would be no 'weight' force or gravitational force acting on an object. Astronauts and their spaceship in deep space would be truly weightless.

When we stand on the ground we feel the upwards contact force or reaction force of the ground supporting us. This feeling tells us that our feet are pressing against the ground. If the ground is taken away from under our feet so that we are falling through the air then we have no experience of any contact force acting on us and so we feel 'weightless'.

You feel 'weightless' while you are in the air after jumping off a springboard or a trampoline and astronauts are given training under 'weightless' conditions in an aircraft by imitating similar conditions. These are examples of *apparent weightlessness*. In fact while you are being pulled towards the earth you must have a weight.

On board the space shuttle, in orbit around the earth, astronaut Commander Thomas K Mattingley II experiences apparent weightlessness.

The astronaut who is in a spaceship orbiting the earth is in a permanent state of free fall towards the earth. The spaceship is also falling towards the earth with the same acceleration which means that there can be no contact force between the astronaut and the spaceship. As he feels no support or contact force from the spaceship around him he experiences apparent weightlessness all the time. He does however actually have a weight $W = mg$ and a free fall acceleration g both directed towards the earth.

Assignments

Describe and explain   the apparent changes in your weight which you experience in a lift as it sets off (a) upwards and (b) downwards. What would you feel if the ropes supporting the lift broke?

Describe   an experience you have had when you were moving in a circular path such as at a fair ground. Explain the forces which were acting on *you*.

Explain   how a spin-dryer works

Try question 8.25

Questions 8

Assume $g = 10\,N/kg$ or $10\,m/s^2$

1 An Intercity 125 train travels from London to York in 2 hours. If the distance is 300 km, find the average speed of the train in (a) km/h and (b) m/s.

2 How far will a walker travel if he walks for 6 hours at an average speed of 1.2 m/s?

3 A girl runs at a constant speed of 5.0 m/s round a running track. How long will it take her to run 1 kilometre?

4 A girl rides a bicycle which records the distance travelled in metres. On a short ride she records the distance every 10 seconds and obtains the following readings:

time/s	0	10	20	30	40	50	60	70	80	90	100
distance/m	0	20	40	60	80	140	200	260	280	280	280

Plot a distance–time graph of her ride.
What was her speed during the first 40 s?
What was her speed between 40 s and 70 s from starting?
What happened after 80 s?

5 A swimmer is 240 m from the beach when he realises that the tide is carrying him out to sea. If the velocity of the tide is 0.5 m/s away from the beach and he can swim at a maximum speed of 0.8 m/s, calculate (a) his maximum resultant velocity towards the beach and (b) the shortest time in which he can reach the beach.

6 A spaceship travelling in space at a constant velocity of 800 m/s launches a probe towards a distant planet. If the probe is launched with a velocity relative to the spaceship of 600 m/s in a direction at right angles to the flight path of the spaceship, calculate the resultant velocity of the probe.

Figure 8.39

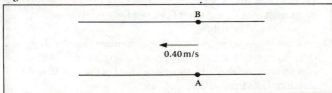

7 A river that is 40 m wide flows at 0.4 m/s in the direction shown in fig. 8.39. A man sets out from A in a rowing boat heading in the direction AB. His speed through the water is 0.80 m/s.
   Determine (a) the time taken to reach the far bank, and (b) the distance from B at which he reaches the bank.   (C)

Figure 8.40

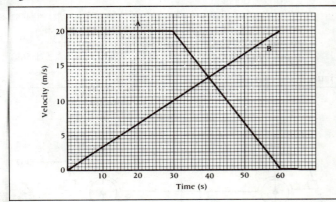

8 The velocity–time graph in fig. 8.40 shows part of the motion of two cars A and B.

a) Describe the motion of car A between 0 and 60 seconds.
b) Write down the time when both cars have the same speed.
c) Calculate i) the difference in the speeds of the cars at 22 s, ii) the acceleration of car B, iii) the distance car A has travelled between 30 s and 60 s.
d) Car A has travelled a greater distance than car B in 60 s. How can you tell this from the graph?   (W 88)

9 Fig. 8.41 represents the velocity–time graph for a lift in a department store
a) Briefly describe the motion represented by OA, AB, BC on the graph.
b) Use the graph to calculate (i) the acceleration of the lift, and (ii) the total distance travelled by the lift.   (JMB)

Figure 8.41

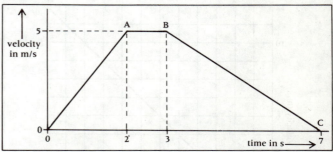

10 A minibus was driven, from rest, in a straight line and came to rest again after travelling a distance of 600 m. Accompanying the driver there were three persons whose duties were to record the speed of the bus and the distance travelled at suitable time intervals. In addition to the instruments on the dashboard of the bus which gave (i) its speed, in km/h, and (ii) the distance travelled, in 0.1 km divisions, a stopwatch was also available.

Figure 8.42 *Diagram of dashboard*

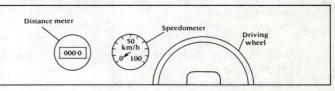

a) Describe, *in detail*, the instructions you would give to each observer on the bus to enable them to obtain and record these measurements. State any precaution you would take to ensure accurate readings.
b) The table gives readings obtained from one such experiment. The speed has been converted to m/s. Plot a graph of the *velocity* of the bus against *time*.

Velocity of the bus in m/s	0	5	10	15	15	15	7.5	0
Distance reading in km	0	0	0.1	0.2	0.3	0.5	0.5	0.6
Time in seconds	0	10	20	30	40	50	55	60

c) With the aid of the graph describe, *in detail*, the movement of the minibus. The values of any acceleration and deceleration should be calculated and stated.
d) From the graph, or otherwise, calculate the *displacement*, in metres, of the minibus at times of 10 s, 20 s and 30 s from the start of its motion and explain clearly why the values you have obtained for the displacement do **not** all agree with the distances recorded in the table.   (NISEC 88)

11 Mary pulls her little brother Brian on a sledge. She pulls with a force of 120N and the friction force between the sledge and the ground is 20N.
a) What is the resultant accelerating force on the sledge?
b) The graph in fig. 8.43 shows the possible motion of the sledge as a result of the action of the accelerating force.
Using the information on the graph,
i) find the speed of the sledge after 3 seconds,
ii) find the speed of the sledge after 5 seconds.
iii) The world record for the 100m sprint is just under 10 seconds. Using this information, comment on your answer to (b)(ii). (Speed = distance/time.) (NEA [B] 88)

Figure 8.43

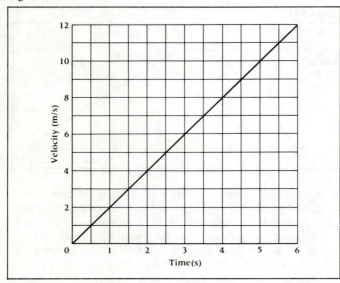

12 Fig. 8.44 is a graph showing the velocity of a body (in m/s) plotted against the time (in s).
a) Describe the motion of the body over (i) the region DE, (ii) the region EF.
b) What is the acceleration of the body over the region marked OB on the graph?
c) What is the distance travelled by the body over the region marked CD?
d) The distance travelled by the body over the region OB is 100m and over the region BC is 200m. What is the average velocity over the region OC? (Joint 16+)

Figure 8.44

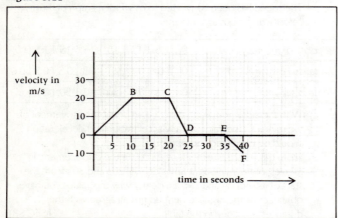

13 The author of a text book starts to describe the motion of an object thrown horizontally from a cliff in the following way 'The *horizontal velocity* and the *vertical velocity* of the body's motion are *independent* and . . .'.
a) i) Explain the meaning of the three phrases printed in italics.
ii) State how the two velocities mentioned would vary as the body falls.
b) If the body described above is thrown horizontally at 15m/s and the vertical cliff is 80m high, calculate
i) the time taken for the body to reach the ground and
ii) the distance from the foot of the cliff to where the body strikes the ground.
(Assume that the acceleration of free fall (g) = 10m/s^2 (N/kg).)
(JMB)

14 a) Name the **two** vertical forces acting on a man falling through the air. b) Explain why a man with a parachute falls, through the air, more slowly than a man without one. (W88)

15 If a force of 240N is applied to an object of mass 8kg and there are no frictional forces opposing the motion, calculate its acceleration. If the object starts from rest, find its velocity after 4s.

16 Calculate the force needed to give a mass of 5kg an acceleration of (a) 2m/s^2, (b) 10cm/s^2 and (c) 5km/s^2.

17 A force of 12N is used to move a box of mass 20kg along the ground. If there is a constant frictional force opposing the motion of 4.0N, what will be the acceleration of the box?

18 A trolley of mass 1.5kg is pulled by an elastic cord and is given an acceleration of 2m/s^2. Find the frictional force acting on the trolley if the tension in the elastic cord is 5N.

19 Calculate the momentum of an object if
a) its mass is 4.0kg and its velocity is 8.0m/s,
b) its mass is 500g and its velocity is 3km/s,
c) a force of 20N is applied to it for 6s and it moves from rest,
d) its mass is 2.0kg and it falls from rest for 10s (assume g = 10m/s^2 or 10N/kg).

20 A front-wheel drive car is travelling at **constant velocity**. The forces acting on the car are shown in the diagram in fig. 8.45. Q is the force of the air on the moving car and P is the total upward force on both front wheels.
a) Explain why i) P = 4000N, ii) Q = 400N.
b) Calculate the mass of the car.
c) The 400N driving force to the left is suddenly doubled.
i) Calculate the resultant force driving the car forward.
ii) Calculate the acceleration of the car.
iii) Draw a sketch graph showing how the velocity of the car changes with time. (Start your graph just before the driving force is doubled.)
d) i) Passengers in a car are advised to wear a safety belt. Explain, in terms of Newton's laws, how a safety belt can reduce injuries. ii) What other design feature in a car can offer protection in a crash? (W88)

Figure 8.45

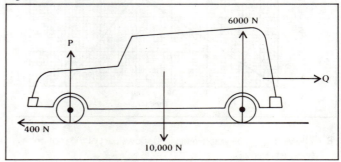

Figure 8.46

GARDEN
EQUIPMENT

21 This question is about a road accident involving a car and a van in a head-on collision.

Fig. 8.46 shows the situation before the vehicles crash. The car and the van were travelling in opposite directions along a straight road where the speed limit is 60 miles per hour (26 m/s). They are involved in a head-on crash which locks the vehicles together and brings them to rest on the spot. The drivers were wearing seat belts and no-one was seriously hurt. The police have the job of working out what happened. They know that the van (of mass 2000 kg) was travelling at a speed of 15 m/s because this vehicle was fitted with a tachometer. But they will have to do some calculations to find the speed of the car (of mass 1000 kg).

a) Calculate the momentum of the van (in kg m/s) before the collision.

b) Explain how you can use momentum to show that the car must have been speeding and calculate the speed of the car.

c) Speedometers have to be accurate to within 10%. Allowing for this, could the police prosecute either driver for speeding? Explain your reasoning.

d) In the collision, the van comes to rest in 0.5 seconds. Calculate i) the deceleration of the van, ii) the force on the van while it is stopping. (SEG [ALT] 88)

22 A sprinter of mass 70 kg accelerates from rest. The speed–time graph of his motion is shown in fig. 8.47.

a) Calculate the acceleration of the sprinter at the start of the race.

b) How far did the sprinter travel in 8 seconds? c) Find the kinetic energy of the athlete after 4 seconds. (SEB SPEC 90)

Figure 8.47

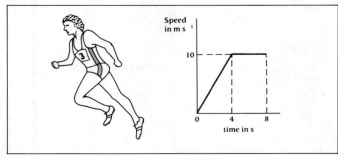

Speed
in m s⁻¹

10

0   4   8
time in s

Figure 8.48

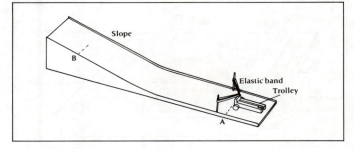

Slope

B

Elastic band
Trolley

A

23 A trolley is pulled back against an elastic band, as shown in fig. 8.48. It is then released from position **A**. The trolley runs along the flat

surface and then up the slope to **B**.

a) The trolley comes to rest at **B**. What type of energy does it have at **B**?

b) The trolley then runs back from **B**. It is stopped briefly by the elastic band stretching again. Sketch a graph of *speed* against *time* for the trolley *from the moment it leaves* **B** until it comes to rest again. (MEG Nuffield 88)

24 The kinetic energy of a moving car is transformed during the braking of the car. The distance over which this energy transformation occurs is called the **braking distance**. The **stopping distance** is greater than the braking distance because of the additional **thinking distance**. This is the distance the car travels at constant speed while the driver reacts to the need to brake.

a) The table below is based on the Highway Code and gives the thinking, braking and stopping distance for a 1000-kg car.

Speed/ km/h	Thinking distance/m	Braking distance/m	Stopping distance/m
50	9	14	23
80	15	38	53
110	21	74	95
160	30	155	185

i) Draw on graph paper a graph of **braking** distance (*y*-axis) against speed (*x*-axis).

ii) Is it sensible to extend (extrapolate) this graph to the origin? Give a reason for your answer.

iii) If a graph of **thinking** distance against speed is drawn it is a straight line through the origin. Why are thinking distance and speed related in this way?

b) When this 1000-kg car is travelling at 20 m/s (72 km/h) i) calculate the kinetic energy of the car, ii) read from the graph the braking distance for this speed, iii) calculate the average braking force during the braking from this speed.

c) A typical car is of length 4.2 m. How many car lengths would it be suitable for a driver in such a car travelling at 110 km/h on a motorway to leave between himself and the car in front. (LEAG 88)

25 A spacecraft of total mass 1 000 kg is travelling round the earth in a circular orbit of radius 12 000 km at constant speed. The gravitational field strength at that distance from the earth's centre is 3 N/kg.

a) What is meant by *gravitational field strength*?

b) The astronaut in the spacecraft has weight and yet he feels weightless. Explain.

c) Does the spacecraft need to produce a force from its rockets to keep it moving at constant speed? Explain.

d) How big is the force towards the centre of the earth acting on the spacecraft?

e) Calculate the speed of the spacecraft.

f) If the spacecraft had been in a circular orbit of twice the radius, it would have experienced a gravitational force only $\frac{1}{4}$ as large. How would its speed have compared with its speed in the first orbit? Explain your answer.

g) A part of the spacecraft is fired forward at 50 m/s relative to the remainder. If the mass of this part is 200 kg, what would be the change in the speed of the remainder?

h) When a spacecraft returns to earth, it requires a *heat shield* to disperse the heat produced.

i) Why is heat produced on re-entry?

ii) State three properties it would be desirable for the material of the heat shield to have, and give a reason in each case. (Nuffield)

9

Some properties of matter

Matter is made up of tiny particles called atoms and groups of atoms joined together called molecules. These particles are much too small to see with the human eye, even when aided with an optical microscope. It is just possible to 'see' large molecules using an electron microscope which uses electrons to 'see' with instead of light. So how small are molecules? How do molecules form solids, liquids and gases? How do molecules behave in each of these states?

A field on microscope picture of a platiniun crystal magnified × 700 000

9.1
MOLECULES IN MOTION

First we shall look for direct evidence of the size of a molecule and for evidence of its kind of motion.

Estimating the size of a molecule

If you look in puddles of water by the roadside or in a garage forecourt after rain you may see rainbow colours in a thin layer of oil floating on the water surface. When some kinds of oil are allowed to spread out on the surface of water they will continue to spread until eventually a thin film forms, which is just one molecule of oil thick. So we can estimate the size of an oil molecule by calculating the thickness of an oil film on water.

We need an oil-free water surface to carry out this experiment. A metal food tray coated in wax provides a suitably large and clean surface, fig. 9.1 (see appendix D4).

● *Fill the tray until it is overflowing with water. If necessary, level the tray with wedges.*

● *Place the two waxed booms across the tray side-by-side in the centre. Slide them towards the edges of the tray and leave them there.*

Figure 9.1 *Estimating the size of an oil molecule*

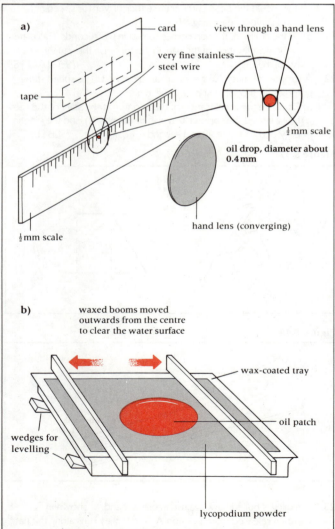

This process should clear the surface of the water and is needed after each attempt at measuring an oil patch.
- *Lightly sprinkle the water surface with a fine powder such as lycopodium. The water is now ready to receive an oil drop.*
- *Using a fine loop of wire attached to a card, dip it into a sample of oil (suitable oils include oleic acid and olive oil).*
- *Make sure that all the oil picked up by the wire loop is collected together in a single drop at the end of the loop.*
- *Hold the drop in front of a $\frac{1}{2}$ mm scale and, viewing it through a hand lens, estimate its diameter.*

You may be able to guess to the nearest $\frac{1}{10}$ mm, but only an estimate of size is needed for this experiment. NB, the drop should not be greater than $\frac{1}{2}$ mm in diameter otherwise it will spread over an area too large for the tray.
- *Now touch the water surface with the oil drop until the oil spreads rapidly over the surface.*

The area covered by oil can be seen as it pushes the powder away leaving a roughly circular and clear patch of oil. This oil film, which is one molecule thick, is called a **monomolecular layer**.
- *Measure the diameter of the oil patch with a metre ruler.*
- *Clear the surface with the booms and repeat the experiment to obtain several sets of results.*

Calculating the thickness of the monomolecular layer
Using specimen results:
diameter of oil drop $d_1 = 0.4$ mm $= 4 \times 10^{-4}$ m
diameter of oil patch $d_2 = 0.2$ m $= 2 \times 10^{-1}$ m
The volume of the oil drop V is given by

$$V = \frac{4}{3}\pi\left(\frac{d_1}{2}\right)^3 = \frac{4}{3}\pi\left(\frac{4}{2} \times 10^{-4}\,\text{m}\right)^3$$
$$V = 3 \times 10^{-11}\,\text{m}^3 \text{ (one significant figure only)}$$

and the area of the oil patch

$$A = \pi\left(\frac{d_2}{2}\right)^2 = \pi\left(\frac{2}{2} \times 10^{-1}\,\text{m}\right)^2$$
$$A = 3 \times 10^{-2}\,\text{m}^2$$

Volume of oil in patch = area of patch × thickness of patch
= volume of oil in the drop, V

$\therefore$ thickness of oil patch $= \dfrac{V}{A}$

$$= \frac{3 \times 10^{-11}\,\text{m}^3}{3 \times 10^{-2}\,\text{m}^2} = 1 \times 10^{-9}\,\text{m}$$

The rough value obtained of 10^{-9} m or 1 nm is an estimate of the length of an oil molecule, because in the monomolecular layer of oil the molecules stand up on end. A more accurate value of the length of an oil molecule is about 2×10^{-9} m but the figure to remember as a guide to the size of molecules is 1 nm or 10^{-9} metres.

Brownian motion
We might ask: do molecules have motion? We cannot see molecules moving because they are too small, but we can look for motion of larger particles which may be caused by the movement of molecules. Some evidence that air molecules move around randomly is provided by the effect called Brownian motion. This effect, which can be seen in both liquids and gases, is easily demonstrated with small smoke particles in air using a simple piece of apparatus like that shown in fig. 9.2.

Figure 9.2 *Brownian motion in a smoke cell*

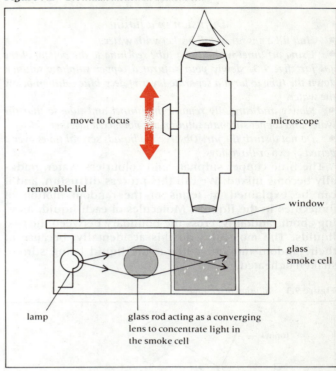

- *Remove the lid from the apparatus and hold a drinking straw upright in the glass smoke cell.*
- *Light the top end of the straw so that smoke passes down the inside of the straw into the smoke cell.*
- *Quickly replace the lid to trap the smoke inside the cell.*
- *Place the apparatus under a microscope and adjust the microscope to focus on the smoke particles inside the glass smoke cell.*

It is best to start with the microscope objective lens (the one at the bottom) very close to the smoke cell and then to raise the microscope slowly until the smoke particles come into focus.
- *Watch the smoke particles carefully and describe their motion.*

Air molecules have motion
The bright specks seen dancing about in a jerky, erratic or random way are the smoke particles brightly illuminated by the concentrated light. They do not often collide with each other, but rather appear to be knocked about by some other invisible particles in the smoke cell.

We believe that the motion of the smoke particles is evidence that air molecules also are moving. The smoke particles, which are large enough to be seen under the microscope, are also small enough to be knocked about by the fast moving air molecules. The jerky, erratic movement of the smoke particles, known as Brownian motion, shows that air molecules move in all directions with a range of speeds and kinetic energies. We describe the motion of the air molecules as **random motion**.

The same effect can be seen with pollen grains or sulphur particles suspended in a liquid. So we also have evidence that the molecules of a liquid have random motion within the liquid. The theory that molecules all have some kind of motion and kinetic energy is called the **kinetic theory of matter**.

Investigating diffusion

Diffusion in a liquid

- *Half fill a gas jar or tall beaker with water.*
- *Using a funnel with a long tube reaching to the bottom of the gas jar, fig. 9.3, slowly pour saturated copper sulphate solution down the tube to form a separate layer of deep blue solution below the water.*
- *Slowly and carefully remove the funnel and tube so that the water and copper sulphate solution are not mixed together.*
- *Do not disturb the jar. Observe the liquids several times over a period of two or three days.*

The blue copper sulphate and colourless water gradually become mixed. We call this process **diffusion**, and it can be explained in terms of the random motion of molecules in the liquids. Molecules of each liquid, moving about randomly, cross the boundary between the two liquids. The molecules do this accidentally, because of their random motion, without the liquid being stirred, shaken or heated.

Figure 9.3   *Diffusion in a liquid*

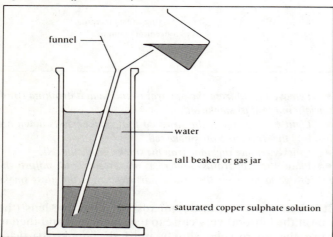

Diffusion into air

- *Close all the windows and doors in the room to cut down the draughts. Then open a bottle of scent in one corner of the room.*
- *If each person puts a hand up when the scent is first detected you can watch the spread of the scent across the room. How long does it take for the scent to reach the far corner of the room?*

The molecules of scent spread around the room by diffusion, and we can detect them by our sense of smell, but we cannot see them because they are too small and too few.

A model of diffusion

- *Arrange a number of coloured marbles at one end of a tray and some clear or white marbles at the other end, fig. 9.4.*

The two colours of marbles represent the molecules of two different substances which are initially separated.

- *Shake the tray in an erratic way so that the marbles begin to move about with random motion.*
- *What do you notice happening?*
- *Is it possible to make the marbles sort themselves out again by continuing to shake the tray?*

Figure 9.4   *A model of diffusion*

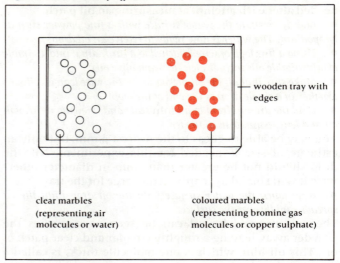

By the process of diffusion the molecules (represented by marbles) become completely mixed up. Diffusion allows molecules of any gas or liquid to invade the whole of the space available to it.

We find evidence that both diffusion and Brownian motion occur in liquids and in gases. This is good evidence that the molecules of both liquids and gases are in restless motion with randomly changing directions and speeds.

Assignments

Describe some everyday examples of diffusion.

Draw what you think is a typical path of a smoke particle in a smoke cell. It helps if you watch one particle for a while.

Construct a random walk diagram for a 'molecule'.

Use a die and the special isometric paper shown in fig. 9.5. Each time you throw the die, move the 'molecule' in the direction indicated in the figure. In this way the random sequence of numbers from the die will generate a random walk for the 'molecule'. You can repeat the process for several 'molecules' on the same sheet of paper.

Figure 9.5   *Random walk*

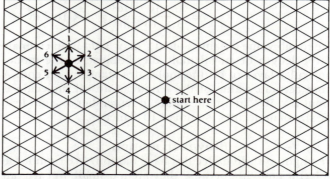

Explain

a) the difference between the motion of smoke particles you observed in a smoke cell and the motion which you would expect to see if it was caused by a draught or by convection.

b) what you would expect to see in a smoke cell if the smoke particles were much larger.

Try Questions 9.1 to 9.4

9.2
SOLID, LIQUID AND GAS, THE THREE STATES OF MATTER

Gases are 'springy', solids are 'stretchy' and liquids form drops. These simple ideas give us clues about the different natures of the three states of matter. If you put your finger over the end of a bicycle pump you can feel the 'spring' in the gas as you try to compress it. Watch the water slowly dripping from a tap or running down a window pane and you can see how water likes to form drops. Stretch a spring and you can feel the forces in the solid pulling against you. To help us explain and understand these properties we shall look at some models of a gas, a liquid and a solid.

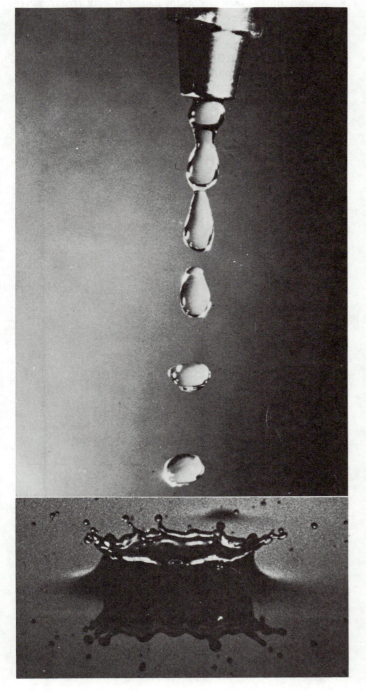

A model of a gas

● *Put about twenty marbles in a tray with vertical sides.*

● *Shake the tray in an erratic way keeping the tray flat on the bench top.*

● *Watch the path of one particular marble. (It helps to have one marble a different colour from the rest.)*

● *Describe its motion over a period of a minute or two, fig. 9.6. You should notice changes of speed as well as direction.*

Figure 9.6 *The random path of a marble*

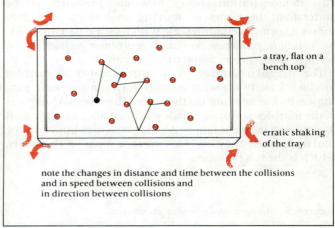

● *Listen to the sounds of collisions. Note the difference in sound when marble collides with marble and marble collides with the walls of the tray.*

● *What changes do you see and hear when*
a) *you shake the tray more violently or more gently?*
b) *you change the number of marbles in the tray to half as many or twice as many?*
c) *you crowd the marbles into one corner of the tray with a short ruler and then, while continuing to shake the tray, you remove the ruler again, fig. 9.7?*

Faster moving marbles with more kinetic energy represent molecules of a gas at a higher temperature. They collide with each other and with the walls of the tray more often and, having more energy, make more noise.

Adding more marbles to the tray is like putting more air in a car or bicycle tyre. Now marbles collide with the walls of the tray more often and, as we shall see, this is why more air in a tyre increases the pressure.

Figure 9.7 *Compressing a gas*

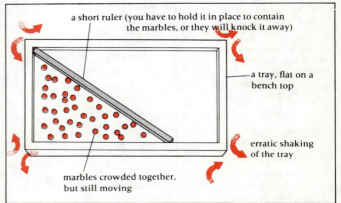

Crowding marbles in one corner of the tray is equivalent to compressing a gas into a smaller volume. Just as a compressed gas tries to expand again, the marbles press the ruler away and quickly reoccupy the whole tray when it is removed. Compare this with the air squashed in a bicycle pump as you hold your finger over the outlet end. You can feel the 'spring' in the gas as the elastic potential energy stored in the gas tries to expand it.

How molecules produce gas pressure

This demonstration shows how gas pressure can be understood in terms of moving molecules. Fig. 9.8a shows a spring weighing scale with its scale pan inverted.

● *Watch the pointer on the scale as a stream of marbles, or ball bearings, is poured over the inverted scale pan.*

The irregular impacts of a large number of marbles produce a fairly constant average force on the scale pan, shown by the reading on the scale. The irregular impacts of the marbles simulate elastic gas molecules bouncing off the walls of their container. As the molecules bouce off the container walls they exert an average force on the walls, which is the gas pressure.

Figure 9.8 *Moving molecules produce gas pressure*

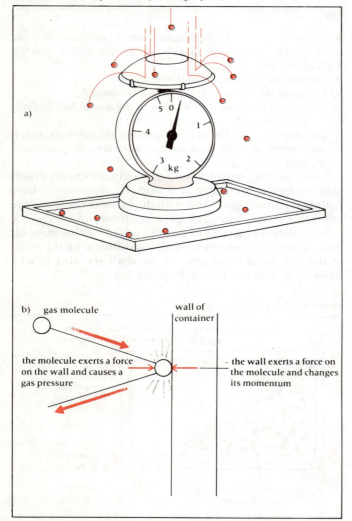

From Newton's third law of motion we know that the force exerted by the molecules hitting the wall is equal and opposite to the force exerted by the wall on the molecules, fig. 9.8. From Newton's second law of motion we know that the force exerted is equal to the rate of change of momentum of the molecules. So when the gas molecules bounce off the walls of their container and their direction is reversed they also have a change of momentum. This change of momentum of the gas molecules is the cause of the force on the container walls and is what we describe as the gas pressure.

A model of a liquid

● *Cover about a quarter of the tray with marbles, fig. 9.9.*
● *Slightly tilt the tray and again agitate it with erratic movements.*
● *Observe the movement of the marbles, particularly in the space above those which are jostling each other.*

The marbles which are close together represent the liquid state. These marbles are able to move around, but are mostly confined to the lower section of the tray.

A few energetic marbles get thrown out of the 'liquid' region into the space above. These marbles represent molecules which have 'evaporated' and entered the 'vapour' or 'gas' region. They return to the 'liquid' region and are replaced by other fast moving marbles.

● *Notice how the marbles become thinned out near the 'liquid' surface.*
● *Notice how the 'liquid' marbles occupy much less space than the 'gas' marbles.*

Figure 9.9 *A model of a liquid*

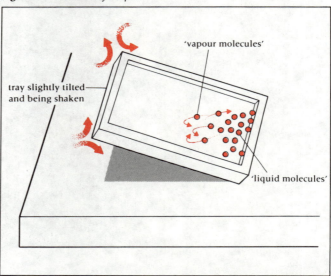

Investigating surface tension effects

Why do liquids form drops? Why does water wet some surfaces but run off others? How can a pond-skater rest on the surface of water without sinking? Why does water rise up a capillary tube but mercury is pushed down to a lower level in a capillary tube?

● *Try some of the experiments shown in fig. 9.10.*
● *Watch carefully what happens to the liquid surface in each experiment. Can you detect any forces pulling the liquid surfaces?*

Floating a needle

- *Rest a steel needle on a small piece of blotting or filter paper and gently place it on the surface of some clean water.*

After a while the soaked paper sinks leaving the needle floating on the surface. This is surprising because steel, being much denser than water, should sink.

- *Look carefully at the water surface around the floating needle, fig. 9.10a. What do you notice?*

The surface is depressed and stretched like an elastic skin.

Pond-skaters or water striders (see photo) have a coating of fine water-repelling hairs that clothe their undersides and legs and prevent them getting wet. The water does not wet the water striders, and its surface becomes 'stretched' and 'dimpled' rather than broken. This stretched elastic surface provides a stronger upwards force to support the water striders (likewise the steel needle), than would be provided simply by the upthrust from displaced water.

Soap film experiments

- *Tie a thread across a wire loop in various ways. Two possibilities are shown in fig. 9.10b.*
- *Dip the whole wire loop into a liquid detergent solution and remove it slowly.*
- *Prick part of the soap film with a needle.*
- *What do you notice about the shape of the cotton thread?*
- *What does the remaining soap film always try to do?*

The soap film has two liquid surfaces, one on each side. Each surface is under tension and is trying to contract. The loose thread is always pulled into the shape which gives the remaining soap film the minimum surface area.

Bubbles formed by soap films trap a certain volume of air. Their spherical shape provides the soap film with the minimum surface area for a given volume of air.

Figure 9.10   *Some surface tension effects*

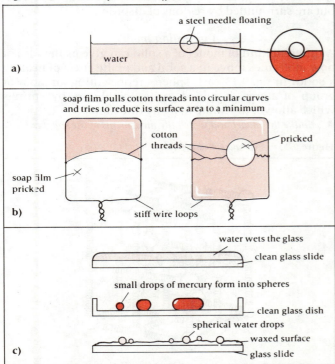

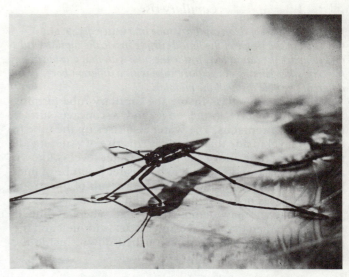

A pond skater can rest on the surface of water without breaking the surface. You can see the dimples where the surface is 'stretched' by the insect's feet. Pond skaters 'row' themselves, or skate, across the surface of water at high speed using their long middle legs.

Forming liquid drops

- *Place a few drops of water from a dropping pipette onto a clean glass slide and also onto a waxed or greased glass slide.*
- *Watch a demonstration with a few small drops of mercury in a clean glass dish.*
- *Compare the results, fig. 9.10c.*

Adhesive and cohesive forces

Two kinds of forces act on a liquid molecule:

Cohesive forces are those forces which attract liquid molecules to each other. Cohesive forces make liquid molecules cling together and are responsible for the formation of liquid drops and surface tension effects.

If liquid molecules are spaced further apart than their normal separation the cohesive forces between them try to bring them closer together. This is what happens at the surface of a liquid; it makes the surface behave like a stretched skin and we call the effect surface tension. If, however, the molecules of a liquid try to come too close together there is a very strong repulsive force between them. This force makes liquids almost incompressible.

Adhesive forces are forces which arise between liquid molecules and the molecules of their container. Adhesive forces may make a liquid stick to or 'wet' the surface of its container.

For example, water wets a glass surface because the adhesive forces between the water molecules and the glass molecules are greater than the cohesive forces between the water molecules themselves. By waxing the glass surface we reduce the adhesive forces and allow the cohesive forces between water molecules to pull the water into spherical drops. This principle is seen at work when fabric is treated with a wax-like substance to make it water-repellent.

The stronger cohesive forces between mercury molecules form spherical drops of mercury even on a clean glass surface. The weak adhesive force between mercury and glass means that mercury does not 'wet' glass.

Capillary rise

- *Dip a length of clean capillary tube into water, fig. 9.11d.*
- *Look at the shape of the liquid surface inside the capillary tube and compare the liquid level inside the tube with that outside it.*
- *Try other lengths of capillary tube with a different bore, i.e. a different size hole through the middle.*

Water wets the glass inside the capillary tube forming an upwards curve at the edge and a concave surface known as a **meniscus**. The water also rises up the tube to a higher level inside than outside. This rise of water inside the capillary tube above the outside water level is known as capillary rise and is found to be greater in a tube of finer bore. Capillary rise is thought to be a result of the strong adhesive forces between the water and glass molecules.

Your teacher may demonstrate that when a tube is dipped into mercury a convex meniscus is formed and the mercury level is depressed, or pushed down, inside the tube below the outside liquid level.

Figure 9.11 *Capillary rise*

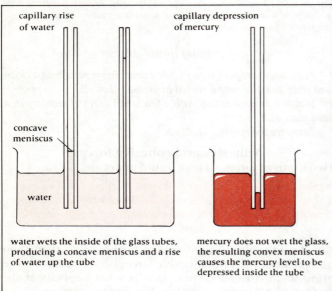

water wets the inside of the glass tubes, producing a concave meniscus and a rise of water up the tube

mercury does not wet the glass, the resulting convex meniscus causes the mercury level to be depressed inside the tube

Models of a solid

Crystalline solids

In a crystalline solid the molecules are arranged in regular patterns giving it a permanent, fixed shape. When a particular solid crystallises out from the liquid state the crystals so formed have a characteristic shape. Many different crystal shapes exist and these depend on the packing arrangement of the molecules.

Models of crystalline solids can be built using marbles or polystyrene spheres. Unless the spheres are glued together a 'fence' or tray with edges is needed (or you can use books) to stop the spheres rolling away. Building a close-packed structure:

- *Lay out a 4 × 5 array of spheres as shown in fig. 9.12a.*
- *Stack a 3 × 4 array on top followed by two more layers.*
- *Look at the sloping faces of the pyramid, the spheres are closely packed in a hexagonal array, fig. 9.12b.*
- *Can you see that the spheres are more closely packed in the planes of the sloping faces than they are in the horizontal planes?*

This difference in arrangements of the atoms is called anisotropy and gives the crystal different properties in different directions.

Figure 9.12 *Arrays of atoms in a close-packed structure*

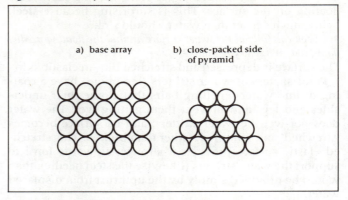

a) base array

b) close-packed side of pyramid

Other kinds of solids

Powders and amorphous materials are also solids. An amorphous material is one without a definite shape. Rapid cooling often produces an amorphous structure. Glass is such an 'amorphous' material; its molecules set in the random orientation they had in the liquid state. In a glassy material there is no long-range ordering of the atoms. In contrast, crystals, which are the most ordered structure possible, are formed when a liquid cools slowly allowing the particles of the liquid time to arrange themselves into an ordered pattern. Powders are formed when crystalline materials are crushed. This results in the atoms in each grain being aligned randomly and producing no order throughout the powder.

Metals and many rocks are described as polycrystalline solids. They contain crystalline grains of different materials or combinations of atoms set within the overall structure. The grains have ordered structures as in crystals but are surrounded by regions of disorder.

A vibration model

A different kind of model of a solid can give us the idea of how molecules move in a solid and of the effect of heating a solid. In fig. 9.13 each sphere, representing an atom or group of atoms, is joined to its neighbours by springs which allow it to vibrate in many directions.

- *Shake the model and watch the movements of the spheres.*

Figure 9.13 *A vibration model of a solid*

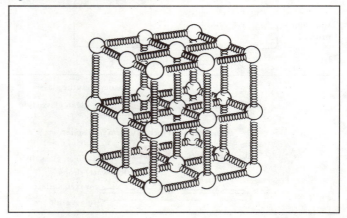

As the spheres vibrate they continually exchange their kinetic energy of motion with the elastic potential energy stored in the connecting springs. If we shake the model harder the spheres move faster and with larger amplitude. This is what happens when a solid is heated; the heat energy added appears as extra vibrational kinetic energy of the molecules. As the speed and amplitude of vibration increases a temperature is reached where the bonds between the atoms (the springs in our model) break and the solid melts.

The springs represent the bonds caused by the forces of attraction and repulsion between the atoms due to the electric charges of their nuclei and electrons.

We can see that it is possible for a solid to keep its overall regular shape while the individual atoms vibrate about fixed positions within the solid.

Stretching materials

The attractive forces between the molecules in a solid provide its characteristic elastic or stretchy properties. When we stretch a solid we are very slightly increasing the spacing of its molecules. The tension we can feel in a stretched spring is due to all the forces of attraction between the molecules in the spring.

Stretching a spiral spring

- *Arrange a stand to hold a millimetre scale close to a hanging spiral spring as shown in fig. 9.14.*
- *Attach a pointer to the end of the spring and take a scale reading of the pointer for an unstretched, unloaded spring.*
- *Hang a slotted mass hanger on the end of the spring and take a series of scale readings as slotted masses are added to the hanger, increasing the stretching force or load.*
- *Record your readings in a table, using the headings in table 9.1.*

Table 9.1

Mass on hanger m/kg	Stretching force mg/N	Scale reading /mm	Extension of the spring /mm	Force / extension $\dfrac{\text{N}}{\text{mm}}$

- *Calculate the stretching force using $F = mg$, where $g = 10$ N/kg.*
- *Calculate the increase in length or extension of the spring by subtracting the inital length or scale reading for the unloaded spring from all the loaded readings.*
- *Calculate for all the readings the value of the ratio: stretching force/extension.*
- *Plot a graph of extension against stretching force, fig. 9.15.*

The ratio stretching force/extension is very nearly constant and we can write:

$$\text{stretching force} = \text{constant} \times \text{extension}$$

The graph of extension against the stretching force is a straight line showing that the extension of a spiral spring is directly proportional to the stretching force. In other words, if the stretching force is doubled the extension is doubled and so on.

Figure 9.14 *Stretching a spring*

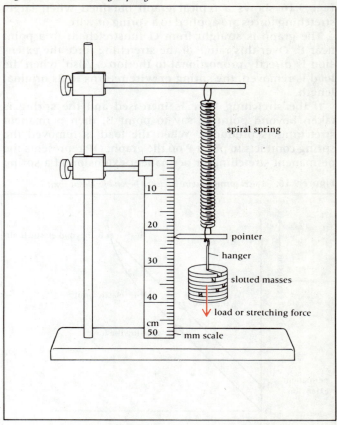

spiral spring

pointer

hanger

slotted masses

load or stretching force

cm
mm scale

Figure 9.15 *Graph of extension against stretching force for a spring*

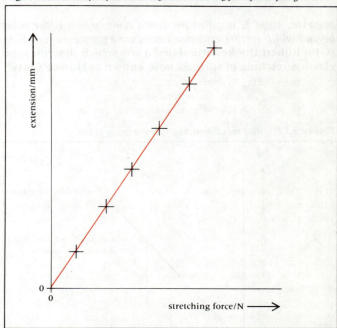

extension/mm

stretching force/N

This relation can be expressed as:

extension ∝ stretching force

The same kind of result is obtained if a straight steel wire is stretched. But what happens if we go on stretching a spring or a wire further and further?

The elastic limit and Hooke's law

Fig. 9.16 shows a typical graph obtained when larger stretching forces are applied to a spring or wire.

The graph is straight from O (unstretched) to a point near E. Over this range of the stretching force the extension is directly proportional to the force. Also, when the load is removed, the spring or wire returns to its original length.

If the stretching force is increased and the spring is taken beyond point E, say to point B, then permanent stretching is produced. When the load is removed the spring contracts to point P on the graph. OP represents the permanent stretching or permanent extension of a spring

Figure 9.16 *A steel spring or wire stretched beyond its elastic limit*

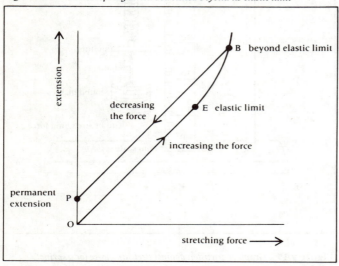

or wire. *Point* E *is called the elastic limit, which is the point beyond which further extension causes some permanent extension (OP).* Robert Hooke formulated a law which describes the elastic stretching of springs, now known as **Hooke's law:**

Provided the stretching force does not extend a spring beyond its elastic limit, the extension of the spring is directly proportional to the stretching force.

Figure 9.17 *Using the calibration graph of a spring balance*

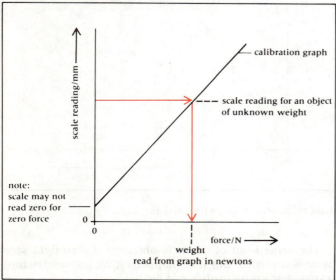

Using a spiral spring as a spring balance

A graph showing how a spiral spring stretches as the load hanging on it is increased can be used to find the weights of objects and is called a **calibration** graph. The calibration converts scale readings at the end of the spring into weights in newtons. Fig. 9.17 shows how the weight of an object is found using the calibration graph. When a spring balance is manufactured the scale readings in millimetres are converted into the required force units. In a physics laboratory the scale should be calibrated in newtons.

Worked example
Measuring weight with an elastic spring

When a load of 12 N is applied to a steel spring it produces an extension of 80 mm without exceeding the elastic limit of the spring. Calculate the weight of an object which, when hung from the same spring, produces an extension of 60 mm.

Using the relation:

$$\text{stretching force} = \text{constant} \times \text{extension},$$

we can find the constant of the spring:

$$\text{spring constant} = \frac{\text{stretching force}}{\text{extension}}$$

$$= \frac{12\,\text{N}}{80\,\text{mm}} = \frac{12\,\text{N}}{0.08\,\text{m}} = 150\,\frac{\text{N}}{\text{m}}$$

In other words, a force of 150 N would produce an extension of 1 m if the elastic limit was not exceeded. The extension produced by the object of unknown weight is 60 mm = 0.06 m. So the weight of the object W is

$$W = \text{the stretching force}$$
$$= \text{spring constant} \times \text{extension}$$
$$\therefore W = 150\,\frac{\text{N}}{\text{m}} \times 0.06\,\text{m} = 9.0\,\text{N}$$

Answer: The weight of the object is 9.0 newtons.

Stretching elastic bands and metal wires

A suitable arrangement for stretching lengths of wire, rubber or nylon thread is shown in fig. 9.18.

● *Clamp the wire at one end, attach a slotted mass hanger at the other end and attach a pointer to give a reading on a mm scale.*
The mm scale measures the change in length or extension of the wire rather than its total length.

Using long lengths (2 or 3 m) of fine copper wire (e.g. 32 SWG) allows a much greater extension to be observed.

● *Take readings of the pointer's position on the scale both as the masses are added and also as they are removed one by one.*
● *Record your readings in a table and calculate the extension.*
● *Plot a graph of extension against stretching force.*

Figure 9.18 *Stretching threads and wires*

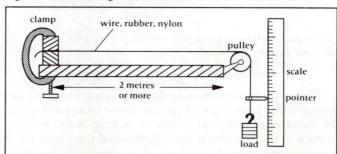

Figure 9.19 *Extension graph for rubber*

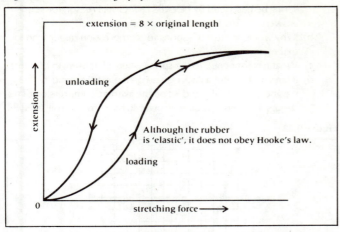

Fig. 9.19 shows a typical graph for a rubber or elastic band. Note that:

a) at no part of the graph does rubber obey Hooke's law, as no part of it is straight.

b) the extension is greater when unloading the masses than when loading. This effect is known as hysteresis and causes energy to be turned into heat in the rubber as it is stretched.

Figure 9.20 *Extension graph for copper wire*

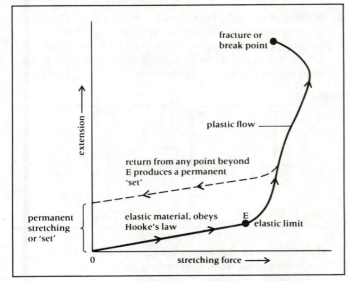

If a copper wire is stretched until it breaks the complete graph is shown in fig. 9.20. Beyond the elastic limit E the wire undergoes plastic flow or plastic deformation. When the region beyond E is large, as it is for copper, the material is described as ductile.

Mechanical properties of a solid

- *Strength*. A strong material needs a big force to break it. Strength often depends on whether the material is stretched, compressed or twisted.
- *Stiffness*. A stiff material does not stretch, bend or 'give' very much and is rigid.
- *Elasticity*. When the force is removed an elastic material goes back to its original size and shape after being pulled, bent or twisted.

- *Plastic* materials stay permanently deformed or stretched even when the force is removed.
- *Ductile* materials can be pulled out into wires and hammered or rolled into thin sheets without breaking. Metals are ductile.
- *Brittleness*. A brittle material breaks suddenly just beyond its elastic limit. Brittle materials also break suddenly when given a sharp knock, but even such fragile materials as glass have some elasticity before they break. Thin glass fibres are very flexible but also brittle and fragile. Pottery and cast iron are brittle.

Assignments

Study table 9.2. Make a list of the experiments you have seen or done and the observations you have made which give evidence for each of the properties in the table. Make sure you know which experiments support which ideas.

Table 9.2

	Solids	Liquids	Gases
Movement of molecules	limited to vibrations about a fixed position	vibrations and random movements throughout the liquid but molecules cling together (cohesion)	independent random motion spreading to fill the space available
Spacing of molecules	close packed usually in a regular formation giving high density	usually slightly further apart than in a solid, also has a high density	far apart giving a low density
Volume and shape	fixed volume and fixed regular shape, solids return to their original shape when stretched or deformed to a certain extent	fixed volume but no fixed shape, liquids take the shape of their container and are almost incompressible	neither volume nor shape are fixed, gases are compressible

Find out

a) Why do the bristles of a brush spread out while they are dipped in a beaker of water, but stick together when they are taken out again?

b) What happens to a floating needle when you add a few drops of detergent to the water? Can you think of an explanation for this? What has this effect to do with washing things in detergent solution?

c) Where does the dampcourse go round your house or school buildings? How do you think a dampcourse works? (Bricks are usually porous and the ground is damp. What effect is responsible for rising damp?)

d) Dip an inverted funnel in a detergent solution and remove it slowly so that a soap film is formed across the wide end of the funnel. What happens? Can you explain this effect?

Calculate, using the estimated size of a molecule of about 10^{-9} m:

e) the number of molecules that could be lined up end to end in one metre,

f) the number of these molecules that might fit into a cubic metre. (There are in fact on average about 10^{28} molecules per m^3 in both solids and liquids and about 10^{25} molecules per m^3 in gases under normal pressures and temperatures.)

Try questions 9.5 to 9.8

Questions 9

1 In an experiment to estimate the size of a molecule of olive oil, a drop of the oil, of volume $0.12\,mm^3$, was placed on a clean water surface. The oil spread into a patch of area $60 \times 10^3\,mm^2$. Use these figures to estimate the size of a molecule of olive oil pointing out any assumption you make.   (O & C)

2 Some perfume is spilled in a corner of a room in which the doors and windows are closed to stop draughts. The scent slowly spreads by *diffusion* to all parts of the room.
   a) By writing about molecules, explain how *diffusion* occurs.
   b) Why does the scent spread slowly even though the molecules move fast?   (MEG 88)

3 This question is about the ways in which molecules move. You are given three descriptions. One description refers to the molecules of a solid. One description refers to the molecules of a liquid. One description refers to the molecules of a gas.
   Select either **solid** or **liquid** or **gas** for each description.
   i) close packed, moving to and fro about fixed positions
   ii) widely separated, moving freely and randomly
   iii) close packed, jostling and changing places with neighbouring molecules   (LEAG 88)

Figure 9.21

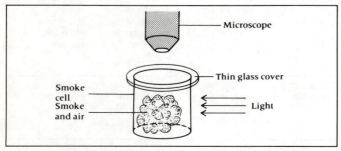

4 Fig. 9.21 shows apparatus used to observe the behaviour of smoke particles in air.
   a) Why is light shone into the container?
   b) Why are smoke particles very suitable for use in this experiment?
   c) Describe what you would see when looking through the microscope into the smoke cell.
   d) What does the experiment tell you about the behaviour of the air molecules in the cell?
   e) What difference, if any, would be seen in the motion of the smoke particles if a weaker light was used?   (NEA [B] 88)

5 Jim takes a spring off an old mattress and decides to experiment with it at school. He sets up the apparatus shown in fig. 9.22 and loads the spring with 5N weights, recording the length of the spring at each loading. He obtains the results given in the graph in fig. 9.23.

Figure 9.22

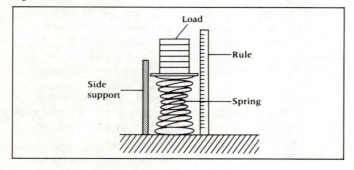

a) Over which range of load values can the behaviour of the spring be described by Hooke's law? Give a reason for your answer.
b) Why would Jim test the spring for compression rather than for extension?
c) What has happened to the spring when it has a length of 1 cm?
d) If Jim's mother put a packing case on the mattress and it weighed 300N, what would be the **average** compression of the springs if the case lay directly over 30 of them?   (NEA [B] 88)

Figure 9.23

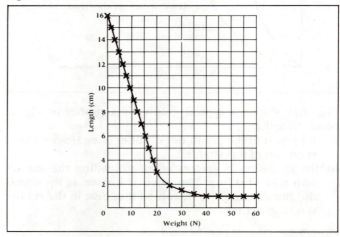

6 The diagrams in fig. 9.24 show how the length of a spring changes when loaded. Use the information given in the diagrams to calculate i) the extension produced by the 200N load, ii) the load that produces a 1 cm extension, iii) the load X, iv) the length of the spring when a 120N load is attached to it.   (W88 part)

Figure 9.24

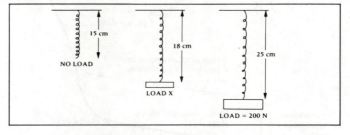

7 a) Sketch an extension–force graph for a rubber band.
   b) Does the rubber band obey Hooke's law? Explain how this is shown on the graph.
   c) On the same axes add graphs for:
      i) a rubber band of half the length;
      ii) two rubber bands of the original length joined together in parallel (side by side).
   d) Explain both of these graphs.

8 a) What is meant by the *density* of a substance?
   b) A spring balance has a maximum reading of 10N and the length of the calibrated scale is 20cm. A rectangular metal block measuring 10cm by 3cm by 2cm is hung on the balance and stretches the spring by 15cm.
      Calculate
      i) the weight of the block;
      ii) the mass of the block;
      iii) the density of the metal from which the block is made.
   (Assume that the acceleration of free fall is $10\,m/s^2$.)   (JMB)

10

Heat and matter

We feel heat when the sun shines and when we sit near a fire. A hot drink or a hot water bottle contains heat. Heat flows along the metal handle of a pan placed on a fire or a hot stove and will burn our hands unless the handle is insulated. Heat causes ice to melt, water to boil and many things to expand.

All these effects of heat have been observed and studied for many centuries, but only in 1842, when J. P. Joule started a series of famous experiments, was the true nature of heat discovered. He converted measured amounts of mechanical energy and electrical energy into heat and found that the amount of heat produced was always in proportion to the amount of energy converted.

Today the unit of energy named after him, the joule, is the universal unit used for all forms of energy including heat energy.

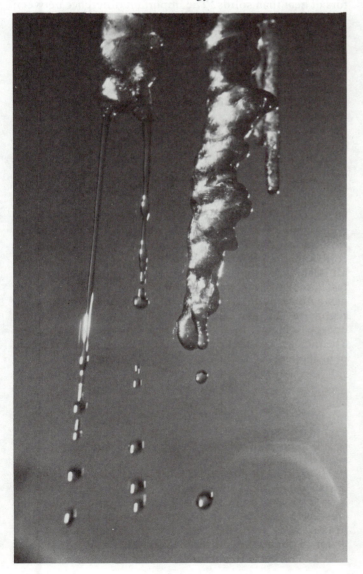

10.1
HEAT ENERGY GETS AROUND

Heat energy has several interesting ways of travelling. Sometimes it uses molecules to help it get around but it can also travel alone as radiant heat.

Conduction of heat

The flow of heat through solids

Rest some rods of various materials across the top of a tripod, fig. 10.1. The rods should be of similar length and thickness to make a fair comparison of the different materials.

● Attach a small nail or matchstick to one end of each rod with vaseline.

● Now heat the other end of all the rods at the same time and watch what happens.

Figure 10.1 Comparing rates of heat conduction of different materials

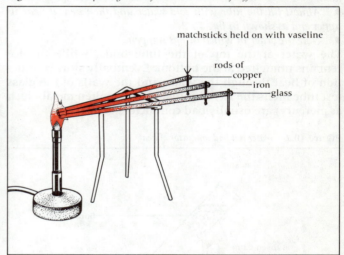

Heat flows along the rods at different rates and the matchsticks drop off after varying lengths of time. Heat flows through the material of the rods without any flow of the materials themselves. The flow of heat through a material without any flow of the material is called **conduction of heat**.

The metal rods in our first experiment conduct heat quite quickly and are good **conductors** of heat. Other solid materials such as glass, wood and plastic conduct heat very slowly and are bad conductors of heat or **insulators**.

We find that all metals are good conductors of heat compared with other materials, silver and copper being particularly good conductors. This suggests that the mechanism of heat conduction in metals might be different from other materials.

All materials conduct some heat by passing energy, fairly slowly, from molecule to molecule. It works something like this. As a material receives heat energy the molecules near the source of heat begin to vibrate more energetically. In other words, the energy which we call heat is present in the material as the kinetic energy of the vibrating molecules. Each vibrating molecule can pass on some of its kinetic energy by 'bumping' into its neighbours and making them vibrate more violently as well.

This passing along of energy from molecule to molecule involves no flow of the molecules themselves and is the basic mechanism of conduction of heat in all materials.

A metal, however, contains *free electrons* which can move independently through the metal material. We shall hear more about these electrons and the role they play in the flow of electric currents in the next two chapters. When a metal is heated these free electrons move faster with more kinetic energy. These fast moving, energy-carrying electrons spread, or diffuse, themselves into the cooler parts of the metal and then transfer their kinetic energy to metal molecules by collisions with them. By this process heat energy is transferred very quickly from hot regions to cooler regions of a metal.

Water is a poor conductor of heat

● *Wedge a piece of ice at the bottom of a test tube so that it cannot float.*
● *Almost fill the tube with cold water and then heat it near its upper end as shown in fig. 10.2.*
● *Note the order in which things happen.*

The water at the top of the tube boils, while the ice remains unmelted at the bottom. Eventually slow conduction of heat through the water and the walls of the glass test tube melts the ice. Liquids (except molten metals such as mercury) are usually bad conductors of heat.

Figure 10.2 *Water is a bad conductor of heat*

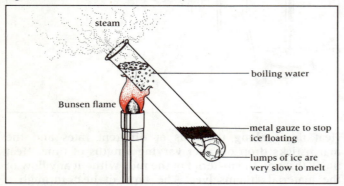

Air is a very poor conductor of heat and is an important insulator

When we feel cold we put on warm clothes. Clothes do not supply heat to our bodies, so how do they make us feel warm? Warm clothes usually contain a lot of trapped air. Pockets of air are held between the fibres of woollen materials and in holes like those in string vests. In a similar way, birds fluff up their feathers in winter to trap more air around them to keep them warm.

Trapped air which cannot be blown away by draughts or carried upwards by convection currents (see below) forms a very good insulator. So wrapping ourselves in clothes keeps us warm, simply by insulating our bodies and preventing the heat produced in our bodies from escaping.

Many materials which trap large amounts of air are used as insulators today. Fibreglass and expanded polystyrene are used as insulators in houses; both these materials owe their good insulating properties to the large amount of trapped air they contain.

A bird can improve its insulation by fluffing up its feathers, so trapping air between them.

Why are liquids and particularly gases poorer conductors of heat than solids? The important difference between the molecules in a solid and those in a liquid or gas is the way they are linked together. If you imagine the molecules in a solid rod as a row of people standing side by side and holding hands, then the heat energy flows along the row as follows. The first person receives the heat energy which makes him shake about. He passes this energy of vibration and movement on to the next person through their linked arms. The second person, now shaking about, passes this kinetic energy on to the next person, again helped by their linked arms, and so on along the row. In time the energy reaches the end of the row.

In a model of a liquid the people are wandering about and although they sometimes hold hands in groups, there is no long chain or continuous linking together of hands, so it is more difficult for them to pass energy on.

In a model of a gas the people are far apart and are running about madly so that they meet only in chance collisions. In a gas there is no linking of the molecules at all. Conduction of heat energy can only happen when a fast-moving energetic molecule from a hot region in the gas collides with a slow-moving molecule from a cooler region in the gas and hands over some of its kinetic energy, which is then carried back into the cooler region. The increase in kinetic energy in the cooler region is a gain of heat energy which raises the temperature of the gas.

Convection

Convection currents in water

● *Fill a beaker with cold water almost to the top. When the water is still, drop in a few small crystals of potassium permanganate near one side, fig. 10.3.*
● *Using a small flame, gently heat the beaker just below the crystals and watch what happens.*

Purple streaks can be seen rising with the water above the crystals (a) which are then carried down the far side of the beaker away from the heat (b). The whole body of the water is circulating in the beaker. This flow of water is called a **convection current**.

Figure 10.3 *Convection currents in water*

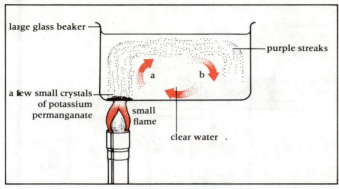

Figure 10.4 *Convection currents in air*

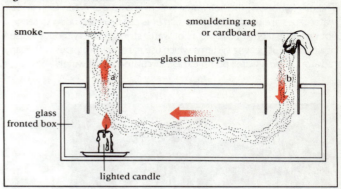

Convection currents in air

The apparatus shown in fig. 10.4 can be used to show convection currents in air.

● *Position a lighted candle under one chimney and then stuff some smouldering rag or cardboard into the top of the other chimney.*

● *Compare what happens when the candle is alight with what happens when it is not.*

Although we cannot see the air moving we can feel a draught of warm air rising out of the candle chimney and we can see smoke being carried down the first chimney (b) and up the candle chimney (a). This flow of air is also called a convection current.

What causes a convection current?

Convection currents occur when a liquid or gas rises above a source of heat. When the water which is at the bottom of the beaker and in close contact with the heat source receives some heat energy it expands. This expansion is caused by the molecules, now moving faster with more kinetic energy, pushing each other further apart. The expanded water is less dense than the surrounding water and so rises above the cooler and denser water around it.

In fig. 10.3 the arrow labelled (a) shows warmer, expanded, less dense water rising and the arrow labelled (b) shows cooler, denser water sinking. The arrows in fig. 10.4 show the same things happening in air.

Convection currents are a flow of liquid or gas caused by a change in density, in which the whole medium moves and carries heat energy with it.

So heat can flow in liquids and gases by means of convection currents.

Note that warm clothes also keep us warm by preventing convection currents carrying warm air away from the surface of our bodies.

Natural convection in the air

Breezes and winds are often caused by one region of air being heated while a neighbouring region remains cool. For example, air rises above large towns and industrial estates, which are hotter than countryside.

At the coast there is often a temperature difference between the land and the sea. The water in the sea hardly changes its temperature between night and day, but the land becomes much hotter than the sea during the day's sunshine and cools down more than the sea during the night. As air is heated by the hot land and rises, it is replaced by sea breezes in the day time. The warmer air rises above the sea at night and draws cooler air off the land forming land breezes.

Gliders and hang-gliders use rising convection currents of warm air known as *thermals* to give them extra height so that they can stay up for longer.

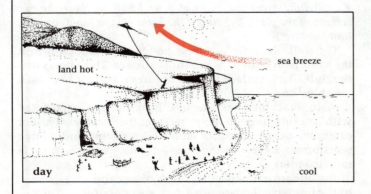

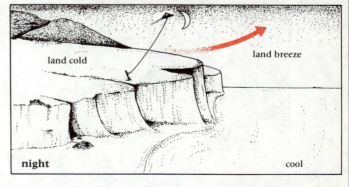

Natural convection in water pipes

The diagram shows the hot water system in a house. When fairly wide water pipes are used, convection currents can be relied upon to carry hot water up from a boiler to a hot water storage tank without the use of a pump. But heating systems which use narrow bore pipes with many radiators need a water pump to help circulate the hot water.

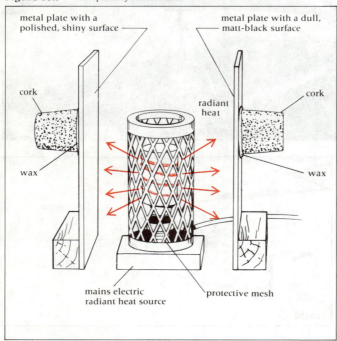

The hot water tank is insulated with a thick layer of air-filled foam material, called lagging.

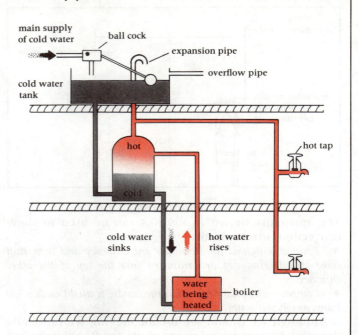

Radiant heat

Absorption of radiant heat

● *Position a radiant heat source midway between two metal plates, as shown in fig. 10.5.*
One plate is painted matt black on the side facing the heat source and the other plate has a polished shiny surface (or a white painted surface) facing the heat source. Each metal plate has a cork fixed with wax to its reverse side.
● *Switch on the electric heater and watch what happens.*

Figure 10.5 *Absorption of radiant heat*

(The red glow of the heater element shows that it is switched on but the red glow is *not* the radiant heat. The red glow is visible light whereas the radiant heat is invisible.)

The cork on the black surface falls off first. This demonstration shows that the black surface absorbs radiant heat more quickly than the shiny or white surface. (The cork behind the shiny surface does eventually fall off.)

You feel cooler if you wear light-coloured, or shiny clothes in the hot summer. These clothes are poor absorbers of radiant heat energy from the sun. If you wear black or dull and dark coloured clothes, or sit inside a dark coloured motor car, the greater absorption of radiant heat energy will make you much hotter.

Emission of radiant heat

● *Mount a thick sheet of copper with one surface painted dull black and the other highly polished in a metal clamp and stand.*
● *Heat it with several bunsen burners to make it very hot.*
● *Remove the burners.*
The whole sheet should be at about the same temperature because copper is a good conductor of heat.
● *Carefully bring the backs of your hands up near the two surfaces of the plate, fig. 10.6. (The backs of your hands are very heat-sensitive.)*

The dull black surface feels hotter even though it is at the same temperature as the shiny surface. This shows that the dull black surface is emitting more radiant heat than the shiny one.

Many machines need to lose heat and are often fitted with cooling fins to help radiate the heat away. For example, a car radiator, a motor bike engine and a large transformer have cooling fins for this purpose. Heat is radiated more quickly if
a) the fins are painted a dull black colour and
b) the fins have the largest surface area possible.

Figure 10.6 *Emission of radiant heat*

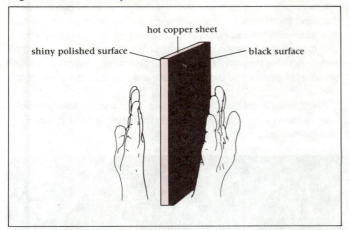

Figure 10.7 *Reflection of radiant heat*

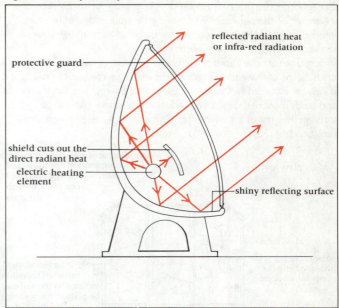

The flow of air between the cooling fins on this Honda bike, together with the emission of radiant heat, prevents this overheating.

A larger surface area also increases heat loss by convection because more air molecules come into contact with the hot surface. The fins are made vertical to allow convection currents to rise.

To get more heating in a room, one method is to increase the size of the radiators and so increase the surface area from which radiant heat can be emitted. In a small room how could you get more heat to radiate from the same size radiator?

Reflecting radiant heat

An electric fire or radiant heater uses a shiny metal reflector behind the hot element, fig. 10.7. We can test whether radiant heat is reflected by the shiny reflector by shielding the element so that direct radiant heat cannot reach us. With a shield in place we can still feel the radiant heat coming from the reflector.

We conclude that the shiny surface is a poor absorber and a good reflector of radiant heat energy.

A fire-fighting suit and a space suit are sometimes covered in a shiny metallic surface. What reasons can you think of for using these materials in each of these cases?

Evidence of the nature of radiant heat

a) Radiant heat is absorbed by all objects and surfaces causing a temperature rise, but dull black or matt surfaces absorb it most quickly.

b) Radiant heat is also radiated by all objects and surfaces causing a temperature fall and dull black surfaces are the best radiators.

c) When someone walks between you and a warm fire the sensation of heating on your skin is immediately stopped at the moment the fire is hidden and just as quickly returns when the person has passed by. This tells us that the radiant heat travels very quickly, in fact almost instantaneously.

d) The radiant heat energy coming from the sun must have travelled through space to reach us, so this form of heat does not need a medium to travel in or molecules to carry the heat energy.

e) Radiant heat energy can also be reflected in the same way that light can, and is often emitted by sources, such as the sun and fires, which also emit light radiation at the same time.

This evidence and much more all leads to the conclusion that radiant heat energy is part of the **electromagnetic spectrum** of radiation. This is a family of many kinds of radiation all of which have certain common properties which link them together. Radiant heat energy is called **infra-red radiation** and belongs next to red light in the family spectrum.

We shall hear more about the electromagnetic spectrum in chapter 23.

The greenhouse effect

Greenhouses are used to help certain plants grow better by providing a warmer air temperature. In summer greenhouses do not need an internal source of heat because they are able to trap enough solar radiation to keep them very warm inside.

Sunshine (or solar radiation) contains radiation of many different kinds. Some of that radiation is the light which we can see, but much of it is invisible infra-red radiation, or radiant heat energy. The sun is very hot and sends most of this infra-red radiation in a form which can easily pass through the glass of a greenhouse. (This is short wavelength infra-red radiation.)

Once inside the greenhouse this infra-red radiation is absorbed by the plants and the soil making them warmer. The warm soil and plants now also emit infra-red radiation, but, since the soil is cool compared with the sun, this radiation is different (it has a much longer wavelength) and cannot pass through the greenhouse glass. In this way solar radiation becomes trapped inside the greenhouse and causes its temperature to rise.

Another application of the greenhouse effect and the absorption of solar radiation is the solar panel, used in some houses to heat water. This is described on p 429.

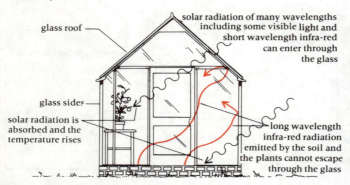

Inside the hot house at Kew Gardens.

The vacuum flask

To keep a drink hot inside a flask, heat losses by all three processes must be reduced to a minimum.

Conduction is totally prevented through the sides of the flask by the vacuum between the double glass walls of the bottle. The cork or plastic stopper contains a lot of trapped air which is a bad conductor of heat.

Convection is also totally prevented by the vacuum and can cause heat loss through the top of the flask only while the stopper is removed.

Infra-red radiation is more difficult to prevent because it can travel through the vacuum between the double glass walls. The radiation loss of heat is greatly reduced by the two silver coatings on the glass walls of the bottle. The outwards-travelling radiation is reduced at the outside silvered surface of the first glass wall because this shiny surface is a bad emitter of infra-red radiation. The inner surface of the second glass wall is also silvered and being a poor absorber and a good reflector of infra-red radiation sends some of it back into the hot drink.

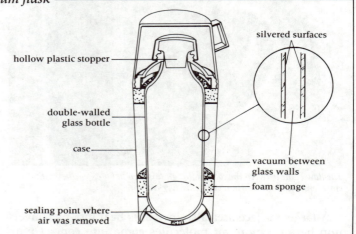

Assignments

Make yourself a convection current detector. Cut a piece of card as shown in fig. 10.8 and bend the segments to form blades. These must all slope the same way. Hang the card on a thread from its centre and hold it over possible sources of convection currents.

Why does it turn?

This is a simple design for a turbine.

Figure 10.8 *A convection windmill*

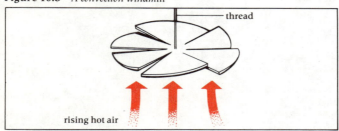

Find out

a) Why can miners in coal mines not use a naked flame in a lamp? How was the risk overcome in the Davy safety lamp?

b) Why does a stone or concrete floor feel very cold to bare feet in winter, but if it is covered with cork tiles (at the same temperature) it feels comfortably warm?

c) What are the base and handles of your cooking pans at home made of? Explain why you think these materials were chosen.

Explain

d) What causes draughts along the floor in your sitting room when the fire is on?

e) Why does a hot-air ballon rise up in the air?

f) Why, when you open the door of a refrigerator, does the cold air come out at the bottom?

g) How do greenhouses get hot in summer?

h) How can a vacuum flask be used to keep cold drinks cold in hot weather?

i) Why is it a good idea to paint houses white in hot countries?

Try questions 10.1 to 10.7

10.2
EXPANSION OF SOLIDS AND LIQUIDS

The increase in size of objects when they get hotter is called expansion. This expansion can be a cause of problems in the construction of machines and buildings, but design engineers have often found ingenious ways both of allowing for the expansion of materials and of making positive use of it.

Demonstrating expansion of a solid

The metal bar in fig. 10.9a will just fit into the gap in the gauge when both the bar and the gauge are cold. Similarly the metal ball will just pass through the ring when both are cold, fig. 10.9b.

● *Heat the bar and the ball over a Bunsen burner flame and then test the fit again.*

Figure 10.9 *Demonstrating expansion of a solid*

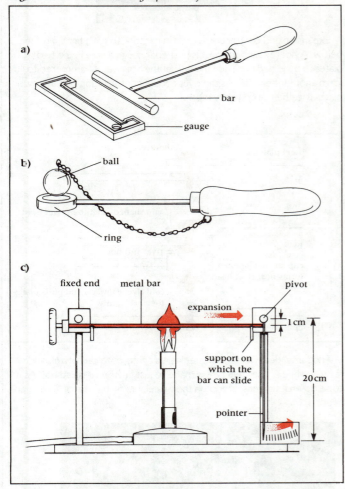

With the rise in temperature of the bar and the ball comes an increase in size: an expansion. The gauge shows that the bar has increased in length, which is called **linear expansion**. The ring shows that the diameter of the ball has increased in all directions. The expansion in area of a solid is known as **superficial expansion** and the expansion in volume is called **cubical expansion**.

We need the gauge and the ring to show that there has been any expansion, because the change in size is so very small. Fig. 10.9c shows one method of making the small

linear expansion of a metal rod easier to see by magnifying the expansion.

A metal bar is mounted so that one end of it is fixed and cannot move; the other end is in contact with a long pivoted pointer. If the length of the pointer is 20 times the distance of the end of the bar from the pivot then the movement of the tip of the pointer will be 20 times greater than the movement of the end of the rod. So when the metal bar is heated and expands, the pointer magnifies the linear expansion 20 times, making it visible.

Testing the force of expansion and contraction

● *To test the force of expansion, fit the tensioning nut and cast-iron bar in the inside positions in the bar-breaking apparatus as shown in fig. 10.10, positions E.*
● *Before heating the steel rod, turn the tensioning nut hand tight so that there is no room for expansion.*
● *Now heat the steel rod with one or more Bunsen burners and watch the cast-iron bar.*
● *To test the force of contraction fit the tensioning nut and cast-iron bar in the outside positions C.*
● *Heat the steel rod first and tighten the tensioning nut while the flames are still on the rod.*
● *Now when there is no room for contraction, remove the heat and again watch the cast-iron bar.*

Figure 10.10 *Both the forces of expansion and contraction are very strong in metals*

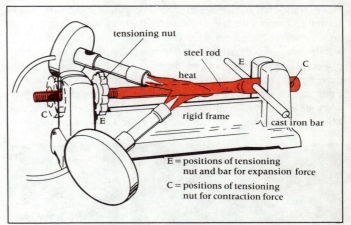

E = positions of tensioning nut and bar for expansion force

C = positions of tensioning nut for contraction force

Molecules provide the forces

Although the cast-iron bar is brittle, it requires a large force to break it. When a solid is heated, its molecules gain extra energy and vibrate more violently and need more room for movement. The molecules try to push their neighbours slightly further away, against their mutual attraction. So heating a solid slightly increases the distance between the molecules and causes expansion in all directions. When a solid has no room to expand, its molecules, in trying to make more space between themselves, produce the force of expansion.

When a hot solid cools down, but is not allowed to shrink, its molecules are held too far apart so that they pull on their neighbours producing a tension in the material. This tension, or force of contraction, is the same force which tries to restore a stretched spring or metal wire to its unstretched state while a stretching force or load is acting on it.

Expansion and contraction, precautions and uses

Although the expansion of a short length of metal is too small to see without magnification, the expansion of a steel or concrete bridge in a hot summer is too great to be ignored. For example, a 100 metre span of steel will increase in length by as much as 5 cm from a cold winter's day to a hot summer's day.

To allow room for expansion in a bridge an expanding joint must be made in the road surface. The concrete or steel span is usually mounted on rollers or rockers, which allow it to move without straining the vertical support pillars.

Railway tracks have been bent and seriously damaged during a very hot day where the gap allowed for expansion was too small. Modern railway track is laid and welded together in lengths of about 1 km. Several precautions are taken to allow for the temperature changes between winter and summer. Concrete sleepers are used which can withstand large forces caused by the temperature changes. At the ends of welded lengths of track a tapered joint is used so that some movement is possible in the track without the jarring caused to trains by a gap. Some lengths of track are heated when they are welded together which leaves the track in tension when it cools down. Then on a hot day, when the track 'expands', all that happens is a reduction of the tension.

This technique is also used in constructing the stainless steel pipes which carry the cooling gases or liquids into the hot core of a nuclear reactor. While the reactor is being built the pipes are welded together under tension, so that when the reactor begins to operate and gets hot in its core, the pipes will not expand and cause damage. The rise in temperature of the pipes only reduces the tension in them.

Pipelines in the chemical industry which carry liquids and gases over long distances must have flexible expansion joints built in them at regular intervals.

Expansion and contraction is used in rivetting to get a tight joint, see below. A hot (expanded) rivet is pushed through a hole in the two plates to be joined (a). Then the end of the hot rivet is hammered to form another head (b). As the rivet cools it contracts and pulls the two plates together more tightly (c).

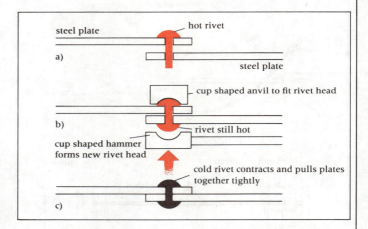

The same idea is used to fit steel tyres tightly onto cartwheels and railway wheels. The steel tyre is made to just fit when it is red hot. As it cools down it tightens its grip on the wheel. No fixing nails or screws are needed.

In this British Rail tyre-heating hearth, the steel tyre is heated by gas burners until it has expanded enough for the wheel to be fitted into it.

The bimetallic strip

A bimetallic strip is made of two strips of different metals, e.g. brass and iron, welded or rivetted together. When cold the double strip is straight, fig. (a). As it is heated the brass expands more than the iron and so the brass forms the outside of a curve and the iron the inside, fig. (b).

Bimetallic strips are used in thermostats and many other mechanical switching circuits, but increasingly these are being replaced by electronic circuits with no moving parts. Can you work out how the bimetallic strip in fig. (c) works as a thermostat? It should switch a heater on when the temperature falls.

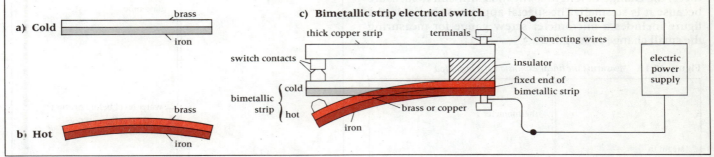

Linear expansivity

When a steel bridge gets hotter it expands. The change in the length of the bridge is called its linear expansion. The engineer who designed the bridge needed to know exactly how steel expands in order to do his calculations to predict the expansion of the bridge. The measure of the tendency of a particular material to expand is called its **expansivity**. We have seen that brass expands more than iron when heated by the same amount (bimetallic strip), so brass has a higher expansivity than iron. The lengthways expansivity of a material is called its **linear expansivity** and is given the symbol α (alpha).

The linear expansion or change in length Δl of a steel bridge depends on three things;
a) the length of the bridge l,
b) the change (or rise) in temperature $\Delta \theta$ and
c) the linear expansivity of steel α.

When the temperature of a length l
of steel rises by $\Delta \theta$, it expands by: Δl

When the temperature of a *unit length*
of steel rises by $\Delta \theta$, it expands by: $\dfrac{\Delta l}{l}$

When the temperature of a unit length
of steel rises by *one degree* it expands by: $\dfrac{\Delta l}{l \times \Delta \theta}$

This is the linear expansivity of steel, α,

$$\alpha = \frac{\Delta l}{l \times \Delta \theta}$$

The linear expansivity of a material is defined by this formula.

Linear expansivity α is the increase in length of a unit length of the material for each degree rise in temperature.

$$\text{linear expansivity} = \frac{\text{linear expansion}}{\text{original length} \times \text{temperature rise}}$$

We can see that the units of linear expansivity will be:

$$\frac{\text{metres}}{\text{metres} \times \text{degrees C}}$$

which simplifies to per degree C or per kelvin (one kelvin (1 K) is one degree Celsius).

Some values of linear expansivities are given in table 10.1. Note how small all the values are. This should not be surprising because these figures give the expansion of only 1 metre of the material for only 1 degree temperature rise. For example $\alpha = 12 \times 10^{-6}$/K for concrete means that the expansion of 1 metre of concrete is only 12×10^{-6} m or 0.000 012 m when the temperature rises by 1 K.

Table 10.1

Material	linear expansivity, $\alpha \Big/ \dfrac{1}{\text{K}}$
concrete	12×10^{-6}
iron	12×10^{-6}
brass	19×10^{-6}
aluminium	26×10^{-6}
glass	9×10^{-6}
nylon	100×10^{-6}

Calculating linear expansion Δl

By rearranging the formula for linear expansivity we obtain a formula which can be used to calculate the expansion of things like bridges and railway lines.

$$\begin{pmatrix} \text{linear} \\ \text{expansion} \\ \text{or change} \\ \text{in length} \\ \text{(of an object)} \end{pmatrix} = \begin{pmatrix} \text{linear} \\ \text{expansivity} \\ \text{(of its} \\ \text{material)} \end{pmatrix} \times \begin{pmatrix} \text{original} \\ \text{length} \end{pmatrix} \times \begin{pmatrix} \text{temperature} \\ \text{rise} \end{pmatrix}$$

$$\Delta l = \alpha \, l \, \Delta \theta$$

Worked Example
Calculating linear expansion

Calculate the linear expansion of a concrete bridge of span 100 m when the temperature rises by 20°C. The linear expansivity of concrete is 1.2×10^{-5}/°C.

Using the formula $\Delta l = \alpha \, l \, \Delta \theta$

$$\Delta l = 1.2 \times 10^{-5} \left(\frac{1}{°C} \right) \times 100\,\text{m} \times 20\,°C$$

$$\therefore \Delta l = 2.4 \times 10^{-2}\,\text{m}$$

Answer: The concrete span will increase in length by 2.4×10^{-2} m or 0.024 m or 2.4 cm.

Measuring linear expansivity

Fig. 10.11 shows some apparatus for measuring the linear expansion of a material in the form of a rod or bar. There are three quantities to be measured, the original length of the rod, its rise in temperature and its change in length. Of these, the change in length is the most difficult to measure because it is so small. The special apparatus shown in the figure includes a micrometer screw gauge for measuring the small change in length Δl.

Figure 10.11   *Measuring the linear expansion of a metal rod*

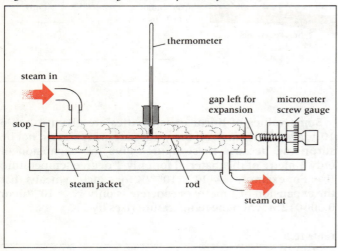

- *Measure the original length l of the rod using a metre ruler.*
- *Fit the rod inside the steam jacket and fit the thermometer in its socket.*
- *Screw up the micrometer so that there is no gap at either end of the rod and take a reading of the micrometer scale x_1.*
- *Note the initial temperature of the rod θ_1.*
- *Unscrew the micrometer to leave room for expansion of the rod and pass steam through the jacket for a few minutes.*
- *Screw up the micrometer again and take a second reading of the micrometer scale x_2.*
- *Note the final temperature of the rod θ_2.*
- *Calculate the change in length $\Delta l = x_2 - x_1$ measured by the micrometer.*
- *Calculate the rise in temperature $\Delta \theta = \theta_2 - \theta_1$.*
- *Check that both l and Δl are in the same units.*
- *Calculate the linear expansivity using*

$$\alpha = \frac{\Delta l}{l \times \Delta \theta}$$

Demonstrating the expansion of liquids

- *Fill a glass flask with coloured water and fit a stopper with a long glass tube so that there is no air in the flask and the water rises a short way up the tube, fig. 10.12. (The glass tube should not extend below the stopper.)*
- *Heat the flask and watch the level of the water in the tube. What happens in the first few seconds of heating? Can you explain this?*
- *To compare the expansion of water with the expansion of other liquids such as alcohol or ether, arrange identical flasks in a water bath so that the different liquids are heated equally and there is less risk of fire with any inflammable liquids.*

We find that in the first few seconds of heating the liquid level drops. Glass is a bad conductor of heat, so at first the

Figure 10.12   *Demonstrating the expansion of water*

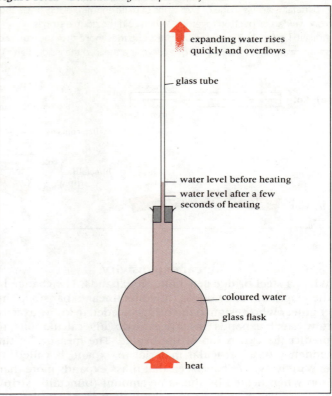

glass flask expands and its volume inside increases. The liquid, which has not started to expand yet, drops to fill the extra volume inside the flask. Once heat reaches the liquid it expands rapidly up the tube and over the top. This shows that the cubical or volume expansion of a liquid is very large. *Liquids expand much more (in volume) than solids do.*

The expansion of the liquid which we see is called its *apparent* expansion. The *real* expansion is actually greater than the observed apparent expansion, because of the expansion of the liquid's container which takes up some of the liquid's expansion.

The expansion of water

Most liquids contract steadily as they cool, and contract further on reaching their freezing point. Water contracts as it cools down from 100°C to 4°C. However, between 4°C and 0°C water behaves unusally in that it expands as it gets colder.

When water freezes its volume increases by about 8%, which is a much larger increase in volume than occurs between 4°C and 0°C. The changes in the volume of water are shown in fig. 10.13a. Fig. 10.13b shows how the density of water changes with temperature. The maximum density occurs at +4°C (minimum volume).

When a pond is freezing over, the densest water at 4°C remains at the bottom of the pond. The less dense (but lower temperature) water, between 3°C and 0°C, floats in layers above it, fig. 10.14. The water on the surface is frozen, but floats because it is less dense than the water below it. The different density layers stop convection currents spreading the heat.

Figure 10.13 *The irregular expansion of water*

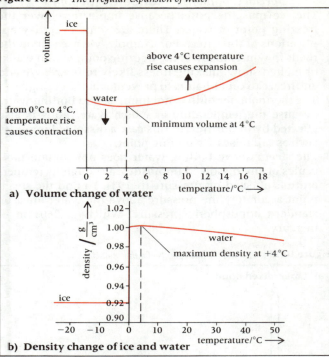

a) **Volume change of water**

b) **Density change of ice and water**

Figure 10.14 *Water temperatures in a frozen pond*

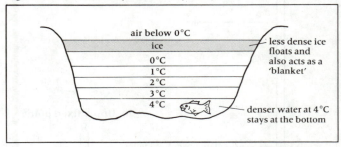

Ice is a bad conductor of heat so that the layer of ice on the top of a pond acts like an insulating blanket and slows down further loss of heat from the water below.

Aquatic animals and plants make use of this phenomenon, by living in the liquid layers when the water freezes over in the winter.

Assignments

Find out

a) What method is used to allow for the expansion of a local bridge near your home or school?

b) What allowance is made for expansion of water in a hot water system of a house (see p. 170).

Explain why

c) A concrete road surface has pitch-filled joints between the slabs of concrete.

d) It is easier to remove a tight metal lid from a glass jar after running hot water over it.

e) A glass bottle cracks when boiling hot water is suddenly poured inside it.

Remember

f) The formula for linear expansion $\Delta l = \alpha l \Delta \theta$

g) The meaning of linear expansivity α.

Try questions 10.8 to 10.12

10.3
HEAT AND TEMPERATURE

Heat is a form of energy which when absorbed by an object makes it hotter and when lost by an object leaves it colder. Words such as hot, warm, tepid, cool and cold tell us about the 'hotness', or temperature, of an object. These words are not very precise, so when we need to be more accurate about the temperature of an object we use scales of temperature with graduations called degrees.

Temperature scales

Temperature is measured with an instrument called a **thermometer**. All thermometers have a scale on them which we read to find temperature, called a temperature scale.

To fix a temperature scale on a thermometer we choose two easily obtainable temperatures, such as the temperatures of boiling and freezing water, and give them numbers. These two temperatures are called the upper and lower **fixed points** of the temperature scale. We then divide the temperature range between the two fixed points into a number of equal parts called **degrees**.

On a centigrade scale there are 100 equal graduations or 100 degrees between the upper and lower fixed points.

The Celsius scale

On the Celsius scale the lower fixed point is the temperature of melting pure ice, known as the **ice point**. The ice point is fixed at zero degrees, written 0°C. The upper fixed point is the temperature of steam just above boiling water, known as the **steam point**. The steam point is fixed at 100 degrees, written 100°C. So, as there are 100 degrees between the two fixed points on the Celsius scale, this is a centigrade scale.

The absolute, or kelvin, scale

Temperatures exist which are much colder than the freezing point of ice, 0°C on the Celsius scale. Experiments suggest that there is a limit to how cold things can get.

At a temperature of −273°C all the heat energy has been removed from any substance. We call this lowest possible temperature **absolute zero**.

A new temperature scale is now used which has the zero of its scale at this absolute zero of temperature. This scale is called the **absolute scale** or **kelvin scale**, after Lord Kelvin who devised it.

One division on the kelvin temperature scale is called a kelvin and is exactly equal to one division or degree on the Celsius scale. It follows that there are 100 kelvins between the ice point and steam point of water.

As the scale divisions on the two scales are equal, 1 kelvin = 1 Celsius degree.

Absolute zero is	0 K	or	−273 °C
The ice point is	273 K	or	0 °C
The steam point is	373 K	or	−100 °C

Note that we write 273 K without a ° (degree) sign. Note also that capital K is the symbol for kelvins and small k is the symbol for kilo (= 1000).

Converting temperatures

If T is the temperature on the kelvin or absolute scale and θ is the temperature on the Celsius scale, then we can see that the relation between T and θ is given by:

$$T/\text{K} = \theta/°\text{C} + 273$$

or more simply

$$T = \theta + 273$$

To convert from a Celsius temperature to a kelvin temperature add 273, and to convert from a kelvin temperature to a Celsius temperature subtract 273.

Temperature changes

We can use either ΔT *on the kelvin scale or* $\Delta \theta$ *on the Celsius scale to mean a change of temperature. Because 1 kelvin division = 1 Celsius degree,* ΔT *always equals* $\Delta \theta$.

For example, on the kelvin scale a rise from 280 K to 300 K is written

$$\Delta T = 300\,\text{K} - 280\,\text{K} = 20\,\text{K}$$

On the Celsius scale the same temperature rise from 7 °C to 27 °C is written

$$\Delta \theta = 27\,°\text{C} - 7\,°\text{C} = 20\,°\text{C}$$

A temperature rise of 20 K = a temperature rise of 20 °C.

Worked example
Converting temperatures between the kelvin and Celsius scales

Convert (a) 37 °C to a temperature on the kelvin scale, (b) 200 K to a temperature on the Celsius scale.

a) Using $T = \theta + 273$ we have

$$T = 37 + 273 = 310\,\text{K}$$

b) Similarly

$$\theta = T - 273 = 200 - 273 = -73\,°\text{C}$$

Answer: 37 °C is equivalent to 310 K on the kelvin scale and 200 K is equivalent to −73 °C (73 degrees below zero) on the Celsius scale.

Finding the fixed points on a mercury thermometer

- *Freeze some pure (distilled) water.*
- *Crush the ice into small, roughly pea-sized pieces and fill a funnel with them.*
- *When the ice begins to melt (and has warmed up to 0 °C) inset the bulb of a thermometer so that it is covered with ice, fig. 10.15a. This should cool all the mercury to 0 °C.*
- *When the mercury stops shrinking, mark the stem of the thermometer at the mercury level.*

This is 0 °C, the lower fixed point, or ice point, on the Celsius scale.

- *Now arrange the thermometer inside a flask so that its bulb is just above the surface of boiling water, fig. 10.15b.*
- *When the mercury stops expanding, mark its level on the thermometer stem.*

This is 100 °C, the upper fixed point, or steam point, on the Celsius scale.

- *If you are making a thermometer, divide the distance between these two fixed points into 100 equal parts, marked as a scale along the stem.*

Special conditions for accurate fixed points

1. The ice must be pure, because impurities lower the freezing point of water. There are some everyday applications of this effect. For example, we use salt on the roads in winter to lower the freezing point of rain water, or melted snow, making it less likely to freeze. We add antifreeze to car radiators to prevent water freezing.
2. The thermometer bulb is not immersed in boiling water because the temperature at which water boils is also affected by impurities. In this case a dissolved impurity such as salt raises the boiling point.
3. The steam above boiling water does not contain molecules of any dissolved impurities, although its temperature is affected by pressure (p 193). To find the steam point accurately the pressure above the water must be standard atmospheric pressure, which is 760 mm of mercury.

Figure 10.15 *Finding the fixed points of a thermometer*

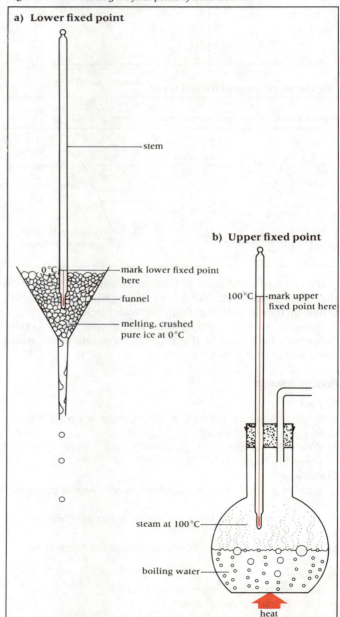

a) **Lower fixed point**

stem

0 °C — mark lower fixed point here

funnel

melting, crushed pure ice at 0 °C

b) **Upper fixed point**

100 °C — mark upper fixed point here

steam at 100 °C

boiling water

heat

Some important temperatures

As well as being familiar with the relation between the Celsius and kelvin scales it is useful to have some idea of where some important temperatures fit on these scales. Fig. (a) shows some everyday temperatures between the fixed points. There are also many important temperatures outside the 0°C to 100°C range. Some of these are shown in fig. (b). Note how the upper part of this diagram has been squashed together (contracted).

a) The fixed points and some everyday temperatures on the Celsius scale

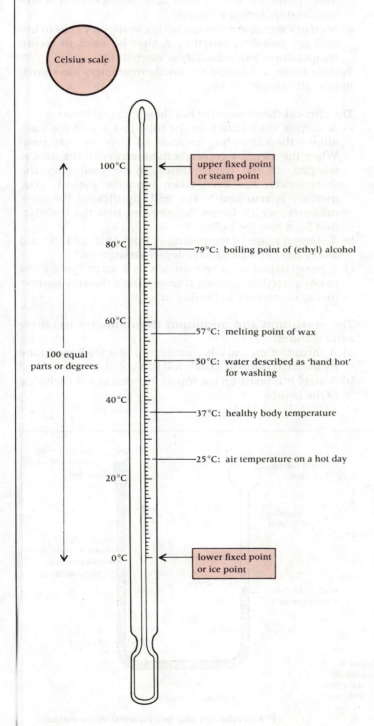

Celsius scale

- 100°C — upper fixed point or steam point
- 79°C: boiling point of (ethyl) alcohol
- 57°C: melting point of wax
- 50°C: water described as 'hand hot' for washing
- 37°C: healthy body temperature
- 25°C: air temperature on a hot day
- 0°C — lower fixed point or ice point

100 equal parts or degrees

b) The absolute, or kelvin, scale of temperature

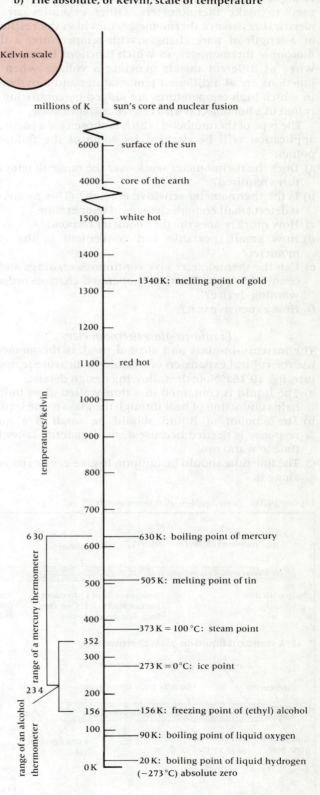

Kelvin scale

- millions of K — sun's core and nuclear fusion
- 6000 — surface of the sun
- 4000 — core of the earth
- 1500 — white hot
- 1340 K: melting point of gold
- 1100 — red hot
- 630 K: boiling point of mercury
- 505 K: melting point of tin
- 373 K = 100°C: steam point
- 273 K = 0°C: ice point
- 156 K: freezing point of (ethyl) alcohol
- 90 K: boiling point of liquid oxygen
- 20 K: boiling point of liquid hydrogen (−273°C) absolute zero

temperatures/kelvin

range of a mercury thermometer

range of an alcohol thermometer

Types of thermometer

There are many different types of thermometer. Any property of a material which changes with temperature can be used to indicate or measure temperature. For example, the expansion of solids, liquids and gases are all used to make thermometers. Other examples are: an electrical resistance thermometer, in which the resistance of a length of wire changes with temperature; a thermocouple thermometer, in which junctions between two wires of different metals generate a voltage when the junctions are at a different temperature; and a pyrometer in which high temperatures are judged by comparing the colour of a hot object with a reference colour scale.

The type of thermometer which is chosen for a particular application will be decided by some of the following points:
a) Does the thermometer work over the range of temperatures required?
b) Is the thermometer sensitive enough? (This means can it detect small enough changes of temperature.)
c) How quickly does the thermometer respond?
d) How small, portable and convenient is the thermometer?
e) Can the thermometer give continuous readings and be connected to an electrically operated chart-recorder or warning device?
f) How expensive is it?

Liquid-in-glass thermometers

The mercury-in-glass and alcohol-in-glass thermometers use the cubical expansion of a liquid to measure temperature, fig. 10.16a. Note the following design details.
a) The liquid is contained in a thin-walled glass bulb to help conduction of heat through the glass to the liquid.
b) The amount of liquid should be small if a quick response is needed because a small quantity takes less time to warm up.
c) The fine tube should be uniform to give even expansion along it.

d) Making the tube finer increases the sensitivity of the thermometer. The same cubical expansion of the liquid will move further along a finer tube allowing up to $\frac{1}{10}$ degree divisions to be used instead of 1 degree divisions.
e) The space above the liquid is evacuated during manufacture to prevent a high pressure of the trapped air when the liquid expands a lot.
f) The maximum ranges of liquid-in-glass thermometers depend on the choice of liquid. Mercury is suitable for most purposes but alcohol can be used to a lower temperature before it freezes.
g) Mercury thermometers are fairly cheap, very easy to use and are portable, but they cannot be used to record temperatures automatically or electrically.

Special forms of liquid-in-glass thermometers are shown in figs. 10.16b and 10.16c.

The **clinical thermometer** has these extra features:
a) A narrow constriction in the tube just above the bulb allows the expanding mercury to force its way past. When the mercury contracts the mercury in the tube is trapped, so the temperature can be read after the thermometer has been taken from the patient. (The mercury is returned to the bulb by flicking the thermometer, which forces the mercury past the constriction back into the bulb.)
b) A limited range of temperatures from 35°C to 42°C and a very fine tube give $\frac{1}{5}$ or $\frac{1}{10}$ degree sensitivity.
c) A pear-shaped cross section acts as a magnifying glass in one direction, making it easier to see the very narrow thread of mercury in the fine tube.

The **maximum and minimum thermometer** has these extra features:
a) A thread of mercury is pushed round a U-shaped tube by the expansion of a bulb full of alcohol.
b) A steel pin floats on the top of the mercury at each side of the U-tube.

Figure 10.16 *Some liquid-in-glass thermometers*

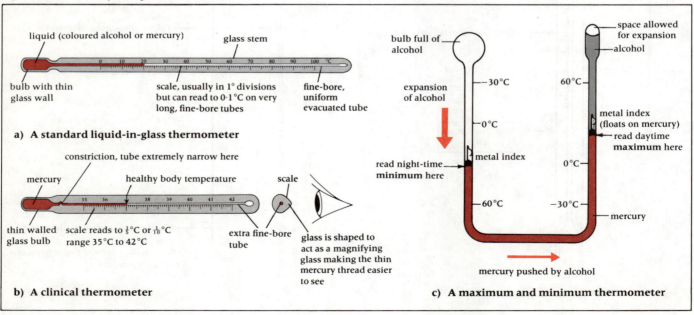

c) As the mercury is pushed (by the expansion of the alcohol) up the right side of the U-tube, it pushes the steel pin with it. When the mercury is pulled back down again (by contraction of the alcohol and surface tension) the pin is left behind, held in the tube by a fine spring.

d) The lower end of the pin (which touched the mercury) gives a reading of the highest, or maximum, temperature reached. As a temperature indicator, the steel pin is often called an index.

e) The mercury rising up the left side of the U-tube (as the alcohol contracts) pushes the other steel pin or index up this side of the tube until the lowest temperature is reached.

f) The lower end of the index on the left side of the U-tube gives a reading of the lowest or minimum temperature reached.

g) The two indices are returned to the mercury surfaces to reset them for the next day by using a magnet.

Thermocouple thermometers

If two different metals are joined in an electric circuit and one wire junction is cold and the other is hot, a small electric current is generated in the circuit, fig. 10.17. The current (or voltage) increases as the temperature difference between the two junctions increases. So if one junction is kept at a fixed cold temperature such as 0°C, then the other junction can be used as a small probe to measure temperatures above 0°C.

Figure 10.17   *A thermocouple thermometer*

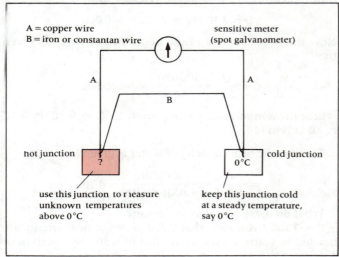

The main advantages of thermocouple thermometers are the following:

a) The wire junction can be very small and needs very little heat to warm it up. This means that it responds very quickly to temperature changes and that it can be used in very small or precise locations.

b) The output of this thermometer is an electrical signal (a current or a voltage), which can be used to operate electrical equipment capable of giving warnings of sudden temperature changes, or of keeping continuous records of temperatures.

c) By choosing particular pairs of metals A and B, temperatures up to about 1 500°C can be measured.

Heat capacity and specific heat capacity

The heat capacity of an object depends on its mass

An electric kettle holds about 1.6 kg of water and a hot water tank may hold about 160 kg of water, that is about 100 times as much, fig. 10.18.

Heating 1.6 kg of water from 20°C to 60°C in the kettle with a 2.5 kW heating element takes about 2 minutes.

Heating 160 kg of water for the same temperature rise in the hot water tank with a 5 kW immersion heater takes about 100 minutes.

Figure 10.18   *Heat capacity depends on the mass of substance heated*

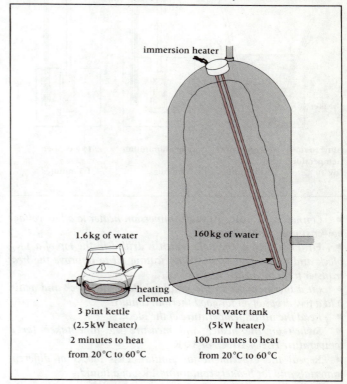

As the 5 kW immersion heater heats twice as quickly as the 2.5 kW kettle, it will take only half as long to produce the same amount of heat. So if the kettle takes 2 minutes to heat 1.6 kg of water, the immersion heater would take only 1 minute to heat 1.6 kg. The immersion heater takes 100 minutes to heat 160 kg of water so we can see that the amount of heat needed is directly proportional to the mass of water.

These figures agree with our experience of waiting for kettles to boil and hot water tanks to heat water for baths. We know that it takes much longer to heat up the water in the tank than it does to boil a kettle. This is because the water in the tank has a much larger heat capacity. In other words, it requires much more heat energy to raise its temperature.

The first factor which affects the heat capacity of an object is its *mass*. The mass of water in the tank requires 100 times more heat because its mass is 100 times greater than the water in the kettle.

The heat capacity of an object is directly proportional to its mass.

The heat capacity of an object depends on its material

We can compare the heat required by different materials by heating equal masses of them with the same electric heater. We use 1 kg samples in the form of cylindrical blocks, fig. 10.19.

Figure 10.19   *Different substances have different heat capacities*

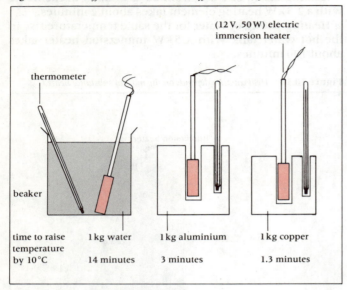

time to raise temperature by 10 °C

1 kg water	1 kg aluminium	1 kg copper
14 minutes	3 minutes	1.3 minutes

- *Connect a 12 volt, 50 watt immersion heater to a low voltage power supply set at 12 volts.*
- *Fit the heater in the hole which is drilled in the top of a 1 kg block and add a few drops of lubricating oil to improve the heat transfer from the heater to the block.*
- *Fit a thermometer in the smaller hole in the block and again add a few drops of oil for good thermal contact.*
- *Read the initial temperature of the block.*
- *Switch on the heater and measure the time taken for a temperature rise of 10 °C, or 10 K*
- *Repeat the experiment for similar 1 kg blocks of different materials and for beakers containing 1 kg of a liquid.*

We find that different materials take different times to warm up by 10 K, which shows that they need different amounts of heat and therefore have different heat capacities. Aluminium needs more than twice as much heat as copper, and water needs about 11 times as much heat as copper, to produce the same temperature rise in the same mass of substance.

The heat capacity of an object depends on what it is made of.

Heat capacity and specific heat capacity

Heat capacity refers to a whole object and we define it as follows:

The heat capacity C of an object is the heat energy needed to raise its temperature by 1 kelvin (1 degree).

Specific heat capacity refers to 1 kg of a substance. We use the word *specific* in physics to mean *per unit mass* which in SI units is *per kilogram*, or *for each kilogram*.

The specific heat capacity c of a substance is the heat energy needed to raise the temperature of 1 kg of the substance by 1 kelvin (1 degree).

We use the symbol Q to stand for a quantity of heat energy to distinguish it from mechanical work W. Both are measured in joules, however.

These definitions can also be written as formulas:

$$\frac{\text{heat capacity}}{(\text{of an object})} = \frac{\text{heat energy}}{\text{temperature rise}}$$

or

$$C = \frac{Q}{\Delta T}$$

The units of heat capacity C are therefore

$$\frac{\text{joule}}{\text{kelvin}} \quad \text{or} \quad \text{J/K}.$$

$$\frac{\text{specific heat capacity}}{(\text{of a substance})} = \frac{\text{heat energy}}{\text{mass} \times \text{temperature rise}}$$

or

$$c = \frac{Q}{m\,\Delta T}$$

The units of specific heat capacity c are

$$\frac{\text{joule}}{\text{kilogram} \times \text{kelvin}} \quad \text{or} \quad \text{J/(kg K)}.$$

We can use the water in the electric kettle, fig. 10.18, to illustrate these two ideas:

The heat energy supplied by the kettle is given by:

$$\text{energy} = \text{power} \times \text{time}$$

where the power of the kettle is 2.5 kW = 2 500 W, and the time of heating is 2 minutes = 120 seconds.

$$\therefore Q = 2\,500\,\text{W} \times 120\,\text{s} = 300\,000\,\text{J}$$

Now the heat capacity of the kettle full of water is given by:

$$C = \frac{Q}{\Delta T} = \frac{300\,000\,\text{J}}{40\,\text{K}} = 7\,500\,\text{J/K}$$

where the temperature rise ΔT, from 20 °C to 60 °C, is 40 °C or 40 kelvin (40 K).

The specific heat capacity of water is given by:

$$c = \frac{Q}{m\,\Delta T} = \frac{300\,000\,\text{J}}{1.6\,\text{kg} \times 40\,\text{K}} = 4\,700\,\frac{\text{J}}{\text{kg K}}$$

What do these two answers mean?

$C = 7\,500$ J/K means that 7 500 joules of heat energy are needed to warm up the kettle full of water by 1 kelvin or 1 degree Celsius.

$c = 4\,700$ J/(kg K) means that 4 700 joules of heat energy are needed to warm up each 1 kg of water by 1 kelvin.

Note that the correct value for water is 4 200 J/(kg K), but we would expect a high value from the kettle data because a lot of heat is lost to the metal kettle and surrounding air during heating.

Table 10.2 gives some useful values of specific heat capacities of substances.

Table 10.2

Substance	Specific heat capacity $c \Big/ \dfrac{\text{J}}{\text{kg K}}$
water	4200
aluminium (alloy)	880
copper	380
ice	2100
nylon	1700
glass	670
lead	126
marble	880

The relation between heat capacity and specific heat capacity

If we multiply the formula for specific heat capacity c by mass m we get:

$$mc = m \times \frac{Q}{m\Delta T} = \frac{Q}{\Delta T} = C$$

So we have:

$$C = mc$$

This confirms that the heat capacity C of an object depends on the mass m of the object and the substance it is made of.

Calculating the heat energy needed to warm things up

By rearranging the formula for specific heat capacity we obtain a formula for the heat energy needed Q:

$$Q = cm\,\Delta T$$

We can see that the quantity of heat energy needed depends on the substance, the mass and the temperature rise. The following example shows how this formula is used.

Worked Example
Calculating heat energy

A hot water tank contains 160 kg of cold water at 20°C as shown in fig. 10.18. Calculate (a) the quantity of heat energy required to raise the temperature of the water to 60°C and (b) the time this will take using a 5 kW electric immersion heater.

The temperature rise

$$\Delta T = 60\,°C - 20\,°C = 40\,°C \text{ or } 40\,K$$

the specific heat capacity of water

$$c = 4200\,\text{J/(kg K)}$$

Using $Q = cm\,\Delta T$

we have

$$Q = 4200\,\frac{\text{J}}{\text{kg K}} \times 160\,\text{kg} \times 40\,\text{K} = 26\,880\,000\,\text{J or } 27\,\text{MJ}$$

Using power = energy/time and rearranging we have:

$$\text{time} = \frac{\text{energy}}{\text{power}} = \frac{27\,\text{MJ}}{5\,\text{kW}} = \frac{27 \times 10^6\,\text{J}}{5 \times 10^3\,\text{W}} = 5.4 \times 10^3\,\text{s}$$

Answer: (a) The energy required is 27 megajoules and (b) the time taken for heating is 5400 seconds (90 minutes). (Allowing for the extra heat, and therefore extra time, needed for heating the copper tank and adjoining pipes and also for some heat loss to the surrounding air, it would be reasonable if the 90 minutes calculated here increased to about 100 minutes in practice.)

Measuring the specific heat capacity of a substance

The specific heat capacity of a solid (in the form of a block) or a liquid can be measured by using the apparatus shown in fig. 10.19.

- *Connect a 12 volt immersion heater to a 12 volt d.c. power supply as shown in the circuits of fig. 10.20 (see also p 249).*
- *Find the mass m of the substance by weighing.*
- *Wrap the metal block or beaker in heat insulating material, called lagging, and cover the beaker with a lid.*
- *Record the initial temperature θ_1 before switching on the heater.*
- *For circuit (a) record the initial reading on the joulemeter and then switch on the heater.*
- *For circuit (b) start the clock as you switch on the heater and record the current reading I (in amperes) and the voltage reading V (in volts).*
- *When the temperature has risen by about 10°C, switch off the heater. (If the temperature rise is too large then too much heat will be lost.)*
- *For circuit (a) record the final reading on the joulemeter.*
- *For circuit (b) stop the clock and read it.*
- *Note the highest temperature reached θ_2 after switching off the heater. (Stir a liquid before reading θ_2.)*
- *Calculate the temperature rise $\Delta T = \Delta\theta = \theta_2 - \theta_1$ and the energy supplied as heat Q.*

In circuit (a) Q is found from the difference between the two joulemeter readings (in joules).

In circuit (b) use $Q = ItV$ where t is the time in seconds (p 251).

The specific heat capacity can be calculated using:

$$c = \frac{Q}{m\Delta T}$$

Figure 10.20 *Circuits for measuring the energy supplied to an electric heater*

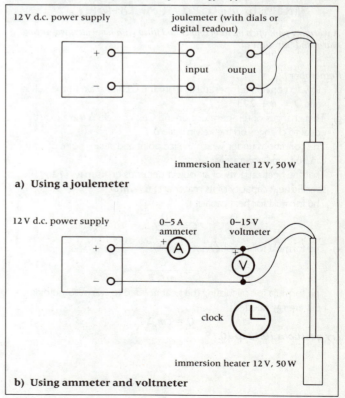

a) **Using a joulemeter**

b) **Using ammeter and voltmeter**

Assignments

Explain

a) Why should the glass wall of a thermometer bulb be thin?

b) Why should the volume of liquid in a thermometer bulb be small?

c) Why is water a good substance to use for cooling a car engine?

d) Why are bottles of hot water good for heating beds.

e) How does the very large heat capacity of the water in the sea affect the temperature of (a) the sea between day and night and (b) the land near the sea in summer and winter?

A warming pan, often made of copper and filled with hot water, was used to warm and 'air' the bed.

Remember

f) How to convert temperatures from the Celsius scale to the kelvin scale, $T = \theta + 273$

g) That changes of temperature on the Celsius scale $\Delta \theta$ are exactly equal to changes on the kelvin scale ΔT

h) The conditions under which the ice point and steam point of water can be found accurately.

i) That the heat capacity of an object depends on its mass m and the specific heat capacity of its material c: $C = mc$

j) The formula for heat capacity:

$$C = \frac{Q}{\Delta T}$$

k) The formula for specific heat capacity:

$$c = \frac{Q}{m \Delta T}$$

l) The formula for calculating the heat needed to warm something up: heat energy needed:

$$Q = cm \Delta T$$

Try questions 10.13 to 10.19

10.4
LATENT HEAT

An iceberg can survive many weeks floating on the sea before it all finally melts. It takes much longer to turn a kettle full of boiling water into steam than it does to bring cold water to the boil in a kettle. These examples show that a lot of heat energy is needed to change ice into water and water into steam. This heat energy which changes the state of a substance is called latent heat.

Ice remains in the glacier while flowers bloom in the sunshine.

Change of state

Latent heat means hidden heat. The heat which changes ice into water is *hidden* in the sense that when the ice melts it is no hotter than before it received the heat. The latent heat turns ice at 0°C into water at 0°C. So latent heat changes the state of an object without causing any rise in temperature. Fig. 10.21a shows the changes of state which are possible for water.

Figure 10.21 *Changes of state*

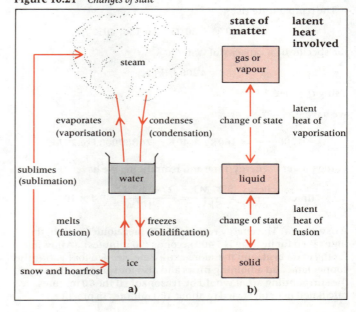

When ice *melts* to form water the change of state is called **fusion**.

When water *evaporates* to form water vapour or steam the change of state is called **vaporisation**.

When steam *condenses* to form water the change of state is called **condensation**.

When water *freezes* to form ice the change of state is called **solidification**.

It is also possible for a change to occur directly from water vapour to ice, or from ice to water vapour. This change of state which misses out the liquid state is called **sublimation**. White hoarfrost, which forms on plants and on the ground, is water vapour from the air turned directly into ice crystals without becoming drops of dew first. Sometimes the hoarfrost can turn back into water vapour in the air without melting first. Sublimation is a two-way change of state.

Frost appears on a window when water vapour in the air turns into these beautiful ice crystals by losing its latent heat to the cold window.

Latent heat is involved with each change of state, fig. 10.21b.

Latent heat of fusion given to a solid melts it; latent heat of fusion removed from a liquid freezes it.

Latent heat of vaporisation given to a liquid evaporates it; latent heat of vaporisation removed from a vapour (or gas) condenses it.

Evaporation requires heat energy and causes cooling

● *Set up a small beaker containing some ether as shown in fig. 10.22a.*

● *Draw a stream of air through the ether, using a pump to increase the rate of evaporation. (Do not suck the ether vapour by mouth.)*

● *Watch the glass beaker and feel it from time to time.*

Ether is very volatile, i.e. it evaporates rapidly. Water vapour condenses from the air on the outside of the beaker showing how cold the glass has become. Eventually the water freezes on the sides of the beaker and between the beaker and the sheet of wood.

● *Try lifting the beaker.*

Figure 10.22 *Evaporation causes cooling*

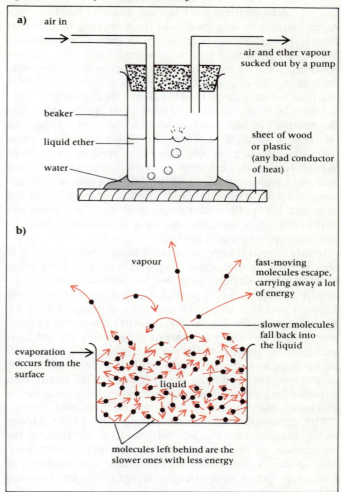

This experiment shows that as ether evaporates it needs heat energy for its latent heat of vaporisation. This heat is removed from the remaining liquid ether and from everything around it, which causes cooling. If the surrounding water is cooled to 0°C then further removal of heat takes away its latent heat of fusion and freezes it.

In evaporation the faster molecules escape

Fig. 10.22b shows how evaporation occurs at the surface of a liquid. Molecules reaching the surface may either escape or fall back into the liquid. Only the faster molecules actually get away, because only they have enough energy to escape from the attraction of the liquid molecules, which pulls slower ones back into the liquid.

The general effect of the faster molecules being able to escape and the slower molecules remaining in the liquid is the lowering of the average energy of the molecules in the liquid. A liquid with slower molecules with less kinetic energy is a cooler liquid. So evaporating molecules leave a cooler liquid behind.

When heat energy is given to a liquid to help it evaporate, more molecules in the liquid are given enough kinetic energy to escape. So the latent heat of vaporisation of a liquid is the extra energy required by the molecules of the liquid to enable them to escape from the liquid surface.

The refrigerator
A heat pump which uses latent heat

A refrigerator makes food cold by removing heat energy from it. Heat is a form of energy, so it cannot be destroyed, but it can be transferred or pumped from one place to another. So the basic idea is to pump heat energy from food inside the refrigerator and give it to the air outside. A machine which pumps heat is called a **heat pump**.

Once the food and air inside a refrigerator is colder than the outside air, heat tends to conduct back in through the refrigerator cabinet to warm the inside up again. (Heat naturally flows from hot places to cold places.) Even though refrigerator cabinets are well insulated, some heat is always leaking back in again. For this reason refrigerators keep switching on from time to time to pump the heat out again. Electric refrigerators have a compressor pump, which is switched on by a thermostat when the inside temperature gets too high.

Inside a refrigerator an evaporating liquid is used to remove heat in the same way as the evaporating ether (p 185) removed heat from its surroundings. A volatile liquid called *Freon* is circulated inside a refrigerator through a closed circuit of pipes by a pump, fig. 10.23. Evaporation of the liquid Freon occurs inside the evaporator, which has several loops of pipe inside the refrigerator cabinet, usually inside the freezing compartment. The pump aids evaporation by pumping vapour out of the evaporator and reducing the pressure. The latent heat required for the Freon to evaporate is removed from the air and food inside the refrigerator making them cold.

The compressor pump compresses the Freon vapour inside the condenser pipes, which are at the back of the refrigerator on the outside and surrounded by air. Pressurising the vapour helps it to condense and causes it to give up its latent heat of vaporisation to the surrounding air.

The flow of liquid Freon back into the evaporator is controlled by a valve. Increasing the flow will allow heat to be removed more quickly and so make the refrigerator colder.

Latent heat and specific latent heat of fusion

The latent heat of fusion of an iceberg is the heat energy needed to melt *all* of it. None of this heat causes any temperature rise of the water formed from the ice.

The latent heat of fusion L of a solid object is the heat energy required to change it from solid to liquid without any temperature change.

Again using specific to mean *per unit mass* or *for each kilogram of mass*, the specific latent heat of fusion of the ice in an iceberg is the heat energy needed to melt each kilogram mass of it without raising its temperature.

The specific latent heat of fusion l of a solid substance is the heat energy required to change 1 kg of it from solid to liquid without any temperature change.

Writing these definitions as formulas we have:

$$\text{latent heat of fusion (of an object)} = \text{heat energy needed to melt all of it}$$

or:

$$L = Q$$

where L and Q are both quantities of energy measured in joules.

$$\text{specific latent heat of fusion of a substance} = \frac{\text{heat energy}}{\text{mass}}$$

or:

$$l = \frac{Q}{m}$$

and the units of l are $\dfrac{\text{joule}}{\text{kilogram}}$ or J/kg.

Latent heat and specific latent heat of vaporisation

The latent heat of vaporisation of a kettle full of boiling water is the heat energy needed to turn *all* the water at 100°C into steam at 100°C. Similarly the *specific* latent heat of vaporisation of water is the heat energy needed to turn 1 kg of boiling water into steam without any temperature rise.

Experiments show that the specific latent heat of fusion of a substance and its specific latent heat of vaporisation are quite different amounts of heat energy. For example, it takes 340 000 J to melt 1 kg of ice and a much larger amount of heat energy 2 300 000 J, to convert 1 kg of boiling water into steam.

The definitions and formulas for vaporisation are very similar to those for fusion. In symbols, the formulas are identical.

The specific latent heat of vaporisation l of a liquid substance is the heat energy required to change 1 kg of it from liquid to vapour without any temperature change.

Figure 10.23 *A refrigerator*

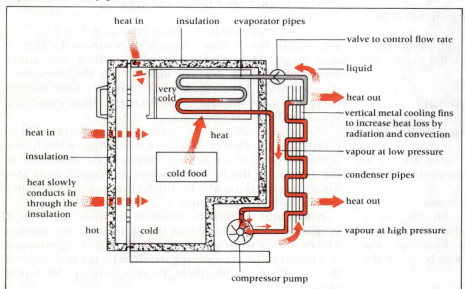

Calculating the heat energy required to melt or evaporate things

By rearranging the formula for specific latent heat, we obtain a formula for the heat energy Q:

$$Q = lm$$

Now we have two formulas for heat energy. Note how they are used:

$Q = cm\Delta T$: this heat energy causes a change of *temperature* ΔT,

$Q = lm$: this heat energy causes a change of *state*, but no temperature change.

Worked Example
Calculating latent heat of fusion

An ice lolly has a mass of 100 g. If the specific latent heat of fusion of ice is 340 000 J/kg, calculate the heat energy needed to melt the ice lolly. (Assume the ice is at 0°C and no temperature rise occurs.)

Using the formula for a change of state:

$$Q = lm$$

and $m = 100\,g = 0.1\,kg$ we have

$$Q = 340\,000\frac{J}{kg} \times 0.1\,kg = 34\,000\,J$$

Answer: 34 kilojoules of heat energy are needed to melt the lolly.

Worked Example
Calculating latent heat of vaporisation

The kettle in fig. 10.18, which contains 1.6 kg of water, is left switched on. After starting to boil, how much heat energy will be used in turning all the water to steam, and how long will it take for the 2.5 kW kettle to boil dry? The specific latent heat of vaporisation of water is 2.3 × 10⁶ J/kg.

Using the formula for a change of state: $Q = lm$ we have

$$Q = 2.3 \times 10^6 \frac{J}{kg} \times 1.6\,kg = 3.7 \times 10^6\,J$$

Using the relation:

$$\text{power} = \frac{\text{energy}}{\text{time}} \quad \text{or} \quad P = \frac{Q}{t}$$

and rearranging we have

$$t = \frac{Q}{P} = \frac{3.7 \times 10^6\,J}{2.5 \times 10^3\,W} = 1.5 \times 10^3\,s$$

Answer: The heat energy needed is 3.7 megajoules and it will take 1500 seconds or 25 minutes for the kettle to boil dry.

Worked Example
Heat capacity and latent heat

Calculate the heat required to convert 5 kg of ice at −20°C into steam at 100°C. The specific heat capacities of water and ice are, respectively, 4 200 and 2 100 J/(kg K); the specific latent heat of fusion of ice is 340 000 J/kg, and the specific latent heat of vaporisation of water is 2.3 × 10⁶ J/kg.

Fig. 10.24 shows the four steps to this question. Four separate quantities of heat energy are needed for the four different changes that occur.

a) Warming up the ice. From −20°C to 0°C is a temperature rise $\Delta T = 20\,K$. The heat needed for this temperature rise is

$$Q = cm\Delta T$$

where c is the specific heat capacity of ice, giving

$$Q = 2\,100\frac{J}{kg\,K} \times 5\,kg \times 20\,K = 210\,000\,J$$

b) Melting the ice. The heat needed for this change of state is

$$Q = lm$$

where l is the specific latent heat of fusion of ice, giving

$$Q = 340\,000\frac{J}{kg} \times 5\,kg = 1\,700\,000\,J$$

c) Warming the water. From 0°C to 100°C is a temperature rise $\Delta T = 100\,K$. The heat needed for this temperature rise is

$$Q = cm\Delta T$$

where c is the specific heat capacity of water, giving

$$Q = 4\,200\frac{J}{kg\,K} \times 5\,kg \times 100\,K = 2\,100\,000\,J$$

d) Evaporating the water. The heat needed for this change of state is

$$Q = lm$$

where l is the specific latent heat of vaporisation of water, giving

$$Q = 2.3 \times 10^6 \frac{J}{kg} \times 5\,kg = 11.5 \times 10^6\,J$$

The total heat energy needed is the sum of all four amounts of heat energy:

a)	210 000 J
b)	1 700 000 J
c)	2 100 000 J
d)	11 500 000 J
	15 510 000 J

Answer: The total heat energy required is about 16 megajoules.
It is interesting to note that most of the heat energy is used for the change of state from water to steam. This is the reason why it is so expensive to obtain fresh water from salt water by evaporation.

Figure 10.24

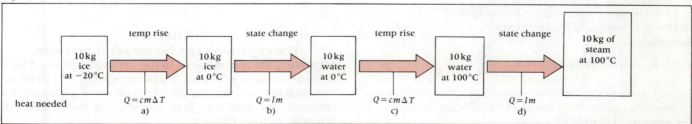

Measuring the specific latent heat of vaporisation of water

An estimate of the specific latent heat of vaporisation of water (also called the specific latent heat of steam) can be made using the apparatus in fig. 10.25. The energy measuring circuits are the same as those in fig. 10.20.

- *Connect the electric immersion heater to its power supply and measuring circuit.*
- *Nearly fill a beaker with hot water, cover it with a lid and weigh it, m_1. Fit the beaker into a lagging jacket.*
- *Insert the heater through the hole in the lid and heat the water until it just begins to boil. Remove the lid to allow steam to escape.*
- *At the same moment read the joulemeter (circuit a); or start the clock and read the ammeter and voltmeter (circuit b).*
- *After 15 minutes switch off the heater and replace the lid.*
- *Read the joulemeter again (circuit a) or stop the clock and record the time of heating (circuit b).*
- *Remove the heater and lagging.*
- *Weigh the beaker again with its lid, m_2.*

The following specimen results show how to calculate the specific latent heat of steam (using circuit b).

m_1	= initial mass of water (+beaker + lid)	312 g = 0.312 kg
m_2	= final mass of water (+beaker +lid)	294 g = 0.294 kg
m	= mass of water converted into steam	$m_2 - m_1$ = 0.018 kg
t	= time of heating	15 minutes = 900 s
I	= current reading from ammeter	= 4.0 A
V	= voltage reading from voltmeter	= 12.0 V

The heat energy supplied by the heater is given by:

$$Q = ItV = 4.0\,\text{A} \times 900\,\text{s} \times 12.0\,\text{V} = 43\,200\,\text{J}$$

(If a joulemeter is used Q is the difference between the two meter readings.)

The specific latent heat of steam is given by

$$l = \frac{Q}{m} = \frac{43\,200\,\text{J}}{0.018\,\text{kg}} = 2\,400\,000\,\frac{\text{J}}{\text{kg}} = 2.4\,\text{MJ/kg}$$

Figure 10.25 *Measuring the specific latent heat of steam*

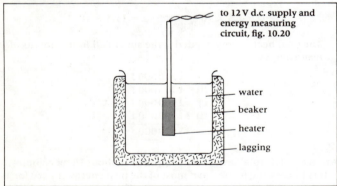

to 12 V d.c. supply and energy measuring circuit, fig. 10.20

water
beaker
heater
lagging

Assignments

Explain

a) How does a refrigerator keep food cold?

b) Why does wearing wet clothes make you feel cold?

c) How does sweating help keep you cool?

Remember

d) The formula for specific latent heat: $l = \dfrac{Q}{m}$

e) The formula for calculating the heat needed to change the state of something (without any temperature rise): $Q = lm$

Try questions 10.20 to 10.23

THE GAS LAWS

When a gas is heated, its molecules gain extra kinetic energy and move about at greater speeds. We detect this change as a rise in temperature of the gas. The increase in energy of the gas molecules may cause both the volume and the pressure of the gas to increase. The gas laws describe the experimental evidence about the ways in which the temperature, pressure and volume of a gas are related to each other.

A three-dimensional model of a gas

Small ball bearings in violent motion inside a perspex tube can be used to represent gas molecules in random motion, fig. 10.26. A piston striking a flexible rubber sheet at the base of the tube feeds energy to the ball bearings. The speed of this piston is controlled by the speed of the electric motor which drives it.

Half cover the base of the tube with small ball bearings.

Fit a freely sliding plug made of expanded polystyrene inside the tube and cover with the loose lid.

Switch on the motor so that the plug rises up the tube and is held there by the impacts of the ball bearings, fig. 10.26a.

A model of a gas expanding when the temperature rises

By adjusting the voltage of the power supply, gradually increase the speed of the motor and watch the gas model. Figs. (a), (b) and (c) show how the plug P rises higher up the tube as the piston speed increases. Since the weight of plug P is constant, its pressure on the 'gas' and the 'gas pressure' supporting it are also constant.

The increasing piston speed gives more energy to the 'molecules' and they move faster. The 'gas' is therefore at a higher temperature. We can see that *increasing the temperature of the 'gas' at constant pressure makes it expand*, as is shown by plug P rising up the tube.

Increasing the pressure of a model gas at constant temperature

With the motor running at its highest speed, representing a constant temperature, fig. 10.26c, note the height of plug P. Remove the lid and place a cardboard disc D (of similar weight to plug P) on the top of plug P. Watch what happens to the 'gas'.

Adding the cardboard disc D doubles the weight of P and represents a doubling of the gas pressure. We find that the volume of the gas is reduced by about half, fig. 10.26d

Add two more cardboard discs to double the 'gas pressure' again, fig. 10.26e.

Increasing the pressure at constant temperature reduces the volume of the 'gas'. This is an inverse relation.

Charles' law
The expansion law for constant pressure

Increasing the piston speed in the gas model showed how a temperature rise causes a gas to expand. The relation between the volume and the temperature of a gas under a constant pressure was first investigated by the French scientist Jacques Charles.

Figure 10.26 *A three-dimensional gas model*

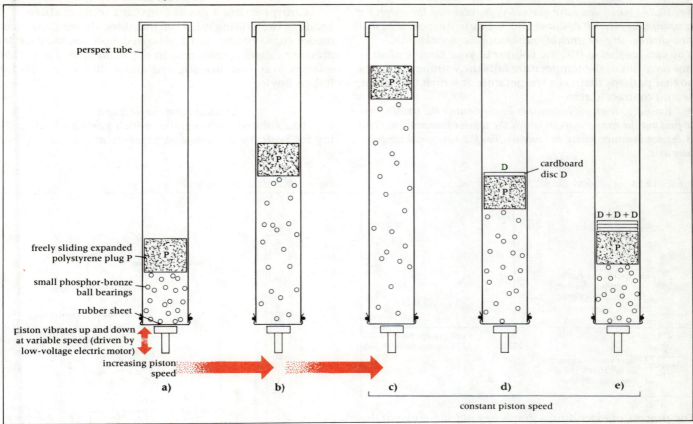

perspex tube

P

freely sliding expanded
polystyrene plug P

small phosphor-bronze
ball bearings

rubber sheet

piston vibrates up and down
at variable speed (driven by
low-voltage electric motor)

increasing piston
speed

cardboard
disc D

D

D + D + D

a)   b)   c)   d)   e)

constant piston speed

Investigating Charles' law

Charles' law, like the other gas laws, applies to a *fixed mass
of gas*. The relations between the temperature, pressure
and volume of a gas describe the behaviour of a constant
number of gas molecules. When the gas is heated or
compressed no molecules must be added or allowed to
escape.

We can investigate Charles' law using a fixed mass of air
which is trapped in a capillary tube by a drop of acid (see
appendix D6).

● *Fit the capillary tube on a 30 cm ruler using rubber bands so
that its sealed end lines up with the zero on the scale, fig. 10.27.*

● *Assemble the rest of the apparatus as shown in fig. 10.27.*

● *Heat the water bath and read the length of the air column at
different temperatures. The heating should be done slowly and the
water stirred so that the temperature of the air will be the same as
the thermometer reading.*

● *Record your readings in a table.*

Temperature/°C
Length of air column/cm (a measure of gas volume)

Figure 10.27 *Investigating Charles' law*

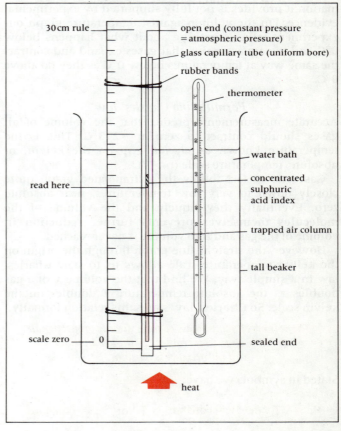

30 cm rule

open end (constant pressure
= atmospheric pressure)

glass capillary tube (uniform bore)

rubber bands

thermometer

water bath

read here

concentrated
sulphuric
acid index

trapped air column

tall beaker

scale zero

0

sealed end

heat

• *Plot a graph of volume against temperature as shown in fig. 10.28. When you have marked your results on the graph for temperatures between 0°C and 100°C they will appear to be in a straight line. Draw a straight line through the plotted points.*

You can see that at 0°C the volume of your sample of air is not zero, but as the temperature falls the volume decreases so that perhaps there is a temperature at which its volume would contract to zero.

• *Extend your graph backwards by continuing the straight line to find out the temperature at which the volume becomes zero. This is the temperature where the line touches the horizontal temperature axis.*

Figure 10.28 *A graph for Charles' law*

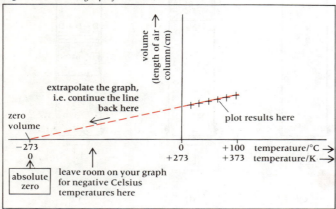

The process of extending a graph beyond the range of the experimental results is called **extrapolation**. This process should always be used with caution because the information it provides is not fully supported by experimental evidence. On the volume against temperature graph our experiment tells us nothing about what happens below 0°C and we are assuming that gases expand and contract the same way at temperatures below 0°C as they do above 0°C.

Formulating Charles' law

Accurate measurements predict that the volume of all gases should contract to zero at −273°C. This is the temperature known as *absolute zero* on the kelvin, or absolute, temperature scale (p 177).

Gases do in fact follow the extrapolated graph quite closely down to very low temperatures, near absolute zero. Eventually they liquefy and the volume of the molecules themselves prevents further reduction in volume of the gas and zero volume is never reached.

However, the straight-line graph through the origin on the kelvin temperature scale allows us to state Charles' law in a simple way. We find that the volume V of a gas doubles as the absolute temperature T doubles on the kelvin scale. So **Charles' law** states this relation formally:

The volume of a fixed mass of gas is directly proportional to its absolute temperature (on the kelvin scale) if the pressure is constant.

Stated in symbols we have:

$$V \propto T \quad \text{or} \quad V = \text{constant} \times T \quad \text{or} \quad \frac{V}{T} = \text{constant}$$

Boyle's law
Compressing a gas at constant temperature

Loading the sliding plug in the three-dimensional gas model (figs. 10.26d and e) showed how an increase in pressure causes a reduction in the volume of a gas. This relation was first investigated about 300 years ago by Robert Boyle.

Investigating Boyle's law

Fig. 10.29 shows an apparatus which gives a direct reading for both the volume and pressure of a fixed mass of gas.

Figure 10.29 *Investigating Boyle's law*

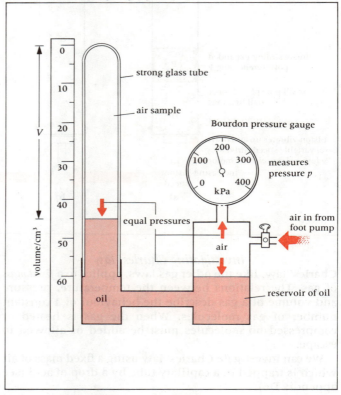

A sample of air is trapped in a strong glass tube by a column of oil. The oil is supplied from a reservoir where it can be pressurised using air from a tyre pump. The pressure above the oil in the reservoir is read directly on a Bourdon pressure gauge. The pressure above the oil in the reservoir is transmitted through the oil to the trapped air in the glass tube. The Bourdon gauge reads the actual pressure of the air including the atmospheric pressure. So when no air has been pumped into the reservoir the gauge reads about 100kPa, which is normal atmospheric pressure at sea level.

• *Gradually increase the pressure p of the air sample and record several readings for p and the volume V, read from the vertical volume scale, as in table 10.3.*

Table 10.3

Pressure p/kPa	Volume V/cm³	$p \times V$ /(kPa cm³)	$\dfrac{1}{p} \Big/ \dfrac{1}{\text{MPa}}$
110	45	4950	9.1
100	50	5000	10.0

• *Look at the results obtained for p and V, what do you notice?*
• *Calculate p × V and put the values in the third column of your table of results.*
We see that as the pressure *p* increases the volume *V* decreases. The values of *p* × *V* are almost constant. These figures suggest that the volume *V* is *inversely proportional* to the pressure *p*.
• *To test this relation, calculate values of* 1/*p* *for each of your readings, column 4 in the table.*
Using an example given in table 10.3, the units can be simplified as follows:

If *p* = 110 kPa (kilopascals), we can write this as
 p = 0.11 MPa (megapascals).
$$\therefore \frac{1}{p} = \frac{1}{0.11\,\text{MPa}} = 9.1\,\frac{1}{\text{MPa}}$$

• *Plot a graph of V/cm³ against* $\dfrac{1}{p}\Big/\dfrac{1}{\text{MPa}}$, *as shown in fig. 10.30.*

Figure 10.30 *A graph for Boyle's law*

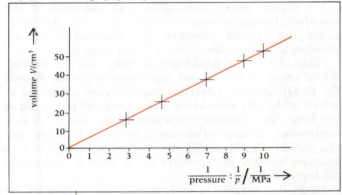

The straight-line graph through the origin gives us the relation between *V* and *p*. We see that the volume doubles when 1/*p* doubles, and the volume is three times greater when 1/*p* is three times greater, and so on. It follows that if *V* is proportional to 1/*p*, it is inversely proportional to *p*. This relation is stated as **Boyle's law**:

The volume of a fixed mass of gas is inversely proportional to its pressure if the temperature is constant.

Stated in symbols we have:

$$V \propto \frac{1}{p} \quad \text{or} \quad V = \frac{\text{constant}}{p} \quad \text{or} \quad pV = \text{constant}$$

The pressure law
Heating a gas at constant volume
A third gas law can be investigated by keeping the volume of a fixed mass of gas constant. A gas is heated but not allowed to expand. As a result of the molecules hitting the walls of their container with higher speeds, the pressure of the gas increases.

The relation between pressure *p* and the absolute temperature *T* of a gas is found to be very similar to the Charles' law relation. If we extrapolate the pressure graph back to zero pressure we find that the pressure of a gas might also become zero at −273 °C. The pressure doubles when the temperature *T* on the absolute, or kelvin, scale is doubled, which gives us the **pressure law**:

The pressure of a fixed mass of gas is directly proportional to its absolute temperature if its volume is constant.

Stated in symbols we have:

$$p \propto T \quad \text{or} \quad p = \text{constant} \times T \quad \text{or} \quad \frac{p}{T} = \text{constant}$$

The gas equation
We can combine the three relations between *p*, *V* and *T* from the three gas laws into one gas equation. We have:

$$\frac{V}{T} = \text{constant} \quad \text{(Charles' law)}$$
$$pV = \text{constant} \quad \text{(Boyle's law)}$$
$$\frac{p}{T} = \text{constant} \quad \text{(pressure law)}$$

which we combine to give the **gas equation**:

$$\frac{pV}{T} = \text{constant}$$

This formula can be used to solve problems about gas pressures and volumes when all three of the quantities *p*, *V* and *T* change at the same time. If the initial values are p_1, V_1 and T_1 and after a change in the gas they become p_2, V_2 and T_2, providing the mass of the gas is constant we can write:

$$\frac{p_1 V_1}{T_1} = \frac{p_2 V_2}{T_2}$$

Rules for using this formula:
a) The mass of gas must be constant.
b) The temperatures T_1 and T_2 must be given in *kelvins*.
c) The units in which *p* and *V* are calculated must be the same on *both* sides of the equation (but any convenient units such as mm of mercury and cm³ may be used).

Worked Example
Using the gas equation

In a chemistry experiment 240 cm³ of oxygen gas are collected. The temperature of the room is 20 °C and the atmospheric pressure, read on a barometer, is 770 mm of mercury. Calculate the volume of gas at standard temperature and pressure. (Standard temperature is 0 °C = 273 K and standard atmospheric pressure is 760 mm of mercury.)

We are given:
 p_1 = 770 mm of mercury
 p_2 = 760 mm of mercury
 T_1 = 20 + 273 = 293 K
 T_2 = 0 + 273 = 273 K
 V_1 = 240 cm³
 V_2 = ?
Rearranging the gas equation gives

$$V_2 = V_1 \times \frac{p_1}{p_2} \times \frac{T_2}{T_1}$$

Remembering that the temperatures must be in kelvins, and that the units of p_1 and p_2, and V_1 and V_2 must be the same, we find

$$V_2 = 240\,\text{cm}^3 \times \frac{770\,\text{mm of mercury}}{760\,\text{mm of mercury}} \times \frac{273\,\text{K}}{293\,\text{K}} = 227\,\text{cm}^3$$

Answer: The volume of the gas at standard temperature and pressure would be 227 cm³.

Worked Example
Compression reduces volume and raises temperature

A bicycle pump holds 60 cm³ of air when the piston is drawn out. The air is initially at 17°C and 1.0 atmospheres pressure. Calculate the pressure of the air as it is forced into the tyre if compression reduces its volume to 15 cm³ and raises its temperature to 27°C.

We are given:

p_1 = 1.0 atmospheres
p_2 = ?
V_1 = 60 cm³
V_2 = 15 cm³
T_1 = 17 + 273 = 290 K
T_2 = 27 + 273 = 300 K

Rearranging the formula gives:

$$p_2 = p_1 \times \frac{V_1}{V_2} \times \frac{T_2}{T_1}$$

Remembering the rules about the units of *T*, *V* and *p*:

$$\therefore p_2 = 1.0 \text{ atmospheres} \times \frac{60\,\text{cm}^3}{15\,\text{cm}^3} \times \frac{300\,\text{K}}{290\,\text{K}}$$

$$\therefore p_2 = 4.1 \text{ atmospheres}$$

Answer: The air enters the tyre at 4.1 times atmospheric pressure.

Note: Questions in which one of the three quantities is constant can be solved the same way. For instance, if the temperature had remained constant, the formula for calculating p_2 would simplify by cancelling T_1 and T_2 to give:

$$p_2 = p_1 \times \frac{V_1}{V_2}$$

Vapours

Evaporation

There is always some vapour above the surface of a liquid because liquid molecules are continually escaping from the liquid surface. If the liquid is in an open container so that the vapour molecules can escape completely, then eventually, under the right conditions, all the liquid may evaporate and become vapour, fig. 10.31a.

Experiments show that the rate of evaporation is increased by:
a) an increase in the temperature of the liquid,
b) an increase in the surface area of the liquid,
c) a draught or wind which blows over the surface.

We can explain these observations as follows:
a) at a higher temperature more molecules in the liquid are moving fast enough to be able to escape from the surface,
b) a larger surface area gives more molecules a chance to escape since more of them are near the surface,
c) a draught which carries the vapour molecules away from the liquid surface stops them returning to the liquid and makes it easier for more molecules to escape.

Vapour pressure

When a liquid is placed in a closed container such as a bottle with a stopper, the space above the liquid fills up with vapour. Liquid molecules evaporate into the space above the liquid, whether it contains air or not. (The difference was seen in the bromine diffusion experiment, p 158. Evaporation into a vacuum is very rapid; evaporation into an air-filled space is much slower.)

The vapour molecules collide with the walls of the container, bounce off and cause a pressure. This pressure, called the **vapour pressure**, is additional to the air pressure already present (due to the air molecules).

Molecules in the liquid are escaping into the vapour, while some in the vapour are returning to the liquid, fig. 10.31b. A state of **dynamic equilibrium** is reached, in which molecules are leaving and re-entering the liquid at the same rate. *Dynamic* describes the fact that there is a continuous exchange of molecules between the liquid and the vapour. *Equilibrium* is used to mean that there is a constant number of molecules in the vapour state above the liquid surface.

When the number of molecules of the vapour in the space above the liquid has reached its maximum the space is said to be **saturated**. Any vapour which is trapped in a space above its liquid will become saturated unless all the liquid evaporates before saturation is reached. The pressure of a vapour in a closed space above its liquid is called its **saturated vapour pressure (s.v.p.)**.

A saturated vapour is one which is in a dynamic equilibrium with its own liquid and which produces a saturated vapour pressure on the walls of its container.

Figure 10.31 *Evaporation*

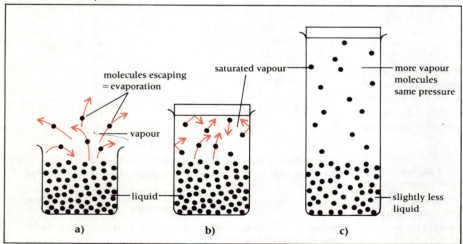

Saturated vapours do not obey the gas laws

If we increase the space above a liquid we leave more room for evaporation, fig. 10.31c. The effect of this increased volume is that more molecules are able to leave the liquid and become vapour, until they set up a new dynamic equilibrium.

The vapour in the new dynamic equilibrium will have the same saturated vapour pressure as before but it will contain more molecules. The changes are as follows:

a) increased volume of vapour,

b) no change in saturated vapour pressure,

c) a small reduction in the volume of liquid.

An increase in temperature raises the saturated vapour pressure but the relation between the s.v.p. and its temperature T is not the proportional relation of Charles' law, which gases follow.

Vapours behave like gases and obey the gas laws providing they do not become saturated.

Evaporation and boiling

Both evaporation and boiling are processes which convert a liquid into vapour. We can tell the difference between the two processes by looking for bubbles forming inside the liquid, which indicates that the liquid is boiling. Another important difference is that boiling occurs only when a certain definite temperature called the *boiling point* is reached. Table 10.4 summarises these differences.

Table 10.4

Evaporation	Boiling
occurs at the surface of liquids	bubbles form within the liquid
occurs at all temperatures	occurs at a definite temperature called the boiling point

When a liquid is heated, the saturated vapour pressure increases. At a particular temperature (the boiling point) the s.v.p. becomes equal to the external atmospheric pressure. At this temperature liquid molecules have enough energy to form bubbles of vapour inside the liquid. In other words, the s.v.p. is strong enough to 'blow up' bubbles inside the liquid without the atmospheric pressure on the surface crushing them. These bubbles, containing saturated vapour, burst at the liquid surface causing the familiar 'boiling' effect. This process is shown in fig. 10.32.

Figure 10.32   *Boiling water*

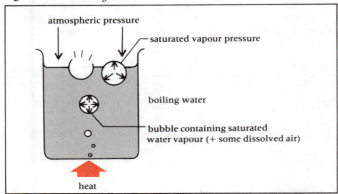

The effect on the boiling point of an increase in pressure (teacher demonstration)

Heat some water in a strong round-bottomed flask as shown in fig. 10.33a. When the water is boiling squeeze the rubber tube on the steam outlet with a clip or pliers and continue heating until the temperature on the thermometer has shown a *small* rise above 100°C. Quickly release the pressure and stop heating.

As more steam is produced inside the sealed flask the pressure on the water surface rises.

The effect of a pressure increase is a rise in the boiling point of water

Figure 10.33   *The effect of pressure on the boiling point of water*

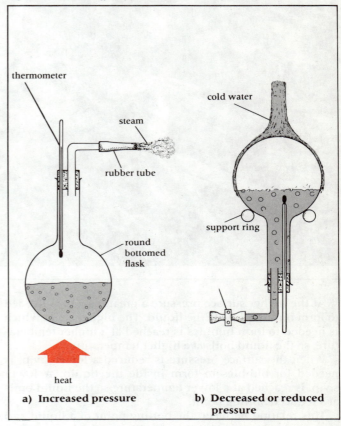

a) **Increased pressure**   b) **Decreased or reduced pressure**

The effect on the boiling point of a decrease in pressure (teacher demonstration)

Drive all the air out of the flask by boiling the water for several minutes. Then stop heating and quickly make an air-tight seal on the tube with a clip.

Invert the flask and support it in a ring over a sink. (Protect all observers with a safety screen between them and the flask.)

Run cold water over the flask and watch the water inside the flask, fig. 10.33b.

The steam above the water condenses and produces a partial vacuum, that is a reduced pressure. The water starts to boil at temperatures well below 100°C. The lower the pressure the lower the boiling point falls.

The effect of reducing the pressure on the surface of a liquid is to lower its boiling point.

The pressure cooker

By increasing the surface pressure on the liquid inside a closed cooker we can raise its cooking temperature. Temperatures of up to 120°C can be used to cook food much more quickly and economically than at 100°C. The pressure can be adjusted by changing the weight placed on the steam-escape valve on the top of the cooker lid. A larger weight produces a higher steam pressure inside and therefore a higher cooking temperature. Steam escapes from the valve when the selected cooking pressure has been reached.

The reverse effect occurs when food is cooked at high altitudes, since air pressure decreases with altitude. A mountaineer has difficulty in cooking an egg in boiling water because the water boils at a temperature much lower than 100°C. For example, in a high-altitude country such as Tibet, water boils at about 90°C.

The picture shows a modern pressure cooker being used to help with the cooking at a high altitude. Why do you think the jug has been placed on the pan lid alongside the pressure cooker?

At the higher surface pressure a greater s.v.p. is needed to form bubbles inside the liquid. The higher s.v.p. which is needed to form bubbles is reached at a higher temperature, so the liquid boils at a higher temperature.

When the surface pressure is reduced a lower s.v.p. is needed for bubbles to form inside the liquid. A lower s.v.p. is reached at a lower temperature so the liquid boils at a reduced boiling point.

This relation between the boiling point of a liquid and its s.v.p. leads to a definition of boiling point:

The boiling point of a liquid is the temperature at which its saturated vapour pressure becomes equal to the external pressure on its surface.

Assignments

Explain

a) When you pump air into a tyre to blow it up, why does the air inside the tyre not obey the gas laws during pumping.

b) Why does water vapour condense on the bathroom windows when you have a bath.

c) Why does dew form when the air temperature drops at night.

Remember

d) the gas equation:

$$\frac{pV}{T} = \text{constant}$$

e) The relation:

$$\frac{p_1 V_1}{T_1} = \frac{p_2 V_2}{T_2}$$

f) The gas laws. Can you remember which quantity is constant for each gas law? Can you remember what to plot on the graph which tests each gas law?

Use the experimental results given in table 10.5 to estimate the temperature at which the volume of the gas would contract to zero volume if it obeyed Charles' law down to very low temperatures.

Table 10.5

Temperature/°C	Length of air column/cm (represents gas volume)
0	18.3
25	20.0
50	21.6
75	23.5
100	25.0

Try questions 10.24 to 10.30

Questions 10

1 The sunlight which shines onto a greenhouse as shown in fig. 10.34 ncludes three regions of the electromagnetic spectrum **A**, **B** and **C**. **B** and **C** pass through the glass but radiation, **A**, does not.
   a) Name the three radiations.
   b) The plants in the greenhouse give out radiation which cannot escape and so the temperature of the greenhouse increases. What is this radiation?
   c) In the evening the insides of a greenhouse may be seen to have water droplets on them. Why does this happen?
   d) How could this be prevented?   (NEA [B] 88)

Figure 10.34

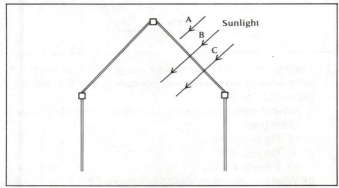

2 a) Describe and explain, with clear diagrams, why light breezes occur on an otherwise calm day at the seaside and these are in different directions at night and day.
   b) Describe and explain an experiment to show that water is a poor thermal conductor.
   c) What molecular movements occur in the process of thermal conduction?   (S)

3 Fig. 10.35 shows the metal shade and bulb of an electric reading lamp.
   Redraw the diagram and add arrows to indicate the convection currents in the air inside the shade when the lamp is in use.
   State the processes by which heat is transferred from the bulb filament to the shade.
   Why does the shade eventually reach a steady temperature?   (C)

4 Two similar cans are partly filled with equal quantities of paraffin. Each holds a thermometer, is covered by a lid, and stands on a wooden bench at the same distance from a radiant heater, as shown in fig. 10.36. One can has a dull black surface, the other a bright silver surface.

The following temperatures are recorded:

Time in minutes	0	1	2	3	4	5
Temperatures in °C, dull black	19	21	23	25	27	29
Temperatures in °C, bright silver	19	20	21	22	23	24

Figure 10.36

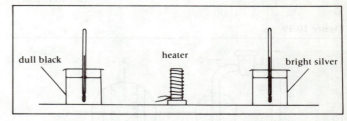

dull black   heater   bright silver

   a) Why can we say that the cans are not heated by conduction?
   b) Why can we say that the cans are not heated by convection?
   c) By what process are the cans heated?
   d) Do equal quanities of heat fall upon the two cans?
   e) Why is there a difference between the rates of heating of the two cans?
   f) Explain an advantage of using paraffin rather than water in the cans.   (O)

5 a) Describe an experiment which you would carry out to show how the nature of a surface affects the heat radiated from that surface in a given time.
      State any precautions which you would take and state your findings for two named surfaces.
      How would you then show that the surface which was the better radiator was also the better absorber of radiation?
   b) As the surface of a pond freezes it is found that each equal increase in the thickness of the ice takes longer to form, even when the air above the ice remains at the same temperature. Explain why this is so.
   c) In the experiment shown in fig. 10.37 the ice remains intact for several minutes as heating progresses. Explain how this can be so.   (L)

6 Fig. 10.38 shows an electric hot plate being used by a cook to heat a saucepan containing water.
   a) By what method is heat transferred through the saucepan to the water?
   b) Explain how a convection current is created in the water.
   c) The cook is rather proud and keeps the saucepan including the base highly polished. Explain any advantages and disadvantages which arise from keeping the saucepan highly polished.
                                                        (Joint 16+)

Figure 10.35

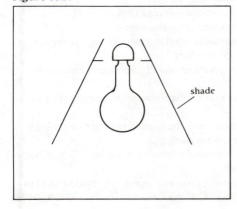

shade

Figure 10.37

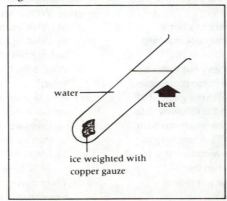

water

heat

ice weighted with
copper gauze

Figure 10.38

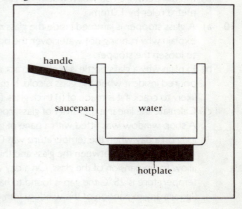

handle

saucepan   water

hotplate

7 Fig. 10.39 is a simple diagram of a vacuum flask with an enlarged view of the part in the circle.
   a) What is the material of the items labelled A and C?
   b) What types of transfer of heat energy are reduced or prevented by the items marked B, C and D?
   c) Explain why A is effective in reducing heat transfer.
   (Joint 16+, part)

Figure 10.39

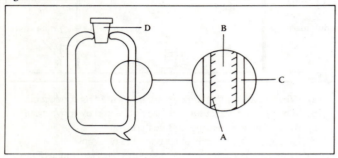

8 Fig. 10.40 shows a device which switches off a domestic appliance when it reaches the desired temperature.
   a) What is this device called?
   b) Name two domestic appliances which make use of this device.
   c) What property does material Z possess?
   d) What materials could be used for X and Y?
   e) Explain why bending occurs when the device is heated.
   (Joint 16+, part)

Figure 10.40

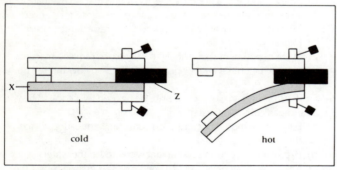

cold   hot

9 Using the values of linear expansivities given in table 10.1, calculate the following:
   a) the expansion of a steel (iron) girder caused by a temperature rise from −10°C to +30°C, if its length is initially 25 metres,
   b) the temperature rise which would increase the length of a brass metre ruler by 1.0mm.

10 a) A glass stopper is jammed inside the glass neck of a bottle. Explain why running hot water over the bottle neck may help to loosen the stopper.
   b) Explain why a glass bottle is likely to crack if very hot water is poured inside it when the glass is cold. Is the bottle more or less likely to crack if it is made of (i) thick glass, (ii) Pyrex glass?
   c) Calculate the linear expansivity of glass from the following data. A shop window was fitted with a pane of glass which was 6.0m long on a day when the temperature was 0°C. A small gap of 1.35mm was left between the glass and the window frame to allow for expansion of the glass. On a day when the temperature is 25°C the gap is found to have just closed. (Ignore any expansion of the window frame.)

11 Fig. 10.41 represents a section of a pond.
   a) State the probable temperature of the water at the two positions X and Y as shown in the diagram.
   b) What fact concerning the density of water do these temperatures indicate?   (AEB)

Figure 10.41

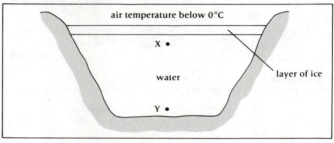

12 The graph in fig. 10.42 illustrates the changes in volume which occur when a fixed mass of pure ice initially at −10°C is heated from −10°C to 20°C.
   a) What changes in volume are indicated by the following sections of the graph?
      i)   The section AB
      ii) The section BC
      iii) The section CD
   b) What can you deduce about the density at D?   (AEB)

Figure 10.42

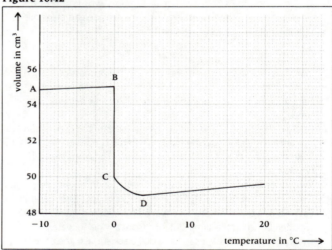

13 You are trying to measure the boiling point of water using a thermometer as shown in fig. 10.43.
   a) i)   What does a thermometer measure?
      ii) What liquid is found in this thermometer?
      iii) What liquid would be used in a thermometer to measure a boiling point of about −50°C?
   b) What is the boiling point of water i) in degrees Celsius, ii) in kelvin?
   c) A nurse may use a clinical thermometer to measure a patient's temperature.
      i)   Give **two** reasons why this thermometer is more suitable than the one used in part (a).
      ii) For a healthy person the instrument will read 37°C. Change 37°C to kelvin.
   d) Write down the name of an instrument that could be used to measure up to 1500°C as in a furnace.   (NEA [A] 88)

14 a) You are provided with an uncalibrated thermometer as shown in fig. 10.44. The mercury level is about one quarter of the way up the stem at room temperature. Describe how you would calibrate the instrument by marking two fixed points and then use it to determine room temperature.
Explain why, in the instrument shown in fig. 10.44,
  i) the glass surrounding the bulb is thin even though this makes it fragile,
  ii) the mercury level will not immediately rise to its final steady level when the thermometer is placed in a warm liquid.
b) Explain why
  i) an alcohol-filled thermometer might be preferred to a mercury-filled one by an Arctic explorer,
  ii) in a clinical thermometer the bulb is not quite full of mercury at room temperature. (L)

15 In a simple experiment to determine the specific heat capacity of aluminium a 50 W immersion heater is inserted in a 2 kg block of aluminium, which also holds a thermometer, as shown in fig. 10.45.
a) How much heat is supplied by the heater every second?
b) How much heat is supplied by the heater in 5 minutes?
c) Why would the temperature of the block be measured more accurately if a little oil was poured into the thermometer hole?
d) It is found that the temperature of the block rises 8 K in 5 minutes. Neglecting heat losses, calculate from this the value of the specific heat capacity of aluminium. (O)

16 When 1 kg of water was heated for 5 minutes with an immersion heater the water's temperature rose by 30°C.
To the questions that follow answer **either**: greater than 30°C; **or** less than 30°C; **or** equal to 30°C.
What would be the temperature rise if the same heater heated
  i) 2 kg of water for 9 minutes?
  ii) 1 kg of paraffin for 5 minutes (given that the heat capacity of the paraffin is less than that of the water)? (W 88)

17 Here is an extract from an advertisement for 'Sun-wave' solar panels:
Flat solar panels of area approximately 1 m² are fixed to the roof. The front glazing is 4 mm glass and each collector contains 7 m of copper tubing which is painted matt black. The tubing is backed with an aluminium foil reflector and fibre glass is used as insulation. A copper waterway is incorporated which circulates water (with the aid of a pump) from the solar panels to the water tank and back again. The short waves from the sun pass through the glass into the collector, hitting the absorber surface, and changing into long waves that are then trapped in the collector (producing a 'greenhouse effect').

a) i) Why are **copper** pipes used?
   ii) Why are the copper pipes painted **black**?
   iii) Why is **insulation** needed in the back of the panel?
   iv) What is meant by the 'greenhouse effect'?
   Before deciding to purchase 'Sun-wave' solar panels, a householder may decide to find the savings that would be made. This could be done by considering the costs of heating the water with a conventional electric immersion heater.
   Here is some data:

Solar energy received by the panel on an average day = 126 MJ
Efficiency of the solar heating system = 20%
Specific heat capacity of water = 4200 J/(kg K)
Temperature of water entering tank = 5°C
Temperature of water required by householder = 35°C
1 kilowatt-hour = 3.6 MJ
Cost of one kilowatt-hour of electricity = 8 pence
Number of days per year for which panels can be used = 200

b) Calculate
   i) the quantity of energy provided on an average day by the sun that can be used to heat the water;
   ii) the rise in temperature of the water required by the householder;
   iii) the mass of water than can be heated;
   iv) the number of kilowatt-hours of electricity which would be required to heat this mass of water;
   v) the cost of electricity that would be needed to heat the water. (SEG 88)

18 Explain the following:
a) An iron gate feels cold to the touch, but its wooden gatepost feels comparatively warm.
b) Cloudless nights in winter are often frosty.
c) The cavities between the walls of some modern houses are filled with plastic foam.
d) The contents of certain cooking pots may continue to boil for a short time after removal from the source of heat.
e) Two identical thermometers, left hanging side-by-side in the laboratory, may indicate different temperatures if the bulb of one of the thermometers has been covered with cotton wool moistened with water. (O & C)

Figure 10.43

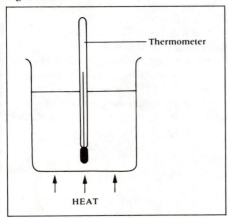

Figure 10.44

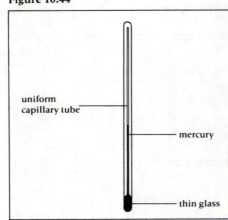

Figure 10.45

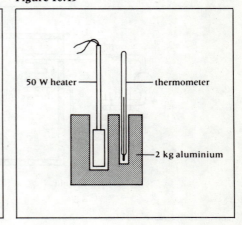

19 a) Fig. 10.46 shows part of a household hot water system.
  i) Why is pipe A connected between the top of the boiler and the top of the storage tank?
  ii) Why is pipe B connected between the bottom of the boiler and the bottom of the storage tank?
  iii) What is the function of pipe C?
  iv) Suggest, with reasons, what might be added to the hot water system above to make it more efficient.

Figure 10.46

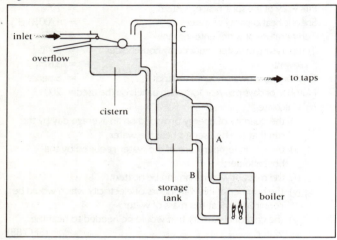

b) The temperature of the water inside an aquarium can be controlled by a thermostat which switches an electric heater on and off. Draw a diagram showing how this may be done using a bimetallic strip. (Your diagram must clearly show the construction of the bimetallic strip.)
  How may different constant temperatures be achieved using your arrangement?
c) The hot water tap of a bath delivers water at 80°C at a rate of 10 kg/min. The cold water tap of the bath delivers water at 20°C at a rate of 20 kg/min.
  Assuming that both taps are left on for 3 minutes, calculate the final temperature of the bath water, ignoring heat losses.   (L)

20 a) Fig. 10.47 shows the essential parts of a simple refrigerator circuit.
  i) Explain carefully what is happening in the pipes at A, stating clearly why this achieves the desired result.
  ii) Why is A situated at the top of the refrigerator?
  iii) Why are the fins metal, and what is their purpose?

Figure 10.47

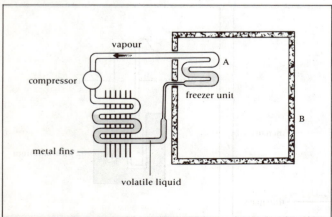

  iv) Name another machine in which metal fins are used to achieve the same purpose.
  v) The walls of the cabinet, B, are sometimes filled with crinkled aluminium foil. Why?
  vi) A person who leaves open the refrigerator door to cool the room on a hot day in summer will not succeed. Give the reasons for this.
b) Explain why a chest type deep freezer (lid at top) is thought to be more efficient than an upright type (door at side).   (L)

21 A heater supplying energy at a constant rate of 500W is completely immersed in a large block of ice at 0°C. In 1320 s, 2.0 kg of water at 0°C are produced. Calculate a value for the specific latent heat of fusion of ice.   (C)

22 A 50W heating coil is totally immersed in 100 g of water contained in an insulated flask of negligible heat capacity.
a) If the temperature of the water is 20°C when the heater is switched on, how long would it take for the water to boil?
b) After the water has been boiling for 15 minutes it is found that the mass of water in the flask has decreased to 80 g. Assuming no external heat losses, calculate a value for the specific latent heat of vaporisation of water.
  [Assume that the specific heat capacity of water = 4200 J/(kg K).]
     (JMB, part)

23 a) What is meant by the statement that the specific heat capacity of water is 4200 J/kg°C?
b) i) A 50W heater is totally immersed in water which is contained in an aluminium pot. The heater is switched on and the initial and final temperature of the water noted over a given period of time. From the following, determine a value for the specific heat capacity of water, showing clearly how you obtain your answer.

Mass of water   = 2.5 kg
Initial temperature of water   = 20°C
Final temperature of water   = 28°C
Time for which heater was operating = 30 minutes

  ii) Suggest two reasons why the value for the specific heat capacity calculated above differs from the true value given in (a). (Assume that all readings have been taken accurately and that the heater has been tested to give 50W precisely).
c) What is meant by the specific latent heat of vaporization of water?
d) Water (boiling point 100°C) is kept boiling in a kettle with an electric heating element and the following measurements taken:

Current flowing through element   = 10 A
Potential difference across element = 250 V
Initial temperature of water   = 100°C
Mass of kettle   = 0.5 kg
Initial mass of kettle and water   = 2.7 kg
Final mass of kettle and water   = 1.9 kg
Time for which the water boils   = 12 minutes

  i) Determine the power rating of the element.
  ii) Determine the specific latent heat of vaporization of water from the measurements taken, showing clearly, how you obtain your answer.
  iii) What is the temperature of the water at the instant the element is switched off?   (NISEC SPEC)

Figure 10.48

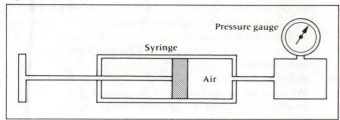

24 Fig. 10.48 illustrates an apparatus in which a fixed mass of air was compressed in a calibrated syringe, which was approximately half full of air at atmospheric pressure and a temperature of 17°C. Corresponding values of volume and pressure of the trapped air are shown in the table.

pressure (kPa)	50	60	75	90	105	120
volume (m³)	0.00048	0.00040	0.00032	0.00027	0.00023	0.00020
$\frac{1}{volume}$ (m^{-3})		2500		3704		5000

a) Complete the table by calculating values for $\frac{1}{volume}$. Some of the values have been entered for you.

b) On graph paper plot a graph of pressure on the y-axis against $\frac{1}{volume}$ on the x-axis.

c) What relationship between pressure and volume of the trapped air can be deducted from your graph? Explain your answer.

d) If the temperature of the air was increased to 27°C, what volume would be occupied by the air at a pressure of 100 kPa? (NEA [A] 88)

25 Fig. 10.49 shows an apparatus which can be used to investigate how the volume of air changes with its temperature while its pressure remains constant.

Figure 10.49

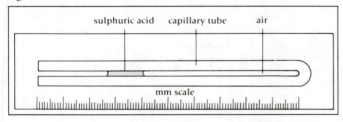

a) What measurement would you make which would represent the volume of the trapped air? Explain your answer, stating the assumption you are making.

b) Describe briefly how you would measure the temperature of the air in the capillary tube. State two precautions you would take to ensure that your temperature measurement was as accurate as possible.

c) Why is the air enclosed by a plug of concentrated sulphuric acid?

d) The table below contains corresponding values of volume and temperature obtained in such an experiment.

Temperature/°C	20	40	60	80	100
Volume/mm³	58.5	63.0	66.5	71.0	75.0

Plot a graph of volume on the y-axis against temperature on the x-axis.

Use your graph to obtain values for (i) the volume of trapped air at 0°C, (ii) the temperature at which the volume of the trapped air would become zero.

e) What is the significance of the temperature in (d)(ii)? What assumption about the behaviour of the gas did you make when you calculated the temperature in (d)(ii)?

f) What conclusion might be drawn from the results of this experiment? (JMB)

26 a) Explain what is meant by the absolute zero temperature. Describe a simple experiment which you could perform in your school laboratory which would allow you to estimate the value of absolute zero on the Celsius scale of temperature. Sketch the apparatus which you would use, list the observations you would make and show how these observations would be used to arrive at the final result. What result would you expect?

b) A motor car tyre contains a fixed mass of air. The pressure of the air was measured as 200 kN/m² above atmospheric pressure when the air temperature was 17°C. After a high-speed run, the air pressure in the tyre was measured again and was found to be 230 kN/m² above atmospheric pressure. What was the new temperature of the air in the tyre if its volume remained constant? (Atmospheric pressure on the day was 100 kN/m².) (JMB)

27 Explain why a bubble of air increases in volume as it rises from the bottom of a pond to the surface. If the volume as it just reaches the surface is double that at the bottom of the pond, estimate the depth of the pond. (Assume that the water temperature is uniform. Take the pressure at the water surface to be 10^5 N/m² (10^5 Pa) and the density of water to be 1 000 kg/m³.)

28 A pressure cooker with a weight on the needle valve is heated until the water boils.

a) Is the temperature of the water greater than, less than or equal to 100°C?

b) If the needle valve sticks what will happen?

c) A climbing expedition to Mount Everest finds that a pressure cooker is essential to cook its food effectively. Explain why this is so. (Joint 16+, part)

29 State what changes, if any, take place in the following.

a) The melting point of ice when salt is added to the ice.

b) The volume of water when it changes to ice.

c) The boiling point of a liquid when the pressure on the liquid is reduced.

d) The density of molten wax (which contracts when it solidifies). (AEB)

30 Using the *simple kinetic theory*, give explanations of each of the following.

a) Ether, placed in a dish by an open window, is found to be at a lower temperature than its surroundings.

b) If an 'empty' aerosol can is left in strong sunlight it may explode.

c) Energy must be supplied in order to melt ice even though the temperature of the resulting water is no higher than that of the ice.

d) When lighting a gas cooker an explosion is less likely if the match is struck first and then the gas turned on than if the gas is turned on first and the match then struck. (L)

PRACTICAL INVESTIGATIONS AND PROBLEMS TO SOLVE I

(More investigations and problems can be found on pages 378 and 379.)
For each of these problems and investigations, decide:
- *what you are going to measure;*
- *what equipment you will need to measure it;*
- *how to record the measurements and*
- *how to process the measurements to draw conclusions.*

Then try out your ideas. If they don't work very well, try changing the equipment you are using or even redesigning your experiment. Answer the following questions:
- *Is what you are trying to measure going to give you useful information?*
- *Is the equipment capable of measuring what you want to know?*
- *What do your measurements tell you? Have you reached a conclusion?*

1 Squash players spend the first part of their time on court warming up both themselves and the ball. It is suggested that temperature affects the bounciness of a squash ball. Your task is to design and carry out an experiment to investigate the effect of temperature on the bounciness of a squash ball.

2 The strength of paper changes when it is wet. This can be a problem when the paper wrapping around a parcel gets wet. Your problem is to design and carry out an experiment to investigate how the strength of wrapping paper changes as it becomes wet. Is there a time delay?

3 Parachutes are used to give falling objects a much lower and safer speed of descent. A maximum downwards speed, called the terminal velocity, is soon reached. Design a simple parachute which can be used to land a 100 g brass weight. Investigate how the size of the parachute affects its terminal velocity.

4 An experiment for a fine day. You will need at least one bicycle with gears.

It is suggested that the acceleration of a cyclist depends upon which gear she uses. Design an investigation to test this hypothesis. You must consider carefully what you are going to measure and what steps you will take to limit the effects of other factors which may also change for acceleration.

5 Design and build a bridge using only paper and glue. The bridge must span a gap of 40 cm and be capable of supporting a load at its centre. You will need to design paper beams so that the paper has greater strength than it would have if it were flat. Test your bridge by placing a load at its centre. Gradually increase the load until the bridge collapses.

What was the maximum load your bridge could support?
What was the weakest feature of your design?
How could the design of your bridge be improved?

Problem 5

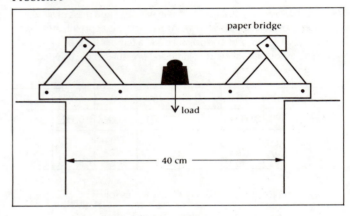

6 It is noticed that when two plane mirrors are set facing each other at an angle a number of images of a small object can be seen in the two mirrors.
Investigate how the number of images that can be seen depends upon the angle between the two mirrors.

7 The formula for the time period of a simple pendulum (a bob swinging on the end of a fine string) does not include a symbol for the amplitude of the swing. Investigate whether the time period of a simple pendulum is independent of the amplitude of the swing.

8 You are provided with sheets of corrugated cardboard, fibreglass and carpet felt. You are asked to investigate which of these three materials is the best thermal insulator. One method you might try is to see how good each material is at keeping a hot drink hot.

Problem 3

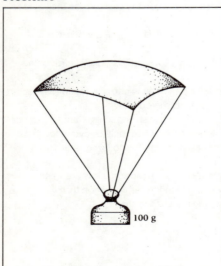

Problem 6

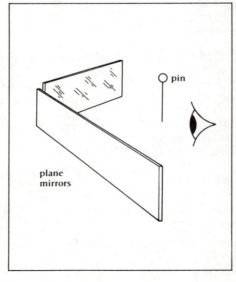

Problem 7

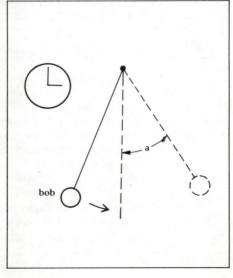

9 Problem 9 shows what happens when rays of light are affected by some objects. The objects have been hidden behind screens.
a) Study the way the light is affected in each diagram and then redraw each of the diagrams to show what you think is behind the screen.
b) Explain carefully the reason for the difference between diagrams (i) and (ii). (NEA [B] SPEC)

Problem 9

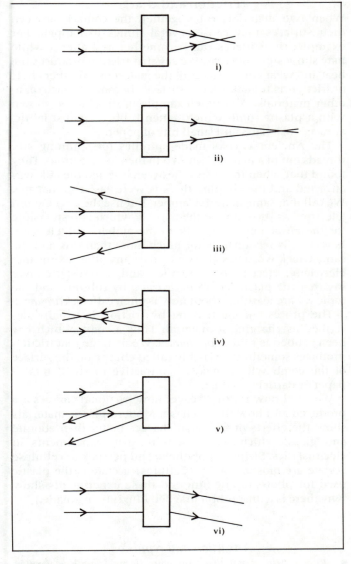

10 The claim made in an advertisement is shown in problem 10 below.
   Design an experiment that you could carry out in a laboratory to test the truth of this claim. Include in your description:
a) the apparatus you would need,
b) how you would use this,
c) what you would measure and record,
d) how you would interpret your results to test the claim. (SEG [ALT] 88)

Problem 11

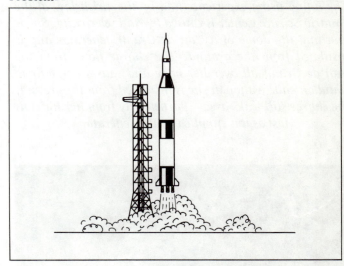

11 Information about the Saturn V rocket, used by NASA to place space probes in orbit, is given below.

Mass of rocket and fuel at lift-off	= 4000 tonnes
Mass of fuel at lift-off	= 3500 tonnes
Constant thrust produced by rocket engine	= 50 000 000 N
Rate of burning fuel	= 10 tonnes/s
Upward acceleration at lift-off	= 2.5 m/s^2

(1 tonne = 1000 kg)

a) Explain how the Saturn V rocket is able to rise from the launch pad. You must refer to exhaust gases in your explanation.
b) i) Show that the force producing the acceleration at lift-off is 10 000 000 N.
   ii) Why does the rocket engine need to produce a thrust much greater than 10 000 000 N at lift-off?
c) Suggest a reason why the acceleration of the Saturn V increases as it rises from the launch pad.
d) How long will the fuel last, assuming a constant burning rate throughout the launch? (SEB 1990)

Problem 10

11

Static electrons

*A*ll matter contains vast numbers of electrons. When a few of them get transferred from the surface of one object to the surface of another, they produce what is known as static electricity.

The hair-raising experience had by the girl below at the Ontario Science Center is caused by static electricity. She is touching the dome of a Van de Graaff generator and is insulated from the ground. The charge flows from the Van de Graaff all over her body and causes her hair to stand on end. Although she is insulated from the ground, the charge still leaks away (particularly from her hair) as fast as it is supplied by the generator.

11.1
CHARGE

Why do some clothes tend to cling to us when we move about, or crackle when we undress? Why do gramophone records and photographic films attract dust particles? The easy answer is to say that static electricity is the cause.

Friction and charge

When two materials rub together, the contact between their surfaces may cause several things to happen. For example, the surfaces can become hot and after a while may show signs of wear. We say that friction produces the heat and wears the surfaces of the materials. Another effect of friction is to make some surfaces become 'attractive' to other materials. A simple example of this can be shown with a plastic comb which, when rubbed against fabric, attracts dust, hairs and small bits of paper.

The Ancient Greeks made spindles for spinning silk threads out of a material known to them as 'electron'. They found that when these spindles were rubbed the silk was attracted and the clinging threads were easier to manage. We call that same material 'amber' but use the Greek word 'electron' as the name of the particle which is responsible for the attraction of the silk to the amber. The electron, however, is responsible for much more than this and the same Greek word has also given us many other words like electricity, electronics, electrified and electrostatics. We say that the plastic comb is *electrified* by rubbing and the topic we are learning about now is often called *electrostatics*.

The process of electrifying by friction is usually described as charging a material. Thus a comb which has been rubbed is said to be *charged*. We imagine that friction produces something invisible called **charge** on the surface of the comb which makes it attractive to small bits of paper or particles of dust.

We shall now try to discover how frictional charges are produced and how they behave. Modern plastic materials show the effects of frictional charges better than ebonite and glass, which were used in early experiments in electrostatics. Strips of polythene and perspex or cellulose acetate are most suitable. (Cellulose acetate is the plastic used for photographic film and these experiments show why there is a dust problem when film is being made.)

Attraction and repulsion

● *Rub a strip of polythene on your sleeve or with a woollen duster then hold it near some small pieces of paper on the bench top and notice what happens.*
● *Repeat the test with small pieces of aluminium cooking foil or small metallised polystyrene balls (expanded polystyrene balls coated with aluminium paint).*
● *Repeat with a strip of cellulose acetate or perspex.*

The charged strips of polythene and cellulose acetate both attract the bits of paper and these may cling on to the strips for some time. The metallised polystyrene balls jump rapidly up and down between the bench top and the charged strips, being first *attracted* to the strips and then thrown off or *repelled*. We notice then that charged materials can produce both attraction and repulsion effects.

Two kinds of charge

- *After rubbing it, support a polythene strip* A *in a stirrup as shown in fig. 11.1a.*
- *Charge another polythene strip* B *by rubbing it and then slowly bring it close to one end of strip* A *without touching.*
- *Test the other end of* A *and notice what happens.*
- *Now charge two strips of cellulose acetate,* C *and* D*, by rubbing them with a duster and repeat the test with* C *in a stirrup and* D *brought near it, fig. 11.1b.*
- *Finally, investigate what happens when the charged cellulose acetate strip* D *is brought near to the suspended charged polythene strip* A*, and similarly when the charged strip* B *is brought near to suspended strip* C*, fig. 11.1c.*

As the figure shows, when the two charged strips are made of the same material they repel each other but when the two different materials are charged and brought near they attract each other. We have to conclude that *two different kinds of charge are produced by friction* on these two different materials.

These two kinds of charge which produce opposite effects are called **negative charge** and **positive charge**. Rubbed polythene gains a negative charge and rubbed cellulose acetate (perspex and glass) gains a positive charge.

Since both the polythene strips have a negative charge and both acetate strips have a positive charge, the tests give the following results:

Two negatively charged strips repelled each other.

Two positively charged strips repelled each other.

One negatively charged strip and one positively charged strip attracted each other. (These are referred to as **unlike charges.**)

We find that
a) like charges always repel each other,
b) unlike charges always attract each other,
c) only two types of charge exist.

No material has ever been found which attracts both positive and negative charges.

Figure 11.1 *Two kinds of charge*

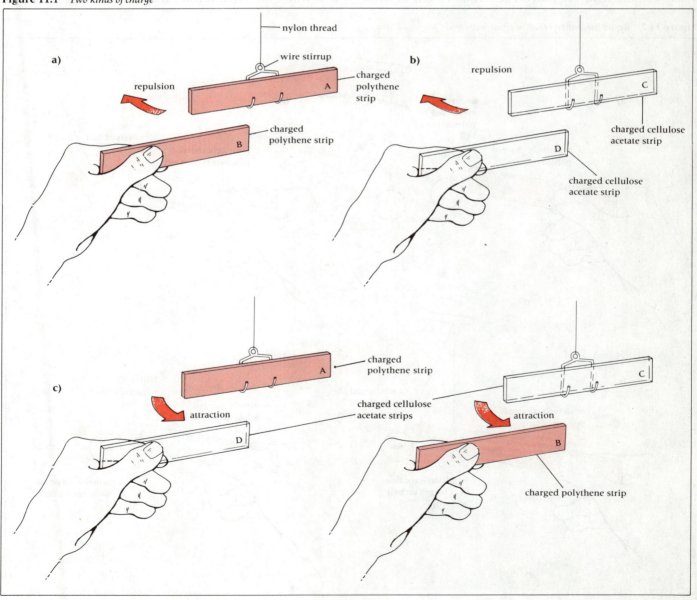

Charged or neutral?

• *Hang a metallised polystyrene ball on a nylon thread and make no attempt to charge it.*

• *First bring a negatively charged polythene strip near to the ball but do not let it touch, fig. 11.2a.*

• *Now bring a positively charged acetate strip near to the ball, again without touching it, fig. 11.2b.*

In both (a) and (b) the uncharged ball is attracted to the strip whether the attracting charge is positive or negative.

• *Allow the metallised ball to touch the negatively charged polythene strip and repeat the tests (a) and (b).*

This time the results are different as shown in (c) and (d). The ball is now repelled by the polythene strip suggesting that it has the same negative charge as the strip. Apparently, when the ball touched the strip it picked up some of its negative charge as a result of the contact. As we would expect, the positively charged cellulose acetate strip attracts the ball, which now has an unlike or opposite charge.

The metallised polystyrene ball had no charge before it made contact with the polythene strip. We can describe this uncharged object as being **neutral**. This evidence shows that some neutral objects are attracted to charged objects. It follows that when attraction occurs we cannot be sure whether the attracted object is charged or neutral, since in both cases attraction may occur.

However, if repulsion occurs then we have reliable evidence that the repelled object has the same charge as the one repelling it. *Only repulsion can confirm that an object is charged*.

We have seen that a metallised ball can receive charge from a polythene strip when they touch. This *charging by contact* explains why the metallised balls jump off the charged polythene strip immediately after being attracted to it. As soon as contact occurs the balls pick up some of the charge on the strip and then the two like charges cause repulsion.

So we have an explanation for repulsion, but some questions remain. How does friction between two surfaces sometimes leave them charged and why is an uncharged or neutral object attracted to a charged object? We might ask, where does charge come from?

Figure 11.2 *Repulsion confirms that an object is charged*

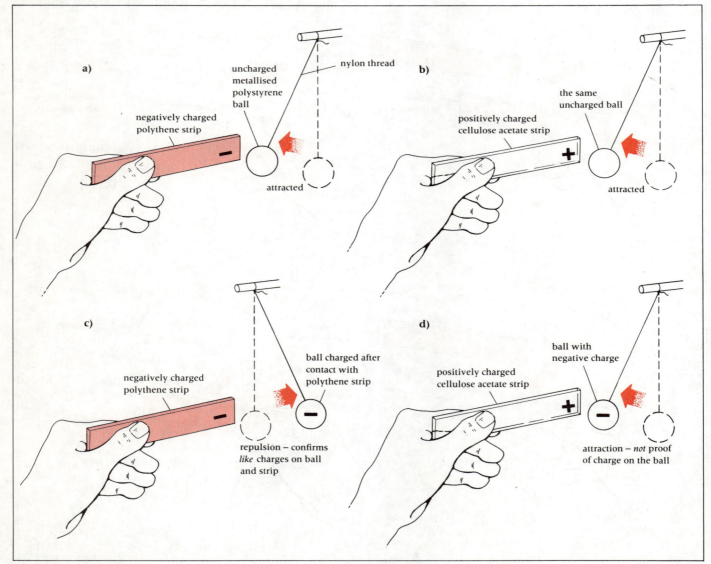

Where does charge come from?

In chapter 17 we shall look in more detail at the structure of atoms and the very small particles of which they are built. For now the following ideas will help us to account for the behaviour of charged objects and materials.

Three important particles are known to be the basic building blocks of all atoms. They are called **protons**, **neutrons** and **electrons**. Fig. 11.3a shows a model of how these particles are thought to form an atom. In this example they form a helium atom. The protons and neutrons are held together in a very small space at the centre of the atom, called its **nucleus**. The electrons exist in orbits around the central nucleus but, compared with the sizes of these particles, the orbits are so far out that the atom is mostly empty space.

Figure 11.3   *Charged particles in atoms*

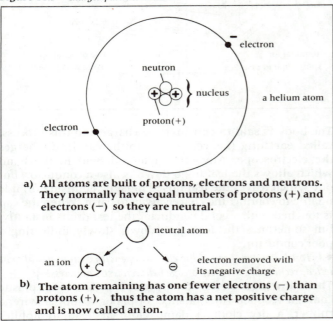

a) **All atoms are built of protons, electrons and neutrons. They normally have equal numbers of protons (+) and electrons (−) so they are neutral.**

b) **The atom remaining has one fewer electrons (−) than protons (+),   thus the atom has a net positive charge and is now called an ion.**

An atom is very small but the particles of which it is built are much smaller. The electron in particular has a mass which is only about $\frac{1}{2000}$ of the mass of the smallest atom, hydrogen, and an even smaller fraction of all the larger atoms.

In some materials one or two electrons can break away from each atom and move about comparatively freely between the atoms. In such materials these mobile electrons are the key to how they conduct electricity.

Electrons have been found by experiment (p 347) to have a negative charge which cannot be removed from them. Each electron always has exactly the same quantity of charge. This fixed quantity of charge is known as *the charge of the electron*. It seems that charge does not exist on its own, but is something which belongs to certain kinds of particles. Protons have an equal quantity of the opposite kind of charge to electrons, i.e. they have a fixed positive charge. This is why it is thought that atoms (which normally have an equal number of protons and electrons of opposite charge) are found to be neutral.

The third kind of particle in the atom, the neutron, as its name suggests has no charge, that is, it is neutral.

Fig. 11.3b shows the idea of an electron (−) being removed from a neutral atom to leave it positively charged. The atom has a net positive charge because it now has one proton (+) more than the number of electrons remaining in the atom. An atom which is charged as a result of losing (or sometimes gaining) an electron is known as an **ion**. Remember, fig. 11.3b does not show what an ion really looks like; it is just a 'model' to help us understand how they behave.

How do objects become charged?

We can now explain what happens when two materials are rubbed together. Electrons, which are quite loosely attached to the atoms at the surface of one material, are removed by the friction process and deposited on the surface of the other material. As the rubbing transfers electrons from one material surface to the other, the negative charge of the electrons is transferred with them. We can summarise the effects of friction as follows.

a) Materials such as polythene gain extra electrons on their surfaces and become negatively charged, fig. 11.4. The cloth or duster used to rub the polythene must have lost these electrons and consequently gained an equal and opposite positive charge.

Figure 11.4   *Charging by rubbing*

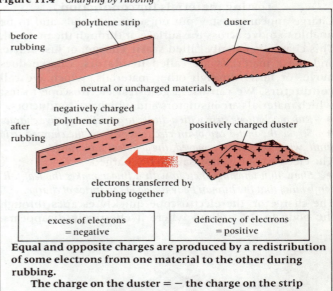

Equal and opposite charges are produced by a redistribution of some electrons from one material to the other during rubbing.
     The charge on the duster = − the charge on the strip

b) Materials such as cellulose acetate, perspex and glass have electrons removed from their surfaces when they are rubbed. These materials become positively charged, while the duster used gains the electrons with their equal negative charge.

c) Charge is never made or destroyed by friction; it is only transferred from one material to another, i.e. it is redistributed. Charge transfer does not occur on its own, but rather results from the transfer of electrons from which the negative charge is inseparable.

d) Objects are made of neutral atoms and are therefore normally neutral themselves.

An object becomes negatively charged when it gains an excess of electrons. Similarly, when an object has some electrons removed, the deficiency of electrons makes the object positively charged.

The gold-leaf electroscope as a charge detector

We have seen that repulsion between two objects can be a reliable test that they are both charged. The gold-leaf electroscope is a simple instrument which uses repulsion to indicate when something is charged. The diagram shows one type of modern instrument which is housed in a wooden case with glass windows to protect it from draughts.

The vital part of the electroscope is a very thin and flexible strip or 'leaf' of gold foil which is attached to a metal plate. (Real gold foil was used in early electroscopes because gold is a good conductor and is very malleable, i.e. it can be rolled or beaten into very thin sheets. Nowadays a cheaper and stronger leaf made of aluminised plastic is often used.) The gold leaf and metal plate are connected by a metal rod, which passes through the centre of a perspex insulator, to the metal cap on the top of the electroscope. (The perspex insulator stops charge given to the metal cap from spreading onto the case and leaking away.)

When the metal cap receives some charge it spreads down to both the metal plate and the gold leaf. Since both the metal plate and gold leaf receive the same kind of charge, they repel each other and the very light gold leaf rises away from the metal plate. We say that the leaf diverges. The charge can be removed from the cap and leaf by connecting them to the case of the electroscope. Touching the cap and case together makes a good enough connection because, as we shall see, both the wooden case and a hand are conductors of charge.

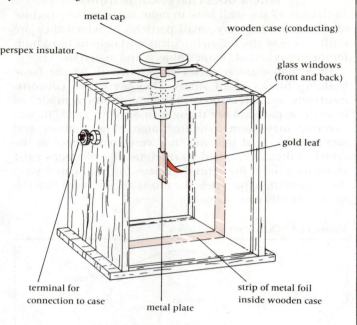

Testing materials for conduction

Charge appears to stay put on some materials and to be unable to move across the surface or through the material. This charge is usually called **static charge**, or just **static**, and these materials are called **insulators**. Charge does move, or flow, through other materials, which we call **conductors**. We can use a gold-leaf electroscope to test which materials are insulators and which are conductors.

● *Rub a polythene strip with a duster to give it a negative charge.*
● *Stroke the metal cap of an electroscope with the charged strip until, when the strip is removed, the gold leaf stays raised.*
The electroscope is now said to be charged.
● *Show that touching the cap with a finger makes the leaf fall, confirming that the human body is a good conductor of charge.*
The charge on the electroscope quickly escapes through the body to the ground where it effectively disappears.

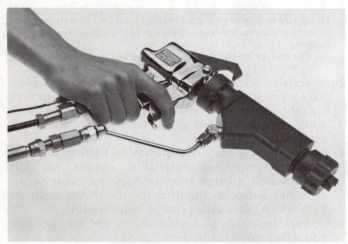

A fine needle at the tip of this electrostatic paint spray gun is charged negatively to 75 kV and gives all the small droplets of paint a negative charge. If the object to be sprayed with the paint is made positively charged then the paint is attracted to it covering it more evenly on all sides and saving paint.

The body is said to **conduct** the charge and this process, called **earthing** (i.e. connecting to the earth), discharges the electroscope. Any other material held in the hand which allows the leaf to go down is also a conductor, but one which leaves the leaf raised must be an insulator; it is unable to conduct the charge away to earth. When the cap is touched with a good conductor the leaf drops instantly, but sometimes the leaf goes down slowly indicating a poor conductor.

● *Touch the metal cap of the electroscope with each material to be tested, recharging the electroscope between tests, as necessary.*
The following materials can be tested: a glass rod, a plastic comb, a wooden ruler, a sheet of paper, a copper wire, a rubber, a dry cloth, a damp cloth, a 'lead' (graphite) pencil, a wax candle, a strip of polythene and a strip of cellulose acetate or photographic film.
● *Make a list of materials tested and record whether they are good conductors, poor conductors or insulators.*

Charging by contact, charge sharing

When a strip of polythene is rubbed it becomes charged. Extra electrons (of negative charge) stay on the surface of the polythene at the points where they become attached by the rubbing. Because the polythene is an insulator the charge cannot flow off it through the hand holding it. In contrast, a metal object, being a conductor, must be insulated from all conducting materials around to prevent charge flowing from it. For this reason all metal objects used in electrostatics have insulating handles or stands. The charge on a conductor is able to spread itself all around the surface. This happens because the repulsion between the like charges of the electrons tends to push them away from each other.

When a charged object comes into contact with an uncharged object what happens depends upon whether each object is made of conducting or insulating material.

A charged insulator touching an uncharged insulator
- Rub a polythene strip A with a cloth and test it on an electroscope to show that is is charged.
- Test another polythene strip B, that has not been rubbed, to make sure that is is not charged.
- Touch the two strips together as in fig. 11.5a then, after separating, test them both again.

The charge remains on A but is not transferred to B. *Charge is not shared between two insulators* unless they are rubbed together so that most of their surfaces make close contact.

A charged insulator touching an uncharged conductor
The metal cap of an electroscope is a convenient conductor to use because the gold leaf will give an indication of whether the cap and leaf have gained any charge.
- Bring a charged polythene strip near to (or even lightly touching) the metal cap of the electroscope so that the leaf rises, then take the strip away again.

Figure 11.5 *Charge sharing*

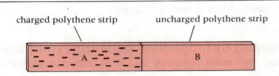

charged polythene strip   uncharged polythene strip

a) An insulator in contact with another insulator; no charge sharing

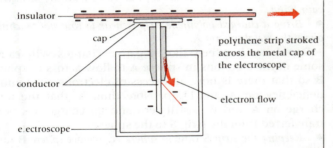

insulator

cap

polythene strip stroked across the metal cap of the electroscope

conductor

electron flow

electroscope

b) An insulator in contact with a conductor: charge is collected from the surface of the insulator

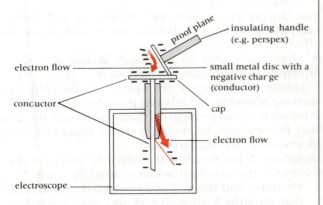

proof plane

insulating handle (e.g. perspex)

electron flow

small metal disc with a negative char ge (conductor)

conductor

cap

electron flow

electroscope

c) A conductor in contact with another conductor: charge flows and is shared

The leaf drops fully showing that no charge has been transferred to the cap. It is, in fact, quite difficult to charge the cap of the electroscope by contact with a charged insulator like the polythene strip.
- Stroke the polythene strip across the cap so that most of the surface of the strip comes into close contact with it, fig. 11.5b.

This time, the leaf may stay partly raised after the strip is taken away and you may have succeeded in leaving some charge on the electroscope. *Charge can be shared between an insulator and a conductor but the charge will not flow off the insulator, rather it has to be collected or picked up from the insulator's surface.* Charging an electroscope by contact with an insulator is not a reliable or recommended method.

A charged conductor touching an uncharged conductor
To demonstrate this we need a charged conductor with an insulating handle. A convenient tool is one called a **proof plane** (shown in fig. 11.5c), which is used in electrostatics for sampling charge in various places.
- Charge a proof plane by contact with a charge generator (see later) or by picking up charge from an insulator which has been rubbed.
- Touch the cap of an uncharged electroscope and then remove the proof plane noting what happens.
- Test the proof plane on another uncharged electroscope to see if any charge remains.

The leaf of the first electroscope stays raised when the proof plane is removed and the second electroscope also receives charge, but not as much; the leaf diverges less. The excess electrons on the proof plane are able to flow from the plane to the cap and leaf of the electroscope because they are all made of conducting metal. The process is one of sharing the excess electrons around, spreading them out by mutual repulsion so that charge is gained by the electroscope, but some also remains on the proof plane. The second electroscope confirms this. *Charge flows between conductors in contact giving them both a share of the same charge.*

To protect computer files from the effects of static electricity and dust, this mat with a sticky surface is used to remove dust and conduct away static charge from people handling the files.

Testing the sign of the charge on an object

● *Give a negative charge to a gold-leaf electroscope by stroking the cap with a negatively charged polythene strip, until enough charge has been transferred to diverge the gold leaf to about 45° from its plate, i.e. about half raised as in fig. 11.6a.*

● *Observe what happens to the gold leaf as you bring first a negatively charged polythene strip, and then a positively charged cellulose acetate strip, slowly up to the electroscope cap.*

● *Now bring your hand slowly towards the cap of the electroscope, fig. 11.6d.*

● *Repeat these tests with a positively charged electroscope.*

Figure 11.6 *Testing for the sign of a charge*

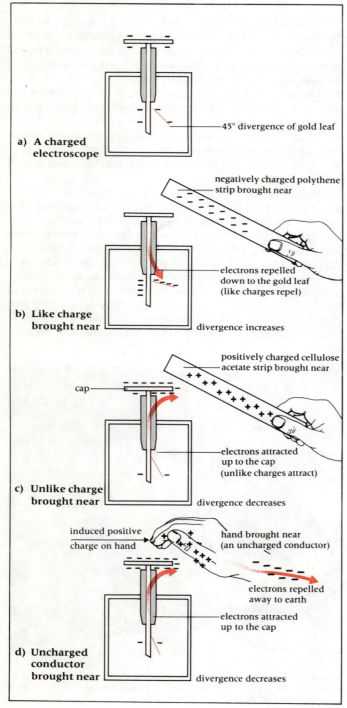

a) **A charged electroscope**

— 45° divergence of gold leaf

— negatively charged polythene strip brought near

— electrons repelled down to the gold leaf (like charges repel)

b) **Like charge brought near**

divergence increases

— positively charged cellulose acetate strip brought near

cap —

— electrons attracted up to the cap (unlike charges attract)

c) **Unlike charge brought near**

divergence decreases

induced positive charge on hand —

hand brought near (an uncharged conductor)

— electrons repelled away to earth

— electrons attracted up to the cap

d) **Uncharged conductor brought near**

divergence decreases

The results shown in fig. 11.6 indicate that *only an increase in the divergence of the gold leaf can be relied upon as proof of a particular charge on an object*. This is the same as the conclusion that only repulsion between two objects can be taken as proof that they are *both* charged. The increase in the divergence of the gold leaf is caused by an increase in the repulsion, which in turn is caused by an increase of charge on the leaf and plate. However, a decrease in the divergence of the gold leaf may be caused either by an object with the opposite charge to the electroscope or by a conductor with no charge at all.

The uncharged conductor behaves as if it has the opposite charge to the cap of the electroscope. This is caused by a flow of electrons in the uncharged conductor which produces what is called an **induced charge** as shown in fig. 11.6d.

When the electroscope is given a positive charge exactly the same results are obtained, but in this case the charges and direction of electron flow are the opposite to those shown in fig. 11.6.

An increase in the divergence of the gold leaf shows that an object has the same charge as the electroscope.

Charging by induction: separating conductors

● *Place two metal objects* A *and* B *in contact as shown in fig. 11.7a.*

Metal spheres on insulating stands are usually used for this investigation but a pair of metal cans insulated from the bench by a sheet of glass or polythene will work just as well.

● *Ensure that the spheres are uncharged by touching them with your hand.*

● *Bring a charged polythene strip* S *near to (but not touching) one of the metal spheres* A.

Repulsion between like negative charges will cause some of the electrons in sphere A to flow across to sphere B so that there is now an excess of electrons on B and a deficiency on A, fig. 11.7b. Note though, that the total charge on A and B is still zero and no charge has been transferred from the strip S to the spheres.

● *Keeping the strip* S *close to sphere* A, *remove sphere* B *some distance away without touching the metal sphere itself.*

The extra electrons on sphere B are now unable to return to sphere A.

● *Remove the charged strip.*

The charges on the spheres will now spread over their conducting surfaces if they are far enough apart not to affect each other.

● *Test these charges on a positively charged electroscope, to identify the sign of the charge on each sphere.*

Sphere A causes the divergence of the gold leaf to increase, showing that it is also positively charged. Sphere B causes the divergence to decrease and its negative charge may be confirmed by testing it with a negatively charged electroscope. We conclude that:

a) Sphere A has received the opposite charge to that on strip S, (positive in this case, caused by a deficiency of electrons) and sphere B has received the same charge as that on strip S (negative in this case, caused by an excess of electrons).

b) The charges on spheres A and B are equal as well as opposite.

Figure 11.7 *Charging by induction*

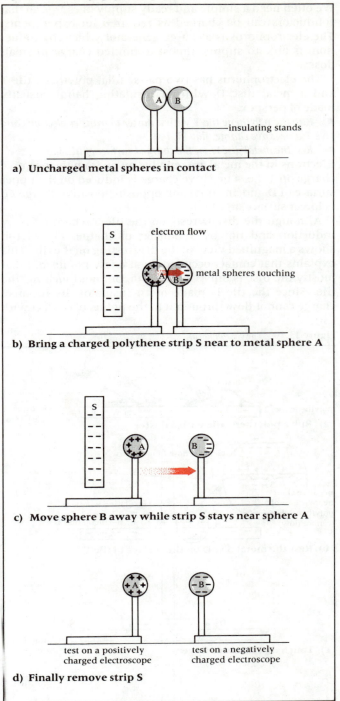

a) **Uncharged metal spheres in contact**

insulating stands

b) **Bring a charged polythene strip S near to metal sphere A**

electron flow

metal spheres touching

c) **Move sphere B away while strip S stays near sphere A**

d) **Finally remove strip S**

test on a positively charged electroscope

test on a negatively charged electroscope

c) No charge was transferred from the strip S and it keeps all its charge.

d) This method of charging conductors without contact is called **charging by induction** and the charges on the spheres are said to be induced charges.

Note how the induction process involves a *remote* effect of the charge on strip S. Charging by induction always involves 'action at a distance'. The negative charge on strip S applies a *repulsive force* to the negatively charged electrons some distance away in the metal spheres, causing them to flow from A to B.

Charging by induction: earthing a conductor

Initially the insulator conductor A has no charge, fig. 11.8a.

- *Bring a charged strip S up to, but not touching, the conductor* A.

Repulsion between the negative charge on strip S and the negative electrons in the conductor A causes the conductor's electrons to flow away from S, fig. 11.8b. The deficiency of electrons produces a positive charge on the near side of A. The excess of electrons produces an equal, negative charge on the far side of A.

- *Earth conductor A by touching with a finger*.

The excess of electrons on the far side of the conductor is now able to flow even further away from the repelling charge on strip S. The electrons are conducted through the body and 'disappear' in the earth.

- *Remove the finger so that the escaped electrons cannot return*.

This leaves conductor A with a permanent deficiency of electrons, or a positive charge.

- *Remove the charging strip* S, *which has retained all its negative charge*.

Now the positive charge spreads over the surface of conductor A, fig. 11.8c.

We conclude that conductor A has received, by induction, a charge opposite to that on the charging strip S. When a single conductor is charged by induction and earthed during the process, it always receives the opposite charge to the one used to induce it.

Figure 11.8 *Charging by induction*

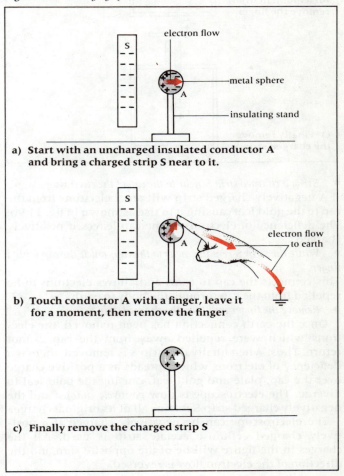

electron flow

metal sphere

insulating stand

a) **Start with an uncharged insulated conductor A and bring a charged strip S near to it.**

electron flow to earth

b) **Touch conductor A with a finger, leave it for a moment, then remove the finger**

c) **Finally remove the charged strip S**

Charging a gold-leaf electroscope by induction

The most effective way of charging an electroscope is to use an induction method. The sign of the charge received, the opposite of the inducing one, can be relied upon.

● *Start with an uncharged electroscope and check that its gold leaf is not diverged, fig. 11.9a.*

Figure 11.9 *Charging an electroscope by induction*

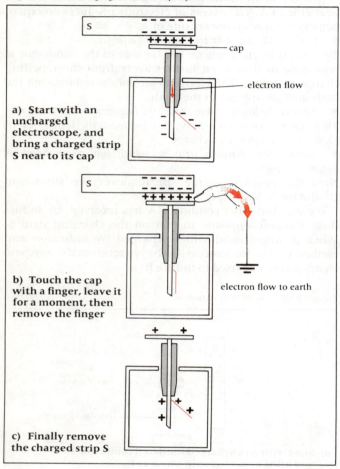

a) **Start with an uncharged electroscope, and bring a charged strip S near to its cap**

b) **Touch the cap with a finger, leave it for a moment, then remove the finger**

c) **Finally remove the charged strip S**

● *Bring a charged strip S near to the cap of the electroscope.*

A negatively charged strip will repel electrons from the cap to the gold leaf causing it to rise as shown in fig. 11.9b. The deficiency of electrons in the cap leaves it positively charged.

● *While the strip S remains near to the cap, touch the cap with a finger.*

This connects the cap to earth and allows electrons to be repelled from the cap through the body to earth.

● *Remove the finger and then the charged strip S.*

Once the earth connection has been removed, the electrons which were repelled away from the cap cannot return. Thus, when finally the strip S is removed, there is a deficiency of electrons which spreads as a positive charge over the cap, plate and gold leaf, causing the gold leaf to diverge. The electroscope is now *positively charged* and the negatively charged strip S retains all of its original charge.

The electroscope can be negatively charged if a positively charged cellulose acetate strip is used. All the charges in the figure will be of the opposite sign and the direction of the electron flow is reversed.

The electrophorus

We often need a simple and ready supply of charge so that conductors can be charged as required for experiments. The electrophorus is a charge generator which, by induction, is able to supply almost unlimited charge in small doses.

The electrophorus has two parts, a flat polythene tile T and a metal disc D with an insulating handle usually made of perspex.

● *Rub the polythene tile T with a duster to transfer electrons and give T a negative charge, fig. 11.10a.*

● *Rest the metal disc D on the charged surface of tile T.*

Electrons in the metal are repelled away from the negative charge on T, i.e. a negative charge is induced on the upper surface of D and an equal and opposite, positive charge on is lower surface, fig. 11.10b.

Although the disc D rests on the tile T, charging is by induction and not by a transfer of charge. Fig. 11.10b shows a magnified view of the disc resting on the tile. This explains that contact occurs only at a few points and that mostly the disc is only very near to, but not touching, the tile. Since the tile is made of an insulator, its negative charge cannot flow through it to the points of contact with

Figure 11.10 *The electrophorus*

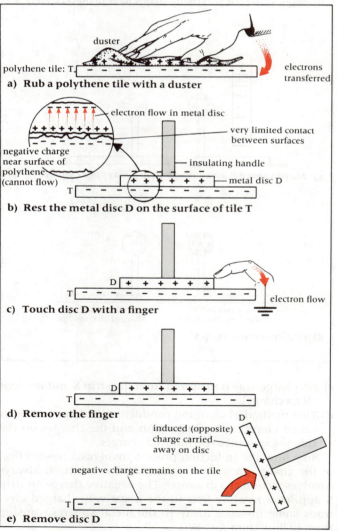

a) **Rub a polythene tile with a duster**

b) **Rest the metal disc D on the surface of tile T**

c) **Touch disc D with a finger**

d) **Remove the finger**

e) **Remove disc D**

the disc. The negative charge remains static on its surface. (If, however, the surface of the polythene tile is damp, then the charge can flow through the conducting film of water. This is why these experiments do not work well in damp or humid conditions.)

● *Briefly touch the metal disc D with a finger to earth it.*
The electrons repelled to the top of the disc are now able to escape to earth through the finger and body, leaving a deficiency of electrons on the disc.

● *Remove the finger before lifting disc D off the tile.*
Since D is insulated, electrons cannot return to it and it will remain positively charged. The negative charge on the tile remains and can be used over and over again to charge the disc positively by the induction process described.

Thus it would, at first, appear that at no cost we have an unlimited supply of positive charge since the charge on the tile is not being 'used up'. However, each time the positive charge on the disc is given to, or shared with, another conductor some electrons are returned to the disc. Thus the flow of electrons to earth is reversed as the disc loses or shares its positive charge. The electrons are never lost or destroyed. The abundant supply of charge from the electrophorus does not mean that is has a limitless source of energy either. Energy is provided each time work is done separating the disc D from the tile T.

The charge distribution on the surface of a conductor

Conductors of various shapes can be used to show how the distribution of charge on a surface is affected by the shape. Two conductors which show the important features of surface charge distribution are the spherical and pear-shaped conductors, fig. 11.11.

● *Charge each of the conductors by touching them with the positively charged disc of an electrophorus.*
The positive charge is able to spread over the surfaces, since they are conductors. (It may be easier to think of positive charge flowing over the conductor, but actually all the charge is moved by electrons flowing from the conductors to spread out the deficiency of electrons of the electrophorus.)

● *Investigate how the positive charge has spread itself over the surface by taking samples of the surface charge with a proof plane.*
When the metal disc of the proof plane is placed on the

surface of the conductor it forms a part of that surface and acquires the charge present at that position on the surface. When the proof plane is removed it takes this small sample of the surface charge with it.

● *Test the charge on the proof plane by sharing the charge with an uncharged electroscope.*
The amount of charge on various parts of the surface can be compared by taking different samples and testing them on the electroscope. A greater divergence of the gold leaf is produced when there is more charge on the proof plane.

The amount of charge collected on the proof plane is a measure of the charge on a certain area of the conductor's surface (an area equal to the area of the proof plane's disc). This charge sample gives a measure of the charge density on the surface of the conductor.

Charge density is defined as the quantity of charge per unit area of a conductor's surface.

The results show that the *charge density is greatest where a surface is most sharply curved.* Thus flat surfaces have a low charge density compared with curved surfaces and the charge density is greatest at corners, edges and points on conductors, i.e. where the curvature is greatest. Fig. 11.11 shows how the charge is evenly distributed on the surface of a spherical conductor, but the charge density is greatest at the pointed end of the pear-shaped conductor.

It should be remembered that this discovery applies only to the surfaces of conductors and not at all to insulators, on which the charge cannot flow to establish any particular distribution.

Investigating hollow conductors

The inside surface of a hollow conductor

● *Charge a hollow metal sphere by touching it with the charged disc of an electrophorus.*
● *Lower a proof plane through the small opening in the sphere taking care not to touch its outside surface with the proof plane.*
● *Make contact with the inside surface of the sphere to sample any charge that may be there, fig. 11.12.*
● *Remove the proof plane and test it on an uncharged electroscope.*
No charge is found on the inside surface of the hollow sphere.

Figure 11.11 *The distribution of charge on the surface of a conductor*

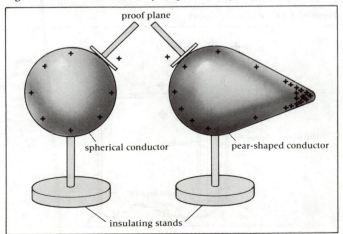

Figure 11.12 *There is no charge on the inside of a hollow conductor*

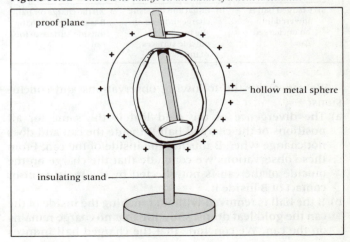

Faraday's ice-pail experiment

Faraday investigated the effect of lowering a charged ball inside a hollow metal can. The can he used was one kept for storing ice, hence the name of this experiment. We can repeat the experiment using a hollow metal can set on the cap of an electroscope. Ideally the can should either be tall, or have a small opening at the top, as shown in fig. 11.13.

● *Charge a small metal ball B by contact with a charged electrophorus and lower it into the metal can, being careful not to touch the can.*

Once the ball is inside the can, the gold leaf of the electroscope diverges.

● *Move the ball around inside the can without them touching and observe the divergence of the gold leaf.*

● *Remove the ball and note whether any charge has been left on the electroscope and can.*

● *Again lower the charged ball into the hollow can, this time allow it to touch the inside surface of the can, watching the gold leaf as contact is made.*

● *Remove ball B and test it on another electroscope.*

Figure 11.13 *The ice-pail experiment*

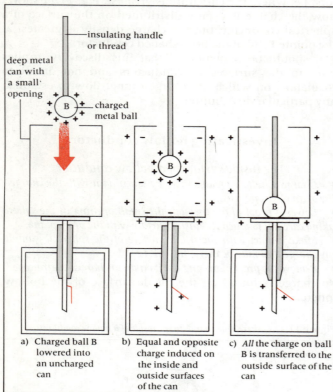

a) Charged ball B lowered into an uncharged can

b) Equal and opposite charge induced on the inside and outside surfaces of the can

c) *All* the charge on ball B is transferred to the outside surface of the can

We can make the following observations and conclusions:

a) The divergence of the gold leaf is the same for all positions of the charged ball B inside the can and does not change when B touches the inside of the can. From these observations we conclude that the charge on the outside of the can is not affected by the movement or contact of B inside it.

b) If the ball is removed without touching the inside of the can the gold leaf drops, showing that no charge remains on the can. We conclude that the charged ball induces

equal and opposite charges on the outside and inside surfaces of the can, which neutralise each other again when the ball is removed. No charge is transferred from the ball to the can.

c) When the charged ball is removed after contact with the inside surface of the can, it is found to be totally discharged. All the charge on B is transferred to the can. Since as there is now no charge on the ball, there is no charge on the inside surface of the can either. The positive charge on B exactly neutralises the induced negative charge on the inside surface of the can. We now understand that the three charges (on B and the inside and outside surfaces of the can) are all equal in magnitude. The effect of touching the inside of the hollow can with the charged ball is to transfer all of the charge from the ball to the outside of the can.

Note that usually when a charged conductor touches an uncharged one, the charge is shared, but when contact is made on the inside of the uncharged conductor all the charge is transferred.

At all times the total, or net, charge inside the hollow can is zero. All charge given to a conductor goes to its outside surface.

This important principle is used in the Van de Graaff generator.

Experiments with a Van de Graaff generator

Producing a spark

● *Connect a small metal sphere S on a conducting stand to the base B of the Van de Graaff generator as shown in fig. 11.14.*

With a continuous supply of charge to the dome D, a spark is produced between D and S. When the distance between D and S is only one or two centimetres the spark may be continuous. For larger distances (10 cm or more) the spark will be intermittent since time is required between sparks for sufficient charge to accumulate on the dome again.

An important idea, which we shall learn more about in the next chapter, is that charge flows round closed paths or *circuits*. In fig. 11.14 charge accumulates on the dome D, then jumps across to S, flows down the conducting stand and through the connecting wire to the base of the generator. Here it is transferred from the base by friction to the belt, which finally completes the circuit by carrying the charge up to the dome.

Figure 11.14 *Producing a spark*

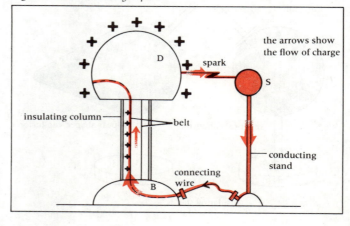

The Van de Graaff generator

The discovery that charge delivered to the inside of a hollow conductor is all transferred to its outside surface led to the invention of a very efficient static charge generator. Van de Graaff built a machine in which charge is continuously delivered to the inside of a hollow metal dome by means of a rotating rubber belt.

In a simple school machine, friction between the rollers and the belt provides charge on both rollers. The sign of the charge depends on the material of the roller. A metal comb B which is connected to the metal base of the generator (which in turn is earthed) becomes positively charged by induction. The negatively charged roller repels electrons from the comb to earth leaving an induced positive charge on the comb. This positive charge is sprayed onto the outside of the belt by the action of points (see below). As the belt is made of an insulating material, the charges it receives remain fixed on its surface and are carried up to the dome.

As the positive charge on the belt arrives inside the dome it induces a negative charge on a metal comb C. This comb is mounted close to the belt and is connected to the inside of the dome. An equal and opposite positive charge is induced on the outside surface of the dome. Comb C 'sprays' electrons onto the belt, thus losing the induced negative charge and neutralising the positive charge on the belt. By this process the positive charge is transferred from the belt to the outside of the dome. The positive charge continues to accumulate on the outside surface of the dome until it begins to leak away by conduction through the air. (Positive charge can be carried through the air away from the dome by charged air molecules (ions), or by particles of dust which jump off the dome taking positive charge with them.) When the leakage of charge equals the rate of delivery of charge by the belt, the dome has accumulated the maximum amount of charge it can hold. Note that if the positions of the polythene and perspex rollers are exchanged the dome becomes negatively charged. Can you explain why?

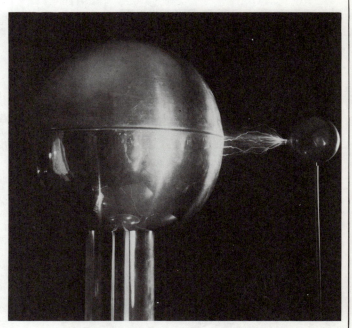

The action of points

A wire 'windmill', known as Hamilton's mill, mounted on an insulating stand or fitted to the top of the dome of a Van de Graaff generator can be used to demonstrate the action of points on a conductor.

Connect the 'windmill' to the dome with a wire to provide a continuous supply of charge, as in the arrangement shown in fig. 11.15. The 'windmill' rotates in the direction away from its points.

When charge is supplied to the metal wires of the 'windmill' most of it accumulates at the points of the wires, where the curvature is very high. The resulting very high charge density at the metal points produces a strong repulsion between the positively charged points and any positively charged particles or ions nearby. Thus positive ions in the air near the metal points (together with others produced in the action) are 'blown' away by the repulsive force, causing what is sometimes called an **electric wind**. Consequently there is a *reaction* force acting on the wire points which moves them in the opposite direction, as shown.

Figure 11.15 *The action of points*

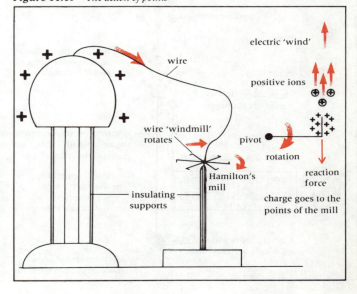

Blowing a candle flame with an 'electric wind'

● *Connect a pin or needle to a Van de Graaff dome and support it on an insulating stand a centimetre or two from a candle flame, fig. 11.16.*

The flame is blown by the 'electric wind' away from the point of the pin by the stream of positively charged ions.

Free electrons in the air will also be attracted to the point of the pin and flow in the opposite direction along the wire back to the dome of the generator. This sometimes causes part of the flame to be attracted to the pin.

Figure 11.16

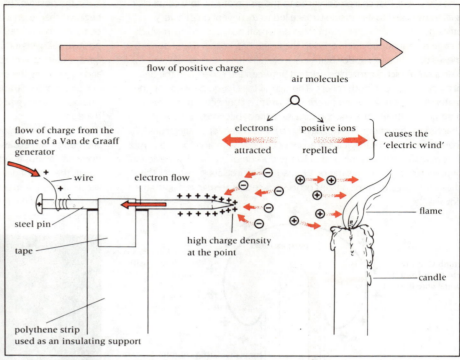

flow of positive charge

air molecules

electrons — attracted   positive ions — repelled   } causes the 'electric wind'

flow of charge from the dome of a Van de Graaff generator

wire

electron flow

high charge density at the point

flame

candle

steel pin

tape

polythene strip used as an insulating support

The lightning conductor

The diagram illustrates how a lightning conductor helps to protect a tall building from being struck by lightning. The lightning conductor is a very thick copper strip which connects some sharp metal points fitted above the highest part of a building to a large metal plate buried deeply in the damp earth below the building. The conductor provides a path for electrons to flow easily in vast numbers from the top of the building to the earth.

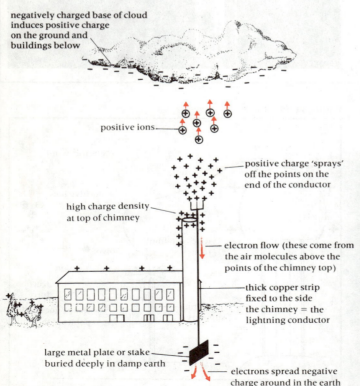

negatively charged base of cloud induces positive charge on the ground and buildings below

positive ions

positive charge 'sprays' off the points on the end of the conductor

high charge density at top of chimney

electron flow (these come from the air molecules above the points of the chimney top)

thick copper strip fixed to the side the chimney = the lightning conductor

large metal plate or stake buried deeply in damp earth

electrons spread negative charge around in the earth

The steady leakage of positive charge towards the clouds from the points and the flow of electrons (from the air) down the lightning conductor to earth helps to prevent a large build-up of charge on the highest parts of the building. The alternative to a steady discharge from the points and through the conductor is a sudden discharge in the form of a lightning strike. The very large and sudden flow of charge that occurs in lightning has enough energy to do serious damage to buildings.

When charge leaks from a lightning conductor or sharp points on any metal object you can sometimes see a faint glow of light and hear a hissing sound. This glow gives the process the name **corona discharge** because light surrounds the top of the pointed conductor like a crown. You can sometimes hear the hissing sound near the Van de Graaff generator and on overhead electricity power lines when charge is leaking across the insulators.

Corona discharge can be seen around the insulators in the photograph. These insulators are being tested at an ultra-high voltage by the Central Electricity Generating Board.

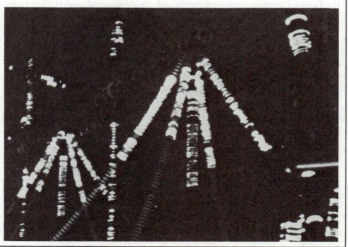

An electrostatic precipitator

An electrostatic precipitator removes smoke and dust from the waste gases going up the chimneys of factories and power stations. The diagram shows how a precipitator works. The wire grid is kept highly charged so that a continuous corona discharge occurs between the grid and the earthed metal plates. This discharge involves a continuous stream of ions which attach themselves to the dust particles in the gas going up the chimney. The charged dust particles are now repelled from the wire grid and attracted to the earthed plates where they become deposited. These plates are tapped from time to time so that the dust and smoke particles fall down the chimney and are removed at the bottom.

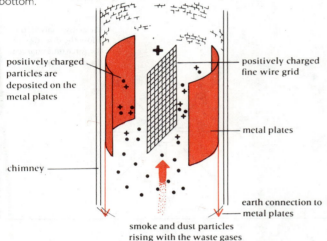

positively charged particles are deposited on the metal plates

positively charged fine wire grid

metal plates

chimney

earth connection to metal plates

smoke and dust particles rising with the waste gases

This factory chimney is in use in both photographs, but in the lower one the electrostatic precipitator is switched on.

Revealing finger prints on surfaces

A concealed finger print on the surface of a paper object such as a bank note, or a plastic object such as a handbag, can be revealed using a charged powder. A metal plate with a coating of a fine powder (e.g. silicon carbide) is given a high positive charge from a 10 000 volt supply. The specimen is connected to the negative terminal of the supply. The powder becomes positively charged and is repelled from the metal plate towards the specimen. When the powder strikes the specimen, particles stick only to the tacky ridges of the finger print. Elsewhere on the specimen the particles lose their positive charge, pick up a negative charge and are repelled back to the plate below. The finger print in the photograph was made visible by this electrostatic process at the University of Manchester Institute of Science and Technology. The finger print, which was on paper, was invisible before the process was used. The same kind of idea is used in most modern photocopiers, where a dark powder or 'toner' is attracted to charged places on a metal plate and is then transferred onto paper.

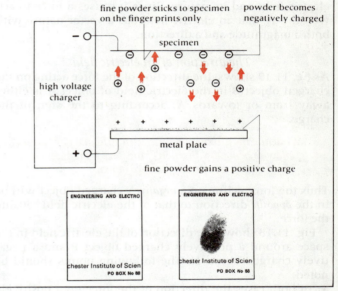

fine powder sticks to specimen on the finger prints only

powder becomes negatively charged

specimen

high voltage charger

metal plate

fine powder gains a positive charge

Assignments

Remember

a) Like charges repel, unlike charges attract.

b) Conductors share the same charge by contact.

c) A conductor charged by induction will have the opposite charge to that on the inducing agent.

d) All the charge delivered to the inside surface of a hollow conductor transfers to its outside surface so that the total charge inside the conductor remains zero at all times.

Draw

e) the diagrams of figs. 11.7, 11.8 and 11.9 showing how a positively charged strip of cellulose acetate could be used to induce the opposite charge on the metal spheres or electroscope. Show the direction of electron flow on your diagrams.

Explain

f) How does a balloon, rubbed on your sleeve, stick to a wall or ceiling? (This is an induction effect.)

g) Why may a damp or dusty dome on a Van de Graaff generator prevent it from working properly?

h) What happens when you bring a charged plastic object such as a comb close to the stream of water from a slow-running tap? (Another induction effect.)

Try questions 11.1 to 11.6

11.2
ELECTRIC FIELDS AND FORCES

We have seen that a charged object can affect other objects nearby without touching them. This action at a distance can be explained by what is called the electric field of the charged object.

The idea of an electric field

In the space around a charged object A we can detect various effects. For example another charged object B may move away from or towards A, fig. 11.17. Such effects are the result of a force which acts on any charge which comes into the region of influence of the charged object A. We call this influence around a charged object its **electric field**.

Any charge entering an electric field has a force acting on it and that action reveals both the existence of the electric field and its nature. Since it causes a force to act, we can say that an electric field has a vector nature with both a magnitude and a direction.

The direction of an electric field

As fig. 11.17 shows, the direction of the force acting on the charged object B in the electric field of object A is either away from or towards A, according to the sign of the charge.

We define the direction of the electric field at a particular place as being the direction of the force it produces on a positively charged object.

Thus the force acting on a *negatively* charged object will be in the *opposite* direction to that of the electric field causing the force.

Fig. 11.18 shows the direction of the electric field in the space around a positively charged object P and a negatively charged object N. The following points should be noted:

a) In both cases the direction of the force on a positively charged object is the same as the electric field direction, but the force acting on a negatively charged object is in the reverse direction to the field.

b) The basic rule that like charges repel and unlike charges attract is in agreement with these diagrams.

c) The electric fields around the objects P and N are symmetrical in three dimensions.

Figure 11.17 *A force acts on charges in the space around charged object A*

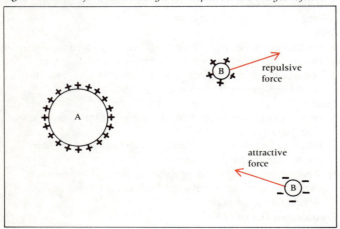

Figure 11.18 *The electric field in the space around a charged object*

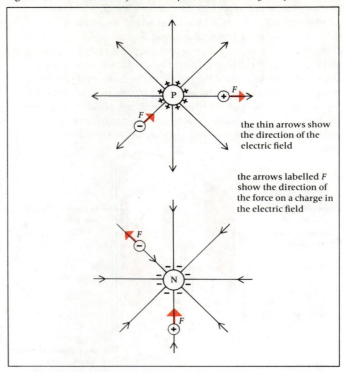

the thin arrows show the direction of the electric field

the arrows labelled *F* show the direction of the force on a charge in the electric field

d) Any electric field can be represented by lines like those drawn in fig. 11.18. These lines are usually called **field lines,** or **lines of force.**

e) We imagine that electric field lines (or lines of force) always start at a positive charge and end at a negative charge.

f) As we would expect from Newton's third law of motion, there will always be an equal and opposite (reaction) force acting on the charged object producing the electric field.

Making the shape of an electric field visible

An electric field exists between any two charged objects. We can make suitable objects from various shapes of metal wire and plates, which when charged are called **electrodes.** Some simple shapes of electrode which can be used are shown in figs. 11.19 and 11.20.

Assemble a pair of these metal electrodes in a shallow glass dish so that they are just covered by a layer of an insulating liquid such as castor oil, fig. 11.19. Connect the dome of a Van de Graaff generator to one electrode and its base terminal to the other electrode to earth it. Lightly sprinkle tiny grains or needles of an insulating material onto the surface of the oil. (Grass seed or semolina powder work quite well.)

The needles receive induced opposite charges at their ends and as the electric field between the electrodes causes forces to act on these charges, the needles become aligned in the direction of the electric field. The lines of force and shape of the electric field are made visible as the needles link together forming lines between the two electrodes. Some of the field shapes that can be demonstrated are shown in fig. 11.20. The direction of the electric field is, in each case, from the positive electrode to the negative electrode along the lines formed by the needles.

Figure 11.19 *Showing electric field lines*

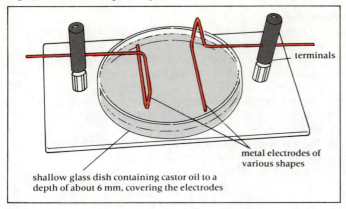

terminals

metal electrodes of
various shapes

shallow glass dish containing castor oil to a
depth of about 6 mm, covering the electrodes

Figure 11.20 *The shapes of some electric fields (the arrows show the direction of the electric fields)*

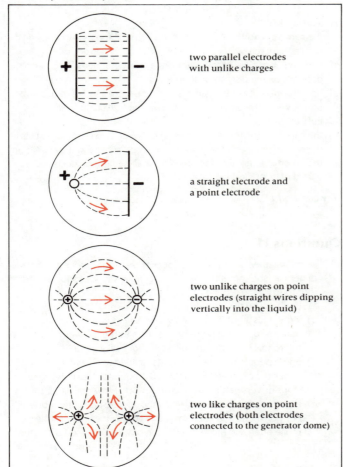

two parallel electrodes
with unlike charges

a straight electrode and
a point electrode

two unlike charges on point
electrodes (straight wires dipping
vertically into the liquid)

two like charges on point
electrodes (both electrodes
connected to the generator dome)

Investigating the force between charged objects

Fig. 11.21 shows an arrangement which can be used to investigate how the force between two charged objects depends both on their separation and on the charge on each object. A charged metal ball B is suspended so that it can be deflected when repelled by a like charge on another ball A. The deflection of ball B indicates the size of the force acting on it rather like the deflection of the gold leaf in an electroscope. A larger deflection indicates a stronger force of repulsion.

Figure 11.21 *The force between charged objects*

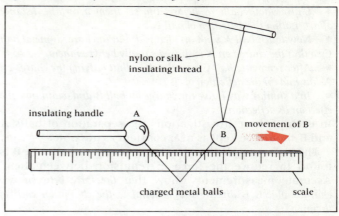

nylon or silk
insulating thread

insulating handle   A

movement of B

charged metal balls   scale

The effect of separation on the force

● *Charge both balls from an electrophorus or Van de Graaff generator.*

● *Bring ball A gradually closer to ball B and notice how the deflection of B changes.*

We find that when A is a long way off a small movement towards B makes very little difference to the deflection of B. As A gets close to B, the deflection of B increases by larger amounts for the same movement of A.

It is clear from this experiment that the force of repulsion increases as the separation decreases, which is an inverse relation. Accurate measurements show that *the force varies inversely with the square of the distance between two charged objects*. This relation, first discovered by Coulomb in 1785, is known as the **inverse-square law** of the force between two charged objects.

This law means that if we double the distance between two charged objects, the force of attraction or repulsion (caused by the electric field of one object acting on the other) becomes four times weaker. Fig. 11.22 illustrates this relation. Two objects A and B are given equal quantities of positive charge Q. When placed a distance d apart there is a force of repulsion F acting on both of them. When their separation is doubled to $2d$, the forces are reduced to $\frac{1}{4}F$, and so on.

Figure 11.22 *The effect of separation on the force between two charged objects*

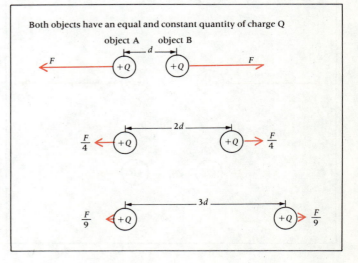

Both objects have an equal and constant quantity of charge Q

object A   object B

F $+Q$ d $+Q$ F

$\frac{F}{4}$ $+Q$ $2d$ $+Q$ $\frac{F}{4}$

$\frac{F}{9}$ $+Q$ $3d$ $+Q$ $\frac{F}{9}$

The effect of the quantity of charge on the force

- *Charge both balls as before and set them a certain distance apart, noting the deflection of ball* B.
- *Now touch ball* A *with an identical, but uncharged, metal ball* C *so that the charge on* A *is shared equally between them.*
- *After removing ball* C *note the effect that halving the charge on* A *has had on the deflection of* B.
- *In a similar way halve the charge on ball* B *and again note the effect on its deflection.*

In each of these investigations the separation of balls A and B should be kept as near constant as possible.

The effect of reducing the charge on either ball A or B is found to reduce the force of repulsion between them. Accurate measurements show that *the force between two charged objects is directly proportional to the charge on each of them.*

In other words, if we double the charge on either of two charged objects we double the force of attraction or repulsion between them. The way in which the force acting on both objects depends on the charge on both of them is shown in fig. 11.23. When a quantity of positive charge Q is placed on both objects A and B they both experience a repulsive force of F. The figure shows how doubling the charge to $2Q$ on one object will double the force to $2F$ on both objects, and so on.

It is important to realise that a change in the charge on just one of the objects affects the force acting on both of them. The two forces will, of course, remain equal and opposite whatever happens to the two charges or the distance between them.

Figure 11.23 *The effect of the charges on two objects on the force between them*

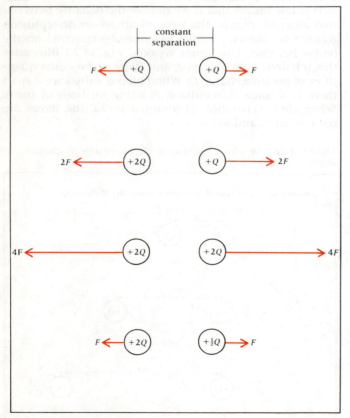

Assignments

Remember

a) An electric field is the region around a charged object in which a force acts on other charged objects.

b) The direction of an electric field is the direction of the force it produces on a positively charged object placed at a particular point in the field.

c) Electric field lines or lines of force go from a positive charge to a negative charge, indicating the direction of the electric field between them.

d) The magnitude of the F acting on a charged object **A** and caused by the electric field of another charged object **B** depends:

inversely on the square of the distance d between them,

$$F \propto \frac{1}{d^2} \ (Coulomb's \ law)$$

directly on both the charge Q_A on object **A** and the charge Q_B on object **B**,

$$F \propto Q_A Q_B$$

Find out

e) the name of another kind of field in which the forces also obey an inverse-square law.

f) how the brightness of the light from a photographer's flash gun depends on the distance from the gun.

g) how the intensity of the gamma radiation from a radioactive source changes with the distance from the source (see chapter 18).

Calculate

h) the change in the force of repulsion between two charged objects if the charge on one of them is made 4 times larger and the charge on the other one remains the same, but their separation decreases to a third of its original value.

Try question 11.7

Questions 11

1 In dry weather, when a driver touches the door before getting out of her car, she sometimes gets an electrical shock as soon as her foot touches the ground.
  a) Explain why this happens.
  b) Why does it not happen in wet weather?
  c) Some people fix a piece of fine chain to the car which drags on the road. How does this prevent shock taking place?
  d) Some people fix a piece of metal shaped as shown in fig. 11.24 under the car body. Even though it does not touch the road, it stops drivers getting an electrical shock. How does it do this?
  e) Describe an experiment which shows that: i) a lightly rubbed, dry polythene rod has a small charge; ii) if the rod is rubbed more often and more quickly, the charge produced is greater.

(Joint 16+, part)

Figure 11.24

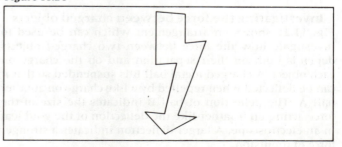

Figure 11.25

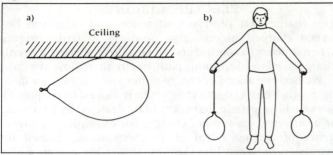

Ceiling

2 This is a question about electric charge. If you rub a balloon on your jumper you can charge it with static electricity. Sometimes it will stick to the ceiling when it is charged, fig. 11.25a.
   a) A balloon which is **negatively** charged has more electrons on it than a balloon which is not charged. i) What are electrons? ii) Why do extra electrons make the balloon **negatively** charged? iii) Where do these extra electrons come from?
   b) Fig. 11.25b shows two negatively charged balloons hanging from threads. i) What would you notice as the balloons are brought together? Give a reason for your answer. ii) What would happen if both balloons were **positively** charged?
   c) A plastic comb can be charged by rubbing it on your sleeve.
      i) How could you tell if a plastic comb was charged?
      ii) It is much more difficult to charge up a **metal** comb by rubbing it on your sleeve. Give a reason for this.   (SEG 88)

Figure 11.26

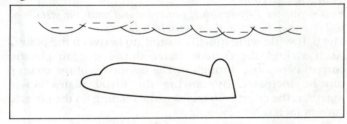

3 An aircraft flies just below a negatively charged thunder cloud. Movement of free electrons causes electrostatic charges to be induced in the aircraft.
   a) Copy fig. 11.26 and show the positions and signs of the induced charges on the aircraft.
   b) Explain, in terms of the movement of electrons, the distribution of the charges you have shown.
   c) What will happen to the induced charges when the aircraft flies away from the cloud?   (LEAG SPEC A part)

4 The police use electrostatic effects to reveal finger prints on paper cheques. This is shown in fig. 11.27.
   a) What is the charge of the powder on the plate?
   b) Why is the powder repelled from the plate?

Figure 11.27

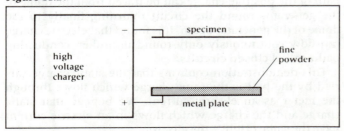

specimen

high voltage charger

fine powder

metal plate

c) The powder remains stuck only to the tacky ridges of the finger print. Why does the powder leave the other parts of the paper and make the finger print visible?

5 You are provided with two small insulated charged conductors of different shape, an electroscope and any other apparatus you may require.
   Describe one experiment, in each case, to show, preferably without discharging the conductors, that they possess (a) opposite charges, (b) unequal charges.   (L, part)

6 a) A student performed the following operations in the sequence described.
      i) Brought a negatively charged polythene rod near to the cap of an uncharged leaf electroscope.
      ii) Touched the cap of the electroscope momentarily with a finger.
      iii) Removed the rod.
   Draw diagrams showing the charge distribution on the cap and the leaf of the electroscope and the position of the leaf after each of the above operations.
   What would be the effect of removing the rod before removing the finger?
   Another student performing the same experiment allowed the polythene rod to rest on the cap of the electroscope in operation (i). State, giving your reasons, whether or not you would expect any marked difference in the results obtained by the two students.
   b) A manufacturer of nylon thread put heavy rubber mats under his spinning machines, which were made of metal, to deaden the noise. The following effects were subsequently noted.
      i) The workers sometimes received an electrical shock when touching the machines. (There was no leak from the mains cable.)
      ii) Small bits of nylon fluff stuck to the thread, but this could be overcome by keeping the air in the workshop moist.
   State and explain the physical reasons for (i) and (ii), and say how you could overcome (i).   (L)

7 Fig. 11.28 illustrates two conductors, A and B. A is mounted on an insulating stand and B, which is very light, is suspended by an insulating thread from P. There are no charges on A or B. A is then charged positively. Redraw and complete fig. 11.28b to show the new rest position of B.
   Show also the charges induced on the conductor B.
   A free positive charge, carried on a very small and very light sphere, is placed between A and the new position B. In which direction will the very light sphere move? Give a reason for your answer.   (C)

Figure 11.28

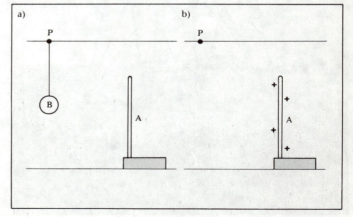

a)   b)

P   P

B

A   A

Electric currents

*A*n *electric current is a flow of electric charge. Now we investigate what makes electric charge flow, how it is measured and what it does.*

12.1
ELECTRIC CIRCUITS

The charge that produces attraction and repulsion between charged objects also produces electric currents when it flows in conductors. To sustain an electric current, charge needs a continuous path, or circuit, to flow around.

We become aware of charge either when it exerts a force on another charged object or when it moves. In chapter 11 we saw how a static charge could be detected by its action, at a distance, on another charge. For example, the divergence of the gold leaf of an electroscope, caused by repulsion between like charges, shows the presence of a static electric charge.

The following experiment shows how the charge which causes repulsion can also be responsible for an electric current when it flows round a continuous conducting path called a **circuit**.

Static charge and electric current
● *Support two metal plates about* 10 cm *apart and well above the bench, by fixing their insulating handles in stands fig. 12.1.*
● *Connect one plate to the dome of a Van de Graaff generator and the other, through a sensitive meter, to the base of the generator. (The meter should be capable of detecting very small electric currents; an instrument called a spot galvanometer is suitable.)*
● *Hang a metallised table tennis ball on an insulating thread so that it is positioned centrally between the two metal plates as shown. (The surface of a table tennis ball can be made into a conductor by coating it with aluminium paint or Aquadag which is a carbon-based paint.)*
● *Operate the Van de Graaff generator and touch the metallised ball on one of the metal plates, then let it go.*
The ball will swing quickly to and fro between the plates. As it swings the current meter shows that an electric current is passing through it. The frequency of the swings can be increased by moving the metal plates closer together; the effect is to increase the reading on the electric current meter.

The positive charge on the Van de Graaff generator dome flows along the wire conductor giving the metal plate A a positive charge. The positive charge on plate A induces an equal and opposite negative charge on plate B by attracting negative electrons to the surface of B.

When the metallised ball touches plate A it receives some positive charge and is immediately repelled from plate A and attracted to plate B, fig. 12.1b. (We could say that in the electric field between the two plates, a force acts on the ball towards B.) On touching plate B the ball gives up its positive charge and collects a negative charge which sends it back towards plate A, fig. 12.1c. Thus each swing of the ball transfers electric charge between the plates.

The swinging ball completes a continuous path, round which electric charge is made to flow by the generator. The path of the positive charge can be traced from the dome of the generator round the circuit returning finally to the dome of the generator again. It is found that electric charge can flow continuously only round unbroken conducting paths called **closed circuits**.

This demonstration confirms that the static charge carried by the ball is the same charge which flows through the meter as an electric current. We believe that static charge and the charge which flows in an electric current have the same origin – the electron.

Figure 12.1 *Charge flows round a circuit*

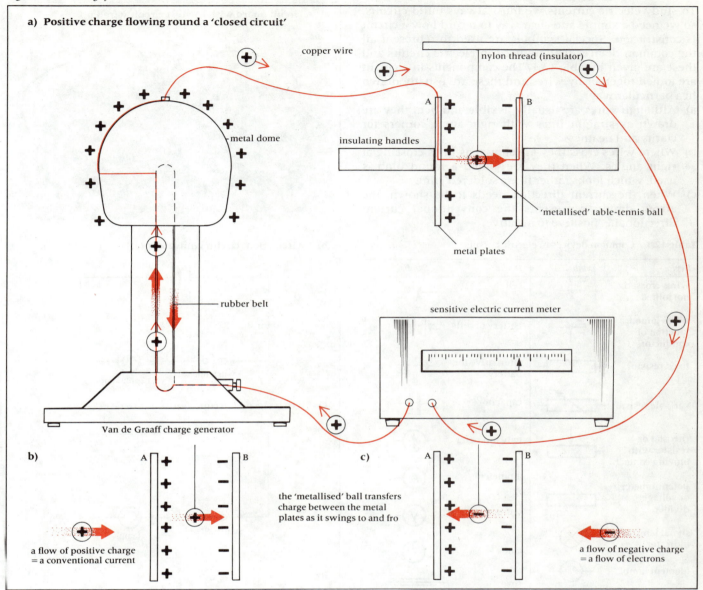

a) **Positive charge flowing round a 'closed circuit'**

copper wire

nylon thread (insulator)

metal dome

insulating handles

A + B −

'metallised' table-tennis ball

metal plates

rubber belt

sensitive electric current meter

Van de Graaff charge generator

b) A + − B

a flow of positive charge = a conventional current

the 'metallised' ball transfers charge between the metal plates as it swings to and fro

c) A + − B

a flow of negative charge = a flow of electrons

Conventional current and electron flow

We think of an electric current as a flow of *positive* charge round a conducting circuit as shown in fig. 12.1a. Arrows are usually drawn on wires to indicate this direction of flow which is called the **conventional current direction**. This means that it is usual to show electric current as flowing from a positively charged point to a negatively charged point (in the direction of the electric field), fig. 12.1b.

For a more accurate description of what happens in an electric current in a circuit, we must remember that charge does not exist on its own but belongs to certain of the very small subatomic particles of which matter is composed, see chapter 17. In particular, metallic conductors, such as copper wires, contain a number of electrons (of *negative* charge, as always), which are able to move through the conductor. Thus electrons drifting through a conductor produce a flow of negative charge in their direction of travel. Negative charge flows from negatively charged

points towards positively charged points. This is the *opposite* direction to the conventional current direction (and opposite to the direction of the electric field), fig. 12.1c.

So what really happens is that the electrons, of negative charge, flow in the wires of fig. 12.1a in the opposite direction to the conventional current (shown by the arrows). When the ball swings from B to A it is transferring excess electrons from B to A. When the ball swings back from A to B the positive charge it carries is really a shortage of electrons, plate A having 'stolen' them. Plate B then replaces the electrons missing from the ball and adds extra electrons to it, thus making the ball negative again.

The early experimenters with electricity thought that current really was a flow of positive charge. Even though we now understand more about conduction in metals, everyone continues to use the conventional current direction, as indeed we shall from now on in this book.

Drawing circuit diagrams

To study electric currents we must always build circuits, so we need a simple and clear way to record how a circuit is constructed. Special symbols are used to represent all the common devices that are used in electric circuits and these are given in table 12.1. The components in a circuit are joined together by wires and these are usually drawn in a particular way:

a) Although wires are usually flexible and bent they are drawn as straight lines with right-angle corners for clarity and neatness.

b) Where wires cross they are shown as lines crossing at right angles. When they are joined a round blob is used, which looks rather like a soldered joint.

c) When the current direction needs to be shown, the arrow drawn represents the conventional current direction, i.e. positive to negative.

Table 12.1   Common devices in electric circuits

Device	Symbol	Device	Symbol
wires crossed, not joined		cell	
wires joined, junction of conductors		battery of cells	
fixed resistor		alternative for battery	
variable resistor		capacitor	
rheostat or resistor with moving contact		galvanometer	
potentiometer or voltage divider		ammeter	
signal lamp or indicator		voltmeter	
filament lamp		clock	
		fuse	
diode or rectifier		switch	
earth or ground		a.c. supply	
heater		coil of wire or inductor	

See pp 348 and 349 for the symbols of some electronic devices.

Testing an electric circuit

Fig. 12.2a shows a type of circuit board which is suitable for investigating basic electric circuits. Connections between components in a circuit can easily be made using copper strips and plug-in wires. The lay-out of the components has the same orderly arrangement as the equivalent circuit diagram.

● *Connect up a circuit as shown in fig. 12.2b including a single 1.5 volt battery (called a cell), a switch and a lamp.*

● *Press the switch to close the circuit.*

When the switch is closed (or 'on') the lamp lights, fig. 12.2c.

Figure 12.2   *Using a circuit board to test an electric circuit*

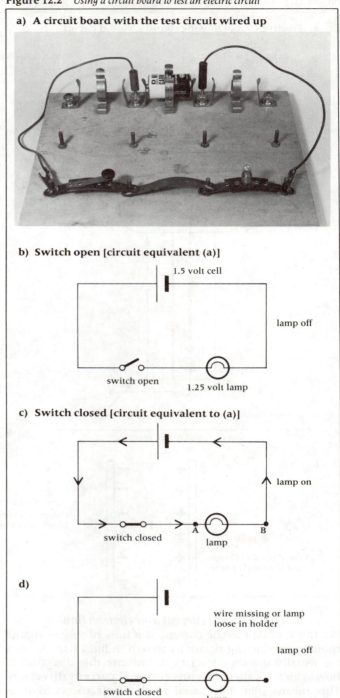

a) **A circuit board with the test circuit wired up**

b) **Switch open [circuit equivalent (a)]**

1.5 volt cell

lamp off

switch open   1.25 volt lamp

c) **Switch closed [circuit equivalent to (a)]**

lamp on

switch closed   A   B
lamp

d)

wire missing or lamp loose in holder

lamp off

switch closed   lamp

● *Investigate whether the lamp will stay lit if the circuit is broken at other points such as A or B, or by disconnecting a wire or by unscrewing the bulb in its holder, fig. 12.2d.*

The lamp lights only when there are no breaks anywhere in the circuit as in fig. 12.2c. The circuit must be closed for the lamp to light.

A **closed circuit**   is one in which there is an unbroken conducting path round which charge can flow continuously. Closing a switch often completes or closes a circuit, which switches a current 'on'.

An **open circuit** is one in which there is a break at one or more points in the conducting path so that there is no current anywhere in the circuit. When a switch is switched 'off' the circuit becomes 'open' and the break in the circuit stops the current.

A **short circuit** acts as a bypass for the electric current by providing an easier or shorter path for it to flow round.

● *Complete or close the circuit again so that the lamp lights.*

● *Now connect a piece of wire between points A and B and note what happens. (Do not leave this wire connected for very long because it will drain the battery very quickly.)*

This piece of wire acts as a short circuit. The current takes the easiest path, avoiding the lamp which now does not light up. The lamp is said to be *shorted out* of the circuit.

Conductors and insulators

We have already seen that some materials, called **conductors**, allow electric charge to flow through them and that others, called **insulators**, do not. We tested materials by touching the cap of a charged gold-leaf electroscope to see whether the charge could leak away through them. Now we can test materials by connecting them in an electric circuit to see whether they will conduct an electric current.

● *Connect up a circuit as shown in fig. 12.3, leaving a gap between points X and Y and using three cells.*

● *Find a good variety of materials for testing.*

Figure 12.3

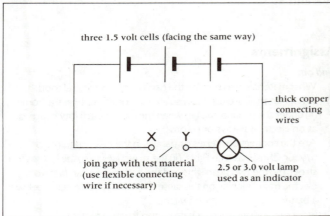

three 1.5 volt cells (facing the same way)

thick copper connecting wires

X Y

join gap with test material (use flexible connecting wire if necessary)

2.5 or 3.0 volt lamp used as an indicator

These materials should include some of the following: nylon, plastic, rubber, graphite ('lead' pencil), water, a salt solution, wood, gold (ring), aluminium (foil), glass (rod or tube), brass, copper and silver.

● *Use the test material to close the gap in the circuit between X and Y and close the switch.*

● *Record the brightness of the lamp for each material tested.*

When the lamp is bright, the material is a good conductor and is passing a large current. When the lamp is dimmer, the material is a poorer conductor and is passing a smaller current. When the lamp is off, very little or no current is flowing and this means that the material being tested is probably an insulator.

Using a lamp as the indicator, we can only show which materials conduct well enough to light it. If we wanted to show much smaller currents in many other materials, we would need to use a more sensitive current detector, such as a milli- or microammeter, in place of the lamp.

Tests like these give the following results:

a) All metals are good conductors of electricity, allowing large electric currents through them.

b) Some materials conduct electricity, but rather less well than metals. These include graphite, some special metal alloys (e.g. constantan and manganin), certain solutions called electrolytes, water (p 264), and the materials used in electronic devices which include germanium and silicon (p 354).

The white porcelain type of insulator on the London Underground insulates the live, current-carrying rail from the ground.

c) It is very difficult to pass any current through some materials. Among these the best insulators are polythene, p.v.c. (used to insulate electric cables), nylon and plastics in general, glass, rubber and natural materials such as quartz (a crystalline mineral), resins and wax.

d) As table 12.2 shows, materials form a spectrum from the best conductors to the best insulators and there is no clear division between them.

Table 12.2   A spectrum of conductors and insulators

Material	Class	
metals like silver, copper, aluminium	good conductors	↑
alloys like manganin and constantan	conductors	increasing
graphite	poor conductor	current with
electrolytes (water)	poor conductors	better
germanium	semiconductor	conduction
silicon	semiconductor	
plastic and rubber	insulator	
quartz	insulator	

Series and parallel connection of conductors

There are two basic ways of joining conductors together, known as **series connection** and **parallel connection**, although many circuits contain combinations of both. The circuits in fig. 12.4 show some possible arrangements of conductors.

● *Using three 1.25 volt lamps as conductors, connect up circuit (a) on a circuit board.*

● *Compare the brightness of the lamps.*

● *Now unscrew any one of the lamps and watch what happens.*

These lamps are said to be connected *in series*.

When lit, the lamps have equal brightness. This shows that the same current flows through all the lamps when they are connected in series. If any one lamp fails or comes loose in its holder the whole circuit is broken and, with no current anywhere in the circuit, all the lamps go off. This happens in some Christmas tree lamp circuits, but not all. Lamps connected in series in a circuit follow one after the other like a series of events.

Figure 12.4 *Connecting in series and parallel*

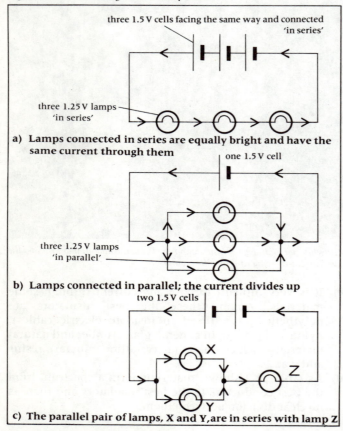

a) **Lamps connected in series are equally bright and have the same current through them**

b) **Lamps connected in parallel; the current divides up**

c) **The parallel pair of lamps, X and Y, are in series with lamp Z**

● *Connect the same three lamps together as shown in circuit (b), using only one 1.5 volt cell in this circuit to avoid burning out the lamps.*

They are now side-by-side or parallel to each other and are said to be connected *in parallel*.

● *Again compare the brightness of the lamps, then unscrew any one of them and note what happens.*

In parallel connection the current in a circuit divides up and only part of it flows in each conductor. When lamps are connected in parallel, if one fails it does not affect the other lamps; less current flows in the circuit as a whole.

Figure 12.5 *The lights at home are connected in parallel with each other so that each one can be switched on independently. When any one switch is closed there is a complete circuit from the fuse box, through the switch and lamp, back to the fuse box.*

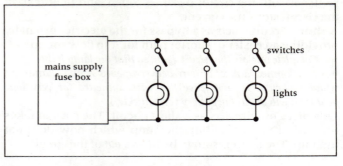

This is the most common arrangement. For example, at home we connect all the lights and electrical equipment in parallel so that they can be switched on and off separately, fig. 12.5.

● *Using three similar lamps as before, connect them to two 1.5 volt cells, as in circuit (c) of fig. 12.4, and again compare their brightness.*

In this circuit a pair of parallel-connected conductors, X and Y, are connected in series with a third conductor Z. The single lamp Z is somewhat brighter than the other two because all the current in the circuit must pass through Z while only half the current passes through each of X and Y.

Assignments

Find out

a) What materials are used for the handles of electricians' tools?

b) What materials are used to insulate electric cables (i) in the home, (ii) under the ground and (iii) when they carry electricity overhead as on electric railways or pylons?

c) Are the connections series or parallel in the following circuits: (i) your Christmas tree lights, (ii) the street lights, (iii) the houses in a street, (iv) the two headlights and the battery of a car, (v) two toy electric trains running on the same track and (vi) an electric bell with a battery and push-button switch?

Draw a circuit diagram for a battery torch with a switch.

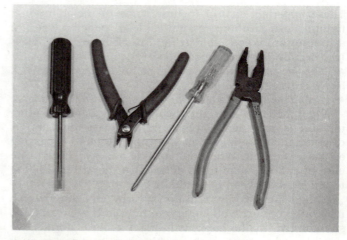

The handles of tools used by electricians, such as screwdrivers and pliers, are insulated in case contact is made with a 'live' wire.

12.2
MEASURING ELECTRIC CURRENTS

The electric current through a particular conductor or machine must not be too large or too small for it to work correctly. For example, too little current may fail to light a lamp or turn an electric motor, but too much current could burn out the filament of the lamp or turn the motor too quickly. So it is important to be able to measure the electric current in a circuit, or through a conductor.

Electric current and charge

To get an idea of the 'strength' or size of an electric current we can compare the flow of electric charge with the flow of water in a river.

We would say that the current was strong if a large quantity of water was flowing quickly down a river. The strength or size of this current could be measured in litres of water flowing past a point in the river in a certain time. We would in fact be measuring the *rate of flow* of the water.

Similarly the 'strength' or size of an electric current is a measure of the *rate of flow of electric charge* past a point in an electric circuit or through a conductor. To use this relation between the size of an electric current and the rate of flow of charge we need symbols and units for these quantities.

The quantity of electric charge Q is measured in **coulombs** (symbol C).

The size or 'strength' of an electric current I is measured in **amperes** (symbol A). In everyday usage we usually say 'amps' but the only correct written abbreviation for ampere is A.

The relation between current I and charge Q

Suppose 24 coulombs of charge pass through a conductor in 8 seconds. The size of the electric current, i.e. the rate of flow of charge through the conductor, will be the charge flowing through in 1 second. The current is 3 coulombs in 1 second or 3 amperes. We see that the relation between current and charge is the following:

$$\text{current} = \frac{\text{charge}}{\text{time}} \qquad I = \frac{Q}{t}$$

Which gives us:

A current of 1 ampere is a flow of charge at the rate of 1 coulomb per second.

This idea of the ampere being 'a coulomb per second' is the one to understand and remember; the formal definition of the ampere, which is given on p 299, is for reference only.

Worked Example
The relation between current and charge

If a charge of 180 C flows through a lamp every 2 minutes, what is the electric current in the lamp?

We have:
$Q = 180\,C$
$t = 2\text{ minutes} = 2 \times 60\,s$

Using $\qquad I = \dfrac{Q}{t}$

gives $\qquad I = \dfrac{180\,C}{2 \times 60\,s} = 1.5\,A$

Answer: The current in the lamp is 1.5 amperes.

The unit of charge: the coulomb

By rearranging the relation between current and charge we get:

$$\begin{array}{ccc} \text{charge} & = & \text{current} \times \text{time} \\ \text{(in coulombs)} & & \text{(in amperes)} \quad \text{(in seconds)} \end{array} \qquad Q = It$$

which gives us the definition of the coulomb:

A coulomb is the charge which flows in 1 second past any point in a circuit in which there is a steady current of 1 ampere.

Worked Example
Charge and current

A battery circulates charge round a circuit for 30 s. If the current in the circuit is 5 A, what quantity of charge passes through the battery?

We have:
$I = 5\,A$
$t = 30\,s$
Using $Q = It$ gives:

$$Q = 5\,A \times 30\,s = 150\,As$$

Now $As = \text{ampere} \times \text{second} = \text{coulomb}$

so: $\qquad Q = 150\,C$

Answer: 150 coulombs of charge pass through the battery. Note that the charge flowing round a circuit passes through the battery as well as the rest of the circuit.

How large is a coulomb?

It is difficult to imagine the vast number of electrons needed to provide just 1 coulomb of charge. Each electron has a minute charge of only 1.6×10^{-19} coulombs, so about 6×10^{18} electrons are needed to make up just 1 coulomb of charge i.e. 6 million million million electrons! In terms of practical quantities of charge which can be stored, a coulomb is a large quantity and we more often use millicoulombs ($1\,mC = \frac{1}{1000}\,C$ or $10^{-3}\,C$) and microcoulombs ($1\,\mu C = 10^{-6}\,C$).

How large is an ampere?

In practical terms the ampere has a convenient size since the currents found in most domestic machines vary from about $\frac{1}{4}A$ in a light bulb to about $10\,A$ in a heater. However, very small currents of milli- and microamperes are common in electronic circuits, while a car battery can briefly pass a current as high as 400 amperes through a starter motor.

Since an ampere is a flow of charge at the rate of 1 coulomb per second, when there is a current of 1 ampere through a conductor it follows that 6×10^{18} electrons must pass through it every second. We should realise, however, that this very large number of moving electrons is still only a small fraction of the total number of electrons present in a conductor.

Measuring current in a series circuit

Ammeters, as their name suggests, are 'amp meters' and measure electric current in amperes. So how do we use an ammeter and where should it be connected in a circuit?

The circuit diagram of fig. 12.6 shows four possible positions for an ammeter in a series circuit.

● *Wire up this circuit, in turn connecting the ammeter in each of the positions shown.*

Figure 12.6 *Measuring current in a series circuit*

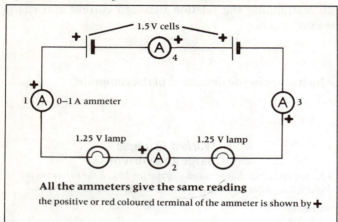

1.5 V cells

1 Ⓐ 0–1 A ammeter

All the ammeters give the same reading

the positive or red coloured terminal of the ammeter is shown by ✚

Note that for its pointer to move the correct way across the scale, the positive (or red coloured) terminal of the ammeter should be connected to the positive terminal of the battery. (This is shown by the + signs in the circuit.)

● *Note the reading of the ammeter when connected in each position.*

In all four positions the ammeter gives the same reading showing that the *current is the same all round a series circuit*. It is even the same through the battery! In this series circuit of a single closed loop there is only one conducting path for the current. So, wherever the ammeter is connected in the circuit, all the current must flow through it, and it will give the correct reading. The ammeter can be connected anywhere in the series circuit.

The current at a junction in a circuit

● *Connect two lamps together in parallel in a circuit as shown in fig. 12.7 so that ammeter 1 reads the current through one lamp, ammeter 2 reads the current through the other lamp and ammeter 3 reads the current returning to the cell.*

● *Note the readings on the ammeters when all three switches are closed and when, in turn, each switch is opened.*

● *More readings can be obtained by using different lamps and more cells in the circuit.*

J is a junction in the circuit. The current entering the junction is measured by ammeters 1 and 2. Ammeter 3 measures the current leaving the junction. The results

Figure 12.7 *Currents at a junction in a circuit*

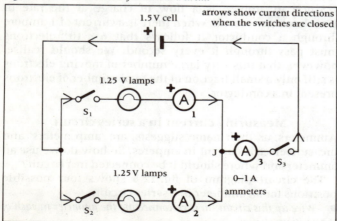

show that the readings of ammeters 1 and 2 always add up to equal the reading of ammeter 3. This discovery is known as **Kirchhoff's first law:**

The total current entering a junction in a circuit must equal the total current leaving it.

This law is used to calculate the currents when conductors are connected in parallel and at junctions, as shown in the examples given in fig. 12.8.

Figure 12.8 *Kirchhoff's first law*

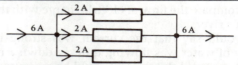

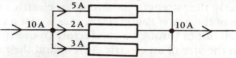

a) The currents in parallel-connected conductors must add up to the total current in the whole circuit. In this example identical conductors pass equal parts of the total current.

b) In this example different conductors pass different currents but the total still equals the current in the full circuit.

Worked Example
Kirchhoff's first law

In fig. 12.9 currents of 5 A and 3 A are entering a junction in a circuit and a current of 2 A is leaving. Find the size and direction of the unknown current x.

Using Kirchhoff's first law:

total current entering junction $= 5\,A + 3\,A = 8\,A$
total current leaving junction $= 2\,A + x$
as these must be equal we have: $2\,A + x = 8\,A$
therefore: $x = 6\,A$

Answer: *x* is a current of 6 amperes leaving the junction.

Figure 12.9

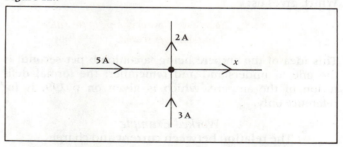

Assignments

Remember

a) The formulas:

$$I = \frac{Q}{t} \quad \text{and} \quad Q = It$$

b) The definition of the unit of charge, the coulomb.
c) The current in a series circuit is the same all round the circuit.
d) Kirchhoff's first law.

Try questions 12.1 to 12.6

12.3
POTENTIAL DIFFERENCE AND THE VOLT

We have seen that charge flows round electric circuits; but what makes it flow, what energy changes take place when it does so? To answer these questions it helps if we first consider what makes water flow.

Water in a pond can remain absolutely still without any currents, while water in a stream rushes restlessly downwards with a strong current. We are familiar with these examples and do not stop to ask why, because we think it is natural for water to behave that way; it is what we expect. Now try answering a few questions:

- *Why does a stream flow downhill?*
- *What effect on the flow of water has the height or steepness of a hill down which it flows?*
- *What energy changes occur as the water descends?*

Of course water flows downhill; it is pulled down by gravity like everything else. We also expect water to flow faster down steeper hills. Water has potential energy at the top of a hill which is mostly converted into heat by the time it reaches the bottom of the hill.

In the still water of a pond there is no current . . .

. . . but in a stream the downhill flow of water produces a strong current.

Charge flows 'downhill'

Now apply this idea to the flow of charge in an electric circuit. We might ask, where is the hill and how high is it? A comparison is shown in fig. 12.10. There are, you might say, high and low levels in electric circuits and charge flows 'downhill' like water. The electrical level or 'height' is called the **voltage** or the **potential**. Using the conventional current direction, (i.e. a flow of positive charge,) 'downhill' is towards the negative terminal of the battery. The battery does the job of pumping charge up to the 'top of the hill' so that it can then flow 'downhill' as a current through the conductors in the circuit. The higher the voltage to which the battery raises the charge, the steeper will be the downward slope and the faster the charge will flow, thus increasing the electric current. At the top of the 'hill' the charge has *electrical* potential energy which is converted into other forms of energy as it flows round the circuit.

Figure 12.10 *Comparison of electrical and water 'hills'*

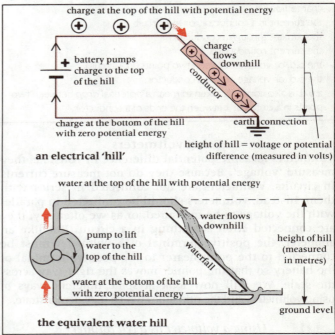

Potential difference, p.d. or voltage

When we state the height of a hill it is given as metres above a certain level, usually sea level or ground level, because what is important is the *difference in height* between the top and the bottom. In the same way, the **potential difference** in an electric circuit between two points is the most important factor.

A difference in water levels causes a current of water to flow downhill, while no difference of levels in a pond results in no flow. In exactly the same way a potential difference in an electrical circuit causes charge to flow 'downhill', and the current depends upon the size of the potential difference.

The words 'potential difference' are often abbreviated to p.d., but its symbol in formulas is *V*. The unit of potential difference is the volt and its symbol is also V (not printed in italics). We often talk about 'voltage' when we really mean 'potential difference in volts'.

Earth potential

Sometimes a point in a circuit is connected to the earth or ground which fixes it at earth potential or zero volts. This is equivalent to using sea level as a reference for heights, sea level being zero height. Earth connections are often made to metal water pipes or metal stakes which go deep into the ground. The symbol ⏚ in a circuit indicates connection to earth or ground. If the negative terminal of a battery is taken to be zero potential, then all points in its circuit have a positive potential difference, or positive voltage above zero.

Using electrical terms correctly

Try to use the correct words when talking about electric charge, current and potential difference. The list below gives some correct ways of using these terms. Avoid the error of talking about a p.d. or voltage 'flowing' or 'going through' a conductor. Similarly a voltage cannot be 'in' a conductor. Ask yourself, how can a 'height difference' flow in or through something? The following phrases illustrate the correct use of terms:

> charge *flows*,
> the current *in* a conductor, or *in* a circuit,
> the current *through* a conductor,
> the current *round* a circuit,
> the p.d. or voltage *between* two points (in a circuit),
> the p.d. or voltage *across* a conductor,
> a p.d. is also correctly taken to mean a *potential drop* between two points in a circuit or between the ends of a conductor.

Using voltmeters

Voltmeters measure potential difference in volts, i.e. they measure 'voltage'. Because they do not measure currents in circuits, voltmeters are not connected in series with them. Instead, voltmeters should be connected in parallel with the voltage to be measured, or as we often say, they are connected *across* something in a circuit. But like an ammeter, the positive terminal of a voltmeter must be connected to the point nearer to the positive terminal of the battery so that the pointer moves the right way across the scale. We will now look at some different ways of using voltmeters and the principles which they demonstrate.

Using a voltmeter as a cell counter

If several similar cells are connected together in series and facing the same way, they can be 'counted' by a voltmeter.

● *Using a circuit board or other cell holder, connect a voltmeter across first one cell, then two and three and so on, fig. 12.11a.*

The voltmeter readings increase in equal steps (of 1.5 V for dry cells) for each additional cell, thus indicating the number of cells. From this investigation we find that *when cells are connected in series, their voltages add up.*

It is helpful to compare this use of the voltmeter with measuring heights. A scale placed alongside a wall could measure its height either in centimetres or in 'rows of bricks'. In fig. 12.11b, the wall is 40 cm or 5 rows of bricks high. In the same way a voltmeter could measure the electrical 'height' or voltage between two terminals of a battery either by a reading in volts or in terms of the number of cells. For example, in the figure, the voltage across the whole battery is either 4.5 volts or it is 3 cells 'high'. Note that in both cases the scale is placed alongside or parallel to the object to be measured.

Figure 12.11 *Comparing a voltmeter with a height scale*

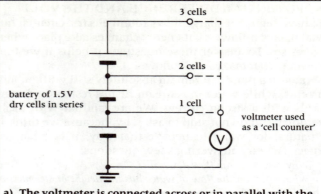

a) **The voltmeter is connected across or in parallel with the battery to be measured. The battery is a total of 3 cells or 4.5 volts 'high'.**

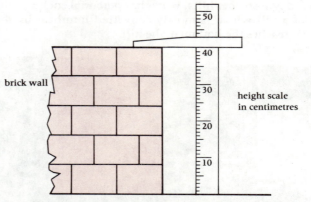

b) **The scale is placed alongside or parallel with the height of the wall to be measured. The wall is 5 rows of bricks or 40 cm high.**

Using a voltmeter in a series circuit

When a voltmeter is used to measure the voltage across a conductor in a circuit or any part of a circuit, it is connected in parallel with it. Fig. 12.12 shows the correct positions for a voltmeter when connected to measure the voltage across conductors in a series circuit. The arrows show the current round the series circuit. Note that it does not flow through the voltmeters.

Figure 12.12 *Connecting a voltmeter to a series circuit*

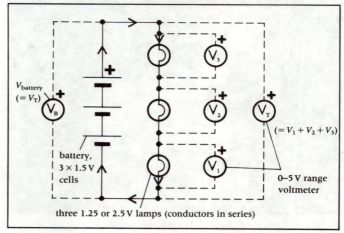

three 1.25 or 2.5 V lamps (conductors in series)

• *Connect up the circuit shown in fig. 12.12, using three 1.5 V cells in series with three similar lamps as the conductors, (lamps rated at 1.25 V are suitable).*

• *Connect a voltmeter (0–5 V range) across each lamp in turn and then across all three lamps and record its readings, V_1, V_2, V_3 and V_T (V_{total}).*

• *Change two of the lamps for others of different ratings, say 2.5 V, and repeat the voltmeter readings.*

The voltmeter readings show that in both cases:

$$V_1 + V_2 + V_3 = V_T$$

The voltages across each of the three lamps connected in series add up to the total voltage across all three of them, V_T. Thus in general we can say that: *the voltages across conductors connected in series in a circuit add up.*

• *Now connect the voltmeter across the terminals of the battery and note its reading, V_B.*

We find that $V_B = V_T$, i.e. the voltage across the battery terminals is always equal to the total voltage across the conductors in series with it. These two positions of the voltmeter are measuring the p.d. or voltage between the same two levels in the circuit and must give the same reading. Notice that one terminal of each voltmeter (reading V_B and V_T) is connected to the 'top' of the circuit with only a connecting wire between their joining places, and the other terminal of each voltmeter is connected to the 'bottom' of the circuit, again with only a wire between their joining places. When there is only a piece of wire between two places in a circuit they are at the same potential or voltage and amount to the same place in the circuit. It follows that *the voltage across the terminals of a battery always equals the total voltage across the components in its circuit.*

Using a voltmeter in a parallel circuit

• *Reduce the battery to only one 1.5 V cell as shown in fig. 12.13, (this is to avoid burning out the lamps).*

• *Connect three lamps in parallel to the battery.*

• *Connect the voltmeter in each of the positions shown and note its readings: V_B across the battery and V_C across the conductors, i.e. the lamps.*

We find that $V_B = V_C$, i.e. the voltmeter reads the same in both positions again. One end of each lamp is connected to the same low potential level in the circuit and the other end of each lamp is connected to another common, but higher, potential level. Thus the p.d. or voltage between these levels is the same for all the lamps and equals the voltage between the terminals of the battery. We can always say that *the p.d.s or voltages across conductors connected in parallel are equal.*

Figure 12.13 *Connecting a voltmeter to parallel conductors*

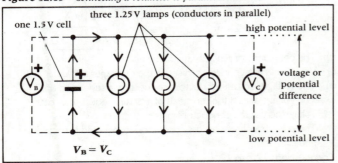

Voltage, work and energy

To understand the energy changes in a circuit it helps to remember that the flow of charge is really a flow of electrons. When electrons are pumped round a circuit by a battery they transport energy from one point in the circuit to another. But at no point in a circuit does the gain or loss of energy by the electrons result in a change in their kinetic energy. Any change in kinetic energy would mean a change of speed, and a change of speed of the electrons would mean a change of current; but we know this does not happen. (The current is the same all round a series circuit.) It follows that the energy change of the electrons as they flow round a circuit must be a change of their *potential energy*, see fig. 12.14.

The cell gives its own stored chemical energy to the electrons. It does this work by raising their electric charge to a higher potential or voltage. Thus the electrons provide the means of passing the electrical potential energy round a circuit to places where it can be converted by a conductor into other forms of energy such as heat and light. The work the cell does in raising the charge to a higher voltage is equal to the energy converted in the rest of the circuit by the conductors. This idea is shown in fig. 12.14. Part (a) shows that the energy converted in the cell and the conductor are equal. Part (b) shows charge Q being lifted up a voltage V by a cell and then flowing through a conductor where it drops an equal voltage V as it returns to the cell at the zero potential level. Part (c) shows the equivalent electric circuit as it is usually drawn.

Figure 12.14

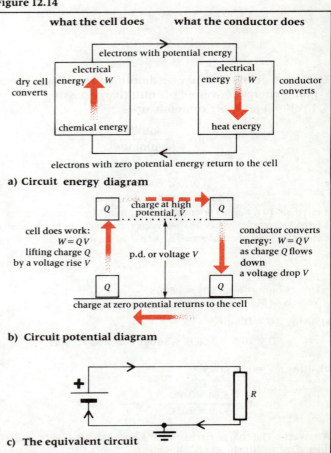

The link between voltage and energy is shown by parts (a) and (b). The work W done by the cell depends on both the quantity of charge Q lifted and the voltage V by which the charge is raised. The amount of electrical potential energy W gained by the charge is given by $W = QV$. An equal amount of energy is converted to other forms as this charge flows through the conductor. In fact we use the amount of energy converted in the conductor for each unit of electric charge that passes through it, to define the p.d. across it.

The voltage or p.d. between the ends of a conductor is equal to the energy converted from electrical to other forms per unit electric charge flowing through it.

Suppose that 24 joules of heat energy are produced in a conductor as 8 coulombs of charge flow through it. The amount of energy converted when 1 coulomb of charge flows through it is therefore 3 joules, i.e. the energy converted per unit of charge flowing through the conductor is 3 joules per coulomb. This energy conversion requires a potential difference of 3 volts between the ends of the conductor. From this example we can see that the relation is:

$$\text{p.d.} = \frac{\text{energy}}{\text{charge}} \quad V = \frac{W}{Q}$$

The unit of potential difference, the volt, is defined in the same way:

The p.d. between the ends of a conductor is 1 volt if 1 joule of energy is converted from electrical to other forms when 1 coulomb of charge flows through it.

These two definitions may be simplified as follows:
p.d. = energy converted per unit charge passing,
1 volt = 1 joule per coulomb, or

$$\text{volts} = \frac{\text{joules}}{\text{coulombs}}$$

Worked Example
Charge, energy and p.d.

A current of 10 A flowing through an electric heater for an hour converts 8.64 MJ of electrical energy into heat energy. Calculate (a) the total charge circulated through the heater and (b) the p.d. across the heater.

We have:

$I = 10\,\text{A}$
$t = 1\,\text{hour} = 60 \times 60\,\text{s} = 3\,600\,\text{s}$
$W = 8.64\,\text{MJ} = 8.64 \times 10^6\,\text{J}$

a) using $Q = It$ gives

$$Q = 10\,\text{A} \times 3\,600\,\text{s} = 36\,000\,\text{C} = 3.60 \times 10^4\,\text{C}$$

b) using $V = \dfrac{W}{Q}$ gives

$$V = \frac{8.64 \times 10^6\,\text{J}}{3.60 \times 10^4\,\text{C}} = 240\,\frac{\text{J}}{\text{C}} = 240\,\text{V}$$

Answer: The charge circulated in 1 hour is 3.60×10^4 C and the p.d. across the heater is 240 volts.

Worked Example
Energy, charge and current

A battery circulates 60 C of charge round a circuit.
a) If the p.d. across a lamp in the circuit is 12 V, how much energy is converted into heat and light by the lamp?
b) If the charge is circulated at a constant rate of flow for 20 s, what is the current in the circuit during this time?

We have:
$Q = 60\,\text{C}$
$V = 12\,\text{V}$
$t = 20\,\text{s}$

a) Using $W = QV$ gives

$$W = 60\,\text{C} \times 12\,\text{V} = 720\,\text{J}$$

(coulombs × volts = joules)

b) Using $I = \dfrac{Q}{t}$ gives

$$I = \frac{60\,\text{C}}{20\,\text{s}} = 3\,\frac{\text{C}}{\text{s}} = 3\,\text{A}$$

(coulombs per second = amperes)

Answer: The energy converted is 720 joules and the current in the circuit is 3 amperes.

Electromotive force

In a circuit there needs to be a source of energy which enables charge to be pumped or forced round a circuit. There are a great variety of sources of electrical energy available for use in circuits (see chapter 13), but here the only source we shall refer to is a 'cell'.

A cell produces an electromotive force (or e.m.f. for short), E, which is responsible for the pumping action of the cell. Although it is helpful to think of the cell as forcing charge round a circuit, the e.m.f. of a cell is not strictly speaking a force. The e.m.f. of a cell is, in fact, a measure of its ability to do the work necessary to pump charge up to a higher voltage. This work was shown in fig. 12.14.

The electromotive force or e.m.f. of a cell (or other source) is equal to the energy converted to electrical form by the cell per unit charge passing through it.

Since the e.m.f. of a cell is also measured in volts, the e.m.f. can be described as *the total number of joules per coulomb available from the cell.*

Assignments

Remember
a) The formulas:

$$W = QV \quad \text{and} \quad V = \frac{W}{Q}$$

b) The definitions of p.d., e.m.f. and the volt.
c) When cells are connected in series their voltages add up.
d) The voltages across conductors connected in series add up.
e) The voltage across the terminals of a battery equals the total voltage across the conductors in its circuit.
f) The p.d.s or voltages across conductors connected in parallel are equal.

Try questions 12.7 to 12.9

12.4
RESISTANCE

Resistance to motion is an everyday experience. We are aware of the effort needed to move ourselves through water, or the need for an engine to keep a car moving, even on a level road. The cause of this resistance is the friction between materials or moving parts, but what do we know about the resistance to a flow of current in a conductor? Resistance there must be, because a battery or generator is needed to drive the current. Resistance arises in all components of a circuit, but while the wires that connect components should have as little resistance as possible, the resistance in a lamp plays a vital role. The resistance is where the electrons give up the potential energy they carry from the battery. For example the resistance of a lamp causes the electrical energy to change into other forms, such as heat and light. If a lamp filament had no resistance, no energy change could occur in it and it would not light up.

The resistance of a conductor depends on two things:
a) its dimensions,
b) the material of which it is made.

A thin wire has a higher resistance than a thick wire; a long wire has a greater resistance than a short one. Again, thinking of water flowing through a pipe, we should expect more resistance when it flows through a long, thin pipe than through a wide, short one.

The material the conductor is made of affects its resistance. A good conducting material has more 'free' electrons than a poorer one, and if these electrons can flow easily through the material its resistance is lower.

Measuring resistance by a simple method

The resistance of a conductor opposes the current in a circuit, so to investigate its resistance we must measure the current through it. We must also measure the voltage which causes charge to flow through the conductor. To do this we require a circuit in which we can measure both the voltage across a conductor and the current through it.

A simple form of such a circuit is shown in fig. 12.15. (A length of thin wire made of a high-resistance alloy called constantan is suitable as the conductor; use a length of 1 metre of insulated wire of diameter 0.4 mm, or standard wire gauge 28 or 30.)

Figure 12.15 *A simple circuit to measure resistance using an ammeter and voltmeter*

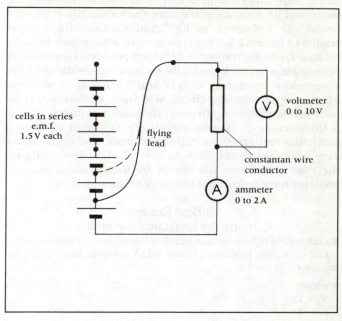

● *Connect the conductor to one cell.*

● *Measure the current through the conductor and the voltage across it.*

● *Increase the number of cells connected in the battery so that the current in the circuit increases.*

● *Record readings of both meters as each extra cell is added.*

You should be able to see a pattern of regular increases for both current and voltage.

● *Divide the voltage by the current for each pair of readings giving a value for V/I. Table 12.3 shows an ideal set of readings.*

The constant value obtained for *V/I* is called the **resistance** of the conductor.

Table 12.3 Voltage and current readings for a constantan wire conductor

V/volts	0	1.5	3.0	4.5	6.0	7.5	9.0
I/amperes	0	0.2	0.4	0.6	0.8	1.0	1.2
$\dfrac{V}{I} = R$/ohms	–	7.5	7.5	7.5	7.5	7.5	7.5

Resistance and the ohm

The relation between current *I* and p.d. or voltage *V* is the basis of the definition of resistance.

The resistance of a conductor is the ratio of the voltage across it to the current through it.

$$\text{resistance} = \frac{\text{voltage}}{\text{current}} \qquad R = \frac{V}{I}$$

We can use the units of p.d. and current, the volt and the ampere, to define the unit of resistance, called the **ohm** (Ω). (The symbol Ω is the last letter of the Greek alphabet 'omega'.)

The ohm is the resistance of a conductor through which the current is 1 ampere when the p.d. between its ends is 1 volt.

How large is an ohm?

In the experiment with a length of constantan wire, we measured its resistance. We used the formula $R = V/I$, and found the resistance to be 7.5 ohms from the sample results in table 12.3. This type of wire was chosen because it has a fairly high resistance which is easy to measure. A wire made of a good conductor such as copper should have a very low resistance, only a small fraction of an ohm. Most components of a circuit will have a resistance typically of a few ohms up to many thousands of ohms.

Conductors in circuits which are designed to have a particular resistance are called **resistors**, and their values vary from less than 1 ohm to several $M\Omega$, depending on their use. Note how the words are used: conductors and resistors have *resistance* to the current through them.

Worked Example
Using the resistance formula

If a current of 4 A flows through a car headlamp when it is connected to a 12 V car battery, providing a voltage of 12 V across the lamp, what is its resistance?

We have:

$V = 12\,V$

$I = 4\,A$

Using $R = \dfrac{V}{I}$ gives

$$R = \frac{12\,V}{4\,A} = 3\,\Omega$$

(volts/amperes = ohms)

Answer: The resistance of the lamp is 3 ohms.

Worked Example
Using $V = IR$

What voltage would be needed to drive a current of 0.2 A through a torch lamp of resistance 22.5 Ω?

We have:

$I = 0.2\,A$

$R = 22.5\,\Omega$

Using $V = IR$ gives

$$V = 0.2\,A \times 22.5\,\Omega = 4.5\,V$$

Answer: The voltage across the lamp should be 4.5 volts.

Ohm's law

The resistance of most conductors is found to vary with temperature. For metals R increases at higher temperatures while for semiconductors R usually decreases at higher temperatures. George Ohm made measurements of the resistance of metal wires and discovered that the ratio V/I remained constant so long as the wire was kept at a constant temperature. In 1826 he announced his now famous law.

The current through a metallic conductor, maintained at constant temperature, is directly proportional to the potential difference between its ends.

$$I \propto V$$

Conductors with a constant value for the ratio V/I are said to obey Ohm's law and are described as **ohmic conductors**.

A model for Ohm's law

In fig. 12.16 a tank contains water which is flowing out through a horizontal pipe at its base.

- *Set up a tank like this in which the water level inside the tank can be seen and compared with the jet of water flowing out of the pipe.*

Figure 12.16 *Ammeter and voltmeter shown in equivalent positions on a water flow diagram. The water current is directly proportional to the height difference, h.*

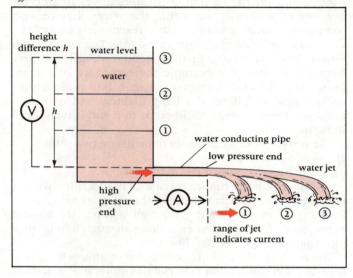

- *Compare the level of the water or its height h above the pipe with the range of the jet for several values of h.*

There is a *pressure difference* between the ends of the pipe which is proportional to the height of water above the pipe. The rate of flow of water or 'water current' is indicated by the range of the water jet beyond the end of the pipe. The slowest flow ① reaches the shortest distance and the fastest flow ③ reaches the farthest.

Now compare this flow of water with a flow of electric charge. We see that the water height difference h corresponds to V, the electrical p.d. as measured by the voltmeter. The 'water current' as shown by the range of the jet is equivalent to the electric current as measured by the ammeter. The water height difference h is the cause of the pressure which pushes the water out through the pipe. Similarly an electrical potential difference V is the cause of the flow of electric charge through a conductor.

The water model shows that the 'water current' is directly proportional to the height difference h. This is the same relationship as the one described by Ohm's law, i.e. the electric current I is directly proportional to the potential difference V.

Assignments

Remember

a) The formulas:

$$R = \frac{V}{I} \quad \text{and} \quad V = IR$$

b) The definition of resistance and the ohm.

c) Ohm's law.

Try questions 12.10 to 12.15

12.5
PRACTICAL RESISTORS AND CIRCUIT CONTROL

Types of fixed resistor

Fixed resistors are made of a variety of materials which are suitable for different uses. The cheapest type of resistor is made by baking in a kiln a composition of carbon black and a binding material, and is surrounded by a ceramic tube, fig. 12.17a. The resistance value of these resistors is not very accurate and tends to vary, but they are still widely used in electric circuits. In fact they are so common in radio circuits that they are often called 'radio resistors'. The resistance value of these cylindrical resistors is given by four coloured bands painted round them. Table 12.4 shows how these coloured bands give the value and the accuracy of the resistor. Another resistance code is given in table 12.5, overleaf.

Since carbon composition resistors are rather inaccurate and variable, they are not suitable for some circuits. Another type of carbon resistor is made by coating a ceramic rod with a thin film of carbon and cutting a helical groove in the carbon film to give the resistor a more accurate and constant value. Even better stability is achieved by coating a ceramic rod with tin oxide instead of carbon. In each case, the resistor is protected by an outer coating of lacquer or resin, or a ceramic tube.

The best quality resistors are wire-wound, fig. 12.17b. The wire usually used is enamelled nichrome alloy which has a very high resistance. Apart from increased accuracy and stability, wire-wound resistors can be made to conduct large currents without being damaged by the heat produced in them.

Sometimes we need a very accurate resistor for use in a measuring instrument, and these are known as **standard resistors**, fig. 12.17c. The wire in these resistors is usually made of constantan or manganin alloys whose high resistances vary very little when their temperature changes.

Figure 12.17 *Some fixed resistors*

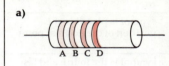

a)

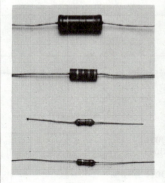

The resistance value of many types of fixed resistor is given by the coloured bands, A B C D, which are painted round the cylinder. The resistance value is found by using the resistor colour code given in Table 12.4

The physical sizes of the resistors (shown here on millimetre graph paper) vary according to the current they can conduct and to the maximum rate of production of heat (i.e. power) which they can stand without being damaged. In order of size the resistors shown here have maximum power ratings of 2 watts. 1 watt, 0.5 watts and 0.25 watts. In the same order, they are of the following types: carbon film, carbon composition, tin oxide and another carbon film resistor.

b)

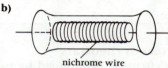

nichrome wire

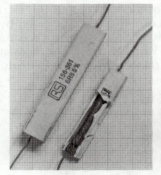

Wire-wound resistors are usually made of nichrome wire covered with a protective enamel coating. The maximum power rating of these resistors may be 10 or more watts.

c)

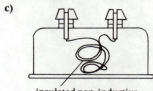

insulated non-inductive coil of constantan wire

A laboratory standard resistor is made very accurately by carefully adjusting the length of a piece of constantan wire. The photograph shows a 1 ohm standard resistor whose resistance value is accurate to within 0.2% of its stated value.

The circuit symbol for all fixed resistors.

Table 12.4 Resistor colour code

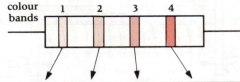

Colour	Value	Value	Multiply by	Tolerance
Black	0	0	1	red = ± 2%
Brown	1	1	10	gold = ± 5%
Red	2	2	100	silver = ± 10%
Orange	3	3	1 000	no band = ± 20%
Yellow	4	4	10 000	
Green	5	5	100 000	This gives the
Blue	6	6	1 000 000	maximum error
Violet	7	7		in the value of
Grey	8	8	not	the resistor.
White	9	9	used	

Examples:
Brown, black, red = 1 0 00 = 1000 ohm = 1 kΩ (± 20%)
Orange, orange, orange = 3 3 000 = 33 000 ohm = 33 kΩ
Green, blue, green = 5 6 00 000 = 5 600 000 ohm = 5.6 MΩ
Brown, green, black, gold = 1 5 and no noughts = 15 ohm (± 5%)

Table 12.5 The new resistor symbol code.

Symbol	Meaning	Use
R	ohms	symbol is placed in the
K	thousand ohms	position of the decimal point
M	million ohms	
F	± 1%	symbol is the fourth and last
G	± 2%	and gives the tolerance or
J	± 5%	accuracy of the value
K	± 10%	
M	± 20%	

Examples:

Symbol		Meaning
1 R O	=	$1.0\,\Omega$
33 R	=	$33\,\Omega$
K 56	=	$0.56\,k\Omega$ or $560\,\Omega$
1 K O	=	$1.0\,k\Omega$ or $1000\,\Omega$
27 K	=	$27\,k\Omega$ or $27000\,\Omega$
M 10	=	$0.10\,M\Omega$ or $100000\,\Omega$
1 M O	=	$1.0\,M\Omega$ or $1000000\,\Omega$
M 33 K	=	$0.33\,M\Omega$ or $330000\,\Omega \pm 10\%$
68 R K	=	$68\,\Omega \pm 10\%$
R 22 J	=	$0.22\,\Omega \pm 5\%$

Variable resistors or rheostats

The current that flows in a circuit is determined by two factors, the e.m.f. or voltage of the battery and the total resistance of the circuit. The current can be increased by reducing the circuit resistance, or reduced by increasing the resistance. By including a variable resistor or rheostat in a circuit we can vary or control the current in the circuit.

The circular type of rheostat shown here with its back removed has a curved, wire-wound track which may be from about 2 cm to 4 cm in diameter. A knob is fitted to the other end of the central shaft.

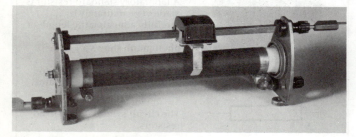

The much larger laboratory rheostat can conduct currents of several amperes without being damaged.

The design of a rheostat depends largely on the size of current which it has to control, because the current must pass through it and will cause a heating effect as it does. Fig. 12.18a shows a small circular type of rheostat. This design has a slider on a rotating arm which connects terminal B with a circular conducting track. A knob, connected to the arm, is turned in order to vary the resistance. The conducting track is made of a thin carbon film or, for larger currents, a coil of fine constantan wire. The length of the track through which the current flows determines the amount of resistance in the circuit. Although the device has three terminals, A, B and C, when it is used as a rheostat only two terminals are needed. If terminals A and B are used, then the minimum resistance is when the slider is round at A and the maximum when it is turned fully clockwise at C. Note that the resistance between terminals A and C is fixed and cannot be varied.

The straight form of wire-wound rheostat, shown in fig. 12.18b, is usually used in laboratory experiments where currents of several amperes are involved. Notice that the slider takes the current from the wire coils and passes it through a solid brass rod to the terminal B at the top. Again note that using terminals A and C provides only a fixed resistance.

Figure 12.18 *Variable resistors or rheostats*

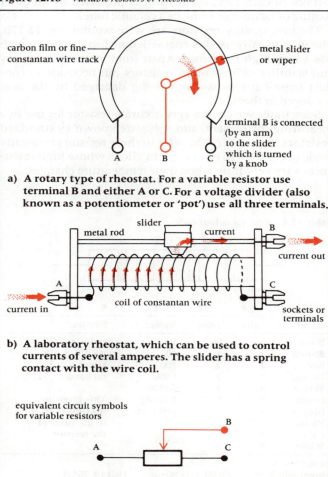

a) **A rotary type of rheostat. For a variable resistor use terminal B and either A or C. For a voltage divider (also known as a potentiometer or 'pot') use all three terminals.**

b) **A laboratory rheostat, which can be used to control currents of several amperes. The slider has a spring contact with the wire coil.**

Using a rheostat to vary the current in a circuit

● *Wire up the circuit of fig. 12.19 and see how the rheostat varies the current in the circuit. This is shown by both the brightness of the lamp and the reading of the ammeter.*

We notice the following points:

a) The rheostat is connected in *series* with the circuit and the current to be varied or controlled.

b) The rheostat cannot reduce the current in the circuit to zero (to do that it would require an infinitely high resistance). When the rheostat is set at its maximum resistance, the ammeter still indicates some current even though the lamp is not lit.

Figure 12.19 *Circuit using a rheostat to vary the brightness of a lamp. Variation of circuit resistance controls current. There is zero resistance from the rheostat and therefore maximum current, when B slides to A.*

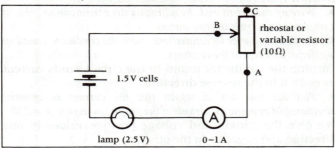

Using a voltage divider

● *Wire up the circuit shown in fig. 12.20 using the same device as was previously used as a rheostat.*

When used as a voltage divider, terminals A and C act as its 'input' side and terminals A and B provide its 'output'. The battery is connected to the input of the voltage divider and a lamp and ammeter are connected to its output.

● *Compare how the voltage divider controls the current through the lamp with the way the rheostat did in the circuit of fig. 12.19.*

The current through the lamp and ammeter can be varied as before but now the current can be reduced to zero by the voltage divider. The voltage divider varies the voltage or p.d. supplied to the lamp and so varies the current through it. The voltage divider also takes the fixed voltage of the battery and divides it into two parts which is why we call it a voltage divider.

Figure 12.20 *The voltage divider working as a smoothly variable battery or source of variable voltage*

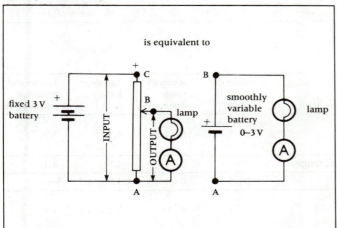

This can be demonstrated using the circuit of fig. 12.21. The input of 4.5 V from the battery is divided into two parts by the voltage divider. These are measured by two voltmeters which show that the two parts add up to the input voltage. So the output of the voltage divider between terminals A and B is a part of the divided input voltage. The size of this output part can be chosen by moving the slider connected to terminal B from A to C. When the slider is moved to end A the output voltage is zero. As it is moved towards C, the output smoothly increases from zero to the full input voltage.

A voltage divider is also known as a **potential divider** or a **potentiometer**. Potentiometers or voltage dividers are more common in electric circuits than rheostats. For example most volume, brightness and tone controls use voltage dividers. We can easily tell if a device is being used as a voltage divider because all three of its terminals will be connected to different points in a circuit and it will probably have four wires connected to it.

Figure 12.21 *The voltage divider divides a fixed voltage into two parts, which when added equal the fixed voltage. The division is smoothly variable.*

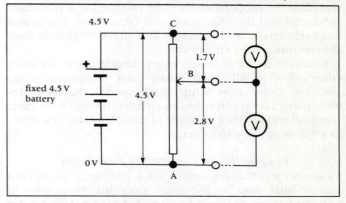

Assignment

Look at the circuit of fig. 12.22. It is an advanced circuit for a light-sensitive switch. Decide how each resistor, labelled A to E, is being used in this circuit, but do not try to explain how the circuit works. Find a resistor which is being used as:

a) a fixed resistor,

b) a rheostat or variable resistor,

c) a potential divider.

(Look carefully at resistor B; two of its terminals are joined together and do not connect to different points in the circuit.)

Figure 12.22 *A circuit for a light-sensitive switch. How is each resistor (A to E) being used?*

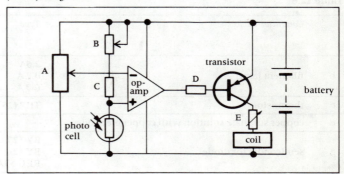

12.6
CHARACTERISTICS

The characteristic of a circuit component or device tells us about the way it conducts electric current and shows how it may be used in an electric circuit.

Ohm's law and the characteristic of a conductor

If we measure the current I through a conductor for various values of the p.d. or voltage V across it and calculate the value of its resistance R, we usually find that each pair of values for V and I give a slightly different value for the resistance R. The most common reason for this change in the resistance of a conductor is the change in temperature produced by the heating effect of the electric current in it. Because of the variation of resistance with temperature, conductors are unlikely to obey Ohm's law unless their temperatures are kept constant.

When we plot a graph of the current I through a conductor, against the voltage V between its ends, the shape of the curve obtained is known as the **characteristic** of the conductor. A straight-line graph (of positive gradient, through the origin) would show that the current was directly proportional to the voltage for a particular conductor and therefore it obeyed Ohm's law. The nearer its characteristic is to a straight line, the more closely does the conductor obey Ohm's law.

However, as we shall see, many conductors do not obey Ohm's law at all, even when their temperatures are constant. Conductors which have special characteristics have particular applications in circuits. (Perhaps the most common and striking example of this is the use of a diode as a one-way valve in a circuit.)

Finding the characteristic of a conductor

To obtain a full characteristic for a conductor we need a voltage that can be increased smoothly from zero to positive values. This can be supplied by a potentiometer or voltage divider circuit as shown in fig. 12.23.

● *Choose the voltage of the battery and the ranges of the ammeter and voltmeter to suit the particular conductor or device to be tested. Some suitable values are given in table 12.6.*

When in doubt about the size of the current, it is wise to start with a high-range ammeter, say 0–10 A or 0–1 A. If the current is found to be small, drop down to more sensitive ranges. This avoids overloading a sensitive ammeter with a large current and damaging it.

● *Since damage could be caused to the conductor by passing too much current through it, connect a protective resistor in series with*

Figure 12.23 *Circuit to find conduction characteristics*

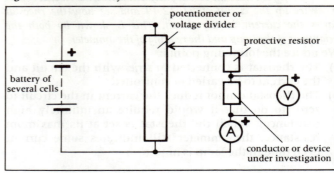

it to limit the current through it. (Table 12.6 suggests suitable protective resistors.)

● *Wire up the circuit and, by adjusting the potentiometer, obtain a set of readings from the two meters.*

● *By changing over its connections, turn the conductor round in the circuit and repeat the readings.*

Turning the conductor round in the circuit sends current through it in the reverse direction.

● *Plot the two sets of results (for the current in opposite directions) in opposite quadrants of the graph, as shown on p 237.* We give the current and voltage positive values in one direction and negative in the other.

Some devices have a high resistance in one direction and a low resistance in the other. In these cases, the low-resistance direction (which allows the larger current through) is called the **forward** direction. The high-resistance direction is called the **reverse** direction.

Assignments

Plot a graph

Table 12.7 gives a set of results obtained for a filament lamp in one direction only.

Plot a graph of I against V and draw the characteristic as a smooth curve through the points.

Up to what value of the current does the lamp obey Ohm's law?

What happens to its resistance above this value of the current?

Try question 12.16

Table 12.7 Voltage and current readings for a filament lamp

V/volts	0	0.5	1.0	1.5	2.0	2.5	3.0	3.5
I/amperes	0	0.04	0.08	0.12	0.15	0.175	0.19	0.20

Table 12.6

	Conductor/device	Type	Voltmeter range (volts)	Ammeter range (amperes)	Supply or battery (volts)	Wirewound protective resistor (ohms)
b	filament lamp	2.5 V 0.2 A 0.3 A	0 – 5	0 – 1	4.5	3.9
c	thermistor	TH7 (25 to 1 Ω range)	0 – 5	0 – 1	4.5	2.2
d	copper sulphate solution with copper electrodes	—	0 – 5	0 – 1	3.0 or 4.5	—
g	semiconductor diode	BY 127 BY 126 REC 53A	0 – 1	0 – 1	3.0 or 4.5	3.9

— *Some important characteristics* —

a) Ohmic conductor

(Here constantan wire) The straight-line characteristic is symmetrical in both directions and passes through the origin of the graph. This conductor closely obeys Ohm's law: $I \propto V$ and R is constant.

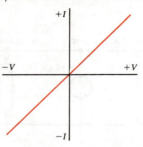

b) Filament lamp

At low currents the characteristic may be fairly straight but as the current rises, producing more heat, the temperature rise increases its resistance. So at a particular voltage where the filament temperature has risen, the current value is lower than it would be for an ohmic conductor.

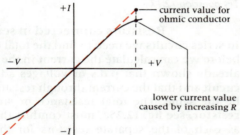

c) Thermistor

A thermistor is a resistor made of a semiconductor material. The resistance of a thermistor changes rapidly as its temperature changes and usually a rise in temperature causes a fall in its resistance. A falling resistance allows the current to increase more rapidly than it would in an ohmic conductor.

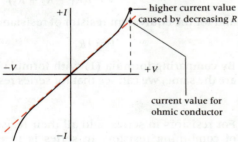

d) Thermionic diode

(Conduction is through a vacuum by electrons emitted from a heated metal electrode.) This device does not conduct at all in the reverse direction, and although part of the characteristic in the forwards direction is roughly ohmic, the current reaches a maximum or saturation value.

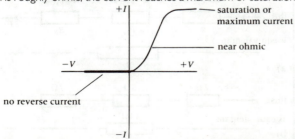

e) Ionic solution or electrolyte

(Conduction is by means of ions.) Different characteristics are obtained for different electrolytes, but the electrodes (i.e. the metal terminals where the current enters and leaves the solution) determine whether the graph goes through the origin. Conduction through the electrolyte itself is ohmic and the graph is straight, but when certain electrodes are used conduction does not begin until the applied voltage reaches a certain value. The electrodes produce an opposing voltage (effectively a cell) which has to be overcome by the applied voltage. Two examples are given. Copper sulphate solution with copper electrodes is ohmic, but dilute sulphuric acid with carbon electrodes requires 1.7 V in either direction before conduction begins (p 266).

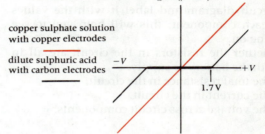

f) Gases

(Conduction is by the movement of ionised gas molecules and electrons.) At low voltages conduction is near ohmic but the current is very small. As the voltage is increased, the speed of the ions increases and more reach the electrodes and so the current increases. Over a wide range of voltages the current is very small and nearly constant. This **saturation current**, as it is called, is constant because the ions are arriving at the electrodes as fast as they are being produced and no change in voltage can make them arrive any faster. The current increases again at very high voltages when the gas begins to conduct in the form of a spark or discharge.

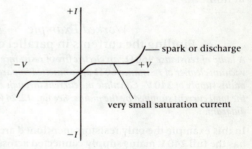

g) Semiconductor diode or p-n junction (silicon)

(Conduction is by electrons and 'positive holes'.) This device is the semiconductor replacement for the thermionic diode and its characteristic shows a lot of similarities. However, a very small current flows in the reverse direction and, apart from a slow start to the current rise, it is nearly ohmic in the forwards direction (p 342).

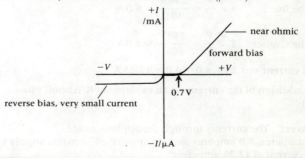

12.7
CIRCUIT CALCULATIONS

The aim of this section is to give help with the solving of numerical questions about electric circuits. Before starting on this work, refresh your memory about series and parallel connection of conductors (p 224). Then read through this section following the worked examples and trying the selected questions at each stage. There is a pattern or system in the way many circuit problems are solved and it is vital to understand the correct method and sequence of using the formulas.

The following stages form a plan for solving circuit problems.

1) Draw a circuit diagram and label it with the values given for each component; this will help you to see what is going on.
2) Decide whether the resistors in the circuit are all in parallel with the battery or supply.
3) Calculate the total resistance in the circuit.
4) Calculate the current in the circuit.
5) Calculate the voltage across circuit components.

Parallel circuits

When a circuit diagram has been drawn, if all the resistors are in parallel across the supply and if the voltage across them is known, then the current through each one can be calculated separately using $I = V/R$. There is no need to find their total resistance, unless this is asked for. If, however, the resistance of the supply or battery has to be included, then this is in series with the other parallel resistors and the total circuit resistance must be calculated first.

Parallel circuits are very common as is shown in the first worked example which is about the parallel circuits used at home.

Worked Example
Finding the currents in parallel circuits

A lamp of resistance 960 Ω, an electric fire of resistance 30 Ω and a vacuum cleaner of resistance 60 Ω are connected in parallel across the mains supply of 240 V. Calculate the current through each appliance and the total current supplied by the mains. See fig. 12.24 for the circuit diagram.

In this example the only resistors mentioned are in parallel. Each has the full 240 V mains supply connected across it and can be used independently. The current through each resistor can be calculated separately and there is no need to find the total circuit resistance.

For the lamp $\quad I = \dfrac{V}{R} = \dfrac{240\,\text{V}}{960\,\Omega} = 0.25\,\text{A}$

For the fire $\quad I = \dfrac{V}{R} = \dfrac{240\,\text{V}}{30\,\Omega} = 8.0\,\text{A}$

For the cleaner $\quad I = \dfrac{V}{R} = \dfrac{240\,\text{V}}{60\,\Omega} = 4.0\,\text{A}$

Total current = 0.25 + 8.0 + 4.0 = 12.25 A

(The addition of the currents is an example of Kirchhoff's first law.)

Answer: The currents through the appliances are 0.25 amperes, 8.0 amperes and 4.0 amperes. The mains supply a total current of 12.25 amperes.

Figure 12.24 *Conductors connected in parallel across the mains each receive the same p.d. of 240 volts. The total current from the mains, 12.25 A equals the sum of the separate currents through the parallel conductors.*

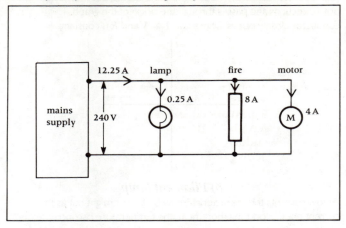

Assignment

Try question 12.17

Resistors connected in series

In series circuits we need to find the total circuit resistance before we can calculate the current in the circuit. We have already shown that p.d.s or voltages add up in a series circuit and that the current through resistors in series must be the same. The total resistance or single **equivalent resistor**, see fig. 12.25a, must conduct the same current I as each of the separate resistors for the same applied voltage V. Adding up the voltages:

$$V = V_1 + V_2 + V_3$$

We can use $V = IR$ so for each resistor

$$V = IR_1 + IR_2 + IR_3$$
$$\therefore \quad V = I(R_1 + R_2 + R_3) \qquad (1)$$

and for the equivalent resistor of resistance R we have

$$V = IR \qquad (2)$$

By comparing formula (1) with formula (2), since V and I are the same, we can see that for series resistors

$$R = R_1 + R_2 + R_3$$

For resistors in series, add all their resistances. The effect of combining resistors in series is to increase the total circuit resistance, and to decrease the current throughout the circuit.

Figure 12.25 *Equivalent resistors (series)*

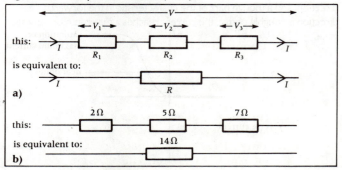

Worked Example
Resistors connected in series

If three resistors of 2 Ω, 5 Ω and 7 Ω are connected together in series, what value single resistor could replace them and allow the same current to flow?

See fig. 12.25b. Using the formula:

$$R = R_1 + R_2 + R_3$$

gives: $\qquad R = 2\,\Omega + 5\,\Omega + 7\,\Omega = 14\,\Omega$

Answer: A 14 ohm resistor could replace them all.

Resistors connected in parallel

We have seen that the p.d. or voltage across parallel connected conductors is the same. Referring to fig. 12.26, the voltage across the single equivalent resistor of resistance R must also equal the voltage across each of the parallel resistors. From Kirchhoff's first law we know that the current I divides up so that:

$$I = I_1 + I_2 + I_3$$

The current in each resistor is given by $I = V/R$

$$\therefore \; \frac{V}{R} = \frac{V}{R_1} + \frac{V}{R_2} + \frac{V}{R_3}$$

As all the voltages are equal we can divide through the equation by V and so we get the formula for resistors connected in parallel:

$$\frac{1}{R} = \frac{1}{R_1} + \frac{1}{R_2} + \frac{1}{R_3}$$

Figure 12.26 *Equivalent resistors (parallel)*

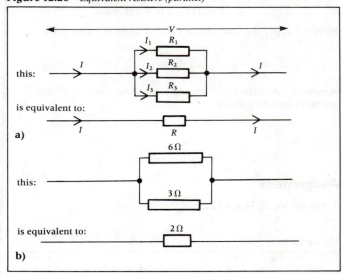

a)

b)

For resistors in parallel, add the reciprocals of the resistances to give the reciprocal of the combined, equivalent resistance. Connecting conductors in parallel provides more ways for the current to flow. This makes it easier for the current to flow, and makes the combined resistance less than each separate resistance. Resistors connected in parallel have a *smaller combined resistance* and conduct a larger total current than they do separately.

Some people prefer to use an alternative form of this formula for only two resistors in parallel. It is easier to use (if you are not using a calculator with a reciprocal key) and

it gives the combined resistance rather than its reciprocal, but it only works for two resistors. By rearrangement, the formula

$$\frac{1}{R} = \frac{1}{R_1} + \frac{1}{R_2} \qquad \text{becomes:} \qquad R = \frac{R_1 R_2}{R_1 + R_2}$$

This is the alternative formula for two parallel resistors.

Worked Example
Resistors connected in parallel

If a 6 Ω resistor is connected in parallel with a 3 Ω resistor, what is their equivalent combined resistance?

See fig. 12.26b. Using the formula:

$$\frac{1}{R} = \frac{1}{R_1} + \frac{1}{R_2}$$

gives: $\qquad \dfrac{1}{R} = \dfrac{1}{6\,\Omega} + \dfrac{1}{3\,\Omega} = \dfrac{1+2}{6\,\Omega} = \dfrac{3}{6\,\Omega} = \dfrac{1}{2\,\Omega}$

$$\therefore \; R = 2\,\Omega$$

Alternatively using:

$$R = \frac{R_1 R_2}{R_1 + R_2}$$

gives: $\qquad R = \dfrac{6\,\Omega \times 3\,\Omega}{6\,\Omega + 3\,\Omega} = \dfrac{18\,\Omega^2}{9\,\Omega} = 2\,\Omega$

Answer: A resistor of 2 ohms is equivalent to the two parallel resistors.

Assignment

Try questions 12.18 and 12.19

Mixtures of series and parallel resistors

When there is a mixture of series and parallel connections of resistors we need to know which ones to combine together first. In arrangement (a) in fig. 12.27 we can say that R_2 is in parallel with *only* R_3 and that R_1 is in series with *both* R_2 and R_3. In arrangement (b) we can say that R_2 is in series with *only* R_3, but together they are in parallel with R_1. Start by combining the resistors which are the *only* ones joined in a particular way. The following examples will help to explain how.

Figure 12.27 *Series and parallel mixtures of resistors*

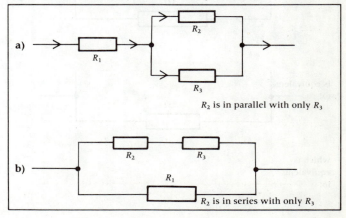

a)

R_2 is in parallel with only R_3

b)

R_2 is in series with only R_3

Worked Example
Mixed series and parallel combinations of resistors

Find the equivalent resistance of the two arrangements of resistors shown in fig. 12.28.

a) The $6\,\Omega$ resistor is in parallel with *only* the $3\,\Omega$ resistor, therefore we combine the parallel pair first. In a previous example we found that $6\,\Omega$ and $3\,\Omega$ in parallel is equivalent to $2\,\Omega$. Then the series pair combine to give:

$$R = R_1 + R_2 = 3\,\Omega + 2\,\Omega = 5\,\Omega$$

Answer: The equivalent resistance is 5 ohms.

b) The $2\,\Omega$ resistor is in series with *only* the $4\,\Omega$ resistor, therefore we combine the series pair first. Again using the formula: $R = R_1 + R_2$ gives:

$$R = 2\,\Omega + 4\,\Omega = 6\,\Omega$$

Then the parallel pair combine, and we know that $6\,\Omega$ and $3\,\Omega$ in parallel are equivalent to $2\,\Omega$.

Answer: The equivalent resistance is 2 ohms.

Assignment

Try questions 12.20 and 12.21

Figure 12.28

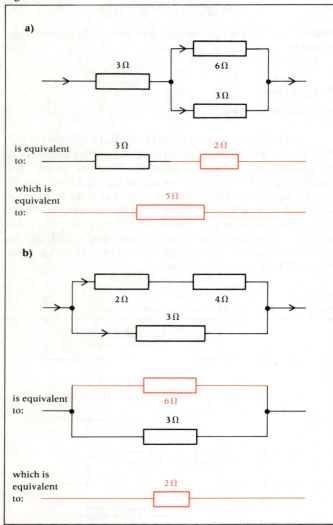

Calculating the current in a series circuit

We have already looked at the calculation of the current in a parallel circuit, but now we must learn how to calculate the current in a circuit which has some or all of its resistors connected in series. Since every bit of resistance in a series circuit reduces the current in the whole circuit, the current in a series circuit is calculated from the total applied voltage and the total resistance in the circuit (including any resistance in the supply or battery itself, p 273).

At this stage we must be clear about how we can use the formulas

$$V = IR \qquad R = \frac{V}{I} \qquad I = \frac{V}{R}$$

These can be used for individual resistors, for groups of resistors or for whole circuits, but their use in each case must be consistent. This means that the values of R, V and I used in a formula must all apply to the same thing, e.g. a single resistor. The next two worked examples show how to use the formulas correctly.

Worked Example
The current in a series circuit

Calculate the current in the circuit of fig. 12.29.

This is a series circuit so we first find the total circuit resistance. For the total circuit

$$R = R_1 + R_2 + R_3$$
$$\therefore \quad R = 1\,\Omega + 2\,\Omega + 3\,\Omega = 6\,\Omega$$

Now we apply the formula for current to the whole circuit:

$$I = \frac{V}{R} = \frac{12\,\text{V}}{6\,\Omega} = 2\,\text{A}$$

Answer: A current of 2 amperes will flow through all the resistors and the battery.

Assignment

Try questions 12.22 and 12.23

Figure 12.29

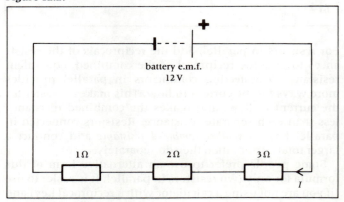

Calculation of current
in parallel branches of a series circuit

Fig. 12.30a shows the current I in a series circuit dividing up through two parallel resistors of resistance R_1 and R_2. We know from Kirchhoff's first law that: $I = I_1 + I_2$. Also, resistors in parallel must have the same voltage V across them. From the formula:

$$I = \frac{V}{R}$$

we can write for R_1 and R_2:

$$I_1 = \frac{V}{R_1} \quad \text{and} \quad I_2 = \frac{V}{R_2}$$

dividing the first equation by the second we get:

$$\frac{I_1}{I_2} = \frac{V}{R_1} \times \frac{R_2}{V}$$

cancelling V we get:

$$\frac{I_1}{I_2} = \frac{R_2}{R_1}$$

As fig. 12.30b shows, the current I divides up so that, in proportion, there is more current through the smaller resistor and less current through the larger one. The currents in parallel resistors are in the ratio I_1/I_2, which is the inverse of the ratio of their resistances, R_2/R_1. If the parallel resistances are equal then the current divides up equally, with half of the total circuit current going through each resistor. When there are more than two parallel resistors, the current through each one must be calculated from the formula $I = V/R$, where V is the voltage across the group of parallel resistors.

Figure 12.30 *Currents in parallel conductors*

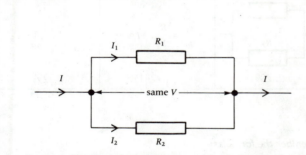

a) Current dividing through two parallel branches of a series circuit, $I = I_1 + I_2$

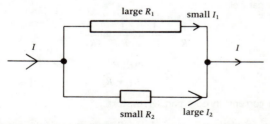

b) For parallel resistors the currents are in the inverse ratio of the resistances: $\dfrac{I_1}{I_2} = \dfrac{R_2}{R_1}$

Worked Example
The currents through resistors
connected in mixed series and parallel arrangements

A parallel pair of resistors of values $4\,\Omega$ and $12\,\Omega$ are together connected in series with another resistor of value $3\,\Omega$ and a battery of e.m.f. $24\,V$. Calculate the current through each resistor.

As we work through the solution of this question, see how we follow the steps suggested at the beginning of this section. The first step is to draw a circuit diagram because this helps us to 'see' what is happening, fig. 12.31. We see that there are some series connections in this circuit and so we must find the total circuit resistance before we attempt to find any currents.

The $12\,\Omega$ resistor is in parallel with *only* the $4\,\Omega$ resistor, so we combine these two first. For the parallel pair

$$R = \frac{R_1 R_2}{R_1 + R_2} = \frac{4\,\Omega \times 12\,\Omega}{4\,\Omega + 12\,\Omega} = \frac{48\,\Omega^2}{16\,\Omega} = 3\,\Omega$$

Then in series we have $3\,\Omega$ and the equivalent $3\,\Omega$ therefore total circuit resistance $= 3\,\Omega + 3\,\Omega = 6\,\Omega$. Applying the formula to the full circuit we have

$$I = \frac{V}{R} = \frac{24\,V}{6\,\Omega} = 4\,A$$

Thus the current through the single $3\,\Omega$ resistor must be 4 amperes because the full circuit current flows through it. The current of $4\,A$ divides up unequally through the parallel pair of resistors so that

$$\frac{I_1}{I_2} = \frac{R_2}{R_1} = \frac{12\,\Omega}{4\,\Omega} = \frac{3}{1}$$

$$\therefore \quad I_1 = 3 I_2$$

The total current of $4\,A = I_1 + I_2$, which is easily divided up so that $I_1 = 3 I_2$, giving $I_2 = 1\,A$ and $I_1 = 3\,A$.

Answer: The current in the $3\,\Omega$ resistor is 4 amperes (I), the current in the $4\,\Omega$ resistor is 3 amperes (I_1) and the current in the $12\,\Omega$ resistor is 1 ampere (I_2).

Figure 12.31

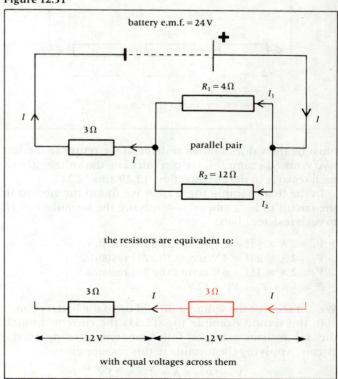

Calculating the voltage across circuit components

One of the most important things to understand about a series circuit is that the full e.m.f. or voltage of the supply is not connected across each of the resistors in the circuit. Instead, the full e.m.f. E is divided up so that:

a) The voltages across the separate resistors add up to the full e.m.f. E applied to the circuit:

$$E = V_1 + V_2 + V_3$$

b) The voltage across each resistor is directly proportional to its resistance: $V \propto R$.

In fig. 12.32 the current through all the series resistors is I. Applying the formula $V = IR$ to each separate resistor we get:

$$V_1 = IR_1 \qquad V_2 = IR_2 \qquad V_3 = IR_3$$

or for the same current we can say: $V \propto R$. The largest resistance will have the greatest voltage across it because that is where the most energy will be converted in the circuit as the charge is circulated. The greater the resistance, the more energy is needed to get the charge through it. (Remember the definitions of p.d. and the volt, p 230.)

When the current in a circuit or in a parallel branch of a circuit has been calculated then the voltage across any resistor can be found. But the current and resistance values used to calculate the voltage across a particular resistor

Figure 12.32

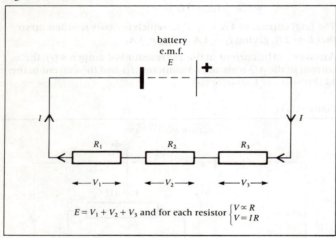

$E = V_1 + V_2 + V_3$ and for each resistor $\begin{cases} V \propto R \\ V = IR \end{cases}$

must be the values for *that* resistor. If we return to the last two worked examples, we can calculate the voltage across each resistor in the circuits (figs. 12.29 and 12.31).

In the first example (fig. 12.29) we found the current in the circuit to be 2 amperes. Applying the formula $V = IR$ to each resistor gives:

$V_1 = 2\,\text{A} \times 1\,\Omega = 2\,\text{V}$ across the $1\,\Omega$ resistor,
$V_2 = 2\,\text{A} \times 2\,\Omega = 4\,\text{V}$ across the $2\,\Omega$ resistor,
$V_3 = 2\,\text{A} \times 3\,\Omega = 6\,\text{V}$ across the $3\,\Omega$ resistor.
$E = V_1 + V_2 + V_3 = 12\,\text{V}$.

We can see that the voltages do add up to the battery e.m.f.

In the second example (fig. 12.31) the current through the $3\,\Omega$ resistor was the full 4 amperes flowing in the circuit. Applying the formula to this resistor gives

$$V = IR = 4\,\text{A} \times 3\,\Omega = 12\,\text{V}$$

Since the parallel pair of resistors also had an equivalent resistance of $3\,\Omega$ it should not be surprising to find that the 24 volt e.m.f. divides equally across the two series $3\,\Omega$ resistors. To check this we can calculate the voltage across each of the parallel pair. For the $12\,\Omega$ resistor, the current through it was 1 A:

$$V = IR = 1\,\text{A} \times 12\,\Omega = 12\,\text{V}$$

Similarly for the $4\,\Omega$ resistor, with a current of 3 A:

$$V = IR = 3\,\text{A} \times 4\,\Omega = 12\,\text{V}$$

Assignments

Study the summary table 12.8. Turn to the page references to revise anything you cannot remember or understand. Look for patterns in the information in the table.

Table 12.8 Comparing currents and voltages for series and parallel resistors

	Current	Voltage or p. d.
resistors in series	is the *same* in each resistor	divides up in *direct proportion* to the resistances
	$I_1 = I_2$	$\dfrac{V_1}{V_2} = \dfrac{R_1}{R_2}, \quad V \propto R$
(p 238)	(p 226)	(p 242)
resistors in parallel	divides up in the *inverse ratio* of the resistances	is the *same* across each resistor
	$\dfrac{I_1}{I_2} = \dfrac{R_2}{R_1}, \quad I \propto \dfrac{1}{R}$	$V_1 = V_2$
(p 239)	(p 241)	(p 229)

Remember the formulas

a) For series resistors:

$$R = R_1 + R_2 + R_3$$

b) For parallel resistors:

$$\frac{1}{R} = \frac{1}{R_1} + \frac{1}{R_2} + \frac{1}{R_3}$$

$$R = \frac{R_1 R_2}{R_1 + R_2}$$

c) For currents through parallel resistors.

$$\frac{I_1}{I_2} = \frac{R_2}{R_1}$$

Try questions 12.24 to 12.27

12.8
AMMETERS AND VOLTMETERS AT WORK

Ammeters and voltmeters are used in a great variety of places. For example, they can be seen in industry, in hospitals, in power stations and in recording studios. We are now going to see how these instruments can be adapted for so many uses.

Different scales can be used

One may at first think that the only instruments which are ammeters or voltmeters are those with scales labelled in amperes or volts. In many uses the scale does read amperes or volts, but in fact there are far more cases where the scale reads something entirely different – yet the instrument is still operating as a current detector or voltage scale. The majority of pointer-and-scale instruments have the same basic design, in which a coil turns between the poles of a magnet when a current flows in the coil. These moving-coil instruments which detect or measure electric current are called **galvanometers**. It is the great variety of ways in which a galvanometer can be used that has made it so common and so valuable.

Instrumentation is about measuring physical quantities, such as pressure and temperature, and converting the readings into a convenient form for display (e.g. a pointer on a scale), or for operating a control system (called automation). What has made electrical instrumentation so successful? The first reason is that many physical quantities can be represented accurately by an electric current or voltage. We call the current or voltage produced an **electrical signal**. Variations in the electrical signal are converted to movements of the pointer of a galvanometer. These pointer movements show how the current through the galvanometer is changing and also represent the changes in the physical quantity being measured. The second factor is that the meter can be any distance away from where the measurements are being made. This is very important when measurements have to be made in dangerous or inaccessible places. For example, temperature can be measured inside a furnace and displayed on a galvanometer scale far away in a control room.

The invention of more devices, called **transducers**, which can convert measurements of many physical quantities into electrical signals, has expanded the use of galvanometers and the scope for instrumentation. The use of light-emitting diodes and liquid crystals in digital displays has made possible a new generation of instruments which can give data about a wide variety of things in an easily read form. The following are a few examples of how transducers connected to current or voltage measuring instruments provide this variety of information.

A microphone is a transducer which converts sounds into an electrical signal, which in turn can be displayed on a sound-level meter. This is sometimes labelled 'signal level' and the scale reads in decibels (dB). The fuel gauge on a car indicates the level of petrol in the tank by a current, which is read on a scale labelled from empty to full. The temperature gauge also tells us how hot the engine is by a current reading. In this case the instrument is an electrical thermometer with a scale in degrees, or labelled from cold to hot. Many cameras now have a built-in light meter which helps to get the exposure

A wide variety of instruments is needed in the cockpit of Concorde.

correct by measuring the light entering the camera. The transducer used here is a light-sensitive cell which converts the light into an electrical signal to be displayed in the viewfinder. The instrument which indicates the correct shutter speed, or which aperture to use, is actually reading a voltage which is proportional to the brightness of the light. In the cockpit of an aircraft there is an array of instruments which indicate a variety of pressures, temperatures, levels and speeds which are almost all in the form of electrical signals measured by galvanometers, but each with a quite different scale. Sensors or transducers are positioned all over the aircraft in fuel tanks, in engines and outside, but the information is all carried as electric currents to the cockpit, so the pilot can check everything throughout the flight. Similarly, in the control room of any large industrial process a large number of instruments will tell the controller whether the machinery or process is working correctly. While the instruments will mostly be moving-coil galvanometers or digital voltmeters, their scales read values of very different physical quantities.

The ideal ammeter and voltmeter

When we use an ammeter to measure the current in a circuit it ought not to change that current. If the ammeter has any resistance at all, when we connect it in a circuit it will increase the total circuit resistance and so reduce the circuit current. Ideally, therefore, an ammeter should have zero resistance.

Unfortunately, a moving-coil galvanometer used as an ammeter cannot have zero resistance, because the current must flow through its coil. This fine wire coil always has some resistance.

Voltmeters are connected in parallel with a part of a circuit. The ideal voltmeter should have infinitely large resistance so that no current flows through it. Any current flowing through a voltmeter means an increase in the total circuit current, which will affect the voltmeter reading. Moving-coil voltmeters cannot have infinite resistance, however, because they are based on the moving-coil galvanometer which needs a small current through the moving coil to make it turn. It follows, then, that all moving-coil ammeters and voltmeters fall short of ideal instruments because of the need for an operating current to flow through their coils.

Electronic (digital) instruments are superior to moving-coil galvanometers, because their electronic circuits are designed to make them nearer to the ideal. This is possible because their circuits only need an extremely small current to operate. In particular, electronic voltmeters are now built with resistances of several million ohms. This makes them very accurate, because they take almost zero current from the circuit.

Another type of instrument called the oscilloscope can also be used as a voltmeter because it also has the advantage of having a very high resistance, p 350.

Table 12.9 gives a summary of the ideal ammeter and voltmeter and how they are used.

Table 12.9 The ideal ammeter and voltmeter

	Ammeter	Voltmeter
Resistance of meter	zero	infinitely large
Connection to circuit	in series with the current to be measured, should not change the current at all.	in parallel with, or across the voltage to be measured, should not pass any current at all.
Circuit diagram with meter	zero R	infinite R, zero I

Reading larger current ranges using shunts

A galvanometer can be converted to read larger currents by dividing up the current so that a small fraction of it flows through the galvanometer. The rest of the current flows through a parallel resistor. This parallel resistor, called a **shunt**, will have to have a resistance smaller than the galvanometer so that the larger fraction of the current will flow through the shunt.

We can compare a shunt with a bypass road for a town, as in fig. 12.33. The traffic is divided up so that the smaller fraction goes through the narrow streets of the town while the larger fraction goes round the wide, fast bypass road. To persuade drivers to use a bypass it must have a low resistance to traffic flow; it should be wide, fairly straight and have very few roundabouts.

Similarly, the shunt which takes the larger fraction of the current round the galvanometer will be made of a thick piece of wire of low resistance, so that the current can flow easily. The wire used for the shunt will probably

Figure 12.33 *Comparison of a shunt with a bypass road*

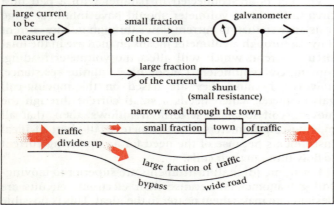

be manganin alloy, because its resistance will change very little when it is heated by the current. The exact resistance of the shunt depends on the resistance of the galvanometer and the size of the current to be measured. The next example shows how the shunt resistance is calculated.

Worked Example
The resistance of a shunt for an ammeter

We require an ammeter for a current range of 0 to 1.0 amperes, to be suitable for measuring the current through a filament lamp in an experiment to find its characteristic. We have available a galvanometer of resistance $100\,\Omega$ which gives full scale deflection (FSD) for a current of $1.0\,\text{mA}$.

The maximum current that can flow through the galvanometer is $1\,\text{mA}$ and we need to allow $1\,\text{A}$ to flow without damaging the instrument. Calculate the resistance of the shunt required.

$1.0\,\text{mA}$ is $\frac{1}{1000}$ of the maximum current to be measured. We therefore connect a shunt in parallel with the galvanometer so that $\frac{1}{1000}$ of the current flows through the galvanometer and $\frac{999}{1000}$ (the rest) of the current flows through the parallel resistor or shunt. This is shown in fig. 12.34a.

For conductors in parallel the current divides up in the inverse ratio of the resistances:

$$\frac{\text{current through shunt}}{\text{current through galvanometer}} = \frac{999\,\text{mA}}{1\,\text{mA}}$$

the current through the shunt is $999 \times$ galvanometer current,

$$\therefore \frac{\text{resistance of shunt}}{\text{resistance of galvanometer}} = \frac{1}{999}$$

$$\therefore \text{resistance of shunt} = \frac{\text{resistance of galvanometer}}{999}$$

$$= \frac{100\,\Omega}{999} = 0.1001\,\Omega$$

Answer: We have calculated that a shunt of resistance $0.1001\,\Omega$ is required, and that when connected in parallel with the galvanometer it will divide up a current so that only $\frac{1}{1000}$ of it flows through the galvanometer. Fig. 12.34b shows how this arrangement is equivalent to a 0–1.0 A ammeter.

Figure 12.34

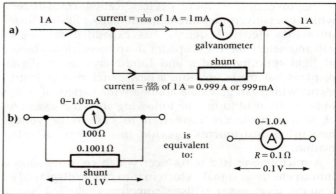

Reading larger voltages using multipliers

We have seen that an ideal voltmeter would have an infinitely large resistance and no current through it. The moving-coil galvanometer, however, requires only a small current to work it and so its maximum resistance is limited by this current. A voltmeter is given the resistance which will just conduct the required working current when connected across the maximum voltage to be measured. We increase the voltmeter's resistance by connecting a large resistor in *series* with the galvanometer. This large series resistor, used for a voltmeter, is called a **multiplier**, to distinguish it from the small parallel resistor used for an ammeter, which we call a shunt. When using a multiplier the galvanometer scale should be multiplied by a given number to convert it to the higher voltage scale.

Fig. 12.35 shows the arrangement for connecting a multiplier in series with a galvanometer to convert it into a higher range voltmeter.

Figure 12.35   *Converting a galvanometer into a voltmeter*

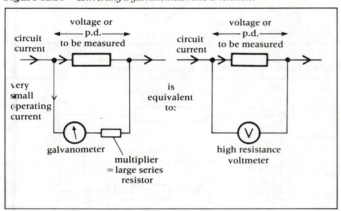

Worked Example
Converting a galvanometer into a voltmeter

Using the same basic galvanometer as before with a resistance of 100 Ω and requiring 1.0 mA for FSD, calculate the resistance of the multiplier required to convert it to read up to 10 volts.

When connected across the maximum 10 V to be measured the maximum current of 1.0 mA should flow through it producing FSD, see fig.12.36. The voltmeter requires a total resistance given by:

$$R = \frac{V}{I} = \frac{10\,V}{1.0 \times 10^{-3}\,A} = 1.0 \times 10^4\,\Omega = 10\,000\,\Omega = 10\,k\,\Omega$$

As the galvanometer already has a resistance of 100 Ω, the extra series resistance needed for the voltmeter is

$$10\,000\,\Omega - 100\,\Omega = 9\,900\,\Omega.$$

Answer: A multiplier of resistance 9 900 Ω connected in series with the 100 Ω galvanometer will provide a voltmeter of total resistance 10 000 Ω which will read a full scale deflection of 10 volts.

Figure 12.36   *Converting a galvanometer into a voltmeter*

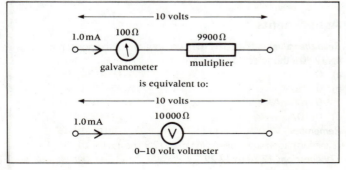

Comparing the conversions of ammeters and voltmeters

It is helpful to compare how we convert a galvanometer to read current scales and voltage scales, so that you clearly recognise the basic differences. Table 12.10 summarises the main points which you should know. The following general points are also worth noting:

a) A galvanometer can be converted to read only *larger* ranges of currents or voltages, thereby making it *less* sensitive. No arrangement of series or parallel resistors can make it more sensitive (read smaller ranges).

b) Connecting shunts in parallel with a galvanometer always *reduces* the resistance (which is a good thing for an ammeter). Connecting multipliers in series always *increases* the resistance (good for a voltmeter).

Table 12.10   Shunts and multipliers

conversion to	Ammeter	Voltmeter
The resistor used is called a	shunt	multiplier
The resistor has	very low resistance	very high resistance
The resistor is connected	in parallel with the galvanometer	in series with the galvanometer
The arrangement	low *R*	high *R*
Equivalent symbol	A	V

Multimeters

When an ammeter or voltmeter is designed for a particular application, a shunt or multiplier can be built into it and a scale fitted which reads the range of values required. For example, some cars are fitted with an ammeter which indicates the current flowing to or from the battery. This needs to be a centre-zero instrument with a range from −50 to +50 A and should be fitted with a scale showing these values. When no current flows the pointer reads zero in the centre of the scale. When the battery is discharging, the pointer moves to the left reading negative values, and when on charge, to the right indicating positive values.

While this instrument is specially designed for a particular job, it would not be suitable for most other applications. The electrical or electronic engineer who makes frequent measurements of electrical quantities requires an instrument which can be adapted quickly to read a wide variety of ranges of several quantities. The most common quantities to be measured are current, voltage (both direct and alternating) and resistance. An instrument which combines several ranges of current, voltage and resistance is usually called a **multimeter**.

To provide several ranges of current and voltage measurements, the multimeter has a set of built-in shunts and multipliers which can be selected by using multi-position switches. The basic circuits for a multimeter are shown in fig. 12.37.

Figure 12.37 *Multimeter circuits*

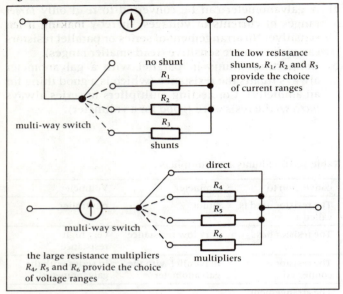

the low resistance shunts, R_1, R_2 and R_3 provide the choice of current ranges

the large resistance multipliers R_4, R_5 and R_6 provide the choice of voltage ranges

Some electronic multimeters have digital readouts which display values that are correct for the range selected; this makes these the easiest to read. Moving-coil multimeters have a pointer and a choice of scales; the user has to select the scale which is most suitable for the range being used and multiply the scale reading by a conversion factor. Fig. 12.38 shows a typical multimeter display where all current and voltage values are read on one of the top two scales. For example, on a 0–1 A range the 0–100 scale would be used, dividing the reading by 100. On a 0–25 V range the 0–250 scale would be used, dividing the scale reading by 10. The pointer in fig. 12.38 would read as follows:

a) selected 0–1 A range, the reading is:

$$\frac{70.0}{100} = 0.70\,\text{A}$$

b) selected 0–25 V range, the reading is:

$$\frac{175}{10} = 17.5\,\text{V}$$

The multimeter as an ohmmeter

In order to measure resistance the multimeter needs an internal cell. This cell sends a small current through the resistance to be measured and displays the current on the scale which reads in ohms. So the multimeter, used as an ohmmeter is really a sensitive ammeter.

Because of variation in the e.m.f. of the internal cell, the instrument needs a zero adjusting control for the resistance range. The zero of the instrument is set when the test leads are joined together providing a short-circuit of zero ohms resistance. The multimeter used as an ohmmeter is shown in fig. 12.39.

Figure 12.38 *A multimeter scale. The top two scales read both current and voltage, the lower scale reads resistance from right to left.*

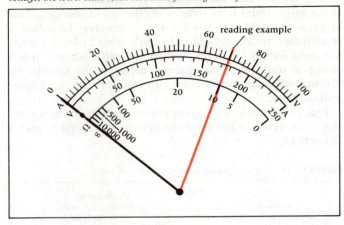

Figure 12.39 *The multimeter used as an ohmmeter*

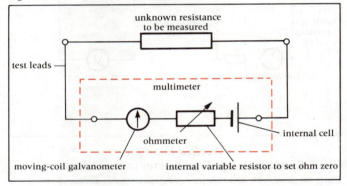

Figure 12.40

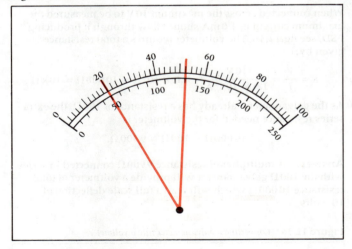

Assignments

Read the values indicated by the two positions of the pointer in fig. 12.40 if the selected ranges are:
a) 0–1 V,
b) 0–250 V,
c) 0–100 mA,
d) 0–10 A.

Remember

The information summarised in tables 12.9 and 12.10.

Try questions 12.28 to 12.29

12.9
RESISTIVITY

As part of their daily work, engineers in the telecommunications and power industries must select suitable electric cables to carry electric currents for many different purposes. Selection is difficult because there is a great variety of cables to choose from.

One important factor to be considered is the resistance of the conductors in the cable. We shall now investigate how the resistance of a conductor is affected by the type of metal it is made of and by its length and thickness.

Investigating how resistance depends on shape

For this investigation a special 'resistivity putty' is available from Unilab Ltd. As this putty is rather messy it should be handled with polythene gloves or rolled between sheets of paper. However, it is non-toxic and can be removed with soap and water.

Resistivity putty can be rolled or patted into various shapes to investigate how the resistance of a conductor depends on its length and cross-sectional area. Fig. 12.41 shows a simple method of making connections to the putty using two coins and gives the circuit diagram to use for measuring its resistance.

Figure 12.41 *Investigations using resistivity putty.*

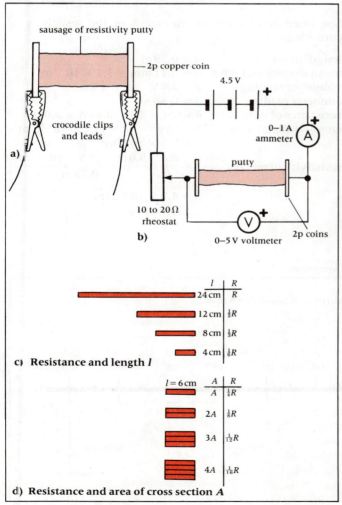

- *Roll about 50 g of the putty into a long, fairly uniform sausage about 24 cm long.*
- *Measure its resistance as shown in fig. 12.41b.*
- *Cut the sausage into various shorter lengths, say 12, 8 and 4 cm as shown in fig. 12.41c.*
- *Measure the resistance of each length of putty and record your results in a table.*

By comparing the results or by plotting a graph of resistance against length, we see that the resistance of the conductor increases proportionally with its length. Increasing the length of a conductor is like connecting resistors in series; their resistances add up.

The resistance R of a conductor is directly proportional to its length l; that is R ∝ l

- *Cut a long sausage of putty into several equal lengths. Stick them side by side to increase the cross-sectional area by various known factors, fig. 12.41d.*
- *Measure the resistance of the sausage each time it is made fatter and record results to show how the resistance changes as the cross-sectional area increases.*

We see that the resistance of the conductor decreases as its cross-sectional area increases. When the area is doubled the resistance of the conductor is halved.

The resistance R of a conductor is inversely proportional to its cross-sectional area A; that is R ∝ 1/A

By selecting a conductor with a particular thickness or cross-sectional area we can control resistance. For example, the filament of a lamp is made of very fine wire to give it more resistance than the rest of the circuit. The wires which carry the very large current from a car battery through the starter motor are made of very thick wire because they must have a very low resistance to conduct the very large current required.

The filament of an electric lamp, being made of a coil of very fine wire of small cross-sectional area, has a high resistance.

Resistance and resistivity

We have already investigated the conduction of electricity through a variety of materials. We know that some materials are good conductors, some poor conductors and others insulators; there is in fact a spectrum from the best conducting materials to the best insulating materials. The resistance of a conductor depends on the material it is made of as well as its dimensions. We call the resisting property of a material its **resistivity**, symbol ρ (the Greek letter 'rho'). Conductors and resistors have a resistance; the materials of which they are made have a resistivity.

When we compare the resistivities of different materials, we actually compare the resistances of conductors made from a standard size specimen of the materials. This gives us a definition of resistivity.

The resistivity of a material is numerically equal to the resistance of a specimen of unit length and unit cross-sectional area.

It follows that:

the resistance R of a conductor is directly proportional to the resistivity ρ of its material; that is R ∝ ρ

A formula for resistivity

We now know how the resistance of a conductor depends on each of the three quantities, *l*, *A* and *ρ*. If we combine these relations we get

$$R \propto \frac{\rho l}{A}$$

By the definition of resistivity, which makes $\rho = R$ when $l = 1$ metre and $A = 1\,\text{m}^2$, we have

$$R = \frac{\rho l}{A}$$

We can rearrange this to give the formula for resistivity

$$\rho = \frac{RA}{l}$$

We can see that the units of resistivity will be

$$\frac{\text{ohms} \times \text{metres}^2}{\text{metres}}$$

which simplifies to ohms × metres or Ω m.

Fig. 12.42 shows the very wide spectrum of resistivities of some important materials. The values given are for room temperature. The resistivities of most materials decrease as temperature rises, with the notable exception of metals, whose resistivities increase at higher temperatures.

Measuring the resistivity of a metal

The resistivity of a metal or alloy is so very small that the resistance of a specimen of length 1 metre and cross-sectional area 1 metre² cannot be measured. Just think of its size! Instead we measure the resistance of a long length of thin wire made of the particular metal, and calculate the resistivity.

A suitable specimen for this experiment would be a metre of constantan wire of diameter about 0.3 mm (SWG 28, 30 or 32). For this wire use a 0–1 A ammeter and a 0–5 V voltmeter with a 4.5 V battery.

- *Measure the resistance of the specimen length of wire using an ammeter and voltmeter as we did before, p 231.*

We also need accurate measurements of the length and cross-sectional area of the wire. Of course, its length *l* can be measured very accurately by laying it flat along a metre ruler, but measuring its cross-sectional area *A* is more difficult. We need a micrometer screw gauge which reads to an accuracy of 0.01 mm to measure the diameter of the wire, from which we can calculate its cross-sectional area.

- *Measure the diameter d at various points along the wire and in different directions across the wire in case it is slightly oval.*
- *From these readings, calculate a mean diameter in millimetres and convert it to a value in metres.*
- *Calculate the cross-sectional area by using the formula*

$$A = \tfrac{1}{4}\pi d^2 \quad (\text{in m}^2)$$

- *Finally calculate the resistivity of the metal by using the formula*

$$\rho = \frac{RA}{l} \quad (\text{in } \Omega\text{m})$$

Specimen results for measurement of the resistivity of constantan:

length of wire $\quad l = 0.79\,\text{m}$
mean diameter of wire $\quad d = 0.31\,\text{mm} = 3.1 \times 10^{-4}\,\text{m}$
voltmeter reading $\quad V = 4.4\,\text{V}$
ammeter reading $\quad I = 0.88\,\text{A}$
resistance of wire $\quad R = V/I = 4.4\,\text{V}/0.88\,\text{A} = 5.0\,\Omega$
cross-sectional area $\quad A = \tfrac{1}{4}\pi d^2 = \tfrac{1}{4}\pi \times (3.1 \times 10^{-4})^2\,\text{m}^2$
$\quad = 7.5 \times 10^{-8}\,\text{m}^2$
resistivity of metal $\quad \rho = \frac{RA}{l} = \frac{5.0\,\Omega \times (7.5 \times 10^{-8})\,\text{m}^2}{0.79\,\text{m}}$

$$\therefore \quad \rho = 4.7 \times 10^{-7}\,\Omega\,\text{m}.$$

Figure 12.42 *The spectrum of resistivities of some important materials at room temperature*

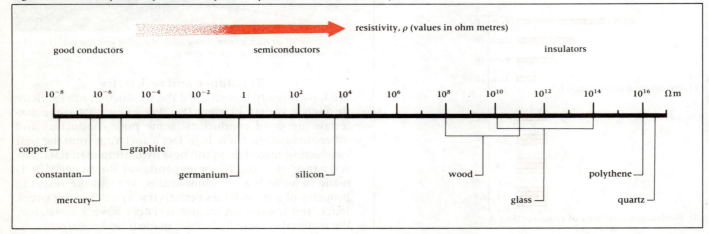

Worked Example
Calculating the resistance of a cable

Calculate the resistance of an aluminium cable of length 20 km *and diameter* 4.0 mm. *The resistivity of aluminium is* $3.0 \times 10^{-8} \, \Omega \, m$

First convert all the units of dimensions to metres;

length $= 20 \times 10^3 \, m = 2.0 \times 10^4 \, m$
diameter $= 4.0 \times 10^{-3} \, m$

Now using the formula $A = \frac{1}{4} \pi d^2$, we can find the cross-sectional area,

$$A = \frac{1}{4} \pi \times (4.0 \times 10^{-3})^2 \, m^2 = 1.3 \times 10^{-5} \, m^2$$

Using the formula $R = \rho l / A$, we can find the resistance

$$R = \frac{(3.0 \times 10^{-8} \, \Omega \, m) \times (2.0 \times 10^4 \, m)}{1.3 \times 10^{-5} \, m^2} = 4.6 \times 10 \, \Omega = 46 \, \Omega$$

Answer: The resistance of the cable is 46 ohms.

Worked Example
Calculating the length of wire required for a shunt

A shunt is required for an ammeter. It should be made from manganin wire of diameter 0.71 mm *and of resistivity* $4.5 \times 10^{-7} \, \Omega \, m$; *the shunt resistance required is* 0.05 Ω. *Calculate the exact length of this manganin wire which will have a resistance of* 0.05 Ω.

First convert the units of the diameter d to metres;

$$d = 0.71 \, mm = 0.71 \times 10^{-3} \, m = 7.1 \times 10^{-4} \, m$$

Using the formula: $A = \frac{1}{4} \pi d^2$ gives

$$A = \pi \times (7.1 \times 10^{-4})^2 \, m^2 = 4.0 \times 10^{-7} \, m^2$$

By rearranging the resistivity formula we get:

$$l = \frac{RA}{\rho} = \frac{0.05 \, \Omega \times 4.0 \times 10^{-7} \, m^2}{4.5 \times 10^{-7} \, \Omega \, m} = 4.4 \times 10^{-2} \, m$$

Answer: The length of manganin wire required for the shunt is 44 mm.

Assignments

Measure the resistivity of 'resistivity putty'. Mould about 50 g of the putty into a uniform bar which is an easy shape to measure. A square section bar about 10 cm long would be suitable. Try to make it uniform by patting it into shape with flat pieces of wood. Measure its length and both its thickness dimensions. Using the same circuit as in fig. 12.41, measure the resistance of the bar of putty. Now calculate the resistivity using the formula $\rho = RA/l$. (As a check, ρ is about 40 m Ω m.)

Remember
a) The formula for resistivity:

$$\rho = \frac{RA}{l}$$

b) The rearranged formula for resistance:

$$R = \frac{\rho l}{A}$$

c) The units of resistivity: ohm metre.
d) The definition of resistivity.
Try questions 12.30 to 12.33

ELECTRICAL ENERGY AND POWER

Calculating the energy conversion in an electrical circuit can tell us how much heat an electric fire will supply, how much work an electric motor will do in a certain time or even the kinetic energy gained by an electron in a cathode ray tube. This section is concerned with the formulas for electrical energy and power. We shall begin with an experiment to investigate the heating effect of an electric current.

An electric drill or power tool converts electrical energy into mechanical energy to do mechanical work.

The heating effect of an electric current

This effect was first investigated by James Joule and is often referred to as the Joule heating effect. The production of heat depends on time, resistance and current. To show this, we can pass an electric current through a coil of resistance wire immersed in water so that the electrical energy converted in the wire will all become heat energy in the water. If we use a thermally insulated beaker, the rise in water temperature will be directly proportional to the heat energy supplied to it. Therefore, the rise in water temperature can be used to show how the heat produced depends on time, resistance and current.

A commercial immersion heater could be used to investigate how the heat produced depends on the current and time. But because we want to vary the resistance of the heater too, we need to make several simple heating coils from various lengths of resistance wire.

● *Make four coils of wire of lengths* 10, 20, 30 *and* 40 cm, *by winding the wire round a pencil. (Constantan wire of diameter* 0.3 mm *(30 SWG) or nichrome wire of diameter* 0.4 *or* 0.5 mm *(or 26 SWG) would be suitable.) The longest coil should have a resistance of about 2 to 3 ohms.*

● *Connect the ends of the largest coil to two screw terminals or crocodile clips supported across the top of a beaker or metal calorimeter.*

The coil should be almost entirely immersed in water, but not touching the sides or bottom of the container. Fig. 12.43 shows a suitable arrangement. You will also need lagging for the beaker, a celsius thermometer and a clock which reads in seconds. The mass of water and hence the thermal capacity of the water and beaker should remain constant throughout the experiments. About 0.2 kg of water should be enough to cover the heating coil in a 250 cm³ beaker.

Figure 12.43 *Investigation of the heating effect of an electric current*

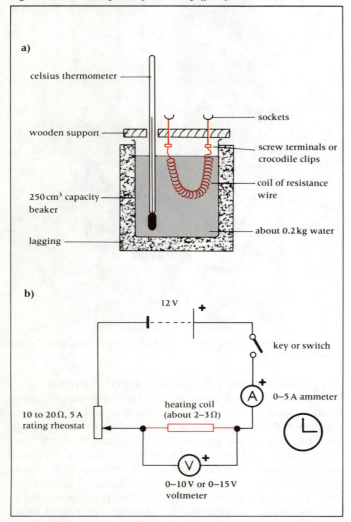

Figure 12.44

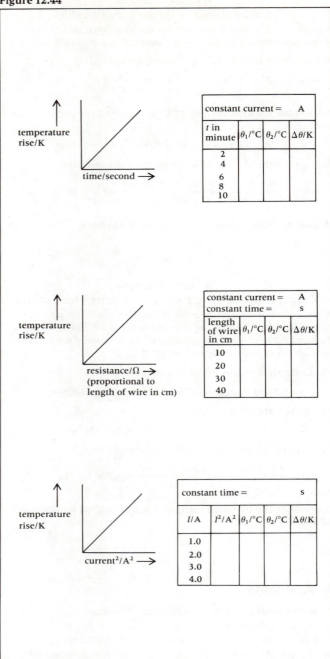

How does the heating effect depend on time?
- *Using the largest coil, set up the circuit shown in fig. 12.43b.*
- *Adjust the rheostat to pass a steady current of 3.0 or 4.0 A.*
- *Check that the current does not change during the experiment but if it does, use the rheostat to correct any variation.*
- *Record the initial temperature θ_1 and start timing as soon as the switch is closed.*
- *Stir the water gently with the thermometer and note the temperature θ_2 about every two minutes for about ten minutes.*
- *Tabulate the value of time t, temperatures θ_1 and θ_2, and calculate the temperature rise $\Delta\theta = \theta_2 - \theta_1$.*
- *Plot a graph of temperature rise $\Delta\theta$ against time t, fig. 12.44.*
How does the heat produced (proportional to the temperature rise) depend on the time?

How does the heating effect depend on resistance?
The same current must flow for the same time through several coils of different resistances. It would be possible to connect the four coils used above in series in a single circuit so that the current and time are identical. However, this would mean using four beakers of equal size and thermal capacity, containing equal masses of water. Also, because the combined resistance of the coils in series would be about 6 ohms, a higher voltage supply, 24 or 30 V, would be needed to produce a satisfactory temperature rise.

A simpler experiment can be carried out using the same beaker refilled four times with the same mass of cold water (about 0.2 kg) and using each of the coils in turn.

- *Use the rheostat to adjust the current to exactly the same value each time. The maximum convenient current should be used, say 3.5 or 4.0 A. (The size of the current will be limited by the largest resistance coil, which should therefore be used first.)*
- *Record the initial and final temperatures for each coil and calculate the temperature rises $\Delta\theta$.*
- *Plot a graph of temperature rise against length of wire in the coil, fig. 12.44.*

The heat produced is directly proportional to the temperature rise, and the resistance of the wire is directly proportional to its length. What, then, is the relation between the heat produced and the resistance of the coil?

How does the heating effect depend on current?

- *Using the largest coil each time, pass currents of 1.0, 2.0, 3.0 and 4.0 A in turn, through it for the same length of time. For each value of the current, refill the beaker with the same mass of cold water. Record the initial and final temperatures and calculate the temperature rises.*

The relation between the temperature rise and the current should be clear. What happens to the temperature rise each time the current is doubled?

- *Plot a graph to confirm this relation, as in fig. 12.44.*

A formula for the heating effect of an electric current

The results of the three experiments show that the electrical work W converted into heat energy is directly proportional to:

the time t,
the resistance R,
the square of the current I^2.

These results, which are sometimes called **Joule's heating laws**, can be combined to give $W \propto I^2Rt$.

Today we define electrical quantities so that the formula for the heating effect of an electric current becomes

$$W = I^2Rt$$

If we make various substitutions in this formula we get several useful alternative relations.
Electrical energy converted: $W = I^2Rt$

Using $R = \dfrac{V}{I}$ $W = I^2\dfrac{V}{I}t = IVt$ (or ItV)

Using $Q = It$ $W = QV$

Using $I = \dfrac{V}{R}$ $W = \dfrac{V^2}{R^2}Rt = \dfrac{V^2t}{R}$

Collecting these formulas together we have (with W in joules):

$$W = QV = ItV = I^2Rt = \frac{V^2}{R}t$$

Worked Example
Calculating electrical energy converted

If a current of 4.0 A is passed through a thin cable for 1.0 hours and its resistance is 20 Ω, how much electrical energy will be converted to heat energy in the cable?

We are given
  $I = 4.0\,A$
  $R = 20\,\Omega$
  $t = 1.0 \times 60 \times 60\,s = 3\,600\,s$
Using electrical energy converted, $W = I^2Rt$ gives

$$W = (4.0\,A)^2 \times 20\,\Omega \times 3\,600\,s = 1.15 \times 10^6\,J = 1.15\,MJ$$

Answer: The electrical energy converted into heat energy is 1.15 megajoules.

Electrical power

We all realise that the electric motor in a washing machine is more powerful than the one in a hair dryer. We also know that heat is produced more quickly in an electric fire than it is in a light bulb. Power is the rate of energy conversion. An electric fire is more powerful than a light bulb because it converts electrical energy into heat and light energy more quickly.

$$\text{power} = \frac{\text{energy converted}}{\text{time}} \qquad P = \frac{W}{t}$$

The unit of power is the **watt** (W)

$$1\ \text{watt} = 1\ \text{joule per second.}$$

We can obtain the formulas for electrical power from the electrical energy formulas by dividing by time:

$$P = \frac{W}{t} = \frac{ItV}{t} = \frac{I^2Rt}{t} = \frac{V^2t}{Rt}$$

$$P = IV = I^2R = \frac{V^2}{R}$$

P is in watts. The formula $P = IV$ is used a lot by electricians because it allows them to calculate the current that will flow through cables to appliances of known power requirements. The size of the current determines the type of cable and fuses that are used. In everyday use this formula is quoted in terms of the units:

$$\text{watts} = \text{amperes} \times \text{volts}$$

The electrical input power of this drill is given as 350 watts.

Energy and power

The work done by electrical machines and appliances can be compared with that done by the mechanical machines we studied in chapter 7. In all cases we saw that, when the power of a machine is known, the work it does or the energy it converts can be found from the relation:

$$\text{energy converted} = \text{power} \times \text{time} \qquad W = Pt$$

For example, an electric motor does work as it converts electrical energy into motion energy, and an immersion heater is an electrical appliance which converts electrical energy into heat energy.

Worked Example
Energy and power

A 5 kW immersion heater is used to heat water for a bath. If it takes 40 minutes to heat up the water, how much electrical energy is converted into heat energy?

We have:

$$P = 5\,\text{kW} = 5 \times 10^3\,\text{W}$$
$$t = 40\,\text{minutes} = 40 \times 60\,\text{s} = 2\,400\,\text{s}$$

Using $W = Pt$ gives

$$W = 5 \times 10^3\,\text{W} \times 2\,400\,\text{s} = 1.2 \times 10^7\,\text{J} = 12\,\text{MJ}$$

Answer: The electrical energy converted is 12 megajoules.

Electrical energy into mechanical energy

All electric motors convert electrical energy into mechanical energy, fig. 12.45a and we can measure the energy in both forms. The electrical energy W fed into a motor can be calculated from

$$W = Pt = I^2 Rt = ItV$$

If the motor is used to lift a mass m a vertical height h by winding up a rope round an axle, the potential energy E_p given to the mass can be calculated from $E_p = mgh$. The potential energy gained by the mass E_p is a measure of the output of useful work from the motor. This work will always be less than the electrical energy supplied to the motor. We can account for most of the wasted energy by saying that it is converted into heat energy. Can you think of two important ways in which energy is wasted during this conversion and suggest ways of reducing this energy loss?

The energy equation for an electric motor is

$$\begin{array}{ccc} \text{electrical energy} \\ \text{input} \end{array} \rightarrow \begin{array}{c} \text{useful work} \\ \text{output} \end{array} + \begin{array}{c} \text{heat energy} \\ \text{lost} \end{array}$$

$$Pt = mgh + \text{heat energy lost}$$

Worked Example
The efficiency of an electric motor

An electric motor, rated at 500 W, was used to raise a load of mass 16.0 kg a vertical height of 125 m. If it took the motor 50.0 s to raise the load, calculate (as a percentage) how efficient the motor was in converting electrical energy into mechanical energy ($g = 10.0\,\text{N/kg}$).

The mechanical work done by the motor is

$$mgh = 16.0\,\text{kg} \times (10.0\,\text{N/kg}) \times 125\,\text{m} = 2.00 \times 10^4\,\text{J}$$

The electrical energy supplied to the motor is

$$Pt = 500\,\text{W} \times 50.0\,\text{s} = 2.50 \times 10^4\,\text{J}$$

The efficiency of energy conversion, as a percentage is

$$\frac{\text{mechanical work output}}{\text{electrical energy input}} \times 100\%$$

$$\text{efficiency} = \frac{2.00 \times 10^4\,\text{J}}{2.50 \times 10^4\,\text{J}} \times 100 = 80.0\%$$

Answer: The efficiency of the electric motor was 80.0%.

Electrical energy into heat energy

Wherever an electric current flows through a resistor, a quantity of electrical energy W (equal to Pt or $I^2 Rt$) will be converted to heat energy, fig.12.45b. We can measure the heat energy produced when it heats a mass of material which is thermally insulated from its surroundings.

In chapter 10 we learnt that the heat energy gained by an object of mass m and specific heat capacity c can be calculated using the formula:

$$\text{heat energy gained} = cm\Delta T,$$

where ΔT is the temperature rise. (When m is in kg, c in J/(kg K) and ΔT in K, the heat energy is measured in J.)

The energy equation for an electric heater is

$$\begin{array}{ccc} \text{electrical energy} \\ \text{input} \end{array} \rightarrow \begin{array}{c} \text{heat energy gained} \\ \text{by an object} \end{array} + \begin{array}{c} \text{heat energy} \\ \text{lost} \end{array}$$

$$Pt = cm\Delta T + \text{heat energy lost}$$

The measured heat energy output is less than the electrical energy input because some heat always escapes. When this relation is used in electrical methods of measuring specific heat capacities, steps must be taken to keep the heat loss to a minimum.

Figure 12.45

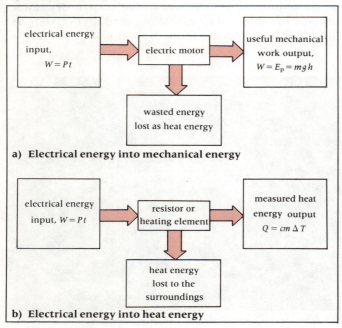

a) **Electrical energy into mechanical energy**

b) **Electrical energy into heat energy**

Worked Example
The heating effect of an electric current

When an electric immersion heater is connected to the 240 V mains supply, a current of 20 A flows through it. If the water tank contains 50 kg of cold water at 20°C, calculate the temperature of the water after half an hour. Assume the tank itself has no heat capacity and is perfectly lagged. The specific heat capacity of water $c = 4.2$ kJ/(kg K).

We use:

$$\frac{\text{electrical energy}}{\text{input}} = \frac{\text{heat energy gained}}{\text{by water}} + \frac{\text{no heat}}{\text{loss}}$$
$$Pt = (IV)t = cm\Delta T$$

where ΔT is the rise in temperature of the water which is to be found.

$P = IV = 20\,\text{A} \times 240\,\text{V} = 4800\,\text{W}$
$t = 30 \text{ minutes} = 30 \times 60\,\text{s} = 1800\,\text{s}$
$m = 50\,\text{kg}$
$c = 4.2\,\text{kJ/(kg K)} = 4.2 \times 10^3\,\text{J/(kg K)}$

$\therefore \quad 4800\,\text{W} \times 1800\,\text{s} = 50\,\text{kg} \times [4.2 \times 10^3\,\text{J/(kg K)}] \times \Delta T$

$$\therefore \quad \Delta T = \frac{4800 \times 1800}{50 \times 4.2 \times 10^3}\,\text{K} = 41\,\text{K} = 41°\text{C}$$

Answer: The water temperature will reach $20°\text{C} + 41°\text{C} = 61°\text{C}$.

The joulemeter

The electrical energy converted by an appliance can always be calculated from the readings of an ammeter and voltmeter connected in the usual way. Sometimes it is convenient to have an instrument which gives a direct reading of energy in joules. Such an instrument is called a **joulemeter**. These instruments are used in every house to measure the electrical energy supplied; there they are called electricity meters. The household meter records the energy converted in kilowatt-hours, which are sometimes called 'units' of electricity. The kilowatt-hour is a large unit of energy (= 3.6 MJ), and is a suitable size for measuring the large quantities of energy used in a modern household.

Joulemeters can be used in electric circuits to measure the energy supplied to an appliance. The circuit connections are shown in fig. 12.46. The two terminals labelled 'supply' are the input to the meter and the two labelled 'load' are the output from the meter to the appliance. The appliance which converts electrical energy is often called the **load**. Joulemeters are now available with a digital readout as well as the traditional dials.

Figure 12.46 *Connection of a joulemeter in a circuit*

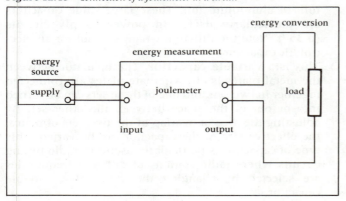

Worked Example
Current and power

An electric fire has three bars, each rated at a power of 1 kW. Calculate the current that will flow through the fire when it is connected to the 240 V mains supply.

The three bars are connected in parallel and their powers will add in the same way as their currents. We have:

total power $\qquad P = 3\,\text{kW} = 3 \times 10^3\,\text{W}$

$$V = 240\,\text{V}$$

Using $I = P/V$ gives

$$I = \frac{3 \times 10^3\,\text{W}}{240\,\text{V}} = 12.5\,\text{A}$$

Answer: The total current through the fire is 12.5 amperes.

Worked Example
Current, power and energy

A torch bulb is labelled 2.5 V, 0.3 A. Calculate the power of the bulb and the energy converted in 10 minutes.

We have
$I = 0.3\,\text{A}$
$V = 2.5\,\text{V}$
$t = 10 \text{ minutes} = 10 \times 60\,\text{s} = 600\,\text{s}$
Using $P = IV$ gives

$$P = 0.3\,\text{A} \times 2.5\,\text{V} = 0.75\,\text{W}$$

Using $W = Pt$ gives

$$W = 0.75\,\text{W} \times 600\,\text{s} = 450\,\text{J}$$

Answer: The power of the bulb is 0.75 watts and it converts 450 joules of electrical energy in 10 minutes.

Assignments

Remember
a) The electrical energy formulas:

$$W = Pt = IVt = I^2 Rt = \frac{V^2}{R}t = QV$$

b) The electrical power formulas:

$$P = \frac{W}{t} = IV = I^2 R = \frac{V^2}{R}$$

Try questions 12.34 to 12.36

12.11
CAPACITORS

Electric circuits contain many different kinds of devices. One of these, called a **capacitor**, is very common and can be used in a variety of ways. We find capacitors in computers, televisions and all electronic circuits, as well as certain electric circuits such as car ignition circuits and some electric power tools. Without capacitors electronic circuits would be very limited.

In this circuit of an old oscilloscope we can see a large electrolytic capacitor in a cylindrical metal can on the left and two much smaller 0.1 μF polyester capacitors on the right. Notice also the three thermionic valves.

Practical capacitors

Although capacitors are connected in electric circuits, they cannot conduct electric charge in the way that resistors and other circuit components do. Rather, they have the ability to store charge, and this makes capacitors particularly useful and quite different from all other devices.

Any object which can hold or store electric charge is a capacitor, but most objects can store so little that they would be of little use as practical devices. Even the dome of the Van de Graaff generator could be called a capacitor, but it does not store very much charge. The first capacitors that were made were very large and yet stored very little charge. Now several different types of capacitor have been invented which can store much more charge in less space, but even these cannot sustain a current for very long in a circuit.

Two metal plates

As we shall demonstrate later, the ability of an object to store charge can be greatly increased by bringing another conductor near to it. So practical capacitors are usually formed from two metal plates held close together, but not touching. When such a capacitor stores charge, one of the metal plates becomes positively charged with a charge of $+Q$. The other becomes negatively charged with an equal charge of $-Q$. The total charge on the capacitor is effectively just Q, because a charge Q flows round a circuit when the capacitor is discharged.

To prevent the positive and negative charges meeting and neutralising each other, the two plates must be separated by an insulating material, fig. 12.47. This material is sometimes just air, but it has been found that other insulating materials, such as various plastics, help a capacitor to store more charge when they are used to separate the plates. Materials which do this are called **dielectrics**.

Figure 12.47 *A simple capacitor*

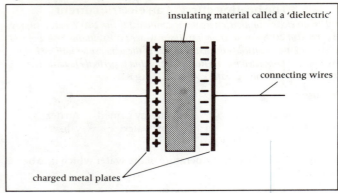

It follows that capacitors do not conduct charge, for if they did they would fail to keep the two opposite charges separate and would therefore not be able to store any charge. Capacitors which begin to conduct or become 'short-circuits' are a common cause of faults in circuits and such capacitors must be replaced.

Examples of practical capacitors

Practical designs of capacitors must pack two plates closely together with a layer of dielectric insulator between them. Some examples are shown in fig. 12.48.

a) This is a common design of a metal foil capacitor with plastic insulation. It consists of two lengths of metal foil separated by sheets of insulating material which are all rolled up tightly into a cylinder and encapsulated for protection. Each length of foil is connected to one end plate and to its wire connecting lead. The insulating material or dielectric may be sheets or films of polyester, polycarbonate, polystyrene, waxed paper etc. The capacitor type takes its name from the insulating material used, e.g. a 'polyester capacitor'. These capacitors come in a variety of shapes and sizes according to the application.

b) This is an **electrolytic capacitor**, which is different from most other types of capacitor because it is polarised. This means that one lead must always be connected to a positive supply terminal and the other to a negative terminal. Although this limits their applications, electrolytic capacitors have the advantage of being able to store much more charge than other types of capacitor of the same physical size. Electrolytic capacitors get their name from the electrolyte (ammonium borate) which fills the space between two rolls of aluminium foil. These large-value capacitors are particularly useful in power supply circuits (p 357). Note the different circuit symbol for an electrolytic capacitor.

c) This is a **variable capacitor**. The insulator between the metal plates is air. Alternate plates are connected together in two sets and one of these sets can be turned into or out of the space between the other set. By adjusting the area of overlap of the two sets of plates, the effective size of the capacitor can be varied. This type of capacitor is particularly useful in radio tuning circuits where radio stations of different frequencies are selected by changing the value of a variable capacitor.

Figure 12.48 *Practical capacitors*

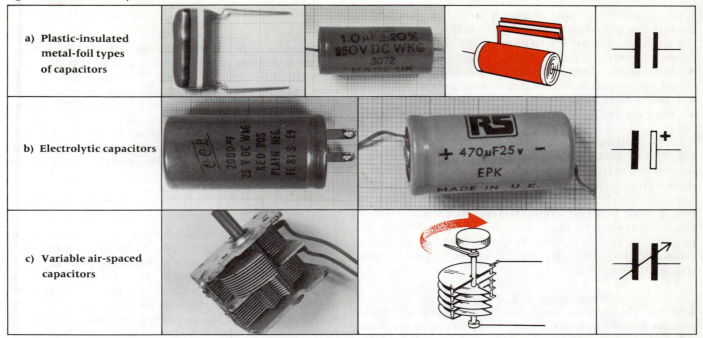

a) **Plastic-insulated metal-foil types of capacitors**	
b) **Electrolytic capacitors**	
c) **Variable air-spaced capacitors**	

Capacitor sizes

A capacitor stores charge. The measure of its ability to store charge is called its **capacitance**, symbol C. The unit of capacitance is the **farad**, symbol F. In practical terms a farad is a very large unit and most capacitors have much smaller values than 1 farad. Common sizes of capacitors are made in smaller units:

microfarads, $\mu F = 10^{-6}\,F$ (1 millionth of a farad),
nanofarads, $nF = 10^{-9}\,F$,
picofarads, $pF = 10^{-12}\,F$.

Electrolytic capacitors are made with capacitances in the range $0.1\,\mu F$ to about $100\,000\,\mu F$ but most other types cover a much lower range from a few pF to a maximum of about $10\,\mu F$.

Charging and discharging a capacitor

A charging circuit

An uncharged capacitor has two plates with neither a surplus nor a deficiency of electrons and no potential difference between them. The simplest way of charging a capacitor is to connect its two leads to a cell or battery. Fig. 12.49a shows a circuit which can be used for charging a capacitor from a cell and suggests suitable values for the components.

- *Connect up this circuit noting the following points:*
- a) The capacitor C is an electrolytic capacitor and its positive-labelled end must be connected towards the positive terminal of the cell in the circuit.
- b) The $+$ terminals of the ammeters must also be connected towards the positive terminal of the cell.
- c) A resistor R of about $100\,\Omega$ resistance is included in the circuit to limit the current to a safe value for the ammeters.
- *Note what both meters indicate when switch S_1 is closed.*
- *Open and close S_1 again and explain what you observe.*

Figure 12.49 *Charging and discharging a capacitor*

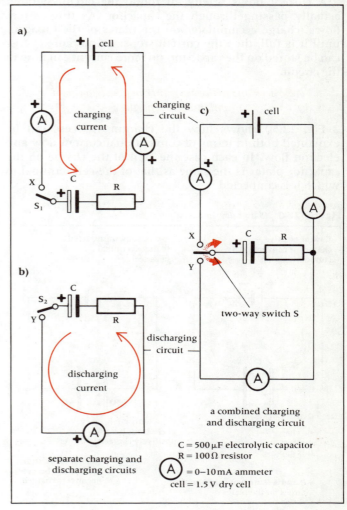

a combined charging and discharging circuit

$C = 500\,\mu F$ electrolytic capacitor
$R = 100\,\Omega$ resistor
$\text{A} = 0{-}10\,mA$ ammeter
cell = 1.5 V dry cell

separate charging and discharging circuits

A discharging circuit

If the capacitor is storing charge, it should be possible to get this charge to flow again in a circuit, that is to *discharge* the capacitor. Circuit (b) (without a cell) can be used to investigate this.

- *Disconnect the charged capacitor and protective resistor R from the cell and reconnect them to another ammeter and switch S_2.*
- *Observe the ammeter when S_2 is closed.*
- *Open and close S_2 again and explain what you observe.*

A combined charging and discharging circuit

Fig. 12.49c shows a combined circuit which uses a two-way switch to change over from the charging circuit to the discharging circuit without the need for rewiring. Notice how the two circuits remain separate because no connection is made between terminals X and Y. The capacitor is either connected to the cell to be charged (switch in position X) or it is connected to the discharging circuit (switch in position Y), but it cannot be connected to both circuits at the same time.

Observations and conclusions

1 In the charging circuit when S_1 is first closed, both ammeters give a flick in the same direction and then return to a zero reading. Opening and closing the switch again gives no further response.

A current flows briefly all round the circuit without actually passing through the capacitor. As this current flows, charge accumulates on the plates of the capacitor until it is full, then the current stops. As no more charge can be stored on the capacitor, no more current can flow in the circuit.

No permanent direct current can flow in a circuit containing a capacitor.

2 Fig. 12.50 shows how the charging process can be explained both in terms of conventional current flow and electron flow. In each case the sign of the charge on the capacitor plates is the same as that of the cell terminal to which it is connected.

Figure 12.50 *Charge flow in a capacitor circuit*

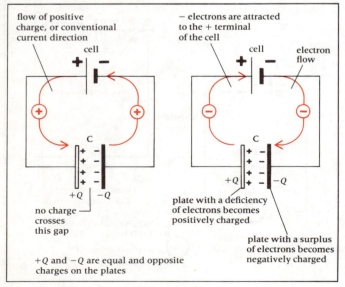

flow of positive charge, or conventional current direction

cell

no charge crosses this gap

$+Q$ and $-Q$ are equal and opposite charges on the plates

− electrons are attracted to the + terminal of the cell

cell

electron flow

plate with a deficiency of electrons becomes positively charged

plate with a surplus of electrons becomes negatively charged

When a charge Q is circulated round a circuit the capacitor stores a charge Q, and has equal and opposite charges of $+Q$ and $-Q$ on its plates.

3 When is a capacitor 'full'? As opposite charges accumulate on the plates of a capacitor a potential difference is established between the plates. We can explain this by saying that the cell takes a quantity of positive charge from the negative plate of the capacitor and 'pumps' it up to a higher voltage or potential and then delivers it to the positive plate of the capacitor. When the cell has filled the capacitor to the same voltage as its own e.m.f., no more charge can be pumped round the circuit and the capacitor is fully charged.

Once the voltage across the capacitor is the same as the e.m.f. of the cell the current in the circuit stops.

4 When switch S_2 is first closed a current flows briefly round the circuit of fig. 12.49b. A charged capacitor will discharge when there is a conducting path or circuit connected between its plates. We can explain this by saying that a current flows from the positive plate to the negative plate until both plates are neutralised. In fact the excess electrons on the negative plate flow round to the positive plate to replace the missing electrons there.

5 Since capacitors are able to deliver an electric current to a circuit, they must store energy as well as charge. That energy is converted to other forms in a circuit as a capacitor discharges.

Measuring the charge stored in a capacitor

We know that the quantity of charge Q flowing round a circuit is given by

$$\text{charge} = \text{current} \times \text{time} \qquad Q = It$$

So it should be possible to calculate the charge stored in a capacitor by measuring the current I and the time t taken to charge it.

Slowing down the charging process

To make it easier to measure the current in the circuit we slow down the charging process by making the resistance in the circuit much larger. If we increase the resistance in the circuit of fig. 12.49 from $100\,\Omega$ to $100\,k\Omega$ ($\times 1\,000$), the current in the circuit will be $1\,000$ times smaller. So it should take $1\,000$ times longer for the capacitor to charge.

- *Set up a circuit as shown in fig. 12.51 to charge a large electrolytic capacitor C.*
- *Set the variable resistor at its maximum value of $100\,k\Omega$.*
- *Close the switch S and start the clock at the same moment.*
- *Observe the reading on the ammeter for about 2 or 3 minutes.*

We notice that the current in the circuit decreases and gradually approaches zero, although it takes several minutes before there is no detectable current. The shape of the graph of current against time, shown in fig. 12.52a, is known as **exponential**. This is a special kind of curve and has the same features as the graph of the decay of a radioactive substance, p 394. In this example the value of the current is found to be reduced by about half for every 35 s that pass. So after 35 s the current has fallen to $50\,\mu A$, after 70 s it has fallen to $25\,\mu A$, after 105 s it has fallen to $12.5\,\mu A$ and so on; never, in theory, reaching zero!

Figure 12.51 *Charging a capacitor slowly*

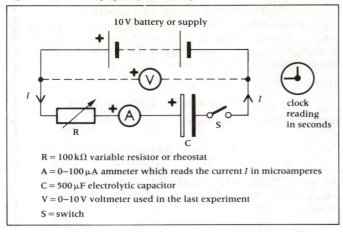

R = 100 kΩ variable resistor or rheostat

A = 0–100 μA ammeter which reads the current *I* in microamperes

C = 500 μF electrolytic capacitor

V = 0–10 V voltmeter used in the last experiment

S = switch

Figure 12.52 *Capacitor charging graphs*

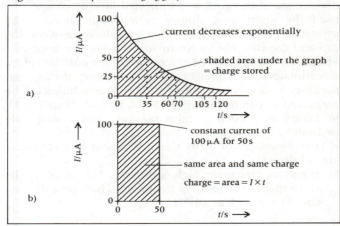

As the current is changing all the time it is difficult to find the charge stored on the capacitor. The area under the graph is equal to the stored charge because

$$area = current \times time = charge$$

but the area under a curve is also difficult to measure.

Charging at constant current

If the current can be kept constant while the capacitor is charged it is possible to calculate the charge stored using the relation, charge = current × time. This can be done as follows.

● *Using the circuit of fig. 12.51 again, set the rheostat to its maximum resistance of 100 kΩ and close the switch S as the clock is started.*

● *Gradually reduce the resistance of the rheostat to keep the charging current steady at 100 μA.*

● *As the rheostat reaches zero resistance the current will quite suddenly drop to zero. At this moment stop the clock.*

Fig. 12.52b shows how this charging process appears on a graph. The capacitor receives the same charge whether the current varies or is kept constant, so the areas under the two graphs which represent the stored charge are equal. In fig. 12.50b the charge stored is calculated as follows. Using $Q = It$ we have

$$Q = 100 \,\mu A \times 50 \,s = 5000 \,\mu C = 5.0 \,mC$$

Comparing different capacitors

● *Repeat the measurement of the time taken to charge at a constant current of 100 μA for several capacitors of different capacitances.*

● *Calculate the charge Q stored on each capacitor and compare these results with their capacitances C.*

Some typical results are given in table 12.11.

Table 12.11 Typical results for the charge stored in different sizes of capacitor charged to the same voltage

Battery voltage V/V	Capacitance C/μF	Current I/μA	Time t/s	Charge stored = It Q/mC
10	1000	100	105	10.5
10	470	100	52	5.2
10	220	100	24	2.4
10	100	100	11	1.1

The results show that:

The charge stored is directly proportional to the capacitance of the capacitor; Q ∝ C

This relation can be compared with beakers of different sizes filled with water to the same level, fig. 12.53. The larger beakers hold proportionally more water when filled to the same level.

Figure 12.53

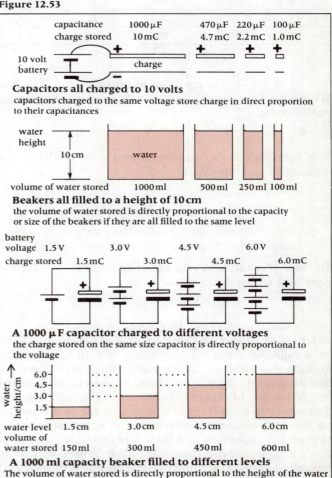

Capacitors all charged to 10 volts
capacitors charged to the same voltage store charge in direct proportion to their capacitances

Beakers all filled to a height of 10 cm
the volume of water stored is directly proportional to the capacity or size of the beakers if they are all filled to the same level

A 1000 μF capacitor charged to different voltages
the charge stored on the same size capacitor is directly proportional to the voltage

A 1000 ml capacity beaker filled to different levels
The volume of water stored is directly proportional to the height of the water

Changing the voltage across the capacitor

• *Using one particular capacitor, charge it to various different voltages and repeat the measurements of time taken to fully charge it at a constant current as before. The output voltage of the battery can be varied by using a potentiometer as in fig. 12.20, or by using a different number of cells in the battery.*

• *Measure the voltage used to charge the capacitor by connecting a 0–10 V voltmeter across the supply as shown in fig. 12.51.*

• *Calculate the charge Q stored on the capacitor and compare it with the voltage V used to charge it.*

Typical results are given in table 12.12, and these show that:

The charge stored is directly proportional to the voltage across the capacitor: $Q \propto V$

This relation can be compared with the volume of water contained in a beaker when it is filled to different levels. The higher levels correspond to higher voltages or p.d.s. The volume of water is directly proportional to the height of the water, fig. 12.53b.

Table 12.12 Typical results for the charge stored in a capacitor when charged to different voltages

Capacitor voltage V/V	Capacitance C/μF	Current I/μA	Time t/s	Charge stored $= It$ Q/mC
1.5	1000	100	16	1.6
3.0	1000	100	30	3.0
4.5	1000	100	47	4.7
6.0	1000	100	61	6.1

The capacitor formula

We can combine the relations $Q \propto C$ and $Q \propto V$ to produce the formula:

$$Q = CV$$

This formula can be rearranged and used to define the capacitance of a capacitor:

$$C = \frac{Q}{V}$$

The capacitance of a capacitor is the charge stored on its plates when there is unit potential difference between the plates.

The unit of capacitance, the **farad** (symbol F), is therefore equal to 1 coulomb per volt.

Calculating the charge stored on a capacitor

In tables 12.11 and 12.12 we calculated the charge stored on a capacitor from measurements of I and t using the formula $Q = It$. If we now examine these results we see that the formula $Q = CV$ could also be used to calculate the charge we would expect to be stored on the capacitor. The differences in the values of Q found by the two formulas are due to the differences between the actual capacitances of practical capacitors and their labelled values, used in $Q = CV$. (Electrolytic capacitors have values which lie typically between the limits -10% to $+50\%$ of its stated value.)

Worked Example
The charge stored on a capacitor

A capacitor is labelled with the capacitance value 470 μF and is charged to a p.d. of 10 V. (a) Calculate the charge stored on the capacitor according to its labelled value. (b) If experimentally it is found to store 5.2 mC of charge, calculate its actual capacitance.

a) Using $Q = CV$ we have

$$Q = 470\,\mu\text{F} \times 10\,\text{V} = 4700\,\mu\text{C} = 4.7\,\text{mC}.$$

b) Using $C = Q/V$ we have

$$C = \frac{5.2\,\text{mC}}{10\,\text{V}} = 0.52\,\text{mF} = 520\,\mu\text{F}$$

Answer: The stored charge should be 4.7 mC according to the labelled value. The experimental value of the capacitance is 520 μF.

The gold-leaf electroscope as a voltmeter

When we used a gold-leaf electroscope in chapter 11, the divergence of the gold leaf indicated electric charge. We used the instrument simply as a charge detector. By connecting a potential difference or voltage between the cap and the case of the electroscope it may be used as a voltmeter. The amount of divergence of the gold leaf gives an estimate of the voltage connected between the cap and the case. A scale is sometimes fitted so that voltages can be measured or compared, fig. 12.54. A p.d. of about 1000 V to 2000 V is needed to raise the gold leaf to full-scale deflection.

The advantages of using this instrument as a voltmeter are as follows:

a) It has an extremely high resistance. This is really the resistance of the perspex insulator, which on a dry day could be as high as $10^{16}\,\Omega$.

Figure 12.54 *The gold-leaf electroscope can be used as a voltmeter*

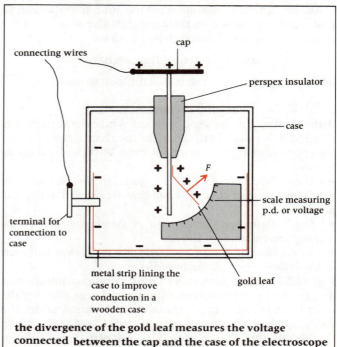

the divergence of the gold leaf measures the voltage connected between the cap and the case of the electroscope

b) It can measure high voltages. Each division on its scale might represent about 100 V. The scale is, however, non-linear; the divisions do not represent equal increases in voltage.

c) It has a very small capacitance. With a capacitance of only a few picofarads it takes very little charge away from any charged conductor or capacitor to which it is connected. Thus when it is connected across a charged capacitor to measure the voltage, it does not significantly reduce the reading by taking charge away.

d) The gold leaf diverges whichever way round a voltage is connected to the cap and case.

Using a gold-leaf electroscope to investigate a parallel-plate capacitor

As most capacitors are formed in some way from two parallel metal plates or strips of metal foil, it is helpful to know how the capacitance of such a capacitor is affected by its physical size and shape.

Two suitable metal plates can be made from sheet aluminium or aluminium foil glued to a firm backing and fixed to a stand. They should have·a fairly large area, say at least 20 cm × 20 cm.

● *Stand plate A on a polythene tile to insulate it from the bench and connect it to the cap of a gold-leaf electroscope.*

● *Make sure that the connecting wire is insulated or does not touch the bench.*

● *Charge the plate by connecting it to a Van de Graaff generator or by connecting a high-voltage supply between the plate and the case of the electroscope, i.e. 'earth'.*

It is important to understand that as long as this plate is not touched its charge remains constant during the experiments.

The gold leaf should have a large divergence indicating a high voltage between its cap and case. As the case rests on the bench it is, effectively, earthed. There are three separate investigations to try, as shown in fig. 12.55.

Bringing an earthed plate up to a charged plate

● *Connect the second metal plate B to the case of the electroscope.* Both plate B and the electroscope case are earthed and the electroscope measures the p.d. between the two plates.

● *Starting with plate B a long way off from A, slowly move plate B towards A and observe the electroscope voltmeter. Do not allow anything to touch the charged plate A.*

● *Then slowly move plate B away again.*

As plate B approaches, the divergence of the gold leaf reduces indicating a drop in voltage. When B is moved away again the leaf rises once more showing that the voltage returns to its original value and the charge remains on A. From the relation $C = Q/V$ we can see that if the charge Q is constant, then $C \propto 1/V$;

The change in capacitance C is inversely proportional to the change in voltage V.

This idea can be illustrated by a water model. Fig. 12.56 shows how a constant volume of water fills a larger beaker to a lower level. This is equivalent to a constant quantity of charge filling a larger capacitor to a lower p.d. or voltage.

Figure 12.55 *Investigating a parallel-plate capacitor*

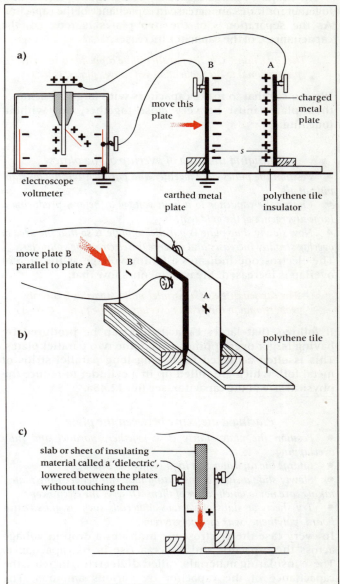

Figure 12.56 *The same volume of water fills different size beakers to different levels*

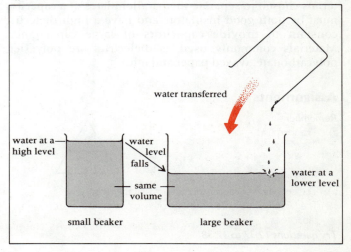

The drop in voltage caused by bringing the plates closer together indicates an increase in capacitance of the capacitor. As the separation *s* of the two plates is reduced, the capacitance *C* of the capacitor increases.

The capacitance C of the parallel-plate capacitor is inversely proportional to the separation s of its two plates: C ∝ 1/s

We can see that to make capacitors with large capacitance their plates must be very close together, but without touching.

Changing the area of overlap of the plates

● *Position the plates close together and parallel, but not overlapping at all.*
● *Charge the capacitor to a high voltage as before, producing a large divergence of the gold leaf.*
● *Now slowly slide plate B parallel to plate A so that the area of overlap steadily increases and observe the electroscope voltmeter.*
The electroscope indicates a falling voltage as the area of overlap is increased. Measurements show that.

The capacitance C of a parallel-plate capacitor is directly proportional to the area of overlap A of the plates, C ∝ A

It follows that larger capacitances can be produced by having larger overlapping areas of the two parallel plates. This is often achieved by having long parallel strips of metal foil, which are rolled up in a cylinder to reduce the physical size of the capacitor, see fig. 12.48a.

Placing a dielectric between the plates

● *Position the plates fairly close together, parallel and fully overlapping.*
● *Charge the capacitor to a high voltage as before.*
● *Slowly slide a sheet of insulating material between the plates, taking care not to touch either of them. Observe the electroscope.*
● *Try sheets or slabs of various materials such as glass, cardboard, polythene, wax and polystyrene.*
In every case the electroscope indicates a drop in voltage across the capacitor and hence a rise in its capacitance. These insulating materials, called **dielectrics**, increase the capacitance of the capacitor by various amounts. The number of times the capacitance is increased by a particular dielectric is a property of that material known as its **dielectric constant**, or **relative permittivity**. The materials chosen to separate the parallel plates of a capacitor must be both good insulators and have a high dielectric constant to provide capacitors of large capacitance. Materials commonly used as dielectrics are polyester, polycarbonate, waxed paper and mica.

Assignments

Remember
a) The formulas:

$$Q = CV \qquad C = \frac{Q}{V}$$

b) The definition of capacitance and its unit the farad.
c) The factors which affect the capacitance of a parallel-plate capacitor.

Try questions 12.37 to 12.38

Questions 12

1 In the circuit of fig. 12.57 if ammeter A_1 reads 0.5 A, what will be the reading on ammeters A_2 and A_3?

Figure 12.57

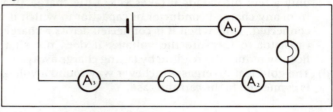

2 A current of 6 A flows through a conductor for 2 minutes. What is the total charge which passes through the conductor?
3 How long will it take for a total charge of 960 C to pass through a conductor if a steady current of 4 A is flowing?
4 What current could circulate a total charge of 3 600 coulombs round a circuit in 20 minutes?
5 The circuit of fig. 12.58 shows the sizes and directions of the currents, at a junction J, in an electric circuit. Calculate the size and direction of the current recorded by the ammeter in the wire JX.

Figure 12.58

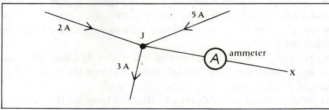

6 Fig. 12.59 shows a circuit containing three identical ammeters A_1, A_2, A_3. State whether the following statements are **true** or **false**:
 i) A_2 gives the reading of the current through the cell;
 ii) the reading of A_3 is the current through the 6R resistor;
 iii) the reading of A_1 is less than the reading of A_3;
 iv) the reading of A_3 is equal to the reading of A_2 minus the reading of A_1. (W 88)

Figure 12.59

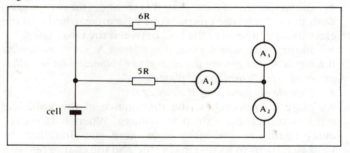

7 If a 12 volt car battery circulated 500 coulombs round a circuit, how much energy did the battery supply?
8 If 660 joules of work was done by a battery in moving 110 coulombs of charge through a lamp, what was the voltage across the lamp?
9 1200 coulombs flowing through a heater converted 14.40 kJ of energy into heat. If the current was 4.80 A, calculate the time for which the heater was switched on and the voltage across the heater.

10 A torch bulb is labelled 3.5 V, 0.3 A; what is its resistance when working?

11 What is the potential difference or voltage required to produce a current of 2.5 A through a conductor of resistance 12.5 Ω?

12 A 12 volt car battery must send 240 amperes through a starter motor to turn the car engine. What must be the maximum resistance in the circuit containing the starter motor and the battery?

13 The mains electricity supply of 240 volts is connected to the following appliances. If they have the resistances stated, calculate the current which will flow through each.
a) vacuum cleaner, R = 40 Ω;
b) electric shower, R = 16 Ω;
c) refrigerator, R = 120 Ω.

14 The mains voltage of 240 V supplies a current of 12.0 A to an electric kettle. What is the resistance of the kettle's heating element as the current flows through it?

Figure 12.60

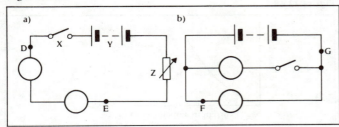

15 Fig. 12.60 shows two circuits (a) and (b), which have been set up using similar components.
a) The circles represent identical lamps. Draw the correct symbol for a lamp.
b) What do the circuit symbols labelled X, Y and Z represent?
c) Which circuit, (a) or (b), shows lamps connected in parallel?
d) Add a meter to circuit (a) so that the potential difference across component Z can be measured.
e) The switches are now closed. What difference is there, if any, in the current at points i) D and E, ii) F and G?
f) i) In which circuit will the lamps be brighter? ii) In which circuit can one lamp be switched off and the other remain on?
g) What effect does adjustment of component Z in circuit (a) have on the lamps?   (NEA [A] 88)

16 A sealed box with two terminals may contain one of the following: a wire coil, a filament lamp, a diode.
a) Draw a circuit diagram to show how you would measure the current – voltage characteristic of the device in the box. Do not attempt to state the ranges of the instruments you would use.
b) Briefly describe the experiment you would perform to obtain the current – voltage characteristic.
c) Sketch the current – voltage graphs you would expect to obtain for each device. Label your sketches clearly.   (JMB, part)

17 The resistance of each of the four side lights on a car is 9.6 Ω and the resistance of each of the two headlights is 3.0 Ω. Calculate the current supplied by the battery when all these lights, which are connected in parallel, are switched on, if the battery voltage is 12 V.

18 Fig. 12.61 shows four combinations of resistors. Find the total or equivalent resistance for each combination.

19 One 4 ohm and two 2 ohm resistors are available. All three are to be connected together in two different arrangements such that the total (resultant) resistance is (a) less than 2 ohm, (b) more than 4 ohm but less than 8 ohm. Draw a diagram of each arrangement and calculate the total (resultant) resistance in each case. (JMB, part)

Figure 12.61

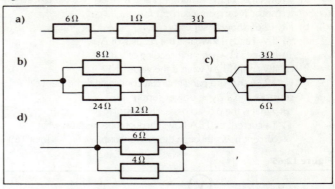

20 Fig. 12.62 shows two combinations of resistors in a mixture of series and parallel connections. Calculate the equivalent resistance of each combination.

Figure 12.62

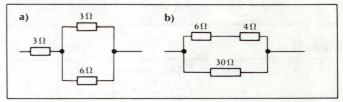

21 Fig. 12.63 shows how three resistors P, Q and R are connected together. If one resistor has a resistance of 2 Ω, another has a resistance of 3 Ω and the third has a resistance of 4 Ω, find which resistor should have which resistance for the combination shown to have (a) the minimum resistance and (b) the maximum resistance.

Figure 12.63

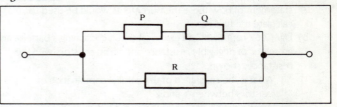

22 Three resistors of values 2 Ω, 3 Ω and 5 Ω are connected in series with a battery of voltage 6.0 V. Calculate the current through each of the resistors, and the current through the battery.

23 A current of 2.4 A flows through a resistor of 2.0 Ω resistance. If the battery in the circuit has a voltage of 12 V, what other resistance is there in the circuit?

24 Calculate the reading of each of the ammeters in the circuit of fig. 12.64.

Figure 12.64

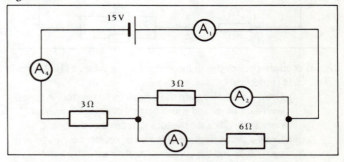

25 a) Write an equation relating *V*, *I* and *R* where *V* is the p.d. across a resistor *R* carrying a current *I*.

b) Fig. 12.65 shows a series circuit.
   i) If V_1 reads 4V what is the reading of A_1?
   ii) If $X = 3\,\Omega$ what is the reading of V_2?
   iii) If V_4 reads 12V what is the value of *Y*?

c) Fig. 12.66 shows a parallel circuit containing three identical resistors and a 2 volt power supply.
   i) What is the p.d. across each resistor?
   ii) If a current of 0.5 A flows through one resistor, what current flows through the power supply?   (Joint 16+)

Figure 12.65

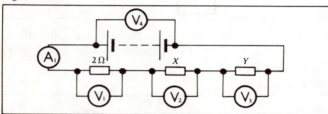

Figure 12.66

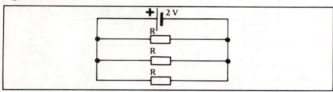

26 a) What is meant by the *resistance* of an electrical component? Describe an experiment to measure the resistance of a length of resistance wire. Your account should include a circuit diagram, a list of measurements you would make, a clear statement of how these measurements would be used to calculate the final result.

b) Fig. 12.67 represents an electrical circuit containing a battery, an ammeter of negligible resistance and three resistors with the resistances shown.
   i) What is the resistance of the parallel combination of resistors between Y and Z?
   ii) What is the resistance of the circuit between X and Z?
   iii) Assuming that the battery has negligible internal resistance, what reading would you expect on the ammeter?
   iv) What is the potential difference between X and Y?
   v) What current would flow through the 3 ohm resistor? (JMB)

Figure 12.67

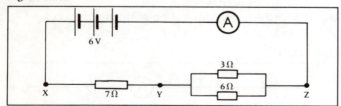

27 a) You are provided with three resistors, of values $1\,\Omega$, $3\,\Omega$ and $6\,\Omega$ respectively.
   i) Draw a circuit diagram showing all three resistors in series with each other and with a battery of negligible internal resistance. Calculate the total resistance of the circuit.
   ii) Draw a circuit diagram of the three resistors in parallel with each other and the combination in series with a battery.

Calculate the total resistance of the circuit.
   iii) Draw a circuit diagram showing the $3\,\Omega$ and the $6\,\Omega$ resistors in parallel with each other, and this combination in series with the $1\,\Omega$ resistor and the battery. Calculate the total resistance of the circuit. If the p.d. across the battery terminals is 6V calculate the current through the 6 ohm resistor.

b) Fig. 12.68 is a graph of a set of readings of the p.d. across a filament lamp plotted against the current through the lamp.
   i) Draw a labelled circuit diagram of the apparatus you would use to obtain such a set of readings.
   ii) Explain why the graph indicates that the filament of the bulb does not obey Ohm's law.
   iii) As the current increases what happens to the gradient of the graph?
   iv) What can be deduced about the filament of the bulb from your answer to (b)(iii)?   (Joint 16+)

Figure 12.68

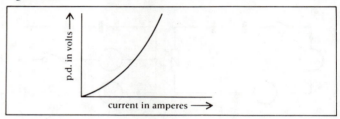

28 Two resistors, one of resistance 4 ohm and the other of unknown resistance, are connected in parallel. This combination is then placed in a circuit and the current passing into the combination is measured for various potential differences across the combination. The results of the experiment are given in the table.

Potential difference/V	1.5	3.0	4.5	6.0	7.5
Current/A	0.75	1.50	2.25	3.00	3.75

a) Draw a labelled diagram of the circuit you would use to perform the experiment. (Do not describe the experiment)

b) i) Plot a graph of potential difference against current.
   ii) From the graph calculate the total resistance of the combination of resistors, explaining clearly how the graph was used.
   iii) Using the resistance of the combination obtained in (ii), calculate a value for the unknown resistance.

c) Describe, with the aid of one diagram in each case, how a moving-coil galvanometer may be converted into (i) a voltmeter, (ii) an ammeter.   (JMB)

29 A moving-coil galvanometer (milliammeter) has a resistance of 5 ohms and will give a full-scale deflection when a current of 0.015 A flows through it. Calculate

a) the potential difference across the meter when a current of 0.015 A flows through it,

b) the value of the resistance which would convert the meter into an ammeter reading up to 3 A, showing how the resistance would be connected to the galvanometer,

c) the value of the resistance which would convert the meter into a voltmeter reading up to 15 V, showing how the resistance would be connected to the galvonometer.   (JMB, part)

30 a) A length of wire has a resistance of about 8 ohm. How would you measure this resistance, using an ammeter and a voltmeter? Draw a labelled circuit diagram. Explain how you would calculate the resistance from your readings.

b) For a particular specimen of wire, a series of readings of the current through the wire for different potential differences across it is taken and plotted as in fig. 12.69a. Describe how the resistance is changing. Suggest the most likely explanation.

c) How would the resistance of a piece of wire change if i) the length were doubled, ii) the diameter were doubled?
The growth of cracks in a brick wall can be detected by measuring the change in resistance of a fine piece of wire. The wire is stretched tightly and fixed firmly to the wall on each side of the crack (fig. 12.69b).

d) If the crack opens slightly, what happens to i) the length of wire, ii) the thickness of wire, iii) the resistance of the wire.

e) In choosing a wire for the element of an electric kettle, A says it should have a high resistance as it is the resistance which causes the heating effect. B says the resistance should be low so that the current is as great as possible. What do you think?

(NEA [B] SPEC)

Figure 12.69

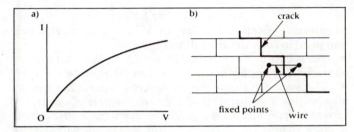

31 Calculate the resistance of a copper cable of length 200 km and cross-sectional area 2.5 mm², if the resistivity of copper is $1.7 \times 10^{-8}\,\Omega$m.

32 If a steel wire of length 132 m and cross-sectional area 2.64 mm² has a resistance of $7.50\,\Omega$, what is the resistivity of steel?

33 An accurate resistor is required to convert a galvanometer into a voltmeter. Calculate the length of manganin wire of diameter 0.032 mm required to make a resistor of value $1.120\,k\Omega$. The resistivity of manganin is $4.50 \times 10^{-7}\,\Omega$m.

34 Describe an experiment to investigate the relationship between the heat produced in a wire and the electric current flowing through it. Your answer should include (a) a circuit diagram, (b) an account of the observations you would make, (c) an account of how you would use these observations to deduce the relationship, (d) a statement of the result you would expect to obtain.

(JMB, part)

35 An electric kettle has a heating element rated at 2 kW when connected to a 250 V electrical supply. Calculate
a) the current that would flow when the element was connected to a 250 V supply,
b) the resistance of the element,
c) the heat produced by the element in 1 minute.   (JMB, part)

36 a) In an experiment to measure the power of a small motor the apparatus is set up as shown in fig. 12.70.
i) Complete the table below naming the meters marked M_1 and M_2 and stating what quantity each measures. Indicate also the polarity $(+/-)$ of the terminals X and Y.
ii) As the rod turns the slack is taken up in the string and the

Meter	Terminals	Name of Meter	Quantity Measured
M_1	X =	....	....
M_2	Y =	....	....

Figure 12.70

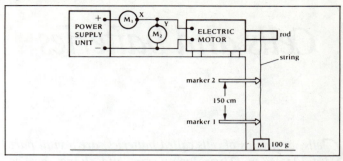

mass M starts to move upwards at a steady speed. M has a mass of 100 g. Determine the tension in the string in newtons. [Take g, the acceleration of free fall as 10 m/s²]

iii) Calculate the change in the potential energy when the mass M rises from marker 1 to marker 2.

iv) If the output mechanical power of the motor is 0.3 W find the time taken for M to move between the two markers, showing clearly how you obtain your answer.

v) Calculate the kinetic energy in Joules of M when it is midway between the two markers.

vi) If throughout the experiment M_1 reads 80 mA and M_2 reads 5.0 V calculate the input electrical power in watts to the motor.

vii) Account for the difference between the mechanical power of the motor, 0.3 W and the input electrical power as calculated in part (vi).

viii) Write down an equation for the efficiency of a motor. Calculate the efficiency of the motor in (a) (vi).

b) A practical industrial motor delivers much more power than a toy motor. This is due to several differences between them. State *two* of these differences.   (NISEC 88)

37 a) Draw a labelled diagram of a parallel plate capacitor.
b) What is the relationship between capacitance, voltage and charge?
A 2 μF capacitor is charged to a potential of 200 V and then disconnected from the power supply.
c) What is 2 μF expressed in farads?
d) What is the size of the charge on each plate of the capacitor?
e) One plate of the capacitor carries a positive charge; the other plate is earthed. Explain why the earthed plate carries a negative charge.   (O)

38 a) A capacitor is a device for storing electric charge.
i) What are the essential features in the construction of a capacitor?
ii) Explain how the charge is distributed when a capacitor is charged.
iii) Describe and explain what happens when the terminals of a charged capacitor are connected by a piece of copper wire.
b) In an experiment with a capacitor, the charge which was stored was measured for different values of charging potential difference. The results are tabulated below.

Charge stored/μC	7.5	30	60	75	90
Potential difference/V	1.0	4.0	8.0	10.0	12.0

i) Plot a graph of charge stored on the *y*-axis against potential difference on the *x*-axis.
ii) Use the graph to calculate the capacitance of the capacitor used in the experiment.   (JMB, part)

13

Cells and batteries

*C*ells, or groups of cells called batteries, are a vital part of life today. Their types, sizes and shapes are as varied as their uses. The photographs show a selection of cells that you might come across:

a) is a wet Leclanché cell;
b) is a zinc–carbon (Leclanché) dry cell;
c) is an alkaline–manganese dry cell;
d) is a rechargeable nickel–cadmium dry cell;
e) is silver oxide button cell;
f) is a mercury oxide button cell and
g) is a lead–acid car battery.

There are many different types of cell

(a) wet Leclanché cell;

(b) dry Leclanché cell;

(c) alkaline–manganese cell;

(d) nickel – cadmium rechargeable cell;

(e) silver oxide button cell;

(f) mercury oxide button cell;   (g) car battery

Types of cell

Traditionally, cells have been divided into primary and secondary types. **Primary cells** use a chemical reaction which converts stored chemical energy into electrical energy in a process that cannot be reversed. These cells are thrown away after their one useful life. **Secondary cells** can be reused, because their chemical reaction can be reversed many times. Secondary cells are storage cells, and store or accumulate the energy fed into them in a charging process. This process sends a chemical reaction in one direction in the cell. The cell can give out the energy when the chemical reaction is reversed in the discharging process. Such cells are therefore also known as **storage cells** or **accumulators**. Both these types of cell use a chemical reaction as the means of supplying electrical energy to a circuit, but there are some cells which involve other direct energy conversions. The **solar cell**, for example, converts light energy directly into electrical energy. A **thermocouple** converts heat energy into electrical energy.

Choosing a cell or battery

In selecting a cell or battery for a particular application, you need to consider the following:
- *The voltage or e.m.f. needed for the device. Note that cells are often connected together in series to increase the available e.m.f.*
- *The amount of charge that the battery can store. The capacity may be quoted in ampere hours, i.e. the number of amperes of current × hours that it can supply. Note that when a large current is taken from a cell it generally will become discharged more quickly and have a smaller effective capacity.*
- *The physical size of a battery of suitable capacity.*
- *Whether the battery needs to be portable and if so, whether a wet cell containing liquid would be safe.*
- *If it is a primary cell, its expected life*
a) *on the shelf, i.e. when it is not being used and*
b) *in use for the application concerned.*
- *If it is a rechargeable (secondary) cell, the number of times it can be recharged and the equipment needed.*
- *The price. Prices vary greatly according to the size and capacity of the battery. Generally, rechargeable cells are much more expensive to buy than primary cells and you must also consider the cost of special charging devices for the rechargeable cells.*

Depending on the application, different factors will be more important than others. For example, watches and cameras need very small batteries and so the choice of a button cell is the most important consideration. But to operate the starter motor on a car requires a very large capacity battery which can deliver up to 400 amperes of current for a short time. The most suitable type of battery for this purpose is still a wet lead–acid battery. The technology of cells is being developed all the time and there is still a need for new designs. One major problem, not yet solved, is the design of a battery that can store enough energy to drive an electric car over a reasonable distance before recharging is necessary and that is not too large and expensive. Another need is for small, button cells, to have longer working lives.

Table 13.1 gives information about various kinds of cells and batteries which are available today. You need to be able to use rather than memorise all this information.

Table 13.1   Cell data

Type of cell	Electrodes	Electrolyte	Voltage and internal resistance	Size and shape	Applications and comparisons
Leclanché wet primary, 1868	+ carbon − zinc (rod)	ammonium chloride solution with manganese (IV) oxide to depolarise	1.5 V, several ohms	large glass jar containing a porous pot	was used for bells and telephones, now replaced by dry cells, not portable
Zinc–carbon dry Leclanché sealed primary 1890	+ carbon − zinc (can)	same as wet Leclanché but electrolyte is set in a paste	1.5 V, about 0.5 Ω	a variety of common small sizes, mostly cylindrical	torches, radios etc., where current demand is small or intermittent, limited shelf life, some develop leaks
Alkaline– manganese dry cell	+ carbon − zinc (powders)	potassium hydroxide, manganese (IV) oxide is compressed and mixed with the carbon	1.6 V, less than 0.5 Ω	same shapes and sizes as above but more expensive	cassette players & motorised toys, applications involving heavy continuous use needing larger currents
Nickel–cadmium rechargeable & sealed, 'nicad' dry cell	+ nickel oxide − cadmium	potassium hydroxide	1.3 V, then very steady at 1.25 V, very low resistance	same shapes as zinc–carbon but much more expensive, also button cells	shavers, calculators, flash guns, cassette players and motorised devices which make heavy current demands, needs special charger
Mercury oxide or silver oxide dry cells	+ mercury(II) oxide or + silver oxide − zinc	alkaline electrolyte	1.35 V	button cells	calculators, watches, hearing aids, cameras, used where size and weight are important but current demand is low
Lead–acid, wet cell, secondary rechargeable	+ lead(IV) oxide (brown) − lead (grey)	dilute sulphuric acid	2.0 V, less than 0.01 Ω	large, box-shaped, usually heavy	cars, used where very large currents and high power is needed, can supply 400 amps
Silicon solar cell, photovoltaic	silicon: + p-type − n-type	none	0.5 V in bright sunlight	small crystal slices, 1 or 2 mm thick	watches, calculators, photographic exposure meters, power for satellites

For example, you should be able to:
- find which cells are secondary, i.e. rechargeable cells;
- find out which cells are dry cells and fully portable;
- find out the voltage or e.m.f. of a particular cell;
- identify poisonous substances used in certain cells;
- choose a cell suitable for a specified application;
- compare the charge-storing capacity of different cells; and, given prices for different equivalent-size cells:
- select a 'best buy' for a particular application.

Basic ideas about cells and batteries

Electrodes

The electrodes are the metal terminals through which electric current leaves and returns to a cell. In the external circuit to which a cell is connected, negative electrons leave from the negative electrode and flow towards the positive one. Conventional current is said to flow from the positive terminal as is shown in fig. 13.1.

E.m.f. or voltage

The voltage of a cell, called its *electro-motive force* or *e.m.f.*, depends mainly on the two different metals used for its electrodes. Other factors, such as exhaustion of the electrolyte and a defect known as polarization can cause the voltage to fall. Most cells have an e.m.f. which is fairly constant for most of their useful life but which gradually falls as they become exhausted or 'flat'. For example, a zinc–carbon dry cell starts life with an e.m.f. of 1.55 V but this soon falls to a fairly steady 1.5 V.

Figure 13.1   *Electron and conventional current flow from a cell*

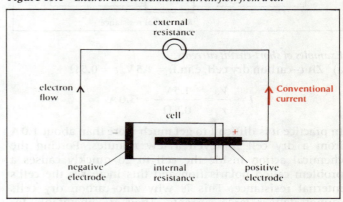

Capacity

The capacity of a cell is a measure of how much electric charge it can store. The capacity of a cell is not a fixed size because it depends on several factors, for example:
- its physical size;
- the rate at which it is discharged: the capacity is reduced if it is discharged quickly at high currents;
- the temperature and age of the cell.

The charge delivered by a cell or battery is given by:

$$\text{charge} = \text{current} \times \text{time}$$

Although electric charge is usually measured in coulombs, to distinguish it, battery capacity is usually measured in the special units of **ampere hours (Ah)**.

The capacity of a battery can be calculated using:

$$\text{battery capacity} = \text{amperes} \times \text{hours}$$

The capacity of a lead–acid car battery is as large as 40 ampere hours (Ah), while a D-size nickel–cadmium cell can hold 4 Ah and a button cell typically only 0.25 Ah. This means that, for example, a D-size nicad cell could supply 0.5 A for 8 hours (0.5 A × 8 h = 4 Ah).

Internal resistance

The electric current delivered to a circuit by a cell or battery also flows through the battery itself. Conduction inside a cell is by means of the movement of charged atoms or groups of atoms called ions in the electrolyte. There is some resistance to the flow of these ions which gives a cell an internal resistance.

The short-circuit current

The internal resistance of a cell limits the current that can flow in its circuit, fig. 13.2. The maximum current a cell can supply is its short-circuit current, obtained by shorting its terminals with a thick piece of copper wire. The resistance R of a thick wire is almost zero so that the maximum current round the circuit will be limited only by the internal resistance r of the cell, since $I = V/r$.

Figure 13.2 *Internal and external resistance in a circuit*

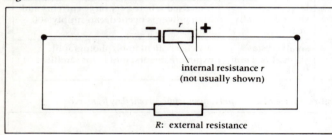

Examples of short-circuit currents:

a) Zinc–carbon dry cell, e.m.f. = 1.5 V, $r = 0.5\,\Omega$

$$I = \frac{V}{r} = \frac{1.5\,\text{V}}{0.5\,\Omega} = 3.0\,\text{A}$$

In practice it is difficult to get much more than about 1.0 A from a dry cell for even a few minutes. Forcing the chemical action inside the cell to go quickly causes a problem called polarisation and this increases the cell's internal resistance. This is why zinc–carbon dry cells cannot deliver large currents. They are unsuitable for applications which need sustained large currents and power, for example, driving electric motors.

b) Lead–acid wet storage cell, e.m.f. = 2.0 V, $r = 0.01\,\Omega$

$$I = \frac{V}{r} = \frac{2.0\,\text{V}}{0.01\,\Omega} = 200\,\text{A}$$

It is possible to take very large currents from lead–acid cells but it can be dangerous. Unless they are very thick, the connecting wires become very hot and can cause burns or even fire. The cell itself can also be damaged. Currents as large as several hundred amperes are needed to operate the starter motor in a car. To obtain such large currents a cell or battery with a very low internal resistance must be chosen.

Measuring the e.m.f. of a cell

The e.m.f. of a cell or battery can be measured directly across its terminals only when it is not supplying a current, i.e. when it is in an 'open circuit'. This is because as soon as a current flows, the voltage between the terminals of a cell falls. We must use a voltmeter with a very high resistance because this will allow only a very small current to flow from the cell.

The terminal voltage of a cell

When a cell supplies a current to a circuit, why does its terminal voltage fall? Fig. 13.3 shows a cell of e.m.f. $E = 1.5\,\text{V}$ and internal resistance $r = 0.5\,\Omega$ connected to an external resistance $R = 2.5\,\Omega$.

Figure 13.3 *On load, the terminal voltage of a cell falls*

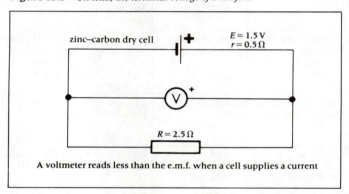

A voltmeter reads less than the e.m.f. when a cell supplies a current

The total circuit resistance = $R + r = 2.5\,\Omega + 0.5\,\Omega = 3.0\,\Omega$

The current in the circuit $I = \dfrac{E}{R + r} = \dfrac{1.5\,\text{V}}{3.0\,\Omega} = 0.5\,\text{A}$

So what will the voltmeter in fig. 13.3 read? Note that it is connected across both the external resistor and also across the terminals of the cell. The terminal voltage of the cell must be the same as the voltage across the external resistor. Using $V = IR$ for the external resistor:

$$V = IR = 0.5\,\text{A} \times 2.5\,\Omega = 1.25\,\text{V}$$

So the voltmeter reads 1.25 V and this is the terminal voltage of the cell and not its e.m.f. of 1.5 V.

The 'lost' volts

What has happened to the missing 0.25 volts? They are 'lost' across the internal resistance of the cell. Using $V = Ir$ for the internal resistance of the cell:

$$V = Ir = 0.5\,\text{A} \times 0.5\,\Omega = 0.25\,\text{V}$$

This does not mean that the voltage across the terminals of the cell will be 0.25 V if we measure it. This voltage cannot be measured directly but is 'lost' from the e.m.f.:

$$\text{terminal voltage} = \text{e.m.f.} - \text{'lost' volts}$$
$$V = E - Ir$$

The equation shows that as the current flowing from the cell increases, the 'lost' volts will also increase and so the terminal voltage falls further. This effect can be noticed if you are listening to the radio in a car. As the starter motor is operated, causing the terminal voltage of the car battery to drop significantly, the radio fades.

Internal resistance calculations

Remember the following:

- *when you calculate the current in a circuit you must include the internal resistance of a cell or battery in your calculation.*
- *the voltage across the terminals of a cell equals the voltage across the external resistors, not the 'lost' volts across the internal resistance of the cell and not the e.m.f. of the cell.*
- *the terminal voltage of a cell is only equal to its e.m.f. when it is not supplying a current.*

Worked Example
Internal resistance and circuit current

A battery has an e.m.f. of 6.0 V and an internal resistance of 1.0 Ω. How much current will flow through an 11.0 Ω resistor connected across its terminals? If a voltmeter is also connected across the battery terminals while this current flows, what will it read?

The circuit diagram is shown in fig. 13.4. The current in the circuit depends on the total resistance $R + r$

$$R + r = 11.0\,\Omega + 1.0\,\Omega = 12.0\,\Omega$$

For the whole circuit, the current

$$I = \frac{E}{R + r} = \frac{6.0\,V}{12.0\,\Omega} = 0.5\,A$$

The voltage across the battery terminals equals the voltage across the external resistor. So, for the external resistor,

$$V = IR = 0.5\,A \times 11.0\,\Omega = 5.5\,V$$

This value could also be found by calculating the 'lost volts' across the internal resistance r and subtracting from the e.m.f. E

$$\text{'lost volts'} = Ir = 0.5\,A \times 1.0\,\Omega = 0.5\,V,$$
$$\therefore \quad \text{the terminal voltage} = E - \text{'lost volts'}$$
$$= 6.0\,V - 0.5\,V = 5.5\,V$$

Answer: The current through the 11.0 Ω resistor is 0.5 A and the voltmeter across the battery terminals will read 5.5 V.

Figure 13.4

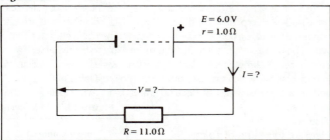

$E = 6.0\,V$
$r = 1.0\,\Omega$
$I = ?$
$V = ?$
$R = 11.0\,\Omega$

Cells in batteries

When cells are joined together to form a battery, two factors are affected by the way the cells are connected. These are the e.m.f. of the battery and internal resistance of the battery. Fig. 13.5 shows three typical arrangements of cells that are used.

a) The cells are connected in series. Their e.m.f.s and internal resistances add up to give a battery with total e.m.f. = $3E$, and total internal resistance $3r$. So this arrangement produces a battery of increased e.m.f., but the current it can supply is limited by a greater internal resistance.

b) The cells are connected in parallel. This arrangement reduces the internal resistance of the battery. Like all parallel resistors, the internal resistances of the cells have a smaller combined resistance. In this case the total internal resistance is $\frac{1}{3}r$. However, connecting the cells in parallel neither increases nor decreases their e.m.f. When side by side like this the cells do not help each other to increase the energy given to the charge which is to be driven round a circuit; but the cells do share the work so they will last longer. So the parallel cells form a battery of total e.m.f. E and total internal resistance $\frac{1}{3}r$.

Cells should not be left connected in parallel when not in use because, if one cell has a slightly higher voltage than another, the stronger cell will send a current in the reverse direction through the weaker cell and the cells will become exhausted. For this reason it is recommended that some types of cell, like nicad storage cells, should never be connected in parallel.

c) A series and parallel arrangement is combined to provide a battery of increased e.m.f. without increased internal resistance. As in (a), each parallel branch has e.m.f. $3E$ and internal resistance $3r$. But the combined resistance of the parallel branches is reduced to just r.

Figure 13.5 *Cells in batteries*

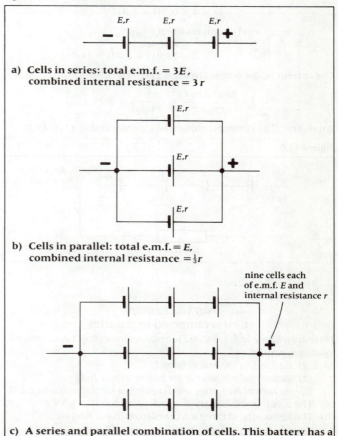

a) Cells in series: total e.m.f. = $3E$, combined internal resistance = $3r$

b) Cells in parallel: total e.m.f. = E, combined internal resistance = $\frac{1}{3}r$

nine cells each of e.m.f. E and internal resistance r

c) A series and parallel combination of cells. This battery has a higher e.m.f. without a higher internal resistance. Total e.m.f. = $3E$. Combined internal resistance = r

The six 1.5 volt cell layers, connected together in series, give this battery a total e.m.f. of 9 volts.

Worked Example
Cells connected in series

Four cells of e.m.f. 1.5 V and internal resistance 0.5 Ω are connected in series. What current will they drive through an external resistor of 22.0 Ω?

The circuit diagram is shown in fig. 13.6. The total e.m.f. of the battery is

$$4E = 4 \times 1.5\,V = 6.0\,V$$

and the total internal resistance is

$$4r = 4 \times 0.5\,\Omega = 2.0\,\Omega$$

$$\therefore \quad \text{total circuit resistance}$$
$$= R + 4r$$
$$= 22.0\,\Omega + 2.0\,\Omega = 24.0\,\Omega$$

The current in the whole circuit is given by

$$I = \frac{\text{total e.m.f.}}{\text{total } R} = \frac{6.0\,V}{24.0\,\Omega} = 0.25\,A$$

Answer: The current through the external resistor is 0.25 A.

Figure 13.6

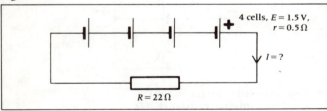

4 cells, $E = 1.5\,V$, $r = 0.5\,\Omega$

$I = ?$

$R = 22\,\Omega$

Worked Example
Cells connected in parallel

Three dry cells of e.m.f. 1.5 V and internal resistance 0.6 Ω are connected together in parallel.
Find (a) the e.m.f. of the battery formed,
(b) the internal resistance of the battery formed and
(c) the current the battery could supply to a lamp of resistance 2.8 Ω.
(a) The e.m.f. of the battery = e.m.f. of each cell = 1.5 V.
(b) The resistance of three 0.6 Ω resistors in parallel is:

$$R = \frac{0.6\,\Omega}{3} = 0.2\,\Omega = \text{internal resistance of the battery.}$$

(c) Total circuit resistance = $2.8\,\Omega + 0.2\,\Omega = 3.0\,\Omega$
The current in the circuit will be:

$$I = \frac{V}{\text{total } R} = \frac{1.5\,V}{3.0\,\Omega} = 0.5\,A = \text{current through the lamp.}$$

Assignments

Data research and processing:
Carry out a 'value-for-money' comparison of a range of different kinds of batteries.
- Select a particular size of cell which is commonly found in shops, say the large single 1.5 V cell often used in torches; it has several size numbers: D, SP2, HP2, R20. Or alternatively, use the small 9 V battery called a PP3 or AA size.
- Find out the price of each kind of cell or battery of the same size. Look for the following kinds:
a) standard quality zinc–carbon,
b) HP (high power) or top quality zinc–carbon (e.g. Ever Ready Silver Seal),
c) alkaline–manganese (e.g. Duracell, Ever Ready Gold Seal),
d) nickel–cadmium rechargeable.
- Decide how you could compare the following qualities of these cells:
1. Cell capacity in ampere hours.
2. Shelf-life.
3. Steadiness of cell voltage over a period of continuous use.
- Carry out suitable tests to obtain the necessary data.
- Make a 'value-for-money' comparison of the cells. Have you included all the relevant costs in your calculations?

Estimation
- Estimate the power of a car and hence the energy needed for a car to travel 100 miles. (Use energy = power × time.)
- Assume that a car could be fitted with an electric motor of the same power, and that it is to be powered by a 12 V battery. Calculate the current which the motor will take from the battery.
- Estimate the capacity of the battery needed to take the car for the whole 100 mile journey without a recharge of the battery.

Make a lemon cell
Insert two thin strips or rods of different materials into a lemon or other citrus fruit, as in fig. 13.7. Connect a voltmeter to these two electrodes using leads with crocodile clips. Electrodes which could be used are zinc, copper, tin, lead, carbon, iron and aluminium. The best voltmeter range would be 0 to 2 volt, but 0 to 5 volt is more likely to be available. If a centre–zero voltmeter is available this will show clearly which electrode in each case has which sign, or polarity. Complete a table of the e.m.f. and polarity for each pair of electrodes and decide which pair gives the largest e.m.f.

Figure 13.7   *A lemon cell*

Find out
Here are some uses of a battery:
  kitchen wall clock,
  portable hedge cutter,
  electric milk float,
  satellite energy storage from solar cell source,
  human heart pacemaker.
For each of these answer the following three questions:
a) What type of battery is used?
b) Why is this type of battery suitable?
c) What possible future improvements in battery design would be an advantage in this application?

Try questions 13.1 to 13.4

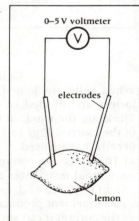

0–5 V voltmeter

V

electrodes

lemon

Questions 13

1 The table gives information about four types of rechargeable battery.

Information	Nickel–Cadmium		Lead–Acid	
	dry	liquid	dry	liquid
Number of times it can be charged/discharged	500	1000	250	250
Working life (years)	10	10–25	4	5
Highest working temperature (°C)	45	45	45	60
Amount of electricity stored (ampere hours)	15	1500	10	200
Fraction of charge lost in a month if unused	3/10	3/10	3/10	2/10

Use the information in the table to answer the following questions.
a) Which type of battery can i) store the most charge, ii) keep its electricity most effectively?
b) In which way do the dry batteries differ most from the liquid batteries?
c) You are given a choice of any of the four types of battery. State, giving **three** reasons, which type you would select for use in a transistor radio. (NEA [A] 88)

The silicon solar cell

A solar cell is quite different from the other cells we have described because it has no internal energy store. It takes in light energy and converts it immediately and directly into electrical energy. It is likely that solar cells will make an increasing contribution to our supply of electricity. Solar cells are now used to recharge the batteries in watches and space satellites, to measure the amount of light entering a camera and, in some sunny countries, as a local source of electricity.

Many materials are capable of converting light energy into electrical energy, but certain semiconductor materials are more efficient than others. The semiconductor which has been used in the space programme as a **photovoltaic cell** is silicon. Photovoltaic means that the cell produces a voltage or p.d. between two terminals when light shines on it.

In bright sunlight a silicon photovoltaic cell can produce a voltage of about 0.5 volts. A current of 20 to 30 mA can be collected from a cell of surface area 1 cm². Like other cells, they can be connected in series to increase the voltage, and in parallel to increase the current available.

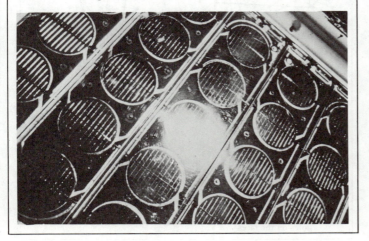

2 When a car battery of e.m.f. 12 V is connected to the car electric starter motor its terminal voltage falls to 10 V. If the resistance of the starter motor and connecting wires is 0.04 Ω, calculate the internal resistance of the battery and the current it supplies to the starter motor.

3 A cell of internal resistance 0.8 Ω is connected in series with ammeters A_1 and A_2, and a lamp, as shown in fig. 13.8. A voltmeter V_1 is connected across the cell and a voltmeter V_2 is connected across the lamp. The reading of A_1 is 0.3 A and that of V_2 is 1.2 V.
(Assume that the ammeters have negligible resistances and that the voltmeters draw negligible currents.)
a) Find the resistance of the lamp.
b) Find the reading of V_1.
c) What is the reading of A_2?
d) Find the electromotive force (e.m.f.) of the cell.
e) What current would the cell deliver if short-circuited by connecting a copper wire across its terminals? (O)

4 An electric circuit contains a lamp, a battery, a rheostat, two ammeters and two voltmeters, as shown in fig. 13.9.
a) How will the readings of ammeters A_1 and A_2 compare? Explain.
b) What difference will it make to the reading of A_1 if the resistance of the rheostat is increased? Why?
c) What difference will it make to the reading of V_1 if the resistance of the rheostat is increased? Why?
d) (It may be assumed that the voltmeters draw negligible current and that the ammeters have negligible resistances.)
What is the resistance of the lamp when A_2 reads 0.3 A and V_2 reads 1.5 V?
At the same time, V_1 reads 1.3 V. If the e.m.f. of the battery is 3 V, what are the 'lost' volts and the internal resistance of the battery? (O)

Figure 13.8

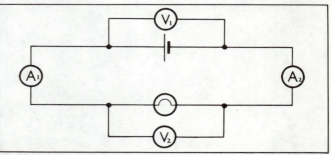

Figure 13.9

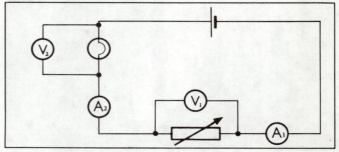

14

Magnetism and electromagnetism

*W*e live on a planet which is a large electromagnet. Many of the machines we use every day depend on the magnetic effects of electric currents and magnetic materials. Before we explore the variety of electromagnetic machines and try to understand how they work, we shall first investigate the properties of magnetic materials and the laws which describe how magnetism and electricity interact.

The aurora borealis or northern lights (shown below), which are sometimes seen in northern skies, are caused by the action of the earth's magnetic field on charged particles or ions travelling down through the atmosphere towards the magnetic pole.

14.1
MATERIALS AND MAGNETS

Certain materials are strongly affected by magnetism. When we investigate the magnetic nature of these materials we find some common properties and some important differences between them.

The simplest way of finding out which materials are strongly affected by magnetism is to test them with another material which is already magnetised. Such a material, known to keep its own magnetism for a long time, is called a **permanent magnet**.

● *Place a selection of objects made of different materials on the bench top for testing. It is important to find samples of as many different metals and alloys (mixtures of metals) as possible.*

● *Bring a permanent magnet up to each material in turn and note whether you can see or feel anything happening.*

● *Make a list of strongly magnetic materials (those attracted to the magnet) and non-magnetic materials (those quite unaffected by it.)*

The special class of materials which, like iron, are strongly affected by magnetism are called **ferromagnetic** materials (ferrum is Latin for iron). The magnet attracts anything containing iron, from a lump of iron ore to a steel ball bearing or a safety pin. However, there are only two other metals which are strongly attracted by a magnet and they are cobalt and nickel. There are also many special alloys containing iron, nickel or cobalt mixed with other metals which have useful magnetic properties.

Investigating the properties of magnets

All magnets have two poles

● *Roll a magnet in some iron filings or small pins and see where and how they stick to the magnet. (You can use plasticine to remove the iron filings from the magnet afterwards.)*

Figure 14.1   *Iron filings or pins show where the poles are on some magnets*

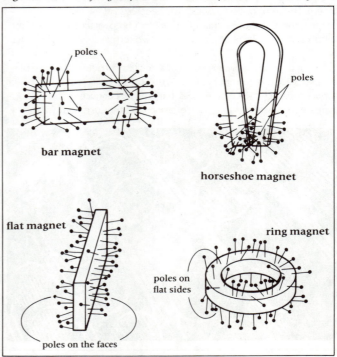

Some examples are shown in fig. 14.1. There are always two places on a magnet to which magnetic materials are attracted. These are called the **poles** of the magnet. The poles are near the ends of a bar or horseshoe shaped magnet but some magnets, made for special applications, have poles in unsuspected places.

A suspended magnet always settles
with its poles pointing the same way

It is important to place this experiment well away from all objects containing iron such as steel pipes or girders in the frame of a building and also not too near to electric cables or other magnets.

● *Suspend a bar magnet using cotton thread from a wooden (non-magnetic) support so that it is balanced horizontally and is free to turn, fig. 14.2.*

The magnet swings or oscillates slowly about a particular direction until it comes to rest with its poles always pointing exactly the same way. One pole of the magnet always points towards a place at the northern end of the earth and so we call it a **north-seeking pole**, or just the **north pole** (N pole) of the magnet. Similarly the other pole of the magnet, called the **south-seeking pole** or the **south pole** (S pole) always points towards the southern end of the earth. This discovery is used in the magnetic compass to help sailors and travellers find their direction. A steel needle or pointer is magnetised with a N pole at the pointing end, or is attached to a bar magnet, mounted so that it can turn freely and point to the magnetic north of the earth. Any other desired direction can be found from a scale on the compass.

Figure 14.2 *A suspended bar magnet*

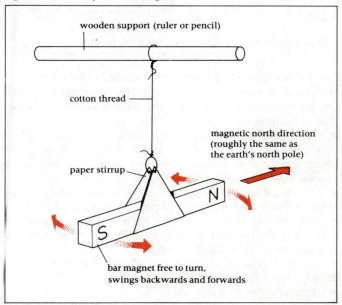

Table 14.1

Pole of suspended magnet A	Pole of hand-held magnet B	Action
N	N	repulsion
S	N	
N	S	
S	S	

Forces between magnets

● *Suspend a bar magnet A horizontally so that it can turn freely as before and label its poles N and S.*
● *Bring the north pole of another similarly labelled bar magnet B slowly towards one pole of magnet A, fig. 14.3.*
● *Repeat with the other pole of magnet A and then again, bringing the S pole of the hand-held magnet B towards each pole of magnet A.*
● *Record the action as shown in table 14.1*

The force rule we discover is very similar to the one we found for two objects with electric charges (p 203).

Like poles repel and unlike poles attract

Figure 14.3

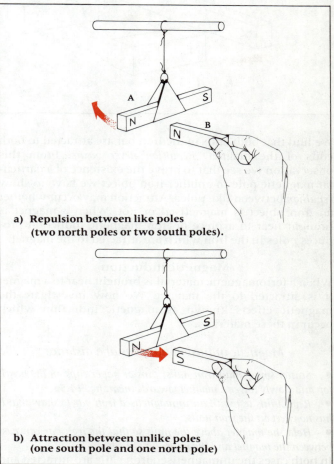

a) **Repulsion between like poles**
   (two north poles or two south poles).

b) **Attraction between unlike poles**
   (one south pole and one north pole)

Testing magnetic poles

● *Carry out a similar investigation to the last one, but this time suspend a bar of iron or some large iron nails in the paper stirrup and label the ends X and Y for identification, fig. 14.4.*
● *Record the action as in table 14.2.*

Table 14.2

End of suspended iron bar	Pole of hand-held magnet	Action
X	N	attraction
Y	N	
X	S	
Y	S	

Figure 14.4 *Testing an iron bar*

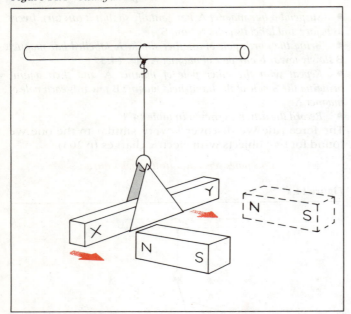

We find that both ends of the iron bar are attracted to both ends of the magnet. *Repulsion never occurs*. From this observation we see that to prove the existence of a particular magnetic pole in another iron object we have to show *repulsion* between like poles. Attraction may occur whether an iron object is magnetised or not. When a magnet is brought near to a piece of iron magnetic induction produces poles in the iron which are attracted to the magnet.

Magnetic induction

When a ferromagnetic material is brought near to a magnet it is attracted to the magnet. We now investigate the magnetic effects known as magnetic induction which occur in these materials.

Magnetic induction happens at a distance

- *Scatter some small iron nails, pins or paper clips on the bench top and slowly lower a magnet towards them, fig. 14.5a.*
- *Repeat this test with an unmagnetised iron bar to show that it has no effect on the iron nails.*
- *Hold the magnet above the nails so that the iron bar can pass between the magnet and the nails, fig. 14.5b.*

In both cases the unmagnetised iron nails are attracted and move while they are still some distance away.

The force of attraction acts from a distance.

For attraction to occur between the iron bar and the nails the bar must become magnetised while under the influence of the permanent magnet above it. We say that there is **induced magnetism** in the iron bar. In fact there is also induced magnetism in the iron nails and this happens before they are attracted to the magnet or the iron bar. Fig. 14.5c shows how the induced magnetic poles are always arranged so that unlike poles result in attraction. *Magnetic induction never results in repulsion.* Note that there is always a *pair* of induced poles. The unlike pole is induced nearer to the permanent magnetic pole causing the magnetic induction.

Figure 14.5 *Action at a distance*

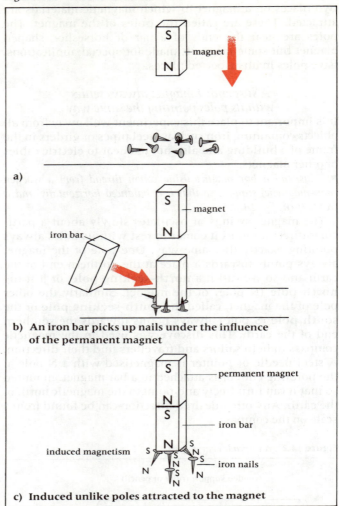

a)

b) An iron bar picks up nails under the influence of the permanent magnet

c) Induced unlike poles attracted to the magnet

Permanent and temporary induced magnetism

We can demonstrate an important difference between two kinds of ferromagnetic materials by the magnetic induction experiment shown in fig. 14.6.

- *Select some small objects such as nails or paperclips made of iron, and some other objects of a similar size made of steel. Steel objects may be more difficult to find but pen nibs or small safety pins are usually suitable.*
- *Attach a fine cotton thread to the first iron nail and first steel nib so that they can hang from a wooden pencil or ruler.*
- *Now pick up these objects with a bar magnet as shown and carefully add more nails and nibs in a chain until no more will stay attached by the induced magnetism.*
- *Carefully separate the top nail and top nib from the magnet allowing the chains to hang on the cotton threads.*
- *Remove the magnet completely.*

The iron nails very quickly lose their induced magnetism and drop off the chain, but the steel nibs retain their induced magnetism and continue hanging in a chain without the magnet. The steel nibs become permanently magnetised themselves. It may also be found that more iron nails than steel nibs could be picked up in a chain by the magnet. It is usually harder to magnetise steel than iron and so the magnetism induced in steel nibs may be weaker.

Ferromagnetic alloys like steel which are harder to magnetise are called *hard* magnetic materials. Those which are easier to magnetise such as the iron used in nails and paper clips are called *soft* magnetic materials, often referred to as soft iron. (The words 'soft' and 'hard' originally referred to the physical hardness of the metal, but they are now used in this magnetic sense to describe the ease with which a material can be magnetised.) Table 14.3 summarises these properties.

Table 14.3

	Soft magnetic materials	Hard magnetic materials
Examples	soft iron	steel
Can be magnetised	very easily	less easily
Induced magnetism is	temporary	permanent

Figure 14.6

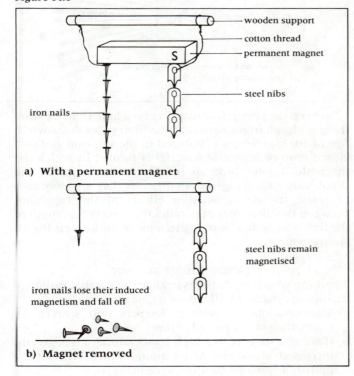

a) **With a permanent magnet**

b) **Magnet removed**

Special magnetic materials

Both soft and hard ferromagnetic materials are needed for many different applications in electrical machines today and much research has been done to find materials with improved magnetic properties, table 14.4.

At one end of the magnetic hardness spectrum new materials have been invented which are magnetically very hard and these make extremely powerful permanent magnets. Two types of material are used. One is an alloy of iron which contains carefully controlled proportions of various metals. The other is a ceramic material manufactured by a process similar to that used for making pottery. In this process, very high pressure as well as heat is applied to form a solid from powders of various metal oxides (including iron oxide). The heated metal oxides, called **ferrites**, form a very hard and brittle solid called a **sintered** magnetic material.

Table 14.4 Special magnetic materials

Type of magnetic material	Names	Composition	Magnetic properties
alloys for permanent magnets	Alcomax Alnico Ticonal	iron alloys with various amounts of aluminium, nickel, cobalt and copper	very hard; very strong permanent magnets
ceramic permanent magnets	Magnadur	powders of various metal oxides formed into a solid by the application of heat and high pressure	very hard; the strongest permanent magnets
soft iron alloy for electro-magnets and transformer cores	stalloy	96% iron 4% silicon	very soft, easily magnetised, temporary induced magnetism easily lost or reversed
nickel-iron alloy used for magnetic shielding	mumetal	74% nickel 20% iron 5% copper 1% manganese	

At the other end of the spectrum, very soft magnetic materials are made from alloys of iron and nickel. These have important uses where the induced magnetism must disappear or change its direction very quickly. These magnetic materials are also very easy to magnetise making powerful temporary magnets. Applications, which include electromagnets, transformer cores and magnetic shielding, will be described later.

Magnetising a steel bar by magnetic induction

A simple method of making a magnet from a steel bar is shown in fig. 14.7. (A steel knitting needle or screwdriver can be used.)

• *Stroke the bar with one pole of a permanent magnet so that the pole passes along the bar in the same direction many times. Between strokes the magnet should be raised high above the bar.*

• *Test the poles induced in the steel bar by bringing them close to a magnetised compass needle. The pole which repels the N pole of the compass will also be a N pole.*

We find that *the pole produced at the end of the bar where the stroke ends is of the opposite kind to the one used on the permanent magnet.* This old method produces only rather weak magnetism in steel and these days magnets are made electrically (p 305).

Figure 14.7 *Making a magnet by magnetic induction*

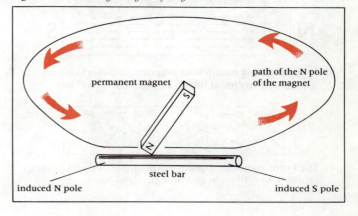

A theory of magnetism

If a magnetised steel knitting needle is cut into pieces, each short length remains magnetised and becomes a magnet with a new N and S pole, fig. 14.8a. If we could cut each small magnet into even smaller and smaller lengths we can imagine that they would still be magnetised. We can suppose that the original magnet was made up of many very small magnets lined up and joined together by their unlike poles in long chains.

Individual magnets are often called **magnetic dipoles** because they always have two poles. Magnets are not found as single magnetic poles or 'monopoles'. We call the line joining the two magnetic poles of a magnet its **magnetic axis**. The idea of tiny magnetic dipoles lined up and linked together inside a large magnet is shown in fig. 14.8b.

It is easy to imagine that the tiny magnetic dipoles are randomly arranged inside an unmagnetised bar, fig. 14.8c. When we magnetise a steel bar by stroking it the tiny magnetic dipoles in the bar are attracted to the stroking pole. Using a N pole, all the S poles of the dipoles are attracted to it and are turned in the direction of the path of the N pole so that they end up with their magnetic axes along the bar and their S poles attracted to the end where the stroking pole left the bar.

Figure 14.8

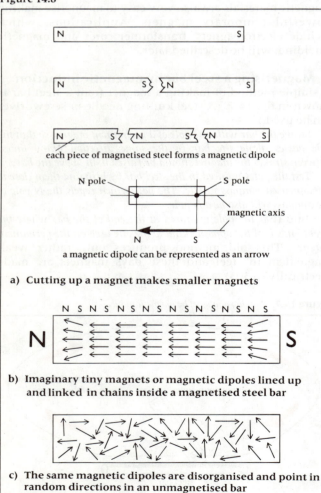

a) **Cutting up a magnet makes smaller magnets**

b) **Imaginary tiny magnets or magnetic dipoles lined up and linked in chains inside a magnetised steel bar**

c) **The same magnetic dipoles are disorganised and point in random directions in an unmagnetised bar**

Storing magnets

If we hang a number of steel knitting needles or pins on the end of a magnet they become magnetised with like poles at the same ends. The mutual repulsion between all these like poles causes the needles to splay out, fig. 14.9.

The same effect occurs inside a magnet near its ends, fig. 14.8b. The magnetic dipoles at the ends of a magnet splay out. Over a long period of time this repulsion can break down the parallel arrangement of the dipoles and gradually destroy the magnetisation of the magnet. It is this effect which quickly destroys the magnetisation of soft magnetic materials.

Figure 14.9   *Repulsion between like poles makes magnetic dipoles splay out*

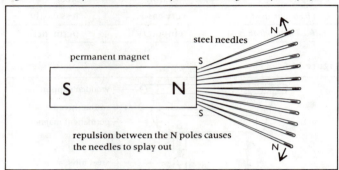

To keep the strength of magnets for a long time we store them with soft iron **keepers** across their poles as shown in fig. 14.10. Magnetism is induced in the soft iron and so a closed loop of magnetic material is formed in which the magnetic dipoles link up in closed chains. When the dipoles are linked together in this way they are more able to resist the demagnetising effects of the repulsion between the like poles of parallel dipoles. The strength of the links can be felt by trying to remove the keepers from a strong magnet.

Demagnetising magnets

Anything which tends to disarrange the parallel magnetic dipoles in a magnet will reduce its magnetism.

a) Storing a magnet without keepers will weaken its magnetism over a period of time.

b) Heating a magnet to a high temperature causes greatly increased vibrations of its atoms which will totally destroy any magnetisation of the material.

c) Dropping and knocking magnets can disarrange the tiny magnetic dipoles.

d) An alternating current in an electromagnet can be used to demagnetise magnets (p 305).

Figure 14.10   *Soft iron keepers are used when magnets are stored to help them stay strongly magnetised*

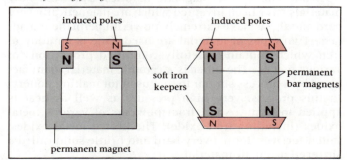

Magnetic domains

The magnetism of materials is thought to originate with the electrons of atoms and all kinds of atoms show some magnetic effects although mostly these are very weak. In ferromagnetic materials the individual atoms act as minute magnetic dipoles or 'atomic magnets'. Groups of neighbouring atoms affect each other so that they naturally set themselves with their magnetic axes parallel, i.e. with all their N poles pointing the same way. Such a natural grouping of 'atomic magnets' or 'atomic dipoles' with parallel magnetic axes is called a **magnetic domain**.

These magnetic domains have been made visible under very powerful microscopes. They are regions within a single crystal of a magnetic material, but each domain is still very much larger than the individual atoms.

In an unmagnetised material the magnetic domains are arranged so that the atomic dipoles tend to link up in closed loops as shown in fig. 14.11a. The arrows represent individual atomic dipoles (with a N pole at the pointed end), but there are really a very large number of these in each magnetic domain.

When a steel bar is magnetised the domains change shape as their atomic dipoles are realigned so that most of them have their axes in the same direction in the material, fig. 14.11b. In magnetically hard materials this realignment of the dipoles within the domains is quite permanent.

There is a maximum level of magnetisation for a material called **magnetic saturation**. This happens when the atomic dipoles in all the magnetic domains have been realigned with their magnetic axes parallel and pointing in the same direction.

In the case of a soft magnetic material, when the cause of the induced magnetism is removed the mutual repulsion between like poles of the parallel dipoles quickly disarranges them again so destroying the magnetisation of the material.

Figure 14.11 *Magnetic domains*

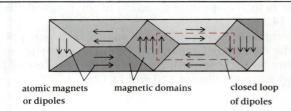

atomic magnets or dipoles   magnetic domains   closed loop of dipoles

The atomic dipoles link up the domains in the closed loops.

a) **Unmagnetised ferromagnetic material.**

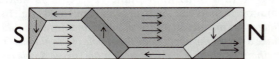

The majority of the atomic dipoles are aligned with their axes along the bar producing a N pole at one end and a S pole at the other.

b) **Magnetised ferromagnetic material.**

Some uses of permanent magnets

a) A magnet is used to reset the metal index in a maximum and minimum thermometer (p 180).
b) A magnet is fixed to the oil drain plug at the bottom of car engines and gearboxes. Small splinters of metal are picked up by the magnet from the oil to prevent their damaging moving parts of the engine.
c) Freezer and refrigerator doors are fitted with a magnetic strip to keep the door closed. The plastic seal around the door has a flexible strip inside which is magnetised. This strip is impregnated with powdered metal oxides which can be permanently magnetised.
d) Letters with magnets in them can be attached to metal notice boards and so can be rearranged very easily.
e) Magnetic ink is used on cheques so that machines in banks can read the cheque number, the account number and the amount of money paid and then automatically feed the information into the bank's computer.
f) Small ring magnets called ferrite cores are used in some computers as a magnetic memory. The rings are magnetised either clockwise or anticlockwise and these two magnetic states represent the binary digits 1 and 0.
g) Magnets are sometimes used to remove iron objects from people. For example an iron splinter can be removed from an eye or a swallowed pin might be removed without the need for surgery.

Many other uses of permanent magnets involving electrical machines are described in the next chapter.

Oily steel splinters, which have been removed from the oil circulating in a car engine, can be seen clinging to this magnetic drain plug.

Assignments

List as many uses of permanent magnets as you can find at home, at school, in the car or anywhere else that you know about. Can you think of any toys that use magnets?

Find out what a *lode stone* is.

Explain how you would distinguish between three metal objects, one made of brass, one made of iron and the third one a magnet. You may use only a plotting compass.

Remember
a) Like poles repel and unlike poles attract.
b) Repulsion between two iron objects is the only proof that they are both magnetised.
c) An opposite magnetic pole is induced in iron nearest to the magnetising pole of a permanent magnet.
d) Soft magnetic materials are easy to magnetise but quickly lose their induced magnetism. Hard magnetic materials are harder to magnetise but stay magnetised a long time.
e) Magnets, particularly small ones, are often referred to as magnetic dipoles because they always have two magnetic poles, one N and one S, some distance apart.

14.2
MAGNETIC FIELDS AND FORCES

Magnets have been found to affect other magnets and iron objects nearby. We can feel a force between two magnets when we hold them close together. These 'action at a distance' effects are caused by the magnetic forces which act in the region around a magnet called its **magnetic field**.

The idea of a magnetic field

A magnet can pick up an iron nail or turn the magnetic needle of a compass while it is still some distance away. There is a magnetic force in the space around a magnet which moves these objects. An iron bar held a short distance away from a magnet becomes temporarily magnetised without touching the magnet. The magnetism is said to be induced in the iron by the magnetic field of the permanent magnet. The result of the induced magnetism is that the iron bar is attracted to the magnet.

The magnetic field around a magnet is the region in which forces act on other magnets and on magnetic materials by inducing magnetism in them.

The forces acting in a magnetic field, like all forces, are vectors and have both magnitude and direction. Since all magnets are dipoles, both the poles of a magnet cause a force to act on the magnetic materials in its field.

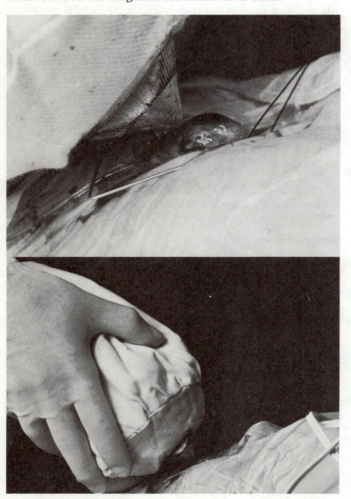

A magnet being used to remove a steel splinter from an eye at Moorfields Eye Hospital.

Finding the direction of a magnetic field

The magnetic field points in different directions at different places around a magnet. The simplest way of showing the direction is to place a small compass needle in the field. The needle points in the direction of the magnetic field. A compass needle, however, has two magnetic poles and magnetic forces will act on both poles to line it up with the magnetic field. Fig. 14.12a shows how we can find the direction of the magnetic force on a single magnetic pole which is free to move in a magnetic field. By using a very long magnet, like a magnetised steel knitting needle, we can keep one pole far enough away for it not to be affected by the magnetic field being investigated.

● *Use a cork to float the magnetised needle with its N pole at the same level as the bar magnet as in fig. 14.12a.*

● *Place the floating needle near the N pole of the magnet and watch what happens.*

The needle moves round in a curved path from the N pole to the S pole of the bar magnet. It moves in the direction of the force acting on it, so this path shows the direction of the magnetic field of the bar magnet, fig. 14.12b.

We define the direction of a magnetic field at a particular place as being the direction of the force it produces on a 'free' magnetic north pole.

But remember that single magnetic poles do not really exist!

The lines we draw to show the direction of a magnetic field are called **magnetic field lines**. As the magnetic forces acting on magnetic materials are directed along these lines they are often called **lines of force**. It follows that:

a) The magnetic field points out of or away from the N pole of a magnet and into or towards its S pole, fig. 14.12b. (Compare with an electric field which goes from a + charge to a − charge.)

b) The force on a N pole placed in a magnetic field is in the same direction as the magnetic field. The force on a S pole is in the opposite direction to the magnetic field, fig. 14.12c.

c) A compass needle aligns itself so that its N pole points in the direction of a magnetic field. The opposite forces acting on its poles turn the needle into the line of the magnetic field, figs. 14.12d and 14.12e.

Making maps of magnetic fields using a plotting compass

A plotting compass is a simple magnetic compass. A small magnet is supported between two glass faces so that it can turn freely in a horizontal plane. The case is made of a non-magnetic material such as plastic or brass, fig. 14.13a.

● *Place a bar magnet on a sheet of paper and draw round it.*

● *Make a dot on the paper near the N pole of the magnet ①.*

● *Position the plotting compass so that the curved S pole end of its needle surrounds the dot.*

● *Make the next dot ② near the N pole end of the plotting compass needle.*

● *Now move the plotting compass so that its S pole is over dot ② and mark another dot ③ near its N pole. This position is shown in fig. 14.13b.*

Figure 14.12   *Finding the direction of a magnetic field*

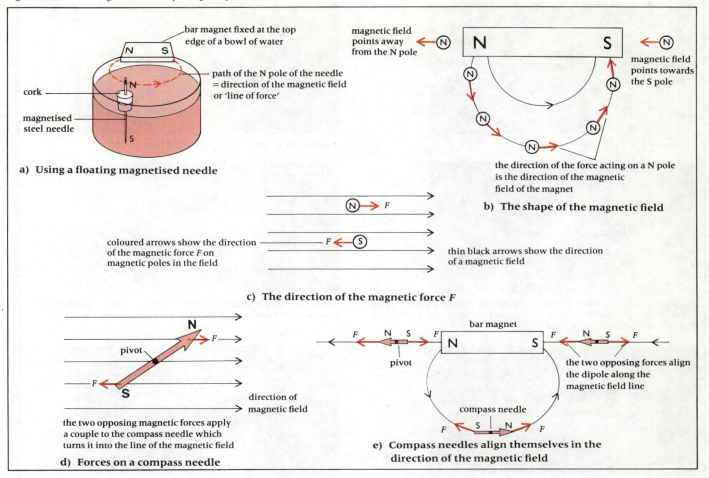

a) **Using a floating magnetised needle**

b) **The shape of the magnetic field**

c) **The direction of the magnetic force *F***

d) **Forces on a compass needle**

e) **Compass needles align themselves in the direction of the magnetic field**

Figure 14.13   *Using a plotting compass*

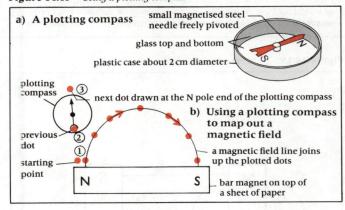

a) **A plotting compass**

b) **Using a plotting compass to map out a magnetic field**

- Continue to plot points in the direction indicated by the N pole of the compass needle until you reach the S pole of the magnet.
- Join up the dots to show the magnetic field line.

A complete map of the magnetic field of the magnet can be made by plotting field lines from several different starting points around the N pole of the magnet. Note that:

a) The magnetic field lines go from the N pole to the S pole and their direction is shown by the needle of the plotting compass.

b) The magnetic field lines never cross each other if the plotting is done accurately; check that the compass needle is always turning freely.

c) Although this method is rather slow it can show the map of a magnetic field quite a long way from a magnet where the magnetic field is weak. A plotting compass can also map the earth's magnetic field which is very weak.

Making maps of magnetic fields using iron filings

- *Arrange a sheet of cardboard or transparent perspex over the top of a magnet.*

Non-magnetic materials such as wooden blocks or aluminium rings can be used to support the sheet so that it is slightly raised above the magnet, fig. 14.14.

- *Using a pepper pot, sprinkle iron filings thinly and evenly over the top of the sheet and then gently tap the sheet with a pencil.*

Figure 14.14   *Using iron filings to show the shapes of magnetic fields*

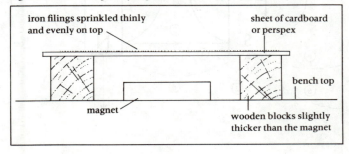

Features of some magnetic fields

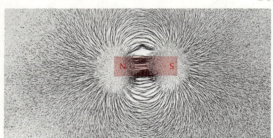

The magnetic field of a single bar magnet.

A pair of bar magnets in line with unlike poles together.

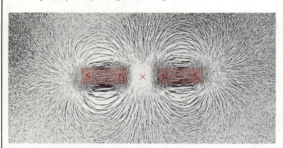

A pair of bar magnets in line with like poles together. There is a position called a neutral point (marked X) between the two like poles where the two magnetic fields cancel or neutralise each other.

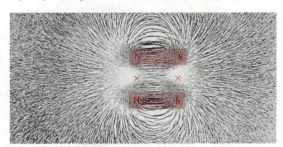

A pair of bar magnets side by side with like poles together. There are two neutral points (marked X) midway between the opposing poles.

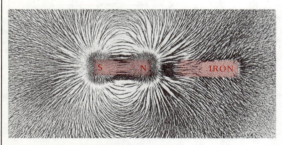

Induced magnetism produced in an iron bar placed in the magnetic field of a permanent magnet. The iron bar draws the magnetic field lines towards it and concentrates the magnetic field through the iron, so producing induced magnetism in it.

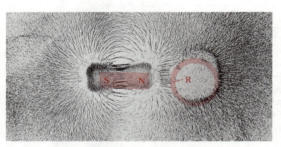

Magnetic shielding by a soft iron ring R. The magnetic field lines are attracted to the soft iron ring and induce magnetism in it, but the region enclosed by the ring is shielded from the magnetic field on the magnet. Soft iron boxes are sometimes used to shield sensitive electrical instruments from magnetic fields.

The iron filings become magnetised by magnetic induction in the magnetic field, forming small magnetic dipoles. Tapping the sheet allows them to move and turn to line up with the direction of the magnetic field. As the filings have to be magnetised, this requires a strong magnetic field. Best results are obtained with small powerful bar magnets such as ones made of Ticonal. If however the sheet is too close to the magnet the iron filings tend to slide in towards the magnet. Permanent records of iron filing field maps can be made by spraying the filings with a fixer such as hair lacquer.

Drawing magnetic field maps

Fig. 14.15 shows the basic structure of a typical magnetic field map. This is not a complete map, rather it shows how to draw one by planning the main features first. The order to work in is as follows:

① Draw the magnet shape(s) and label the magnetic poles.

② A line of force will leave the N pole and one will enter the S pole at the end of each magnet.

③ Lines of force will curve round from the N pole to the S pole on both sides of each magnet.

Figure 14.15 *Drawing a magnetic field map*

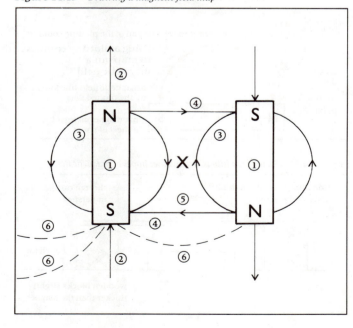

④ When there are two magnets a line of force will reach across from the N pole of one magnet to the S pole of the other if unlike poles are nearby.

⑤ Look for any places where either lines of force pass in opposite directions or like poles of two magnets are nearby and mark an X between them for a neutral point.

⑥ Draw in extra lines of force to complete the map.
Note that:

a) Arrows should be drawn to show the direction of the field lines (N to S).

b) The magnetic poles are not at the very ends of a bar magnet.

c) Lines of force should never cross each other.

d) The magnetic field is strongest where the lines of force are closest together.

Assignments

Remember that the direction of a magnetic field is:

a) the direction of the force it produces on a 'free' magnetic N pole at a particular place,

b) the direction in which a magnetised compass needle points,

c) always from a N pole of a magnet towards a S pole.

Map or draw some more magnetic fields:

d) the magnetic field of a single pole of a bar magnet (this can be found by standing a magnet on end underneath a sheet of card or perspex, fig. 14.16a),

e) the magnetic field between two flat magnadur magnets attached to an iron yoke, fig. 14.16b (the lines of force between these poles are parallel and equally spaced giving what is called a *uniform* magnetic field),

f) a pair of bar magnets arranged as in fig. 14.16c,

g) the field of a small bar magnet in the earth's magnetic field, arranged with its N pole pointing either magnetic N or magnetic S.

Describe

h) in your own words the method you have used to map the magnetic fields.

i) in your own words what you understand by the term *a magnetic field*.

Figure 14.16 *Some more magnetic fields to draw or map*

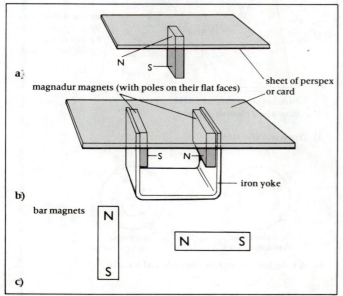

14.3
ELECTRIC CURRENTS HAVE MAGNETIC EFFECTS

When in 1819 Oersted first discovered the magnetic field around an electric current in a wire, he made the first important step in understanding the link between magnetism and electric charge. In the following experiments we investigate the magnetic field patterns produced when an electric current flows in wires of various shapes. It is now known that all moving charges, whether in a wire or not, have magnetic fields around them.

Oersted's experiment

● *Connect a thick copper wire to a low-voltage power pack which can supply several amperes of current. If an accumulator is used a rheostat of 5 or 10 A rating should be included in the circuit to protect it, fig. 14.17.*

● *Before switching on the current, place a plotting compass at various places around the wire and note the direction of the magnetic field (the direction of the N pole of the needle).*

● *Now switch on the current and place the plotting compass above and below the wire and note the direction of the magnetic field.*

● *What happens when the current direction is reversed?*

There is clearly a magnetic field around the wire which is not caused by the wire itself. The direction of this magnetic field depends on the direction of the current and the position around the wire.

Figure 14.17 *Oersted's discovery*

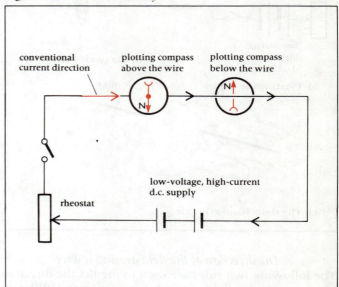

Ampere invented a rule to predict the direction of the compass needle when it is placed near to a wire carrying an electric current. It is known as *Ampere's swimming rule*

Imagine you are swimming along the wire in the direction of the current and you are facing the compass needle, then the N pole of the needle will turn towards your left hand.

If the wire is above the needle you will be swimming face-down and if the wire is below the needle you will be swimming on your back. Test this rule on fig. 14.17.

The magnetic field pattern due to a current in a straight wire

● *Support a stiff card horizontally with a thick straight copper wire passing vertically through its centre, fig. 14.18a.*

● *Using the same circuit as for Oersted's experiment, pass a large current through the wire vertically downwards.*

● *Place a plotting compass at various positions around the wire and note the direction of the magnetic field.*

● *Note the effect of reversing the current.*

● *Now sprinkle iron filings thinly and evenly over the card.*

● *With the current on, tap the card gently until the iron filings show the magnetic field pattern.*

These experiments show that there is a magnetic field which goes in concentric circles round a wire carrying a current.

Figure 14.18 *The magnetic field round a wire*

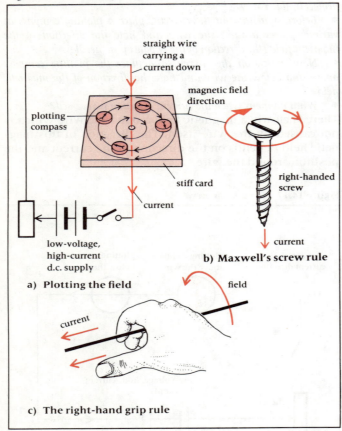

a) **Plotting the field**

b) **Maxwell's screw rule**

c) **The right-hand grip rule**

The direction of the field around a wire

The following two rules are used to predict the direction of the magnetic field around wires in many different situations. You should be able to use one of them and to express it in your own words.

Maxwell's screw rule   If a right-handed screw is turned so that it moves forwards in the same direction as an electric current, its direction of rotation gives the direction of the magnetic field due to the current, fig. 14.18b.

The right-hand grip rule   If a wire carrying a current is gripped with the right hand and with the thumb pointing along the wire in the direction of the current, the fingers point in the direction of the magnetic field around the wire, fig. 14.18c.

The magnetic field due to the current in a flat coil

● *Make a flat circular coil by winding ten or more turns of wire round a cylindrical former such as a length of plastic pipe or a cardboard tube of diameter about 4 cm or more.*

● *Push the turns of wire close together and slide them from the former onto a piece of stiff card as shown in fig. 14.19a.*

● *Connect the coil to a low-voltage, high-current supply.*

● *Sprinkle iron filings over the card both inside and outside the coil.*

● *Switch on the current and tap the card until the magnetic field pattern shows up.*

● *Use a plotting compass to find the direction of the magnetic field at various points around the coil.*

It is helpful to compare the magnetic field around each side of the coil where it passes through the card with the field of a single straight wire.

● *Note the direction of the current at each side and apply a direction rule to check the direction of the magnetic field.*

Note that the circular lines of force around the wires become squashed together inside the coil. The magnetic field is stronger inside the coil.

A rule for the magnetic poles of a coil

The magnetic field points out of the coil in fig. 14.19a towards us in the same way that the lines of force point out of the N pole of a magnet. So the face of the coil we see with the current flowing anticlockwise round it produces a N pole. The rule for remembering this is shown in fig. 14.19b. The pole is given by the letter which points in the same direction as the current. A coil will always have the opposite pole on its other face because the current direction is reversed when seen from the other side.

Figure 14.19 *The magnetic field due to a flat circular coil carrying a current*

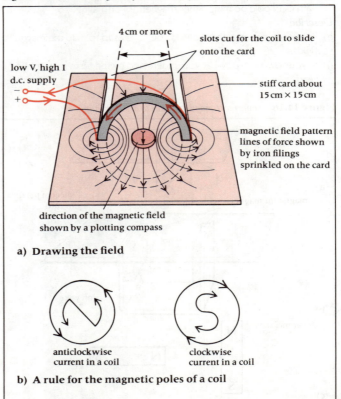

a) **Drawing the field**

anticlockwise current in a coil

clockwise current in a coil

b) **A rule for the magnetic poles of a coil**

The magnetic field pattern due to a solenoid

- *Make another coil, this time stretching out the turns to form a long coil, called a solenoid, about 10 cm long and 4 cm or more in diameter.*
- *Mount the solenoid on a stiff card or hardboard as shown in fig. 14.20a.*
- *Again using plotting compass and iron filings, obtain a map of the magnetic field showing both its shape and direction.*

Note that:

a) The direction of the magnetic field inside is opposite to the direction outside the solenoid.

b) The clarity of the iron filing pattern shows that the magnetic field is stronger inside the solenoid than outside.

c) This magnetic field map is similar to that of a bar magnet, fig. 14.20b.

The solenoid with an iron core

We have already seen that magnetism is induced in a soft iron bar when it is placed in the magnetic field of a permanent bar magnet (p 282). The solenoid has a magnetic field which is very similar to that of a bar magnet and we can fit an iron bar inside the solenoid where its magnetic field is strongest. A bar of soft iron inside a solenoid is called its **core**. The magnet produced by a current in a solenoid with a soft iron core is called an **electromagnet**. When the current in the solenoid is switched off the temporary magnetism in the soft iron core quickly disappears. Consequently, the magnetism of an electromagnet can be switched on and off.

Figure 14.20 *The magnetic field due to a solenoid*

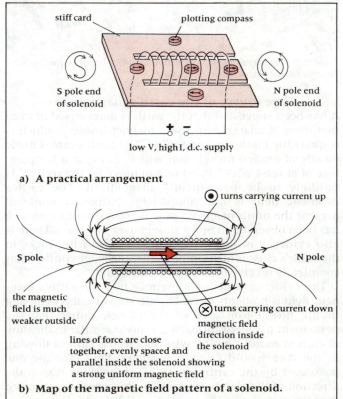

a) **A practical arrangement**

b) **Map of the magnetic field pattern of a solenoid.**

Investigating the strength of an electromagnet

We can investigate how several factors affect the strength of an electromagnet using the practical arrangements shown in fig. 14.21. The factors are: the current in the solenoid, the number of turns of wire on the solenoid and the shape of the iron core.

- *Clamp a soft iron nail or iron rod, about 10 cm long in a wooden holder as shown in fig. 14.21.*
- *Wind 10 turns of insulated copper wire tightly together round the nail and connect to a circuit containing a switch, a 10Ω rheostat, a 0 − 5 A ammeter and a low-voltage, high-current d.c. supply.*
- *Switch on the current and adjust the rheostat to give a particular current, say 2 A, through the solenoid.*
- *Dip the end of the electromagnet into a beaker full of small iron nails, panel pins or paper clips by lifting the beaker up to the electromagnet.*
- *Remove the beaker and count the number of nails attached to the electromagnet.*
- *Switch off the current and note that most, if not all of the nails fall off the electromagnet.*

Figure 14.21 *Testing electromagnets*

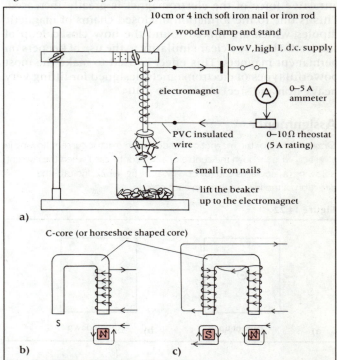

The current and number of turns in the solenoid

- *Repeat the experiment for various different values of the current using the same number of turns on the solenoid.*
- *Keeping the current constant by readjusting the rheostat each time, repeat the experiment for several different numbers of turns of wire on the solenoid.*

The strength of an electromagnet increases as the current is increased and as the number of turns is increased. After a certain current and number of turns is reached, further increases in current or turns do not make the electromagnet any stronger. Can you explain this in terms of the magnetic dipoles in the magnetic domains of the iron? (This is an example of magnetic saturation.)

The shape of the iron core

- *Replace the straight iron nail or rod with a U-shaped iron core (sometimes called a C-core or a horseshoe-shaped core).*
- *Wind a single coil of 10 turns on one side of the C-core as shown in fig. 14.21b, and investigate the strength of this electromagnet compared with the straight one using the same number of turns and the same current.*
- *Wind a second coil as shown in fig. 14.21c.*

Note how the windings on the two sides must go opposite ways round to produce a pair of opposite poles on the two ends of the electromagnet.

- *What reasons can you think of that might explain why the electromagnets made as in (b) and (c) are stronger than the straight one shown in (a)?*

The C-core electromagnet is more powerful than a straight one for the obvious reason that both poles can be used to attract or lift iron objects; but it is actually much more than twice as powerul as a single pole. As the magnetic poles get closer together the magnetic field between them becomes stronger and so is capable of inducing stronger magnetism in the iron objects it picks up. When the gap between the poles is closed by the soft iron object the attractive force of the electromagnet is greatly increased. This is due to the formation of closed chains of magnetic dipoles which are linked round the now closed loop of soft iron. There is a clear similarity to the use of keepers on permanent magnets. This effect is used to make the most powerful types of electromagnets designed for lifting very heavy iron and steel objects, see p 304.

Assignments

Draw the map of the magnetic field due to an electric current flowing in a vertical wire through the centre of a horizontal card when the current is flowing (a) upwards and (b) downwards, fig. 14.22. Indicate the direction of the lines of force in each case.

Figure 14.22

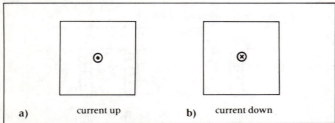

a) current up   b) current down

Find out how electromagnets are used in the world around you.
Remember

c) a rule for predicting the direction of the magnetic field round a straight wire carrying a current,
d) a rule for predicting the magnetic poles on the faces of a coil carrying a current.

Try questions 14.1 to 14.4

14.4
THE EARTH AS A MAGNET

For many centuries sailors and explorers have used the magnetic compass to help them navigate. Very recently we have discovered that some migrant birds have magnetic sensors in their heads which help to guide them using the earth's magnetic field. We have also found out that the earth's magnetism forms a protective barrier against some of the charged particles of radiation that reach earth from space. In 1958 an American space satellite, Explorer 1, discovered the radiation belts, now known as the Van Allen zones, which surround the earth. These are composed of the charged particles of cosmic and solar radiation which have been trapped by the earth's magnetic field. We are only just beginning to appreciate some of the ways in which the earth's magnetic field affects life on earth.

These birds have magnetic sensors in their heads which help them find their direction by using the earth's magnetic field.

The source of the earth's magnetism

It has been suggested that the earth is magnetised or even that there is a large permanent magnet inside it which is responsible for the magnetic field. The earth's core is made mostly of molten nickel–iron which, being at a temperature of at least 2 200°C and free to move by convection, is unlikely to be permanently magnetised. The earth's magnetic field is not completely stationary; while at present the north magnetic pole is in northern Canada, it has been observed to move slowly over the years. There is also evidence of magnetisation in iron-bearing rocks in the earth's crust which suggests that the magnetic field has completely reversed direction occasionally.

The evidence leads us to believe that the earth's magnetic field is probably caused by electric currents circulating in the nickel–iron core of the planet, rather than by permanent magnetisation. By a process similar to the flow of current round a coil of wire, an electric current flowing in the core would cause a magnetic field like the one possessed by the earth, fig. 14.23. If for some reason the direction of the electric currents in the core were reversed then the direction of the magnetic field would reverse.

Figure 14.23 *The magnetic field of the earth, explained by electric currents in the core*

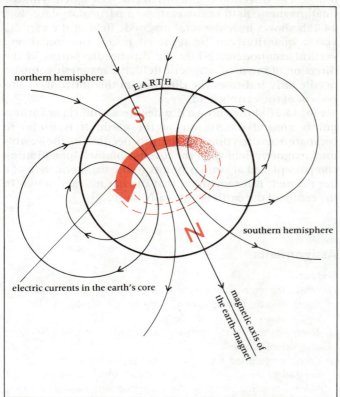

Figure 14.24

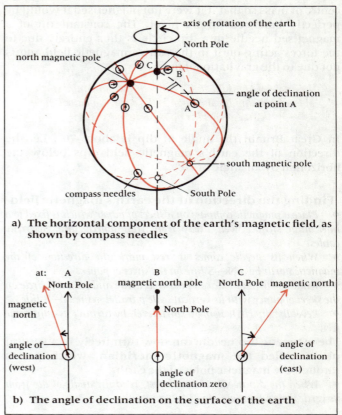

a) **The horizontal component of the earth's magnetic field, as shown by compass needles**

b) **The angle of declination on the surface of the earth**

Magnetic needles in the earth's magnetic field

The magnetic compass

We use a magnetic compass to find the direction of the *horizontal* component of the earth's magnetic field. The compass needle is pivoted so that its N pole can turn freely in a horizontal circle, i.e. in a plane parallel to the surface of the earth.

Magnetic compasses all over the surface of the earth point towards one point on the earth's surface called the **north magnetic pole**, fig. 14.24. At present this point is in northern Canada, some 1 000 km from the true North Pole. (The true North Pole is the place in the Arctic which lies on the axis of rotation of the earth.)

The angle of declination

Because the magnetic north pole and the true North Pole are in different places on the earth's surface, from most points on the surface of the earth, there is an angle between the directions of the two poles.

The angle at a particular point on the earth's surface between the direction of the true North Pole and the magnetic north pole is called the angle of declination.

Fig. 14.24 shows how, at point A, the direction of magnetic north is west of true North. Great Britain is in a position similar to point A and the angle of declination at the present time is about 8° west. From certain positions on the surface of the earth (point B) the two poles are in line and the angle of declination is zero. From other positions (point C) the direction of magnetic north is found to be east of true North.

The dip circle and the angle of dip

The lines of force in the earth's magnetic field pass through the earth and enter and leave the ground at an angle. To find the direction of these lines of force we use a magnetic needle which is pivoted horizontally so that it can turn in a vertical circle. The magnetised needle and circular scale, called a **dip circle**, is used to measure the **angle of dip** of the earth's magnetic field, fig. 14.25.

Figure 14.25

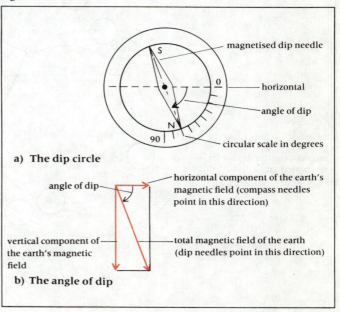

a) **The dip circle**

b) **The angle of dip**

The pivot of the needle must pass exactly through its centre of mass so that if it were not magnetised it would be perfectly balanced on its pivot. The constant tilt of a magnetised needle in a dip circle is thus entirely due to the forces acting on it in the earth's magnetic field and is not due to the gravitational field.

The angle of dip is the angle between the horizontal surface of the earth and the direction of the earth's magnetic field at a particular point on its surface.

In Great Britain the angle of dip is about 70°, i.e. the direction of the earth's magnetic field dips below the horizontal by an angle of 70°.

Finding the direction of the earth's magnetic field

- *Place a magnetic compass on a sheet of paper which is fixed to a horizontal surface well away from iron objects and electricity cables.*
- *When its needle comes to rest mark the direction of the magnetic north by a N − S line on the sheet of paper.*
- *Remove the magnetic compass well away and set a dip circle on the sheet of paper with its vertical scale parallel to the N − S line.*
- *Level the dip circle using a spirit level, by turning its adjustable feet.*

The magnetic dip needle can now turn freely in a vertical plane, called the **magnetic meridian**, which passes through the magnetic poles of the earth.

- *When the dip needle comes to rest, read the angle of dip from its vertical circular scale, fig. 14.25a.*

The direction of the earth's magnetic field has now been found. The direction of the dip needle is aligned with the total magnetic field of the earth at a particular place. Fig. 14.25b shows how the total magnetic field of the earth (a vector quantity) can be resolved into a horizontal and vertical component. While the dip needle points in the direction of the total magnetic field, the magnetic compass needle only indicates the direction of the horizontal component of the earth's field.

Fig. 14.26 shows how a dip needle would set at various points around the surface of the earth. It is useful to compare the directions of the dip needles in the earth's total magnetic field with the results obtained for a plotting compass placed at similar points in the magnetic field of a bar magnet. But remember that the bar magnet drawn in the centre of the earth is not really there.

Assignments

Find out

a) What is the angle of declination at the present time at the place where you live? This is given on Ordnance Survey maps, but the angle is slowly changing over the years.

b) When did the magnetic compass last point to the true North Pole from where you live?

c) If you are interested in space satellites and space research, find out more about the magnetic field around the earth. You will find information about this in many modern space and astronomy books.

Figure 14.26   *A section through the earth's magnetic field*

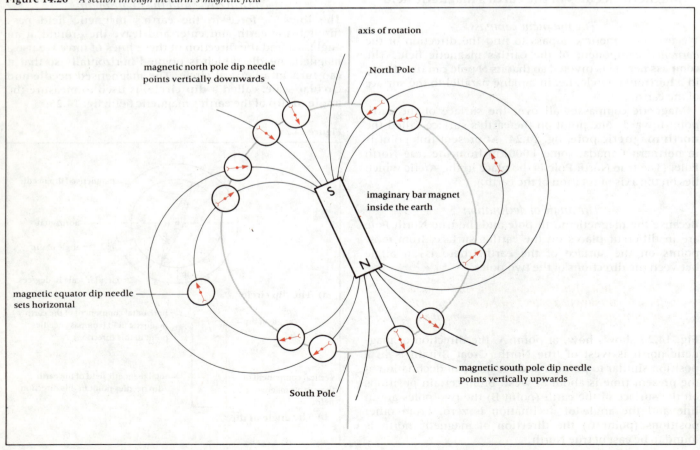

axis of rotation

magnetic north pole dip needle points vertically downwards

North Pole

imaginary bar magnet inside the earth

magnetic equator dip needle sets horizontal

magnetic south pole dip needle points vertically upwards

South Pole

14.5
ELECTRIC CURRENTS IN MAGNETIC FIELDS

We have seen that electric currents cause magnetic fields around them so it is not surprising to find that when a wire carries an electric current through another magnetic field the two magnetic fields interact and produce some interesting effects. The forces between the two magnetic fields can move wires and turn coils which carry an electric current. We call the motion produced the **dynamic effect** or **motor effect** of an electric current.

Figure 14.27 *The force acting on a wire carrying a current in a magnetic field*

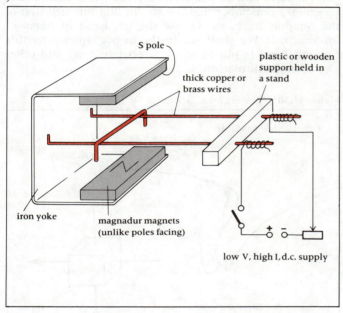

S pole

plastic or wooden support held in a stand

thick copper or brass wires

iron yoke

magnadur magnets (unlike poles facing)

low V, high I, d.c. supply

A current-carrying wire in a magnetic field

● *Attach two magnadur flat magnets to an iron yoke with unlike poles facing to provide a strong magnetic field, fig. 14.27.*

● *Mount two stiff, straight lengths of copper wire or brass rod parallel and horizontal on an insulating support as shown.*

● *Sit a length of copper wire or brass rod across the parallel wires so that it can move smoothly and freely along them. (The bent-over ends stop the wire falling off as it moves.)*

● *Connect a low-voltage, high-current supply to the ends of the parallel wires so that a complete circuit is formed.*

● *With the magnets in place, switch on the current and watch what happens.*

● *Repeat the test with the current direction reversed and then the direction of the magnetic field reversed.*

We notice that the wire is thrown or catapulted out of the magnetic field of the magnets. When either the current or the magnetic field direction is reversed the direction of the movement of the wire is reversed.

An alternative method of showing the catapult force which acts on a wire carrying a current is shown in fig. 14.28a. A wire 'swing' is hung between the poles of a large U-shaped magnet. (The Eclipse 'Major' magnet, a very powerful Alcomax permanent magnet, is ideal for this demonstration.) When a current is passed through the wire it swings out sideways. We notice that:

The wire is thrown out of the magnetic field as if the field had been stretched like a piece of elastic in a catapult which is suddenly let go.

The catapult force acts sideways or at right angles to both the current direction and the magnetic field direction.

The second point is illustrated in fig. 14.28b.

Figure 14.28 *The catapult force acting on a swinging wire*

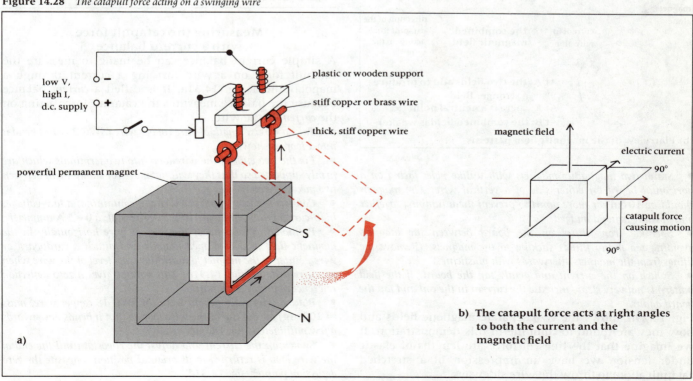

low V, high I, d.c. supply

plastic or wooden support

stiff copper or brass wire

thick, stiff copper wire

powerful permanent magnet

S

N

a)

magnetic field

electric current

90°

catapult force causing motion

90°

b) **The catapult force acts at right angles to both the current and the magnetic field**

The shape of the field producing the catapult force

The magnetic fields of the permanent magnets and the current in the wire are combined when they occupy the same region. We can set up the arrangement shown in fig. 14.29a to obtain a picture of what happens when these two magnetic fields are put together.

Figure 14.29   *The field producing the catapult force*

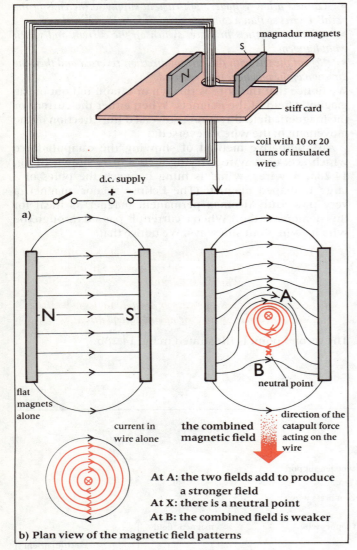

a)

b) **Plan view of the magnetic field patterns**

At A: the two fields add to produce a stronger field
At X: there is a neutral point
At B: the combined field is weaker

• *Stand two magnadur magnets with unlike poles facing on a horizontal board on either side of a vertical wire. The magnets should be 10 cm or more apart to prevent them jumping together and possibly being damaged.*

• *Sprinkle iron filings on the board between the magnets avoiding too many filings sticking to the magnets. (Remove the filings from the magnets afterwards with plasticine.)*

• *Switch on the current and gently tap the board. If the field pattern is not very clear, increase the current in the coil and tap the board again.*

Fig. 14.29b shows the two separate magnetic fields and how they are found to combine in this demonstration. If we imagine that the lines of force are lengths of elastic under tension we have an impression of a stretched catapult about to throw the wire sideways.

Fleming's left-hand rule, the motor rule

The direction of the catapult force can always be predicted by drawing the combined magnetic field. However, J. A. Fleming devised a rule for quickly predicting the direction of the catapult force, fig. 14.30.

Hold the thumb and first two fingers of your *left* hand at right angles to each other. Point your first finger in the direction of the magnetic field (N to S pole) and your second finger in the direction of the current. Your thumb now points in the direction of the thrust. (We use the word *thrust* for the catapult force so that we can link it to the thumb by the letters *th*.)

If you have difficulty in remembering which hand to use you might remember that we are finding the direction of the catapult force, so we use the left hand in *Fleming's left-hand rule*. We shall see in the next chapter how this catapult force is put to use in electric motors and other electromagnetic machines.

Figure 14.30   *Fleming's left-hand rule*

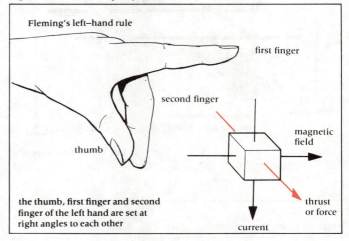

Measuring the catapult force with a current balance

A simple current balance can be made to measure the catapult force on a wire carrying a current through a magnetic field, fig. 14.31a. It is called a *current* balance because the balance measures the catapult force acting on the *current* in the wire.

• *Take a 60 cm length of bare copper wire (SWG 24) and bend it into a loop as shown.*

• *Fix the two ends of the wire loop into two terminals which are firmly supported so that they will not move (or clamp them between two pieces of wood).*

• *Connect the loop of wire to a circuit containing: a low-voltage, high-current d.c. supply; a 10 Ω, 5 A rheostat; a 0 – 5 A ammeter.*

• *Position the free end of the loop of wire horizontally in the magnetic field of a U-shaped magnet and attach a cardboard or wire pointer to the magnet to mark the exact level of the wire when no current flows, fig. 14.31b. This pointer gives a zero deflection reference point for the balance.*

• *Place a wire rider (about 2 cm of SWG 18 copper wire, mass ≈ 200 mg) on the end of the wire loop so that it bends downwards a few millimetres, fig. 14.31c.*

• *Switch on the current and adjust the rheostat until the end of the wire loop is returned to its original position, opposite the zero deflection pointer, fig. 14.31d.*

Here we can say that the two forces are in equilibrium:

the catapult force F = the weight of the wire rider mg
    (upwards)   (downwards)

Note that the catapult force must act upwards to balance the downwards weight of the rider. The directions of the magnetic field, current and catapult force are shown in fig. 14.31a.

● *To find the value of the catapult force weigh the rider to find its mass m.*

The force, given by mg, is in newtons if m is in kg and $g = 10\,\text{N/kg}$.

Figure 14.31 *A simple current balance for measuring the catapult force*

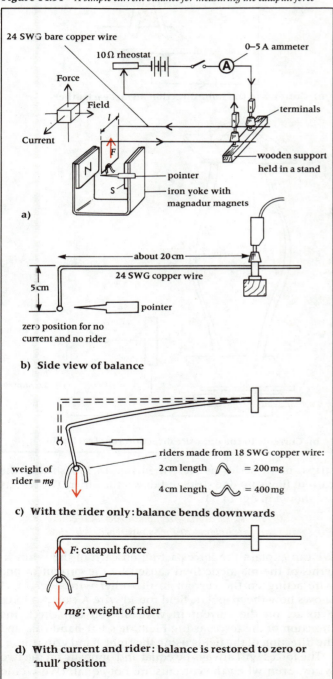

a)

b) **Side view of balance**

riders made from 18 SWG copper wire:
2 cm length = 200 mg
4 cm length = 400 mg

c) **With the rider only: balance bends downwards**

d) **With current and rider: balance is restored to zero or 'null' position**

Investigating what affects the strength of the catapult force

The current balance can be used to investigate how the catapult force F depends on various factors.

The current I in the wire loop

● *Make several wire riders of lengths* 1 cm, 2 cm, 3 cm, *etc. (masses ≈ 100 mg, 200 mg, 300 mg ...).*
● *Place each rider on the balance and find the current necessary each time to return the balance to zero deflection.*

The catapult force F is directly proportional to the current I.

The length l of the horizontal part of the wire which lies in the magnetic field

● *Make several loops of wire with horizontal sections of lengths l equal to 1, 2, 3 and 4 cm.*
● *Starting with a rider of mass 100 mg on the 1 cm length of wire, a 200 mg rider on the 2 cm length and so on, show that the same current will restore the balance to zero deflection each time.*

The catapult force F is directly proportional to the length of wire l in the magnetic field.

The angle θ between the current direction and the magnetic field direction

● *Using a short length of wire l in the magnetic field, adjust the current so that a 200 mg rider balances the catapult force when the wire is at right angles to the magnetic field, fig. 14.32a.*
● *Now turn the magnets so that the angle θ between the current direction and the magnetic field is reduced gradually from 90° to 0°, fig. 14.32b.*

Figure 14.32 *Turning the wire at different angles to the magnetic field (views from above)*

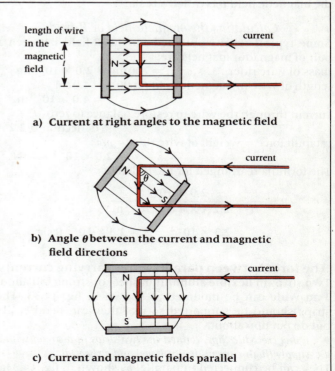

a) **Current at right angles to the magnetic field**

b) **Angle θ between the current and magnetic field directions**

c) **Current and magnetic fields parallel**

As the angle θ is reduced the wire loop gradually bends downwards showing that the catapult force (upwards) is getting weaker. If, when $\theta = 0°$, the current is switched off it makes no difference to the deflection of the loop, showing that the catapult force is now zero. The catapult force F is greatest when the magnetic field is at right angles to the current and reduces to zero when the field and the current are parallel, fig. 14.32c. Accurate measurements show that:

The catapult force F is proportional to sin θ.

Note that the catapult force always stays perpendicular to both the magnetic field and the current whatever the value of the angle θ between them.

The strength of the magnetic field

By using different magnets it is possible to show that the catapult force F is greater in stronger magnetic fields. We can use this force to measure the strength of a magnetic field. The symbol B is used to represent the value of the magnetic field causing the catapult force. Sometimes a magnetic field is referred to as a B-field, as a kind of abbreviation.

A formula for the catapult force

Combining these results gives us a formula for the catapult force F.

$$F = B\left(\begin{array}{c}\text{a measure of the}\\\text{magnetic field}\end{array}\right) \times I\,(\text{current}) \times l\left(\begin{array}{c}\text{length}\\\text{of wire}\end{array}\right) \times \sin\theta$$

If we keep the current and the field perpendicular so that the force has its maximum value, i.e. $\sin\theta = 1$, we have:

$$F = BIl$$

If F is in newtons, I in amperes and l in metres, the unit of the magnetic field B is called a **tesla**.

Calculating the magnetic field or B-field

Some typical results obtained for the B-field between a pair of magnadur magnets are:

mass of wire rider . . . . = 200 mg = 2.0×10^{-4} kg
length of wire in magnetic field (at 90°) = 4.0 cm
$$= 4.0 \times 10^{-2} \text{m}$$
current through the wire to restore balance to zero
deflection = 2.2 A
catapult force = weight of wire rider = mg
$$= 2.0 \times 10^{-4} \times 10 \text{ N}$$
The formula rearranged for the B-field is:

$$B = \frac{F}{Il} = \frac{2.0 \times 10^{-3}\,\text{N}}{2.2\,\text{A} \times 4.0 \times 10^{-2}\,\text{m}}$$

$$= 2.3 \times 10^{-2}\,\frac{\text{N}}{\text{Am}} = 2.3 \times 10^{-2}\,\text{tesla}$$

The force between parallel wires carrying currents

Two strips of flexible aluminium sheet or strong foil about 1 cm wide can be mounted as shown in fig. 14.33. The strips should be supported so that they can bend easily but do not flop about.

● *Using crocodile clips, connect the two strips in a circuit with a d.c. supply (high-current, low-voltage) and a rheostat.*

They can be connected in parallel as shown in fig. 14.33a, so that the current flows in the same direction in the two

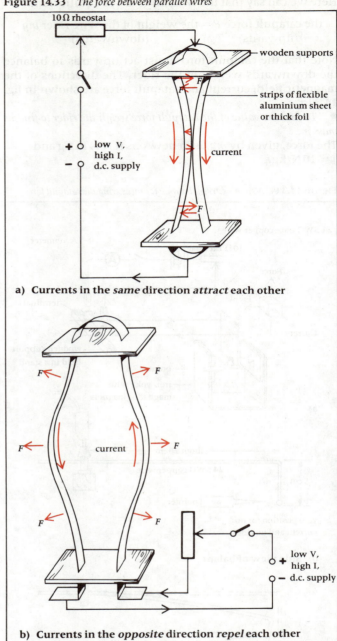

Figure 14.33 *The force between parallel wires*

a) Currents in the *same* direction *attract* each other

b) Currents in the *opposite* direction *repel* each other

strips, or in series as shown in fig. 14.33b so that the current flows up one strip and down the other. The results are shown in fig. 14.33.

Currents in the same direction attract each other and currents in opposite directions repel each other.

We can explain the forces acting on each of the wires in terms of the magnetic field caused by the current in one wire acting on the current in the other wire. Fig. 14.34 shows how the magnetic field round wire A causes a force F to act on the current in wire B. You can check the direction of the forces using Fleming's left-hand rule and the direction of the fields using the right-hand grip rule.

The forces will always be equal and opposite on the two wires, even when the currents are not equal. (We would expect this from Newton's third law.)

Figure 14.34 *The magnetic fields around two parallel wires*

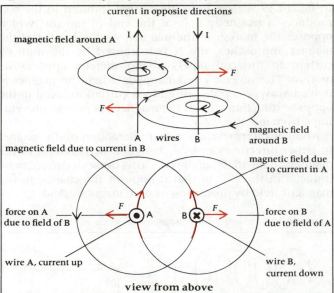

The definition of the ampere

The force between two parallel wires carrying a current is used to define the SI unit of current, the ampere. While the full explanation of this definition requires work beyond the scope of this book, the definition is given here for reference. (It is not usually expected that you will learn this definition, but rather that you understand the experimental basis of it.)

The ampere is that constant current which, if maintained in two straight parallel conductors of infinite length, of negligible circular cross section, and placed 1 metre apart in vacuum, would produce between these conductors a force equal to 2×10^{-7} newton per metre of length.

Assignments

Find the direction of the catapult force in each of the examples in fig. 14.35. When possible do this first by sketching the shape of the field, then check your result using Fleming's left-hand rule.

Remember the things that affect the force on a wire carrying a current and the conditions under which it is strongest.

Try questions 14.5 to 14.7

Figure 14.35 *Finding the direction of the catapult force*

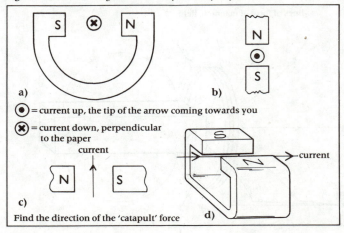

14.6
ELECTROMAGNETIC INDUCTION

We have already seen examples of induced electric charge and induced magnetism. There is a third kind of induction in which electric currents are induced in wires by magnetic fields; this is called electromagnetic induction.

Cutting lines of force with a wire

• *Fit two magnadur magnets to an iron yoke with unlike poles facing to form a strong U-shaped magnet, fig.14.36. The lines of force cross horizontally between its poles from N to S.*
• *Connect a loop of copper wire in series with a sensitive galvanometer ('spot' galvanometer) or centre-zero milliammeter.*
• *Move the wire up and down between the poles of the magnet and watch the meter.*

Figure 14.36 *'Cutting' a magnetic field*

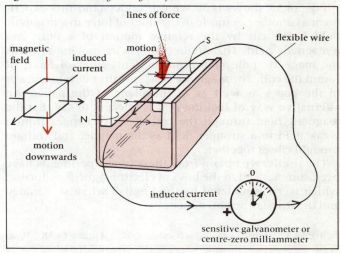

We notice that an electric current flows through the meter. This current must have been induced by the magnetic field, because there is no current source in the circuit itself. As the wire moves through the magnetic field a force acts on the electrons in the wire. This produces the current. This effect is known as the **dynamo** or **generator effect**.

• *Investigate which way the induced current flows when the wire is moved (a) upwards and (b) downwards.*

When a conventional current flows into the terminal of the meter which is labelled + or coloured red, the pointer or spot moves to the right.

• *Investigate how the induced current is affected by the speed with which the wire is moved.*
• *What happens when the wire is moved horizontally across the gap from N pole to S pole, so that it slides in between the lines of force?*
• *Try reversing the magnetic poles and also moving the magnet instead of the wire.*

The *direction* of the induced current depends on both the direction of the motion of the wire and the direction of the magnetic field. But it makes no difference whether the wire moves or the magnet moves because it is their *relative motion* which causes an induced current.

The *magnitude* of the induced current depends on how quickly the magnetic field lines (or lines of force) are cut by the wire. If the wire moves parallel to the field and does not cut the lines of force no current is induced.

Electromagnetic induction using a coil and a magnet

For this experiment a coil with a large number of turns is needed.

- *If a suitable coil is not available make one by winding about 50 turns of thin insulated wire round a tube of diameter* 3 cm *or more.*
- *Connect the coil in series with a sensitive meter, fig. 14.37.*
- *Move the N pole of a bar magnet in and out of the coil and watch the meter.*
- *Investigate how the direction of the induced current depends on the pole of the magnet used and the direction of motion of either the magnet or the coil.*
- *Investigate how the magnitude of the induced current depends on (a) the speed of the relative motion, (b) the strength of the magnet and (c) the number of turns on the coil. (Wind extra turns on your coil, but remember that extra turns means extra resistance in the circuit.)*

Fig. 14.38 shows how we can understand this experiment as another example of the lines of force in a magnetic field being cut by the relative motion of a wire and a magnetic field. The single coil of wire moving towards the magnetic pole of the magnet cuts lines of force all round the coil. The magnetic field spreads out into and out of the paper as well as in the plane of the paper. An alternative way of looking at this effect is to say that the magnetic field through the coil of wire changes from a weak field to a stronger field as the magnet and coil are brought closer together.

The results we obtain from these two experiments have been summarised in the laws of electromagnetic induction which were formulated by the English scientist, Faraday and the Russian scientist, Lenz.

The direction of the induced current, Lenz's law

In fig. 14.39 we see how the current induced in the coil produces a magnetic pole at the end of the coil which opposes the motion of the magnet. As the N pole of the magnet approaches, the N pole caused by the induced current in the coil repels it and thereby opposes its motion. The reverse effect happens when the magnet is moved away; a S pole caused by a current induced in the opposite direction attracts the magnet's N pole so preventing it from moving away.

Another interpretation is that the motion of the magnet in either direction causes a change in the strength of the magnetic field affecting the coil. This causes a current to be induced in the coil so that it opposes the change in the magnetic field by producing its own magnetic field.

Figure 14.39 *The induced current opposes the motion*

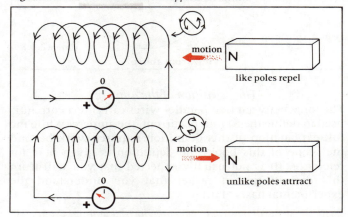

Figure 14.37 *Moving a magnet into a coil*

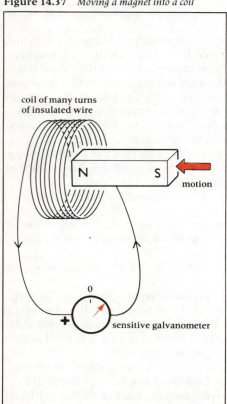

Figure 14.38 *Moving the coil cuts lines of force*

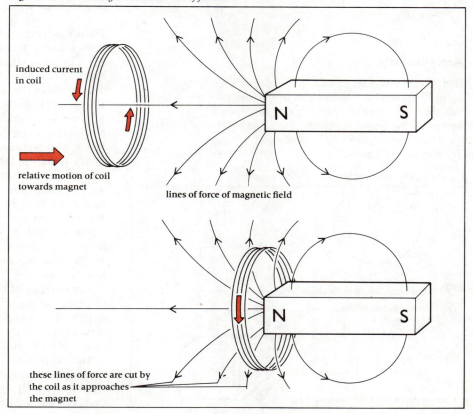

From observations of this kind we obtain **Lenz's law**, which applies to all induced currents:

The direction of an induced current is such as to oppose the change causing it.

In the examples we have seen so far the change has involved motion of a wire, coil or magnet. However, in machines such as transformers the magnetic field changes without any movement and still causes induced currents.

Lenz's law and conservation of energy

When a change or movement causes an induced current to flow in a circuit, energy has been supplied to that circuit. This energy appears as heat in the wires of the circuit and as kinetic energy of a moving coil in a meter which detects the current. For energy to be conserved and not created from nothing by induction, the energy given to the circuit must be supplied when the magnet or coil is moved.

In the case of the magnet being moved into a coil, fig. 14.38, the person moving the magnet does work pushing the magnet into the coil or pulling it out again afterwards. This work supplies the necessary energy for the current in the coil. Work requires a force to work against and in this example it is provided by the repulsion between the like poles as the magnet is pushed into the coil.

We can see that Lenz's law, about the induced current opposing the change, follows from the need to work against a force in order to do work to supply the energy. The direction of the induced current can easily be predicted by using Fleming's right-hand rule or dynamo rule.

Fleming's right-hand dynamo rule

If we compare this dynamo or generator effect with the very similar motor effect we can see that one effect is the mirror image or reverse of the other. In the motor effect a current in a coil causes it to turn but in the dynamo effect the turning of a coil produces a current. We used Fleming's *left*-hand rule for the *motor* effect. A similar *right*-hand rule can be used to predict the direction of an induced current in the *dynamo* effect, fig. 14.40.

Hold the thumb and first two fingers of your right hand at right angles to each other. Point your first finger in the direction of the magnetic field and your thumb in the direction of motion of the wire. Now your second finger points in the direction of the induced current.

Figure 14.40   *Fleming's right-hand rule*

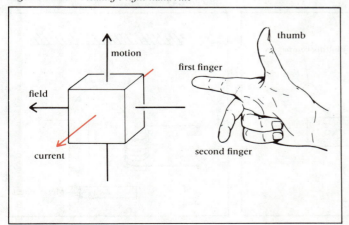

Faraday's law of electromagnetic induction

All experiments in which electromagnetic induction occurs show that some kind of a change is necessary. A magnet placed inside a coil and left there causes no electromagnetic induction at all. A wire held still in a constant magnetic field experiences no induced currents or forces.

Any *change* in the magnetic field or *movement* of a wire or coil which causes the lines of force to be cut results in an induced e.m.f. A potential difference or voltage appears between the ends of the wire or coil. If this wire or coil is connected in a closed circuit the change also produces an induced current.

Faraday was the first to observe how the *magnitude* of the induced e.m.f. depended on the *rate of change* causing it. In our experiments, when a magnet was moved quickly into a coil a large induced current was detected. When it was moved slowly a proportionally smaller induced current flowed. The same was seen with the speed of movement of the wire cutting the lines of force of a magnet.

The experiments also show that when a more powerful magnet is used the magnitude of the induced current increases. We can think of a more powerful magnet as having more lines of force between its poles and so when a wire moves through the field it cuts lines more often. Again it follows that the magnitude of the induced e.m.f. depends on the rate of cutting of the lines of force. **Faraday's law of electromagnetic induction** can be stated as:

The magnitude of the induced e.m.f. between the ends of a wire (or coil) is directly proportional to the rate at which it cuts magnetic lines of force.

Alternatively, it is just as correct to think of the magnetic lines of force moving and cutting the turns of wire in a coil (fig. 14.38), where the coil is stationary and the magnet moving. Similarly we can say that *the induced e.m.f. is proportional to the rate of change of the magnetic field through a coil.*

The number of turns on a coil

The turns of wire on a coil, wound from a continuous length of wire, are connected in series. When magnetic lines of force cut the coil an e.m.f. is induced in each separate turn of the coil. Each turn of wire acts as a source of e.m.f. like a cell. Being connected in series, the e.m.f.s of all the turns add up so that the total e.m.f. between the ends of the coil is the sum of the e.m.f.s induced in all the turns, fig. 14.41.

The induced e.m.f. induced in a coil is directly proportional to the number of turns.

Figure 14.41   *The e.m.f. induced in a coil is proportional to the number of turns on the coil*

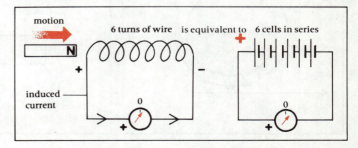

The magnitude of the induced current

The induced current detected in our experiments by a sensitive meter would not be present if there was no closed circuit. The meter, connected to a wire (fig. 14.36) or a coil (fig. 14.37), provides a closed series circuit of low resistance in which an induced current can flow. If we took away the meter and repeated the experiments with a moving wire or coil in an open circuit no induced current could flow. But would electromagnetic induction still occur?

We can answer this question by asking another. In the circuits containing the meter, in which an induced current was detected, where was the source of e.m.f. that drove the current round the circuit?

The wire or coil which cuts magnetic lines of force is the source of the e.m.f. and, like a cell or battery, there is an e.m.f. between the two terminals whether it is connected to a closed circuit or not. The induced e.m.f. does not depend on being connected to a closed circuit.

The magnitude of the induced current does however depend on the resistance of the circuit in which the e.m.f. is induced. For this reason we state Faraday's law in terms of the induced e.m.f. rather than the induced current. A variety of ways of inducing e.m.f.s and currents are described in the next chapter.

Assignments

Explain
a) What is the difference between the magnetic effect and the motor effect of an electric current?
b) In what way is the dynamo effect the reverse of the motor effect?
c) What are the differences between the three kinds of induction: magnetic induction, induction of electric charges and electromagnetic induction?
d) How is Lenz's law a special case of the law of conservation of energy?

Remember
e) The laws of electromagnetic induction stated by Lenz and Faraday.
f) That the magnitude of an induced e.m.f. depends on the number of turns on a coil as well as the rate of cutting of the magnetic lines of force.
g) That the magnitude of an induced current also depends inversely on the total resistance in the circuit.

Use Fleming's right-hand rule on figs. 14.36 and 14.38.
Try questions 14.8 and 14.9

Questions 14

1 Fig. 14.42 shows two bar magnets placed so that unlike poles are facing each other.
   a) Redraw fig. 14.42 and sketch the magnetic field around and between the magnets.
   b) Name a suitable material from which these magnets could be made and state why it is suitable.
   c) In a magnetic tape recorder, the recording is made by magnetising the tape by different amounts in different places. The mechanism is shown in fig. 14.43.
      i) What material is suitable for A?
      ii) Why is this material suitable?
      iii) On which side of the head, X or Y, is the tape magnetised?
      (Joint 16+)

Figure 14.42

2 Fig. 14.44 shows some plotting compasses around a bar magnet. One of the compass needles has been drawn in. Copy the figure. Use your knowledge of the magnetic field to:
   a) label the poles of the magnet;
   b) draw the other compass needles in the blank circles.
   (LEAG SPEC A)

3 Fig. 14.45 shows an electromagnet made by a pupil in the laboratory. The electromagnet is to pick up and release a metal object.
   a) Name a suitable material for part X.
   b) Why is it made from this material?
   c) The electromagnet will just lift a metal object of mass 0.15 kg. What will be the least force exerted by the magnet to do this? The strength of the gravitational field at the earth's surface, g, can be taken as 10 N/kg.
   d) Name a metal which the magnet will not attract.
   e) State **two** changes which the pupil could make so that a heavier metal object could be lifted by the electromagnet. (NEA [B] 88)

4 Many small pieces of soft iron, identical in size and shape are held just below end A of the arrangement shown in fig. 14.46.

   Describe and account for what happens when the switch is closed and then opened again. Compare these results with what would happen if pieces of hard steel of identical size and shape were used instead of the soft iron.

   Suggest a value for the current in a practical electromagnet which is used to lift heavy sheets of mild steel. (L, part)

Figure 14.43

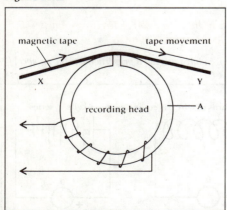

Figure 14.44

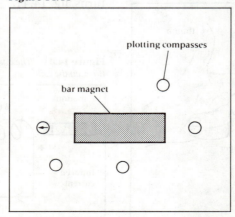

Figure 14.45

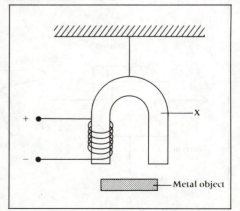

Figure 14.46

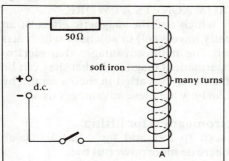

Figure 14.47

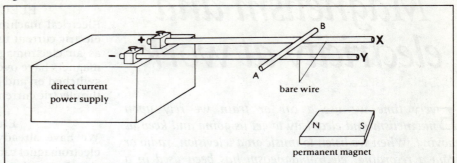

5   The apparatus shown in fig. 14.47 may be used to cause the wire AB to move along the rails X and Y.
    Show on a copy of the diagram, the best position for the permanent magnet to achieve this.
    With your arrangement, state the direction in which the wire AB would move.   (AEB)

6   a) Fig. 14.48a shows a wire XY carrying a steady electric current in the direction shown. Sketch magnetic field lines, in the horizontal plane indicated, to show the pattern of the magnetic field produced by the current in this plane. Mark clearly the direction of the field. Ignore the effects of the earth's magnetic field.
    b) With the current switched off a short piece of unmagnetised soft iron, AB, is placed alongside the wire in a horizontal plane, as shown in diagram (b). The current is switched on again. State clearly any effects produced in the soft iron by the current.
    c) Diagram (c) shows a second wire, OQ, parallel and close to XY, carrying a current in the opposite direction to the current in XY. Indicate by means of an arrow on a copy of diagram (c) the direction of the *magnetic field* due to the current in XY, at the point P, which is the point where OQ passes through the horizontal plane. Draw a second arrow to show the direction of the *force* acting on OQ at P due to the current in XY.
    d) Suggest any factors which you consider may affect the magnitude of the force on OQ at P.   (C)

7   A wide horseshoe magnet is placed on the pan of a sensitive balance. A fixed piece of wire, connected to a switch and power supply, passes midway between the poles of the magnet. The apparatus is illustrated in fig. 14.49. (a) In what direction is the force on the wire when the switch is closed? (b) State whether the reading on the balance will increase, decrease or remain the same and explain your answer.   (Joint 16+, part)

8   Fig. 14.50 shows a hollow coil C about 200 turns mounted on a wooden base. G is a centre-zero galvanometer. When the switch is closed the galvanometer needle is deflected to the right. Suppose the coil is connected to the galvanometer as in fig. 14.51 and the N pole of a long bar magnet is
    a) lowered quickly into the coil,
    b) left at rest on the base of the coil,
    c) moved quickly in and out of the coil about twice per second.
    Describe and explain the resulting movements of the galvanometer needle in each case.   (L, part)

9   State two laws of electromagnetic induction, one law relating to the size of the induced e.m.f. (Faraday's law) and the other relating to its direction (Lenz's law).
    Describe a simple experiment which illustrates the truth of one of these laws. Explain how the law can be deduced from the observations made in the experiment.   (JMB, part)

Figure 14.48

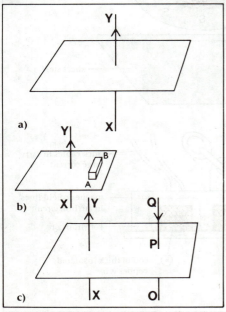

a)

b)

c)

Figure 14.49

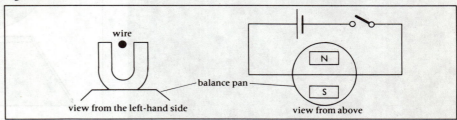

view from the left-hand side   view from above

Figure 14.50

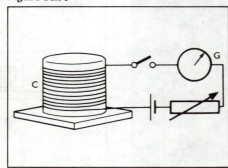

Figure 14.51

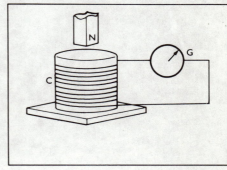

15

Magnetism and electricity at work

Every time we use a car or train we rely upon magnetism and electricity to get us going and keep us moving. When we listen to music on a television, radio or from a recording, electromagnetism has been used in a variety of ways before the sound reaches our ears. Many machines such as record players, washing machines, hairdriers and power tools would not be much use without their electric motors. We can see that many jobs are made easier, travel and communication is made quicker and generally life is made more interesting and enjoyable by the ingenious ways in which magnetism and electricity have been put to work for us.

15.1
ELECTROMAGNETS AT WORK

Electrical machines which use the magnetic effect of an electric current usually have a coil or solenoid which acts as an electromagnet. The main advantages that electromagnets have over permanent magnets is that they can be switched on and off and can be varied in their strength by the electric current. So how are these advantages used?

Electromagnets for lifting

We have already seen (p 291 and fig. 14.21c) that an electromagnet can be made more powerful by:
a) using a soft-iron core inside the coil or solenoid,
b) having a large number of turns on the coil,
c) passing a large current through the coil,
d) using a core shape which brings both magnetic poles close together,
e) forming a closed loop of soft iron from the electromagnet core and the iron object being lifted so that there is an 'iron circuit' for the magnetic field lines.

Fig. 15.1 shows a design for a lifting electromagnet which uses all these ideas to make it very powerful.

Electromagnets are particularly useful in industry for
a) lifting steel plates and slabs which are difficult to handle or attach hooks and chains to,
b) lifting large quantities of loose iron and steel, e.g. scrap metal,
c) for separating ferromagnetic metals from others such as copper, brass, aluminium and lead.

Figure 15.1 *An electromagnet for lifting*

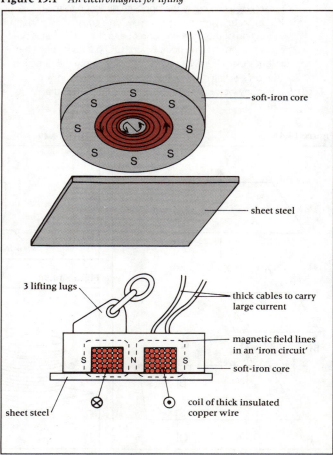

An electromagnet can be used to lift and sort scrap iron and steel. Objects containing any of the ferromagnetic metals, iron, nickel, cobalt or one of their alloys such as steel, are lifted by the electromagnet. Objects made of other metals are left behind.

Making permanent magnets

A solenoid with a large number of turns is required.

● *Connect the solenoid in series with a switch and a supply of direct current capable of passing several amperes through the solenoid, fig. 15.2.*

A bar of steel, a weak magnet that needs remagnetising or a steel knitting needle can be used.

● *Place the steel object fully inside the solenoid and switch the direct current on and off.*

A high current is needed only for a short time to produce the maximum level of magnetisation. Longer times will not increase the strength of the magnet but may overheat the solenoid.

Figure 15.2   *Making a permanent magnet electrically*

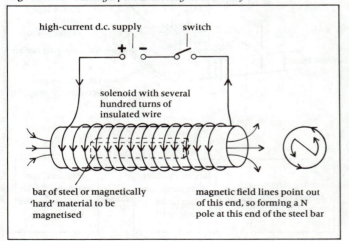

high-current d.c. supply   switch

solenoid with several hundred turns of insulated wire

bar of steel or magnetically 'hard' material to be magnetised

magnetic field lines point out of this end, so forming a N pole at this end of the steel bar

Demagnetising, removing magnetism

● *Use a similar solenoid to the one used for magnetising a steel bar but this time connect it to a supply of alternating current, fig. 15.3. (A transformer stepping the mains supply down to 12 V and capable of supplying about 8 A is suitable.)*

● *Place the magnetised object fully inside the solenoid and switch on the alternating current.*

● *With the current still on, slowly pull the object out of the solenoid and remove it well away from the magnetic field.*

● *Check to see whether the material is completely demagnetised by trying to pick up some unmagnetised pins. If it is not completely demagnetised, repeat the process.*

The alternating current from the mains supply will flow in both directions round the solenoid 50 times every second, so the magnetic field inside the solenoid will actually change direction 100 times every second. As the steel object is slowly removed from this magnetic field the strength of the field will gradually weaken. The effect is to reverse the magnetisation of the object repeatedly, but at a slightly weaker strength every time the field reverses. Thus the strength of the magnetisation gradually climbs down step by step every time the magnetic field reverses and the object has moved further out of the field.

Figure 15.3   *Demagnetising*

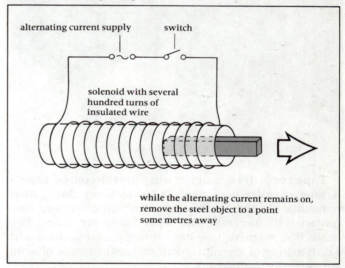

alternating current supply   switch

solenoid with several hundred turns of insulated wire

while the alternating current remains on, remove the steel object to a point some metres away

A bank of electromagnets used to move sheet steel.

Fig. 15.4 shows a device which is used to demagnetise the recording and erase heads of a tape recorder or cassette player. This special kind of cassette has a pair of electromagnets which come into close contact with the two tape heads. A magnetic field generated by the electromagnets enters the metal of the tape heads and magnetises them. The electromagnets are supplied with a high frequency alternating current, which an electronic circuit causes to fade away. As the alternating current fades, the reversing magnetic field also gets weaker and disappears. This process thereby reduces any magnetism in the tape heads to zero. This is an important procedure for keeping a tape recorder in its best condition, because when permanent magnetism occurs in the tape heads it causes a background hiss to be added to all tapes which pass across them.

Figure 15.4 *Demagnetising tape recorder heads*

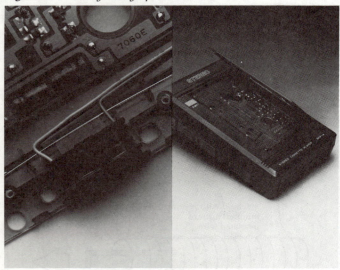

Tape recorders, magnetising metal-coated tape

A tape recorder uses two, and sometimes three, small horseshoe shaped electromagnets which come into close contact with the magnetic tape. These are called tape heads. The magnetic tape has a strong plastic backing with a thin coating of a ferromagnetic material, capable of being permanently magnetised.

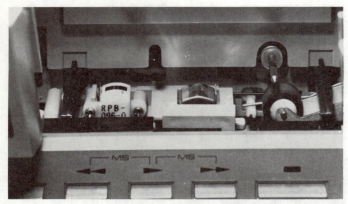

The magnetic tape, driven by the capstan and rubber pinch wheel, is pulled across the erase head and the record–playback head, which can be seen on the left and in the centre of the picture. The poles of the electromagnets show up as dark strips on the faces of the heads, but the narrow gaps between the poles are too fine to see.

The tape is driven at a constant speed past the narrow gap between the magnetic poles of each electromagnet. The first head, called the **erase head**, is used only during recording to demagnetise the tape. The erase head removes anything which may have been previously recorded and also any permanent magnetism in the tape. A high-frequency alternating current (about 85 kHz) is passed through the erase head coil. The tape is demagnetised as it passes across the gap where the magnetic field of the electromagnet is reversing at the same high frequency.

The second head is the **recording head**, fig. 15.5a. (If there is a third head it is used for playback, otherwise the second head is used for both recording and playback.) This head must be made very accurately. For example the gap between the magnetic poles must be only a few microns wide (a micron is a micrometre, or millionth of a metre). The sides of the gap must be parallel and exactly vertical. For stereo recording there are two identical electromagnets mounted one above the other and separated by a magnetic shield of soft iron. The magnetic fields of each electromagnet must magnetise less than a quarter of the width of the tape for each channel, so that their magnetisation does not overlap. The whole head is surrounded by a magnetic shield except at the front where the tape passes.

Varying electric currents are passed through the coil of the electromagnet. These currents represent sound in an equivalent electrical form. The current varies in magnitude as the loudness of the sound varies, and changes direction at the same frequency as the pitch of the sound. The tape passing across the electromagnet gap becomes magnetised and the strength and direction of the magnetisation varies in the same special way as the sound it represents. Fig. 15.5b shows the steps in the process.

The playback process involves the same varying magnetisation of a magnetic tape reproducing the same varying electric current in the coil of the electromagnet in the playback head. This is an example of electromagnetic induction. Note how the whole process is exactly reversed.

Figure 15.5 *A tape recording head*

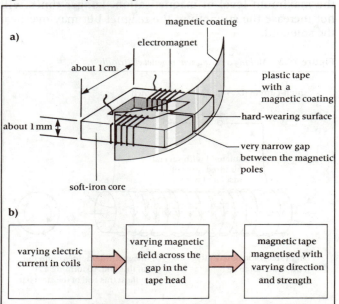

The electric bell

Electric bells and buzzers all use an electromagnet in a simple way to attract a soft-iron bar called the **armature**. For a bell to ring continuously an automatic mechanism is needed to switch the current off and on again rapidly. The circuit in fig. 15.6 shows how this mechanism works in many electric bells. It can be explained in steps as follows:

a) When the push-button switch is closed there is a complete circuit (the contact screw and spring strip are normally closed in the rest position of the armature).

● *Follow the path of the current round the complete circuit and see if you can predict the poles on the electromagnet.*

b) A current flows through the electromagnet, so magnetising the iron core.

Figure 15.6   *The electric bell*

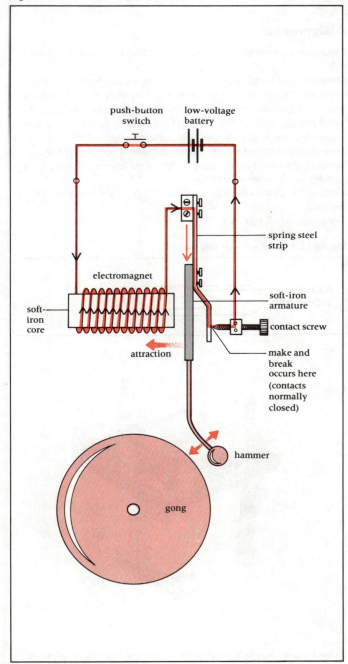

c) The electromagnet induces magnetism in the soft-iron armature, which is attracted to the pole of the electromagnet.

d) The hammer strikes the gong.

e) The spring strip contact moves away from the screw contact and breaks the circuit, switching off the current.

f) The electromagnet, no longer magnetised, releases the armature, which springs back to its starting position.

g) The spring strip contact now touches the screw again, remakes the circuit and switches on the current.

The cycle repeats rapidly from (b) to (g) for as long as the push-button switch is closed. The hammer strikes the gong repeatedly making a continuous ringing sound. The spring strip and screw contacts, where the current is automatically switched on and off, are often called a **make and break** mechanism, because it 'makes' and then 'breaks' the circuit.

Relays

A relay is a switching device which uses an electromagnet. It has two or more completely separate circuits, fig. 15.7. The input circuit (terminals 1 and 2) supplies current to the electromagnet: only a very small current is needed. The electromagnet attracts one end of a soft-iron armature, which is pivoted so that its other end acts as a lever. This lever opens or closes contacts in the second or output circuit, by pushing a spring metal strip. The spring strips and contacts can be arranged so that they are normally open or normally closed and several sets of contacts can be operated by the same armature lever.

The purpose of a relay can vary but some of its advantages and uses are given in the following five points:

a) One circuit can be used to control (switch on or off) another circuit (or several circuits) without any direct electrical connection between them.

b) The input circuit can work on a low-voltage (safe) supply and control another circuit on a high-voltage (dangerous) supply.

Figure 15.7   *A relay*

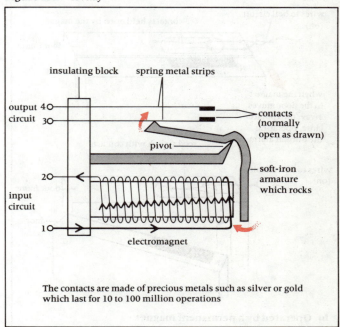

The contacts are made of precious metals such as silver or gold which last for 10 to 100 million operations

c) A small current in the input circuit can switch a large current in the output circuit. For example, in the car ignition circuit, a low current through the ignition switch operates a relay which closes a circuit so passing a very large current through the starter motor (fig. 15.42). Similarly a transistor switch using a very small current can control a relay which switches much larger currents in other circuits (fig. 16.29).

d) The relay can be considered as a mechanical current amplifier when it is used as in (c) above.

e) The relay can be used as a level sensor. Only when the current through the electromagnet reaches a precise level will the armature swing and the output circuit operate. The input current level at which this happens can be made to represent the level of some quantity being measured.

Reed switches

A reed switch contains two or three thin iron strips or reeds inside a sealed glass tube. The contacts on the ends of these reeds can be normally open as in fig. 15.8a, or normally closed as in fig. 15.8b.

The iron reeds are usually unmagnetised, but when magnetism is induced in them they will attract or repel each other depending on their induced poles.

Figure 15.8 *A reed switch*

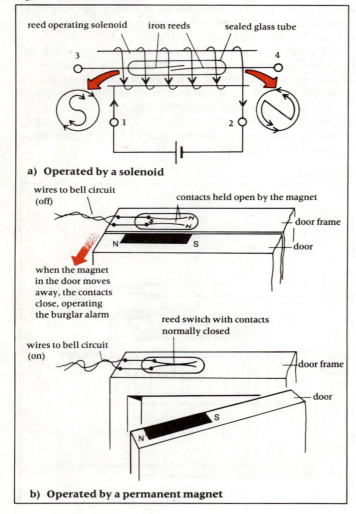

a) **Operated by a solenoid**

b) **Operated by a permanent magnet**

Fig. 15.8a shows how a solenoid can be used to work a reed switch. When used like this it performs exactly like a relay. The input current is supplied to the solenoid (terminals 1 and 2) and the reed switch operates the output circuit (terminals 3 and 4). In this example the normally open contacts become closed when the solenoid magnetises the iron reeds in the same direction so that the opposite poles on the free ends of the reeds attract.

Fig. 15.8b shows how a reed switch can also be operated by a permanent magnet. It provides a simple burglar alarm. The magnet magnetises the two iron reeds in the same direction, but in this case like poles hold the normally closed contacts apart until the door is opened and the magnet moves away. In the absence of the magnet the iron reeds lose their magnetism, close their contacts and switch on the burglar alarm circuit.

Assignments

Make yourself a burglar alarm using a reed switch, magnet and electric bell.

Draw the reed switch shown in fig. 15.9 inside a solenoid. Show the direction of the current in the solenoid and the magnetic poles induced on each of the iron reeds. Which pair of contacts is normally open (when the solenoid current is off) and which pair is normally closed? Explain why this reed switch is called a change-over switch.

Look closely at some tape recorder heads. See if you can see the narrow gap where the tape passes. *Do not* put any metal object like a screwdriver anywhere near them.

Find out if any local industries use electromagnets and if so, how they are used.

Try questions 15.1 to 15.4

Figure 15.9 *A change-over reed switch*

A simple change-over relay which is operated by applying 12 V across the coil.

15.2
APPLICATIONS OF THE MOTOR EFFECT

Coils can be made to twist and turn in different ways in a magnetic field. Electric motors can be made to run on direct current or alternating current and with or without brushes. The amazing variety of ways in which the motor effect is now used is a tribute to the inventiveness of electrical engineers over the last 100 years and to the fascinating nature of these machines.

A moving-coil loudspeaker

By far the most common type of loudspeaker uses a moving coil to cause the motion that reproduces sound waves in the air. Moving-coil loudspeakers are used both in hi-fi equipment, where high-quality reproduction is required, and also in public address systems.

Figure 15.10 *A moving-coil loudspeaker*

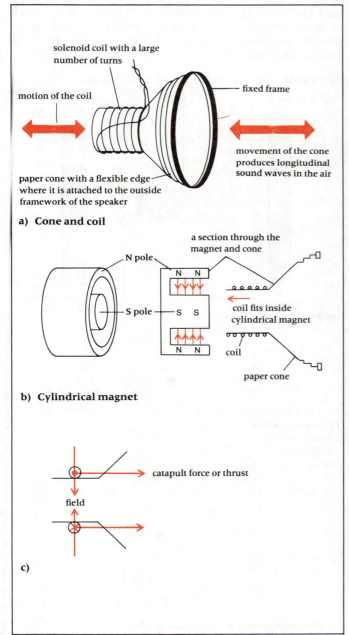

a) **Cone and coil**

b) **Cylindrical magnet**

c)

The moving coil is attached to a flexible paper cone which moves the air molecules to produce sound, fig. 15.10a. The coil fits into a cylindrical magnet which has a centre S pole surrounded by a N pole as shown in fig. 15.10b. The coil is free to move in and out of the cylindrical gap in the magnet where the magnetic field lines point radially inwards.

Rapidly changing currents from an amplifier are passed through the speaker coil and as the current direction reverses, the movement of the coil and cone changes direction. Fig. 15.10c shows the current direction which produces a forwards thrust on the coil. You can check the direction of the catapult force using Fleming's left-hand rule.

A loudspeaker stack at the Notting Hill Carnival. The powerful moving-coil speakers give high quality sound as well as high volume.

Making a simple moving-coil galvanometer

A moving-coil galvanometer is an instrument which detects and measures direct electric currents by the catapult force they produce as they pass through a magnetic field. With a calibrated scale the instrument becomes an ammeter. A simple working model can be made using a kit of parts (a 'Westminster' electromagnetic kit is suitable), or from similar parts collected together. The only special parts are the pair of powerful magnadur magnets with poles on their faces. Fig. 15.11 shows how to build the instrument.

● *Attach the two magnadur magnets with unlike poles facing, to an iron yoke, fig. 15.11a.*
● *Using a wooden frame with a channel or slot cut round its edges, wind 10½ turns of thin (26 SWG) PVC-insulated wire around it so that the wire begins and ends at opposite ends of the frame. Leaving about ½ metre of wire at each end to make a spring, you will need, in total about 3 metres of wire.*

The turns of wire must be tightly and neatly wound on the frame so that it will be able to turn freely between the magnets without catching.

● *At each end of the frame make a spiral spring with the spare wire at the ends of the coil. Starting at the centre, work outwards bending the wire into a flat even spiral, fig. 15.11b.*
● *Fit the frame and coil on the axle, supported as shown in the figure.*
● *Adjust the spiral springs so that the frame is level and fix the ends of the springs to the wooden base with drawing pins.*
● *Attach a pointer.*
● *Connect your galvanometer in a circuit as shown in fig. 15.11c, and test it for different currents in both directions.*
● *Why does the coil turn?*
● *What are the two jobs done by the spiral springs?*
● *What would happen if the springs were omitted?*
● *What would happen if an alternating current was passed through the coil?*

Fig. 15.12 shows how a catapult force acts on both sides of the coil, but in opposite directions because the current is in opposite directions. These two forces produce a *couple* or *turning effect*, which turns the coil clockwise (as seen from the viewing end) against the spiral spring which becomes wound up. The spiral spring, attempting to unwind, produces a turning effect in the opposite direction, i.e. anticlockwise.

Figure 15.11   *A model galvanometer*

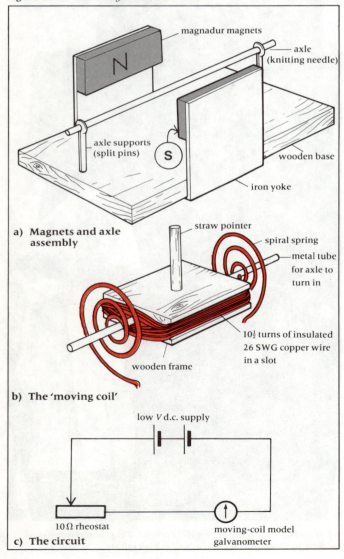

a) **Magnets and axle assembly**

b) **The 'moving coil'**

c) **The circuit**

Figure 15.12   *The catapult forces acting on a model galvanometer coil*

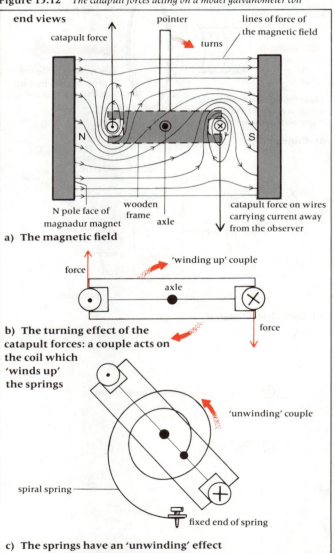

a) **The magnetic field**

b) **The turning effect of the catapult forces: a couple acts on the coil which 'winds up' the springs**

c) **The springs have an 'unwinding' effect**

The coil turns to such an angle that these two turning effects exactly balance each other, i.e. until they are in equilibrium. So if the current was increased, the catapult forces would increase and produce a stronger turning effect which would turn the coil further round until the spring, now more tightly wound up, could balance it. This explains how the galvanometer is able to measure different size currents; that is by winding up a spring more tightly as the current increases.

Compare this idea with the stretching of a spring which we investigated in chapter 9. In both cases there is stretching or twisting which is proportional to a force.

● *Check the directions of the catapult forces in the figures using Fleming's left-hand rule.*

The design of a sensitive moving-coil galvanometer

We have seen that the catapult force on a wire depends on:
a) the current in the wire,
b) the length of the wire and
c) the magnetic field strength.
The force is greatest when the current and the field are at right angles.

A sensitive meter should be able to detect very small currents. It should also be accurate and easy to read, with a scale which has evenly spaced graduations. Such an instrument is shown in fig. 15.13. The design features are:

a) The moving coil has a large number of turns, because this provides a long length of wire in the magnetic field.

b) The coil has a large area because this both increases the length of wire, and also increases the turning effect if the sides of the coil are further from the axle (turning effect = force × perpendicular distance from axle).

c) The magnets used are strong permanent magnets and they are helped by a soft-iron cylinder mounted between the poles as shown in fig. 15.13. The concave shape of the pole faces and the presence of the iron cylinder both increase the strength and also change the shape of the magnetic field. The field in the gap between the poles and the iron cylinder is described as **radial** because it points either towards or away from the centre of the cylinder like spokes of a wheel. This design of magnetic field is used to give a *uniform scale* on the instrument (i.e. one with equally spaced graduations) and to give the greatest sensitivity to small currents.

d) The two spiral springs, which control how far the coil turns, are made from very fine hair springs which have a weak unwinding effect.

e) The pointer is sometimes replaced by a small mirror which reflects a ray of light. As the ray can be much longer than a metal pointer, the movement of the end of the light ray is magnified by its greater length. The sensitivity is also doubled by the fact that the reflected light ray is turned through an angle twice as large as the angle which the mirror (attached to the coil) turns through. Can you explain this? (Hint: see p. 11.)

The sensitivity of a galvanometer is defined as the scale deflection of the pointer or the spot of light produced by a small unit of current. The small unit of current may be 1 milliampere or even 1 microampere.

Figure 15.13   *A moving-coil galvanometer*

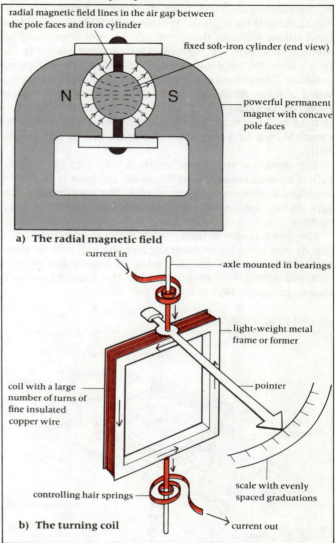

radial magnetic field lines in the air gap between the pole faces and iron cylinder

fixed soft-iron cylinder (end view)

N   S

powerful permanent magnet with concave pole faces

a)   The radial magnetic field

current in

axle mounted in bearings

light-weight metal frame or former

coil with a large number of turns of fine insulated copper wire

pointer

controlling hair springs

scale with evenly spaced graduations

b)   The turning coil

current out

Making a simple direct current motor

In the moving-coil galvanometer current passes into and out of the coil through the springs which also control how far it turns. Without those springs a motor needs another method of supplying current to its rotating coil or **rotor**. In a motor the rotating coil can be called the *rotor* or the *armature*. The advantage of using the name rotor is that this word always refers to the rotating part of any type of motor or generator. The word armature often refers to a part of an electrical machine which does not rotate. For example, we first met the term armature in the electric bell.

The device which supplies the current to the rotor is called a **commutator** and we shall see how it works by building a simple working model. The commutator also acts as an automatic current-reversing switch for d.c. motors.

The same wooden frame with a metal tube through it which was used for making the moving coil of a galvanometer can be used for the rotor of the motor (from a 'Westminster' electromagnetic kit). The finished motor is shown in fig. 15.14.

Figure 15.14   *A model d.c. motor*

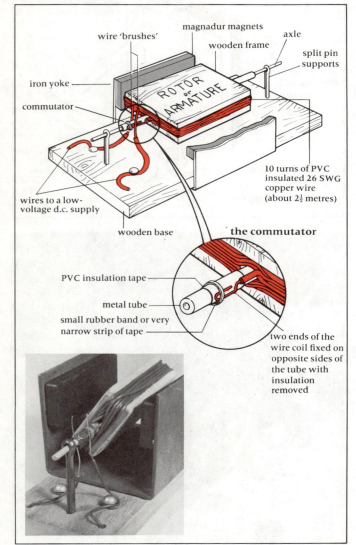

To make the rotor or armature

● *Wind 10 turns of PVC-insulated wire tightly round the frame, pressing them into the slots. Both ends of the coil should finish at the same end of the frame.*

● *Wrap a short length of PVC insulation tape round the metal tube which sticks out of the wooden frame.*

To make the commutator

● *Strip about 2 cm of the insulation from both ends of the wire coil and bend these into two flat U-bends, (see insert detail of the commutator).*

● *Press the two bared wire ends against opposite sides of the insulated tube and fix them there using a small rubber band (cut from rubber tubing) or with a very narrow strip of tape.*

● *Assemble the armature on an axle supported by split pins as shown.*

● *Make two wire brushes by stripping about 2 cm of insulation from two short lengths of wire, which will form the contacts with the commutator as it rotates.*

● *Fix the wire brushes to the base board with drawing pins so that they press against the two sides of the commutator. They should make electrical connection when the armature is horizontal. The pressure of the brushes can be increased by dropping the rotor down to the base board and crossing the brushes above the commutator. When the rotor is raised again the brushes are pushed apart by the commutator and so press against it.*

● *Fit the magnadur magnets on their iron yoke around the armature making sure that the rotor is free to turn.*

● *Connect the brushes to a low-voltage d.c. supply and switch on. Some adjustment of the brushes may be necessary to get the motor running smoothly.*

Some questions about your motor

● *Why should the two halves of the commutator be on the sides of the metal tube, rather than the top and bottom, when the rotor frame is horizontal?*

● *What happens at the commutator when the rotor has turned to the vertical position?*

● *Work out the directions of the current in the coil and the catapult force on each side of the rotor coil when it is horizontal.*

● *What would happen if the direction of the current in the rotor coil did not reverse twice in every revolution?*

Split-ring commutators

In a commercial motor, built to a similar design to your model, the commutator would be made from two insulated halves of a copper ring. Carbon brushes are pressed against it by springs, fig. 15.15. This arrangement gives a good area of contact between the commutator and the brushes and allows current to flow through the rotor coil for most of its revolution. Carbon is chosen for the brushes because it is soft compared with the copper commutator. Worn carbon brushes are easy to replace, but a worn commutator would be expensive to replace.

The two splits in the copper ring are positioned so that the brushes change over from one half to the other half of the ring at exactly the moment when the rotor coil is in the vertical position in fig. 15.14. In general this position is the one in which the rotor would stop and turn no further if the current direction was not reversed. The momentum of the rotor carries it past this stopping position. Then as the current in the coil is reversed by the split-ring commutator, the reversed catapult forces act to turn the rotor through another half revolution. We can see that the *split-ring commutator is an automatic switch*, which reverses the current in the rotor coil twice in every complete revolution.

Figure 15.15   *A split-ring commutator used in d.c. motors*

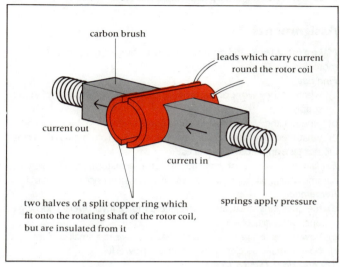

carbon brush

leads which carry current round the rotor coil

current out

current in

two halves of a split copper ring which fit onto the rotating shaft of the rotor coil, but are insulated from it

springs apply pressure

The back e.m.f. in a motor

When the rotor coil turns, its wires cut the magnetic field lines which gave rise to the catapult forces. So at the same time as the motor effect is turning the rotor, the rotor coil is also acting as a dynamo. The direction of the induced e.m.f. in the rotor coil will be such as to oppose the rotation (Lenz's law).

When a motor is running

Here there are two e.m.f.s at work: the *externally applied* e.m.f. of a battery or supply and an *internally induced* e.m.f. due to the dynamo effect. (The induced e.m.f. is less than the applied e.m.f.) These e.m.f.s are in opposition and only the difference between them can actually make a current flow in the rotor coil. The internal opposing e.m.f. is called a *back e.m.f.* because its direction is 'backwards', i.e. opposite to the externally applied e.m.f.

$$\begin{matrix} \text{external} & & \text{internal} \\ \text{applied e.m.f.} & - & \text{induced e.m.f.} & = I \times R \\ \text{(battery)} & & \text{(back e.m.f.)} \end{matrix}$$

I is the current which actually flows in the rotor coil.
R is the resistance of the rotor coil and the external circuit.

When a motor is stationary

This is either the moment when the motor is just switched on, or when the motor has been forced to stop with its power supply still switched on (e.g. when a machine driven by the motor gets jammed, as sometimes happens with power tools).

When the rotor is not turning there is no dynamo effect and with no back e.m.f. the applied e.m.f. is not opposed. The effect of this is to produce a much larger current in the rotor coil. This current can cause overheating and damage to the rotor coil.

Two steps can be taken to prevent damage:
a) A series resistor is included in the circuit with the external e.m.f. and the rotor coil to limit the current while the rotor gains speed. This resistor is switched out of the circuit when the motor reaches a certain speed.
b) A thermal cut-out switch is included in the circuit of the rotor coil so that, if it is stopped with the current still flowing, as soon as the temperature starts to rise the current is cut off. Many electric motors used in power tools are protected by a thermal cut-out switch.

Electric motors which run on alternating current

Alternating current (a.c.) changes its direction of flow very rapidly at a controlled rate. If an electric motor is connected to the mains supply in Britain the current will flow in both directions 50 times every second (the mains frequency is 50 Hz). This means that the current changes direction 100 times every second at a regular rate. It follows that if we use an alternating current for an electric motor the commutator does not need to do the reversing of the current in the rotor to keep it turning, because the current is already reversing itself.

So a motor which runs on a.c. needs a different kind of commutator with two separate complete rings called **slip rings**. Fig. 15.16 shows how these are fitted to the rotating shaft in a commercial design. Each end of the rotor coil is connected to a separate copper ring and the same carbon brush makes continuous contact with the same ring.

Each time the current reverses direction the rotor must have turned half a revolution and be passing through the vertical position. In other words, if the current changes direction 100 times a second, the rotor must make exactly 100 half revolutions every second to keep up with the rate of reversal of the current.

Such a motor is called a **synchronous motor**, because its speed (50 complete revolutions per second) is matched exactly, synchronised, to the frequency of the alternating current. A synchronous motor can only run at the one speed or not at all. This means that these motors will not self-start without special arrangements being built into their design. Once running, if they are slowed down by a load which is too heavy they simply stop.

Synchronous motors have special uses because of their very constant speed of rotation, e.g. they are often used for mains-operated electric clocks which keep very accurate time.

Figure 15.16   *A twin-ring, or slip-ring, commutator used in a.c. motors*

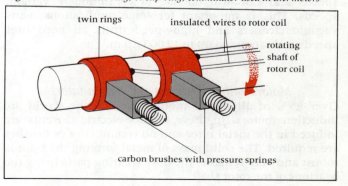

twin rings

insulated wires to rotor coil

rotating shaft of rotor coil

carbon brushes with pressure springs

Different motors for different jobs

Commercial electric motors range in size from the very small, like those which drive model racing cars, to the very large like those used to drive trains. The design of these electric motors also varies in several basic ways.

In choosing an electric motor for a particular job we must think about the following points:
a) Has it got enough power to do the job?
b) Will it run at the correct speed?
c) What will it cost to buy and then to run?
d) How heavy will the motor be which has enough power for the job? What is its power to weight ratio?
e) Can its speed be adjusted, is it continuously variable, or will it only run at one synchronised speed?
f) How reliable is the motor, for example, has it got brushes to wear out? (Reliability and maintenance needs affect the running costs.)

To a great extent the choice of motor type depends on which of these questions is the most important to us. If, for example, the motor must have a variable speed this rules out all the synchronous designs. Or if we need a very robust and reliable motor which will very rarely need maintenance then we are likely to choose a type of motor known as an induction motor which has no brushes or commutator to wear out.

Motors with brushes and commutators

The simple designs which we have looked at, both d.c. and a.c., need a commutator and brushes to supply the current to the rotating coil or rotor. This type of motor can be improved in several ways:
a) The permanent magnets are replaced by electro-magnets. These stationary coils, which provide the magnetic field, are called either the **stator coils** or the **field coils**.
b) Multiple coils are wound on the rotor and the commutator has a pair of contacts for each coil.
c) Instead of a pair of field coils a 4-pole arrangement is sometimes used for the magnetic field.
d) The rotor is made of soft iron. The rotor coils are sunk into the surface of the iron so that the magnetic field passes from the iron core of the stator coils to the iron core of the rotor coils across the narrowest possible air gap. (Magnetic fields are stronger in iron than in air.)
e) A.c. motors can be supplied with 3-phase alternating current (p 317).

These motors have some advantages over induction motors. For example d.c. motors can run at very high speeds and can also operate at continuously variable speeds. This is important for some applications. Fans, vacuum cleaners and high-speed drills all need high speeds and can run at up to 12 000 r.p.m.

Motors without brushes and commutators

Over 95% of all power motors used in the world are induction motors. In these motors electric currents are induced in the metal rotor and no commutator or brushes are required. The solid mass of metal forming the rotor is robust and very reliable, the only wearing parts being the bearings of the rotor shaft.

Their simple and robust design make induction motors a popular choice for most hard working jobs. However these motors do have limitations. Induction motors must have an a.c. supply (only changing currents produce induction effects), and their maximum possible speed is limited by the frequency of the supply. (A 50 Hz supply limits the motor to 3 000 r.p.m.) The principle of the induction motor is explained in the next section.

Assignments

Make a list of tools, and domestic appliances which have electric motors.
Find out
a) whether they work on batteries (a d.c. supply) or the mains (an a.c. supply),
b) whether they have brushes and a commutator,
c) what their electrical power requirements are (often given on a label).
Do not take any motors to pieces!
Explain the meanings of these terms: rotor, armature, stator, field coils, commutator, split ring, slip rings, sensitivity, synchronous, back e.m.f.
Remember
d) the questions to ask about an electric motor if you were choosing one for a particular job,
e) how to make a galvanometer more sensitive to a small current,
f) how to make an electric motor more powerful.
Try questions 15.5 to 15.6

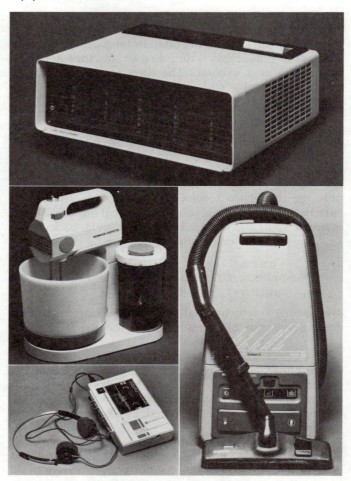

Of these four machines three contain an alternating current electric motor and one contains a direct current motor. Do you know why?

15.3
APPLICATIONS OF THE DYNAMO EFFECT

Sometimes the dynamo effect is used to generate large quantities of electrical power. Equally important are the very sensitive devices, such as moving-coil microphones and record player pick-ups, which convert sound and mechanical vibrations into small varying electric currents.

Making electricity by moving a magnet

We have already seen (fig. 14.37) that an electric current can be induced in a coil by moving a permanent magnet in and out of the coil. This arrangement can be improved (a) by using an iron core for the coil and (b) by rotating the magnet.

● *Wind a coil of about 20 turns of insulated copper wire round an iron C-core and connect the coil to a centre-zero meter, fig. 15.17a.*

● *Move a bar magnet towards and away from the C-core as shown in the figure and watch the meter. Can you explain why the induced current flows in the direction shown in the figure?*

Using the iron C-core greatly increases the strength of the magnetic field, which changes through the coil and induces a larger current.

The most convenient way of producing a continuously changing magnetic field through a coil is to rotate either the magnet or the coil. Fig. 15.17b shows how the magnet can be rotated between the 'poles' of the iron C-core.

● *Attach a small powerful permanent bar magnet to an axle using glue or tape.*

Figure 15.17 *Making electricity by moving a permanent magnet*

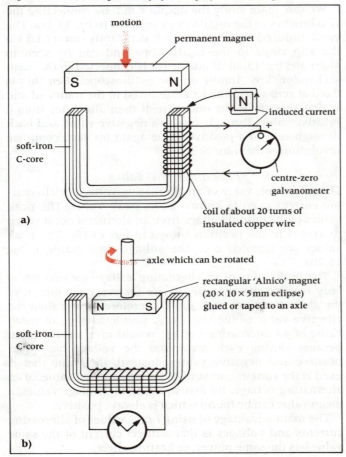

a)

motion
permanent magnet
S   N
N → induced current
soft-iron C-core
centre-zero galvanometer
coil of about 20 turns of insulated copper wire

b)

axle which can be rotated
rectangular 'Alnico' magnet (20 × 10 × 5 mm eclipse) glued or taped to an axle
N   S
soft-iron C-core

The axle can be rotated by hand to demonstrate the effect, although it is difficult to stop the magnet sticking to the iron core unless the axle is mounted so that it is held in the centre of the gap.

● *What do you notice about the induced current recorded by the meter?*

The iron C-core has induced magnetism which reverses as the permanent magnet rotates. So the direction of the induced current in the coil reverses as often as the induced magnetic poles are reversed.

─── *Converting sound to electricity in a microphone* ───

When a varying and alternating current is supplied to a moving-coil loudspeaker its coil moves in and out of a specially designed magnet. If we shout into a loudspeaker and connect its coil to a sensitive meter or an amplifier, we find that it can also be made to work as a microphone. This is an example of the motor effect in reverse acting like a dynamo.

The diagram shows how a moving-coil microphone is constructed. It is very similar to a moving-coil loudspeaker, fig. 15.10, but is considerably smaller and the diaphragm to which the moving-coil is attached is sensitive to the very small pressure changes in the air caused by sound waves. The magnet of the microphone has exactly the same form as in fig. 15.10b. The diaphragm, which moves only a short distance, needs a hole in it to allow the air inside the microphone to flow out and in as the sound waves push it in and out. A moving-coil microphone gives very good quality reproduction of sound.

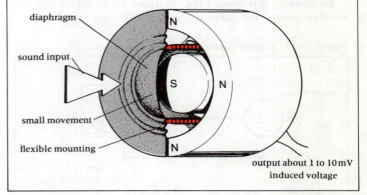

diaphragm
sound input
small movement
flexible mounting
N   S   N   N
output about 1 to 10 mV induced voltage

Generating electricity by rotating a coil

Using a model d.c. motor as a d.c. generator

In the last section a model motor was built which worked on direct current, fig. 15.14. You can now use this model as a dynamo or d.c. generator.

● *Connect the wire brushes to a centre-zero meter instead of a battery.*

● *Spin the rotor coil with your fingers and watch the galvanometer.*

● *What kind of electric current is produced? Does it make any difference if the rotor spins faster or in the opposite direction?*

Your motor is now a dynamo converting the mechanical work you put into it into electrical energy.

The output of a d.c. dynamo

When the dynamo has a split-ring commutator (exactly like the one used for a d.c. motor) it produces direct current. The rotor coil itself actually has an alternating e.m.f. induced in it but the split-ring commutator reverses the direction of the output e.m.f. every time it reverses in the rotor so that a direct current flows from the brushes.

The output of a d.c. generator can be displayed on an oscilloscope by connecting the two wires from the brushes to the Y-plates of the oscilloscope (input terminals), fig. 15.18a. (The output of the model d.c. generator is too intermittent to give a clear trace on the screen. A generator with a split-ring commutator is needed and it should be driven at a constant speed.)

● *Switch on the time base of the oscilloscope to scan across the screen once while the rotor revolves two or three revolutions.*

Fig. 15.18b shows the type of output produced by a good d.c. generator. The oscilloscope works as a voltmeter (displaying a voltage–time graph), so the trace on the screen shows how the e.m.f. of the generator varies. Although the e.m.f. is always in the same direction (always positive, or always negative), it is not constant. The e.m.f. varies from zero to a peak value. As the rotor spins, the wires on the coil cut the magnetic field lines at different rates. The maximum voltage is induced when the wires cut the field lines at right angles, fig. 15.18c. When the wires are moving parallel to the field lines, and not cutting them, the induced voltage drops to zero, fig. 15.18d.

An a.c. dynamo or alternator

Most generators of electricity are designed to produce alternating current. These a.c. dynamos are often called **alternators**.

To convert our model d.c. dynamo to an alternator we need to change the commutator to a slip-ring type.

Figure 15.18 *Investigating the output e.m.f. of a d.c. generator*

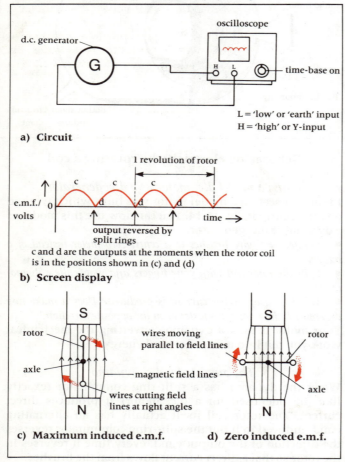

a) **Circuit**

b) **Screen display**

c) **Maximum induced e.m.f.**   d) **Zero induced e.m.f.**

● *Unfasten your split-ring type of commutator and adjust the ends of the wire coil on the rotor so that the two wire ends finish at opposite ends of the rotor.*
● *Insulate both ends of the tube through which the axle passes.*
● *Wind the bare wire ends of the rotor coil round each end of the tube in two compressed coils to form two slip rings.*
● *Fix the wire ends with tape or rubber bands so that they do not unwind, fig. 15.19.*

One wire brush is needed at each end so that the same brush is in continous contact with the same slip ring.

● *Connect the brushes to a centre-zero meter and spin the rotor slowly, watching the meter. What difference do you notice? What happens if the rotor spins quickly?*

Figure 15.19   *A slip-ring commutator for a model alternator*

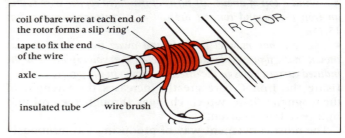

The output of an a.c. generator or alternator

The meter needle swinging to both sides of the central zero shows that the induced current changes direction rapidly, i.e. it alternates. At faster speeds the meter needle and moving coil cannot swing fast enough from side to side, and so stay near to the central zero.

We can again study the output e.m.f. by connecting the two brushes to the oscilloscope input as in fig. 15.18a. The e.m.f. induced in the rotor coil is constantly connected via the slip rings to the oscilloscope and can be seen to alternate, fig. 15.20. If one brush is connected to the earth socket (or 'low' input) on the oscilloscope, then this is fixed at zero volts. The e.m.f. induced in the rotor coil and which appears at the other brush then alternates from a positive value, through zero to a negative value and back through zero to a positive value again for each complete revolution of the rotor coil.

Peak and r.m.s. values

The maximum value of an alternating voltage or alternating current (a.c.) is known as its **peak value**. The peak values of the output voltage from an alternator occur when its coil is in the position shown in fig. 15.18c. The peak values are labelled c on the voltage–time graph in fig. 15.20a.

The peak value of an alternating voltage (or current) is only reached momentarily, twice for every revolution of the alternator rotor coil, and is therefore greater than the effective value of the supply. A simple average or mean value of an alternating voltage would give a zero result, because during each revolution the voltage has equal positive and negative values. Instead, the value that is used is the **root-mean square value**, or r.m.s. value, of an alternating voltage. By first squaring the voltage values, a mean value can be found which is always positive.

The main advantage of using r.m.s. values of alternating currents and voltages is that a direct current of the same value has the same power, or heating effect.

When using the formula:

power = current × voltage

for an alternating supply, this becomes

average power = r.m.s. current × r.m.s. voltage

The values quoted for alternating currents and voltages are usually the r.m.s. values and these should always be used in energy and power calculations for a.c. circuits. The relation between the r.m.s. and peak values of I or V can be shown to be

$$\text{r.m.s. value} = \frac{\text{peak value}}{\sqrt{2}} = 0.71 \times \text{peak value}$$

Fig. 15.20b shows the alternating voltage supplied to houses in Britain. The 'mains' supply has a peak value of ±340 V. So the r.m.s. voltage of the 'mains' supply is 340 V/$\sqrt{2}$ = 240 V.

Practical a.c. generators or alternators

Alternators which are used to generate electricity in the power industry have improved designs for greater efficiency and power output, fig. 15.21a.

Figure 15.20 *Alternating voltages*

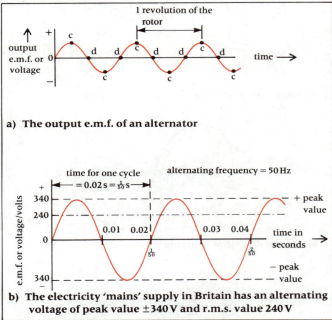

a) **The output e.m.f. of an alternator**

b) **The electricity 'mains' supply in Britain has an alternating voltage of peak value ±340 V and r.m.s. value 240 V**

Figure 15.21 *A practical alternator*

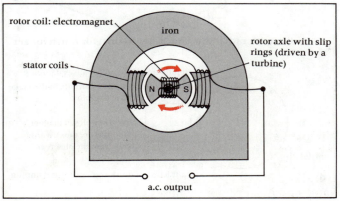

a.c. output

a) The permanent magnet is replaced by a d.c. electro-magnet.

b) The d.c. electromagnet becomes the rotor and works like the rotating permanent magnet in a bicycle dynamo.

c) The generated alternating supply is obtained from the stator coils without this large induced current having to pass through any brushes.

d) A small fraction of the generated output is fed back as a direct current through slip rings to power the electro-magnet of the rotor. This current is much smaller than the main a.c. output from the stator coils and so there is less difficulty in passing it through brushes and slip rings to the rotor. Sometimes a smaller d.c. generator is mounted on the same shaft as the main alternator. The output of the d.c. generator is used to supply d.c. to the electromagnet rotor of the alternator.

Three-phase alternating supplies

The alternating voltage induced in a single coil falls momentarily to zero twice in every cycle. The power available also falls to zero at the same frequency. To provide a more even supply of power, alternators are designed with three sets of coils, which produce three separate supplies of alternating current known as a **three-phase supply**. Three-phase alternators have three pairs of stator coils and three rotor electromagnets set at 120° to each other.

A three-phase supply has three live wires each carrying alternating currents in a circuit. A single common return wire or **neutral** wire may be used for all three circuits, fig. 15.22a. The alternating current in the three phases is timed to change direction at a different moment so that there is always some current flowing in two of the three wires, fig. 15.22b. This gives a steadier supply of power which is more suitable for driving powerful machines.

Figure 15.22 *Three-phase alternating electricity*

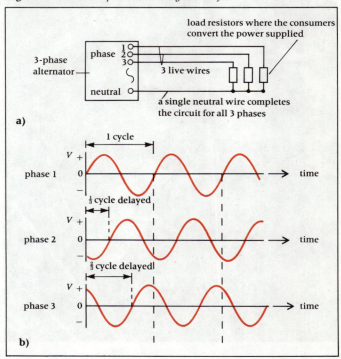

A bicycle dynamo

The diagram shows how the arrangement of a rotating magnet as the rotor and a stationary coil as the stator is used in a practical design for a small bicycle dynamo. The rotor magnet is in the form of a cylindrical permanent magnet with poles on opposite sides of the cylinder. The iron core has concave 'poles', shaped to fit closely round the rotor. An advantage of this type of design is that there are no brushes and no commutator to wear out. What kind of current is produced by this type of dynamo?

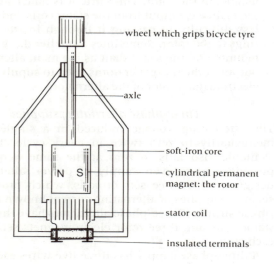

- wheel which grips bicycle tyre
- axle
- soft-iron core
- cylindrical permanent magnet: the rotor
- stator coil
- insulated terminals

Electromagnetic damping

Damping is a word used to describe how movement and vibrations are reduced or slowed down, in the same way that we 'damp down' a fire, or the 'dampers' on a piano stop the strings vibrating. *Electromagnetic damping* can be used to slow down the movement of a conductor as it passes through a magnetic field. It has the special advantage that, unlike a mechanical damping or braking system, there is no mechanical contact with the moving conductor and no brakes to get worn.

Demonstrating eddy current damping

A disc made of a non-magnetic metal such as copper or aluminium can be used to demonstrate the damping or braking effect caused by electromagnetic induction.

- *Support the disc vertically so that it can turn freely on its horizontal axis, fig. 15.23a, and set it spinning.*
- *Bring a magnet up to the disc so that the moving sheet metal cuts through the magnetic field lines of the magnet, fig. 15.23b.*

Figure 15.23 *Eddy current damping*

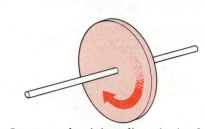

a) **Copper or aluminium disc spinning freely**

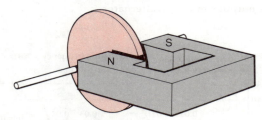

b) **A solid disc stops spinning**

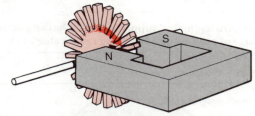

c) **A disc with radial slits continues to turn for much longer**

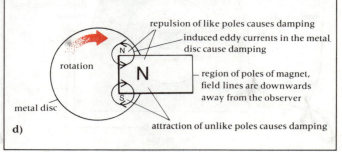

- repulsion of like poles causes damping
- induced eddy currents in the metal disc cause damping
- region of poles of magnet, field lines are downwards away from the observer
- attraction of unlike poles causes damping

rotation

metal disc

d)

The disc quickly stops turning without any direct contact or friction with the magnet. However, another disc with radial slits cut in it, fig. 15.23c, shows a different result.

Cutting the magnetic field lines induces a current in the rotating disc. (You can check the direction of the induced currents shown in fig. 15.23d using Fleming's right-hand rule.) The induced current flows radially outwards (away from the axle) in the disc. But currents only flow in closed circuits. In the disc the induced currents flow back round in a loop outside the influence of the magnet's field. These current loops are called **eddy currents**, because of their similarity with the eddy currents found in air and water, for example a whirlpool.

The damping effect of these induced currents or eddy currents can be explained simply by saying that the direction of the eddy currents will be such that they oppose the motion producing them, and so tend to damp that motion. In other words it is an example of Lenz's law.

Alternatively we can look for an explanation based on forces between magnetic poles as shown in fig. 15.23d. The eddy currents will produce their own magnetic fields as shown and these will experience forces of attraction and repulsion from the poles of the permanent magnet. A N pole (due to the eddy current) approaching the N pole of the permanent magnet is repelled and this causes the motion to slow down. *All* the poles are reversed when seen from the other side and the effect is the same.

The slits cut in the second disc greatly reduce the scope for eddy currents as they cut across the paths of the currents. The slits act like high resistances in the eddy current circuits.

Applications of electromagnetic damping

Emergency brakes

The principle demonstrated with the spinning disc is used for emergency braking on some kinds of vehicles. The permanent magnet is replaced with a large powerful electromagnet. A thick metal disc mounted on a wheel axle passes freely between the poles of the electromagnet. In an emergency (when the mechanical or hydraulic brakes have failed), a large direct current is fed from a battery through the electromagnet providing a strong braking effect on the disc, so slowing down the vehicle.

Dead–beat galvanometers and ammeters

The coil which turns in a moving-coil galvanometer or ammeter is wound on a light-weight metal frame. This frame behaves like a coil of one turn and of very low resistance (a closed loop of metal). As the coil turns due to the catapult forces acting on it (the motor effect), the metal frame also turns and cuts the magnetic field lines between the magnet poles and the iron cylinder (see fig. 15.13). Eddy currents large enough to damp the movement of the coil are induced in the frame because of its very low resistance. The frame is designed so that the amount of electromagnetic damping is just right.

The coil and its pointer turn smoothly and come quickly to rest at the correct deflection without overshooting the reading and oscillating about it. The damping has no effect on the reading because when the coil and frame are not turning there is no induction effect and no damping. This condition is described as **dead-beat**.

When a meter is moved about or transported, a short length of wire of low resistance should be connected between its terminals. This joins the moving coil in a closed circuit of low resistance. The combined damping effect of the eddy currents in the metal frame and the induced current in the closed circuit of the coil is sufficient to stop the coil almost completely from swinging about and turning. This protects it against bumps and jolts which might otherwise damage its bearings, or delicate hair springs. The electromagnetic damping holds the coil and frame in an invisible and 'elastic' grip.

The induction motor principle

An aluminium disc spinning so that it cuts a magnetic field has eddy currents induced in it which oppose its motion relative to the magnetic field and stop it turning, fig. 15.24a.

If the aluminium disc was stationary and the magnetic field was rotated round the disc so that there was the same relative motion between them, then the same eddy currents would be induced in the aluminium disc as before, fig. 15.24b. If the magnetic field continues to rotate at the same speed, then to reduce the relative motion between the disc and the field, the disc must begin to rotate with the field. So the rotating magnetic field drags the disc round with it. It is as if the magnetic field were a kind of fluid flowing round the disc carrying it along with the flow.

In induction motors the rotating permanent magnet shown in (b) is replaced by electromagnets which form the stator coils of the motor. The effect of a rotating magnetic field is created by the alternating currents fed to the stator coils which surround the rotor. The aluminium disc becomes a copper cylinder rotor which has no brushes or commutator. The rotor currents are the eddy currents provided by electromagnetic induction.

Figure 15.24 *The induction motor principle*

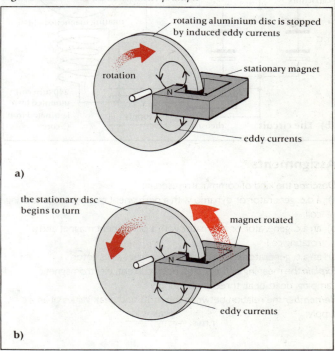

rotating aluminium disc is stopped by induced eddy currents

stationary magnet

rotation

eddy currents

a)

the stationary disc begins to turn

magnet rotated

eddy currents

b)

A simple induction motor using a rotating magnetic field can be demonstrated using the arrangement shown in fig. 15.25a. The rotor R is an aluminium can balanced and pivoted on a vertical needle. (An empty 35 mm film can is suitable for use as a rotor. It will balance better if a slight dimple is made inside at the centre of its base.) Two electromagnets set at right angles to each other provide the rotating magnetic field. The circuit shown in fig. 15.25b is used to produce the rotating magnetic field effect. Coil A is fed with alternating current straight from the low-voltage a.c. supply. Coil B is fed with alternating current which has passed through a large capacitor C. This capacitor causes a phase change of the alternating current, so that the changes in direction of the current in coil B occur at slightly different times (about $\frac{1}{200}$ second earlier) than in coil A. In this way, as the current in coil B alternates ahead of the current in coil A, the effect of a magnetic field rotating from coil B to coil A is created.

In real induction motors the phase change of the current is not usually produced by using a capacitor. An alternative method is to make use of the phase difference available in the three phases of the national grid system.

Figure 15.25 *A model induction motor*

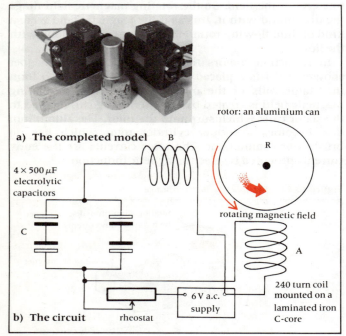

a) **The completed model**

rotor: an aluminium can

R

rotating magnetic field

$4 \times 500\,\mu F$ electrolytic capacitors

C

B

A

240 turn coil mounted on a laminated iron C-core

6 V a.c. supply

b) **The circuit**   rheostat

Assignments

Describe the kind of commutator used in:
a) a d.c. generator or dynamo with a permanent magnet and a rotating coil,
b) an a.c. generator or alternator with a permanent magnet and a rotating coil.
c) an a.c. generator with a d.c. electromagnet as its rotor.

Explain the meaning of the terms: eddy current, electromagnetic damping, dead-beat, three-phase supply.

Remember the relation between the r.m.s. and peak values of an a.c. supply:

$$\text{r.m.s.} = \frac{\text{peak value}}{\sqrt{2}}$$

Try questions 15.7 and 15.8

15.4
ELECTROMAGNETIC INDUCTION WITHOUT MOTION:
INDUCTION COILS AND TRANSFORMERS

Transformers and induction coils provide a means of transferring energy from one circuit to another without any electrical link between them. At the same time the electrical energy can be converted to a higher or lower voltage for different uses. When the voltage of an electric supply is changed it is said to be **transformed**.

Two coils and a direct current

- *Wind about 10 turns of insulated copper wire onto two iron C-cores as shown in fig. 15.26.*
- *Connect coil ① to a low-voltage d.c. supply and include a 0 – 1 A ammeter in the circuit to show that a current is flowing.*
- *Switch on the current.*
- *Connect the second coil ② to a centre-zero meter.*
- *With a steady direct current flowing in coil ① bring the two iron C-cores together so that they form a closed iron loop and watch the meter.*
- *Now separate the C-cores and then bring them together again.*
- *Note what the meter indicates while*
a) *the cores are held apart,*
b) *the cores are moving together,*
c) *the cores are held together,*
d) *the cores are moving apart.*

Coil ① forms an electromagnet. When it is moved so that its magnetic field lines are cut by the turns of wire on coil ②, the induced e.m.f. in the coil ② sends a current through the meter. The effect is exactly the same as when a permanent magnet was moved up to a similar coil, fig. 15.17. When the two coils are held together and a steady direct current flows in coil ①, the meter shows zero induced current in coil ②.

- *Clip the two iron C-cores together and watch the meter while:*
e) *the direct current in coil ① is switched on,*
f) *the direct current in coil ① is switched off,*
g) *the rheostat is used to change the current in coil ①.*

The effect of changing the current in coil ① is the same as moving the coil. When the current changes, the magnetic field which it produces in coil ① also changes, so that its magnetic field lines cut the turns on coil ②. When the current in coil ① is switched on or off the current and the magnetic field change very rapidly. The rapid change of the magnetic field through coil ② induces a large e.m.f. in

Figure 15.26 *For induction to occur with a direct current we need motion, or a changing current*

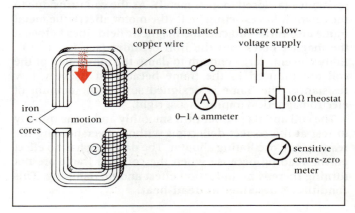

10 turns of insulated copper wire

battery or low-voltage supply

iron C-cores

motion

A

0–1 A ammeter

$10\,\Omega$ rheostat

sensitive centre-zero meter

its turns, which lasts for a short time and shows up as a flick of the pointer on the meter. When the rheostat is used to produce a more gradual change of the current in coil ① the induced e.m.f. in coil ② is much smaller but lasts longer.

Electromagnetic induction can be used to transfer electrical energy from one circuit to another by means of a magnetic field which links the two coils. The two coils are called the **primary coil** and the **secondary coil**.

Only when there is a change of current in the primary coil does an induced e.m.f. appear in the secondary coil.

There are two ways of making the current in the primary coil change continuously.

a) Direct current in the primary coil can be switched on and off rapidly by an automatic make and break mechanism like the one used in an electric bell. Such an electric machine is called an **induction coil**.

b) An alternating current can be used in the primary coil which, because it is continuously changing in both direction and magnitude, is ideal for producing an induced e.m.f. in the secondary coil. This electromagnetic machine is called a **transformer**.

The induction coil

An induction coil produces a high voltage in its secondary coil by electromagnetic induction. The direct current in the primary coil is switched on and off by the make and break mechanism producing the changes of current and changes of magnetic field necessary for electromagnetic induction to occur in the secondary coil, fig. 15.27.

When the current in the primary coil is switched on the induced magnetism in the iron core attracts the soft-iron armature. The moving armature opens a gap between two contacts which breaks the primary coil circuit so switch-

ing off the current. As the induced magnetism fades away the armature springs back, closes the contacts and remakes the circuit so that the current flows in the primary coil again. This cycle of events repeats automatically. The secondary coil is usually wound on top of the primary coil so that it has the same changing magnetic field through it.

Two things help to make the induced e.m.f. in the secondary coil very large.

a) The secondary coil has a large number of turns.

b) The rapid change in the primary current, particularly when it is switched off, causes a very rapid change in the magnetic field through the secondary coil.

We have seen that both these factors will increase the e.m.f. induced between the ends of the secondary coil; in fact e.m.f.s of several 100 kV can be produced. Such a high voltage can be very dangerous as the sparks produced by any voltage greater than 6 kV may generate X-rays in low-pressure discharge tubes. So induction coils can be used only with suitable screening or when the voltage produced is less than 6 kV. (See DES Administrative Memorandum No 2/76 HMSO 1976: *The use of ionising radiations in educational establishments*.)

Making a transformer

● *Wind a primary coil of 10 turns of insulated wire onto one arm of a C-core, fig. 15.28. Connect this coil to a low-voltage a.c. supply (1 V a.c., high-current power unit.)*

● *Wind a secondary coil of 10 turns on another C-core, leaving some extra wire for winding more turns.*

● *Connect the secondary coil in series with a small filament lamp (2.5 V, 0.3 A). How does the brightness of the lamp change when the number of turns on the secondary coil is increased or decreased?*

● *Replace the lamp with a centre-zero meter.*

● *Connect an oscilloscope in place of the meter.*

● *Connect the two ends of the secondary coil to the input terminals of the oscilloscope.*

● *Switch on the internal time base and adjust the control until a wave graph can be seen on the screen.*

● *Again investigate the effect of changing the number of turns on the secondary coil.*

a) Why is the lamp brighter when there are more turns on the secondary coil? What effect do you think more turns on the primary coil would have?

b) Why is the galvanometer unable to give a reading even though a current is flowing through it? (The current is alternating at a frequency of 50 Hz.)

c) What does the vertical height on the oscilloscope graph measure, and how does this quantity depend on the number of turns on the secondary coil? (The oscilloscope displays a voltage–time graph.)

Figure 15.27   *The induction coil*

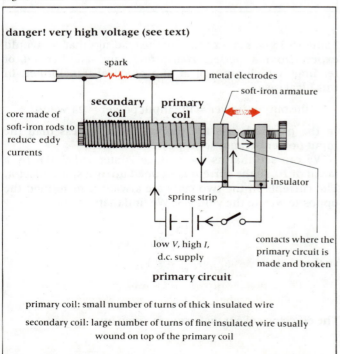

danger! very high voltage (see text)

spark

metal electrodes

secondary coil

primary coil

soft-iron armature

core made of soft-iron rods to reduce eddy currents

insulator

spring strip

low *V*, high *I*, d.c. supply

contacts where the primary circuit is made and broken

primary circuit

primary coil: small number of turns of thick insulated wire

secondary coil: large number of turns of fine insulated wire usually wound on top of the primary coil

Figure 15.28   *Making a transformer*

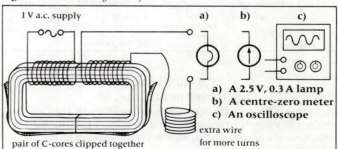

1 V a.c. supply

a)   b)   c)

a) A 2.5 V, 0.3 A lamp
b) A centre-zero meter
c) An oscilloscope

extra wire for more turns

pair of C-cores clipped together

The transformer

The transformer is a very common and very important electrical machine. Every mains-operated television and record player has one. Battery chargers and model train sets depend on transformers; the whole system of distribution of electricity across the country also depends on them. The transformer transfers electrical energy from one circuit to another by electromagnetic induction between two coils. This is sometimes called **mutual induction**, because it takes place between two coils.

While transferring the energy, the transformer enables us to change or *transform* the voltage or e.m.f. to a larger or smaller value. This is the transformer's most important use.

Step-up and step-down transformers

We have found that winding more turns on the secondary coil of a simple transformer increased the brightness of a lamp and produced a larger amplitude graph on the oscilloscope screen. Each turn on the secondary coil has an e.m.f. induced in it, and these e.m.f.s add up to a higher total voltage as the number of turns increases. If the number of turns on the primary coil is increased, the opposite effect occurs and the lamp becomes dimmer. We find that the voltage induced across the secondary coil depends on the ratio of the number of turns on the two coils as follows:

$$\frac{\text{secondary voltage}}{\text{primary voltage}} = \frac{\text{secondary turns}}{\text{primary turns}} \qquad \frac{V_2}{V_1} = \frac{n_2}{n_1}$$

where V_1 = the alternating input voltage to the transformer, the primary voltage,
V_2 = the alternating output voltage from the transformer, the secondary voltage,
n_1 = the number of turns on the primary coil,
n_2 = the number of turns on the secondary coil

Fig. 15.29a shows a *step-up* transformer which changes a voltage to a higher value, i.e. $V_2 > V_1$. This transformer has more turns on the secondary coil than the primary coil in the ratio of the voltage step-up required.

Fig. 15.29b shows a *step-down* transformer which changes a voltage to a lower value, i.e. $V_2 < V_1$. Here $n_2 < n_1$ in the required ratio for the voltage reduction.

Figure 15.29 *The transformer formula*

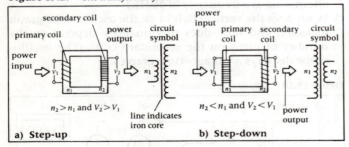

a) **Step-up**   b) **Step-down**

Worked Example
The transformer turns ratio
A transformer is required to step down the mains voltage of 240 V to provide a 12 V supply for an electric toy. If the primary coil is wound with 1 000 turns of wire, calculate the number of turns required for the secondary coil.

Using the formula $\quad \dfrac{n_2}{n_1} = \dfrac{V_2}{V_1}$

we have $\quad \dfrac{n_2}{1\,000} = \dfrac{12\,V}{240\,V}$

$\therefore \quad n_2 = 1\,000 \times \dfrac{12}{240} = 50 \text{ turns}$

Answer: A secondary coil with 50 turns would provide an alternating voltage output from the transformer of 12 V.

Power transfer in a transformer

As with all other machines, it is impossible to get more power out of a transformer than is fed into it. If a transformer was a perfect electrical machine, which wasted no power at all, then the output electrical power from the secondary coil would be equal to the input electrical power to the primary coil.

Fig. 15.30 shows how these two powers can be measured. The ammeters and voltmeters must be special ones suitable for measuring alternating currents and voltages. For each coil the ammeter reads the current in the coil and the voltmeter reads the e.m.f. or voltage across the terminals of the coil. The electrical power is given by:

$$\text{power} = \text{current} \times \text{voltage} \quad \text{or} \quad P = IV$$

Figure 15.30 *Transformer power*

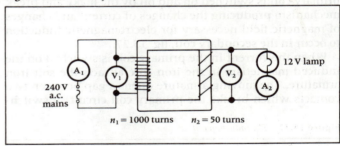

Table 15.1 gives an example of the readings that we would expect from a perfect transformer and another set of readings which are typical of a practical transformer. In this ideal case,

$$\text{the output power} = \text{the input power} = 24 \text{ watts.}$$

In the practical case the output power is less than the input power because some power is wasted.

We can see that as the voltage is stepped down by a factor of 20, so the current is stepped up by a similar factor. The currents in the two coils are always transformed the opposite way to the voltages. We find that:

$$\frac{I_2}{I_1} = \frac{V_1}{V_2}$$

so that $\qquad I_2 V_2 = I_1 V_1$

The ratio of the currents in the two coils is the inverse of the ratio of the voltages across the coils.

The efficiency of a transformer is given by:

$$\text{efficiency} = \frac{\text{output power}}{\text{input power}} \times 100\% = \frac{I_2 V_2}{I_1 V_1} \times 100\%$$

Table 15.1

	Primary coil			Secondary coil		
	I_1/A	V_1/V	input power/W	I_2/A	V_2/V	output power/W
Ideal readings	0.1	240	24	2.0	12	24
Typical readings	0.1	240	24	1.9	11.3	21.5

Worked Example
Transformer power and efficiency

A step-up transformer has 10 000 turns on its secondary coil and 100 turns on its primary coil. An alternating current of 5.0 A flows in the primary coil when it is connected to a 12 V a.c. supply. Calculate:
a) the power input to the transformer,
b) the e.m.f. induced across the secondary coil.
What is the maximum current that could flow in a circuit connected to the secondary coil if
c) the transformer is 100% efficient and
d) the transformer is 90% efficient but produces the same e.m.f. across the secondary coil as the ideal one?

a) Input power = $I_1 V_1$ = 5.0 A × 12 V = 60 W.

b) The secondary coil voltage V_2 is found using:
$$\frac{V_2}{V_1} = \frac{n_2}{n_1}$$

$$\therefore \quad V_2 = V_1 \times \frac{n_2}{n_1} = 12\,V \times \frac{10\,000}{100} = 1\,200\,V$$

c) Output power = input power = 60 W and the output power is $I_2 V_2$, so when the transformer is 100% efficient

$$I_2 = \frac{60\,W}{V_2} = \frac{60\,W}{1\,200\,V} = 0.05\,A$$

d) We use:
$$\text{efficiency} = \frac{\text{output power}}{\text{input power}} = 90\%$$

$$\therefore \quad \frac{\text{output}}{\text{power}} = \frac{90}{100} \times \frac{\text{input}}{\text{power}} = \frac{90}{100} \times 60\,W = 54\,W$$

Now output power = $I_2 V_2$ = 54 W, so when the transformer is 90% efficient

$$I_2 = \frac{54\,W}{V_2} = \frac{54\,W}{1\,200\,V} = 0.045\,A$$

Answer: The power input to the transformer is 60 W, inducing an e.m.f. of 1 200 V across the secondary coil. The maximum current through the secondary coil is 0.05 A at 100% efficiency, and 0.045 A at 90% efficiency.

Designing a transformer to reduce power losses

There are four ways in which power can be wasted in a transformer. To make a transformer as efficient as possible these must be reduced as much as possible within given size and cost limits. Table 15.2 gives the causes of power losses and the steps which can be taken to reduce them, and fig. 15.31 shows an efficient design.

Some general points about transformers are:
a) Well-designed transformers are very efficient, usually better than 99% efficient.
b) Large transformers used in the distribution of electricity handle many hundreds of kilowatts of power. Even if only 1% of this is wasted, a lot of heat is produced. Such transformers are oil cooled.
c) Many transformers have more than one secondary coil. Each coil, with a different number of turns, can provide different voltages for various components in a circuit.

Table 15.2

Cause of power loss in a transformer	Design features which reduce the power loss
heating effect of the current in the copper wires of the coils: lost power in each coil $= I^2R$, where R = resistance of the wire in each coil	thick copper wire of low resistance is used, particularly for the coil carrying the high current at low voltage; the cost of copper wire and the size and weight of the transformer limit the gauge of wire that can be used
heating effect of eddy currents induced in the iron core: lost power = I^2R, where R = resistance of a closed loop in the iron core where eddy currents flow	the iron core is laminated, cutting across the path of any induced eddy currents; the high resistance between the laminations greatly reduces the eddy currents and also the heat they would produce
energy is used in the process of magnetising the iron core and reversing this magnetisation every time the current reverses, also heats the iron core	the iron core is made of very soft iron, which is very easily magnetised and demagnetised by the magnetic field of the primary coil
some of the magnetic field lines produced by the primary coil do not link with the secondary coil, reducing the e.m.f. induced in the secondary coil	the core is designed for maximum linkage between the primary and secondary coils; common method is to wind the secondary coil on top of the primary coil, the iron core must always form a closed loop of iron

Figure 15.31 *In this step-down transformer the primary coil has a large number of turns of a thin gauge of wire and the secondary coil has a much smaller number of turns of thicker wire to carry a larger current.*

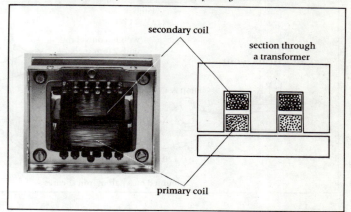

Assignments

Explain
a) Why do transformers require a.c. rather than d.c.?
b) What is the difference between an induction coil and a transformer?
c) What are the ways in which power is wasted in a transformer, and how are these reduced to a minimum?
d) Why should water not be used for cooling a transformer.

Remember the formulas:

$$\frac{V_2}{V_1} = \frac{n_2}{n_1}$$

$$\text{efficiency} = \frac{I_2 V_2}{I_1 V_1} \times 100\%$$

Try questions 15.9 to 15.11

15.5
THE NATIONAL GRID

Power stations which generate electrical energy are connected by cables to all the places where electricity is used. The cables arriving at factories, offices and homes are usually underground in towns and cities, where they are unseen, safe and forgotten. In Britain there is also a network of cables which link all the power stations together so that the demand for electricity can be shared out between them. This network of cables, which are mostly carried overhead on pylons, is known as the national grid.

Building model power lines

Fig. 15.32 shows how working models of electric power lines or transmission lines can be set up in the laboratory.

● *Connect two lamps (12 V, 24 W) to a 12 V a.c. power supply. Lamp A is connected by two short copper wires to the power supply, but lamp B is connected by two long thin wires, fig. 15.32a.*

These two wires represent a long distance of overhead cable on the grid system and should have several ohms resistance, see fig. 15.32c for the practical arrangement. The effect of transmitting the power at the low 12 V level can easily be seen. Lamp B is very dim compared with lamp A.

● *Now connect a transformer at each end of the long wires supplying lamp B.*

A suitable step-up transformer can be made using a 120 turn coil as the primary coil and a 2400 turn coil as the secondary coil. A step-down transformer can be made using similar coils with the larger number of turns on the primary coil, fig. 15.32b. Such a pair of transformers

Figure 15.32 *Model power lines*

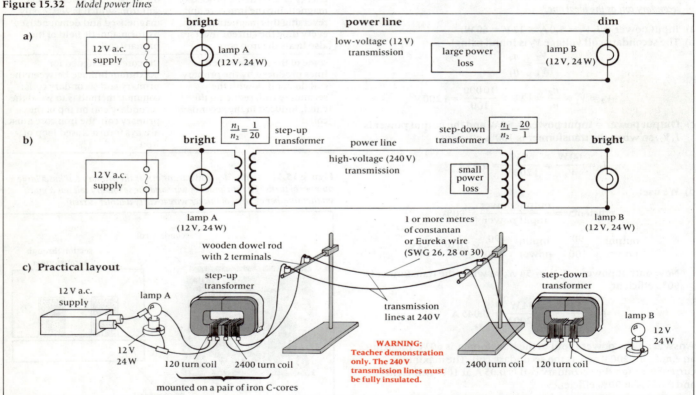

Figure 15.33 *Power loss in transmission lines*

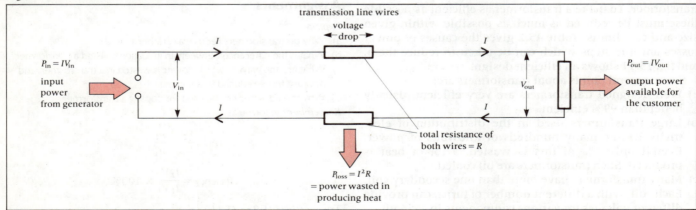

enable the power to pass along the wires at a higher voltage: for the first transformer

$$V_2 = V_1 \times \frac{n_2}{n_1} = 12\,V \times \frac{2400}{120} = 240\,V$$

Warning: Teacher demonstration only. The 240 V transmission lines must be fully insulated.

The effect of transmitting the power at the higher voltage of 240 V is to make lamp B almost as bright as lamp A. The power loss from the power lines is now quite small. The overall efficiency of the high-voltage lines depends on the efficiency of the transformers.

Reducing power loss in transmission lines

Fig. 15.33 shows how the power fed into a transmission line P_{in} divides into two parts. Most of the power P_{out} should reach the end of the line where it is used by a consumer, but some of the power P_{loss} is wasted by the resistance of the wires, producing heat. The power in a wire at any point is given by the formula:

power = current × voltage   or $P = IV$

The power fed into a transmission line is $P_{in} = IV_{in}$, where V_{in} is the voltage connected to the wires from a generator, or transformer, at the beginning of the line.

The power reaching the consumer is $P_{out} = IV_{out}$. The current I is always the same all along a wire, but the voltage drops along the wire as power is lost so that V_{out} is less than V_{in}.

The power loss is due to the heating effect of the current in the wire, which is I^2R, where R is the resistance of the wire in the transmission line. As fig. 15.33 shows,

$$P_{out} = P_{in} - P_{loss}$$

So $$IV_{out} = IV_{in} - I^2R$$

As the same current flows all round the circuit we can divide through by I giving

$$V_{out} = V_{in} - IR$$

where IR is the voltage drop along the wire.

The power loss is kept to a minimum by keeping the voltage drop along the wire as low as possible. As the voltage drop equals IR, it is clear that both I and R should be kept as low as possible.

The resistance of the wires R is kept low by using thick wires with a large cross-sectional area. Aluminium alloys are used because they are cheaper and much lighter than copper.

The same power can be carried in a wire at low current if the voltage is made high. In the example of the model power lines, 24 W of power was supplied. At 12 V the maximum current in the wires is

$$I = \frac{P}{V} = \frac{24\,W}{12\,V} = 2\,A$$

At 240 V the current in the wires is

$$I = \frac{P}{V} = \frac{24\,W}{240\,V} = 0.1\,A$$

When the current is 20 times smaller the power loss as heat in the wires (I^2R) is 400 times smaller. This shows how effective it is to transmit electrical power at a high voltage. These ideas can be summarised as follows:

a) Wires must have a low resistance to reduce power loss.
b) Electrical power must be transmitted at low currents to reduce power loss.
c) To carry the same power at low current we must use a high voltage.
d) To step up to a high voltage at the beginning of a transmission line and to step down to a low voltage again at the end we need transformers.
e) Transformers only work when they are supplied with alternating current.

We can now understand why electrical power is transmitted as *an alternating current at a very high voltage.*

Calculating power loss in transmission lines

Power loss in any wires is due to the heating effect of the electric current in them.

Power loss in wires = I^2R

In this formula, R is the resistance of just the wires themselves, but I is the current in the wires and in the load, which uses the electricity at the end of the power line (i.e. lamp B in the model). To find this current, the *total* resistance in the circuit (including the load, the wires and the power supply) must be calculated.

Worked Example
Power loss in a model power line

In a model of a power line, a 12 V a.c. supply of negligible resistance is connected by wires of total resistance 4 Ω to a lamp of resistance 6 Ω. Calculate:
a) *the current flowing in the wires,*
b) *the power loss in the wires,*
c) *the voltage drop along the power line, and the voltage available to the lamp at the end of the line,*
d) *the power converted in the lamp.*

a) Using $I = V/R$ where V is the supply voltage and R is the total circuit resistance, since the lamp and wires are in series:

$$R = 4\,\Omega + 6\,\Omega = 10\,\Omega$$

the current in the wires and lamp is

$$I = \frac{12\,V}{10\,\Omega} = 1.2\,A$$

b) The power loss in the wires is I^2R, where R is the resistance of the wires:

$$P = (1.2\,A)^2 \times 4\,\Omega = 5.76\,W$$

c) The voltage drop is IR, where R is the resistance of the wires:

$$V = 1.2\,A \times 4\,\Omega = 4.8\,V$$

So the voltage available to the lamp is

$$12\,V - 4.8\,V = 7.2\,V$$

This can be checked by calculating $V = IR$ for the lamp.
d) The power converted in the lamp is IV:
$$P = 1.2\,A \times 7.2\,V = 8.64\,W$$

Answer: The power converted in the lamp is only 8.64 W, compared with the 24 W of power which would be converted if the same lamp was connected directly to the 12 V supply.

Figure 15.34 *The grid system*

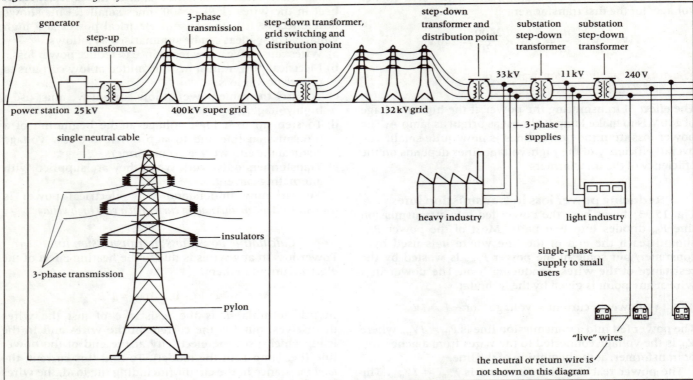

Transformers and the national grid system

The generators at a power station produce a three-phase alternating electric current at a voltage of 25 kV. A step-up transformer immediately increases the voltage to 400 kV for transmission over long distances in overhead cables, fig. 15.34.

The very high voltage keeps the current in the wires low but causes insulation problems. To prevent current leaking to the ground the cables hang on long insulators high above the ground from tall towers called **pylons**, fig. 15.34. Pylons carry cables in sets of three for the three phases of the alternating supply. The single extra cable which links the tops of the pylons is the neutral wire. This much thinner wire completes the circuit back to the generator for all three (or six) cables carried by the pylons.

The **super grid** is a network of transmission lines which link all the major power stations in the country with all the major users. Regional control and switching centres enable power to be sent where and when it is needed, and also allow some stations and lines to be shut down for maintenance work without cutting off the consumers.

A series of transformers and switching stations, known as **sub-stations** step the voltage down from the super grid in gradual steps as shown in fig. 15.34. Industrial consumers who require large amounts of energy receive a three-phase supply at 33 kV or 11 kV. Small users such as houses receive only a single-phase supply at 240 V. The load is spread over the three phases by connecting one house in three to each phase.

The advantage of three-phase transmission lines

We have seen that alternators are designed to generate three separate alternating voltages which are staggered in time to give a more even output of power, p 317. Another advantage of a three-phase supply is that transmission lines require only a single, relatively thin neutral wire to complete the circuit for all three phases. Because of the phase differences, the currents in the three live wires change direction at different times, so at any given time, two currents will be flowing in opposite directions. When the three currents are combined in the neutral wire the total current will be less than the current in each of the single live wires. This means that the neutral wire can be thinner than each of the three separate live wires. This considerably reduces the cost of a transmission line. The neutral wire can be seen linking the tops of the pylons.

When a single-phase alternating voltage is supplied to a house, two equally thick wires must be used for the live and neutral wires to complete the circuit.

Assignments

Find out where your local transformer sub-station is and at what voltage the power arrives there.

Describe how this transformer disposes of the heat produced in its coils and core.

Explain

a) Why is electrical power transmitted at a very high voltage?

b) Why is the current carried in the national grid alternating?

c) Why do pylons have three wires on each side and a single wire joining their tops?

Remember the formulas:

d) power loss in a wire = I^2R,

e) voltage drop along a wire = IR.

Try questions 15.12 to 15.13

15.6
ELECTRICITY AT HOME

Nowadays it is difficult to imagine what life would be like without electricity in our homes and we take its ready availability for granted. Since we need to use electricity safely we have to know about the design of the circuits and safety devices used in houses and how to fit fuses and plugs correctly. To use electricity economically we need to understand the power requirements of different appliances and how to calculate the cost of the electrical energy they use.

The mains supply to houses in Britain

To describe the electricity supplied to our homes we need to know that:

a) it is a single-phase alternating supply,

b) the frequency is 50 hertz, i.e. the alternating current flows in each direction 50 times every second,

c) the peak voltage is ±340 V,

d) the r.m.s. voltage is 240 V.

We have seen that alternating currents are essential to the national grid system, so that transformers can be used to step up and step down the voltage for transmission lines. There are some advantages and some disadvantages in having an alternating supply at home.

Advantages of an a.c. supply

The frequency of the supply is very precisely controlled, so we can use a.c. electric motors which synchronise their rotation to the mains frequency. This gives them a very accurate speed. Such motors are used in electric clocks, tape recorders and record players.

Another advantage of having a.c. at home is that the 240 V supply can be transformed down to a low, safe voltage, say 12 V, for operating toys and electric bells.

Disadvantages of an a.c. supply

Some appliances and electronic devices require a d.c. supply. Examples include electronic equipment such as radios and amplifiers and battery chargers. These devices are fitted with a circuit called a rectifier circuit which converts the alternating supply into a direct one. The action of rectifiers is described in chapter 16.

However, all electric heating appliances including kettles, fires and irons would work equally well on an a.c. or a d.c. supply.

Colour codes for wires in the home

There are two different colour codes in use for wiring in the home. The older British red – black – green code is now being replaced by the European standard brown – blue – green/yellow code. All new appliances are fitted with wires coloured in the European code. The codes are given in table 15.3.

All appliances need two wires to form a complete circuit from the supply through the appliance and back to the supply. These are the **live** and **neutral** wires. The live wire delivers the power at the high voltage which alternates between +340 V and −340 V peak values. The live wire is therefore the dangerous one capable of giving a serious electric shock. Switches in circuits must be fitted in the *live wire* so that switching off disconnects the high voltage from an appliance. Some appliances also have a

Table 15.3 Wire colour codes

Code	Live wire **L**	Neutral wire **N**	Earth wire **E**
wires on *new* appliances (new European standard)	brown	blue	green with yellow stripe
Wires on *old* appliances and permanent fixed cables (old British standard)	red	black	green

third wire known as the **earth wire**, because it is connected to the earth.

Connecting a cable to a three-pin plug

● *Remove the back of the plug and estimate the length of wire needed to reach to the large earth pin E, fig. 15.35.*
Remember that the outer sheath of the cable should be firmly held by the cable grip and care should be taken not to remove too much of the outer sheath.

● *Cut away the outer sheath being careful not to cut into the coloured insulation on the inner wires.*

● *Remove one to two centimetres of the coloured insulation from each of the three inner wires.*
Be careful not to cut off any of the fine strands of wire and leave enough insulation on each wire to protect it right up to the connector.

Figure 15.35 *13 A socket and plug*

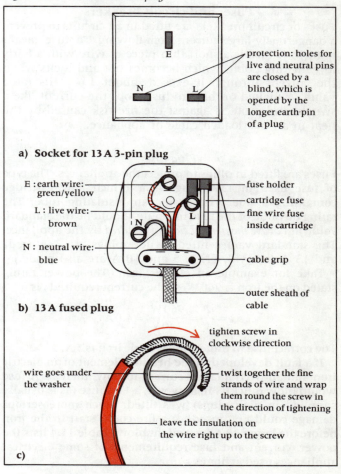

protection: holes for live and neutral pins are closed by a blind, which is opened by the longer earth pin of a plug

a) **Socket for 13 A 3-pin plug**

E : earth wire: green/yellow

L : live wire: brown

N : neutral wire: blue

fuse holder
cartridge fuse
fine wire fuse inside cartridge
cable grip
outer sheath of cable

b) **13 A fused plug**

tighten screw in clockwise direction

wire goes under the washer

twist together the fine strands of wire and wrap them round the screw in the direction of tightening

leave the insulation on the wire right up to the screw

c)

- *Twist the fine strands in each wire together so that no stray strands are left loose inside the plug.*
- *If the plug has wrap-round screw terminals, bend the end of each wire round in the direction of tightening of the screw, as shown in fig. 15.35c.*
- *Fit the wires to the correct terminals according to the colour code, table 15.3.*
- *Tighten the cable grip and test to see that the cable will not pull out.*
- *Check the value of the cartridge fuse and if necessary, replace it with one of the correct value, see table 15.4. Check that the cartridge is tightly held in its holder.*
- *Replace the back of the plug.*

Table 15.4 Examples of power and fuse requirements of some household appliances

Appliance	Power/W	Current/A	Fuse to fit in plug
3-bar electric fire or electric kettle	3 000	12.5	13 A
hairdrier, electric iron or vacuum cleaner	960	4.0	5 A
food mixer or electric drill	480	2.0	5 A or 3 A*
television, refrigerator or freezer	180	0.75	2 A
sewing machine or table lamp	60	0.25	2 A or 1 A* *if available

Fuses and circuit breakers

Fuses or circuit breakers are fitted in all circuits to prevent a dangerously large current from flowing. To 'fuse' means to melt. A fuse is a short thin piece of wire with a fairly low melting point, which becomes hot and melts when the current through it exceeds about 1.6 × its rated value. A melted or 'blown' fuse stops the current like a switch, and protects against the fire risk caused by the heat in an overloaded cable or appliance.

Plug fuses

Fuses are fitted in plugs to protect the appliances. The type of fuse used in 13 A plugs is a 1 inch long cartridge containing a fine wire inside an insulating tube. The rating or value of the fuse in a plug should be the standard value *just above the normal current required* by the appliance. The standard values fitted in new plugs are 3 A (red) and 13 A (brown). (1, 2, 5 and 10 A are also made.)

Take for example an electric iron. The power rating stated on the iron is 960 W, so the current required , is

$$I = \frac{P}{V} = \frac{960\,\text{W}}{240\,\text{V}} = 4.0\,\text{A}$$

The correct fuse to protect the electric iron is 5 A.

If a fault developed in the heating element of an electric iron and allowed a larger current to flow, then a 5 A fuse would blow quickly. If, however, a 13 A fuse (as is usually supplied in new plugs) was fitted, much more serious damage could be caused, or a fire could start in the iron before the larger 13 A fuse might blow. Table 15.4 lists the power, current and fuse requirements of some electrical appliances used at home.

Consumer units or fuse boxes

The fuse in a plug protects an appliance, and the fuses or circuit breakers in the fuse box or consumer unit (see fig. 15.37) protect the permanently wired circuits and the house from a fire risk. For example, the 13 A sockets in a house are all wired in parallel in a ring circuit, (see fig. 15.38). The maximum safe current for this circuit is 30 A, so it should be protected by a 30 A fuse. While there could be six or more sockets in a ring circuit, each rated at 13 A, this does not mean that six electric fires each requiring 12.5 A can be plugged in safely; 75 A (6 × 12.5 A) would very seriously overload the cable.

Consumer units or fuse boxes may be fitted with wire fuse carriers made of porcelain or plastic, cartridge fuse holders or miniature circuit breakers. Some examples of these are shown in the photograph. The most common of these is the porcelain or plastic carrier to which is fitted a length of fine fuse wire. Fuse wire is available in several thicknesses rated at 5, 10, 15 and 30 A to suit the various circuits to be protected. When replacing a fuse wire all traces of the old wire must be removed and the new wire wound round the screw terminals so that it is not under tension when the screws are tightened.

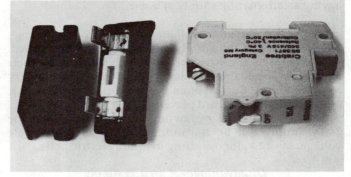

On the left is a 15 A plastic fuse holder and socket. The fine fuse wire is threaded through the porcelain shield and attached by screws to the brass connecting pins. On the right is a 15 A miniature circuit breaker (mcb) which does the same job, needs no fuse wire and is safer. The mcb switches itself off very quickly if the current exceeds 15 A.

Miniature circuit breakers (mcb's)

These are a modern alternative to fuses and have several advantages. They can be reset quickly by a switch or button as there is no melted wire to replace. More important, they can break an overloaded circuit in less than 0.01 s, which is much less time than it takes for a fuse wire to melt. In an mcb the current passes through an electromagnet which becomes more powerful as the current increases. Above the safe current value, the strength of the electromagnet is great enough to force apart some contacts which break the circuit.

Earth leakage circuit breakers (elcb's)

Some modern consumer units have a main switch which will switch off all the circuits in the house in a very short time (say, less than 25 ms) if an earth leakage current of more than about 25 mA occurs. Any current flowing to earth means that a fault has occurred. This includes the potentially lethal event of a current flowing through a person to the ground. When all is well in the household circuits, the current flowing in live and neutral wires will be the same. An elcb detects any imbalance between the currents in these two wires. The two currents are passed through the coils of two opposed electromagnets. Any imbalance of the two electromagnets releases a circuit-breaking contact. An elcb can stop an electric shock so quickly that the risks are greatly reduced. There is also a fire risk when currents flow through earth wires from faulty appliances, and an elcb will cut off the electricity supply when any leak of this sort occurs.

Another name for an elcb is a residual current device or RCD. Socket outlets and portable sockets on extension leads are available fitted with an RCD. These provide protection when we use a particular appliance such as a lawn mower or a hedge trimmer where there is higher than usual risk of electrocution.

The main fuse

The Electricity Board fits the main fuse in a house to protect their electricity meter and the whole house against a total overload. This cartridge fuse, which may be rated 50 A or more, is sealed by the Board and should be touched by no one but the Board's electrician.

Earth wires

The third, green and yellow (or all green) wire in a cable, called the earth wire, protects a person who may touch a faulty or live appliance. Faulty insulation or a loose wire may make an electrical connection between the metal case or an exposed metal part of an appliance and the mains live supply. When someone touches the live metal part, a circuit is made in which a current flows from the mains live supply, through the metal appliance and through the person, completing the circuit via the ground or earth, fig. 15.36.

Figure 15.36 *Protection by an earth wire*

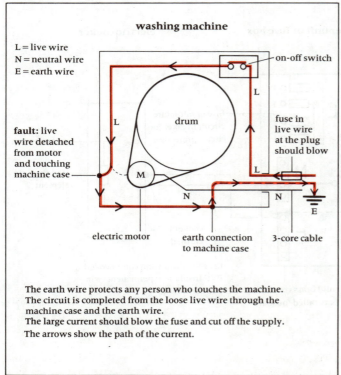

The earth wire protects any person who touches the machine.
The circuit is completed from the loose live wire through the machine case and the earth wire.
The large current should blow the fuse and cut off the supply.
The arrows show the path of the current.

The earth wire is connected to the metal case of an appliance. Earth wires must have a good low-resistance connection to earth, so that when a fault occurs and a current flows through the live wire and the earth wire in series, the fuse in the live wire will blow and cut off the supply. A high-resistance earth wire would not allow enough current to flow for the fuse to blow. In this way a low-resistance earth wire, and a fuse (of the correct value) in the *live* wire act together as a safety device.

Some earth wires are connected to the outer metal armour of the Electricity Board's incoming supply cable, which provides a good connection of low resistance with the ground. Other earth wires are connected to a cold-water pipe near the point where it comes out of the ground or to a long copper stake driven well into the ground.

It is dangerous to use an appliance which needs an earth wire on a 2-pin socket or lighting circuit. If an appliance is supplied with a 3-core cable, it *must* be fitted with a 3-pin plug and used on a power circuit which has an earth wire.

Double insulation

Double insulated domestic appliances are very common now. They should be marked with the symbol ▣ on their specification plate or label. Electric drills, vacuum cleaners, hair driers and food mixers are usually double insulated; so what does this in fact mean?

In practical terms it means that these appliances will not have an earth wire and only require twin cable. If an appliance is totally enclosed in an insulating plastic body or structure so that there is no direct connection between any external metal parts (screws, handles or attachments) and the internal electric components, then there is no risk of the user getting an electric shock in the event of a fault developing inside the appliance. For example, the metal whisks of a food mixer will be mounted in plastic insulating bushes. The attachments to a vacuum cleaner will all be fixed into plastic parts of the casing and any screws will not make contact with the electric motor inside. So, firstly the electric cable is insulated from the internal metal parts of the machine. Secondly the internal metal parts which could become live if a fault developed are also insulated from all the external metal parts which can be touched by the user. This is called double insulation.

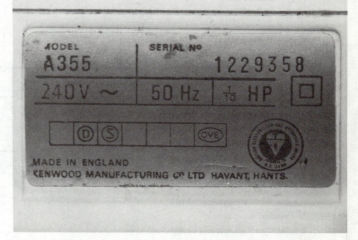

The double insulation symbol, a square inside a square, can be seen on this food mixer label.

Household wiring

Fig. 15.37 shows the arrangement of the wiring installations in a house which are, from left to right:

The incoming cable Power enters the house through an armoured cable which, if it comes underground, may provide an earth connection through its outer metal sheath.

The Electricity Board's main fuse.

The electricity meter. This records the electrical energy converted in the whole house in kilowatt hours (kWh). It may have either dials or a digital readout.

The consumer unit, or fuse box. This does several jobs:

a) It contains the main switch to switch off all circuits in the house. This is a double-pole switch which breaks both the live and neutral wires.

b) It distributes the current to several separate circuits, so it is a junction box.

c) It houses the fuses or mcb's which protect each separate circuit. See table 15.5 for fuse ratings.

The appliances. Only one circuit is shown in fig. 15.37, a cooker circuit. This has its own switch in the live wire and a load resistor which represents the heating elements which convert the electrical energy to heat energy. Each circuit is connected to a separate fuse, with a rating chosen to suit the circuit. The table in fig. 15.37 gives an example of the four circuits provided in a typical modern house. The higher rating fuses should be fitted next to the main switch. In some houses, extra fuses are fitted for extra ring-main circuits, e.g. for an electric shower or a circuit in the garage. When electricity is used as the main source of heating in a house, extra circuits are provided which are controlled by a time clock. The clock switches on the heating circuits at off-peak times (overnight for example)

Table 15.5

Fuse number	Fuse rating	Circuit protected
1	40 A	cooker
2	30 A	socket, ring main
3	15 or 30 A	immersion heater
4	5 or 10 A	lighting

when the electricity is supplied at a cheaper rate. These circuits have a separate meter for the cheaper rate of electricity.

The lighting circuits. Fig. 15.38 shows the important differences between the lighting and power socket circuits in a home. The current in a lighting circuit is quite low, only $\frac{1}{4}$ ampere for each 60 W lamp. So a single twin and earth cable can supply all the lights in a house. In some houses, one cable supplies the upstairs lights and another the downstairs ones. Each lamp is connected in parallel using a junction box at a convenient point along the cable. A separate spur cable supplies each lamp via a single-pole switch connected in the live wire.

The ring main for power sockets. Power sockets are connected along a closed loop or ring of cable which effectively doubles the thickness of wire used because current can reach any socket via two routes. This ring circuit reduces the gauge of wire which has to be used and reduces the risk of overloading the circuit when several sockets are in use. It should be understood that the sockets on a ring still provide parallel circuit connections to each appliance plugged in; the ring circuit is in no way a series circuit. There is no connection between the live and neutral wires except through an appliance.

Figure 15.37 *A house wiring diagram*

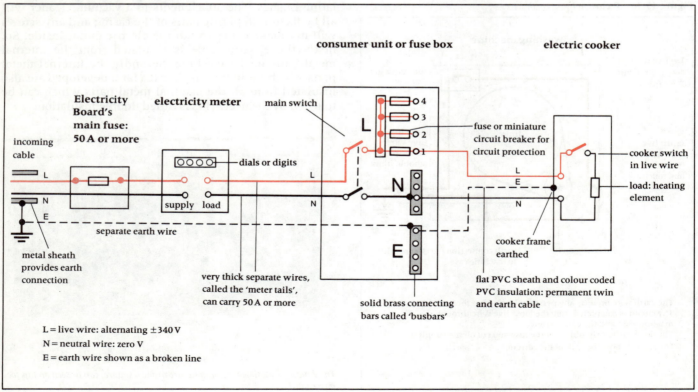

L = live wire: alternating ±340 V

N = neutral wire: zero V

E = earth wire shown as a broken line

Figure 15.38 *House lighting and power socket ring main circuits*

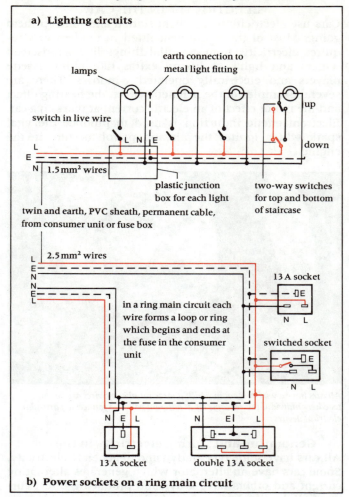

a) Lighting circuits

lamps

earth connection to metal light fitting

switch in live wire

up

down

two-way switches for top and bottom of staircase

plastic junction box for each light

1.5 mm² wires

twin and earth, PVC sheath, permanent cable, from consumer unit or fuse box

2.5 mm² wires

13 A socket

in a ring main circuit each wire forms a loop or ring which begins and ends at the fuse in the consumer unit

switched socket

13 A socket

double 13 A socket

b) Power sockets on a ring main circuit

The cost of electricity

We pay for the energy used or converted in our homes. The amount of energy converted depends on the power of the appliances and the time for which they are switched on:

$$\text{energy converted} = \text{power} \times \text{time}$$

The SI unit of energy, the joule (1 watt × 1 second) is too small for the large amounts of energy used in a modern home. The unit of energy for which we pay is the **kilowatt hour** (kWh):

$$1 \text{ kilowatt hour} = 1\,000 \text{ watts} \times 1 \text{ hour}$$
$$\therefore \quad 1 \text{ kWh} = 1\,000 \text{ watts} \times 3\,600 \text{ seconds}$$
$$= 3.6 \times 10^6 \text{ J} = 3.6 \text{ MJ}$$

Calculating the cost of electricity

The number of 'units' used or converted by an appliance is given by the relation:

$$\frac{\text{number of}}{\text{kWh units}} = \frac{\text{number of}}{\text{kilowatts}} \times \frac{\text{number of}}{\text{hours}}$$

The cost of the electricity used or converted by an appliance is given by the relation:

$$\text{cost} = \frac{\text{number of}}{\text{kWh units}} \times \frac{\text{price per}}{\text{kWh unit}}$$

Worked example
The cost of electricity

How many kWh *units of electrical energy will be used in a day by (a) a* 3 kW *electric fire and (b) a* 60 W *electric lamp? What will each cost if the price of a unit of electricity is* 5 p?

The number of units = kilowatts × hours
a) the fire will use:

$$3 \text{ kW} \times 24 \text{ h} = 72 \text{ kWh}$$

b) the lamp will use:

$$\frac{60 \text{ W}}{1\,000} \times 24 \text{ h} = 1.44 \text{ kWh}$$

The cost = kWh units × price per unit
a) The fire will cost: $72 \times 5 \text{ p} = 360 \text{ p}$,
b) The lamp will cost: $1.44 \times 5 \text{ p} = 7.2 \text{ p}$.

Answer: The electric fire will use 72 units of electricity costing 360 p compared with the lamp, which will only use 1.44 units in a full day, costing as little as 7.2 p.

Calculating an electricity bill

The electricity meter, like a car milometer, records continuously from when it is installed. To calculate the energy used over a certain period (usually 3 months) the reading at the beginning of the period must be subtracted from the reading at the end. Fig. 15.39a shows two readings taken 3 months apart on a digital meter.

$$\text{The number of units used} = 3121 - 2481 = 640 \text{ kWh}$$

(Note that the last figure sometimes gives tenths of a unit and should be ignored.)

Figure 15.39 *Electricity meter readings*

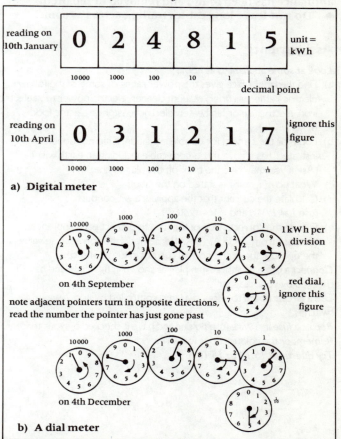

reading on 10th January

0 2 4 8 1 5

unit = kWh

10000 1000 100 10 1 ⅒

decimal point

reading on 10th April

0 3 1 2 1 7

ignore this figure

10000 1000 100 10 1 ⅒

a) Digital meter

1000 100 10
10000
1 kWh per division

on 4th September

red dial, ignore this figure

note adjacent pointers turn in opposite directions, read the number the pointer has just gone past

1000 100 10
10000
1

on 4th December

b) A dial meter

Reading a meter with dials needs more care than a digital meter, because the pointers of adjacent dials turn in opposite directions, fig. 15.39b. The number to read is the one which the pointer has just gone past.

Try to read the dials in the figure and then check your readings with these:

    reading on December 4th: 07928
    reading on September 4th: 07657
    number of units used =   271 kWh.

Some do's and don'ts with electricity

- **Do** switch off and disconnect appliances when not in use, and *always* before starting any repair work.
- **Do** learn how to fit plugs correctly and safely.
- **Do** fit the correct value fuse in 13 A plugs to suit the appliances.
- **Do not** expose any wiring unless it is disconnected or unplugged.
- **Do** replace fuse wire with the correct rating wire, *never* with ordinary wire or things like paper clips.
- **Do not** fit plugs which are damaged, have no cable grip or have a loose fuse.
- **Do not** overload circuits and sockets with too many appliances plugged into multiway adaptors.
- **Do not** take mains appliances, for example hairdriers, into bathrooms where holding with wet hands can be dangerous.
- **Do not** use appliances requiring an earth lead on a 2-pin socket or lighting circuit.
- **Do not** replace a fuse until the fault in the circuit or appliance has been found and removed or repaired.
- **Do not** leave long cables trailing across a room.

Assignments

Look at some appliances at home

a) Find the label which gives the power rating of some of the following: electric kettle, hairdrier, vacuum cleaner, electric power tool, electric iron, electric fire or fan heater, electric toaster or electric food mixer.

b) Look for the symbol for double insulation. If it is not double-insulated, has it got a three-core cable including an earth wire? (Ask if you may take the back off the plug to check.)

c) What power rating is stated on the label?

d) Calculate the current that the appliance will conduct. (use $I = P/V$ and $V = 240 \text{V}$.)

e) Decide the correct value of the fuse to protect each appliance. (Choose from 2 A, 5 A, 13 A.) Remember: the correct fuse value is the one just above the normal current in the appliance.

Connect a cable to a three-pin plug as shown in fig. 15.36 and have it checked by your teacher.

Use a circuit board to wire up a lamp controlled by two 2-way switches.

Repair a fuse in a wire fuse holder and have it checked by your teacher.

Remember the colour codes for wires.

Try questions 15.14 to 15.18

15.7
ELECTRICITY IN THE CAR

Cars use electricity to get them started and to keep them going. Most of the equipment fitted in modern cars requires electricity, from essential things like windscreen wipers and lights to luxury extras like radio-cassette players and electrically operated windows. There are several examples of the magnetic effect, the heating effect and the motor effect of an electric current at work in a car. Electromagnetic induction is used to produce the vital spark which ignites the air and petrol mixture in the engine.

All cars have a wiring loom in which a large number of wires are bound together. Individual wires come off at different points to connect to particular electrical devices.

Generating and storing electricity in the car

All cars have some type of dynamo to generate electricity. Some cars have an alternator which generates alternating current and others have a dynamo producing direct current. All the circuits in a car (which include several electric motors) must be able to work on the direct current supplied by the car battery. So the output of an alternator is also converted to d.c. before it is supplied to the car circuits.

The dynamo or alternator is usually driven by a belt from a pulley on the car engine. So while a car is moving and its engine is turning, electricity is always being generated. When the engine stops the dynamo also stops.

The car battery is used both to start the engine and to top up the supply when the dynamo is not producing enough electricity. While a car is being driven the dynamo usually generates more electricity than the car needs and then the 'spare' electricity is used to recharge the battery. A simplified circuit diagram for a car is shown in fig. 15.40.

The dynamo (Ⓖ for generator) and battery are connected in parallel so that either of them can supply a direct current I to the car circuits. However, when the dynamo is producing 'spare' electricity this flows in the reverse direction back through the battery, recharging it. Some cars are fitted with a centre-zero ammeter reading from -50 A to $+50$ A. When connected in the position shown in the diagram, it indicates a positive current when the battery is being charged and a negative current when the battery is discharging (i.e. supplying the current to circuits in the car). As the battery plays such a central role in the working of a car we shall look at it in more detail.

Figure 15.40 *A simplified car wiring diagram*

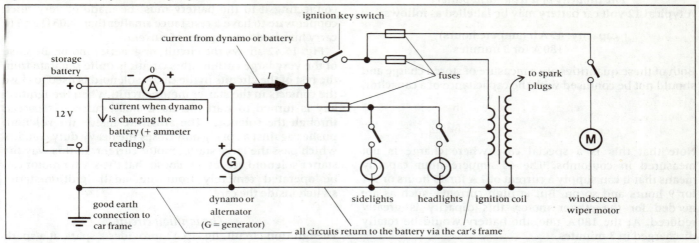

The car battery

Despite the need for regular attention, topping up and charging, and despite their size and weight, lead–acid accumulators are still the most suitable battery for use in a car. A car has many electrical needs which a lead–acid battery can satisfy. It can supply a very large current for a short time or a small current for a long time. It has a very steady e.m.f. and can be recharged many times. These advantages mean that this type of battery will probably be in use for many years to come.

A car battery usually has six 2 V cells connected in series, which gives it a steady voltage of about 12 V. The arrangement of cells is shown in fig. 15.41. It is designed to be able to supply a very large current for a short time, to operate the starter motor. To allow a current of several 100 amperes to flow, the internal resistance of the whole battery must be very low, less than 0.01 ohms. To achieve this the cells must have large-area plates and very small gaps between the plates, so that the electrolyte forms a very narrow conductor of large cross-sectional area (see the section on resistivity, chapter 12). The strong acid solution also helps to make the electrolyte a good conductor by providing a high concentration of ions. Because the plates are close together, separators are needed between them.

When fully charged, the battery voltage will be about 13 volts (6 × 2.2 V), but it is constant at about 12 volts for most of its discharge life. When the voltage falls to about 11 volts (6 × 1.8 V) it is ready for recharging.

Figure 15.41 *Six cells connected in series in a car battery*

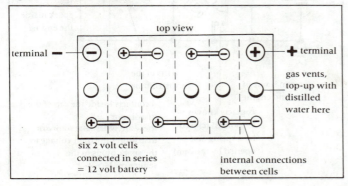

This cut-away picture of a car battery shows how the whole of the space inside is filled with thin flat lead plates of large surface area. These plates are held apart by thin insulating separators and are totally immersed in sulphuric acid.

Care of lead–acid batteries

a) Lead–acid cells must be recharged as soon as their e.m.f. falls to 1.8 V. Leaving them discharged allows the lead sulphate coating to change to a crystalline form which cannot be reconverted to lead and lead(iv) oxide, thus destroying the cells.

b) During charging, and particularly when fully charged, bubbles of hydrogen gas are released from the negative plates. Since hydrogen and the oxygen in the air combine explosively, no naked flame or spark must come near a cell on charge.

c) The state of charge of a cell can be checked using a hydrometer to measure its relative density (see p 102), but we must take care to avoid spilling the acid.

d) Water evaporates from the cells and should be replaced by topping up, using only distilled water.

e) A short circuit across one of these batteries can damage its plates because its very low internal resistance will allow a large current of several 100 amperes to flow. The plates may become bent and the active coating may flake off. This is why excessive use of the starter motor on a car will shorten the life of the battery. Starter motors have very low resistances, almost equivalent to a short circuit and conduct very large currents.

The capacity of a lead–acid battery

A typical 12 volt car battery may be labelled as follows:

capacity: 45 A h (ampere hours)
180 A for 3 minutes

Both of these quantities are a measure of stored charge and should not be confused with the capacitance of a capacitor.

charge = current × time

battery capacity = amperes × hours

Note that this is a special case where charge is not measured in coulombs. The 45 ampere hour capacity means that it can supply a current of 1 A for 45 hours or 5 A for 9 hours and so on. But at high currents, such as are needed for the starter motor, this capacity is greatly reduced. At the 180 A rate, the battery would be totally discharged in 3 minutes.

$$\text{capacity} = \text{amperes} \times \text{hours}$$
$$= 180 \times \frac{3}{60} = 9 \text{ A h}$$

When buying a battery for a car, look at the second, high-current figure which indicates how good it will be at starting the car. In general the higher this figure is, the better the battery.

The starter motor

The starter motor has the difficult job of turning a cold stiff engine when the lubricating oil is at its thickest. To provide the necessary power from a 12 volt battery requires an enormous current. This can be as large as 400 A.

The *maximum power* of a starter motor can be calculated from the formula: power = current × voltage. So, as an example, if a 12 V battery supplies a current of 400 A, the power supplied is given by:

$$P = IV = 400 \text{ A} \times 12 \text{ V} = 4800 \text{ W} \quad \text{or} \quad \text{about 5 kW}$$

Some of this power is wasted as the large current produces heat in the wires.

The *maximum resistance* of the starter motor, the battery and the connecting wires is given by:

$$R = \frac{V}{I} = \frac{12 \text{ V}}{400 \text{ A}} = 0.03 \, \Omega$$

Figure 15.42 *A starter motor circuit*

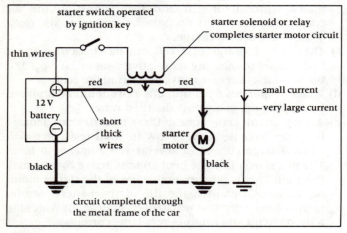

It is clear that the wires inside the motor itself and those connecting it to the battery must be made of very thick copper wire to have a resistance smaller than 0.03 Ω and to carry hundreds of amperes of current.

Fig. 15.42 shows the circuit for a starter motor. Because of the very large current, this circuit is quite separate from the rest of the circuits in the car. A solenoid is used to close the contacts in the starter motor circuit. When the ignition key is turned to start the car a small current is passed through the solenoid. The armature inside the solenoid pushes against a spring and closes the heavy duty contacts which pass the large starter motor current. In this way the starter solenoid acts as a relay so that the starter motor can be operated remotely from the small ignition-starter switch inside the car.

The ignition circuit

The ignition circuit, fig. 15.43, provides a spark at a spark plug inside the engine to ignite the air and petrol vapour mixture. A voltage of several thousand volts is needed to give a satisfactory spark and this has to be produced from a 12 V battery or dynamo.

An induction coil, known as the **ignition coil**, is used to produce the high voltage for the sparks. The primary circuit current in the ignition coil is supplied by the 12 V battery or dynamo. This current is passed through the **distributor** (a device attached to and driven by the engine), where it is switched on and off continuously. The mechanism which makes and breaks this primary circuit is known as **the points**, which are a pair of contacts held closed by a spring and opened by a cam every time a spark is required at one of the spark plugs. The cam, fixed like a collar to the distributor's rotating shaft, has four rounded and protruding corners which in turn push aside one of the contacts to produce a gap, thus breaking the primary circuit. The resulting rapid fall of current in the primary circuit as the points open, and the large number of turns on the secondary coil, together induce a very high voltage

Figure 15.43 *The ignition circuit of a car*

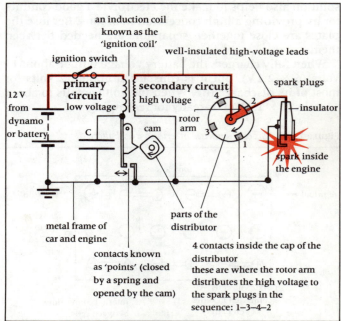

for a brief moment across the secondary of the ignition coil. This high voltage is fed back to the distributor by a well-insulated cable.

A rotating arm or **rotor arm**, fixed on the end of the distributor's shaft, distributes the high voltage to each of the spark plugs in turn. A spark plug is fitted to each cylinder of the engine. (A cylinder is a chamber in which a compressed petrol–air mixture is ignited by the high-voltage spark of the spark plug. The exploding vapour forces a piston to move out of the cylinder and to turn the main shaft of the engine in a manner similar to the way the push of your leg turns the pedals of a bicycle.)

The distributor has to be accurately timed so that the cam opens the points, and the rotor arm distributes the high voltage to a particular spark plug, just at the moment when the air and petrol vapour mixture is compressed and ready for ignition. All this happens very quickly. The points will open and close about 40 times a second when an engine is idling, rising to over 400 times a second when a car is driven at speed.

The capacitor C helps to reduce the spark which occurs across the points as they open. This increases the induced voltage across the secondary coil and reduces the damage done to the points by the spark.

Safety in the car

Many of the electrical devices in a car make it safer to drive. First the car itself must be protected from faults in the electrical circuits.

The fuses

All circuits, except the starter motor's high-current circuit, are protected by a fuse. A typical fuse value is 30 A. The high-current values of the fuses are necessary because the circuits operate on only 12 volts. For example, a single headlamp requires a current of over 4 A to produce 50 W of power.

The fuses are very necessary because of the fire risk in a car. A worn or loose wire may cause a short circuit. The heating effect of the large currents that can be supplied by a car battery can easily set fire to a wire when it is short circuited. Wires can be cut and trapped in accidents and with spilt petrol around there is then a high risk of fire. An excess current in any circuit with a fault should blow the fuse and break the circuit, protecting the wiring, the car and the passengers. The car battery should be disconnected when working on a car to reduce the risk of accidental short circuits, which can be caused by metal tools and disconnected wires. When a car is involved in an accident, switch off the ignition switch.

Safety devices

Many devices in the car give warnings or increase safety. These include lights, direction indicators, horn, windscreen wipers, heated rear windows, temperature and fuel gauges and so on. Some examples of how these devices use electricity are now briefly described.

The lights and heated rear window use the heating effect of an electric current.

The flashing indicators in older cars use a bimetallic strip to switch the lights on and off. In modern cars an electronic circuit operates the flashers.

The windscreen wipers use an electric motor.

The horn uses the magnetic effect of an electric current.

The fuel gauge, which indicates the petrol level in the tank, uses a float to move a sliding contact on a rheostat. This varies the current through an ammeter labelled to read the fuel level.

A temperature gauge uses another ammeter labelled with a temperature scale. The ammeter measures the current flowing through a resistor whose resistance changes with temperature.

The speedometer

The speedometer uses a simple induction motor effect, fig. 15.44. A permanent magnet is rotated by a cable driven from the gearbox of a car. This magnet is close to an aluminium disc but has no direct connection with it. The rotating magnet induces eddy currents in the aluminium disc. These induced currents cause the aluminium disc to turn to attempt to reduce the relative motion between the magnet and the disc. The aluminium disc is, however, controlled by a spring which limits how far it can turn. The faster the magnet rotates the larger the induced eddy currents and the further the aluminium disc turns against the controlling spring. A pointer attached to the disc reads the speed of the car on a scale calibrated in miles per hour or kilometres per hour.

Figure 15.44 *A car speedometer*

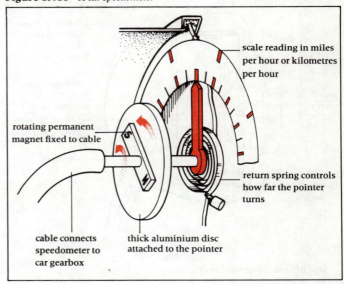

scale reading in miles per hour or kilometres per hour

rotating permanent magnet fixed to cable

return spring controls how far the pointer turns

cable connects speedometer to car gearbox

thick aluminium disc attached to the pointer

Assignment

Make a list of all the electrical devices that are used in a car.

Find out on which effect of an electric current their action depends.

Find out, where possible, their power and current requirements. (A car hand-book will tell you the power of the lamps in the list of replacements.)

Record your findings in a table: like table 15.6.

Try question 15.19

Table 15.6

Device	Action depends on	Power/W	Current/A
starter motor	motor effect	up to 5000	up to 400
brake warning lights	heating effect	21 each	1.75 each

15.8
ELECTRICITY AND THE TELEPHONE

In Britain in 1980 there was about one telephone for every two people and the total number continues to grow every year. Not only is the number of telephones growing but also the range of equipment and facilities available to telephone users is expanding rapidly. However the basic telephone hand-set including the microphone and the earpiece has not changed much for a very long time.

The transmitter, a carbon–granule microphone

Most telephones use a carbon–granule type of microphone. The original patent for a carbon–granule microphone was taken out in 1878 and the design has changed only slightly in over 100 years. A diagram and transmitting circuit of the type currently in use is shown in fig. 15.45.

Sound waves cause a light aluminium diaphragm to vibrate in and out at its centre at the same frequency as the sound waves. The centre of the diaphragm is attached to a hemispherical electrode which protrudes into a shallow chamber full of carbon granules.

Figure 15.45 *A carbon-granule microphone*

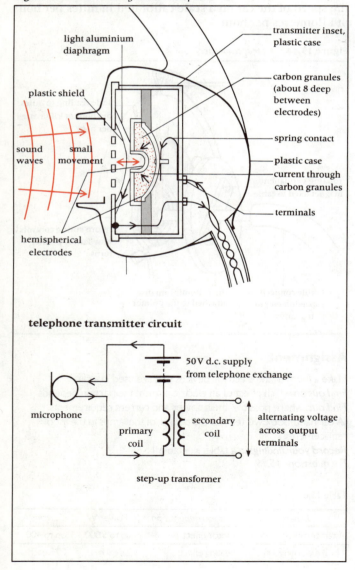

When a telephone is in use, a direct current flows from the telephone exchange to the microphone in the hand-set. This current passes through the carbon granules which fill the space between the two hemispherical electrodes. When a sound moves the diaphragm and attached electrode, the resistance of the contact between the loose granules of carbon changes with the variation of pressure between them. As the resistance of the carbon granules varies, the direct current also varies inversely.

A transformer primary coil in the microphone circuit conducts the direct current, but only variations in this current (caused by sounds) produce an induced voltage in the secondary coil. The transformer has more turns on its secondary coil and steps up the voltage variations. Thus an amplified alternating voltage, which represents the sound vibrations, is transmitted along the telephone wires to the local exchange. There it is redirected to the receiver of another telephone.

Features of the carbon–granule microphone

The carbon–granule microphone has several advantages, which is why it has been in service for so long:
a) It has a simple and robust design.
b) It is cheap.
c) It does not require an amplifier in the telephone. It is the only microphone which produces a large enough alternating voltage to drive an earpiece in another telephone on a local call without any amplification.

However, there are some drawbacks.
d) It has a poor frequency range, i.e. it does not respond to the full range of sound frequencies heard by the human ear.
e) It generates noise, a type of hiss known as *white noise*. This is caused by random movement of the carbon granules.
f) It is less sensitive and gives a lower output voltage on long telephone lines where the resistance of the long lengths of wire reduces the direct current through the microphone.

Replacement microphones of different design are now being considered and tested by telephone engineers.

The telephone earpiece

The telephone earpiece uses a rocking armature type of receiver, fig. 15.46a. Its movement depends on the magnetic effect of an electric current. This type of receiver has a high sensitivity, which makes it suitable for use with the carbon–granule microphone without the need for an amplifier. It is also very reliable, robust and relatively cheap compared with other types.

A flexible non-magnetic diaphragm is attached to one end of a soft-iron armature. A permanent magnet induces magnetism in the core of the electromagnet and in the iron armature. Equal attraction forces hold the armature in a balanced position on its pivot, fig. 15.46b.

When the alternating current, known as the **signal**, arrives in the coils of the electromagnet (the signal coils), the strength of the two poles of the electromagnet vary at the same frequency as the signal. At the instant when one pole is stronger and the other is weaker, the balance of the forces on the armature is upset and the armature rocks in sympathy with the variations of the current in the coils. By this means the diaphragm is set vibrating at the same frequency as the signal. The small in and out movements of the diaphragm produce sound waves in the air which are heard by the listener.

Telephone transmission lines

Telephone cables or transmission lines now carry communications of three main types, voice, data and visual telecommunications (e.g. TV pictures). The rapidly expanding demand for telephone lines has led to many new developments which are aimed at carrying more communications with improved reliability. These include the following:

a) Many conversations can now be carried along the same wire.
b) Ocean cables and satellites are increasing the telephone links around the world.
c) Integrated-circuit electronic systems are replacing the relatively unreliable electromagnetic relays in telephone exchanges.
d) Glass fibres, carrying messages as pulses of light, are replacing copper wires for long distance lines. The glass fibre transmission lines will be cheaper and can carry more communications than a much thicker and heavier multi-stranded copper cable.
e) Signals are increasingly transmitted along lines in digital form rather than analogue form.

The future for telephone systems is an ever-expanding field. Equipment will become more sophisticated with more facilities available to everyone. For example, conversations between three or more people in different places will become a normal event. Telephones will be used to make direct links with computers and data stores. Multi-channel television will be available, transmitted underground along glass fibre telephone lines.

Assignments

Find out

a) Where is your local telephone exchange?
b) Does it use electromechanical relays for switching, or is it a modern electronic exchange?
c) How many telephone lines are connected to your local exchange?
d) How many separate conversations can be carried simultaneously by the same telephone wire?
e) At what speed do telephone signals travel along wires and how long does it take for a signal to reach somewhere like Australia from the UK by satellite, or by land and undersea cable?

Figure 15.46 *An earpiece of a telephone receiver*

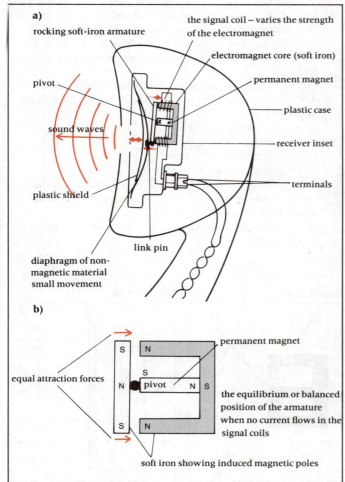

An engineer is seen here working on a British Telecom system X digital subscriber switching system. System X uses digital electronics and microelectronic circuits to carry out all switching. All information from speech to computer data is handled in the form of short electrical pulses. This system can provide automatic alarm calls, can divert calls to other numbers, can bar unwanted calls and can also enable three-way calls to be set up.

Questions 15

1 a) Describe how you would use a solenoid (i) to magnetise a steel bar so that the north-seeking pole is at a marked end of the bar, (ii) to demagnetise a bar magnet. Draw a circuit diagram in each case and say what type of power supplies you would use.

  b) Use the domain theory of ferromagnetism to explain the magnetic process in both the experiments you have described. (JMB)

2 a) What is meant by a *strong* magnet?

  b) Describe, with the aid of a diagram, how you would use a solenoid in an attempt to increase the strength of a magnet. Mark clearly on your diagram the direction of the current and the N pole of the magnet.

  c) Fig. 15.47 illustrates an electric bell operated by a battery. Describe

   i) the magnetic properties of soft iron which make it suitable for the core of the electromagnet and for the armature A.

   ii) the function of the steel strip S.

  Explain why the bell works when connected to a low-voltage a.c. supply instead of the battery. (C)

Figure 15.47

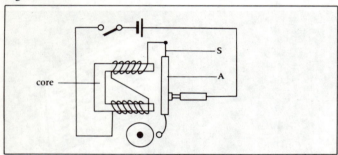

3 Fig. 15.48 shows part of a tape recorder. The tape has a magnetic material coated on one surface. The 'recording head' can magnetise this material.

  a) Suggest a material which would be suitable for part A. b) Why you think the material you suggest would be suitable? c) Should the magnetic material be on side X or Y of the tape?

  d) Suggest how you could magnetise the tape more strongly as it passes over the recording head.

  e) Explain your answer to part (d). (NEA SPEC B)

4 *Read the passage carefully before answering the questions that follow.*

  A reed switch is a switch operated by a magnetic field.

  The switch consists of two metal contacts (called reeds) inside a glass envelope which is filled with an inactive gas to prevent corrosion of the contacts.

  The reeds are made of a metal which contains iron and when a magnet is brought near, the reeds become magnetised, attract one another, making contact to complete the circuit in which the switch is placed. When the magnet is removed the magnetism in the reeds disappears and the springiness of the metal pulls the contacts apart breaking the circuit.

  A reed relay is a reed switch with a coil of wire wrapped around the glass envelope containing the switch. When a current passes through the coil, the coil behaves as a magnet and its magnetic field causes the reed switch to close. The coil circuit is separate from the circuit containing the reed switch and for this reason relays play an important part in electronic switching systems.

  a) Why is the glass envelope filled with an inactive gas?

  b) Explain why the magnetism in the reeds disappears when the magnetic field is removed.

  c) What advantage does a reed relay have over a reed switch?

  d) Give one practical or industrial use of reed relays. (W 88)

5 Fig. 15.49 shows a moving-coil loudspeaker. Copy the shaded part and indicate on your sketch the positon and nature of any magnetic poles. Why is the area of the cone large?

  Describe and explain what happens to the coil and what is heard when

  a) a dry cell is attached to the terminals of the loudspeaker and the current is switched on and then off.

  b) a 50 Hz alternating voltage is applied to the terminals.

   When the loudspeaker is emitting a sound how does the movement of the coil alter if the sound becomes (c) louder, (d) of higher pitch? (L, part)

6 Fig. 15.50 shows a rectangular plane coil ABCD of several turns of wire located in the magnetic field due to the two pole-pieces N and S. The flux lines are parallel to each other and evenly spaced in the region of the coil. The coil is free to rotate on the vertical axle XY.

  When a current is passed through the coil in the direction ABCD the coil starts to turn, and eventually comes to rest. With the aid of diagrams, show:

  a) why the coil begins to turn;

  b) in which direction it begins to turn;

  c) why eventually it comes to rest;

  d) the position in which it comes to rest.

  The effect described is the basis of the moving coil meter and of the d.c. motor. With the aid of diagrams, explain what essential modification is required to produce:

  e) a simple meter (assuming a pointer has already been attached);

  f) a d.c. motor. (O & C)

Figure 15.48

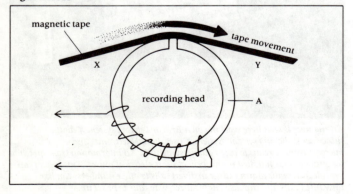

Figure 15.49

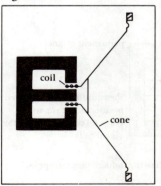

Figure 15.50

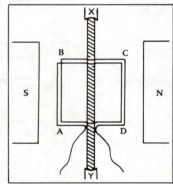

7 Fig. 15.51 shows the structure of a bicycle dynamo.
   a) As the wheels turn the axle, which other part of the dynamo rotates?
   b) What happens in the soft-iron core when the wheels turn?
   c) Where will there be an induced voltage?
   d) How could a larger voltage be produced?
   e) What type of current is produced by this dynamo? Why?
   f) Explain why no current is induced when the bicycle stops.

Figure 15.51

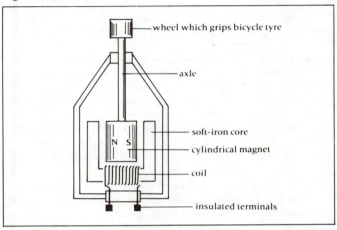

8 a) Fig. 15.52 shows a simple generator. Explain why an e.m.f. is produced between the ends of the coil when it is rotated.
   Answer (i), (ii) and (iii) by drawing three sketch graphs, one below the other on a sheet of graph paper.
   i) Draw a sketch graph showing how the e.m.f. between the ends of the coil varies with time over at least one revolution of the coil. Relate the positions of the coil to the values shown on your graph.
   ii) Draw a sketch graph showing what you would expect if the speed of rotation of the coil were doubled.

Figure 15.52

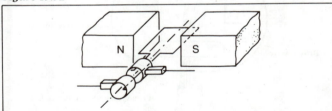

   iii) Draw a sketch graph showing what you would expect if in addition to rotating at twice the speed, the coil contained twice as many turns.
   The output from the generator is found to be unsuitable for charging a car battery. Why is this? What modification to the generator would be necessary to enable this to be done?
   b) A power station generator produces an e.m.f. of 33 000 V at a frequency 50 Hz. The domestic supply is approximately 250 V, 50 Hz. Explain how the output of the power station can be modified for use in the home.   (L)

9 a) What are the advantages of transmitting power at
      i) very high voltage? ii) alternating voltage?
   b) A 6 V, 24 W lamp shines at full brightness when it is connected to the output of a mains transformer, fig. 15.53.
      Assuming the transformer is 100% efficient, calculate
      i) the number of turns in the secondary coil if the lamp is to

work at its normal brightness,
      ii) the current which flows in the mains cables.
   c) Explain whether, and how, the number of secondary turns of the transformer should be altered if
      i) two 6 V lamps in **series** are to work at normal brightness,
      ii) two 6 V lamps in **parallel** are to work at normal brightness.   (W88 part)

Figure 15.53

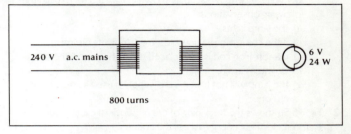

Figure 15.54

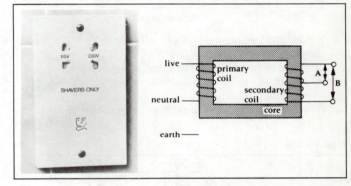

10 A bathroom razor socket contains a transformer. This is to isolate the user from the mains supply, fig. 15.54.
   a) On the right-hand diagram,
      i) add an S to show the correct position for a Switch.
      ii) add an F to show the correct position for a Fuse.
      iii) connect the earth lead to the correct point on the transformer.
   b) The transformer has two outputs, labelled A and B on the right-hand diagram. Complete the table below to show the number of turns and output voltage for each coil.

input voltage	primary turns	secondary turns	A or B	output voltage
230 V	5000	2500		
230 V	5000	5000		(LEAG 88)

11 Fig. 15.55a represents a transformer with a primary coil of 400 turns and a secondary coil of 200 turns.

Figure 15.55

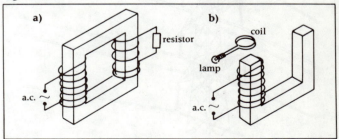

a) If the primary coil is connected to the 240 V a.c. mains what will be the secondary voltage?

b) Explain carefully how the transformer works.

c) Calculate the efficiency of the transformer if the primary current is 3 A and the secondary current 5 A.

d) Give reasons why you would expect this efficiency to be less than 100%.

e) The secondary coil is removed and a small coil connected to a low voltage lamp is placed as shown in fig. 15.54b. Explain the following observations:

   i) the lamp lights,

   ii) if the coil is moved upwards, the lamp gets dimmer,

   iii) if a soft-iron rod is now placed through the coil, the lamp brightens again,

   iv) the lamp will not light if a d.c. supply is used instead of an a.c. one. (L)

12 Fig. 15.56 shows the beginning of the network of cables used to transmit electrical energy across the country.

a) What is the name off this network of cables?

b) The turns ratio of the transformer is:
$$\frac{\text{number of secondary turns}}{\text{number of primary turns}} = \frac{11}{1}$$

   Calculate the output voltage of the transformer.

c) Why is it called a step-up transformer?

d) Why is electrical energy transmitted at high voltage?

e) Why could it be dangerous to fly a kite near to high voltage overhead cables?

f) Why is it safe for a bird to perch on a high voltage overhead cable: (MEG 88)

Figure 15.56

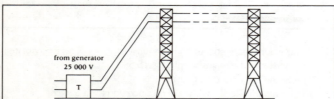

13 Fig. 15.57 shows overhead power lines transmitting electricity at 400 kV.

a) How is the electricity prevented from flowing to earth through the metal pylon?

b) Some people think that underground cables would be better. What are the advantages and disadvantages of underground and overhead cables?

Figure 15.57

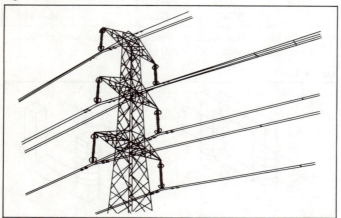

c) The transmission lines carry 200 MW.

   i) Calculate the current in the lines.

   ii) If the total resistance of the transmission lines is 10 Ω, calculate the voltage drop when the lines carry 200 MW.

   iii) Calculate the power loss in the lines.

   iv) What happens to this 'lost' power?

d) Explain why electricity is transmitted at high voltages like 400 kV rather than low voltages like 240 V. (SEG [ALT] 88)

14 a) The following information appears on the rating plate of an electric fire.

> MODEL KG–65
> Electrical Supply   220 V
> Maximum Power   2200 W
> WARNING:
> This appliance MUST be earthed.

   i) Calculate the current which will flow through the element of the fire in normal use.

   ii) How much electrical energy, in Joules, would this fire use in 5 hours at its maximum power?

   iii) If electrical energy costs 7 pence per kWh what is the cost of running this fire for 5 hours at its maximum power?

   iv) Name one form of energy, other than heat, produced by a red-hot electric element.

b) The most common cartridge fuses are 1 A, 3 A, 5 A and 13 A. Which one of these fuses is most appropriate for use with the electrical fire in part (a)? Explain your answer. (NISEC 88 part)

15 Fig. 15.58 shows a plug wired up by a fifth-year pupil and connected to a hair dryer rated at 1000 W, 240 V a.c.

a) Several mistakes have been made. Name any four of them.

b) The earth pin on a plug is longer than the other two pins. Explain why. (NEA [B] 88)

Figure 15.58

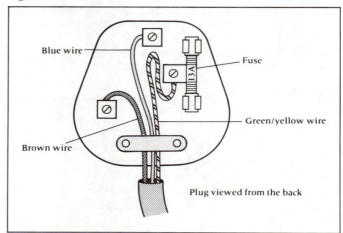

Figure 15.59

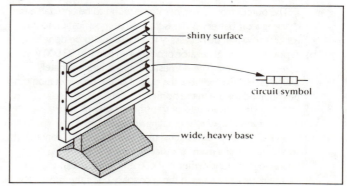

16 Fig. 15.59 shows a radiant electric fire for use on a 240 V mains supply. It has four heating elements, each rated at 0.75 kW.
   a) Two important features of the fire are the shiny surface and the wide, heavy base. Explain the purpose of these.
   b) The four heating elements are controlled by three switches. One of these controls the middle **two** elements, the others control **one** element each.
      i) Redraw and complete the circuit diagram in fig. 15.60 to show how this can be done.
      ii) What is the power rating of the fire?
      iii) The electricity board charges 7p for each kilowatt hour of energy. What is the cost of using the fire from 6 p.m. to 11 p.m.?
      iv) When used on the 240 V mains supply, what is the current drawn by the fire?
      v) What value fuse should be fitted to the plug?   (LEAG 88)

Figure 15.60

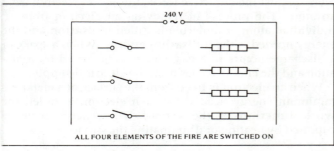

240 V

ALL FOUR ELEMENTS OF THE FIRE ARE SWITCHED ON

17 Electrical energy supplied to your home is measured in kilowatt hours.
   a) What is a kilowatt hour?

Figure 15.61

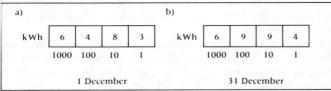

a) kWh	6	4	8	3	b) kWh	6	9	9	4
	1000	100	10	1		1000	100	10	1
	1 December					31 December			

   b) Fig. 15.61a and b shows the readings on an electricity meter at the start and end of December. How many kWh have been used during the month of December?
   c) Below is a list of appliances found in many homes, with their power ratings.

Appliance	Power rating	Time used (hours)	kWh
Television	200 W		
Electric kettle	2 kW		
Immersion heater	3 kW		

A student arrives home in the evening, switches on the television at 7 p.m. and leaves it on until midnight. She makes three pots of tea using the kettle for five minutes on each occasion. She puts on the immersion heater for two hours, has a bath and goes to bed.
   i) Complete the table to show for each appliance
      A) the time for which it was used,
      B) the number of kWh used.
   ii) What will be the total cost if 1 kWh costs 6p?   (NEA [B] 88)

18 Fig. 15.62 shows an *incorrect* attempt to wire three sockets A, B and C to the mains supply. When a mains electric heater is plugged into any one of the three sockets, no current flows in the circuit. When similar mains heaters are plugged into each of the sockets simultaneously, a current flows in the circuit but the heaters give out much less heat than they were designed to do. Explain these observations.

Figure 15.62

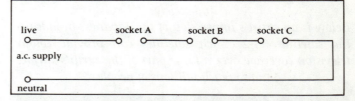

Draw a circuit diagram to show the three sockets correctly wired to mains supply so that the three heaters can operate normally. Include a fuse, and an earth wire, in your circuit diagram.
   Explain clearly why a fuse and an earth wire are used in a mains wiring circuit.   (C part)

19 Fig. 15.63 shows the lighting circuits of a car.
   i) If a break occurred in the wire at B, which 5 lamps would go out? (List them with the smallest number first.)
   ii) Lamp 6 has a broken filament. How does this affect lamps 3, 4 and 5?
   iii) The switches at T, U – V and W are drawn differently because they work differently. Describe briefly how each one works.
   iv) Where must the indicator lights flasher unit be put in the circuit.   (Joint 16+ part)

Figure 15.63

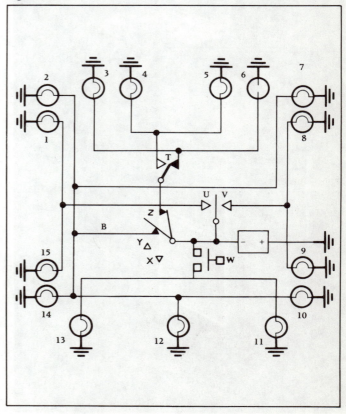

16

Electronics

The world is being changed by electronics. It is causing a revolution in every sphere of life. In this chapter we look at electronics in several different ways to gain a broad understanding of what it is about and of its effects on life and society.

Below is an artist's impression of the Japanese broadcasting satellite BS-2. Such satellites can provide colour television coverage over a large part of the earth's surface and so reach millions of people.

16.1
FREE ELECTRONS

Electrons form an essential part of all atoms and, like atoms, they are too small to see. We can find out what they are like by freeing them from their parent atoms. In 1897 the first confirmation that a very small charged particle existed came from the Cavendish laboratory in Cambridge when J. J. Thomson measured the speed, and the ratio of charge to mass of the electron. Since then a series of experiments have revealed more about the properties of electrons and shown us how they can be used in electronics.

Freeing electrons

Electrons are attached to atoms by the attractive force between their negative charge and the positive charge of the atomic nucleus. For this reason energy is needed to work against the attractive force and free an electron from an atom. The process of removing an electron from an individual atom or molecule is called **ionisation** and the energy needed is the **ionisation energy**. When a spark or a discharge occurs in a gas, electrons are freed by ionisation and the energy comes from the electrical supply.

When an electron is freed from the surface of a metal the minimum energy needed for each electron is called the **work function** of the surface. This energy may be supplied by heating or illuminating the surface.

Free electrons in an ionised gas

When we described how gases conduct electricity (chapter 13), we learnt that gases normally contain a small proportion of ionised atoms and that the missing electrons are free to move through the gas. The number of ionised atoms and free electrons increases when a gas is exposed to ionising radiation, and when a high voltage is used to make it conduct.

For the first experiments carried out by J. J. Thomson, the only available source of free electrons was from an electric discharge in a gas. A stream of electrons, called **cathode rays**, was produced when a high voltage of several kilovolts was applied to two electrodes in a gas at low pressure. Fig. 16.1 shows one of Thomson's original 'cathode-ray tubes'. The rays were named *cathode rays* because they travelled away from the cathode (the negative electrode) in the gas discharge tube. The negatively charged electrons were attracted by the positive anode and accelerated so that they flew through a slit in the anode and travelled on as an invisible ray until they hit the end of the glass tube. Here their kinetic energy was converted into light by fluorescence. The free electrons were provided by gas atoms which became ionised in the discharge between the cathode and anode. These electrons did not originate from the cathode itself.

Thermionic emission

A large number of electrons are free to roam about inside a metal but an electron travelling outwards at the surface is held back by the attractive forces of the atomic nuclei near the surface. When a metal is heated, however, some of its electrons may gain enough thermal energy to escape from its surface. This effect is known as **thermionic emission** (literally thermal ion emission). A more accurate description of the process would be *thermal electron emission*.

Figure 16.1   *J. J. Thomson's 'cathode ray tube', 1897*

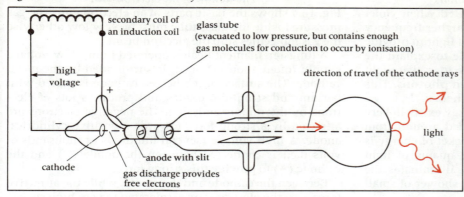

Figure 16.2   *A thermionic valve has an evacuated glass tube containing metal electrodes. One of these electrodes, called the cathode, emits electrons when it is heated to a high temperature. Transistors and other semiconductor devices have now replaced most thermionic valve applications.*

Figure 16.3   *Evidence for photoelectric emission*

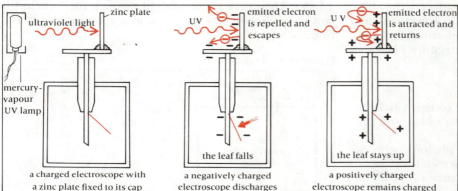

Thermionic emission can only be produced with certain metals, because it occurs at temperatures similar to their melting points. A tungsten filament lamp was found to release electrons from its filament at 2300 K. At this temperature the filament is white hot, produces a lot of heat and has a short life. Such a thermionic source, known as a *bright emitter*, is rarely used.

It has been found that a metal filament coated with oxides of barium and strontium will release lots of thermal electrons at the much lower temperature of 1300 K and will still emit some electrons at 1000 K. These oxide-coated filaments glow a dull red colour.

In most thermionic tubes electrons are emitted from a separate electrode called the cathode. The oxide-coated cathode is heated indirectly by a small wire heater or filament very close to it. The advantage of this arrangement is that the heater circuit is entirely separate from the cathode and does not affect the voltage on the cathode.

The process of thermionic emission of electrons from a heated metal can be likened to the process of evaporation of molecules from a heated liquid. In both cases small particles escape from the surface and become 'free', and energy is supplied by heating. We call the heat energy needed for the release of electrons from a metal surface its **work-function energy.**

The work-function energy is the least energy that must be supplied to remove an electron from the surface of a metal.

If an electron is further below the metal surface, extra energy is needed.

Thermionic emission is important because it is used to produce the stream of electrons from the electron gun which forms a vital part of all cathode-ray tubes. However, since the end of the nineteen sixties, thermionic valves have only been used for certain special purposes as they have largely been replaced by the transistor and the silicon 'chip'. Fig. 16.2 shows a typical thermionic valve.

Photoelectric emission

Electrons may also be freed from the surface of a metal by illuminating its surface with light. It is found that the frequency of the light must be above a certain value.

For most metals the light must be in the ultraviolet part of the spectrum. The effect can be demonstrated using a gold-leaf electroscope as shown in fig. 16.3.

● *Thoroughly clean a zinc plate and mount it on the cap of a gold-leaf electroscope.*

● *Charge the electroscope negatively and then illuminate the zinc plate with ultraviolet light (avoid looking directly at any ultraviolet light source, it can damage your eyes). Watch the electroscope leaf.*

● *Investigate the effect of charging the electroscope negatively again, and inserting an ultraviolet absorbing filter between the lamp and the zinc plate (i.e. a filter which does not transmit UV but allows some violet light through – a sheet of glass does this).*

● *Investigate the effect of charging the electroscope positively and illuminating the zinc plate with UV light.*

We find that the electroscope leaf falls steadily only when the plate is negatively charged and is also illuminated with UV light. Fig. 16.3 shows how, when the plate is positively charged, the emitted electrons are attracted back to the zinc plate and there is no net loss of charge. This confirms that the particles emitted are negatively charged.

The quantum theory and photons

The evidence that electrons are not emitted when violet light is used but are emitted when the higher frequency UV light is used required a new theory of light to provide an explanation. Albert Einstein was able to explain the photoelectric effect using Max Planck's idea that light energy exists in small energy drops called **photons**. This important theory is called the **quantum theory**.

The quantum theory states that all energy exists in very small separate packets, which are handed around from one owner to the next. An individual packet of energy is called a **quantum** (plural quanta). The quanta of light energy are called **photons**. When light illuminates the surface of a metal the photons are like a shower of small energy drops raining on the surface.

The electrons in the metal require a minimum amount of energy, the work-function energy of the surface, to be able to escape. If the energy of a photon is less than the work-function energy of the surface, the electrons will not be freed. The energy W of a photon is given by Planck's formula:

$$W = hf$$

where h is a constant known as Planck's constant (which has the incredibly small value of 6.6×10^{-34} J s), and f is the frequency of the light.

As electrons are only emitted when the photon energy is greater than the work-function energy of a metal surface, it follows that there is a minimum frequency of light for which the emission of electrons will occur. This provides convincing evidence in support of the quantum theory of radiation.

The electron gun

Thermionic emission is used to produce a continuous supply of electrons. An electron gun is then used to form a narrow beam of fast electrons or a 'cathode ray' inside an evacuated tube. Such tubes, known as **cathode-ray tubes** (CRTs), have many applications including the television receiver, cathode-ray oscilloscope (CRO), visual display unit, radar screen, and X-ray tube. We shall use some special CRTs to investigate the properties of cathode rays, or fast-moving electron streams.

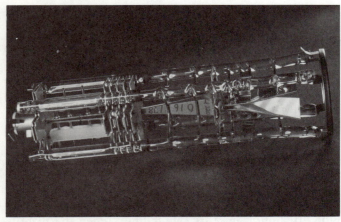

In this electron gun from a CRT you can identify the pairs of X and Y deflection plates by their wedge shapes, which match the increasing deflection of the electron beam as it leaves the gun. All the plates and electrodes are assembled on, and are insulated by, the four glass rods.

The design of an electron gun

Fig. 16.4 shows the basic parts of an electron gun. The gas pressure inside the glass tube must be very low otherwise the gas molecules stop the electron beam.

A tungsten filament heater, operated from a low-voltage supply (often about 6 V a.c.), is surrounded by the metal cathode. The surface of the heated cathode is coated with barium and strontium oxides and produces lots of electrons by thermionic emission. The negative electrons are attracted by a strong force towards a positive cylindrical anode. A high voltage (several hundreds or thousands of volts d.c.) is connected between the anode (+) and the cathode (−) to provide this strong force.

Between the cathode and the anode, while the attractive force acts, the electrons are accelerated to very high speeds ($\frac{1}{10}$ the speed of light for an anode – cathode voltage of 2.5 kV). Then they shoot straight through the hole in the end of the anode and travel on at a constant speed as a cathode ray.

Figure 16.4 *An electron gun*

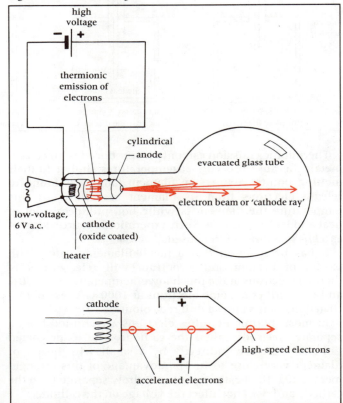

Calculating the energy and speed of electrons from an electron gun

Energy is given to the electrons between the cathode and anode by the high-voltage power supply. The electrical energy supplied W is given by:

$$W = QV \quad \text{or} \quad eV$$

where $Q = e$ is the charge of the electron (-1.6×10^{-19} C), and V is the accelerating voltage between cathode and anode. This energy is converted into kinetic energy E_k of the electron. The speed of an electron v can be calculated using the kinetic energy formula: $E_k = \frac{1}{2}mv^2$.

Worked Example
The energy and speed of electrons in a CRT

Calculate (a) the energy and (b) the speed of electrons fired from an electron gun using an accelerating voltage of 1000 V. The charge e and mass m_e of an electron are:

$$e = 1.6 \times 10^{-19}\,C \qquad m_e = 9.1 \times 10^{-31}\,kg$$

a) Using electrical energy $W = QV = eV$ gives

$$W = 1.6 \times 10^{-19}\,C \times 1000\,V = 1.6 \times 10^{-16}\,J$$

b) Using kinetic energy $E_k = \tfrac{1}{2}mv^2$ we have, by rearranging the formula:

$$v = \sqrt{\frac{2\,E_k}{m}} \doteqdot \sqrt{\frac{2 \times 1.6 \times 10^{-16}\,J}{9.1 \times 10^{-31}\,kg}} = 1.9 \times 10^7\,m/s$$

Answer: The kinetic energy of each electron after acceleration by the electron gun is 1.6×10^{-16} J. The speed of the electron is 1.9×10^7 m/s. (Note how very fast this is compared with the speed of sound at about 330 m/s and how near it is to the speed of light at 3×10^8 m/s.)

Investigating the properties of electron streams

Using the Teltron range of thermionic cathode tubes we can investigate many of the properties of free electrons in electron streams or cathode rays. Fig. 16.5 shows a Teltron tube mounted in the universal stand which holds all the tubes. It is fitted with a pair of coils known as **Helmholtz coils**, which can provide a uniform magnetic field inside the tube when magnetic deflection is required. The instructions given in the following experiments are intended only for demonstrators.

Cathode rays cast a shadow

Mount a Maltese cross cathode-ray tube in the universal stand. Connect power supplies to it as shown in fig. 16.6 but do not switch on. Note that:
a) *The Maltese cross should be connected to the anode.*
b) *The high-voltage supply required is 2 to 5 kV, available from an e.h.t. power unit (extra high tension), and the direct (unprotected) terminals should be used.*
c) *With a directly heated filament, the negative high-voltage connection can be made to either of the two filament terminals on the end of the tube.*

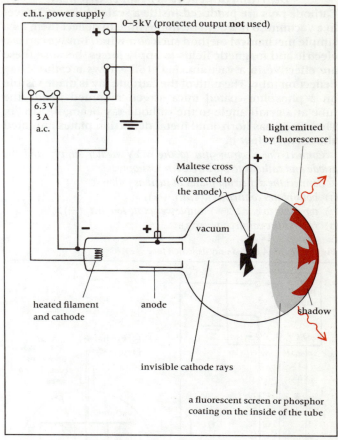

Figure 16.6 *A Maltese cross cathode-ray tube*

Figure 16.5 *A Teltron tube mounted in its stand and fitted with a pair of Helmholtz coils. There is a range of these evacuated tubes with electron guns designed for the study of the properties of electron streams.*

First switch on the filament alone, making the cathode glow brightly.
The light from the cathode casts a shadow of the Maltese cross on the end of the tube. We already know that light travels in straight lines and casts sharp shadows from point sources.
Switch on the e.h.t. supply, slowly increasing the voltage from zero to above 2 kV.
Invisible cathode rays travelling down the tube strike the fluorescent screen and cause green light to be emitted. (The kinetic energy of the electrons is converted into light energy by a process called fluorescence, which takes place in the phosphor coating inside the end of the tube.) The shadow cast by the Maltese cross stays on the end of the tube, but the screen around it now emits green light.
Bring a pole of a bar magnet up to the neck of the tube where the electrons emerge from the anode.
We find that there are two shadows:
a) the shadow cast by light which is not affected by the magnet and
b) the cathode ray shadow which is moved and distorted by the magnet (this is another example of the catapult force).
The fact that cathode rays do cast a shadow shows that they travel in straight lines. As the cathode-ray shadow coincides with the light shadow when there is no magnet, we can also see that there is no significant downwards deflection due to the weight of the electrons. This is due to their very high speed. We have also seen that cathode rays cause fluorescence.

Deflecting cathode rays by an electric field

Cathode rays are produced inside a sealed tube and travel in a vacuum so it would be impossible to deflect them by a simple mechanical method such as a wind. But we can use electric and magnetic fields to apply forces, because these are effective in a vacuum. Fig. 16.7 shows a cathode-ray deflection tube. The path of the cathode ray is made visible on a phosphor-coated mica screen mounted inside the tube at a small angle to the cathode-ray beam, fig. 16.7b. The screen has horizontal metal deflection plates mounted above and below it.

Connect the electron gun to the 6.3 V heater supply and the anode and cathode to the direct e.h.t. supply.

Connect the deflecting plates in turn as follows:
① *both plates connected to e.h.t. (+),*
② *top plate to e.h.t. (+) and lower plate to e.h.t. (−),*
③ *reverse the connection in* ②*.*

Figure 16.7   *A cathode-ray deflection tube*

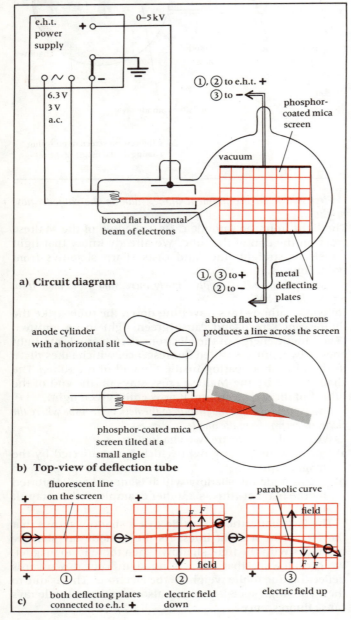

a) Circuit diagram

anode cylinder with a horizontal slit

a broad flat beam of electrons produces a line across the screen

phosphor-coated mica screen tilted at a small angle

b) Top-view of deflection tube

fluorescent line on the screen

parabolic curve

① both deflecting plates connected to e.h.t **+**

② electric field down

③ electric field up

c)

In each case slowly increase the voltage of the e.h.t. supply from zero up to about 3 kV and watch the screen.
Fig. 16.7c shows the results.
In case ① the deflection plates are both at the same voltage as the anode and there is no electric field between them. The cathode ray produces a horizontal line of fluorescence across the screen, showing no deflection when there is no electric field.
In case ② the lower plate is negative, producing a vertical downwards directed electric field (from + to −). The cathode rays are deflected upwards showing that they carry a negative charge. (Force *F* upwards is in the opposite direction to the field, and negative electrons are attracted to the + plate.)
Case ③ confirms the results; by reversing the electric field the deflection is also reversed.

The path of the electrons in the electric field has the shape of a parabola. In case ③ the path is exactly the same as that of a bullet fired horizontally from a gun. The earth's gravitational field causes the downwards force on the bullet, called its weight. In both cases the horizontal motion is at constant speed and the vertical motion is with uniform acceleration due to a constant force.

Cathode rays
a) *can be deflected by electric fields,*
b) *carry a negative charge and*
c) *travel in a parabolic path in an electric field at right angles to their initial direction.*

Deflecting cathode rays by a magnetic field

The same deflection tube can be used to show the effect of a magnetic field on cathode rays. In fig. 16.5 two large circular coils can be seen mounted on the sides of the tube. These Helmholtz coils produce a uniform magnetic field inside the deflection tube when a steady direct current is passed through them. Fig. 16.8a shows how the coils should be connected to a low-voltage supply. The horizontal magnetic field produced by the coils can be varied in strength by changing the current through them.

Connect the electron gun to the e.h.t. supply as before. Connect both deflection plates to e.h.t. (+) so that they produce no deflection. Switch on the filament supply and then increase the e.h.t. voltage until a bright horizontal line of fluorescence is visible on the screen.

Switch on the current through the Helmholtz coils and, using the rheostat, observe what happens as the current in the coils is increased.

Which way is the electron stream deflected and what is the shape of the track? You can use Fleming's left hand rule to work out the charge carried by the cathode rays, fig. 16.8b. You can use a magnetic compass to check the direction of the magnetic field inside the coils. Reserve the current through the coils and check your conclusions.

Cathode rays
a) *can be deflected by magnetic fields,*
b) *carry a negative charge (they travel in the opposite direction to the conventional current as used in Fleming's left-hand rule).*
c) *travel in a circular track in a magnetic field at right angles to the magnetic field,*
d) *turn in a tighter circle (of smaller radius) as the strength of the magnetic field is increased.*

Figure 16.8 *Deflecting cathode rays in a magnetic field*

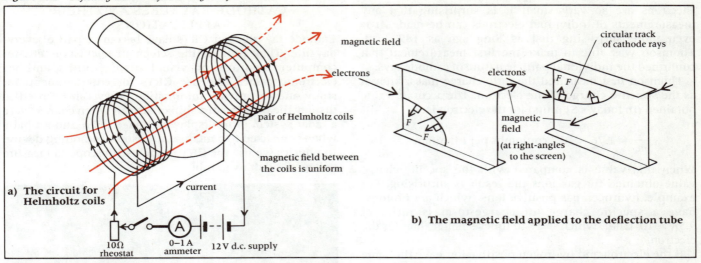

a) **The circuit for Helmholtz coils**

b) **The magnetic field applied to the deflection tube**

Figure 16.9 *The Perrin tube*

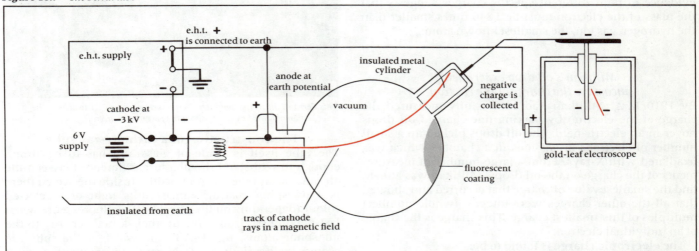

Collecting the charge carried by cathode rays

Perrin was the first to show that the charge carried by cathode rays or electron streams could be collected. In 1897 by a similar experiment, J. J. Thomson was able to show that the charge collected was inseparable from the cathode rays and was a property of them. A modern form of the apparatus used is shown in fig. 16.9, and we can repeat Thomson's experiment.

Mount the Perrin tube in the universal stand and connect the supplies to it as shown in fig. 16.9. Note the following points:

a) *The anode is connected to e.h.t. (+), which is then connected both to earth and to the case of an electroscope.*

b) *The cathode filament circuit, connected to e.h.t. (−), will be at about −3 kV and should be insulated from the bench top by a sheet of glass or polythene. (A 6 V battery can provide a suitable isolated low-voltage supply for the filament.)*

c) *The insulated metal cylinder (sometimes called a Faraday cage) should be connected by a short insulated lead to the electroscope cap to reduce leakage of charge.*

Although a pair of permanent magnets can be used successfully to deflect the cathode rays, more control can be provided by using the Helmholtz coils connected as in the circuit of fig. 16.8. By slowly varying the current in the coils, the beam of electrons can be bent until it enters the collecting cylinder. The position of the beam can be seen from the fluorescent spot at the end of the tube.

Immediately the cathode rays enter the cylinder the gold leaf rises on the electroscope showing a rise in voltage of the cylinder and the collection of charge. When tested, the electroscope confirms that negative charge has been collected. When the cathode rays miss the cylinder no charge is collected; therefore the charge belongs to the cathode rays themselves.

The negative charge is a property of the free electrons and is inseparable from them.

Other properties of cathode rays

Among the properties which will be mentioned again later are these:

a) In the X-ray tube streams of very fast electrons produce heat and X-radiation (p 352).

b) Beta radiation, fast electrons, can penetrate thin aluminium foil (p 393).

c) Fast electrons affect photographic plates.

d) Fast electons in a gas cause ionisation of gas molecules (p 268).

Measuring electrons

Electrons are so very small it is surprising that any measurements of individual electrons can be made. It is even more surprising that as long ago as 1897 J. J. Thomson was able to make the first measurement that confirmed the individual particle nature of cathode rays.

Thomson first measured the ratio charge/mass, known as the **specific charge** of the electron. The accurate value obtained (in today's SI units) for the electron is:

$$\frac{e}{m_e} = 1.76 \times 10^{11} \text{ coulombs per kilogram}$$

When this value is compared with the specific charge value obtained for gas ions the result is surprising. For example, hydrogen gas positive ions (which are protons) have a specific charge, or charge-to-mass ratio, of $9.58 \times 10^7 \text{ C/kg}$, which is 1840 times smaller than for the electron.

If the electron and hydrogen positive ion have the same value of charge (but opposite signs) then the experimental evidence of their different specific charges suggests that the mass of the electron must be 1840 times smaller than the hydrogen gas ion, the smallest known atom!

Millikan's oil drop experiment measures the charge of the electron

By 1910 R. A. Millikan had successfully measured the charge of the electron by making fine charged oil drops hover in an electric field. The oil drops picked up a small number of electrons or gas ions in a chamber which was irradiated with X-rays. From a large number of measurements of the charge on the oil drops, Millikan was able to find the smallest value of charge that occurred and showed that all the other charges were integral (whole number) multiples of this smallest charge. This charge is the charge of an individual electron.

The **electronic charge** is found to be:

$$e = -1.6 \times 10^{-19} \text{C}$$

So we can calculate the mass of the electron, m_e:

$$m_e = \frac{e}{e/m_e} = \frac{1.60 \times 10^{-19}\text{C}}{1.76 \times 10^{11}\text{C/kg}} = 9.1 \times 10^{-31}\text{kg}$$

This mass is so small that is impossible to comprehend. It is so small that it hardly makes any difference to the mass of an atom.

The importance of these experiments was that they proved the existence of the electron and gave a vital clue about the structure of matter.

Assignments

Name three methods by which free electrons can be produced.
Draw a labelled diagram of an electron gun and indicate the voltage supplies it needs.
Remember how the formulas $W = eV$ and $E_k = \frac{1}{2}mv^2$ are used to calculate the energy and speed of the electrons in a cathode ray.
Make a list of the properties of cathode rays of electron streams, and where possible state an observation which supports each property.
Try questions 16.1 to 16.5

CATHODE-RAY TUBES AND THEIR APPLICATIONS

Cathode-ray tubes or CRTs have become a part of everyday life. They provide the screen in television sets, for computer monitors or visual display units and for electronic games. We use CRTs for entertainment, for study and to provide information. They can be used to display the position of a ship or aircraft when connected to a radar system or to display an image of an unborn baby when connected to a medical ultrasound scanning device. In physics we use the CRT in the oscilloscope to measure voltages and draw graphs.

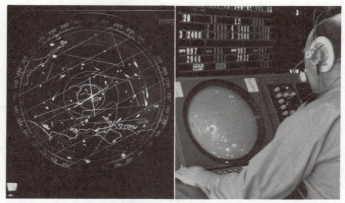

This Ferranti air traffic computer control system uses a CRT to display the information obtained by radar about the positions of aircraft.

The construction of a cathode-ray tube

The tube itself is made of glass and has to be strong enough to withstand the pressure difference between the air outside and the vacuum inside. Inside the screen there is a phosphor coating which emits light of a selected colour (depending on the phosphor) by fluorescence when fast-moving electrons give up their kinetic energy to the phosphor atoms, fig. 16.10. The walls of the tube are coated inside with a black graphite conducting paint which is connected to the final anode (3). The electrons, accumulating on the phosphor coating of the screen, repel each other, and when they reach the conducting graphite paint return to the (+) terminal of the e.h.t. supply so completing the circuit. Connections are usually made to all the electrodes and deflection plates via pins sealed in the end of the tube. See fig. 16.10.

The electron gun

In a CRT the electron gun is improved in the following ways:

a) A control electrode, called the **grid**, is fitted around the cathode. This grid is usually negatively charged compared with the cathode and is used to control the number of electrons in the beam. When the grid is made more negative its greater repulsion allows fewer electrons to reach the anode.

The number of electrons reaching the screen determines the brightness of the light emitted by the screen. Thus the negative voltage of the grid is used as the *brightness* or *brilliance control*.

b) More than one anode is used to accelerate the electrons and to focus them into a narrow beam. The anodes become increasingly positive compared with the

Figure 16.10 *CRT with deflection plates*

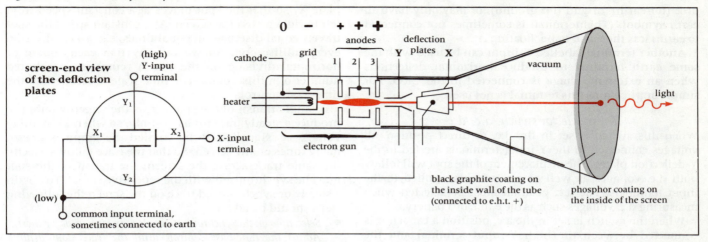

cathode as the electrons pass through them in order 1, 2, 3. The beam is *focused* by making small changes in the voltage of the central cylindrical anode, 2.

The deflection plates

Deflection plates are usually fitted in CRTs designed for use in oscilloscopes. The electron beam is deflected by the electric field between a pair of plates when a small voltage is connected across them. The Y-plates, Y_1 and Y_2 above and below the beam cause a vertical deflection of the beam and the X-plates, X_1 and X_2 at the sides of the beam cause a horizontal deflection. Because of the very high speed of the electrons in the beam, changes of voltage on the deflection plates produce almost instantaneous deflections of the spot of light on the screen.

Two effects combine to produce a continuous trace or line of light on the screen when the beam is deflected quickly by a changing voltage on the deflecting plates.

a) Persistence of vision in the human eye leaves a lingering impression of where the spot of light on the screen has been.

b) When the molecules of the phosphor on the screen absorb the kinetic energy of the electrons striking the screen, some of this energy may continue to be radiated as light for a few seconds afterwards. This effect is known as phosphorescence. Some CRTs are coated with a phosphor which gives several seconds persistence and these are used for display screens in radar and ultrasound scanning. (Electronic memories can now store the data and make an image remain indefinitely on a screen when required.)

Using a cathode-ray oscilloscope, CRO
The following discussion refers to fig. 16.11.

On/off and brilliance
On some CROs these are combined in the same control.
• *Switch on and turn up the brilliance until the bright spot has been found on the screen.*
When in use the brilliance should be set as low as possible for a clear trace to be seen. Excessive brilliance causes the phosphor coating on the screen to deteriorate, also a sharper trace can be obtained at lower brilliance. The brilliance control may also be labelled *intensity*.

Focus
• *Adjust the sharpness of the spot or trace by turning the focus control slowly. Sharpness is also affected by the brilliance.*

Shift controls
The X-shift control moves the spot horizontally by applying a d.c. voltage to the X-deflection plates. (If a positive voltage is applied to plate X_1 in fig. 16.10, the electron beam is attracted to it and the spot of light on the screen is deflected to the left.) Similarly the Y-shift control moves the spot vertically by applying a d.c. voltage to the Y-deflection plates.

• *Use the shift controls to centre the spot. If the spot is not visible, turn both shift controls to the centre of their tracks and then, with the brilliance turned up, turn them slowly in opposite directions until the spot is found.*

Some CROs have a 'trace locate' button which shrinks the full X and Y ranges onto the screen and so reveals where the spot is hiding.

Input terminals
A *pair* of Y-input terminals are provided because the CRO measures p.d.s or voltages between two points in a circuit and, like a voltmeter, requires two leads to connect it in parallel. One terminal, labelled Y-input or high, causes an upwards deflection of the spot when connected to a

Figure 16.11 *A cathode ray oscilloscope (CRO)*

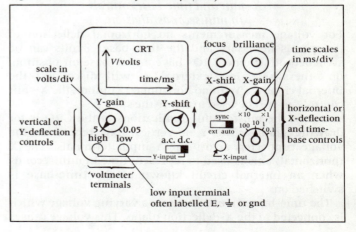

positive voltage. The other terminal may be labelled low, or E (for earth), or gnd (for ground), or may only have an earth symbol. (This terminal is sometimes not connected to earth; it is then said to be 'floating'.)

Another terminal labelled X-input can be used with the same earth terminal to produce horizontal deflections when an external voltage is connected to the CRO. For simple applications this terminal is not used.

Input selector switch: a.c./d.c.
When this switch is set in the d.c. or direct position, all voltages connected to the Y-input terminals are fed to the Y-deflection plates. The displacement of the spot will follow both d.c. voltages as well as a.c. voltages applied to the input terminals. The d.c. position must be selected when measuring d.c. voltages such as the e.m.f. of a battery.

When the switch is set in the a.c. position a capacitor is connected between the high Y-input terminal and the Y-deflection plates which acts as a barrier to any d.c. voltage. Because a capacitor does not conduct direct current, it only allows changing or alternating voltages to reach the deflection plates. The switch should be set in the a.c. position when only a.c. voltage measurements are required.

Some CROs have a central earth or ground position on this switch (a.c. – gnd – d.c.). When the earth position is selected the input terminals are connected together and the vertical position of the spot on the screen represents zero input p.d. or voltage. On CROs without this position, the input terminals can be joined with a short lead to find the zero input position of the spot. The Y-shift control should then be used to set the spot at the centre of the screen or the bottom of the screen. This sets the zero on the vertical voltage scale.

Y-gain in volts/division
The Y-gain control is an amplifier control. An amplifier is connected between the Y-input terminals and the Y-deflection plates so that small voltages can be amplified until they are large enough to deflect the electron beam. The control is calibrated in volts per division. These divisions are usually centimetre or $\frac{1}{2}$ cm squares ruled on the plastic filter fitted in front of the screen. The Y-calibration refers to these squares so that each vertical cm or $\frac{1}{2}$ cm division represents the number of volts selected on the Y-gain control. Using this calibration the CRO can be used as a voltmeter.

X-gain and time-base controls in milliseconds/division
For voltage measurements no horizontal deflection of the spot is necessary and the time-base circuits can be switched off. (Some CROs have a time-base off position, on others selecting 'external X' will disconnect the internal circuits.) With no horizontal (X) input, the X-shift control can still be used to centre the spot.

One of the most useful applications of the CRO is for drawing graphs. The Y or vertical axis represents the voltage connected across the Y-input terminals. The X (horizontal) axis is used to represent time in milliseconds when an internal circuit known as the **time-base** is switched on.

The time-base circuits generate a varying voltage which is connected to the X-deflection plates. This voltage causes plate X_2 to become steadily more positive compared with plate X_1 so that the spot moves at a constant speed from left to right across the screen. At a constant speed the spot travels equal distances in equal times, e.g. a 1 cm division every millisecond. So we can say that each square or horizontal division on the screen represents a selected number of milliseconds. The time scale is selected on the time-base control.

A single transit of the spot across the screen would not produce a steady trace for us to study, so when a repetitive voltage or 'signal' is to be studied the graph is drawn many times on the screen. So that the trace follows exactly the same track across the screen, the transit of the spot must be synchronised with the repetitive signal. To obtain a steady or synchronised trace on the screen the following steps should be taken:

● *Select auto on the synchronisation selector switch (sync for short).*
● *Adjust the time-base controls until the trace stops drifting across the screen.*
● *On some CROs a control labelled 'trigger level' must be turned until the trace locks in a steady position on the screen.*

The number of times the spot travels across the screen from left to right in one second is called the **time-base frequency**. A time-base frequency must be selected which is the same as or less than the frequency of the repetitive signal being studied. The spot returns from the right of the screen to the left of the screen in a very short time known as the fly-back time. During this time the electron beam is shut off by a negative voltage fed to the control grid so that the spot is not visible.

The CRO as a voltmeter
● *Switch off the time base.*
● *Connect the voltage to be measured to the Y-input terminals.*
A suitable range of voltages is shown in fig. 16.12.

Note the important difference between the appearance of a d.c. and an a.c. voltage displayed on the screen. The d.c. voltage only displaces the spot, but the alternating voltage causes it to fly up and down at high speed between its positive and negative peak values tracing out a vertical line on the screen.

The value of a d.c. voltage measurement is found using:

$$\text{d.c. voltage} = \begin{array}{c}\text{displacement of spot} \\ \text{from zero position} \\ \text{(in divisions)}\end{array} \times \begin{array}{c}\text{selected range on} \\ \text{Y-gain control} \\ \text{(in V/division)}\end{array}$$

The value of an a.c. voltage measurement can be given as either a + and − peak voltage or a peak-to-peak voltage which is double the peak value. The peak or maximum value is $\sqrt{2}$ × the r.m.s. value.

$$\text{peak voltage} = \begin{array}{c}\text{length of vertical trace} \\ \text{from zero position} \\ \text{(in divisions)}\end{array} \times \begin{array}{c}\text{selected range on} \\ \text{Y-gain control} \\ \text{(in V/division)}\end{array}$$

$$\text{peak-to-peak voltage} = \begin{array}{c}\text{full length of} \\ \text{vertical trace} \\ \text{(in divisions)}\end{array} \times \begin{array}{c}\text{selected range on} \\ \text{Y-gain control} \\ \text{(in V/division)}\end{array}$$

The input resistance of a CRO (i.e. internal resistance between Y-input terminals) is extremely high, typically several million ohms. This makes the oscilloscope very nearly an ideal voltmeter, since it takes almost zero current from the voltage sources it is measuring.

Figure 16.12 *The CRO as a voltmeter*

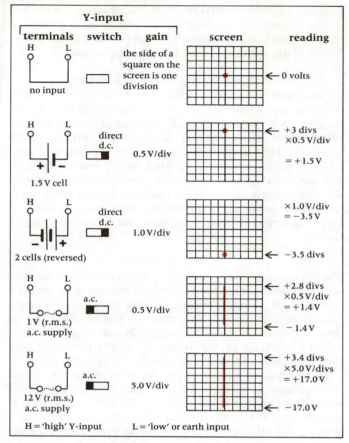

terminals	switch	gain	screen	reading
		the side of a square on the screen is one division		← 0 volts
1.5 V cell	direct d.c.	0.5 V/div		+3 divs ×0.5 V/div = +1.5 V
2 cells (reversed)	direct d.c.	1.0 V/div		×1.0 V/div = −3.5 V ... −3.5 divs
1 V (r.m.s.) a.c. supply	a.c.	0.5 V/div		+2.8 divs ×0.5 V/div = +1.4 V ... − 1.4 V
12 V (r.m.s.) a.c. supply	a.c.	5.0 V/div		+3.4 divs ×5.0 V/divs = +17.0 V ... −17.0 V

H = 'high' Y-input   L = 'low' or earth input

The CRO displaying waveforms

Fig. 16.13 shows various traces on the screen when the time-base is switched on. (a) and (b) show a horizontal trace produced when there is no Y-input.

In case (a) a slow time-base frequency of 1 Hz is selected. The spot will travel slowly across the screen once every second. To select this frequency the time-base control should be set to 100 ms/div:

$$\frac{\text{time to cross a screen}}{\text{10 divisions wide}} = 10\,\text{divs} \times 100\,\text{ms/div}$$
$$= 1000\,\text{ms} = 1\,\text{second}$$

In case (b) the time-base control is set to 10 ms/div. A permanent horizontal line is seen across the screen.

$$\text{time to cross the screen} = 10\,\text{divs} \times 10\,\text{ms/div}$$
$$= 100\,\text{ms} = 0.1\,\text{s}$$

$$\frac{\text{time-base}}{\text{frequency}} = \frac{1}{\text{time to cross the screen}} = \frac{1}{0.1\,\text{s}} = 10\,\text{Hz}$$

In examples (c), (d) and (e) the same Y-input is connected to the CRO. Using a mains step-down transformer, an accurate 50 Hz alternating voltage is available, and the CRO can display the waveform of this varying voltage. The display is a voltage − time graph of the transformer output. The three cases show how the display is affected by selecting different time-base frequencies.

In case (c) the time-base frequency of 50 Hz is the same as the frequency of the displayed voltage waveform. The trace crosses the screen once in every cycle of the waveform.

In case (d) the time-base takes twice as long to cross the screen, so the waveform is displayed twice in each transit.

In case (e) only half of the waveform is drawn during one transit, so the second half of the waveform (broken line) appears in the same place as the first half. When the time-base frequency is too fast the displayed waveform can be confusing. The time-base frequency should be the same as or a simple fraction of the frequency of the waveform to be displayed. A very wide range of a.c. frequencies can be displayed on the screen by selecting a suitable time-base frequency.

Examples (f) and (g) show how the selection of the d.c. input switch position can have a particular use when displaying waveforms. A half-wave rectified voltage is connected to the Y-input terminals (p 355). With the switch set in the a.c. position, case (f), the waveform sets itself partly above and partly below the horizontal centre zero line. This is misleading because, as case (g) shows, when the d.c. position is selected we find that the voltage variations are entirely on or above the zero line; the voltage never becomes negative.

Figure 16.13 *The CRO displaying waveforms*

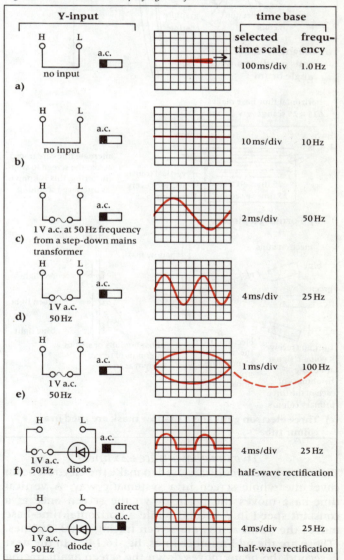

Y-input	time base	
	selected time scale	frequency
a) no input (a.c.)	100 ms/div	1.0 Hz
b) no input (a.c.)	10 ms/div	10 Hz
c) 1 V a.c. at 50 Hz frequency from a step-down mains transformer (a.c.)	2 ms/div	50 Hz
d) 1 V a.c. 50 Hz (a.c.)	4 ms/div	25 Hz
e) 1 V a.c. 50 Hz (a.c.)	1 ms/div	100 Hz
f) 1 V a.c. 50 Hz diode (a.c.)	4 ms/div	25 Hz half-wave rectification
g) 1 V a.c. 50 Hz diode (direct d.c.)	4 ms/div	25 Hz half-wave rectification

Television

Magnetic deflection coils are used

A television set has a CRT which uses two pairs of deflection coils mounted outside the neck of the tube. The magnetic fields of these coils deflect the electron beam by the catapult force. An advantage of magnetic deflection is that a much wider angle of deflection can be achieved. This means that wide screens are possible with relatively short tubes, fig. 16.14a. Can you explain why the X-deflection coils are above and below the neck of the tube? (Think of Fleming's left-hand rule.)

Figure 16.14 *A television CRT*

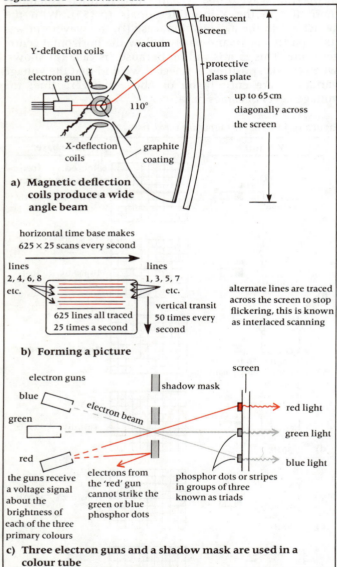

a) **Magnetic deflection coils produce a wide angle beam**

b) **Forming a picture**

c) **Three electron guns and a shadow mask are used in a colour tube**

Two time-bases trace 625 lines

Two time-base circuits are used to make the electron beam cover the whole screen in a systematic way. A vertical time-base moves the beam down the screen once at a constant speed in $\frac{1}{50}$ second while a horizontal time-base moves the beam across the screen tracing out $312\frac{1}{2}$ lines. (These are the odd number lines, fig. 16.14b.) In the next $\frac{1}{50}$ second the beam moves down the screen again tracing

out another $312\frac{1}{2}$ lines (the even number lines). So in $\frac{1}{25}$ second all 625 horizontal lines are traced across the screen.

The reason for creating the picture in this way is to reduce flicker to an undetectable level. The persistence of vision of the eye causes the picture to appear steady.

Brightness variations form a picture

As the electron beam travels across each of the 625 lines covering the whole screen, the brightness is varied. The brightness is controlled by the number of electrons in the beam, more electrons producing greater brightness or intensity. The number of electrons in the beam is controlled by the voltage connected to the control grid. A voltage signal picked up by the television aerial contains the instructions about when and how to vary the brightness as the beam scans the screen. The picture is built up from variations in the brightness along each of the 625 lines.

Colour pictures

The colour of the light emitted by the screen depends only on the choice of phosphor. A colour television screen can simulate all colours by using only the three primary colours of light (p 31). The screen is coated with dots of three different phosphors arranged in a triangular pattern called a **triad**, fig. 16.14c. When electrons strike these dots they each emit one particular colour, red, green or blue.

Three separate electron guns fire electrons at the screen through a mask with accurately drilled holes, one for each triad. The shadow mask (like the Maltese cross) casts a shadow and this prevents electrons from the wrong gun striking the other phosphor dots. The voltage signal from the television aerial contains information about the brightness of each of the three colours and how it should change across the screen. For example, if no electrons were fired from the 'green' gun, but electrons reached the screen from the 'blue' gun and the 'red' gun the screen would appear magenta. Neither the electrons nor the phosphor dots are coloured but any colour can be produced by the mixing of the three primary colours of light emitted by the phosphor dot triads.

X-ray tubes

X-rays are used in medicine and industry for a variety of purposes. The most common application is in the production of X-ray 'shadow images' or photographs, fig. 16.15a. Shadows are cast by solid objects which absorb X-rays.

A point source of X-rays is needed

X-rays travel in straight lines but, unlike light rays, they cannot be focused by lenses. It follows that the shadows cast by X-rays will have sharp edges only if the rays come from a point source. For this reason, a source of X-rays used to produce images must be as small as possible.

The heating problem

X-rays are produced when fast electrons strike a solid target and give up their kinetic energy. However, more than 99% of the kinetic energy of the electron beam is converted into heat energy as the electrons come to rest in the target. This causes a major problem at the target. To prevent the target melting the following steps are taken:

a) The target anode is made of a metal with a high melting point (tungsten, which is often used, melts at 3380 °C).

b) The target area is increased about 100 times by rotating the anode. While X-rays are produced at one small location (about 2 mm × 6 mm) the area of anode in which heat is produced spreads over most of the anode disc, fig. 16.15b.

To reduce the effective area of the source of X-rays so that it acts like a point source, the anode surface is inclined slightly. X-rays are produced in *all* directions, but from the direction of the X-ray window the source appears to have an area of only about 2 mm × 2 mm.

The design of the X-ray tube

The glass X-ray tube is evacuated. Heat is lost from the anode through the vacuum by infra-red radiation. The whole tube is surrounded by cooling oil and then enclosed in a lead jacket, so that X-rays leave only through an X-ray transparent window.

The anode is rotated by an induction motor which requires no connections to the rotor inside the evacuated glass tube.

Figure 16.15 *X-ray tube with a rotating anode*

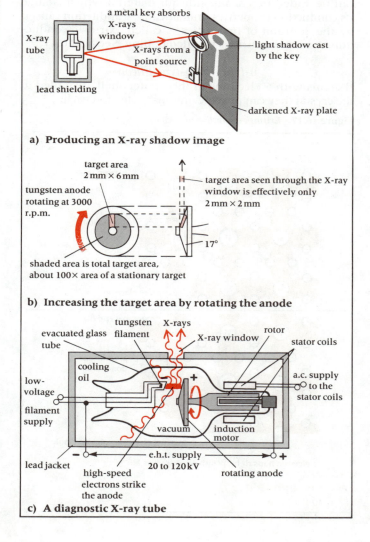

a) **Producing an X-ray shadow image**

b) **Increasing the target area by rotating the anode**

c) **A diagnostic X-ray tube**

A rotating anode X-ray tube. The evacuated glass tube contains the thermionic cathode on the right, and the rotating anode disc attached to the rotor of an induction motor on the left.

Power supplies

a) A low-voltage a.c. supply is required for the stator coils of the induction motor.

b) A variable low-voltage supply is required for heating the filament. By varying the filament temperature the number of electrons emitted by the cathode is controlled. The number of electrons striking the target anode per second determines the intensity of the X-rays.

c) A variable e.h.t. supply is required to accelerate the electrons from cathode to anode. Voltages over the range 20 kV to 120 kV are used to produce X-ray images. The X-rays produced at low voltages, known as **soft X-rays**, have low energy, long wavelengths and poor penetration. The X-rays produced at high voltages, known as **hard X-rays** have high energy, short wavelengths and good penetration. The choice of X-ray energy depends partly on the application and partly on safety considerations.

Assignments

Name four applications of CRTs.

Name and state the purpose of the parts of a cathode-ray tube.

Have a close look at your television screen and look for the phosphor triads and the horizontal scan lines.

Find out at least one industrial application of X-rays.

Use an oscilloscope to measure:

a) the e.m.f. of a simple cell (or a lemon cell, p 268),

b) the e.m.f. of a solar cell,

c) the output of a bicycle dynamo.

Use an oscilloscope to display:

d) the waveform of a musical instrument (p 451),

e) the waveform of the voltage generated in a human heart (if an electrocardiograph amplifier and electrodes are available)

Explain

f) Why should cameras and photographic films not be left inside luggage while passing through the baggage check at airports?

g) Why does a television need two time-base circuits?

h) Why is a point source of X-rays needed to produce X-ray shadow photographs.

i) Why does a television CRT use deflection coils instead of plates.

j) How are full colour pictures produced on a CRT screen from only three types of phosphor dot.

Try questions 16.6 to 16.9

16.3
USING SEMICONDUCTOR DEVICES

A great variety of semiconductor devices are used in electronic circuits both as separate or 'discrete' components and also as parts of integrated circuits or ICs. In this section we shall see how semiconductor materials can be made to conduct in a controlled way and how a few common devices are used in simple circuits.

Semiconductor materials

The group of materials known as semiconductors contain a limited number of free and mobile charged particles. (About one charge carrier per million atoms compared with about one electron per atom of silver or copper in a metallic conductor.) The best known semiconductor material is silicon, of which 'silicon chips' are made. Other semiconductor materials such as germanium and lead sulphide have been used and have special applications. Some new materials such as gallium arsenide and indium antimonide have new properties and roles to play in semiconductor devices.

What makes a semiconductor material special?

The conduction properties of semiconductor materials can be controlled and changed during manufacture and it is this which makes them particularly useful. Semiconductor materials have a natural conduction called intrinsic conduction and an artificial extra conduction due to impurities added in manufacture called impurity conduction or extrinsic conduction.

The crystalline structure of silicon

In the solid state silicon forms a regular, three-dimensional crystalline structure in which each atom has four near neighbours. This structure is represented in a two-dimensional plan in fig. 16.16. There are four electrons in the outer shell of a silicon atom and eight are needed for a stable structure. Bonds, (shaded in red) known as covalent bonds, are formed between neighbouring pairs of silicon atoms. Each atom shares one electron with each of its neighbours so that there are two electrons shared in each bond. Also, each atom has part ownership of eight electrons in its outer shell making it stable (shown by the ring of eight electrons around each atom).

Figure 16.16 *A lattice of silicon atoms form a crystal*

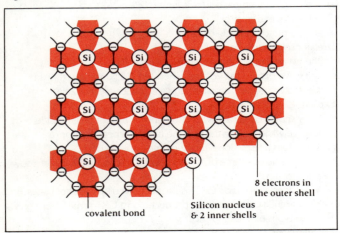

Intrinsic conduction

In a pure crystal of a semiconductor such as silicon, all the outer electrons of all the atoms are tied up in bonds with neighbouring atoms. We would expect no electrons to be 'spare' or free to move through the crystal so that it could conduct electricity. However, as the temperature of a pure silicon crystal increases, the kinetic energy of the vibrating atoms causes some electrons to break free. More electrons are freed as the temperature rises making semiconductors better conductors at higher temperatures.

For every free electron there is a gap or hole in the bonding structure between the atoms of the crystal. Since the atoms are normally neutral, the loss of a negative electron leaves behind a net positive charge at the place where there is a hole. Like free electrons, these positive holes move through the semiconductor material and form part of the electric current in it. A positive hole moves when an electron hops into it. This electron creates a new hole in the bond which it left. Holes and electrons move in opposite directions. The free electrons and positive holes which are formed in pairs provide the charge carriers which are the cause of intrinsic conduction in semiconductor materials.

Extrinsic conduction

During manufacture small amounts of different elements can be added to a semiconductor material which change its conduction properties dramatically. Conduction caused by the addition of impurities is called extrinsic conduction. Adding impurities is called doping. See fig. 16.17.

n-type and p-type silicon

If atoms with 5 electrons in their outer shell are added to the crystal they can provide one spare electron each.

Figure 16.17 *Extrinsic conduction in silicon*

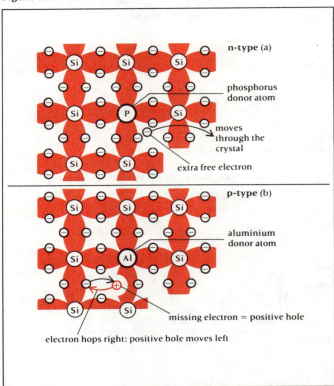

This is because only four electrons are needed to complete the shared bonds with the four neighbouring atoms. Phosphorus, arsenic and antimony atoms are used as donors to dope silicon and make n-type crystals with 'spare' conduction electrons. These 'spare' electrons are free to move through the crystal and form an electric current.

If atoms with only three electrons in their outer shell are used to dope the crystal, for each of these atoms it will have one electron missing in its structure. The missing electrons provide positive holes for conduction, fig. 16.17b. Aluminium, gallium and indium are used to provide these *positive* charge carriers and make p-type material.

The p–n junction diode

A single crystal of silicon can be made in which part of it is n-type material and part p-type. Small pieces of such a crystal with leads attached to the p- and n-type regions are called **p–n junction diodes**. Some examples are shown in fig. 16.18. They are called *diodes* because they have *two* electrodes, two ends to which two wires are attached. p–n junction diodes have special properties caused by the action of the positive and negative charge carriers. These travel in opposite directions when a voltage is applied.

The junction diode conducts in one direction only

The most important property of a p–n junction diode is its one-way only conduction. A diode acts like a one-way valve to electric current.

● *Connect a silicon diode as shown in fig. 16.19a, with the negative end or 'cathode' of the diode (the end which is usually marked) to the negative terminal of the cell.*

The lamp lights, a current flows and this is called **forward biasing**. The diode conducts as electrons and holes are able to cross over the junction between the p- and n-regions and complete the circuit.

Figure 16.18 *Diodes*

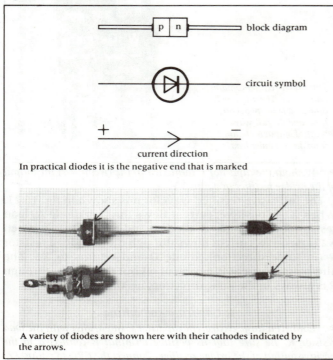

block diagram

circuit symbol

current direction

In practical diodes it is the negative end that is marked

A variety of diodes are shown here with their cathodes indicated by the arrows.

● *Reverse the connections to the diode as shown in fig. 16.19b, where its negative end is joined to the + terminal of the cell.*

This is called **reverse biasing**. No current can flow through the diode because the electrons and holes are attracted to opposite ends of the diode leaving the boundary region without any free charge carriers. The diode now acts as an insulator.

The characteristic of a p–n junction diode

In section 12.6 we found the characteristics of various conductors, which are shown on p 237. The characteristic of a p–n junction diode shows a large current $+I$ in one direction, the forward-bias direction, and a very small current $-I$ in the other, reverse-bias direction. This very small reverse current, perhaps a few microamperes, is due to intrinsic conduction caused by thermal electrons and holes.

Using diodes to convert a.c. to d.c.

Half-wave rectification by a single diode

The one-way conduction property of a diode can be used to convert alternating current into direct current. This process is called **rectification**.

● *Referring to fig. 16.20a (overleaf), connect a resistor R of 100 Ω resistance or more to the 12 V output of a mains step-down transformer.*

This is called a **load resistor** because it represents any device that the transformer might be expected to send a current through. The current through this load resistor will alternate at a frequency of 50 Hz, which is too fast to see on a d.c. ammeter. The voltage across it, which is always directly proportional to the current through it, will follow the same changes of direction at the same times. We can show how the voltage across the load resistor varies by connecting it to an oscilloscope.

Figure 16.19 *p–n junction diodes conduct in one direction only*

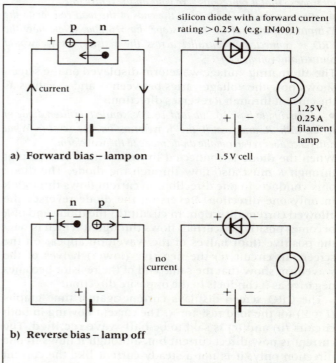

silicon diode with a forward current rating > 0.25 A (e.g. IN4001)

current

a) Forward bias – lamp on

1.5 V cell

1.25 V
0.25 A
filament lamp

no current

b) Reverse bias – lamp off

Figure 16.20 *Half-wave rectification by a diode*

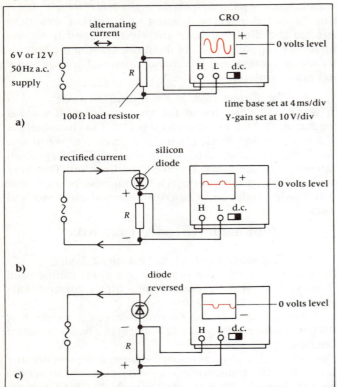

- *Switch on the CRO.*
- *Set the time-base selector at* 4 ms/div *(a frequency of* 25 Hz*) and adjust the Y-gain control to a value which suits the transformer output voltage, say* 10 V/div *for a* 12 V *r.m.s. output.*
- *Select the d.c. or 'direct' position on the Y-input switch.*
- *Before making any connections to the CRO, centre the horizontal trace so that it represents zero volts on the centre line.*
- *Connect the two leads from the ends of the load resistor to the Y-input terminals of the oscilloscope, fig.* 16.20a. *(Note how the CRO is connected in parallel across the resistor, as we would connect any voltmeter.)*

The alternating voltage waveform displayed on the screen shows how the voltage varies between + and − values as the current through R reverses direction.

- *Referring to figs.* 16.20b *and* 16.20c, *connect a silicon diode in series with R and watch the waveforms on the screen. What difference does reversing the diode make to the waveform?*

When the diode is connected in *series* with R any current through R must also flow through the diode. The diode only conducts in one direction, so current flows through R in only one direction. Reversing the diode reverses the allowed current direction. In circuit (b) the diode end of R becomes positive as current flows through it which is why the positive (top) halves of the waveform appear on the screen. In circuit (c) the negative (lower) halves of the waveform show that the same end of the resistor becomes negative as it conducts in the opposite direction.

The CRO screen displays current against time graphs ($I \propto V$) for the load resistor R. The current flowing in both circuits (b) and (c) is said to be **half-wave rectified**. The current is now direct current but, although it flows in one direction only, it is not a steady current like the current from a battery.

A low voltage power unit

Many modern electronic appliances such as cassette players and radios can be powered by either batteries or the mains. Their circuits are designed to work on a low-voltage steady d.c. supply of say 6 V. Fig. 16.21 shows the three changes which must be made to the a.c. mains supply to convert it into a smooth 6 V d.c. supply.

1: Step-down transformer

The 240 V r.m.s. mains is transformed down to 4 V r.m.s. using a transformer with a turns ratio of 60 to 1. The peak output voltage this will provide is given by:

$$peak\ output\ voltage = 4\ V\ r.m.s. \times \sqrt{2} = nearly\ 6\ V.$$

Safety notes:
Students should not work with live mains supplies but if the transformer part of this circuit is housed in an earthed metal box then the rest of the circuit may be investigated using an oscilloscope. The metal core of the transformer must also be connected to mains earth for safety reasons.

- *Set a CRO to display d.c. voltages at* 2 V/div.
- *Test the output of the transformer secondary coil at points XX by connecting the CRO across XX.*

The voltage waveform obtained across XX, as shown on a CRO, is an alternativing sine-wave of amplitude 6 V.

2: Full-wave rectification

A single diode allows only half of an alternating current to flow through the load R, so for half the time the power is cut off. A more even supply of power to the load is provided if the whole of the alternating current is made to flow one way through it. This can be done by using four diodes or a bridge rectifier which contains four diodes. This process is called full-wave rectification.

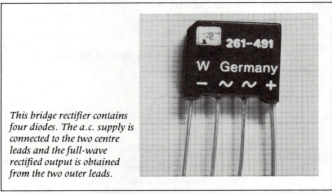

This bridge rectifier contains four diodes. The a.c. supply is connected to the two centre leads and the full-wave rectified output is obtained from the two outer leads.

- *Without the capacitor C connected in the circuit, test the output of the bridge rectifier at points YY.*

The full-wave rectified waveform is shown in fig. 16.21. When X1 is positive, current flows through diode A, load resistor R (from + to −), diode B and returns to X2. When X2 is positive, current flows through diode C, load resistor R (in same direction), diode D and returns to X1. The alternating current is never cut off but the current through R is always in the same direction, i.e. d.c.

3: Capacitor smoothing

- *Connect a large electrolytic capacitor C in parallel with the load resistor R. Be sure that its positive terminal is connected to the positive output of the bridge rectifier. Connect the CRO to points ZZ.*

Figure 16.21 *Converting a.c. to smooth d.c.: a low voltage power unit*

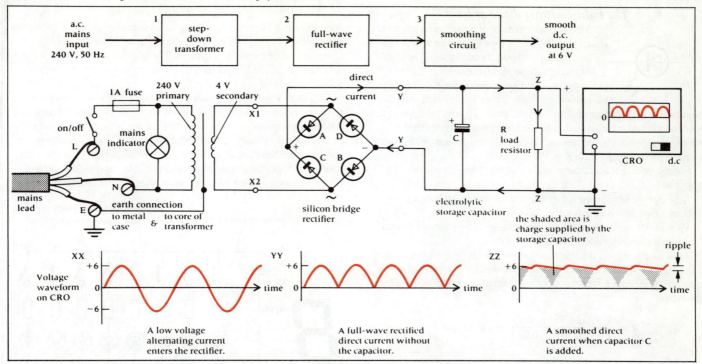

A low voltage alternating current enters the rectifier.

A full-wave rectified direct current without the capacitor.

A smoothed direct current when capacitor C is added.

A large storage capacitor C will make the d.c. output almost smooth. A small variation or ripple, shown by the two arrows on the waveform in fig. 16.21, will remain.

- *Try using different sizes of capacitors.*

The larger capacitors give the greatest smoothing effect. While the voltage is rising to its peak value across both R and C, the capacitor C is charged up and stores energy and charge. While the output voltage of the rectifier drops to zero, the capacitor supplies charge and energy keeping the current through the load R more constant.

If we consider the third graph in fig. 16.21 to be a current against time graph, the shaded areas represent the extra charge supplied from the capacitor.

$$\text{(area} = \text{current} \times \text{time} = \text{charge)}$$

Better stabilisation of the output voltage can be achieved using a zener diode but full stabilisation is very hard to achieve. Integrated circuits are available for this job.

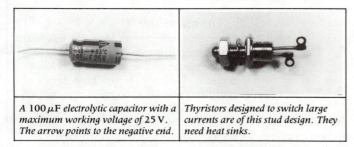

A 100 µF electrolytic capacitor with a maximum working voltage of 25 V. The arrow points to the negative end.	*Thyristors designed to switch large currents are of this stud design. They need heat sinks.*

Power control using a thyristor

A **thyristor** or **silicon-controlled rectifier: SCR** is made rather like a pair of p–n junction diodes connected in series, fig. 16.22a. It has a third connection, called a gate, which controls the main current flow from anode to cathode. If a small positive current (of a few mA) flows into the gate, a much larger current (of several A) is allowed to flow from anode to cathode. In fig. 16.22b:

- *closing switch S_1 turns on or triggers the thyristor.*
- *opening switch S_1 again leaves the thyristor conducting. The thyristor is latched on.*
- *Only disconnecting the power supply or closing switch S_2 (which deprives the thyristor of its current) will turn it off again.*

The thyristor behaves like a relay which stays on once it is switched on. This *latching* feature makes it useful in burglar alarm circuits. It also has the advantage of having no moving parts and of being able to handle large currents such as those needed by an electric motor. A motor would be connected in the position of the load resistor R. Its current gain, i.e. the factor by which the anode current is larger than the gate current, can be 1000 or more. Resistor R_1 limits the gate current to a few mA, but this could come from a sensor or transistor only capable of supplying a small current.

Figure 16.22 *Power switching using a thyristor*

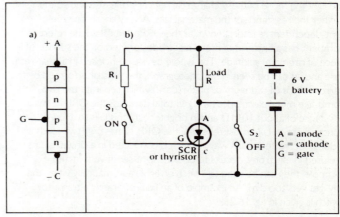

— Useful semiconductor devices —

The light-emitting diode or l.e.d.

Junction diodes made of gallium arsenide and gallium phosphide emit light when a forward-biased current flows through them. The colour of the light depends on the semiconductor material. The main advantage of these devices as indicator lights is their low current and low heat production compared with filament lamps. To limit the current to a low and safe value for an l.e.d., a protective resistor is connected in series with the supply, see below. To calculate the protective resistance needed we use the formula $R = V/I$.

For example, a manufacturer might give the following information about an l.e.d.:

$I_F = 20\,mA$, which means that a typical current in the forwards-biased direction is 20 mA,

$V_F = 1.4\,V$, which means that a safe voltage to be connected across the l.e.d. (when forward biased) is 1.4 volts.

If the supply voltage is 5.0 volts, then there must be a voltage drop of $5.0\,V - 1.4\,V = 3.6\,V$ across the protective resistor as shown in the diagram. The required resistance of the protective resistor is therefore given by:

$$R = \frac{V}{I} = \frac{3.6\,V}{20\,mA} = \frac{3.6\,V}{0.02\,A} = 180\,\Omega$$

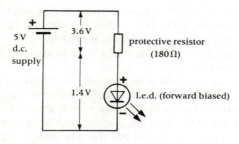

The light-dependent resistor, l.d.r.

Cadmium sulphide is one of several semiconductor materials whose resistance varies with the amount of light falling on it. As light energy is absorbed by this material, electrons and positive holes are released and become available for conduction. The brighter the light the more of these charge carriers are released and so the further the resistance of the material falls. A very common type of light-dependent resistor, known as the ORP12 (a catalogue reference number), has a fine grid of metal electrodes printed on the surface of a piece of cadmium sulphide. The whole device is enclosed in a clear resin case and, as can be seen in the photograph, the resistor is formed by the narrow strips of cadmium sulphide (dark) which separate the two comb-like metal electrodes (light). In total darkness the ORP12 has a resistance of about $10\,M\Omega$ and in brighter light this falls to a few hundred ohms. Like most resistors, the ORP12 has only two leads and it does not matter which way round it is connected in a circuit. L.d.r.s are used in electronic circuits to operate light-sensitive switches. For example, they can be used to switch on lights automatically when the daylight gets too dim. An example of an l.d.r. operating a transistor switch is given in fig. 16.26.

L.e.d. displays

A display can be formed from seven l.e.d. segments as shown below. All the numbers from 0 to 9 can be displayed by selecting different combinations of the seven segments. The diagram also shows a circuit which can be used for operating such a seven segment display with an additional decimal point.

Each l.e.d. segment must have a protective resistor and a switch to turn it on or off. In an electronic device such as a calculator the switching is done electronically, but this circuit can be set up with manual switches to find out how it works. (Note that the circuit shown has a display with common cathodes, i.e. all the negative terminals of the l.e.d.s are connected together to a single pin on the display. Some displays have a common anode and the switches are connected on the cathode side.)

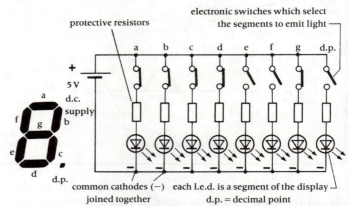

A seven-segment display shown with segments a b c d and g conducting and emitting light. Which number would the display indicate?

The photodiode, or light-sensitive junction diode

All semiconductor junction diodes are sensitive to light, so they are usually encapsulated in a light-proof coating. Junctions with transparent coats have the useful property that as light falls on them their reverse bias resistance drops and allows a larger reverse biased current to flow. Using this effect photodiodes can detect light passing through, for example, the holes in computer punched cards and tapes. They are also used in alarm systems and camera exposure controls. Large-area photodiodes are used in light communication systems.

The thermistor

Thermistors are resistors which are designed to change their resistance in a particular way as their temperature changes. Most thermistors decrease in resistance as their temperature rises. Being made of a semiconductor material, a rise in temperature releases more electrons and positive holes improving the intrinsic conduction of the material and so lowering its resistance. Thermistors are used to control currents in circuits. Being initially cold, they have a high resistance before a current flows in them and so they can be used to prevent high currents through lamp filaments and electric motors at the moment of switching on. Thermistors sometimes operate thermostats and fire alarms as their resistance change can be used to switch electronic circuits when the temperature changes.

The transistor

The transistor is a semiconductor sandwich usually made of silicon. A thin layer of p-type silicon sandwiched between two layers of n-type silicon form an npn transistor, fig. 16.24a. With the types of silicon reversed we have a pnp transistor, fig. 16.24b.

A transistor has three leads connected to the three layers, which are called the **emitter**, **base** and **collector**. The *emitter* emits or sends charge carriers through the thin *base* layer to be collected by the *collector*. In an npn transistor the n-type emitter material sends negative electrons to the collector. In a pnp transistor the p-type emitter material sends positive holes to the collector. In *both* cases however the arrow on the emitter shows the direction of conventional (i.e. positive charge) current flow.

The main current through a transistor is produced by the flow of charge carriers from emitter to collector. When a transistor is 'switched on' a current flows in the direction shown in fig. 16.24 between the emitter and collector. The current in the collector lead is called the collector current.

The base is used to control the collector current through the transistor. The base can switch the collector current on or off and the transistor will act as *an electronic switch*. The base can also supply a small current which is copied and amplified by the collector current; in this case the transistor acts as *a current amplifier*.

Using the transistor as an electronic switch

● *Connect up the circuit shown in fig. 16.25a.*

To do this it is important that you know which lead of the transistor is which. Fig. 16.25b shows how a tag or a flat side on a transistor will identify the leads by their positions.

Figure 16.24 *The transistor*

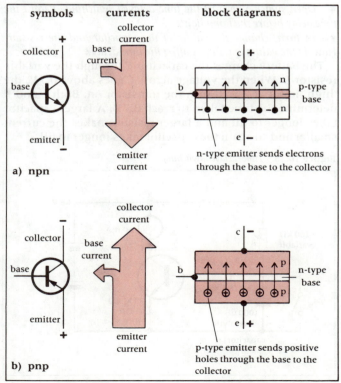

● *Using an npn transistor, connect the emitter to the negative terminal of the battery or low-voltage d.c. power supply.*
● *Connect the collector through a 6 V lamp to the positive terminal of the battery.*
● *Connect one end of a 10 kΩ resistor to the base so that its other end becomes an 'input' which can control the transistor.*
● *Find out what happens when:*
a) *the input is left unconnected,*
b) *the input is connected to the low or zero volt terminal (the negative terminal of the battery),*
c) *the input is connected to the high or +6 V terminal (the positive terminal of the battery).*

When the input is at a low voltage, cases (a) and (b), the lamp is off and no current flows through the transistor. A low input voltage to the base switches the transistor off.

When the input is at a high voltage, case (c), a small current flows through the 10 kΩ resistor R and through the transistor from base to emitter. This is called a base current. This switches the transistor on and allows a large current to flow from collector to emitter which lights the lamp. A high input voltage to the base switches the transistor on, fig. 16.25c.

Figure 16.25 *The transistor as a switch*

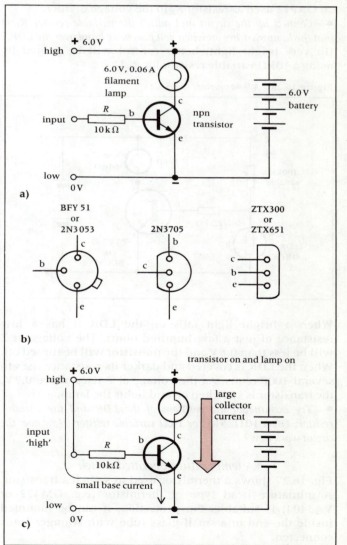

Transistor switching circuits

An npn silicon transistor is switched off when the input voltage between its base (b) and its emitter (e) is below 0.6 V, a **'low'** input. The same transistor is switched on when the input voltage rises above about 0.7 V, a **'high'** input. A voltage divider can be used to control the input voltage to a transistor switch. In the circuits of figs 16.26 and 16.27 the voltage divider is provided by the two resistors R_1 and R_2.

The point marked Z gives the input voltage to the transistor base (b). The voltage at Z is given by:

$$\text{input voltage} = \frac{R_2}{R_1 + R_2} \times \text{supply voltage (6 V)}$$

Resistor R_3 is there to protect the transistor which it does by limiting the base current.

The transistor switch can be operated by light or heat if a suitable sensor is connected in one half of the voltage divider. A change in the light level or temperature alters the resistance of the sensor and causes the input voltage at Z to rise or fall, so switching the transistor on or off.

A light-operated switch

Fig. 16.26 shows a light-dependent resistor (LDR) such as an ORP12 used as resistor R_2 in the voltage divider.

● *Connect up the circuit and adjust the variable resistor R_1 so that the lamp switches on when you pass your hand over the LDR.* [In very bright light, better control will be gained by using a 10 kΩ variable resistor for R_1.]

Figure 16.26   *A light-operated switch*

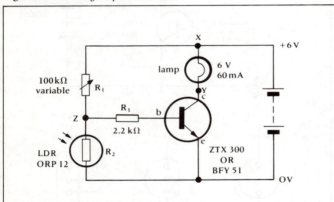

When a bright light falls on the LDR it has a low resistance of just a few hundred ohms. The voltage at Z will be less than 0.6 V and the transistor will be turned off. When the LDR is covered and darker its resistance rises to several 1000 ohms. As the voltage at Z rises above 0.7 V, the transistor is switched on and lights the lamp.

● *Try exchanging the positions of the LDR and the variable resistor. Use a 10 Ω, 5 kΩ or 1 kΩ variable resistor. How does the circuit work now?*

A temperature-operated switch

Fig. 16.27 shows a thermistor used to switch a transistor. A miniature bead type of thermistor (e.g. GM472 or VA3404) is suitable. For protection, it can be mounted inside the end of a small glass tube with stronger leads connected.

Figure 16.27   *A temperature-operated switch*

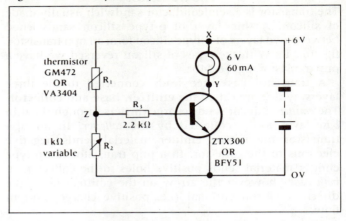

● *Connect up the circuit as shown and adjust the variable resistor R_2 so that the transistor and the lamp are just turned off.*
● *Gently warm the thermistor by hand or by dipping in warm water.*

As the temperature rises, the resistance of the thermistor and the voltage across it falls. The voltage across R_2 will therefore rise causing the transistor to switch on. The lamp, now lit, could be used as a high temperature warning or even a fire alarm.

● *Try connecting the thermistor in the position of R_2. R_1 now needs to be a 10 kΩ variable resistor. Explain how this circuit could be used as a thermostat. [A clue is given in fig. 16.29.]*

Time-delayed switching

● *Connect up the circuit shown in fig. 16.28. Note that the electrolytic capacitor should be connected with its negative terminal to the 0 V line from the battery.*
● *Close switch S to discharge the capacitor.*
Point Z is connected to 0 V and the transistor is off.
● *Open switch S to start the time-delayed switching. Record the time delay before the lamp lights.*
● *In turn, change the values of the capacitor and the resistor. How do the values of C and R affect the time delay?*

The battery charges the capacitor C through the variable resistor R. When the voltage across C goes above 0.7 V, the 'high' input at Z will turn the transistor on. Both C and R determine the time taken to reach 0.7 V. A larger capacitor takes longer to fill and a larger resistor makes the current smaller and so again the capacitor takes longer to fill.

Figure 16.28   *Time–delayed switching*

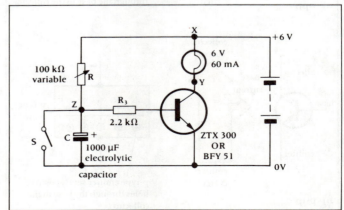

Using a relay as the output stage for a transistor in the circuits of figs 16.26, 27 and 28 the output stage of the transistor circuit was shown as a lamp connected between points X and Y. Other output devices which do not require a large current (such as buzzer) may be connected between X and Y. However, a device which requires a large current (such as a heater or an electric motor) cannot be connected directly because the current would overheat and destroy the transistor.

A relay is a device which allows a small current to control a much larger current, see page 287. The coil of the relay is connected in the control circuit, circuit A, between the points X and Y. When selecting a relay, the coil should be chosen to have a suitable resistance and an operating voltage not greater than the power supply used for circuit A. The switch of the relay is connected in a completely separate circuit in which the large current flows, circuit B. There is no direct connection between these two circuits and they use independent power supplies. In fig. 16.29 the two parts of the relay are shown by the coloured area.

Figure 16.29   *A relay-operated output*

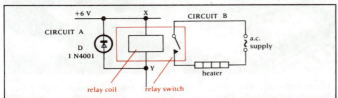

Transistor protection: diode D

The coil of the relay should have a reverse-biased diode D connected across it. When the coil is conducting (current flows from X to Y) the diode is *reverse biased* and so does not conduct. When the current through the coil is falling a large voltage is induced across the coil by electromagnetic induction. The diode is *forward biased* to this voltage and acts as a low resistance short-circuit to it. This prevents damage occurring to the transistor.

Circuit B: a separate circuit

The only connection between circuit B and circuit A is through the magnetic field of the relay coil in circuit A which closes or opens the switch in circuit B. The contacts of the switch may be normally open (NO) or normally closed (NC). Normally open contacts remain open until a current flowing through the relay coil closes them. Normally closed contacts remain closed until the relay is activated.

Any device and any power supply may be used in circuit B but the relay selected must have switch contacts rated high enough for the current needed by the device. For example, circuit B could contain a heater as shown in fig. 16.29. This could be operated from the mains a.c. supply. In this way a thermistor-controlled circuit (like that shown in fig. 16.27 but with R_1 and R_2 exchanged) could be used as a thermostatic switch for an electric heater.

Using a transistor as a current amplifier

This experiment shows in a direct way how a transistor acts as a current amplifier. Transistors used as amplifiers in electronic circuits work the same way.

● *Complete a circuit containing a filament lamp and a battery by holding the two wires in your hand as in fig. 16.30a.*
● *Try wetting your hands and gripping the wires very tightly. Can you get the lamp to light?*

Figure 16.30   *A transistor current amplifier*

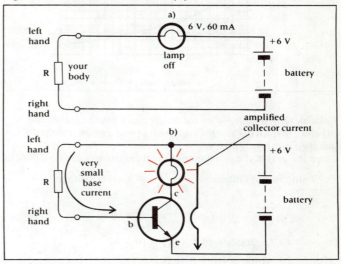

● *Now connect the same lamp and battery in the collector–emitter circuit of a transistor as in fig. 16.30b. This time complete the circuit by connecting the +6 V supply through your body to the base of the transistor.*
● *Again try wetting your hands and gripping the wires tightly.* The resistance of your body will limit the current through it from a 6 V battery to less than 1 mA. This current is far too small to light the lamp on its own. The action of the transistor is to amplify this current a hundred or more times. The amplified current flows through the lamp and lights it. (60 mA is needed to light the lamp brightly, but for some people it may light only dimly.)
● *Why does wetting your hands make the lamp brighter?*

The important feature of current amplification by a transistor is that the collector current is directly proportional to the base current, being controlled by it.

This makes amplification of complicated waveforms possible without changing their shape and character. For example the transistor can be used to amplify the signal from a microphone without changing the sound it represents.

The current amplification or gain (symbol h_{FE}) of a transistor is given by the formula:

$$\text{current gain} = \frac{\text{collector current}}{\text{base current}} \quad h_{FE} = \frac{I_c}{I_b}$$

The value of the current gain of a transistor can be anywhere between 10 and 1000 but a typical value might be 100. A ZTX300 npn transistor, for example, has a current gain in the range 50 to 300.

Worked Example
Voltage divider for a transistor switch

A thermistor and a fixed resistor are connected in a voltage divider in the same positions as shown by R_1 and R_2 in fig. 16.27. The variation of the resistance of the thermistor over a range of temperatures is shown in fig. 16.31.

Figure 16.31 *Variation of the resistance of a thermistor with temperature*

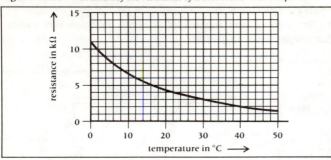

Calculate the resistance of the fixed resistor which would cause the voltage at point Z to reach 0.7 V when the thermistor is at a temperature of 40 °C and the power supply is 10 V.

From the graph, the resistance of the thermistor is $2 \, k\Omega = 2000 \, \Omega$.

$$\text{Using: input voltage} = \frac{R_2}{R_1 + R_2} \times \text{supply voltage}$$

$$0.7 \, V = \frac{R_2}{2000 + R_2} \times 10 \, V$$

$$\therefore \quad (0.7 \times 2000) + (0.7 \times R_2) = 10 \times R_2$$

$$\therefore \quad 1400 = (10 - 0.7) R_2 = 9.3 R_2$$

$$\therefore \quad R_2 = 150 \, \Omega$$

Answer: a $150 \, \Omega$ resistor should be connected in position R_2.

Worked Example
A transistor current amplifier

If the transistor in fig. 16.27b has a current gain of 100 and the lamp was brightly lit by a current of 60 mA, calculate the current through your body and hence your body's resistance.

We use:
$$h_{FE} = \frac{I_c \text{ (collector current and lamp current)}}{I_b \text{ (base current and body current)}}$$

$$\therefore I_b = \frac{I_c}{h_{FE}} = \frac{60 \, mA}{100} = 0.6 \, mA$$

The body's resistance:
$$R = \frac{V}{I} = \frac{6 \, V}{0.6 \, mA} = 10 \, k\Omega$$

Answer: The current through the body is 0.6 mA and the body's resistance is $10 \, k\Omega$.

Note: We simplify this calculation by assuming that the voltage across the resistor in the base circuit is the full 6 volts of the battery. When the transistor is conducting there is in fact a small battery. When the transistor is conducting there is in fact a small p.d. of about 0.7 V across the base–emitter junction (a forward-biased p–n junction). The true voltage across the resistor in the base circuit is nearer to $6.0 \, V - 0.7 \, V = 5.3 \, V$.

Assignments

Build circuits with transistors which:
a) using a thermistor and a relay, will switch on a 12 V immersion heater when the temperature of a beaker of water falls below 40 °C.
b) using an LDR and a relay, will sound an alarm when a beam of light is broken.
c) using an LDR and a SCR or thyristor, will sound an alarm and latch (stay on) when a beam of light is broken.

Try questions 16.10 to 16.19.

16.4
ANALOGUE ELECTRONIC SYSTEMS

Electronic circuits have become more and more complex involving large numbers of components. Groups of components are manufactured and assembled as a package or building block from which much larger and more complex circuits can be built. These packages of components are called **integrated circuits** or **ICs**.

Electronic systems, designed to do a particular job, are built from these building blocks using a range of ICs.

A recording studio uses a large electronic system which is built from many simple building blocks. These include amplifiers, filters, mixers, echo units and a variety of input and output devices.

The detailed circuits of a modern radio receiver are so complicated that only an electronics expert could make sense of them. However, if we show the design of the radio receiver as an electronic system built not from components but from building blocks, it is much easier to understand. See p 368.

The structure of an electronic system

All electronic systems have the three basic building blocks shown in fig. 16.32. In an *analogue* system:
● the **INPUT** block might contain a sensor of light, sound, temperature, pressure or a magnetic field. For example, a tape-recorder head would detect a varying magnetic field and a record-player needle would sense the mechanical vibration cut into a groove.
● the **PROCESSOR** block might contain a current or voltage amplifier, a voltage comparator, a time delay or a latch.
● the **OUTPUT** block might contain a lamp, buzzer, solenoid relay or an electric motor.

Figure 16.32 *The basic building blocks of an electronic system*

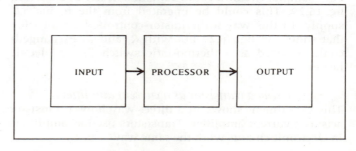

Examples of analogue systems

A burglar alarm is a simple example of an electronic system. There could be several inputs from sensors or switches, for example, a light sensor detecting a broken light beam and a microswitch on a window frame. The inputs might be processed by a transistor switch and a thyristor acting as a latch. The thyristor would also drive an alarm bell as the output device (see fig. 16.33a).

Figure 16.33 *Two analogue systems*

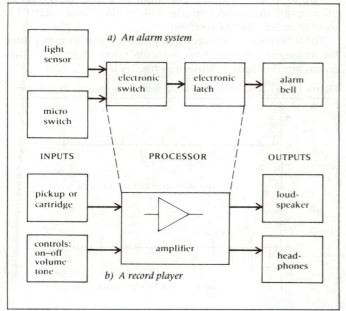

INPUTS   PROCESSOR   OUTPUTS

a) An alarm system

b) A record player

A record player which plays 12-inch discs with grooves cut in them is an example of an analogue system. It also has several inputs: a signal input from the pickup or cartridge, and control inputs including an on–off switch, volume and tone controls. It may have outputs to headphones as well as loudspeakers, (fig. 16.33b).

The information in this system is in the form of an electric current or voltage and the basic process which the system carries out is amplification. So what do we mean by analogue information?

In this example the information stored in the grooves of the disc and converted first into the mechanical vibrations of a needle and then into a small varying electrical voltage, is an analogue signal. The sound, which the signal represents, varies continuously in both frequency (pitch) and amplitude (loudness). For example, the loudness of the sound can have any value on a continuous scale from quiet to loud. The electrical signal is a copy of the sound with all the variation of loudness and pitch it originally had. The information or signal contained in this kind of model is called analogue information.

Analogue information in an electronic system is in the form of a current or voltage which varies smoothly and continuously in the same way as the information which it represents.

A good analogue system does not distort the shape of the signal nor add unwanted features to its waveform. For example, an amplifier should only multiply the amplitude of the signal by a certain factor. This should stretch the waveform vertically but make no other changes to it.

Input and output blocks: transducers and peripherals

All electronic systems need input and output blocks to connect them to the outside world. In computer terminology, devices called **peripherals** provide the means of feeding information into and out of a system; e.g. a keyboard may be the input block or peripheral and a CRT may be the output block or peripheral of a computer.

Since electronic systems handle information in an electrical form, devices are needed which convert different kinds of information into electrical information at the input end. Other devices convert the processed electrical information coming out of the electronic system into a form which we can read, hear or store. Such devices are called **transducers**. Table 16.1 shows some common transducers and how they are used.

The other important input to all electronic systems is the *power supply*. Whenever a signal (i.e. information in an electrical form) is amplified or changed in a way which requires additional energy, a power supply must be fed into the system. Building blocks such as amplifiers which need a power input are called **active components** because they play an active role in changing or enlarging the signal. Other building blocks such as the tuned circuit in a radio receiver which may not require any power input are called **passive components**, fig. 16.34.

Table 16.1 Transducers: input and output devices for analogue systems

Transducer	Application	Information or signal conversion from → to
microphone	sound input	sound-wave → alternating current or voltage
loudspeaker	sound output	alternating current → sound wave
thermistor thermocouple	electronic thermometers	temperature → current variation voltage variation
photodiode l.d.r.	light meters	light variations → current variations
aerial	input and output from radio systems	radio waves ↔ alternating currents
pickup or cartridge	record players	mechanical vibrations ↔ alternating current or voltage
electromagnet	tape recorder	varying magnetisation of tape ↔ alternating voltage
voltmeter	measurements	voltage change → pointer movement

Figure 16.34

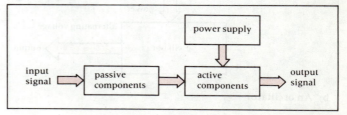

Building blocks for analogue systems

Although a complete system can be very complex, the number of basic building blocks used is quite small. Table 16.2 lists the common blocks and states what they do. We shall now look at two of these building blocks in more detail. The amplifier and the oscillator are used in most analogue systems.

Table 16.2

Analogue building block	What it does
power pack	provides input power for all active components in the system
amplifier	increases the amplitude of a signal
rectifier	converts alternating signals to direct ones
tuner	a circuit which selects one particular frequency
filter	allows only signals of particular frequencies to pass, or amplifies those of certain frequencies more than other frequencies
adder subtractor	combines two signals together
multiplier divider integrator differentiator	makes mathematical changes to the signal, can be used to solve mathematical equations and problems
oscillator or signal generator	generates an alternating voltage of a frequency and wave shape controlled by its internal circuits

The amplifier

The amplifier is an active component of a system and requires a power input, fig. 16.35a. The power supplied is used to increase the amplitude of the input signal. Where this contains several different frequencies, the amplifier should increase the amplitude of all the frequencies by the same proportion. Such an amplifier is said to be **linear**. One of the problems with amplifiers in analogue systems is that to some extent they fail to do this and they may change the wave shape of the signal while increasing its amplitude. In the case of a sound amplifier this **non-linear** amplification changes the quality or 'tone' of the sound. In extreme cases of non-linear amplification the output can sound distorted.

Figure 16.35 Building blocks

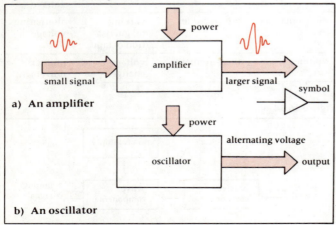

a) An amplifier

b) An oscillator

Oscillators or signal generators

Many systems contain an active component which generates a signal. In an analogue system this building block is known as an **oscillator** or **signal generator**. For example, a radio frequency (r.f.) oscillator generates an alternating voltage at a radio frequency (about 100 kHz to 100 GHz), which is sent to an aerial to be converted to radio waves. Similarly an audio freqency (a.f.) oscillator generates an alternating voltage at an audio frequency (20 Hz to 20 kHz). An a.f. oscillator or signal generator is often used in sound experiments (p 451) and might be used to make electronic music.

The alternating voltages produced by signal generators for analogue systems have various wave shapes for different applications as shown in fig. 16.36.

Figure 16.36   Some examples of voltage waveforms generated by oscillators

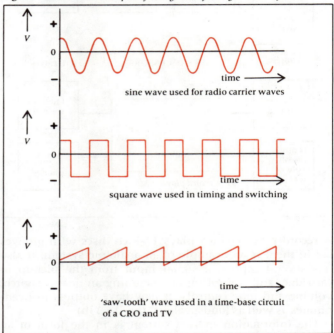

sine wave used for radio carrier waves

square wave used in timing and switching

'saw-tooth' wave used in a time-base circuit of a CRO and TV

Feedback

Biological and mechanical systems use feedback, as well as electronic systems. For example we use feedback to control our movements all the time. Try this experiment.

● *Close your eyes and then, from some distance apart, bring your two index fingers together pointing them so that they should touch at the tips. Try this several times. How often do you miss and why do you miss?*

When your eyes are open you can do this accurately every time. The output of your brain instructs your hands to move, but the movement is never perfectly accurate. The error in the movement is noticed by your eyes. Your eyes supply an input to your brain telling it how to adjust the movement to correct the error. Thus your eyes take information about the brain's output and feed it back into the brain as an input. This process is called **feedback**. Your eyes provide a **feedback loop** for the brain, fig. 16.37a.

Feedback involves all or a fraction of the output of a system (or information about it) being fed back to the input of the system in such a way that its output is controlled or changed.

Negative feedback in electronic systems

When part of the output of an amplifier is fed back to the input so that the output is *reduced*, it is called **negative feedback**.

This is achieved by feeding back a fraction of the output signal in *antiphase* with the input signal to which it is added. This happens if the waveform of the output signal is an inverted version of the input waveform, fig. 16.37b. This apparently foolish action, which reduces the amplification or gain of an amplifier, has many advantages to offer and is used in all practical or 'operational' amplifiers (see section 16.5).

Negative feedback is used to reduce distortion in analogue systems. The more feedback that is used the more 'linear' or accurate is the amplification, but also the smaller is the amplification. Negative feedback can make the gain of an amplifier almost independent of the amplifying device in the circuit.

Positive feedback in electronic systems

When part or all of the output signal of an amplifier is fed back to its input so that it reinforces and increases the input signal, the system has **positive feedback**.

The effect of increasing the input signal is to further increase the output signal which causes an ever rising amplitude at both ends of the amplifier. A limit is reached when the power loss from the system is equal to the power input from the power pack.

This arrangement, using positive feedback, is used to keep electrical oscillations going in oscillators or signal generators. A practical example is shown in fig. 16.37c. If you take the output from a loudspeaker and feed it back into the microphone input of the system it will oscillate or 'howl'.

Analogue systems

Power pack

The block diagram of fig. 16.38a should be compared with the circuit diagram for an l.e.d. display shown on p 348. It is easier to understand what the system does from the block diagram. Blocks can be shown on this type of diagram without understanding how they work. The voltage stabiliser block removes the ripple in the output of the power pack. In larger systems the whole power pack appears as only one block.

Multimeter

The input block contains resistors of various values which can be selected for different measurement ranges, fig. 16.38b. These are connected as shunts or multipliers for current and voltage ranges. Devices which protect the instrument against overloads and misuse are also included in the input stage of the system. The processor block is an amplifier which takes the small input signal and makes it large enough to operate a milliameter or digital display. The readings given by the instrument are produced by the output stage of the system.

Figure 16.37 *Feedback*

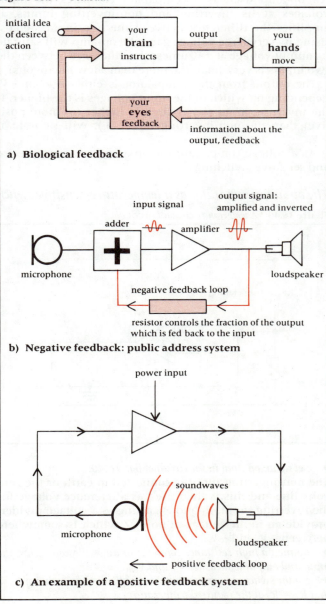

a) **Biological feedback**

b) **Negative feedback: public address system**

c) **An example of a positive feedback system**

Figure 16.38 *Block diagrams of some analogue systems*

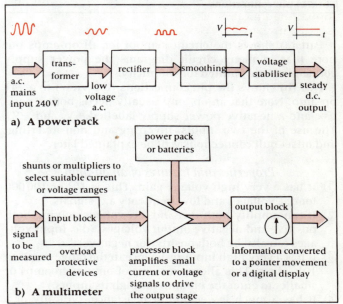

a) **A power pack**

b) **A multimeter**

16.5
THE OPERATIONAL AMPLIFIER OR OP-AMP

An operational amplifier is a high performance amplifier widely used in analogue systems. Even though digital systems are rapidly replacing analogue systems in many applications such as hi-fi and communications systems, the operational amplifier is so versatile that ingenious engineers are continually finding new uses for it. While the basic property of an operational amplifier is its ability to amplify an electrical signal, it can carry out many other circuit operations such as switching and adding. The operational amplifier, or **op-amp** as it is known, is used as a building block for more complex circuits and systems.

The IC Op-amp

An **operational amplifier** or **op-amp** is a complex circuit containing many transistors and resistors wired up to give it specially designed electrical properties. The op-amps most commonly used are constructed as an integrated circuit (IC) within a silicon chip. The chip is sealed inside a protective plastic or ceramic case which is fitted with connecting pins. A commonly used op-amp IC is the 741 which fits an 8-pin d.i.l. (dual in-line) socket. Fig. 16.39 shows the basic connections to a 741 op-amp.

Figure 16.39 *An op-amp symbol and pin connections*

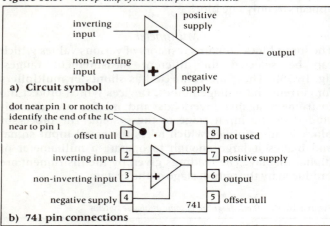

a) **Circuit symbol**

b) **741 pin connections**

Part (a) shows the circuit symbol for all op-amps but note that in many circuit diagrams the power supply connections are omitted for simplicity.

Part (b) shows the pin connections (top view) to a 741 op-amp. Note that an op-amp usually needs both a positive and a negative power supply labelled $+V_s$ and $-V_s$. The use of the two inputs, inverting and non-inverting, and offset null connections will be explained later.

Properties and features of an op-amp

① It has a very high voltage gain. This is 10^5 or 100 000 for d.c. voltages and low frequency a.c. signals.

② It can amplify both d.c. and a.c. voltages giving both positive and negative output voltages, (d.c. input voltages can also be both positive or negative).

③ It has a very high input resistance in the range of 10^6 to 10^{12} ohms. This allows currents of only microamps or smaller to enter the op-amp through its inputs.

④ It has a much lower output resistance, typically of a few hundred ohms. This allows the output to conduct tens or hundreds of milliamps and to operate devices such as relay coils.

⑤ The output voltages cannot exceed the power supply voltages, $+V_s$ and $-V_s$. In fact the maximum output voltage is slightly less than $\pm V_s$. When the op-amp attempts to produce an output greater than $\pm V_s$, the output is said to saturate.

⑥ The basic process which the op-amp performs is the amplification of the voltage difference between its two input terminals. These are labelled + and −. If one of the inputs is held at zero volts or earthed, the other can be used as a single input to the op-amp:

The inverting input (−) changes the sign of a steady d.c. input voltage and inverts the waveform of an alternating voltage. The non-inverting input (+) does not alter the sign of the input voltage nor invert an alternating signal.

The op-amp as a voltage comparator

An op-amp used as a comparator compares the input voltages at its inverting and non-inverting inputs. It amplifies the difference between the two input signals. Using its very large voltage gain, the op-amp will saturate its output voltage at $\pm V_s$ unless the difference between the two inputs is very small (no more than a few microvolts).

The output from the comparator is either $+V_s$ or $-V_s$ depending on which of the input voltages is the higher. If the inverting input (−) is slightly higher (or more positive) than the non-inverting input (+), will saturate at $-V_s$.

As a voltage comparator the op-amp has a very precise and sensitive switching action.

The op-amp comparator as a temperature–sensitive switch

Figure 16.40 *A temperature-operated switch*

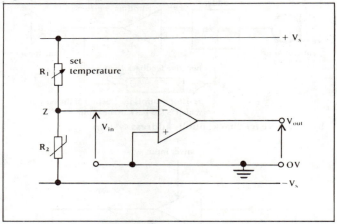

● *Set up an op-amp in the circuit of fig. 16.40.*
The non-inverting input is connected to earth or the zero volts line and this is then used as a reference voltage for the inverting input. R_1 and R_2, acting as a voltage divider, provide an input voltage at point Z which is somewhere between $+V_s$ and $-V_s$.

● *Connect a high resistance voltmeter or an oscilloscope across the input and output to measure V_{in} and V_{out}.*

● *Set the switching temperature T_s by adjusting the value of R_1 so that $R_1 = R_2$ at the required temperature.*

At T_s the voltage at Z will be zero: midway between $+V_s$ and $-V_s$. When the temperature is below T_s, the resistance of the thermistor R_2 will be higher than R_1 making the voltage at Z positive ($> 0\,V$). The comparator compares the voltage at Z with the zero volts at the non-inverting input and when Z is positive its output saturates at $-V_s$.
● *Warm the thermistor so that its temperature rises above T_s.* The resistance of the thermistor will fall to just below that of R_1, causing the voltage at Z to fall below zero.
● *Notice how sharply the op-amp comparator switches its output from $-V_s$ to $+V_s$. There are no in between values of V_{out}.*
● *Use this positive output voltage to operate a relay.*
● *Use the relay to switch on an alarm or operate a cooling system.*

The open-loop gain of an op-amp

As we shall see later, when the op-amp is used as a voltage amplifier it is usually provided with some *negative feedback* to improve its performance. **Feedback** is the name used to describe the passing of some of the output signal back to the input of an amplifier. When the loop in the circuit which allows feedback is not connected it is described as an 'open loop'. So the **open-loop gain** of an op-amp is the gain it has *when there is no feedback*.

Figure 16.41   *Input and output voltages for an op-amp*

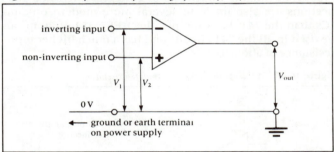

Referring to fig. 16.41, the open-loop gain is given by:

$$\text{open-loop voltage gain } A_{OL} = \frac{V_{out}}{V_{in}} = \frac{V_{out}}{V_2 - V_1}$$

where $V_2 - V_1$ is the voltage difference between the two input terminals which the op-amp amplifies.

Figure 16.42   *The open-loop voltage gain of an op-amp*

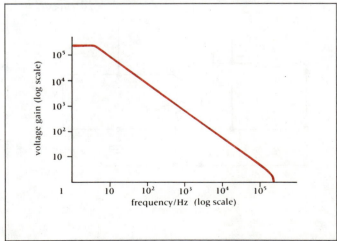

The open-loop gain of an op-amp is not a fixed value. The graph of fig. 16.42 shows how the open-loop gain of a 741 op-amp varies with frequency. We can see that the very high gain of 100 000 falls rapidly as the frequency rises until there is no gain left at about 1 MHz. The open-loop gain given by this characteristic is the maximum gain that the op-amp can deliver. Negative feedback is used to reduce the open-loop gain so that the voltage gain of the amplifier is more constant over a range of frequencies.

An inverting voltage amplifier with negative feedback

As a voltage amplifier the op-amp is usually used with negative feedback as shown in fig. 16.43. Without negative feedback the very high open-loop voltage gain of the amplifier makes it unsuitable for most applications as an amplifier. The output tends to go to either the positive or the negative supply voltage and stay there, i.e. it saturates. Alternatively, the gain varies dramatically with the frequency of the signal being amplified, producing other unwanted effects. With negative feedback the gain of the amplifier is controlled solely by the external feed-back resistor R_f and the input resistor R_{in}. The gain is also constant over a range of signal frequencies and is much lower than the open-loop gain.

Figure 16.43   *An inverting amplifier with negative feedback*

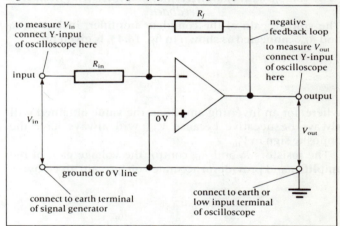

A resistor R_f connected between the output terminal and the inverting input terminal of the op-amp provides negative feedback. The value of R_f determines the fraction of the output signal which is fed back to the inverting input. The feedback signal is in antiphase with (the opposite of) the externally applied input signal. This opposition follows from the action of the op-amp, which makes its output voltage have the opposite sign to that of the inverting input. The output signal is always in antiphase with the input signal at the inverting input. This is explained in fig. 16.44.

In the circuit shown in fig. 16.43, the non-inverting input is connected to the ground or 0 volts line making $V_2 = 0$ volts. The voltage difference between the two inputs of the op-amp is effectively just the voltage applied to the inverting input and so the op-amp works as a single-input inverting amplifier with negative feedback.

Figure 16.44 *The effect of negative feedback*

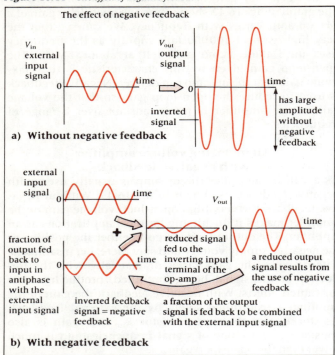

a) **Without negative feedback**

b) **With negative feedback**

The voltage gain of an inverting amplifier with negative feedback

The voltage gain of the complete amplifier, including R_f and R_{in}, connected as shown in fig. 16.43, is given by:

$$\text{voltage gain } A = \frac{V_{out}}{V_{in}}$$

where, for an inverting amplifier, the value obtained will always be negative because V_{out} will always have the opposite sign to V_{in}.

The resistors R_f and R_{in} control the voltage gain of the amplifier and provide an accurate voltage gain given by:

$$\text{voltage gain } A = -\frac{R_f}{R_{in}}$$

This gain is independent of the open-loop gain of the op-amp IC provided that the voltage gain of the amplifier is not greater than the open-loop gain.

Worked Example
Op-amp inverting amplifier

An inverting op-amp has a feedback resistor of $200\,k\Omega$ and an input resistor of $10\,k\Omega$. It receives an input signal of $+0.4\,V$. Calculate the output voltage.

$$\text{voltage gain } A = -\frac{R_f}{R_{in}} = -\frac{200\,k\Omega}{10\,k\Omega} = -20$$

and since:

$$\text{voltage gain } A = \frac{V_{out}}{V_{in}}$$

we have: $V_{out} = V_{in} \times A$

$\therefore \quad V_{out} = 0.4\,V \times -20 = -8.0\,V$

Answer: The output voltage $= -8.0\,V$.

Using an op-amp to amplify d.c. voltages

Even when there is no input voltage connected to either of the op-amp's inputs there is often a small voltage difference of 1 mV or less between its two inputs. This small voltage, known as an *offset voltage*, when amplified can produce an output voltage of up to 1 V. Such a d.c. voltage appearing at the output can, when added to the true output voltage, produce an inaccurate result. In some circuits it can cause the output voltage to drift and go to saturation. Clearly this was a problem which needed a solution when the op-amp IC was being designed.

The offset null pins

The '**offset null**' pins allow the offset voltage to be cancelled out or made **null**. Fig. 16.45a shows how a $10\,k\Omega$ variable resistor is used to apply correcting voltages to the 'offset null' pins, numbers 1 and 5. The resistor, working as a voltage divider, is adjusted until the output is exactly zero when the input is zero (connect both inputs to ground). In an op-amp circuit without negative feedback the output voltage will swing suddenly between the supply voltages when the offset null resistor is correctly adjusted. Fig. 16.45b shows the complete inverting amplifier circuit. All except the input and output connections are usually already made on the op-amp boards and modules available in schools or colleges. The same connections are also made to several more modern op-amp ICs than the 741. For example, the J-FET 351 op-amp can be used in all the 741 circuits and has a much higher input resistance of about $10^{12}\,\Omega$.

Figure 16.45 *A full op-amp inverting amplifier circuit*

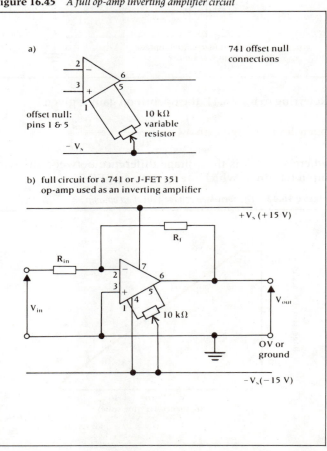

Measuring the voltage gain of an inverting amplifier

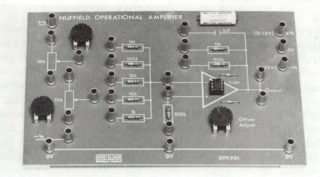

An op-amp board.

- *Connect an op-amp module to its special power supply ($+15$ V, 0, -15 V) or to a pair of 9 V batteries.*
- *Select a signal frequency of 1 kHz on a signal generator and connect the generator output to the input resistor R_{in} of the op-amp.*
- *Connect resistors in the op-amp circuit as suggested in table 16.3. (Some op-amp modules have built-in resistors for R_{in} and R_f, but you will probably need to connect other values from an external resistance substitution box.)*
- *Measure the input and output voltages V_{in} and V_{out} using an oscillosocpe as an a.c. voltmeter, see figs. 16.12 and 16.13, switching on the time-base of the oscilloscope to display the input and output voltage waveforms.*

The waveforms should appear as shown in fig. 16.44. An output waveform like that in fig. 16.46 shows that the output voltage is saturating alternately at the positive and negative supply voltages. This happens when the amplitude of the input voltage is too large and the amplifier is overloaded. The output waveform is said to be **clipped**. To overcome this problem, reduce the amplitude of the signal generator output voltage. The voltage gain of the amplifier should be calculated using values of V_{in} and V_{out} obtained when the output is not saturated.

- *Record the measured voltages in a table like table 16.3.*
- *Calculate the voltage gain expected from the formula $A = R_f/R_{in}$, and the gain measured from the formula $A = V_{out}/V_{in}$.*

Table 16.3

R_{in}/Ω	R_f/Ω	R_f/R_{in}	V_{in}/V	V_{out}/V	$A = V_{out}/V_{in}$
100k	100k	1			
100k	1 M	10			
values	values				
between 1 k	between 1 k				
and 1 M	and 1 M				

Figure 16.46 *The output voltage saturates when the amplitude of the input signal is too large*

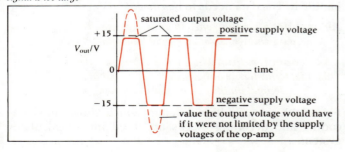

Investigating how the voltage gain of an inverting amplifier depends on the signal frequency

- *Using the circuit of fig. 16.43, select $R_f = 1\,M\Omega$ and $R_{in} = 1\,k\Omega$.*

With these values we would expect a high gain given by

$$A = R_f/R_{in} = 1\,M\Omega/1\,k\Omega = 1000$$

- *Set the signal generator to several frequencies in the range 10 Hz to 100 kHz and at each frequency measure V_{in} and V_{out}.*
- *Record the values as shown in table 16.4.*
- *Calculate the actual voltage gain of the amplifier at each frequency using the formula $A = V_{out}/V_{in}$.*
- *Plot a graph of voltage gain A against frequency f as shown in fig. 16.47.*

Note that both the scales use values which increase in powers of ten. These scales are logarithmic.

The results of the investigation show that, using negative feedback, an op-amp can provide a *constant voltage gain*, but with the following limitations:

a) The voltage gain of an op-amp with negative feedback (solid line on the graph in fig. 16.47) cannot exceed the open-loop gain of the basic op-amp without feedback (broken line on the graph).

b) The output voltage cannot exceed the supply voltages.

The range of frequencies of the input signal for which the voltage gain of the amplifier is roughly constant is called the **bandwidth** of the amplifier. This bandwidth is found to reduce when a higher voltage gain is required from the amplifier.

Within the limitations mentioned a negative feedback amplifier gives a stable and distortion-free amplified output signal.

Assignment

Try questions 16.20 to 16.22

Table 16.4

f/H_2	V_{in}/V	V_{out}/V	$A = V_{out}/V_{in}$
10			
100			
1 k			
10 k			
100 k			

Figure 16.47 *How the gain of an op-amp varies with the signal frequency*

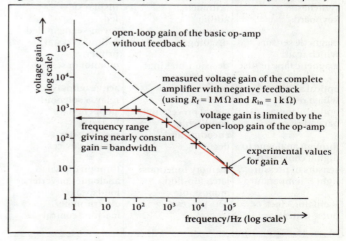

16.6
DIGITAL SYSTEMS

Features of digital systems

Your calculator is an example of a **digital system**. All the information it handles is in a digital form. So what do we mean by digital information?

Any information which has a limited number of values and can change only suddenly from one value to another is digital information.

You can see that the information **INPUT** to your calculator is in a digital form. When your fingers touch the keyboard information is fed in from the outside world.

The **OUTPUT** from your calculator is usually a display or sometimes a printer. The information displayed or printed out is in separate digits. Both the input and output digits are usually numbers in the decimal system.

The **PROCESSOR** in your calculator which processes the data, i.e. does the calculations, also works with digital information but it uses a different number system known as the **binary code**. Fig. 16.48 shows the main blocks in the digital system of your calculator.

All electronic systems have the three basic blocks: **INPUT**, **PROCESSOR** and **OUTPUT**. Some examples of these are listed in table 16.5 below. The calculator also has a **MEMORY** block. The memory block stores permanent programmes as well as numbers used during data processing. The memory block stores all its information in the binary code.

Figure 16.48 *Building blocks in an electronic calculator*

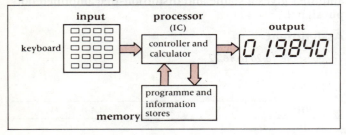

Table 16.5 Building blocks in digital electronic systems

Inputs	Processors	Outputs
switches keyboards	arithmetic: adding, counting, dividing,	displays: led, seven-segment, led & liquid crystal printer,
magnetic sensors which read: magnetic tape or disc	decision-making: logic gates,	monitor or screen
optical sensors which read: optical/compact discs, bar codes	comparator signal processing: amplifying, encoding and decoding	active outputs: relay & solenoid, stepper motor, buzzer, warning devices, control devices
through analogue to digital converters: sensors of pressure, light & temperature	memory functions: latch flip-flop, monostable,	through digital to analogue converters: loudspeaker
oscillator, clock or pulse generator	data processing	electric motor, machine control

Codes used in digital systems

The decimal code

In everyday life we use the **decimal system** with its ten digits 0 to 9. In the decimal number 365, the digit (5) on the right gives the number of units. Its place gives it the lowest value and makes it the least significant digit in the number. The digit (3) on the left gives the number of hundreds and its position makes it the most significant digit in the number 365. The values of the digits increase from right to left in powers of 10: 10^0, 10^1, 10^2 etc. The decimal code is used in digital electronic systems for the convenience of the user only at the inputs and outputs of systems.

The binary code

The simplest digital system uses only two states. The two states can be stored or processed in a variety of forms. In fact any code which has two and only two symbols or states is a binary code. Some examples are given in table 16.6. Think of the table as giving the answer to the question: 'Is the door closed?'

Table 16.6

Answer yes	Closed switch	Magnetised memory	High voltage	Logic 1
Answer no	Open switch	Memory not magnetised	Low voltage	Logic 0

The symbols used when writing a binary code are 0 and 1. These are often called *logic symbols* and are usually referred to as **logic** (or logical) **0** and **logic 1**. When binary-coded information is processed in a digital system, the **logic 1** state is represented by a **'high' voltage**, often 5 V, and the **logic 0** state is represented by **0 V**.

Logic 1 and 0 are used as the two digits in binary arithmetic. In a binary number also, the value of the digits increases from right to left, this time in powers of 2, i.e. 2^0, 2^1, 2^2, 2^3 etc. The decimal values of these digits are therefore: 1, 2, 4, 8 etc.

For example, the binary number 1101 is equivalent to: $(1 \times 8) + (1 \times 4) + (0 \times 2) + (1 \times 1) = 13$ in decimals. The four digits in the binary number 1101 are called **'bits'** and 1101 is a **four-bit binary number**. The 1 on the right is the **least significant bit** or **l.s.b.** The 1 on the left is the **most significant bit** or **m.s.b.**

Encoding and decoding

When a key on a keyboard is pressed the decimal number it represents must be converted into a binary-coded number before the processor of the calculator or computer can handle it. The data needs to be *encoded*.

*An **encoder** is a circuit, often in an IC and usually at the input of a system, which converts data into a binary code.*

*A **decoder** is a circuit which converts binary-coded data into a decimal form. It may also encode the data so that it will show a decimal number on a seven-segment display.*

Fig. 16.49 gives a block diagram which shows how encoding and decoding blocks are fitted into a calculator system.

Figure 16.49 *Encoder and decoder in a calculator system*

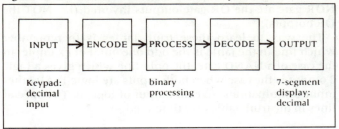

INPUT → ENCODE → PROCESS → DECODE → OUTPUT

Keypad: decimal input   binary processing   7-segment display: decimal

When an analogue signal such as a temperature or sound is to be fed into a digital system it must be converted into a binary-coded digital form. A circuit which does this is called an **analogue to digital converter (ADC)**. Fig. 16.50 shows how this process is used when digital recordings of music are made.

Figure 16.50 *A digital recording system using an ADC*

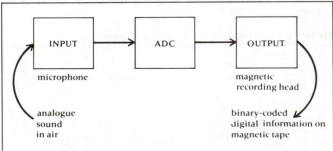

INPUT → ADC → OUTPUT

microphone   magnetic recording head

analogue sound in air   binary-coded digital information on magnetic tape

If we wish to play back a digital recording, say from a compact disc, we need to convert the binary-coded digital information back into an analogue varying signal. The analogue signal, in the form of an alternating voltage, will drive a loudspeaker. The loudspeaker is the output of the system which reproduces the analogue sound. The circuit which converts digital information into an analogue signal is called a **digital to analogue converter (DAC)**.

Fig. 16.51 shows how a continuously varying analogue signal is converted into a digital signal.

Figure 16.51 *Analogue to digital conversion of a waveform*

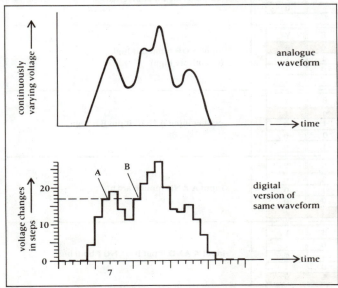

continuously varying voltage

analogue waveform

time

voltage changes in steps

20

10

0

A   B

digital version of same waveform

7   time

The voltage is sampled rapidly in equal short time intervals and is given a value on a digital voltage scale. The voltage can change only in steps according to the digital values allowed on the scale. For example, the seventh sample shown at point A has a digital value of 17. The same value occurs again at point B. The digital value of the voltage at each sample time is converted into a binary code for recording or transmission. If the sampling is not rapid enough some detailed information will be lost in the digital conversion.

Integrated circuits in silicon 'chips'

All of the processing devices, the encoders and digital memories can be built in the form of integrated circuits. These ICs can be made very small and are encapsulated inside plastic or ceramic cases. Fig. 16.52 shows a common package. The silicon 'chip' in which the IC is constructed is a very small piece of silicon perhaps only 5 mm square and less than 1 mm thick. Within that tiny chip of silicon there can be thousands of components. In very large scale integration the number of components in a single chip can approach a million!

Figure 16.52 *The packaging of a silicon 'chip'*

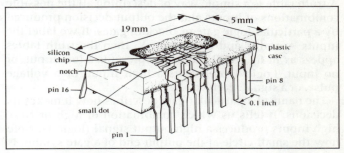

19 mm   5 mm

silicon chip   plastic case

notch

pin 16   pin 8

small dot   0.1 inch

pin 1

Digital logic families

There are two main families of digital electronic ICs. Each family contains a wide range of logic circuits. The two families are constructed using different semiconductor technology. The circuits of the **TTL** family use **Transistor–Transistor Logic**. The other family, **CMOS**, stands for **Complementary Metal Oxide Semiconductor**. In table 16.7 the two families are compared. Systems should be built using ICs from only one family to ensure compatibility.

Table 16.7 Properties and requirements of TTL and CMOS logic ICs

Property	TTL	CMOS
Power requirements:		
Supply voltage, V_s	5 V ± 0.25 V d.c. stabilised	3 V to 15 V d.c.
Power taken	milliwatts	microwatts
Input voltage logic levels:		
'low', logic 0	0 to +0.8 V	0 to 0.3 × V_s
'high', logic 1	+2 V to +5 V	0.7 × V_s to V_s
Unused inputs	behave as if connected to a high voltage & assume logic 1	must be connected to logic 0 or 1 otherwise behave erratically
Switching speed	fast	slower

Logic gates

A variety of building blocks which work like switches or gates are used in digital systems. These basic building blocks are called logic gates. Usually several logic gates are constructed in a single IC and in most cases each gate has two or more inputs.

These blocks are like 'gates' because they have to be 'opened' to let information pass through them and reach their outputs. Each type of gate is 'opened' by a particular combination of information fed to its inputs in binary code. The information, like a key, has to fit the gate. So a logic gate gives an output in the binary code which depends on the binary-coded information given to its inputs. In this way it makes a 'decision': to open or not to open. When the right information is given to the inputs of a gate it decides to 'open' and gives its 'yes' answer as a 'high' voltage signal or logic 1 output. This is like saying 'yes, the key fits'.

There are five basic types of logic gates from which all the more complicated logic or digital building blocks are constructed. These are shown in fig. 16.53.

Truth tables

A truth table is a simple way of describing all the possible combinations of inputs and the output decision produced by a particular logic gate or block of gates. If we label the inputs A and B and the output Y, then the truth tables appear as in fig. 16.53. Logic 0 represents a low input, or no input. Logic 1 represents a high input, or a voltage pulse, or a signal received.

The name of a particular gate describes how it makes its decisions. It tells us which combination of high or NOT high inputs produces a high output signal (logic 1). Note how the small circle at the output end of a gate symbol is the inverting or NOT symbol. For example the outputs of a NOR gate are the OR gate outputs inverted (i.e. NOT the OR outputs).

The way to learn these truth tables is to remember which line in the table is the odd one out. For example, in the case of a NAND gate, the only line which gives a logic 0 output is the case when both inputs are logic 1. All other input combinations give an output of logic 0. The special lines in the truth tables are tinted red.

Testing logic gates

To test a logic gate or combination of logic gates we need an indicator to show the logic state of the output. An l.e.d. is a suitable indicator because it needs only a low current which the outputs of all logic gates can supply. If the output of a logic gate Y is 'high', i.e. at +5 V, it must light an l.e.d. when connected to the input I of the logic level indicator. A simple circuit is shown in fig. 16.54. The method of calculating the series resistance R is given on p 338.

Figure 16.54 *A circuit for an l.e.d. logic state indicator*

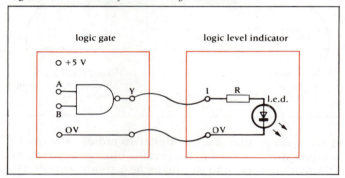

Figure 16.53 *Logic gates*

logic gate	symbol	is equivalent to	truth table			the output is **high**, logic 1 when:
NOT	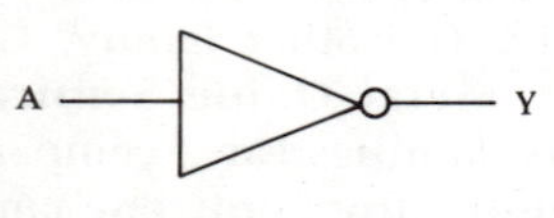 A ▷o Y	INVERTER	input A: 0, 1	output Y: 1, 0		input A is **NOT** high (output is the input inverted)
OR	A, B ⊐ Y	(inclusive) OR	A: 0,0,1,1	B: 0,1,0,1	Y: 0,1,1,1	input **A OR** input B is high (or both are high)
NOR	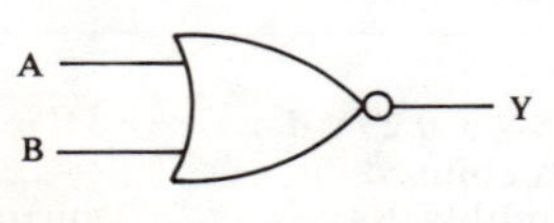 A, B ⊐o Y	OR-NOT	A: 0,0,1,1	B: 0,1,0,1	Y: 1,0,0,0	neither input **A NOR** input B is high
AND	A, B ⊐ Y		A: 0,0,1,1	B: 0,1,0,1	Y: 0,0,0,1	input **A AND** input B are high
NAND	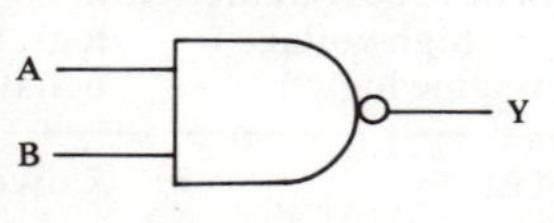 A, B ⊐o Y	AND-NOT	A: 0,0,1,1	B: 0,1,0,1	Y: 1,1,1,0	input **A AND** input B are **NOT** both high

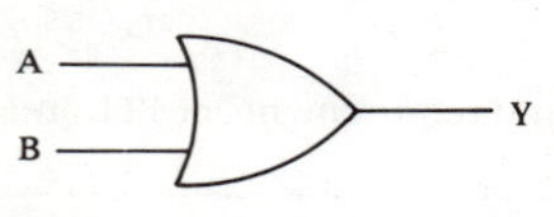

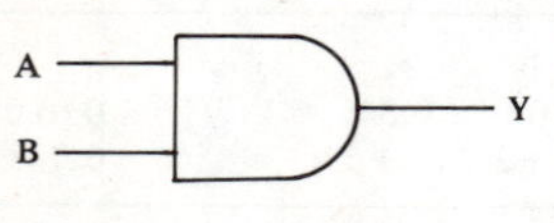

- *Connect a logic gate unit (or module) to a 5 V stabilised d.c. power supply. Make sure that the positive terminal of the supply is connected to the +5 V socket of the logic gate unit.*
- *Connect the output socket Y of the logic gate to the input socket I of a logic level indicator unit. Make sure that the 0 volts or 'ground' line is also connected.*

When the l.e.d. lights it indicates a 'high' output voltage from the logic gate and represents logic 1.

- *Test the gate by connecting each of its inputs A and B to the 'high' voltage level (the +5 V socket) and the 'low' voltage level (the 0 V socket) in turn so that all the four possible input combinations are tried. Do not connect the same input to both +5 V and 0 V at the same time as this would short circuit the power supply.*

If you are using TTL logic ICs (e.g. those with four reference numbers beginning with 74) remember that an unconnected input will behave as if it is connected high.

- *Record the test results in the form of a truth table for each logic gate.*

Combinations of logic gates

Building other logic gates

When logic gates are manufactured on an IC it is easier and more economical to put a large number of the same logic gate on one chip than to make a chip with a particular mixture of gates. For these reasons we find that it is common to build other logic gates using entirely NAND or NOR gates. Fig. 16.55 shows how several logic gates can be built using only NAND gates.

- *Using several NAND gate units and a logic level indicator unit, test the combinations shown in fig. 16.55.*
- *Complete truth tables for each of the combinations showing the logic levels at all intermediate connections.*

Figure 16.55 *Building gates from combinations of NAND gates*

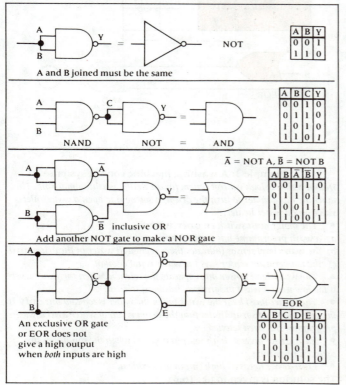

Binary arithmetic

Logic gates are ideal for doing simple binary arithmetic because they use the same binary code. The circuit shown in fig. 16.56 is known as a half-adder. It will add two binary bits (single digits, 0 or 1) and give the answer as a two-bit binary number. The least significant bit (l.s.b.) appears at the output of an **EOR** gate, labelled **SUM**. The most significant bit (m.s.b.) appears at the output of an **AND** gate, labelled **CARRY**. Note that a half-adder has only two inputs, A and B, so it can add only two bits together.

Figure 16.56 *A half-adder*

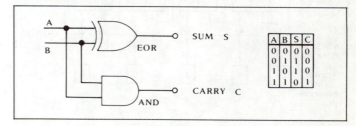

To add two numbers with more than one bit each needs a full-adder for each column of bits. A full-adder has **three** inputs, two as before to add a bit from each number, A and B, and an extra input, **CARRY IN**, to add the carry bit from the previous column. Fig. 16.57 shows how a full-adder can be constructed from two half-adders. Since an EOR gate and an AND gate can be built using NAND gates alone, it is possible to build both a half-adder and a full-adder using only NAND gates. As several full-adders can be assembled in a single IC, we can see how very complex circuits can be constructed using only simple NAND gates.

Addition of two four-bit numbers requires one half-adder to add the two bits in the l.s.b. column and three full-adders to add the three bits (two from the numbers and a carry bit) in each of the other columns, fig. 16.58.

Figure 16.57 *A full-adder*

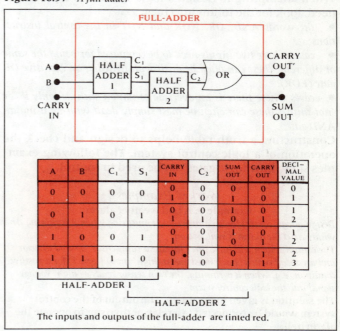

The inputs and outputs of the full-adder are tinted red.

Figure 16.58 *A four-bit adder*

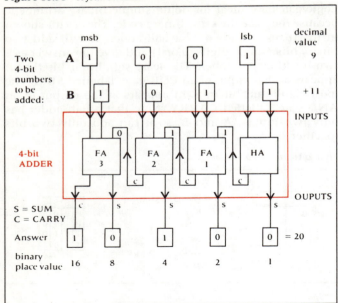

Using the example of adding the two numbers: A 1001 (decimal value 9) and B 1011 (decimal value 11), we can see how a four-bit adder works.

The **half-adder** *(HA)* adds the two least significant bits: 1 and 1 giving a sum 0 and a carry 1.

Full-adder 1 (FA1) adds the next two bits from A and B, 0 and 1, to the carry bit from the half-adder, 1, giving a sum 0 and another carry 1. And so on.

The carry bit from full-adder 3 (FA3) gives a fifth bit in the answer: 10100 (decimal value 20).

This four-bit adder can be constructed in a single IC using only NAND gates.

Logic gates in control

Systems of logic gates are often used to control machines. When attempting to design a logic system for a particular need, look for the following clues:

- *the words AND, OR, NOR and NOT in the control instructions;*
- *cases where two inputs need to be compared for being the same or different; these can often be dealt with by using an exclusive OR gate (EOR);*
- *cases where there is an override condition such as 'only if' or 'not until'; these can often be most simply dealt with by using an AND gate.*

Constructing a truth table helps to design and check the operation of a logic control system. The following examples show how logic gates can be used in control.

Worked Examples

Example 1: A bad weather alarm

*Design a logic control for an alarm which must ring when it is raining **or** windy **and** the alarm is switched on.*

The control system will have three inputs: a rain sensor, a wind detector and an on/off switch. Each must give a 'high' input (logic 1) for a positive condition, e.g. when it is raining, the rain sensor sends a 'high' voltage signal into the logic control sytem.

The solution is given in fig. 16.59. The output of the control system would be connected to a relay or thyristor to operate the alarm bell.

Figure 16.59 *Logic control system for bad weather alarm*

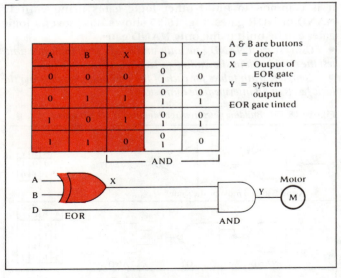

Example 2: A lift motor control system

Design a logic control for a lift motor. The lift operates between the ground floor and the first floor and there are two operating buttons.
The control system must turn the motor on only if:
a) *the lift doors are closed **and***
b) *only one lift operating button is pressed.*
When closed, the doors send a logic 1 to the logic control. When pressed, the operating buttons latch on and send a logic 1 to the logic control.
Drawing a truth table first for the operating buttons, A and B, shows that the logic gate we need is an exclusive OR gate or EOR. It will not give a high output when either no buttons are pressed or both are pressed. This logic function is that of a comparator which gives a high output only when the two inputs are different. See fig. 16.60.

Figure 16.60 *A lift motor control system*

A	B	X	D	Y
0	0	0	0	0
			1	
0	1	1	0	0
			1	1
1	0	1	0	0
			1	1
1	1	0	0	0
			1	

A & B are buttons
D = door
X = Output of EOR gate
Y = system output
EOR gate tinted

Example 3: A washing machine control system

Design a simple logic control system for an electric washing machine. The machine has a motor to turn the drum, a solenoid to open a water inlet valve and a water heater.

- *The motor must switch on **only when**:*
a) *a wash programme has been selected,*
b) *the water level sensor indicates there is enough water in the drum **and***
c) *the temperature sensor says the water is hot enough.*
- *The water level sensor must operate the solenoid when the water level is too low **and** a wash programme has been selected.*
- *The heater must heat the water when the water is too cold but **only if** the water level sensor indicates that the drum is full of water **and** a wash programme has been selected.*
- *Each sensor gives a 'high' output (logic 1) when the conditions are correct for washing.*
- *Each device needs a 'high' input to operate it.*
The solution is given in fig. 16.61.

Figure 16.61   *A simple washing machine logic control system*

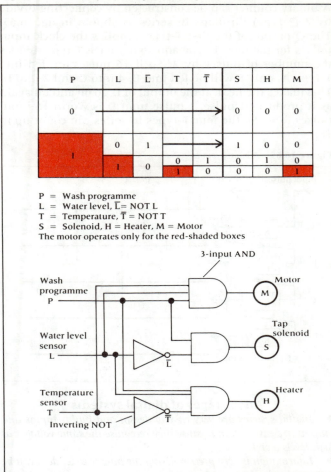

P	L	$\bar{L}$	T	$\bar{T}$	S	H	M
0					0	0	0
	0	1			1	0	0
1	1	0	0	1	0	1	0
			1	0	0	0	1

P = Wash programme
L = Water level, $\bar{L}$ = NOT L
T = Temperature, $\bar{T}$ = NOT T
S = Solenoid, H = Heater, M = Motor
The motor operates only for the red-shaded boxes

In this control system there are three inputs and outputs.

The motor is controlled by all three inputs, all of which must send a logic 1 to a 3-input AND gate before the motor can start.

The solenoid, which opens the water inlet valve, is controlled by both the washing programme and the water level sensor. But the solenoid must switch on if the water level sensor sends a logic 0 indicating that the water level is too low. The logic signal from this sensor must therefore be inverted (a NOT gate) to feed a logic 1 into the AND gate and turn the solenoid on.

The water heater is also controlled by all three inputs. As the water must be heated only when the water is NOT hot enough, a NOT gate must also be included in the line from the temperature sensor.

● *Using kits of logic gates and suitable power supplies and other components, assemble working models of the three control systems described above.* Motors and solenoids will need relays or other devices such as thyristors to drive them. The input sensors must be connected in an electronic switch circuit which will give output voltages equal to logic 0 or 1.

Flip-flops

Flip-flops are devices which can have only two possible outputs between which they switch or 'flip and flop'. The output of a flip-flop made from logic gates can be either logic 0 or logic 1 and it can switch between these. Flip-flops can be made to oscillate, to remember and to count, making them most useful electronic devices.

The SR flip-flop

Figure 16.62   *The NOR gate SR flip-flop*

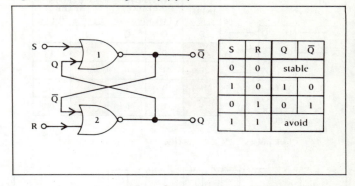

S	R	Q	$\bar{Q}$
0	0	stable	
1	0	1	0
0	1	0	1
1	1	avoid	

Figure 16.62 shows how an SR flip-flop can be made from two NOR gates. The outputs of a pair of NOR gates are fed back into each others inputs. The other two inputs are the S input which SETS the output Q to logic 1 and the R input which RESETS (or clears) the output Q to 0.

The truth table can be interpreted as follows:

If S and R are both 0, Q does not change.
If S = 1, then Q is SET to 1 also.
If R = 1, then Q is RESET to 0.
If S = R = 1: avoid this case.

Note that $\bar{Q}$ is the opposite of Q, i.e. $\bar{Q}$ = NOT Q.

You can check how it works by completing a truth table for each of the NOR gates. For example, if S = 1, then $\bar{Q}$ = 0. If $\bar{Q}$ and R = 0 then Q = 1. And so on.

This flip-flop is also known as a **bistable** because it has two stable output states, i.e. the output can remain indefinitely at either 0 or 1 unless a new logic 1 is applied at R or S to change it. The bistable property of this flip-flop makes it a simple form of electronic memory. It will remember what state its output has been left in even if the logic 1 applied to R or S is removed. A similar flip-flop can be built using two NAND gates.

The D-type flip-flop

The **D** or **DATA flip-flop** is a more useful flip-flop which can also be built using only one type of logic gate, usually NAND gates. The D flip-flop is a basic memory unit often used in silicon-based memory ICs. It is also often used in ICs which divide by 2 or count.

This flip-flop has only one input, the DATA or D input, rather than the two S and R inputs which cause confusion when both receive a logic 1 input. The D flip-flop also has an extra **CLOCK** terminal. The output of the flip-flop cannot change (and will therefore store its output data: a 0 or 1) unless the CLOCK input goes from logic 1 to logic 0. This means that we can control when it should store new data. The data (logic 0 or 1) is supplied to the D or DATA input, but is only stored if and when the CLOCK input goes from 1 to 0. In effect, the CLOCK input controls when we can 'write' data to the memory. Figure 16.63a shows a block symbol and a truth table for the D flip-flop.

What do we mean by a 'CLOCK'? In logic circuits which handle a sequence of data and instructions an electronic clock is needed to synchronise the processing of data. Processing in all computers must be synchronised by clocks.

Figure 16.63 *A D-type flip-flop*

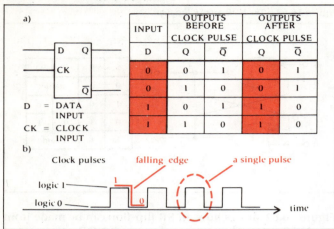

INPUT	OUTPUTS BEFORE CLOCK PULSE		OUTPUTS AFTER CLOCK PULSE	
D	Q	$\bar{Q}$	Q	$\bar{Q}$
0	0	1	0	1
0	1	0	0	1
1	0	1	1	0
1	1	0	1	0

D = DATA INPUT

CK = CLOCK INPUT

b)

Clock pulses   falling edge   a single pulse

logic 1

logic 0

time

The clock generates a series of pulses which change between logic 0 and 1, usually at a high frequency so that processing can be done very quickly, fig. 16.63b. The D flip-flop will store the new data waiting at its D input only at the instant when the falling edge ⌐‾⌐₀ of one such pulse (logic level change 1 to 0) arrives at its CK input.

Timers and counters

Electronic circuits known as **astables** or **crystal oscillators** are commonly used to generate a regular series of clock pulses. For example, a digital watch uses a quartz crystal oscillator to regulate its time-keeping. To keep track of the passing of time or to organise the processing of data in a computer, a pulse counter is needed to count the clock pulses.

The D flip-flop can be used as a clock pulse counter or a counter of any digital information presented as logic level 1 pulses. The D-type becomes a divide by 2 flip-flop when its $\bar{Q}$ output is fed back into its D input, fig. 16.64a. The clock input (CK) receives the pulses to be counted. When $\bar{Q}$ is at logic 1, a 1 is also waiting at the D input for the next clock pulse. On its falling edge, the 1 at D is stored and appears at Q, changing $\bar{Q}$ to 0. By this process, each pulse received at the clock changes the outputs over. This 'toggle' action gives the flip-flop the name of a toggle or T-type. Fig. 16.64b shows how this process divides the number of pulses by two.

Figure 16.64 *T-type flip-flops divide by 2*

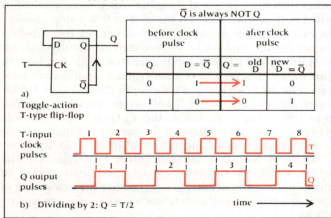

	$\bar{Q}$ is always NOT Q			
	before clock pulse		after clock pulse	
Q	D = $\bar{Q}$	Q = old D	new $\bar{Q}$ = Q
0	1	→ 1	0
1	0	→ 0	1

a) Toggle-action T-type flip-flop

b) Dividing by 2: Q = T/2

A binary counter

A binary counter can be constructed by connecting divide by 2 (T-type) flip-flops in series as shown in fig. 16.65. The Q output of the first T-type supplies the clock input pulses for the next T-type and so on. Each T-type divides the number of pulses by 2. So if 16 pulses are fed into T-type A, only 8 will be fed into B, 4 into C and 2 into D. This makes the l.e.d. at the output of D go on and off again as it produces 1 pulse. 1 pulse from D is worth 16 input pulses to A. So the four T-types in series can count up to 16.

Figure 16.65 *A binary counter*

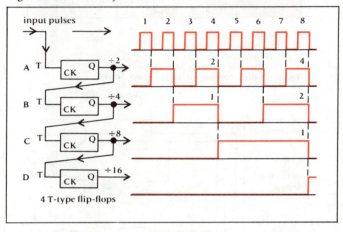

4 T-type flip-flops

Advantages of digital systems

- *Digital systems are very easy to design and build. Inputs and outputs are easy to join because they recognise the same voltages as logic levels 0 or 1.*
- *Information is very precise. Using the binary code, data can be composed only of 0s and 1s.*
- *Information can be stored very easily. Information in the binary code (bits) can be stored in silicon chips and on magnetic discs each storing millions of bits.*
- *Circuits can be fully integrated. Whole circuits containing thousands of logic gates can be built on a single silicon chip. For example, a complete calculator circuit can be built on a single chip.*
- *Digital systems can think faster. The simplicity of logic gates and their small size allows them to change their logic states very quickly. Only a few nanoseconds are needed per step in a sequence of logic changes.*

Assignments

Design systems for the control problems below. In each case work in the following order:
- draw a block diagram of the system;
- if logic gates are to be used, draw truth tables;
- work out the combination of logic gates needed;
- build and test the system.

1. A fire alarm system which will cause an alarm to ring if either a high temperature or smoke is detected.

2. A door bell which will ring only in the day time.

3. A motor driven cooling fan on a desk which switches on only when the air temperature gets too hot and someone is sitting in the chair at the desk.

4. A circuit which rings an alarm when it has counted to ten. Then make the circuit reset itself.

Try questions 16.23 to 16.26

16.7
COMMUNICATION SYSTEMS

Communication systems enable us to transmit information over almost any distance. That information can be speech or music and pictures or data of any kind. The information to be transmitted is called the **signal**.

Corruption and attenuation of the signal

Signals which are transmitted over great distances can become corrupted because of interference which alters or distorts the signal and noise which makes it more difficult to detect or 'hear'. We hear noise as unwanted hiss and crackle in the background. It can be caused by the electromagnetic radiation which is generated by all electrical machines such as motors, transformers and switches. The ratio between the amplitude of the signal and the noise is called the **signal-to-noise ratio**. It is this ratio which must remain large if the signal is to remain clear and not get lost in the noise.

Signals all get weaker as they travel due to energy absorption in cables or fibres or the spreading out of radio waves as they travel. The weakening of a signal is called **attenuation**. The design of a communication system must minimise these problems to deliver good quality signals to the receiver. The receiver may be a person, a computer, a printer or various other controlled machines.

Transmission systems

Figures 16.66, 69 and 70 show the variety of transmission systems used today. They divide into two main categories:

- *Cable and optical fibre links. These include:*
1. Coaxial cables across land and sea.
2. Optical fibres also across land and sea.
- *Radio and microwave links. These include:*
3. Microwave links directly between towers.
4. Satellite microwave links.
5. Very high frequency (VHF) radio and ultra high frequency (UHF) TV broadcasts. These 'space waves' are radio waves which travel directly along the line of sight from transmitting aerial to receiving aerial.
6. High and medium frequency radio waves which are known as 'short waves' and 'medium waves'. These 'sky waves' are reflected by the ionised layers of air in the upper atmosphere called the ionosphere.
7. Low frequency 'long wave' radio waves which travel as 'ground waves'.

Radio and microwave communication systems

A wide band of the electromagnetic spectrum is used for carrying information. Figure 16.67 shows where the different kinds of radio and microwaves fit in the spectrum.

Figure 16.67 *The radiowave and microwave spectrum*

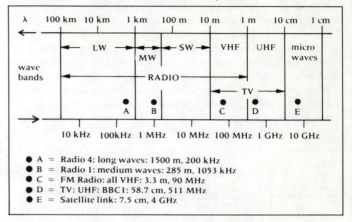

- A = Radio 4: long waves: 1500 m, 200 kHz
- B = Radio 1: medium waves: 285 m, 1053 kHz
- C = FM Radio: all VHF: 3.3 m, 90 MHz
- D = TV: UHF: BBC1: 58.7 cm, 511 MHz
- E = Satellite link: 7.5 cm, 4 GHz

The formula connecting the wavelength λ and the frequency f of a radio wave or microwave is $c = f\lambda$ where $c = 3 \times 10^8$ m/s and is the wave speed. Worked examples are given on page 486.

Modulation

The radio or microwave is called a carrier wave when it carries information. The process by which the carrier wave is made to carry information is called **modulation**.

Modulation is a process in which information is superimposed on a carrier wave, changing its amplitude or frequency, in a way that allows the information to be transmitted and recovered.

Several kinds of modulation are used in communication systems. Some carry analogue signals and others digital.

Amplitude modulation (AM)

The most well known form of amplitude modulation is that which is used in long, medium and short wave radio links. An analogue audio frequency (a.f.) signal representing the sound waves to be transmitted is carried by the radio waves. The analogue signal is added to the radio wave so that its amplitude varies but its frequency does not. This process is described in the box on the next page.

Figure 16.66 *Transmission systems*

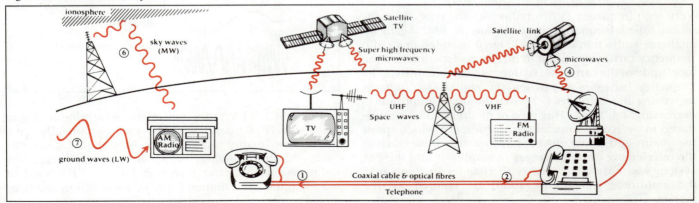

Amplitude modulated (AM) radio transmitter and receiver

Sound waves cannot travel very far through the air, so they are carried 'piggyback' on radiowaves which can travel much further. The top diagram shows how the sound waves are first converted to an electrical analogue signal by a microphone. As the output of the microphone has a small amplitude the first block in the processor must be an amplifier.

An oscillator generates a radio frequency (r.f.) signal, which is then added to the analogue signal by an *adder block* called a **modulator**. This is where the r.f. signal starts to carry the analogue signal. The output of the system is fed to an aerial which acts as a transducer converting the r.f. electrical signal into radiowaves. The radiowaves carry the analogue signal representing the sound waves at the speed of light (3×10^8 m/s).

Another aerial on a receiver some distance away converts a small fraction of the radiowave energy back into an r.f. electrical signal, as shown in the lower diagram. The aerial on the receiver also picks up all the other radiowaves from other stations within range. Many signals of different radio frequencies are fed into the radio receiver system. The first block in the system is a **tuner**, which can be adjusted to select a particular station by matching its own natural conduction frequency to the frequency of the required radio station. At this one frequency a much larger signal is produced in the tuner by the process called *resonance* (p 436). The selected r.f. signal is then amplified by the next block in the system. This is followed by a detector block which separates the analogue signal from the r.f. (carrier) signal by a rectification process. This is called **demodulation**, and is the reverse of the modulation process in the transmitter system, where the analogue signal was added to the r.f. (carrier) signal. The separated analogue signal is amplified before being fed into the output block, which is a loudspeaker. This transducer converts the analogue signal back into sound waves.

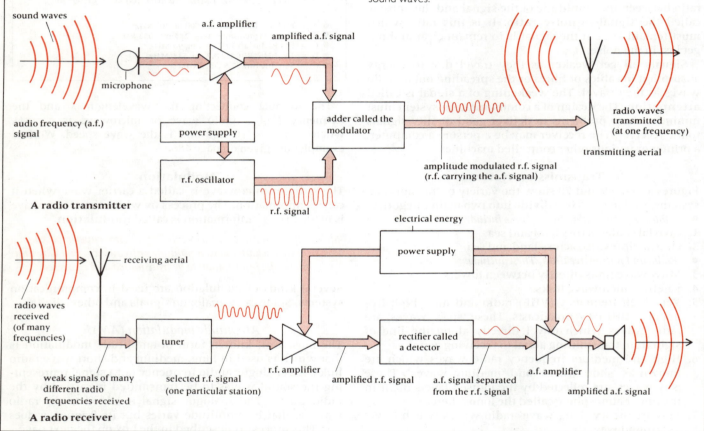

A radio transmitter

A radio receiver

Frequency modulation (FM)

Very high frequency (VHF) radio uses the type of modulation called **frequency modulation** or **FM**. Again an analogue audio frequency (a.f.) signal is added to a radio frequency carrier wave. But in this form of modulation the amplitude of the carrier wave does not change; instead its frequency increases (+) when the signal amplitude is increasing and decreases (−) when the signal amplitude is decreasing, fig. 16.68. This type of modulation was introduced to overcome some of the interference which spoilt the quality and clarity of AM radio systems. Since much of the interference caused changes in amplitude and the FM system was only concerned with changes in frequency, the amplitude interference could be ignored by the system.

Figure 16.68   *Frequency modulation*

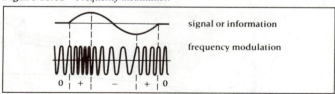

signal or information

frequency modulation

Both AM and FM radio transmissions carrying analogue signals have limitations. AM systems generally give rather poor quality reception but have the advantage of using simple technology which is relatively cheap. FM radio does give clearer reception but the VHF band of wavelengths has a limited range of about 80 km and uses more complex and expensive technology.

Microwave communications

Microwave links are widely used both locally and internationally. Line-of-sight links between towers, often placed on high buildings or on hill tops, are used to carry both telephone and television signals. A concave metal dish on the transmitter is used to send a focussed beam of microwaves to the receiver. At the receiver another concave dish focusses the microwaves on to an aerial at its focus. Over long distances several 'hops' are needed because:

• *the curvature of the Earth limits the distance over which there is a direct 'line of sight' and*

• *the microwaves get weaker with the square of the distance travelled. Therefore they need amplifying at repeater stations.*

Terrestrial microwave links use super-high frequencies of about $10\,\text{GHz}$ ($= 10^{10}\,\text{Hz}$).

Figure 16.69 *Terrestrial microwave links*

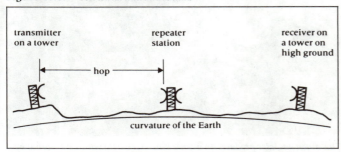

Satellites are used for international communication links. Microwaves are sent up to a satellite at one frequency, say $4\,\text{GHz}$, and are then amplified and converted to another frequency, say $11\,\text{GHz}$, before being sent back down to a receiving station on the ground. The range of frequencies available is between 3 and $30\,\text{GHz}$.

Communication satellites are often placed in a high orbit over the equator. This orbit is called a **geosynchronous** or **geostationary** orbit. In this orbit the satellite orbits the Earth once every 24 hours so that it appears to hover above the same point on the Earth. The high orbit allows a large area of the Earth to be 'seen' by the satellite.

Figure 16.70 *Satellite microwave links*

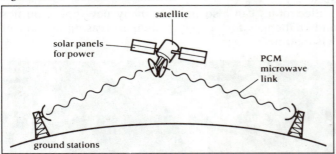

Pulse code modulation

Microwave links carry signals which are **pulse code modulated (PCM)**. PCM is a form of digital modulation in which the microwaves are switched on and off in pulses of varying length. The pulse lengths are arranged in a binary code. It is a simple matter to transmit data which is already in a binary digital code such as that used by computers. The microwave signal is of the form shown in fig. 16.71a. If however an audio analogue signal, fig. 16.71b, is to be transmitted, it must first be converted to a pulse amplitude modulated signal (PAM), fig. 16.71c. The PAM signal is produced by a regular sampling of the analogue signal. The value of each sample, i.e. the pulse height in volts, is then converted to a binary code, fig. 16.71d. Pulses of constant height but varying length are then strung together in series so that they carry the information in a binary code. A pulsed microwave carrying this digital code is said to be 'pulse code modulated'.

Figure 16.71 *Pulse code modulation*

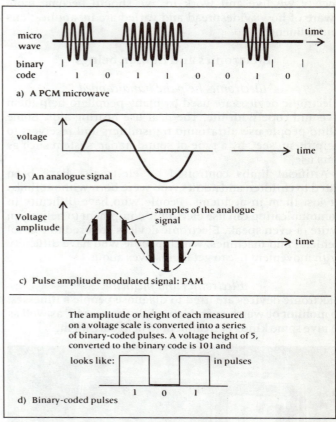

Advantages of digital transmission

Digital coded pulses are very simple signals to transmit. They are either a 1 or a 0. Many digital transmissions use what we would call a telephone system. Such systems can transmit digitally coded data, sound and pictures over long distances. Whether the signal travels in metal cables, optical fibres or as PCM microwaves, the pulses get smaller and lose their sharp edges as the distance increases. But, unlike analogue signals which can only be amplified along with all the added noise, digital pulses can very easily be regenerated to their original form with no unwanted noise added and no information lost. In a digital system it is only necessary to detect the presence or absence of a pulse and not its shape. So digital information is very resistant to corruption by noise and interference.

Assignments

Try questions 16.27 and 16.28

16.8
ELECTRONICS AND SOCIETY

Electronics is now affecting the whole of life in the most dramatic ways. We are surrounded by electronic equipment and machines which are controlled by microprocessors wherever we go. This is equally true at home, in hospital, at the office or in industry. Electronic systems and computers are no longer just for the experts in research, education and defence establishments, but are becoming small enough, cheap enough and versatile enough for everyone to use.

In this brief look at how electronics is affecting the society we live and work in, we should become more aware of how widespread and varied are its applications and influence.

Electronics and human beings

Electronics help the handicapped

Electronic devices are used by many people to help them live and cope with life. The deaf use hearing aids. Some blind people use ultrasound transmitters and receivers to help them 'see', by a type of sound 'radar' system such as bats use.

Artificial limbs controlled by electronics have been fitted to children and adults who were born without limbs or lost them in accidents. People who have difficulty in communicating can use electronic equipment to help them write or even speak. Electronic devices are used to control vehicles and machines, which people who have difficulty with movement use to get themselves about.

Electronics and medicine

Electronic devices are used to diagnose people's illnesses, to monitor or watch over them while they are ill, as well as to give some kinds of treatment to help cure them.

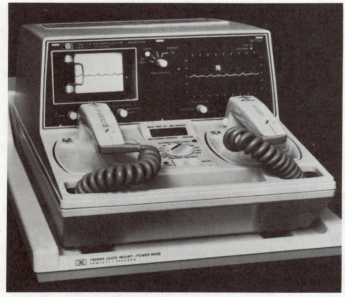

This electronic machine has two uses. It can detect the electrical signals from a patient's heart, display them on a CRT and print them on a paper chart. We call this an electrocardiograph or ECG. The two electrodes are removed from the front panel and placed on the chest of a patient suffering from heart failure. Then a carefully controlled electric shock is delivered via the electrodes to restart the heart.

Electrocardiograph or ECG machines tell a doctor about the working of the human heart, and electroencephalograph or EEG machines about the working of the human brain. Images of parts of people can be made using X-rays, ultrasound, thermography and other scanning systems, which all provide information about the state of the body which the eye cannot tell us. Electronic devices which measure pulse rate, temperature, pressure and other things about the body can be connected up to machines which watch over a seriously ill person and give warnings when anything vital changes or goes wrong.

In some illnesses treatment is given to patients by electronic equipment. Examples include X-ray therapy for some skin diseases and forms of cancer, infra-red heat treatment and laser treatment for some eye disorders.

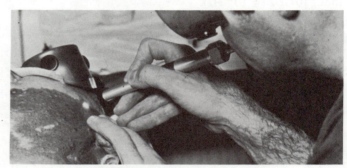

This argon gas laser produces an intense blue/green light which is absorbed by the red blood in the patient's port wine birth mark. The light energy from the laser heats and damages the blood vessels which eventually heal forming lighter coloured tissue under the skin.

Electronics and the nature of work

Many jobs are repetitive, boring or even dangerous. When electronics can be used to control machines and robots can do the repetitive and dangerous jobs, we welcome the improvement in the quality of life for society as a whole. But society must then share out both the available work and the increase in the quality of life, including the new work and the leisure provided by the electronics industry. Everyone should benefit by the electronics revolution and everyone should be able to make a useful contribution to society.

Electronics can also provide many new opportunities and challenges for people in their increasing amount of leisure time.

The jobs that some of these people used to do may now be done by electronic or computer-controlled machines. The sharing of work and the creation of new jobs in the electronics industry can help to restore the balance between workers and jobs.

Electronics and communications

We are all familiar with electronic devices, such as the telephone, radio and television, which provide communication between people. Communication is so important to life and human society that our appetite for new developments in this field seems to grow as fast as the technology.

Telephones are now available with memories and answering machines. Three-way telephone conversations will soon be common. Eventually video phones, where we can watch as well as speak to people, will become cheap enough for us all to use. Perhaps the days of travelling long distances to conferences are nearly over; and the days when electronic conferences with multi-channel video phones, which will be cheaper and much quicker to organise, are nearly here.

Telephones can also be used to send 'electronic mail' by using a typewriter linked to the telephone system. The Post Office now sorts and directs its letters by electronic machines. Televisions are rapidly becoming more than just an entertainment and news service. Our TV screens can now provide 'viewdata' in the form of 'electronic magazines' called 'Ceefax' and 'Oracle'.

Satellites are used to pass information and TV programmes round the world. Satellites use electronics to handle the information and they also require enormous computer power to get them into space in the first place.

Electronics and research

Electronic devices play a vital role in many kinds of scientific research. For example the electron microscope has allowed us to 'see' things which could never be seen by light. We can find out details of the shapes and structure of living cells and even the molecules of which they are built. X-rays have shown us the secrets of crystal structures and help us to understand more about the nature of matter.

While research like this may hopefully lead to benefits, such as cures for human diseases, sadly much research is also directed towards killing people more efficiently. A large amount of research in electronics is concerned with defence and military applications. Electronics is used in watching activities such as radar systems and satellite-based detection of 'enemy' activities. All missile guidance systems depend heavily on electronics and computers. In many countries a large defence budget and the Arms Race together provide both the means and the stimulus for the development of new electronic devices and computer systems.

Many electronic inventions can be used for both peaceful and military purposes. For example, satellites are instruments for communication, for scientific research, for astronomy, for watching and predicting the weather, for finding mineral resources under the ground or spotting disease among crops. But they are also instruments with the potential to fight wars. This is not to say that satellites and other electronic devices are bad for human society or even for human survival, but it does mean that people must find out what they can do and then learn to use them wisely.

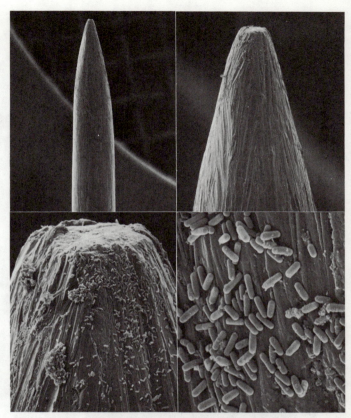

These four pictures, obtained using a scanning electron micrograph, show bacteria on the point of a household pin under magnifications of ×20, ×100, ×500 and ×2500.

Electronics in industry

Process control and automation

The working of any machine or process in industry needs to be watched and controlled. Electronic instruments find out what is happening by watching with their electronic 'eyes'. 'Eyes' or sensors which read temperature, pressure, speed, flow rate, thickness and so on, feed electrical information into automatic control systems to keep the machine working at maximum accuracy and efficiency. If *we* want to know what is going on, the instruments will also display their measurements on dials or digital read-outs.

The computer control system sends out electrical instructions to the machinery to make adjustments where necessary. The whole system is fully automatic, being both supervised and controlled by electronics.

Robots

Some machines are connected to an electronic or computer control system with a memory that can learn. We call these machines robots. They are more than just an automatic machine. Their electronic memories can store all the information about the way a person does a particular job and then repeat the job over and over again, copying exactly what the person did.

A robot can learn to do other jobs as well, providing it is first taught how to do them. A robot will do a job as many times as required, doing it with exactly the same skill and accuracy as the person who taught it. But the robot does not get tired and does not forget.

Computer graphics

Computer graphics equipment provides a visual link between the computer and its user. But more than that, engineers in several industries are now using computer displays to help them design and visualise their products. For example a draughtsman or architect may use computer graphics to help design, draw and change the shape or proportions of a product. An architect can look at a building from different directions and different distances and then alter its shape. All this happens on the screen of a cathode-ray tube.

Machines, buildings, maps and even animated art and cartoons can be drawn and created on a CRT with the aid of computer graphics.

Computers, aided by graphical display on the CRT of a monitor, are now used to design a wide range of things from the largest buildings to the smallest microcircuits.

Microprocessors take over

A microprocessor is a single silicon chip which can carry out all the control and data-handling operations needed for a computer. All computers contain a processor at their heart, where it is referred to as the central processing unit or CPU. Microprocessors have found other uses in controlling and operating all kinds of machines and systems. Fig. 16.72 shows a block diagram of the extra blocks or elements needed to build an electronic system using a microprocessor. The complete system is also called a microcomputer.

A microprocessor to operate traffic lights

A good example of an application for which a microprocessor is ideally suited is the operation of traffic lights. Traffic light systems are all designed to regulate the flow of vehicles and people in a busy street or at a road junction. Since the flow of people and vehicles varies from one place to another and from one time of day to another, the system must be flexible and adaptable.

The first automatic traffic signals had fixed timing and knew nothing about the traffic, or what other signals down the road were doing, or what time of day it was. They just went through a timed sequence of signal colours without any variation.

Traffic signals today can be made to sense and count the number of cars in the main stream of traffic and even the number of cars waiting in access lanes. This kind of information is fed into the microprocessor and stored in a temporary memory. In a permanent memory there are

Figure 16.72 *A microprocessor-controlled system or microcomputer*

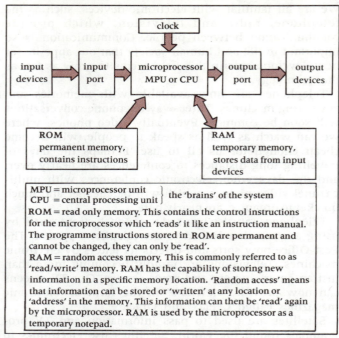

MPU = microprocessor unit
CPU = central processing unit } the 'brains' of the system
ROM = read only memory. This contains the control instructions for the microprocessor which 'reads' it like an instruction manual. The programme instructions stored in ROM are permanent and cannot be changed, they can only be 'read'.
RAM = random access memory. This is commonly referred to as 'read/write' memory. RAM has the capability of storing new information in a specific memory location. 'Random access' means that information can be stored or 'written' at any location or 'address' in the memory. This information can then be 'read' again by the microprocessor. RAM is used by the microprocessor as a temporary notepad.

programmed instructions about how to adjust the timing of the signals to keep the traffic flowing at the optimum rate. This includes information about how to change the timing at rush hours and at weekends. The processor contains the digital logic which controls the sequence of the signals. The whole system requires accurate timing not only for the signals, but also for the controlling and synchronising of information in the microprocessor. The system also needs to know the time of day. For all these purposes an accurate clock or timer feeds timing pulses into the microprocessor.

The output of the whole system is a set of instructions to switches which operate signal lights at the correct times.

Other systems controlled by microprocessors

The range is growing so quickly that we can give only a few examples here to show how many aspects of life are now affected by electronics and microprocessors. All the following machines may now contain a microprocessor to increase their flexibility and range of uses: watches and clocks, calculators, electronic and video games, washing machines, dishwashers and microwave ovens, typewriters, taxi fare meters, vending machines, car ignition systems, bank terminals, pay-roll systems, test instruments, such as digital voltmeters, and supermarket stock control systems.

Assignments

Choose a machine which now uses a microprocessor to improve and control its operation.

Find out what inputs and outputs are connected to the microprocessor.

Explain

a) In what ways has the microprocessor increased the flexibility or capacity of your machine?

b) What controls have been taken over by the microprocessor which previously the user or operator had to supervise?

Try questions 16.29 and 16.30

Draw a block diagram for your machine, like the one shown in fig. 16.72, and name all the parts you have identified.

Write an essay answering as many of the following questions as you can. How do we make sure that everyone shares the benefits (freedom from dull or dangerous jobs, increased leisure time) of the electronics revolution?

How can we make sure that all the work is shared out fairly?

Is doing a job an essential part of life?

Is it necessary to make a rigid distinction between work and leisure?

Do we need to reconsider the idea of 'a full day's work'?

Questions 16

1   Fig. 16.73 shows a simple form of cathode-ray tube (known as a maltese cross tube), which produces a sharp shadow of a cross on a fluorescent screen.

Figure 16.73

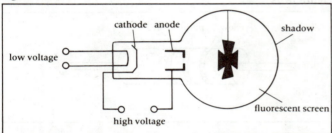

a) Explain what is meant by cathode rays and how they are produced.

b) State the properties of cathode rays, mentioning similarities and differences compared with light.

c) Draw a diagram to show how two plane circular coils carrying electric current (Helmholtz coils) could be used to deflect the shadow on the screen downwards. By means of labels on your diagrams, or otherwise, explain how the deflection is produced.

d) Describe two ways by which the brightness of the screen around the deflected shadow could be increased, and explain what effect each would have on the size of the deflection.   (O)

2   Fig. 16.74 shows a simplified version of a tube designed to illustrate the effect of an electrostatic field on a stream of electrons. Explain carefully

a) how A is made to emit electrons, naming a suitable substance for A and giving a reason for your choice,

b) the purpose of B and how this purpose is achieved,

c) how you would establish the nature of the charge on the electrons,

d) why the air must be removed from the tube,

e) how the path of the electron beam is made visible.   (L, part)

Figure 16.74

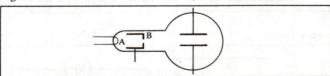

3   Describe, using one simple diagram in each case, what happens when a narrow beam of electrons is passed through (a) a uniform electric field and (b) a uniform magnetic field. What do these results indicate about the charge on an electron?   (JMB, part)

4   A TV tube uses a voltage of 4550 V to accelerate electrons released from its cathode by thermionic emission. Explain:

a) what is meant by thermionic emission,

b) how thermionic emission is produced in the CRT,

c) what assumption you will make about the initial speed of the electrons produced by thermionic emission.

If the charge of the electron is -1.6×10^{-19} C and the mass of the electron is 9.1×10^{-31} kg, calculate:

d) the energy of an electron as it strikes the TV screen,

e) the speed of the electron as it strikes the TV screen.

5   A student performed a Millikan oil-drop experiment several times and obtained the following results for the charge on different oil drops: 3.2×10^{-19} C, 8.0×10^{-19} C, 4.8×10^{-19} C, 1.6×10^{-19} C, 9.6×10^{-19} C, and 1.6×10^{-18} C. Explain how a value for the charge of an electron can be deduced from these results and obtain that value.

In another experiment the student measured the specific charge of the electron and obtained the value 1.75×10^{11} C/kg. Using the results of both his experiments the student was able to estimate the mass of the electron. Repeat his calculation and explain the steps involved in obtaining your answer.

6   Fig. 16.75a and c show the screen, Y-gain and time-base controls from a typical oscilloscope displaying a waveform.

a) What is the setting of the Y-gain control?

b) What is the peak voltage of the waveform?

c) What is the time-base setting?

d) What is the period of the trace?

e) What is the frequency of the waveform?

f) If the time-base setting is altered to 1 ms cm^{-1} and the Y-gain to 2 V cm^{-1}, draw the resultant trace on the graticule shown in b.

(NEA [A] SPEC)

Figure 16.75

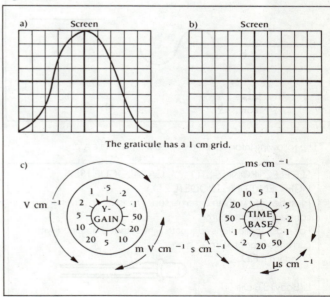

The graticule has a 1 cm grid.

7   Describe two experiments (one for each case), to show how a cathode-ray oscilloscope (CRO) may be used to

a) measure the voltage of a car battery,

b) measure the maximum voltage of a car or cycle dynamo.

Your answers should include labelled diagrams of the circuits you would use and an indication of the values you might obtain. You should not draw or attempt to explain how the CRO works.

(JMB, part)

8 A cathode-ray oscilloscope is adjusted so that the spot is in the centre of the screen. The time-base is switched off. The 'Y' plates (i.e. Y_1 Y_2) are then connected in turn to the power supplies shown in fig. 16.76. Redraw the screen diagrams and mark in what would be seen in each case. Each vertical division on the screen represents 1 volt. (Joint 16+, part)

Figure 16.76

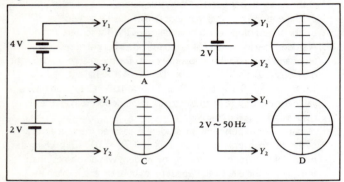

9 A cathode-ray oscilloscope is adjusted so that the time for the electron beam to make one traverse of the screen from P to Q is $\frac{1}{100}$ s and the Y-plate sensitivity is 1 cm ≡ 15 V. The trace on the screen is then as in fig. 16.77a
Figs. 16.77b and c show the traces obtained when the oscilloscope is connected in turn to two voltage sources A and B.
State all the information about the voltage of each source A and B that can be obtained from the traces. (L)

Figure 16.77

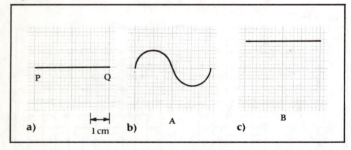

10 The following information is copied from a catalogue.

SEB ELECTRONICS LTD   CATALOGUE 1987
LIGHT-EMITTING DIODE (L.E.D.)
Must be used with an external series resistor or driven from a constant current source.
For most applications adequate light output is obtained with a current of about 10mA.
Standard L.E.D.

L (Body) 8.6 Dia 5.1
Cathode identified by 'flat' on body
Operating current   10mA
Operating voltage   2V

a) Draw a circuit diagram to show how a resistor and a 5.0V d.c. supply can be used to light the L.E.D.
b) Using data from the catalogue, calculate the value of the resistor. (SEB 1990)

11 a) What is the value of the resistor shown in fig. 16.78?
    b) The structure of a silicon crystal may be represented by the diagram of fig. 16.79.
        i) How many bonding (or valence) electrons has each silicon atom?
        ii) Explain why pure silicon has a high resistivity.
        iii) State whether an n-type or a p-type semiconductor results by doping pure silicon using: an element with 5 bonding electrons: an element with 3 bonding electrons. Explain your reasoning. (O)

12 a) A capacitor can store electrical energy. With the aid of a circuit diagram, explain how you would demonstrate this.
    b) Fig. 16.80 shows an electrical circuit containing diodes, capacitors and lamps which has been connected for some time.
        i) State for each lamp whether it is on (lit) or off (unlit).
        ii) Explain your answers. (O)

Figure 16.78

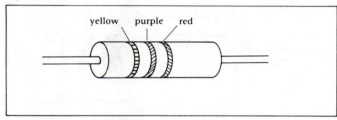

yellow   purple   red

Figure 16.79

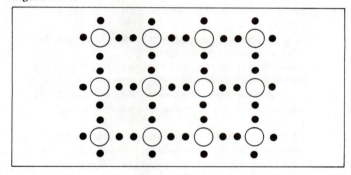

Figure 16.80

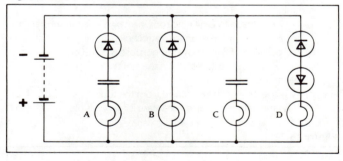

Figure 16.81

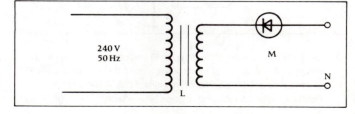

13 Fig. 16.81 shows a circuit diagram which includes a half-wave rectifier.
   a) What is meant by rectification?
      Name the components labelled L and M.
      What is the polarity of N?
   b) Explain the action of component M in causing rectification.
   c) Redraw and add to fig. 16.81 a labelled smoothing circuit and a load resistor.
   d) In fig. 16.82 (A) shows the trace on a cathode-ray oscilloscope when the input to the rectifier is connected to the Y-plates. Assuming that the controls of the oscilloscope are unchanged, copy and complete the figure to show the trace of
      i) the unsmoothed output of a half-wave rectifier,
      ii) the smoothed output of a half-wave rectifier. (Joint 16+)

14 The circuit in fig. 16.83 is to illuminate a warning light (the light-emitting diode) when the temperature falls below a certain value. At what resistance value of the thermistor will the warning light illuminate? (You may assume that the diode and the transistor are both silicon devices.) (L)

Figure 16.84

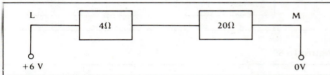

15 Fig. 16.84 shows a simple circuit.
   a) Calculate (i) the resistance between L and M, (ii) the current through the resistors, (iii) the potential difference (voltage drop) across the 4Ω resistor.
   b) Study carefully the circuit of Fig. 16.85.
      i) Name the devices labelled Y and Z.
      ii) A protective device has been missed from the circuit.
         (A) What is this device protecting?
         (B) Add this protective device to the circuit diagram.
      iii) The potential difference across the $10\,k\Omega$ resistor is 5 V. In what state is component Z, component Y, the lamp?
   c) An L.D.R. (light dependent resistor) has a dark resistance of $10\,M\Omega$ and a light resistance of $500\,\Omega$. If the L.D.R. replaces the $2\,k\Omega$ resistor in Fig. 16.85, suggest, giving your reasons, a practical use for this circuit. (NEA [A] 88)

16 Fig. 16.86 shows a light-operated circuit.
   a) Name the component labelled ORP 12.
   b) Which of the following pairs of values gives the resistance of the ORP 12 when the light is as stated?

	A	B	C	D	E	F
Dark	1 kΩ	10 kΩ	100 kΩ	10 MΩ	10 MΩ	10 MΩ
Bright	10 MΩ	10 MΩ	10 MΩ	1 kΩ	10 kΩ	100 kΩ

   c) Compare the current passing through the two lamps when the ORP 12 is illuminated.
   d) When the ORP 12 is shaded, what happens to the current: across the base–emitter junction: across the collector–emitter junction: through the right-hand lamp (top right)?
   e) When the shading is removed from the ORP 12, what happens to: the transistor: the right-hand lamp: the left-hand lamp? (Joint 16+, part)

Figure 16.82

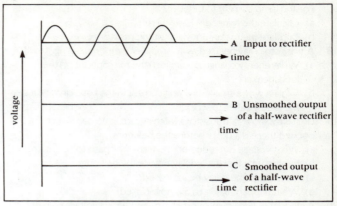

Figure 16.83

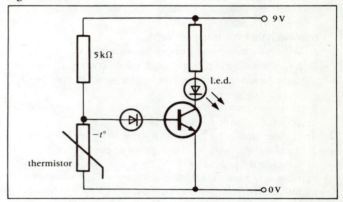

Figure 16.85

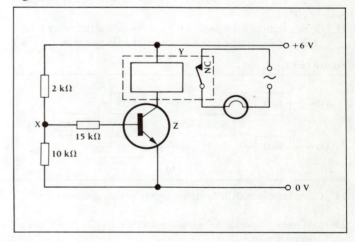

Figure 16.86

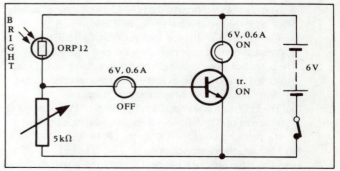

17 The transistor in the circuit of fig. 16.87 has a current gain of 500. For the lamp current to be 50 mA.

    a) calculate the minimum base current,

    b) calculate the maximum value of the resistor *R*.

18 A transistor may be used as an amplifier (see circuit in fig. 16.88).

    a) What energy change occurs in: i) the microphone; ii) the loudspeaker?

    b) Explain why a smaller varying current in the base circuit causes the loudspeaker to work.   (WJEC SPEC part)

19 Fig. 16.89 shows a transistor switching circuit. When switch S is opened there is a delay before the bell rings.

    a) What voltage is needed on the transistor base to turn it on?

    b) Explain why there is a delay before the bell rings.

    c) How could the delay be increased?

    d) What happens if switch S is now closed?

20 A student builds up an inverting op-amp with a voltage gain of -20. The power supply used is ± 15 V.

    a) What will be the output voltage, V_{out}, if the input voltage, V_{in}, is:
        i) $+0.5$ V; ii) -2 V?

    b) What is the maximum possible V_{out}?

    c) What is the minimum V_{in} to give this maximum V_{out}?

    d) If V_{in} is an a.c. signal of amplitude ± 2 V, describe V_{out}.

21 a) State the factors which limit the voltage gain of an operational amplifier using negative feedback.

    b) Give one reason why negative feedback is usually used when an op-amp is used as a voltage amplifier.

22 The power supplies for the op-amp shown in fig. 16.90 are ± 15 volts and the alternating input voltage has a peak value of ± 2.0 V

    a) Calculate the peak value of V_{out}.

    b) How would you expect the peak output voltage to change if the 20 kΩ resistor were to be replaced by a 200 kΩ resistor?

    c) What would be the largest input voltage that could be amplified without saturation occuring if the feedback resistor remains at 200 kΩ?

23 Draw a truth table for each of the systems (a) and (b) shown in fig. 16.91 and identify the logic function which each possesses.   (L)

Figure 16.91

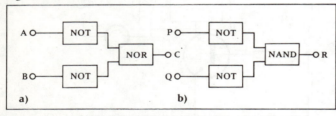

    a)   b)

24 Complete the truth table for the system of logic gates shown in fig. 16.92.   (W 88)

25 a) The circuit symbol for a two-input NAND (NOT AND) gate is shown in fig. 16.93a.

       Also given is the truth table for a two-input NAND gate.

       In the design of logic circuits, the NAND gate is a basic "building block" as in the circuit shown in fig. 16.93b.

       i) Complete the truth table for this circuit, fig. 16.93c.

       ii) What arithmetic operation does this circuit perform?

    b) The output of a logic circuit can be displayed using an LED and associated series resistor, as in the circuit shown in fig. 16.93d.

       i) When the LED is lit, what is the logic state at A?

       ii) The LED has a potential difference drop of 2 V across it and a current of 10 mA flowing through it when it is lit. What is the p.d. across resistor R when the LED is lit?

       iii) Calculate the resistance of R.

       iv) Why is the resistor R needed?   (NEA [A] SPEC part)

Figure 16.87

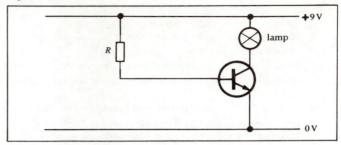

Figure 16.88

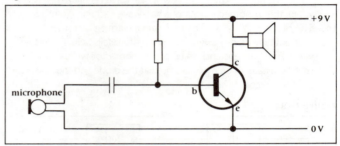

Figure 16.89

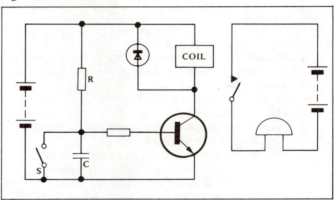

Figure 16.90

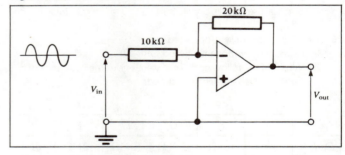

Figure 16.92

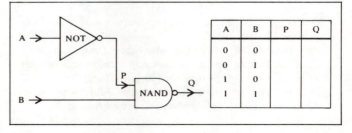

A	B	P	Q
0	0		
0	1		
1	0		
1	1		

Figure 16.93

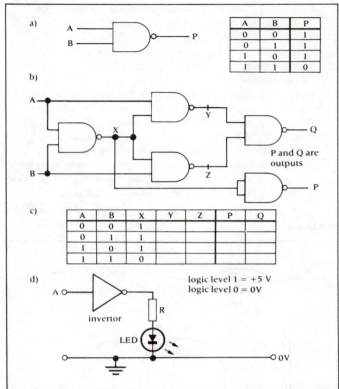

a)

A	B	P
0	0	1
0	1	1
1	0	1
1	1	0

b)

P and Q are outputs

c)

A	B	X	Y	Z	P	Q
0	0	1				
0	1	1				
1	0	1				
1	1	0				

d)

logic level 1 = +5 V
logic level 0 = 0V

invertor

LED

0V

26 A gas-fired central heating boiler has an alarm system which flashes a lamp if the boiler temperature becomes too high or if the flame goes out. The block diagram in fig. 16.94 shows the system.
The temperature sensor gives a logic '1' when it is too hot and a logic '0' when it is normal.
The flame sensor gives a logic '1' if the flame is on and a logic '0' when the flame is out. The lamp is switched on by logic '1'.
a) Name logic block "X".
b) Name the logic block "Y" and explain why it is needed in this system.
c) Name the logic block "W" and explain how it is used to make the lamp flash when the alarm is triggered.
d) Copy and complete the truth table for the alarm system.
(SEB 1990)

Figure 16.94

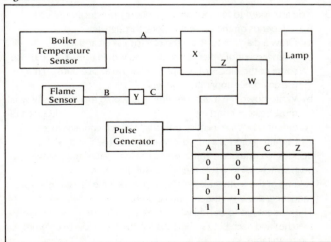

A	B	C	Z
0	0		
1	0		
0	1		
1	1		

27 British viewers saw their first live television broadcast from America in 1962. Signals were beamed from a dish aerial in the USA to the satellite Telstar which amplified and re-transmitted them to a dish aerial in Britain. Viewers were able to see a picture on their TV screens for only 20 minutes at a time. After each 20 minute block of time another 70 minutes had to pass before pictures could be seen again.
When the satellite Early Bird was placed in geostationary orbit in 1965, round the clock TV broadcasting from the USA to Britain became possible.
a) Which satellite had a period of 24 hours for its orbit?
b) What was the period of the orbit of the other satellite?
c) Which satellite had the higher orbit?
d) Explain why:
   i) signals from Telstar could be received for only 20 minute blocks of time;
   ii) signals from Early Bird could be received all of the time.
e) State why dishes were used with the aerials which received signals from the satellites.
(SEB 1990)

Figure 16.95

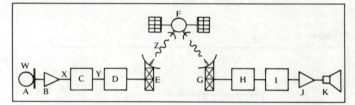

28 The figure shows a block diagram of a satellite radio communication system.
a) Copy and complete the table of the parts of the system and their purpose.
b) State the form of the signal at the points W, X, Y and Z.

Part of system		Purpose
A		Converts sound into electrical signal
B		Increase signal strength
C	Encoder	Converts analogue signal into digital pulses
D	Transmitter	Adds digital pulses to microwave carrier
E		Aims a beam of microwaves at F
F		
G		
H	Receiver	
I		
J		
K	Loudspeaker	

29 Write a paragraph about the importance of television in society for each of these uses:
a) communication, broadcasting news etc.,
b) education, The Open University,
c) advertising,
d) data transmission and display, 'Ceefax' and 'Oracle',
e) monitoring and surveillance.

30 Name one example of an application of the microprocessor which has resulted in the loss of jobs for people. Have any new jobs been created in this application?

PRACTICAL INVESTIGATIONS AND PROBLEMS
TO SOLVE II

(More investigations and problems can be found on pages 200 & 201.)

1 A student designs a simple electrical continuity tester. Her circuit is shown below in problem 1a.
   a) What is the purpose of resistor R?
   b) When leads X and Y are joined together the meter reads its maximum value of 1 mA. Calculate the value of resistor R.
   c) The student decides to use her tester to make two checks on the electric heater shown in problem 1b.
      i) She wishes to find out if the heater is properly earthed.
         A) Where should the student connect leads X and Y?
         B) What will the meter read if the heater is properly earthed?
      ii) She now wishes to check if there is a break in the heating element. She connects leads X and Y to pins L and N of the plug and obtains a zero reading on the meter. She concludes that the element is broken. This is not necessarily true. Give TWO other possible reasons for the meter reading zero.   (SEB 1990)

Problem 1

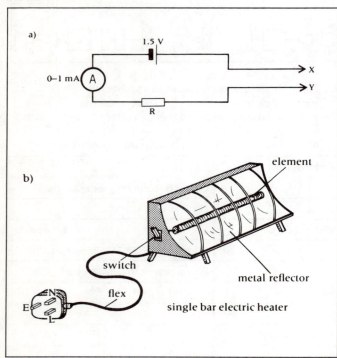

2 This question is about some of the many problems with which an electrician might be faced at work.
   Imagine that you are an electrician and that the following problems and questions are brought to you by some of your customers. In each case describe how you would answer them. Remember, your business depends on satisfied customers.
   In each of the first four cases, explain a possible cause for your customer's trouble and how it might be solved.
   Customer 1 'A lamp in my house has gone out.'
   Customer 2 'The lights downstairs have all gone out'.
   Customer 3 'All the lights in the house have gone out.'
   Customer 4 'All the lights in the houses in my street have gone out.'

The next customer would like a clear wiring diagram.
   Customer 5 'I would like two lamps in the dining room both worked from the same switch.' Draw a suitable wiring diagram.
The next customer is quite bewildered by the problem and you must give this customer a clear, full explanation of how you make the calculation.
   Customer 6 'How much a week will it cost me to run my 1 kW electric fire and 2 kW electric radiator if they are both used for 4 hours each evening?' (1 kWh costs 5p)
The last customer does not even know there is a problem!
   Customer 7 'I should like you to come and fit a mains socket in the bathroom so that I can have the electric fire on when I have a bath.' Explain what the problem is.
                                                              (NEA [B] SPEC)

Problem 3

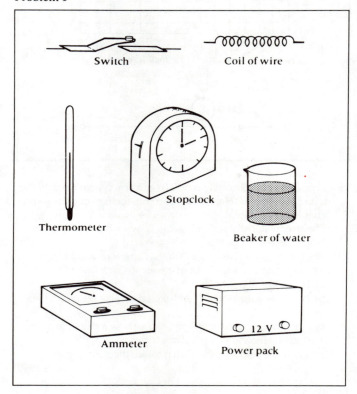

3 You are asked to find out how the energy produced in a coil of wire depends on the current passing through it.
   a) Draw a diagram to show the electric circuit and the arrangement of other apparatus you would use. (You may use some of the pieces of equipment shown above but you may use any other apparatus you wish.)
   b) Write down the measurements you would make, and say what must be kept constant.   (SEG Part 88)

4 A student builds a miniature traffic light which contains three light-emitting diodes, red, yellow and green (R, Y and G). He decides to control the traffic light using logic gates and just two push switches. He wants the switch sequence, shown in the first column of the truth table below, to produce the usual traffic-light sequence, shown in the second column.
   The student has to decide which logic gates to use between the switches and each LED so he adds the third and fourth columns to the truth table.

He wires the circuit so that each switch when pressed 'on' gives a logic I output. To turn on an LED it must have a logic I input.

Switch sequence		Light sequence	Switch output		LED input		
			A	B	R	Y	G
Off	Off	red on	0	0	I	0	0
Off	On	red and yellow on	0	I	I		
On	Off	green on	I	0	0		
On	On	yellow on	I	I	0		

a) Complete the six missing entries in the truth table.
   The diagram for problem 4 shows part of the circuit to control the traffic lights.
   In parts b), c) and d) you have to copy the diagram and draw on the necessary wires and logic gates between the switch outputs, points A and B, and each LED.
b) Compare columns A and R in the truth table. What do you notice? Now complete the diagram between point A and the red LED, leaving room for later connections.
c) Compare columns B and Y in the truth table and hence complete the diagram between point B and the yellow LED.
d) What must be the outputs of the switches for the green LED to be on?
   Now complete the diagram between points A and B and the green LED. (MEG 88)

Problem 4

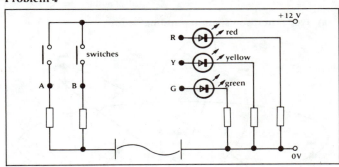

5   You are provided with a 5 V or 6 V power supply, some connecting leads and 10 different lamp boxes. You have no other equipment. The lamps, labelled A, B and C, are identical 3.5 V bulbs connected together by wires which are hidden inside a box. All you can see is the lamps and the terminals, labelled X, Y and Z, to which they are connected inside the boxes. The boxes may be numbered in any order.

Problem 5

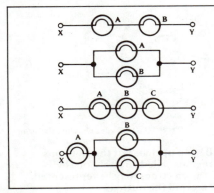

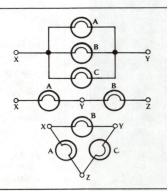

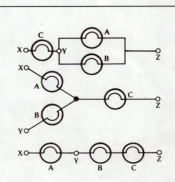

A lamp box

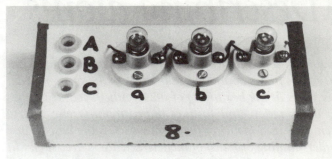

- Your task is to work out the hidden wiring inside the boxes without looking inside them. Ten possible circuits are given below.
- You should start with the simpler boxes which have only 2 lamps and 2 terminals.
- The wires should be connected to the terminals only and never both wires from the supply to the same terminal.
- You may unscrew the lamps if you wish.
- Record the evidence which you obtain from your tests.
- Explain how you arrived at your conclusion about which box is which.

6   Problem 6 has six magnet puzzles for you to solve. Some possible arrangements of magnets, iron bars and a non-magnetic material such as glass are shown below. These are hidden inside identical plastic or cardboard tubes and all feel about the same weight. You are provided with another magnet and a plotting compass.
- Devise some tests to work out what is inside each tube.
- Identify the poles of any magnets.
- Explain how you arrive at your conclusions.

Problem 6

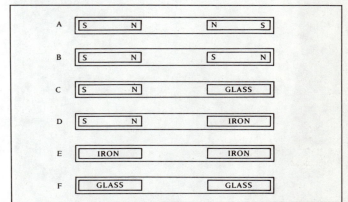

17

Describing atoms

When something is too small to see, finding out what it is like is bound to be difficult. In the space of just a few years at the beginning of the twentieth century, more by imagination and instinct than anything else, a few men gave us a model of the atom which is the basis of our understanding of much of modern science. Three stages in the evolution of models of the atom, shown in fig. 17.1, are described in this chapter.

Sir Ernest Rutherford, who had great intuition about the structure of matter, gave us the idea and the evidence that atoms have a very small central core or nucleus.

17.1
CLUES AND MODELS

The picture of the atom has changed or evolved as new experimental clues and mathematical theories have been pieced together. Table 17.1 gives some of the more important events in the history of experiments, theories and models which help to describe atoms.

The discovery of cathode rays, positive rays and radioactivity around the beginning of the century began to cause doubt about the atom as an indivisible particle. They provided evidence that particles smaller than atoms existed and that these had electric charges. This led both Kelvin and J. J. Thomson independently to the idea that perhaps atoms were solid balls of positively charged matter in which negative electrons were dotted about like currants in a pudding. There needed to be enough negative electrons to make the whole pudding electrically neutral, fig. 17.1a.

Figure 17.1 *Models of atoms*

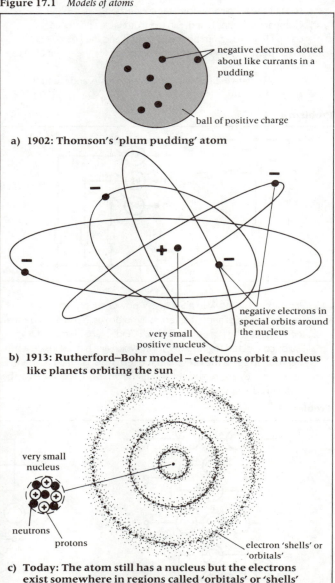

negative electrons dotted about like currants in a pudding

ball of positive charge

a) 1902: Thomson's 'plum pudding' atom

negative electrons in special orbits around the nucleus

very small positive nucleus

b) 1913: Rutherford–Bohr model – electrons orbit a nucleus like planets orbiting the sun

very small nucleus

neutrons

protons

electron 'shells' or 'orbitals'

c) Today: The atom still has a nucleus but the electrons exist somewhere in regions called 'orbitals' or 'shells' around the nucleus. The electrons may be represented as 'waves' or 'clouds' around the nucleus.

Table 17.1 Models of the atom

before 1897	Atoms were thought to be very small indivisible particles.
1897	Joseph John Thomson provided evidence that the cathode rays may come from all elements and that, being particles much smaller than atoms, are common constituents of all atoms.
1902 1907	Lord Kelvin and J. J. Thomson independently suggested a 'plum pudding' atom model.
1906	Ernest Rutherford observed alpha particles passing through a thin sheet of mica without making holes in it. Could alpha particles pass through the atoms themselves?
1911	Geiger and Marsden fired alpha particles through gold foil and found evidence for the nuclear model of the atom, as proposed by Rutherford.
1913	Niels Bohr found the link between atomic spectra, quantum theory and Rutherford's nuclear model of the atom. Bohr's new model had electrons in stable orbits around the nucleus, like planets round the sun.
1932	J. Chadwick identified the neutron as a neutral particle found in the nucleus of atoms, along with the protons.
today	There are now many complicated models involving new theories and particles, but we try to picture the atom with its very small and very dense nucleus, surrounded by a cloud of negative electrons. The position of an electron cannot be precisely pinpointed and it is thought to behave as a wave as well as a particle.

Scattered alpha particles provide a clue

In 1906 Rutherford first noticed that alpha particles (positively charged particles emitted by some radioactive materials such as radium, p 388) passed straight through a thin sheet of mica. There appeared to be no holes made in the mica and it seemed solid enough. A few alpha particles were deflected or scattered from the straight through direction and this interested Rutherford. How could these tiny 'bullets' of matter pass straight through a solid substance without any apparent effect most of the time, and just occasionally get deflected off course?

These first results gave Rutherford the idea that atoms might have a very small central core or **nucleus** with a strong electric charge which would deflect the alpha particles, but only when they came very near to it. This nucleus would be surrounded by electricity of the opposite charge which filled up the rest of the atom, but through which the alpha particles could pass as if it were empty space. At this point Rutherford did not know that the charge of the nucleus was positive and did not have precise experimental data to support his new idea.

The experiments of Geiger and Marsden 1909–1911

When in 1909 Rutherford became Professor of Physics at Manchester University, he put two of his assistants to work investigating the scattering of alpha particles by metal foils. They were H. Geiger and E. Marsden. Rutherford asked them to investigate how the number of alpha particles scattered varied with the angle of deflection. The apparatus they used is shown in fig. 17.2.

The source of alpha particles was a small glass tube which contained radioactive products obtained from radium. In an evacuated metal box a narrow beam of alpha particles rained down on a thin metal foil. Various metals were used, such as platinum, silver and gold. Alpha

particles were detected by a zinc sulphide screen mounted on the end of a microscope. In the dark a very small flash of light or scintillation can be seen when an alpha particle gives up its kinetic energy to the atoms in the zinc sulphide screen. (A similar effect to electrons striking the phosphor coating inside a television screen.)

During their experiments, Geiger and Marsden counted over 100 000 flashes of light while looking down the microscope. They counted alpha particles deflected by angles in the range 5° to 150° from the straight through direction.

Geiger and Marsden did much to confirm the existence of the atomic nucleus and to measure its size and electric charge.

Figure 17.2 *The layout of the experiments of Geiger and Marsden*

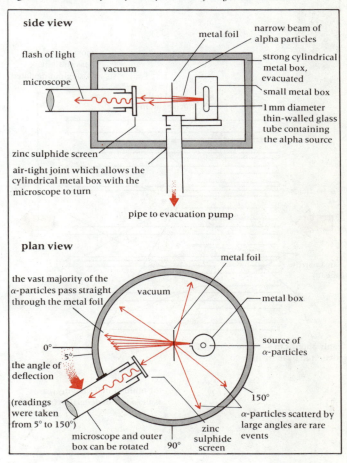

Alphas are scattered backwards

Most of the alpha particles passed straight through the metal foils. The number deflected fell off quickly as the angle of deflection increased. A very small fraction, about $\frac{1}{8000}$ (for platinum), were deflected by angles greater than 90°, i.e. they bounced back towards the source, fig. 17.3. This result was so surprising that Rutherford commented: 'It was about as credible as if you had fired a 15 inch shell at a piece of tissue paper and it came back and hit you.'

The experimental results obtained by Geiger and Marsden showed a good agreement with predictions based on Rutherford's nuclear model of the atom. The model could now be expanded:

a) The nucleus has a positive charge. A deflected alpha particle is repelled by the nucleus of a single metal atom by the repulsion between the like (positive) electric charges of the alpha particle and atomic nucleus.

b) The repulsion obeys the inverse square law of the force between two charged objects (see fig. 11.20) and will become very strong as they approach very close (so explaining the alpha particles which bounce back).

c) The charge on the nucleus of the atom depends on its relative atomic mass. (This was confirmed by the experimental results for different metal foils. We now know that the size of the positive nuclear charge depends on the number of protons in the nucleus of the atom.)

Figure 17.3   *Alpha particle scattering*

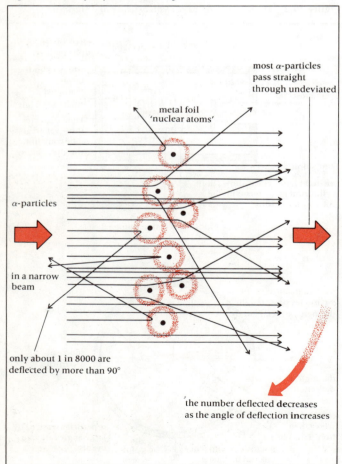

The size of the nucleus

As well as finding out about the positive electric charge on the nucleus, the scattering experiments gave an accurate clue about the size of the nucleus. Rutherford estimated that the closest approach of an alpha particle to a gold nucleus in a head-on collision was about 10^{-14} m. So the gold nucleus must be no larger than this.

The oil–film experiment (p 156) estimates the length of a molecule to be about 0.000 000 001 m or 10^{-9} m. The diameter of a single atom, which varies from one kind of atom to another must be only a few tenths of a nanometre (between 10^{-9} and 10^{-10} m). We now have evidence that the nucleus at the core of an atom is only about 0.000 000 000 000 01 m or 10^{-14} m in diameter. This amazing clue about the atom suggests that the diameter of the nucleus is only $\frac{1}{10000}$ of the diameter of the whole atom. The nuclear atom model is empty except for this very tiny dot in the centre where all its positive charge and almost all of its mass is concentrated, fig. 17.4. The scale of the nucleus compared with its atom can be compared with a walnut at the centre of Wembley Stadium.

Figure 17.4   *Comparing atomic and nuclear sizes*

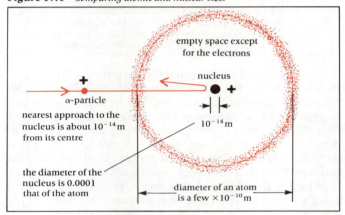

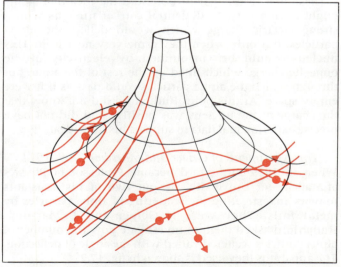

Firing an alpha particle at an atomic nucleus is like rolling a marble up a conical-shaped hill. As an α-particle gets nearer to the positively charged nucleus the force of repulsion increases in the same way as the conical hill gets steeper towards its peak. Particles approaching about head-on are scattered backwards while particles approaching off-centre are deflected only slightly from their incident direction.

If the diameter of the nucleus is $\frac{1}{10000}$ or 10^{-4} of the diameter of the atom, then the volume of the nucleus is roughly $\frac{1}{1000000000000}$ or 10^{-12} of the volume of the atom (volume $\propto$ radius cubed). It follows that if almost all of the mass of the atom (except for the electrons) is concentrated in its nucleus in this very small volume, the matter of the nucleus must be extremely dense.

Atomic spectra give the clue to the electron structure

The Danish physicist Niels Bohr took Rutherford's nuclear model of the atom very seriously and he tried to find solutions to the problems posed by this new model. While Rutherford had been able to estimate the size of the atomic nucleus, there was no theory yet which could explain why the radius of the atom, as a whole, was so much greater than that of the nucleus. It seemed that negative electrons should just fall into the positive nucleus because of the attractive forces. As they accelerated inwards they should emit electromagnetic radiation. Rutherford's nuclear atom needed a fixed atomic radius and some form of stable structure.

In 1913 Bohr saw a formula invented by a Swiss schoolmaster called Johann Balmer. Years later Bohr said: 'As soon as I saw Balmer's formula the whole thing was immediately clear to me.' Balmer had noticed a pattern in the frequencies of the spectral lines from a discharge through hydrogen gas. His formula showed that the frequencies of light emitted from a hydrogen atom fitted a precise mathematical series.

Bohr now imagined the electrons to be in orbits around the atomic nucleus with precise radii that fitted the Balmer formula, fig. 17.1b. However, for Bohr's new model of the atom to work, he had to propose some new rules about what electrons were allowed to do in atoms:

a) Electrons revolve in orbits around the nucleus without emitting radiation. Only certain fixed radii are allowed for these orbits in a stable atom.

b) In these 'allowed' orbits, each electron has a certain amount of energy, which is fixed. Orbits farther away from the nucleus have larger allowed energies.

c) An electron can 'jump' from one orbit of high energy E_2 to another of lower energy E_1 causing the energy difference $(E_2 - E_1)$ to be emitted as one quantum of electromagnetic radiation (a light photon), fig. 17.5. The frequency f of this radiation is given by Planck's formula for the energy of a photon

$$E_2 - E_1 = hf$$

Figure 17.5 *Atomic line spectra explained by the Bohr model of the atom*

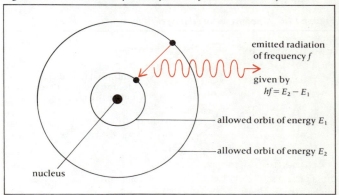

emitted radiation of frequency f

given by
$hf = E_2 - E_1$

allowed orbit of energy E_1

allowed orbit of energy E_2

nucleus

Why don't atoms collapse?

An electron is attracted to the nucleus of an atom by the attraction forces between their unlike charges. The equal and opposite forces between the electrons and the nucleus hold an atom together. If the electrons around the nucleus were stationary there would be nothing to stop them being pulled into the nucleus causing the atom to collapse inwards.

In the Bohr model of the atom each electron is moving with a particular velocity v. A moving electron will, like any other object, travel in a straight line unless a resultant force acts on it (Newton's first law of motion). This force F is provided by the electrostatic attraction, which acts in an inwards or centripetal direction on the orbiting electrons, fig. 17.6. The radius of an electron orbit r is fixed by the requirement that the centripetal force F must provide an inwards acceleration of exactly v^2/r (p 151) to keep the

Imagine a walnut at the centre of Wembley Stadium and you have some idea of how small the nucleus is at the centre of an atom.

electron orbiting at a constant distance r from the nucleus, and so keep the atom stable.

In this way the Bohr atom gives orbiting electrons particular allowed orbits, which are occupied without any energy loss in a stable atom. The Bohr model was successful because it managed to combine the quantum theory (p 344) with Rutherford's nuclear model, and also provided an explanation of the atomic line spectrum of hydrogen.

Figure 17.6 *The force on an orbiting electron*

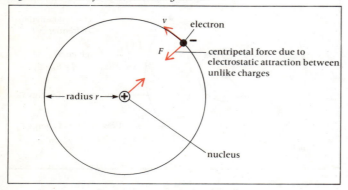

Chadwick finds the neutron

In 1920 Rutherford predicted the existence of a neutral particle in the nucleus of the atom. He proposed the name **neutron** for this new neutral particle, and suggested that it should have a mass very slightly greater than that of the proton.

In 1932 James Chadwick showed that radiation emitted from an experiment in which alpha particles were fired at beryllium must be particles fitting Rutherford's description. The neutron is now known to exist as a basic constituent of matter and the nuclei of all atoms, except hydrogen, contain neutrons as well as positive protons.

Today's model of the atom

We still think of the atom as having a very small nucleus composed of protons and neutrons, but our view of the nature of electrons has changed. The main objection to Bohr's model is that it pinpoints electrons in definite orbits with definite positions. We have evidence that electrons have a wave-like nature similar to light. We believe that the allowed orbits are the regions that can contain a whole number of electron 'waves'. This idea is similar to the way in which sound waves can be seen to fit onto the stretched string of a musical instrument.

The electron structure of the atom is now described by very advanced mathematical theories known as wave mechanics, but we are able to visualise the electrons as existing in cloud-like 'shells', called orbitals, around the nucleus, fig. 17.1c. An electron is more likely to be in a region where the shading is heavier.

Assignments

Make a table

a) comparing the atomic models of Thomson, Rutherford and Bohr.

b) listing the major clues which helped these scientists to think up their models.

c) listing Bohr's proposals for the new rules for electrons in atoms.

17.2
PARTICLES BUILD ATOMS

Now that we have found the three particles which build atoms we can understand how each different kind of atom is constructed, how they form a series of elements and sometimes have variations known as isotopes.

The atomic number, or proton number Z

An atom is normally electrically neutral and since the proton (+) and the electron (−) each have the same amount of charge, there must normally be equal numbers of protons and electrons in atoms. However, atoms can gain or lose electrons and become ions without changing into a different substance or element. So the number of electrons is not a reliable guide to the type of atom. The number of protons in the nucleus of an atom is the only factor which determines to which element an atom belongs.

All the different elements are identified by the number of protons in the nuclei of their atoms. So atoms of a *particular* element are labelled by a number called their **atomic number** or **proton number**, which is given the symbol Z. See table 17.2.

The atomic number or proton number Z of an element is the number of protons in the nucleus of all atoms of that element.

A neutral atom has exactly the same number of electrons as protons.

Neutron number N and mass number A

The nucleus of an atom contains protons and neutrons. These are both referred to as **nucleons**, i.e. particles which belong in the nucleus of an atom. The number of neutrons, known as the **neutron number** N, in a nucleus is similar to the number of protons Z, but is often greater than Z, particularly for large atoms. The main effects of the number of neutrons in the nucleus of an atom are on its mass and its stability. Neutrons, having a mass very similar to that of a proton, make a great difference to the total mass of an atom.

Table 17.2 Elements identified by their proton number Z

Atomic number or proton number Z	Element	Neutron number N	Mass number or nucleon number* $A = Z + N$	Nuclide symbols and isotopes
1	hydrogen	0	1	^1_1H (hydrogen)
		1	2	^2_1H (deuterium)
		2	3	^3_1H (tritium)
2	helium	1	3	^3_2He
		2	4	^4_2He
3	lithium	3	6	^6_3Li
		4	7	^7_3Li
4	beryllium	5	9	^9_4Be
5	boron	5	10	$^{10}_5B$
		6	11	$^{11}_5B$
6	carbon	6	12	$^{12}_6C$†
		7	13	$^{13}_6C$
		8	14	$^{14}_6C$‡

* Heavy type indicates the most common natural isotope.

† Standard nuclide on which the unified atomic mass unit is based.

‡ This isotope provides the carbon–14 'clock', see radioactive dating, chapter 18.

The number of nucleons in the nucleus of an atom is called its nucleon number or mass number A.

This is a whole number and has no units. It is not really a unit of mass, but since the mass of an atom is dependent on the number of nucleons there is a link between the actual atomic mass and the nucleon number A. We can write an equation for the nucleon number:

$$\begin{array}{ccc} \text{nucleon number} & = & \text{proton number or} \\ \text{or mass number} & & \text{atomic number} \end{array} + \begin{array}{c} \text{neutron} \\ \text{number} \end{array}$$

$$A \quad = \quad Z \quad + \quad N$$

Isotopes and nuclides

If two atoms have the same number of protons they are atoms of the same element.

Sometimes two atoms, with the same proton number, have different numbers of neutrons. These atoms are of the same element but they have different nucleon numbers or mass numbers. We call these atoms **isotopes** (a name obtained from two Greek words: *isos* meaning same, and *topos* meaning place). Isotopes occupy the same place or position in the table of elements because they are of the same element.

Isotopes are atoms with the same proton number Z but different nucleon numbers A.

Fig. 17.7 shows the different numbers of neutrons (and hence different nucleon numbers) in the three isotopes of hydrogen. In the case of hydrogen, the isotopes also have special names. Table 17.2 gives the first six elements and

So we would say that hydrogen has three isotopes but when we talk about just one of them, say deuterium, we should refer to the nuclide which has 1 proton and 1 neutron in its nucleus.

Nuclei which have too few or too many neutrons are found to be *unstable*. This means that they are prone to breaking up and shooting out fragments in a process known as **radioactive decay**.

The nuclide symbol

There is a standard notation or way of specifying a particular nuclide which tells us both its proton number and nucleon number, fig. 17.8. The final column in table 17.2 shows how this notation is used. The simplest possible nuclide, a hydrogen atom with only one proton in its nucleus is written $^1_1$H and so on. The number of neutrons in the nuclide can be found by subtracting the two numbers on the nuclide symbol. Since $A = Z + N$, we have the number of neutrons $N = A - Z$.

Worked Example
Using $A = Z + N$

$^{235}_{92}$U *is a nuclide of uranium. How many neutrons and how many protons does it contain?*

By reading the symbol we can see that its proton number $Z = 92$, and its nucleon number $A = 235$.

The number of neutrons $N = A - Z = 235 - 92 = 143$.

Answer: The number of protons is 92 and the number of neutrons is 143 in the nuclide $^{235}_{92}$U.

Figure 17.7 *Isotopes of hydrogen*

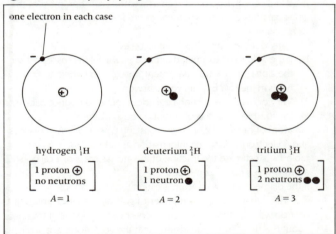

Figure 17.8 *Standard notation for specifying a nuclide*

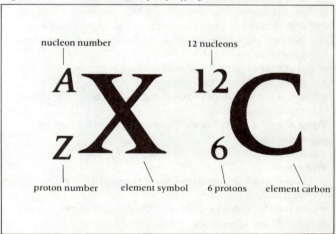

shows that most of them have naturally occurring isotopes. The isotopes are not found in equal proportions, and in most cases nearly all of the atoms are of one particular isotope (indicated by heavy type in table 17.2). For example, 98.9% of carbon atoms have the nucleon number 12.

When we wish to talk about one particular isotope of an element we use the word **nuclide** to refer to atoms with the same nucleon number as well as proton number.

The word nuclide describes a species of atoms of which every atom has the same proton number (atomic number) and the same nucleon number (mass number).

Atomic masses

The mass of an atom is almost equal to the mass of the nucleons it contains. However, as the masses of atoms and their nucleons is very small when expressed in kilogrammes a more convenient unit of atomic mass was devised. This unit is approximately the same as the mass of the proton or the neutron.

The unit of atomic mass chosen, called the unified atomic mass unit, is equal to $\frac{1}{12}$ of the mass of the nuclide $^{12}_6$C (the commonest carbon nuclide).

So the mass of one $^{12}_6$C carbon atom is exactly 12 unified atomic mass units, written 12 u.

One $^{12}_{6}\text{C}$ carbon atom has a mass of $2.0 \times 10^{-26}\,\text{kg}$, so it turns out that

$$1\,\text{u} = \frac{2.0 \times 10^{-26}\,\text{kg}}{12} = 1.7 \times 10^{-27}\,\text{kg}$$

The masses of other atoms expressed in unified atomic mass units are not whole numbers. The masses of the electron m_e, proton m_p and neutron m_n are given in table 17.3. The simplest hydrogen atom composed of 1 proton and 1 electron has a mass

$$m_\text{H} = m_p + m_e = 1.0073\,\text{u} + 0.0005\,\text{u}$$

giving $m_\text{H} = 1.0078\,\text{u}$.

Table 17.3 Particles in the atom

Particle	Symbol	Charge	mass/u*	Relative mass	Location in the atom
electron	e	$-e^\dagger$	0.00055 m_e	—	in orbitals or shells around the nucleus
proton	p	$+e$	1.0073 m_p	$1836\,m_e$	nucleons, i.e. particles found in the nucleus
neutron	n	neutral	1.0087 m_n	$1839\,m_e$	

* The unified atomic mass unit $\text{u} = 1.66 \times 10^{-27}\,\text{kg}$.
† The charge of one electron $= -1.6 \times 10^{-19}$ coulombs.

Assignments

Explain
a) what you understand by the term 'atom'.
b) what you understand by the term 'element'.
c) what all atoms of the same element have in common.
d) what all atoms of a particular isotope of an element have in common, and what makes atoms of one isotope different from those of another.
e) what it is that protons and neutrons have in common that distinguishes them from electrons.
f) what it is that protons and electrons have in common that neutrons do not have.

Write down the number of each kind of particle in the nuclides represented by:
g) ^1_1H h) ^3_1H i) $^{14}_6\text{C}$ j) $^{19}_9\text{F}$ k) $^{235}_{92}\text{U}$

Write down
l) a word which describes the relation between the following nuclides: $^{12}_6\text{C}$ and $^{14}_6\text{C}$.
m) the nuclide symbol for an atom of chlorine (Cl) which contains 17 protons and 18 neutrons.
n) the nuclide symbol for the isotope of uranium which contains 3 neutrons more than the one given in (k) above.
o) the symbol for the nuclide, $\frac{1}{12}$ of which is called 'the unified atomic mass unit'.

Try questions 17.1 to 17.7

Questions 17

1 Here is a list of particles studied in physics.
alpha; atom; beta; electron; molecule; neutron; proton
From the list choose one particle which
a) has a positive charge,
b) is found in the nucleus of an atom,
c) is given out by radioactive materials,
d) is uncharged.   (NEA [A] 88)

2 The following is a list of particles:
proton; neutron; electron; alpha particle; beta particle.
State which **two** of the above
  i) are positively charged,
  ii) are found in the nucleus of an atom,
  iii) occur in equal numbers in all neutral atoms,
  iv) are identical.   (W 88)

3 Fig. 17.9 represents a neutral atom.
a) For this atom, copy and fill in the Table.
b) The diagram represents an atom of lithium. In what way will an atom of another isotope of lithium differ from this one?
                                                       (MEG 88)

Figure 17.9

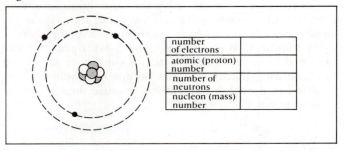

number of electrons	
atomic (proton) number	
number of neutrons	
nucleon (mass) number	

4 A particular atom of neon has an atomic number 10 and a mass number 20. How many protons and neutrons has this atom? Another atom of neon has a mass number of 22. In what ways is this atom similar to and different from the first atom?

5 When a narrow beam of alpha particles from a radioactive source is fired at a very thin gold foil in an evacuated chamber, most of the alpha particles pass straight through the foil without any change of direction.
a) Why does the chamber have to be evacuated?
b) If the gold foil is examined under a powerful microscope after the alpha particle bombardment, would you expect to find any holes in the foil? Explain your answer.
c) Very occasionally an alpha particle is found to bounce back from the foil. How can this observation be explained?

6 An atom of carbon-12 has 6 protons and 6 neutrons in its nucleus and 6 electrons. Using the information about relative masses in table 17.3, estimate the fraction of the atomic mass of a carbon-12 atom which is not in its nucleus.

7 If the diameter of a gold atom is estimated to be about $3 \times 10^{-10}\,\text{m}$, calculate the number of gold atoms which an alpha particle will pass through when it penetrates a gold foil of thickness about $6 \times 10^{-7}\,\text{m}$. (If you assume that the gold atoms are arranged in layers, this will give a rough estimate of the number of layers of atoms in the gold foil.)

In an alpha-particle scattering experiment using this foil, one in 18 000 of the alphas which hit the foil were thrown backwards and did not pass through it. Estimate the proportion of alphas thrown backwards by each layer of gold atoms.

Assuming that the proportion of the alphas thrown backwards by a layer of atoms gives an estimate of the proportion of the area of the layer which is occupied by nuclei, calculate a value for the diameter of a gold nucleus. (The area of the foil occupied by one gold atom, assuming all the atoms are square and are packed close together, is given by $[3 \times 10^{-10}\,\text{m}]^2$. The diameter of the nucleus can be found by assuming that it also has a square area, the square-root of which is the required estimate.)

18
Radioactivity

*N*atural radioactive decay of atoms has been happening on the earth ever since it was formed, over 4000 million years ago. In 1896 Henri Becquerel discovered that some uranium salt emitted radiation which could pass through black wrapping paper and produce a dark silhouette on a photographic plate. Only then did we begin to explore what is now called radioactivity.
Marya Sklodowska was born in Warsaw in 1867. She is remembered as Marie Curie who, working with her husband Pierre, discovered several new elements including polonium and radium which emitted invisible radiation. Marie Curie described these as being 'radioactive' and invented the word 'radioactivity'. She is seen here working in her laboratory. Marie Curie died in 1934 of an illness which was caused by overexposure to radiation.

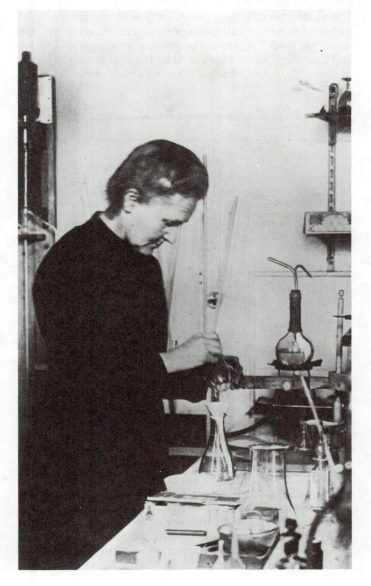

18.1
THREE KINDS OF RADIATION

The radiation emitted by uranium was of unknown nature and only gradually was it found that there were three different types of radiation emitted by radioactive materials. The names alpha (α), beta (β) and gamma (γ), used for the then unknown radiations, are still used today even though we have found out what they are.

Safety precautions

Before starting on experiments which use radioactive materials, teachers and students should be aware of the safety precautions and regulations which apply to these materials. Table 18.1 lists the safety precautions which apply to the use of all radioactive materials. Details of sources suitable for use in schools are given in appendix D10.

Table 18.1 Safety precautions when using radioactive materials

1	Always use forceps or a lifting tool to move a source; never use bare hands.
2	Arrange a source so that its radiation window points away from your body.
3	Never bring a source close to your eyes for examination; it should be identified by a colour or number.
4	When in use, a source must always be attended by an authorised person and it must be returned to a locked and labelled store in its special shielded box (fig. 18.1) immediately after use.
5	No eating, drinking or smoking must take place in a laboratory where radioactive materials are in use.
6	After any experiment with radioactive materials wash your hands thoroughly before you eat. (This applies particularly to the handling of radioactive rock and salt samples and all open sources.)
7	In the UK, students under 16 years old may only handle sources for which special authorisation is **not** required.

Figure 18.1 *Radioactive sources must be handled with tweezers. A cobalt gamma source is seen here being removed from its shielding lead 'castle'.*

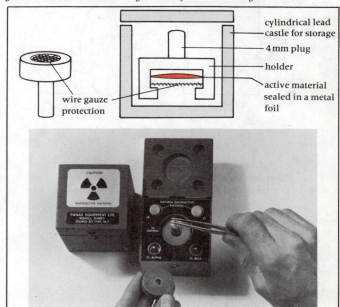

Detecting invisible radiation

The effect on photographic paper

(Radioactive rock samples can be obtained from suppliers without special authorisation. A uranium ore such as pitch blende would be suitable and 'Becquerel plates' with a sample of uranium oxide painted in a recess are also available.)

● *Glue a few crystals of a uranium rock or salt to a small piece of paper about the size of a large coin.*

● *Place the radioactive material on top of a completely light-tight wrapped sheet of photographic paper.*

● *Leave it in a safe place where it will not be disturbed for several days. (The length of time depends on the amount and activity of the radioactive material and the type of photographic paper.)*

● *Unwrap and develop the photographic paper in a dark room.*

● *Anyone who handles the radioactive material should wash their hands thoroughly afterwards.*

The uranium material leaves a dark patch or silhouette on the photographic paper where it has been.

Counting scintillations produced by radiation

This was the method used by Rutherford, Geiger and Marsden to count alpha particles as they were scattered from a thin metal foil (p 381). The alpha particles emitted by radium have kinetic energy which can be converted into tiny flashes of light called **scintillations** when they strike a fluorescent material.

A **spinthariscope** is a scintillation viewer which contains a small sample of radium and a screen coated with a fluorescent substance such as zinc sulphide, fig. 18.2. (This radium source is very weak and requires no special authorisation. It is safe to allow students to use this apparatus.)

● *Take the spinthariscope into a very dark room.*

● *Sit for five or ten minutes until your eyes have adapted to the darkness and become much more sensitive.*

Figure 18.2   *A spinthariscope*

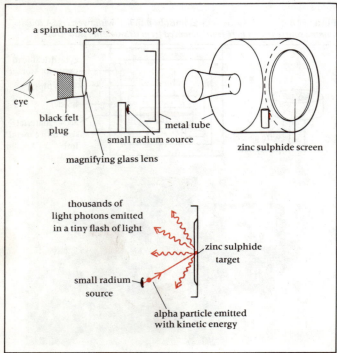

● *Now remove the black felt plug from the eyepiece of the spinthariscope. (If this plug is removed in daylight, the zinc sulphide screen will fluoresce and remain 'foggy' for some time afterwards.)*

● *If you concentrate you can count the tiny flashes of light produced as individual alpha particles strike the screen. Now you can imagine how difficult it was for Geiger and Marsden to count over 100 000 of these scintillations during their experiments!*

The ionising effect of radiation
Teacher demonstration

Charge a gold-leaf electroscope and hold a lighted match very near to its cap.

The gold leaf slowly falls showing that the cap of the electroscope loses charge. The energy released in the flame produces ions and electrons from the surrounding air molecules. Depending on the charge on the electroscope, either the positive ions or the negative electrons will be attracted to the cap and neutralise its charge causing the leaf to fall. Fig. 18.3 shows what happens when the electroscope is positively charged.

Charge the electroscope again and, using forceps, hold a radium source up to the cap in place of the lighted match.

Repeat the test with the opposite charge on the electroscope.

In both cases the electroscope discharges showing that the radiation from the radium source can ionise air molecules and that the effect is not caused by the charges of the alpha or beta particles. Radioactive materials emit ionising radiations.

Figure 18.3   *Ions in the air near an electroscope cause the charge to leak away*

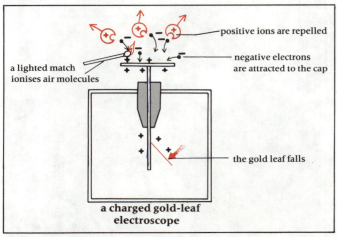

Connect a spark counter to a 0−5 kV e.h.t. supply, (danger: high voltage!) (Spark counters are available from suppliers but a simple device can be made using a length of stiff wire and some wire gauze for electrodes as shown in fig. 18.4.)

Turn up the high voltage supply until sparks jump across the gap between the wire and the gauze. Then turn down the supply voltage slowly until the sparks just stop.

Holding a radium source in forceps, bring it close to the air gap between the wire and gauze until sparks occur again.

At the reduced voltage the air could not conduct. The radiation from the radium source produces extra ions in the air gap allowing it to conduct more easily and produce sparks.

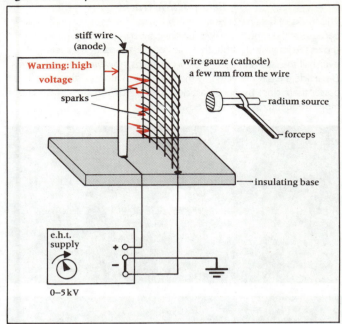

Figure 18.4 *A spark counter*

Making particle tracks visible

We cannot see the radiation emitted by radioactive materials, but by its ionising property we can make the track of a particle visible. As an ionising particle passes through a gas it leaves a trail of positive gas ions along its track. Any water (or alcohol) vapour molecules in the surrounding air are attracted to these positive ions. So the positive gas ions act as centres round which small liquid droplets can form, fig. 18.5.

Air *saturated* with water vapour contains so many water molecules that if any more are added some will try to condense and form liquid water. Air holds less water molecules in the vapour state at lower temperatures. So when saturated air is cooled, it is caught holding too many water molecules and is said to be *supersaturated*. The excess water molecules will condense on any suitable small particles, such as dust particles or positive gas ions which attract them.

The tracks that can be seen as an ionising particle passes through air supersaturated with a vapour, are formed from many thousands of tiny liquid droplets condensing around the trail of positive ions. An alpha particle can form over 100 000 ions in a single track. The vapour trails of high-flying aircraft are formed by a similar process. Vapour condenses on ions produced along the track of the aircraft, and sunlight is reflected from the small liquid droplets that form.

Figure 18.5 *The formation of a water droplet around a positive ion*

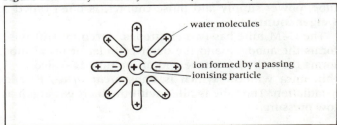

Setting up a cloud chamber

A cloud chamber contains air supersaturated with water or alcohol vapour, fig. 18.6. The conditions which produce clouds in the chamber are also suitable for making vapour trails. To produce these conditions the following steps should be taken:

a) Remove the transparent lid and, using an eye dropper, soak the felt strip at the top of the chamber with water and alcohol. (A 50% ethanol and water mixture works well.) Before replacing the lid and after levelling the chamber a few drops of the liquid can be placed on the metal base plate.

b) Remove the base cover and the foam sponge plug from the underside of the chamber. Spread a layer of crushed dry ice (solid CO_2 obtained from a liquid CO_2 cylinder) evenly over the underside of the metal base plate. Replace the sponge plug and the base cover.

c) If the chamber has no built-in radioactive source, mount an alpha source at a low level on one side of the chamber, facing in a horizontal direction across the base plate.

d) Replace the transparent lid and level the chamber using three thin wedges. (You can tell when the chamber is not level by the drifting of clouds across the base plate.)

e) Position a lamp at the side of the chamber about level with the source, a few millimetres above the base plate.

f) Rub the top of the chamber with a clean cloth. This charges the plastic top and, by attraction, removes dust particles and any existing ions from the chamber.

Now watch carefully through the top of the chamber. Within about a minute the cooling produced by the dry ice will cause the air above the base plate to become supersaturated. The tracks of alpha particles become visible for several minutes while the conditions in the chamber remain right.

Figure 18.6 *A cloud chamber*

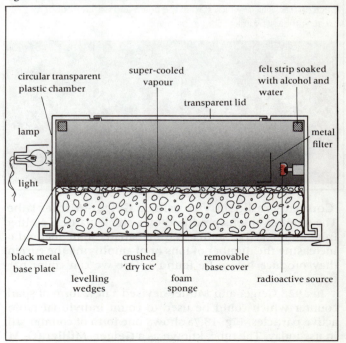

Cloud chamber tracks

Alpha particle tracks

The track left by an alpha particle in a cloud chamber is bold and straight. The tracks are bold because many tiny water droplets form around the many ions produced along the α-particle tracks. The tracks are very straight because the α-particle has a large mass and momentum compared with the electrons which it pulls off gas atoms as it passes them by.

Beta particle tracks

The tracks of beta particles are much fainter than those of α-particles because of their much weaker ionising power. The centre photograph shows the long, straight but thin track of a fast β-particle travelling at over half the speed of light. At slower speeds, however, the very light β-particle is easily deflected by the electrons of atoms in the air as it passes them close by. So β-particles suffer sudden changes of direction more frequently as they slow down.

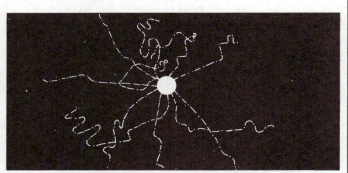

X-ray and gamma-ray tracks

There is no track along the path of an X-ray or gamma photon. However, in an intense beam of X or gamma radiation containing many millions of photons, occasionally a photon is stopped and absorbed by an atom. Then an electron is thrown out of the atom producing a short faint and wandering track similar to that of a slow β-particle. The path of an intense beam of X or gamma radiation can produce many of these electron tracks all of which start in the beam and wander out of it creating the wispy image shown below.

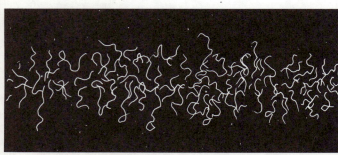

Beta particles in a magnetic field

The very low mass of the beta particle makes it easy to deflect in a magnetic field. The faster particles travel in gradual curves of large radius, while the slower particles travel in tighter circular curves. We can tell that β-particles carry a negative electric charge by the direction in which they are deflected in a magnetic field. The curved paths shown below are caused by a magnetic field which is pointing vertically downwards at right angles to the page.

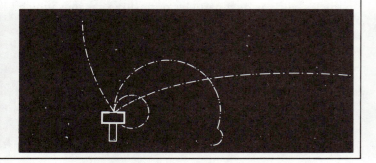

Counting radioactivity using a Geiger–Müller tube

The number of ions produced by a single particle emitted from a radioactive source is quite small. Some method of increasing the number of ions or of amplifying the current they produce has to be found before we can count particles.

In 1928 Geiger and Müller devised a new form of spark counter which could be used to count individual radioactive particles. Fig. 18.7a shows one form of counter still in use today. The tube, known as a **Geiger–Müller (G–M)** tube, is the radiation detector. The whole counter, i.e. tube, power supply and pulse counter, is often called a **Geiger counter**.

The G–M tube has two electrodes. A central stiff wire forms the anode + and the outer cylinder or metal tube forms the cathode −. One end of the tube is sealed by a thin mica window which is transparent to α-, β- and γ-radiation. The tube is filled with the inert gas argon at low pressure.

Figure 18.7 *A Geiger–Müller tube*

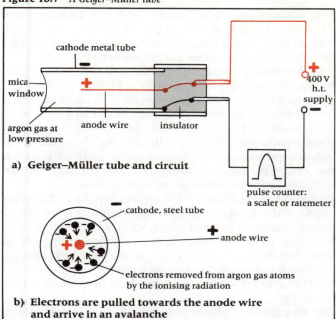

a) **Geiger–Müller tube and circuit**

b) **Electrons are pulled towards the anode wire and arrive in an avalanche**

How the G–M tube works

The G–M tube is a gas amplifier. When radiation enters the tube and produces a few ions, the G–M tube dramatically increases the number of ions to produce a large pulse of charge which is collected by the wire electrode. The high voltage (about 300 to 500 V) connected between its electrodes provides an electric field which pulls the electrons towards the positive central wire electrode, fig. 18.7b. These accelerating electrons, rushing towards the central wire, cause more ions to be formed as they collide with argon gas atoms. An avalanche of electrons arrives at the central wire with a large quantity of charge, which then flows round the electric circuit as a pulse. This is counted electronically.

α- and β-radiation ionise the argon gas by removing electrons from the argon atoms. Some photons of γ-radiation are absorbed by the outer metal tube and knock electrons into the argon gas which then produce more ions and an avalanche of electrons.

The scaler and the ratemeter

These are electronic pulse counters used to count radioactive particles when connected to a G–M tube. A coaxial cable is used to connect the G–M tube to the counter, which supplies the high voltage required for the tube as well as carrying the pulses from the G–M tube to the counter.

The *scaler* gives a reading which is the total or cumulative number of particles counted from the moment it was started. To obtain a count rate, the time of counting must be noted as well.

$$\text{count rate} = \frac{\text{number of counts}}{\text{time of counting}}$$

The *ratemeter* gives a reading which is the count rate in counts per second. This is obtained by electronically averaging the number of counts over a certain time (called the integration time).

Penetration and absorption of radiation
Teacher demonstration

Set up an arrangement shown in fig. 18.8 with a source (see appendix D10) facing the mica window of a G–M tube a fixed distance away.

With an α-source this distance should be only 1 or 2 cm away because of the short range of α-particles in air. (An α-source may overload a G–M tube at close range because of the strong ionising property of α-radiation. In this event a lead or aluminium screen with a small hole in the centre should be placed in front of the source to reduce the amount of α-radiation.) With a γ-source a gap of several centimetres is needed for lead absorbing sheets.

Figure 18.8 *Penetration and absorption of α-, β- and γ-radiations*

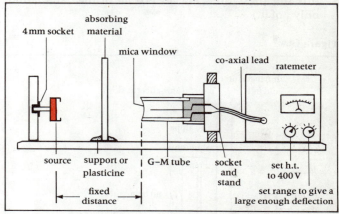

Adjusting the ratemeter

Set the h.t. voltage supply for the G–M tube to about 450 V.

Set the range switch so that the pointer gives a good deflection but is not off the end of the scale.

Taking a background reading

In all radiation experiments where counts or count rates are recorded it is necessary to subtract from all the readings a **background** count rate. The background radiation count is always present due to natural radioactivity in the ground, bricks of buildings and the bombardment of the atmosphere by cosmic radiation.

Remove all sources some distance from the experiment and switch on the ratemeter with its range switch on the lowest range.

The count rate may be so low and erratic that it is difficult to get a steady reading.

● *If the ratemeter has a loudspeaker, when this is switched on you can count the background radiation as the number of clicks you hear in a minute. Do not forget to divide your count by 60 to obtain the background count rate in counts per second.*

Corrected count rate = count rate − background count rate

● *Tabulate the results as in table 18.2.*

Table 18.2

Source	radiation type	Absorbing material	Thickness of absorber	Count rate counts/s	Corrected count rate counts/s

Background count rate = _____ counts/second

The results obtained are summarised in fig. 18.9. The following are the important points:

a) α's and β's have definite maximum ranges. α's are totally absorbed by a thick sheet of paper, or a few layers of human skin. β's are absorbed by a few millimetres of aluminium.

b) γ-radiation has no maximum range. About 1 cm of lead will absorb half of the γ-radiation, another 1 cm of lead will halve it again, allowing $\frac{1}{4}$ of the original radiation to pass through. Thus 5 cm of lead will halve the γ-radiation five times, reducing the level of γ-radiation to $\frac{1}{2} \times \frac{1}{2} \times \frac{1}{2} \times \frac{1}{2} \times \frac{1}{2}$ or $\frac{1}{32}$ of its original value. The thickness of absorbing material which will absorb half of some γ-radiation is called the *half-thickness*. (Note that twice this thickness does not absorb *all* of the γ-radiation, only $\frac{3}{4}$ of it.)

Figure 18.9

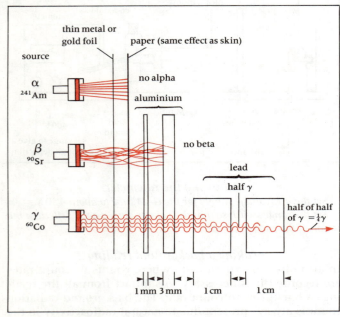

The effect of distance on radiation
Teacher demonstration

Alpha radiation
Use the practical arrangement shown in fig. 18.8, but without the absorbing material, to investigate how far α-particles travel in air.

Move the G–M tube in 5 mm steps away from the source and record the count rate in each position.

Within a few centimetres all the α-radiation has been stopped by the air. The α-particles give up their kinetic energy as they form ions along their track. When all their energy has been used up pulling electrons off air molecules, α-particles pick up two electrons and become helium atoms.

Beta radiation
The range of β-particles in air is difficult to measure, because they can travel several metres before they come to rest. They give up their kinetic energy more slowly as they form far fewer ions along their tracks. A β-particle which has been stopped in its track is an electron.

Gamma radiation
To investigate the effect of distance on γ-radiation it is better to turn the G–M tube with its side facing the γ-source, fig. 18.10a. Measure the distance d from the source to the side of the G–M tube. Suitable values of the distance d are shown in table 18.3.

In each position, take a reading of the count rate.

● *Record a reading of the background radiation count rate.*

The count rate falls off rapidly as the G–M tube is moved away from the source, but it does not fall to zero. With γ-radiation, the drop in count rate is *not* a result of absorption of the radiation by the air, since γ-radiation has only a very weak ionising effect on air molecules. Rather the cause is the 'spreading out' or 'dilution' of the radiation as it gets farther away from its source.

The *inverse-square law* which applies to light and all other parts of the electromagnetic spectrum, also applies to γ-radiation (which is part of that spectrum, chapter 23).

● *To test the inverse-square law for γ-radiation, complete the table of results by calculating $1/d^2$, which is the inverse square of the distance of the G–M tube from the source.*

● *Plot a graph of the corrected count rate against $1/d^2$, fig. 18.10b.*

The straight-line graph confirms that the count rate is inversely proportional to the distance squared.

Figure 18.10 *The effect of distance on gamma radiation*

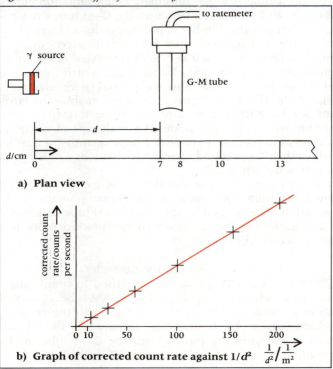

a) **Plan view**

b) **Graph of corrected count rate against $1/d^2$**

Table 18.3

d/cm	7	8	10	13	20	30
d/m	.07					.30
$\dfrac{1/d^2}{1/\text{m}^2}$	204					11
count rate (counts/s)						
corrected count rate (counts/s)						

Fig. 18.11 shows the same idea in another way. It shows that γ-radiation gets weaker as the area it covers gets larger. The radiation is spread over an area A at a distance of 1 metre. At 2 metres it is diluted over an area 4A and at 3 metres over an area 9A. As the area increases with the square of the distance, so the count rate C and the intensity of the radiation decrease according to the inverse-square law with distance.

Here lies the reason why it is so important to handle all sources with long forceps. For example, a source held in the fingers 1 cm away exposes them to concentrated or intense radiation. But the radiation from the same source, held in 10 cm long forceps is reduced to $\frac{1}{100}$ of the intensity, simply as a result of being 10 times further away.

Figure 18.11 *The range of α-, β- and γ-radiations in air*

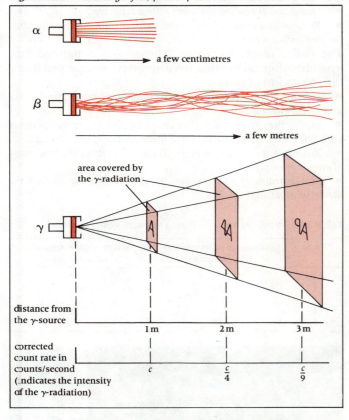

The effect of a magnetic field on radiation
Teacher demonstration

Beta radiation

A β-source with a narrow slit in its aluminium housing can be used to produce a thin beam of β-particles. This is called a **collimated source.**

Arrange this source so that the beam of β-particles enters the mica window of a G–M tube some 10 or 15 cm away, position A in fig. 18.12.

Move the G–M tube to each side and check that most of the radiation is going straight ahead into the tube.

Fit a pair of magnadur flat magnets to an iron yoke with their unlike poles facing. Using a magnetic compass, find out which pole is which and hence which way the magnetic field goes (N to S).

The count rate recorded by the ratemeter will fall significantly, almost down to background level.

Figure 18.12 *Deflection of β-particles by a magnetic field*

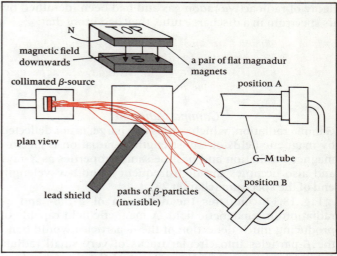

Move the G–M tube to the sides until the count rate increases again, position B. Check that β-particles are not reaching the tube by a direct route by placing a lead shield in the way as shown in fig. 18.12.

The tracks of the β-particles are definitely bent by the magnetic field. This shows that the particles carry an electric charge. Can you confirm (by the direction of bending) that these particles are the negatively charged electrons we believe them to be?

In 1900 Becquerel was the first person to deflect β-particles in an electric field. The information gained from deflecting β-particles in both a magnetic field and an electric field gave an estimate of the speed of β-particles and their charge-to-mass ratio. The results showed that:
a) the speed of β-particles is about half the speed of light,
b) the charge-to-mass ratio e/m, was about the same as for the electron.

More accurate experiments showed that at these high speeds the mass of the β-particles increased, as predicted by Einstein's theory of relativity. When the increase in mass of the β-particles is allowed for, the value of e/m agrees with that of the electron.

Beta radiation is a stream of electrons which are ejected from the nuclei of decaying atoms at speeds between 0.3 and 0.7 times the speed of light.

Alpha radiation

Attempts to deflect α-particles will be unsuccessful unless a very powerful magnet is available. This is because their mass is about 7000 times greater than the β-particles and they are therefore much more difficult to deflect. The positive charge carried by α-particles results in their deflection in the opposite direction to β-particles.

Between 1903 and 1909 Rutherford, working at different times with Geiger and Royds, carried out a series of experiments which identified α-radiation. They carried out very accurate experiments to measure the deflection of α-particles in magnetic and electric fields and found a value for the charge-to-mass ratio Q/m which was the same for all α-particles whatever their speed. They also succeeded in measuring the charge on an individual α-particle, finding it to be twice the electronic charge.

Finally, when helium gas had been collected from the decay of radioactive radon gas and had been identified by its spectrum in a discharge tube, they had proof that:

Alpha radiation is a stream of helium nuclei, which are particles composed of 2 protons and 2 neutrons. Alpha particles are ejected from the nuclei of decaying atoms and eventually become helium atoms by picking up two electrons.

Gamma radiation

Gamma radiation, which carries no charge, is not deflected by magnetic fields of any strength. γ-radiation is electromagnetic radiation and has the same properties as X-rays and also occupies the high-frequency, short-wavelength end of the spectrum (chapter 23).

Fig. 18.13 compares the deflection of α-, β- and γ-radiation in a magnetic field. A magnetic field capable of producing much deflection of the α-particles would bend the β-particles into circular tracks of very small radius. Table 18.4 summarises the natures of the three kinds of radiation emitted in natural radioactive decay.

Figure 18.13 *Radiation in a magnetic field*

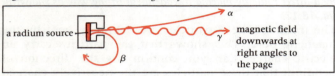

a radium source — magnetic field downwards at right angles to the page

Table 18.4 The nature of α-, β- and γ-radiations

	alpha, α	beta, β	gamma, γ
What it is	a helium nucleus — 2 protons, 2 neutrons	a fast moving electron	electromagnetic radiation of very high frequency f and short wavelength
Charge	the charge of 2 protons: $+2e$	the charge of an electron: $-e$	no charge
Mass	approximately 4u (contains 4 nucleons)	m_e: the mass of an electron about $\frac{1}{7000}$ of the α-particle	zero
Energy	kinetic energy	kinetic energy	photon energy $E = hf$

Assignments

Make a table summarising all the properties of the three kinds of radiation. Include the following in your table: ionising effects, type of tracks in a cloud chamber, range in air, absorption by different materials and the effect of a magnetic or electric field.

Draw a diagram to show the deflection of α-, β- and γ-radiations as they pass between two metal deflection plates in a vacuum. (These are similar to the deflection plates in a CRT.) A high voltage connected across the plates produces the strong electric field between them.

Explain

a) Why are β-particles much more easily deflected than α-particles as they travel both through the air and through magnetic fields?

b) What is the difference between a scaler and a ratemeter?

c) What is meant by background radiation?

d) How are particle tracks made visible in a cloud chamber?

Try questions 18.1 to 18.4

18.2
RADIOACTIVE DECAY OF ATOMS

We cannot tell when the nucleus of a particular unstable atom will break up and hurl out an alpha or beta particle. However, we find that out of a large number of the same kind of unstable atoms a constant fraction of them will decay in a certain time. It is easy to tell which atoms have decayed because they change into a different element.

A model of radioactive decay

Using a large number of dice or small wooden cubes to represent unstable atoms, we can build a model of radioactive decay. The wooden cubes can be made by carefully sawing cubes off a length of 1 cm × 1 cm square-section hardwood. Paint one face of every cube. Several hundred cubes (or dice) are needed for each experiment.

• *Shake and throw all the cubes onto a table top.*

• *Remove and count all the cubes with a painted face upwards (or dice with a 6 upwards).*

The cubes represent atoms. Those with a painted face upwards after each throw have 'decayed' or changed into a different kind of atom.

• *Pick up all the remaining cubes (undecayed atoms) and shake and throw them again.*

• *Repeat the process, each time counting and removing the cubes with coloured faces upwards, until there are very few cubes left.*

• *Record your counts in a table. Table 18.5 shows the beginning of one set of results.*

• *Plot a graph of the number of cubes remaining against the throw number, fig. 18.14.*

Table 18.5 Specimen results from a radioactive decay model

Throw	*Number of cubes removed (decayed atoms)*	*Number of cubes remaining (undecayed atoms)*
start	0	200
1	32	168
2	28	140
3	25	115
4	18	97

In a game of snakes and ladders your fate is decided by the throw of a dice. Chance may lead you up a ladder, or down a snake. The decay of a particular unstable atom is also an unpredictable chance event. In this experiment we use a large number of dice to represent a large number of unstable atoms. At each throw those dice which show a six have 'decayed'.

Figure 18.14

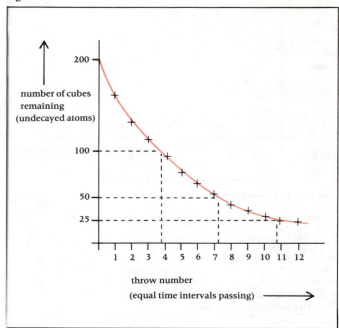

This decay graph has the same special shape that we obtained for the charge flowing out of a capacitor, fig. 12.52. It is called an **exponential curve**. The graph gradually gets less steep and only falls to zero after a very long time. In radioactive decay experiments it is usual to find the time taken for half of a sample of radioactive material to decay. In this model experiment the passing of time is represented by the number of throws.

● *From your graph, find the number of throws which halved the number of cubes. You can find several values for the halving number of throws by halving the number of cubes several times, as shown in fig. 18.14.*

After about $3\frac{1}{2}$ throws the number of cubes has halved, after each further $3\frac{1}{2}$ throws the number of remaining cubes halves again. We find that the number of cubes remaining is halved after the same number of throws $(3\frac{1}{2})$, no matter how many cubes there are left. This is to be expected with six-sided cubes because there is a constant chance $(\frac{1}{6})$ of the painted face turning upwards. Each time the cubes are thrown we expect 1 in 6 of them to 'decay', i.e. to turn out with their painted faces upwards. What we do not know before they are thrown is which particular ones will be the cubes to 'decay'. We can never say about any particular cube that it will 'decay' at the next throw. All we know is that it has a constant chance of $\frac{1}{6}$ of 'decaying' each time it is thrown. The moment when it will actually decay is unpredictable and happens *randomly*.

Half-life

The decay of radioactive atoms is very similar to the 'decay' of the wooden cubes or dice, except that even in a small sample of a radioactive substance there are enormous numbers of unstable atoms. The halving number of throws of the cubes represents the halving time of the unstable atoms, called their **half-life**.

The half-life of a sample of radioactive substance is the time taken for half of the unstable atoms to decay.

However large the sample and however many unstable atoms of a particular kind it contains, *the half-life is constant*. Over a half-life, each atom has a 50% chance of decaying. However, the decay of atoms in a sample appears to be totally random in two ways:

a) we cannot tell *which* particular atoms are going to decay.

b) we cannot tell *when* they are going to decay.

For the very large number of atoms in the sample, the random decay of individual atoms is averaged out to give a constant half-life for the sample as a whole. The value of a half-life is unaffected by all external changes such as temperature, pressure and even chemical reactions.

The half-life of some substances is very long. For example, thorium $^{232}_{90}$Th has a half-life of 14 000 million years. We would have to wait rather a long time to notice any change in a sample of thorium!

A decay series

For us to measure the half-life of a radioactive substance we need one with a half-life of not more than a few minutes. There is a problem with this because such a short half-life means that the substance decays away so quickly we cannot store it. However radioactive substances often decay several times in a series of steps, emitting radiation and producing a new substance at each step, fig. 18.15a. A **parent** substance produces **daughter** and **granddaughter** substances in what is called a **decay series**. Each decay step has a different half-life, and we use this to obtain a substance with a short enough half-life for our experiment.

We store a parent substance with a long half-life, which therefore does not decay away quickly, and obtain from it a daughter product with a short half-life. We need a daughter product which can be separated from the parent easily. One such parent–daughter example which is often used in schools is thorium and radon. Thorium, the parent, is a solid (usually in the form of thorium hydroxide) with a half-life of 14 000 million years. Thorium decays in a long series of steps until it reaches a stable isotope of lead. In the middle of the series, the fifth-generation daughter product is a gas. This gas, radon, can be separated easily from all the other solid substances, and has a convenient half-life of 52 seconds, fig. 18.15b.

Figure 18.15

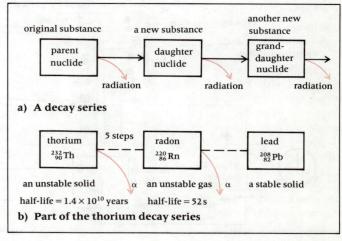

a) A decay series

b) Part of the thorium decay series

Measuring the half-life of radon gas
Teacher demonstration

Fig. 18.16 shows an ionisation chamber used with a direct current (d.c.) amplifier to measure the half-life of radon.

Connect the ionisation chamber to the d.c. amplifier. The metal can is insulated from its mounting but the whole chamber screws directly onto the input of the d.c. amplifier so that the connection between its central negative electrode and the amplifier is as short as possible.

Connect a 9 V battery between the outside can of the chamber and the earth input terminal of the amplifier so that the can is made the positive electrode.

Adjusting the d.c. amplifier

Set the input selector switch to its 'rest' or set-zero position so that the input of its amplifier is short-circuited and the input is zero.

Adjust the set-zero control to give a zero output current on the milliammeter.

Select the highest resistance input, usually $10^{11}\,\Omega$.

Filling the ionisation chamber with radon gas

Connect the polythene bottle radon generator to the ionisation chamber with the two thin rubber tubes.

Release both clips on the tubes.

Squeeze the bottle two or three times until the reading on the milliammeter just goes past full-scale deflection.

Refit the clips on the tubes.

Taking readings and finding the half-life

- *Start the clock as the current reading falls to full-scale deflection.*
- *Take readings of the current every 10 s for about three minutes.*

(Do not dismantle the ionisation chamber. Leave it for half an hour until the remaining radioactivity has decayed to a very low level.)

- *Plot a graph of the current readings against time and draw a smooth curve through the points, fig. 18.17.*
- *Read from the graph the time at which the current had fallen to $\frac{1}{2}$, $\frac{1}{4}$ and $\frac{1}{8}$ of its initial value I_0.*
- *From these times find three values of the half-life of radon gas. How well do they agree?*

The half-life of radon $^{220}_{86}\mathrm{Rn}$ is 52 s and is constant within the accuracy of the experiment.

Figure 18.17

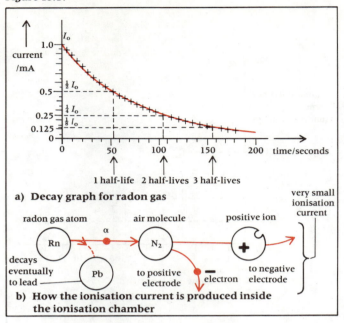

a) **Decay graph for radon gas**

b) **How the ionisation current is produced inside the ionisation chamber**

How the current is produced

Fig. 18.17b shows how a current is produced inside the ionisation chamber. The radon gas atoms decay randomly (half of them in 52 s) emitting α-particles. The α-particles ionise air molecules producing several million ion pairs every second. The central rod (the negative electrode) and the outer can (the positive electrode) collect these ions producing a very weak electric current of about $10^{-11}\,\mathrm{A}$, called the **ionisation current**.

This small ionisation current is fed into the d.c. amplifier which amplifies it about 100 million times until it is large enough to work a moving-coil ammeter.

As the number of radon gas atoms in the chamber gets smaller and smaller they emit fewer and fewer α-particles. These produce proportionally less ions and so a smaller ionisation current flows: *the ionisation current is directly proportional to the number of radon atoms remaining*. The time taken for the ionisation current to reduce to half its initial value is also the time taken for half the radon gas atoms to decay, i.e. the half-life of radon.

Figure 18.16   *Measuring the half-life of radon gas*

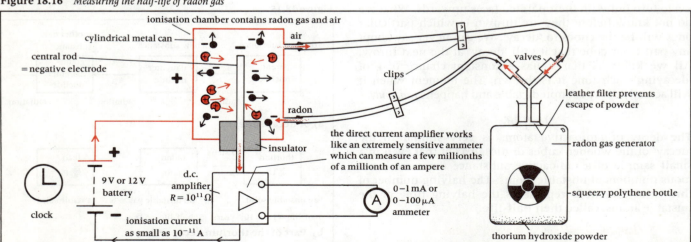

Nuclear changes

When a radioactive atom decays, its nucleus breaks up, throws out an α-particle or β-particle with some energy, and forms a new atom of a different element. We can represent these events as a nuclear equation in which a parent nuclide X changes into a daughter nuclide Y.

Alpha decay

The general equation is

$$
\underset{\substack{\text{parent}\\\text{nuclide}}}{{}_{Z}^{A}X} \rightarrow \underset{\substack{\text{daughter}\\\text{nuclide}}}{{}_{Z-2}^{A-4}Y} + \underset{\substack{\text{alpha}\\\text{particle}}}{{}_{2}^{4}\alpha} + \text{energy}
$$

For example

$$
\underset{\substack{\text{uranium}\\\text{parent}\\\text{nuclide}}}{{}_{92}^{238}U} \rightarrow \underset{\substack{\text{thorium}\\\text{daughter}\\\text{nuclide}}}{{}_{90}^{234}Th} + \underset{\substack{\text{alpha}\\\text{particle}}}{{}_{2}^{4}\alpha} + \underset{\substack{\text{+ gamma photon}\\\text{energy}}}{\text{+ kinetic energy}}
$$

When a nuclide decays by emitting an alpha particle its atomic number or proton number Z decreases by 2 and its mass number or nucleon number A decreases by 4.

This is the result of an α-particle carrying 2 protons and 2 neutrons away. When a nucleus emits an α-particle some spare energy is released as kinetic energy of the particle and some as a γ-photon, fig. 18.18. When the particle is thrown out the new nucleus recoils so that momentum is conserved, but the nucleus has very little kinetic energy because of its low velocity.

The result of α-decay is that a new element is produced with a proton number 2 below its parent in the table of elements. For example, uranium ($Z = 92$) decays by α-emission to form thorium ($Z = 90$) and similarly, radium ($Z = 88$) decays by α-emission to form radon ($Z = 86$).

Figure 18.18 *Alpha decay*

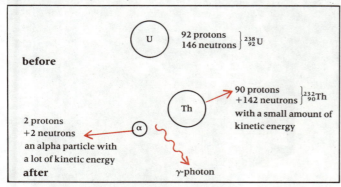

Beta decay

The general equation is

$$
\underset{\substack{\text{parent}\\\text{nuclide}}}{{}_{Z}^{A}X} \rightarrow \underset{\substack{\text{daughter}\\\text{nuclide}}}{{}_{Z+1}^{A}Y} + \underset{\substack{\text{beta}\\\text{particle}}}{{}_{-1}^{0}\beta} + \text{energy}
$$

For example

$$
\underset{\substack{\text{strontium}\\\text{parent}\\\text{nuclide}}}{{}_{38}^{90}Sr} \rightarrow \underset{\substack{\text{yttrium}\\\text{daughter}\\\text{nuclide}}}{{}_{39}^{90}Y} + \underset{\substack{\text{beta}\\\text{particle}}}{{}_{-1}^{0}\beta} + \text{kinetic energy}
$$

To explain how an electron can be emitted from a nucleus which contains no electrons, we have to explain what happens to the nucleons in the nucleus.

Beta emission causes the proton number Z to increase by 1 but the nucleon number A does not change.

Thus if the nucleus gains one proton without a change in the number of nucleons it must have lost a neutron. This change can be explained by the decay of a neutron inside the nucleus producing a proton which remains in the nucleus and a β-particle, which is ejected.

$$
\underset{\substack{\text{neutron}}}{{}_{0}^{1}n} \rightarrow \underset{\substack{\text{proton}}}{{}_{1}^{1}p} + \underset{\substack{\text{beta particle}}}{{}_{-1}^{0}\beta} + \text{energy}
$$

We can see that electric charge is conserved because the $+e$ charge of the proton and the $-e$ charge of the β-particle add up to give no change in the total charge.

Gamma photons

Gamma photons are usually emitted at the same moment as either an α- or β-particle. When a nucleus ejects an α- or β-particle there is often some spare energy produced in the decay process. Decay leaves a nucleus in an excited state, i.e. it possesses more energy than it can comfortably hang on to. Almost instantly the spare energy is released in a quantum, i.e. a single photon of gamma radiation. The size of this quantum puts it into the high-frequency, gamma part of the electromagnetic spectrum.

High-frequency electromagnetic radiation coming from the nuclei of decaying atoms is called gamma radiation. It is emitted in quanta called gamma photons.

What makes a nucleus unstable?

A nucleus seems to be stable when it has the right balance of protons and neutrons. Having too many neutrons or too few neutrons makes it unstable and likely to break up. The unstable nucleus tends to decay by whichever process will bring it nearer to the correct number of neutrons for stability.

Fig. 18.19 shows the trend for stable nuclei. For small nuclei Z and N are about equal. As the nuclei become larger, N becomes gradually larger than Z. For example, the stable nuclide ${}_{6}^{12}C$ has $Z = 6$ and $N = 6$, but the stable nuclide ${}_{82}^{208}Pb$ has $Z = 82$ and $N = 126$.

Figure 18.19 *Stable nuclides have a certain balance of protons and neutrons*

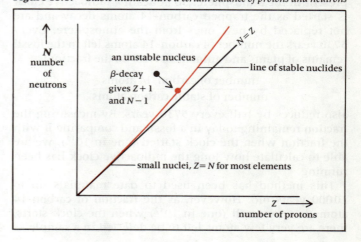

The effect of β-decay, the emission of a negative electron, is to increase Z by 1 and decrease N by 1. This type of decay happens to nuclei which are above the line of stable nuclei and have too many neutrons, because it brings them nearer to that line. (Another type of decay involving the emission of a positron (a positive electron) can occur with nuclei which are below the stable line. This results in a change in Z of -1 and in N of $+1$, which brings the nucleus nearer to the stable line.)

Radioactive dating

Radioactive substances with long half-lives stay around for a very long time. When a particular radioactive nuclide and its decay products become trapped as a rock solidifies or a plant or creature dies, they start a very slow radioactive clock. This clock tells us how much time has passed since the radioactive nuclide was trapped.

One such clock is a **uranium clock**, which starts when rocks are formed. At the end of their decay series uranium atoms form a stable isotope of lead. A measurement of the fraction of uranium atoms which have decayed into lead atoms tells how old the rock is.

For example, after 4500 million years, the half-life of uranium $^{238}_{92}U$, there are the same number of lead atoms as uranium atoms because half of the uranium atoms have decayed. After another 4500 million years there will be 3 lead atoms for every 1 uranium atom. (After 2 half-lives only $\frac{1}{4}$ of the uranium atoms will remain, $\frac{3}{4}$ of them will have decayed to lead.)

Dating using the carbon-14 clock

The carbon-14 clock tells us how long ago something died. Radiation from space converts nitrogen in the atmosphere into the radioactive nuclide of carbon, $^{14}_{6}C$ (or carbon-14). A very small but constant proportion of carbon dioxide in the atmosphere contains these unstable carbon-14 atoms which decay by β-emission with a half-life of 5730 years. The other carbon atoms are the stable isotopes $^{12}_{6}C$ and $^{13}_{6}C$. Only about one carbon atom in 10^{12} is radioactive carbon-14. Living plants take in carbon dioxide (containing this small proportion of radioactive carbon-14) as they grow and animals eat some of the plants. The carbon atoms become trapped in the remains of a plant or animal at the moment of its death. At death all plants and animals contain, trapped in their tissues and bones, the same proportion of carbon-14 as is always present in the atmosphere. So at death a radioactive clock is started as the trapped carbon-14 atoms decay and are not replaced by new ones from the atmosphere. Every 5730 years the number of carbon-14 atoms left in the fossil remains of plants and animals halves. So the fraction:

$$\frac{\text{number of carbon-14 atoms}}{\text{number of stable carbon atoms}}$$

also reduces by half every 5730 years. By measuring the fraction remaining today in a fossil and comparing it with the fraction when the clock started (one in 10^{12}), we are able to calculate how long the radioactive clock has been running.

This method has been used to date materials up to 10 000 years old. However, as the fraction of carbon-14 atoms is very small (one in 10^{12}) when the clock starts, there are very few atoms left to be detected in a sample as old as 10 000 years. Therefore at least a gram of the sample is required in order to obtain a reliable result, and the counting of the small number of decays of carbon-14 atoms inevitably takes several days. So, for example, a whole bone of an old skeleton is needed (and destroyed) in the dating process. Carbon-14 dating has been used successfully with large samples of wood, charcoal, peat and bone, e.g. wooden furniture and papyrus scrolls from ancient Egypt have been dated by this method.

Recently a new technique has been developed which will extend the use of carbon-14 dating to fossil remains which are 10 times older, i.e. up to 100 000 years old, and which can test samples 1000 times smaller, i.e. down to a few milligrams. The radiocarbon accelerator laboratory at Oxford uses a 3 MV ion accelerator to separate carbon-14 ions from the carbon-12 and carbon-13 ions in a mass spectrometer. The accelerator fires the carbon ions at high speed through a powerful magnetic field so that the carbon-14 ions are deflected in a slightly different direction to the other ions. A milligram sample of carbon from a fossil remain 100 000 years old contains as few as 50 atoms of carbon-14. The new technique is capable of measuring this extremely small number of atoms. It is hoped that it will now be possible to date individual fossil seeds and grains, Palaeolithic cave paintings using only fragments and even a single thread from the Turin Shroud.

The 3 MV ion accelerator at Oxford University is used to separate carbon-14 ions from the lighter ions of carbon.

Assignments

Explain the meaning of these terms:
half-life, random decay, parent nuclide, daughter nuclide.
Explain how a radioactive 'clock' can tell us the age of a fossil.
Write a general formula for alpha decay and beta decay.
Try questions 18.5 to 18.8

18.3
RADIATION AND PEOPLE

Every day we are exposed to low levels of background radiation which we are unaware of and can do nothing about. But the artificial and very active sources of radiation, which have many valuable industrial and medical uses, also have considerable risks. When humans and radiation come together the risks and the benefits must be carefully assessed.

Dangerous radioactive materials are handled by remote control and are kept in airtight chambers.

What is ionising radiation?

All radiation which can cause ionisation is dangerous to all animals including humans. Ionising radiation comes in many forms and from many sources. It includes the radiation from natural radioactive substances, i.e. α-, β- and γ-radiation, but it also includes radiation from artificial sources, such as X-ray machines and nuclear reactors.

X-ray machines produce X-radiation which, like γ-radiation, is high-frequency electromagnetic radiation. X-rays have the same properties and the same effects on humans as natural γ-radiation.

Nuclear reactors produce other kinds of radiation which are also harmful to humans. As well as large amounts of γ-radiation and radioactive waste materials, nuclear reactors produce large numbers of neutrons which are released from the nuclei of uranium atoms as they break up in the process called **fission** (p 423). Neutrons fired out of atomic nuclei are also a form of ionising radiation which is dangerous to humans.

So we can see that ionising radiation includes both electromagnetic radiation of high frequency, i.e. X- and γ-rays, and fast moving particles or fragments of atoms, such as α-particles, β-particles, neutrons and protons.

How are people exposed to ionising radiation?

People are exposed to ionising radiation from many sources. The natural radiation to which we are exposed continuously all our lives is called **background radiation**. In the UK about 87% of the average total radiation dose received by a person comes from background radiation.

The pie chart in fig. 18.20 shows where the total dose of radiation comes from. The *internal sources* are radioactive nuclides, such as potassium $^{40}_{19}K$, which are found in our own bodies. The γ-rays and radon gas in the air come largely from naturally occurring radioactive materials such as radium, thorium or uranium in the earth's crust. Even the bricks of which our houses are built emit radiation. Most of the cosmic radiation is absorbed by the earth's atmosphere, but some reaches the ground. However, people in high-flying aircraft and astronauts get a much larger cosmic radiation dose.

Most of the artificial radiation dose is caused by X-rays and radioactive isotopes used in medicine. Perhaps as little as 0.5% of radiation comes from the remaining fallout from atmospheric nuclear bomb tests, and a similarly small amount of radiation exposure comes from industrial sources. However, in the medical and industrial cases, while the *average dose* per person in the country is quite small, the dose received by some individuals is much larger than the background level of radiation. The total dose of radiation received by a patient in hospital or by an industrial worker is very carefully monitored and controlled.

Figure 18.20 *Average radiation dose for people in the UK (source: National Radiological Protection Board)*

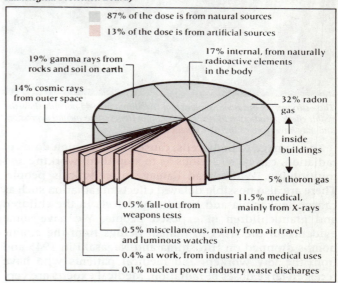

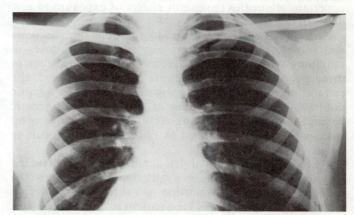

The majority of the extra artificial radiation dose people receive comes from medical uses such as when a chest X-ray picture is taken.

The effects of radiation on people

People are exposed to a variety of radiations which are dangerous. The danger is due to the absorption of energy from the radiation by tissues of the body. As the body absorbs radiation energy, ions are produced which can change or destroy living cells.

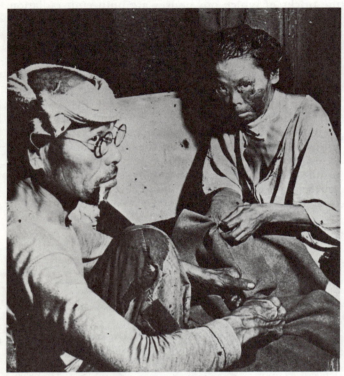

These survivors of the atomic explosion at Hiroshima in 1945 not only show the visible effects of radiation burns, but also suffer from serious mental depression, and may have received invisible genetic damage.

The damage to body cells, caused by very high doses of radiation, can be so serious as to stop them working and multiplying. Widespread damage of cells kills people. There are also possible delayed effects of radiation such as cancer, leukaemia and hereditary defects in the children and grandchildren of exposed people. We have some evidence of the risks to human beings from the atomic bombs dropped on Hiroshima and Nagasaki in 1945 and from the few workers and medical patients who have received large doses of radiation. Medical experts are very uncertain about the long term effects of exposure to low doses of radiation.

Cancers have been observed in atomic bomb survivors and some groups of workers exposed to large doses of radiation. When hereditary defects occur in the children of these people, although the cause cannot be proved, the connection with radiation exposure is beyond reasonable doubt. Accurate predictions of the extent of genetic damage to future generations of children, based mostly on studies of animals exposed to radiation, are difficult to make.

We base our estimates of risk of radiation damage on the assumption that as the dose of radiation gets smaller so the risk gets smaller, in proportion. On this basis we decide what is an acceptable risk for a patient having an X-ray, or for a worker who uses radioactive materials or radiation in his or her job.

For millions of years the background level of radiation is all that nature has had to cope with. Our bodies are highly sensitive to all ionising radiations and cannot repair some of the damage caused by radiation. Damaged cells remain in the body for many years and accumulate. So the effect of radiation can be delayed to an unknown time years in the future.

Fig. 18.21 shows the doses received by people from various sources and how they compare with a fatal dose of above 10 sieverts. Table 18.6 explains the terms and units used to measure radioactivity and the dose a person receives.

Figure 18.21   *Doses of radiation and known effects on humans*

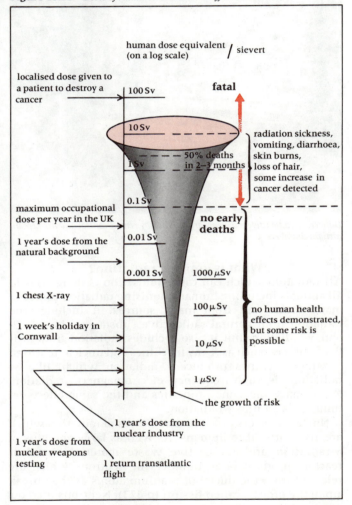

Table 18.6

Quantity	Definition	Unit
activity A (of a source)	number of atoms which decay in 1 second = number of α- or β-particles emitted in 1 second	becquerel (Bq) 1 Bq = 1 decay per second
absorbed dose D (of ionising radiation)	energy absorbed by 1 kg of tissue	gray (Gy) 1 Gy = 1 J/kg
dose equivalent H (of ionising radiation)	absorbed dose × a quality factor Q $H = D \times Q$	sievert (Sv) H in sievert = D in gray × Q

Activity describes how dangerous or active a radioactive source is in terms of the number of particles it emits in 1 second. The activity of a source is measured in the unit **becquerel**, Bq.

The **dose of radiation absorbed** by a person is measured in terms of the energy absorbed by 1 kilogram of body tissue. The unit of absorbed dose D, measured in joules per kilogram, is called the **gray**, symbol Gy.

Different kinds of radiation cause damage to living cells to different extents and so it is necessary to multiply the absorbed dose D by a weighting factor or **quality factor** Q to take this into account. This gives us a **dose equivalent** H, which is the best guide to the effect on a person of a certain dose of a particular type of radiation. Dose equivalent is measured in **sieverts**, symbol Sv.

The quality factor Q is also called the **relative biological effectiveness** of the radiation.

$$\begin{matrix} \text{dose equivalent} & & \text{absorbed dose} & & \text{quality factor} \\ H & = & D & \times & Q \\ \text{(sievert)} & & \text{(gray)} & & \end{matrix}$$

X-rays, γ-rays and β-particles have a quality factor of $Q = 1$, but α-particles are 20 times more dangerous for the same absorbed dose and so are given $Q = 20$.

For practical purposes all you need to know is that the dose equivalent a person has received takes into account the *type* of radiation as well as the *amount of energy absorbed* from the radiation. The examples given in fig. 18.21 show dose equivalents in sieverts.

Precautions for humans

The main risk from α- and β-radiation comes from inside a person. Since α- and β-particles do not penetrate very far into the body, the risk from external sources is quite small. However, care must be taken to avoid radioactive materials being eaten or inhaled from the air. So no eating, drinking or smoking is allowed where any radioactive materials are handled, and disposable gloves and protective clothing are worn. Masks are worn in mines where radioactive dust particles are air-borne.

X-rays and γ-rays can be absorbed deep inside the body, and people exposed to external sources of X- and γ-radiation must be protected as much as possible. The dose a person receives can be limited by:
a) using shielding,
b) keeping a large distance between the person and the source,
c) keeping exposure times as short as possible.

People who work with ionising radiation wear a *film badge* which gives a permanent record of the radiation dose received. The sensitive film is covered with various metal filters through which different kinds of radiation can pass and darken the film behind. The darkness of the film behind the different filters indicates both the amount and type of radiation received. Workers are also checked for radiation contamination by using sensitive radiation monitors before they leave their place of work.

A worker handling radioactive materials may use remote-controlled tools and sit behind a shielding wall made of lead and concrete.

In medical diagnosis, when X-rays are used to produce 'shadow pictures' of bones and internal organs, the radiation dose is kept to a minimum in the following ways:
a) The X-ray beam is restricted to expose only the part of the body where the image is required. Lead shutters or 'beam definers' are used to control where the X-rays come out of the X-ray tube.
b) Aluminium filters are used to absorb unwanted 'soft' X-rays, which are absorbed by the body rather than pass through to the X-ray plate.
c) The exposure time is kept as short as possible.
d) Modern systems use an intensifying screen which makes it possible to obtain an X-ray image from a much lower level of radiation.
e) The number of X-ray exposures is limited for any patient, but *any* exposure is to be avoided for unborn babies and young children.

When radioactive materials are used in medicine they are chosen with care. The radioactive nuclide should have a short half-life so that any material remaining in the body quickly decays away. The radioactive substance used should also have a short **biological half-life**. This is the time taken for the body to get rid of half of the radioactive substance. In other words, the substance should be one which will pass through the body, or be filtered out of the blood as quickly as possible.

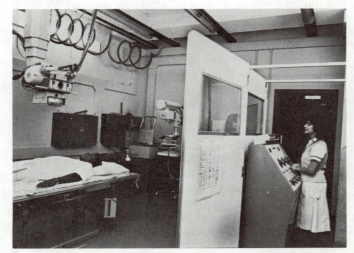

The radiologist stands behind a radiation-absorbing screen while the X-ray machine is operated.

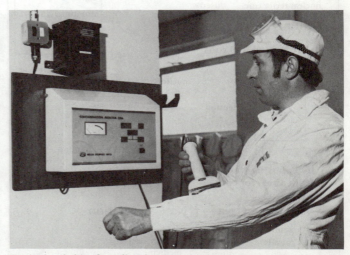

Monitoring clothing for radioactive contamination in a change room at the Springfields Works of British Nuclear Fuels Limited.

Isotopes are produced from used uranium fuel rods

Many radioactive isotopes, which do not exist naturally because of their short half-lives, can be made artificially. A nuclear reactor is one important source of synthetic radioactive isotopes, or **radioisotopes** as they are often called.

Nuclear reactors use fuel rods of uranium to produce lots of heat for power stations but they never completely 'burn up' all the uranium in a fuel rod. The used fuel rods are reprocessed to reclaim the unused uranium for new fuel rods. The fission or breaking up of a uranium atom produces new elements from the fragments of its nucleus. Among these new elements or **fission products** are several useful radioisotopes. When the used fuel rods are reprocessed these radioisotopes can be extracted, fig. 18.22a.

Isotopes are made in nuclear reactors

Some radioisotopes can be made by exposing a natural stable nuclide to a dose of neutrons from the core of a nuclear reactor. Sometimes when a neutron collides with a nucleus in the exposed material it becomes attached to that nucleus. The new nuclide is likely to be radioactive because its extra neutron may make it unstable. An example of this process, known as **neutron capture**, is shown in fig. 18.22b.

A neutron n strikes the nucleus of a stable sodium nuclide $^{23}_{11}$Na. Some energy is released as a γ-photon and a new radioactive nuclide $^{24}_{11}$Na is produced. This radioisotope of sodium, which decays by emitting a β-particle, is used in medicine. Some other important radioisotopes produced in nuclear reactors include cobalt, $^{60}_{27}$Co, americium, $^{241}_{95}$Am, and plutonium, $^{239}_{94}$Pu.

Figure 18.22 *Producing new radioisotopes*

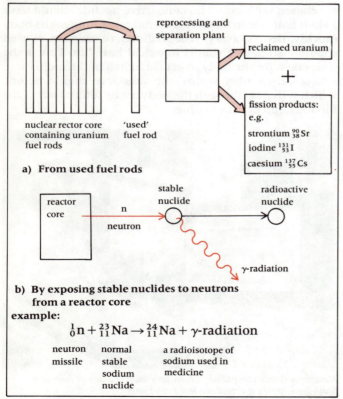

a) **From used fuel rods**

b) **By exposing stable nuclides to neutrons from a reactor core**
example:

$$^{1}_{0}n + {}^{23}_{11}Na \rightarrow {}^{24}_{11}Na + \gamma\text{-radiation}$$

| neutron missile | normal stable sodium nuclide | a radioisotope of sodium used in medicine |

Uses of radioactive isotopes in medicine

We have found many uses for radioisotopes in medicine both in **diagnosis** (finding out what is wrong) and in **therapy** (treating a disease).

A radioisotope and a stable isotope of the same element have exactly the same chemical properties and so they form the same compounds. For example, $^{23}_{11}$Na, the stable isotope of sodium, forms sodium chloride or salt which is very important in the working of nerve cells and the body in general. The radioisotope sodium-24 ($^{24}_{11}$Na) also forms sodium chloride molecules of exactly the same chemical properties. Salt containing some atoms of the sodium-24 isotope is said to be **labelled**. By this we mean that it carries an identifying marker which can be followed wherever it goes.

If a patient eats some of this sodium-24 labelled salt, its progress through the body can be followed. A Geiger–Müller tube or other radiation detector moved over a patient's body shows where the labelled salt has reached and how much salt has been absorbed. While the radioactive salt behaves quite normally in its chemical role, the atoms of the radioisotope sodium-24 will decay while inside the body. They emit γ-rays which pass out of the body to be detected.

It is sometimes necessary to attach radioisotopes to molecules which the body will deliver to specific sites in the body. For example, strontium chloride, labelled with the radioisotope $^{85}_{38}$Sr, is delivered to the bones, because it has similar chemical properties to calcium chloride. If a patient is given a dose of radioactive iodine-131 in solution, it finds its way to the thyroid gland.

Radioisotopes are given to patients internally for various diagnostic tests. One important kind of test is to produce an image of a particular part of the body such as the brain, lungs, liver or bones. An image can be produced when a γ-emitting radioisotope reaches the part of the body to be examined. A gamma-camera is used to detect γ-rays coming from the part of the body which has absorbed the radioisotope. By scanning across the region, a picture can be built up from the emitted γ-rays. Radioisotopes used for producing gamma images include iodine-131 and the important pure gamma source technecium-99 M. (M indicates a special unstable state of the nucleus which decays by emitting γ-radiation only. The state is called 'metastable'.) Technecium-99 M is used for many purposes because it has the following advantages:

a) It emits only γ-rays which mostly escape from the body. No α- or β-radiation is produced to cause local internal ionisation damage.

b) It has a short half-life of 6 hours.

Other diagnostic tests using radioisotopes include dilution tests, absorption tests and leakage tests. For example it is possible to measure the volume of blood or water in a person's body by measuring the dilution of a known volume of a radioistope given to the person The absorption of vitamins and minerals can be measured using radioisotope-labelled samples. Blood leaking into the digestive system can be detected and measured by labelling the blood with a radioisotope.

Radioisotopes are also used to treat various illnesses. Marie Curie was the first person to lend some radium for the treatment of a cancer growth. As early as 1910 needles containing a few milligrams of radium were used to treat cancer. Caesium ($^{137}_{55}$Cs) is now usually used instead of radium to give large local doses of γ-radiation to cancer growths inside the body. Other radioisotopes such as iodine-131, strontium-90, phosphorus-32 and gold-198 are also used for special kinds of treatment.

A nuclide of cobalt, $^{60}_{27}$Co, is widely used to give patients a large dose of γ-radiation for the treatment of internal cancers. A restricted beam of γ-radiation is carefully directed at the cancer site from an external cobalt source. The source, a cube about the size of a sugar lump, may be very radioactive (as high as 10^{15} Bq) and capable of giving a person a lethal dose if their whole body was exposed to its γ-radiation. This dangerous source is heavily shielded and is operated by remote control from behind thick lead and concrete walls.

As well as killing cancer cells, γ-radiation can be used to kill germs or bacteria. This is called **sterilisation**. Instruments and dressings used in hospitals may carry bacteria which could infect a patient and so everything which comes into contact with a patient must be sterilised. Some instruments can be boiled in water to destroy the harmful bacteria but water and heat can damage some materials. Using γ-radiation to sterilise instruments and dressings avoids these problems.

Uses of radioisotopes in industry and agriculture

Many industrial applications of radioisotopes use them as **tracers**. A radioactive substance added to a fluid in a pipeline can be used to measure the flow rate in the pipeline and to find leaks. Gamma radiation can pass out through the walls of a pipeline and so a radiation detector can follow the progress of a radioactive marker as it flows, mixed in with the fluid in a pipeline. Where fluid leaks out underground, the radioactive tracer will also be found to spread out away from the pipeline so showing the position of the leak. Leaks in drainage systems can be found by this method. The isotope of sodium, $^{24}_{11}$Na, is useful in water pipelines and sewage systems because, in the form of sodium chloride, it is soluble in water.

Tracers are used in agriculture to measure the uptake by plants of food from fertilizers in the soil. The phosphate ion in plant food can be labelled with the radioisotope phosphorus-32. As a plant absorbs this labelled food we can trace it to growing parts of the plant and learn about plant growth and improve our use of fertilisers.

The radioisotope cobalt-60 is used to check welds in steel structures and pipelines. Gamma radiation from a large cobalt-60 source placed on one side of a steel structure exposes a photographic plate at the other side, in the same way that X-ray images are produced. A flaw such as a bubble or crack inside a weld on a pipeline would be visible on the exposed film. The advantage of using γ-sources over X-ray equipment is that the γ-source can be used in remote and difficult places without the need for the special high-voltage power supplies required by X-ray equipment.

Another common use for radioisotopes in industry is in the measurement and automatic control of sheet thickness. Sheets of plastic, paper and metal are manufactured to accurately controlled thicknesses. A radioisotope sends radiation through the sheet material as it comes off the production line. α- or β-radiation is used for thin sheets. A radiation detector on the other side of the sheet measures the intensity of the radiation passing through the sheet. Any small variation in sheet thickness produces a change in the count rate of the detector. This information can be fed back to the machinery to adjust the sheet thickness automatically.

A caesium ^{137}Cs radioactive source being used to measure the thickness of a pipe wall.

Assignments

Remember

a) the main sources of background radiation.

b) precautions taken by workers who use radioactive materials and those also taken to limit the dose given to patients when X-ray pictures are taken.

Describe

c) one medical use of radioisotopes.

d) one industrial use of radioisotopes.

Explain

e) the difference between the half-life of a radioisotope and the biological half-life of a radioactive substance given to a patient to swallow or by injection.

f) why α- or β-radiation is more dangerous from sources inside the body than from outside.

g) what is meant by the 'dose equivalent' of a human exposure to radiation, and state the unit in which it is measured.

h) the properties which would make a radioisotope suitable for internal use in medicine.

i) how radioisotopes can be made.

Try question 18.9

Questions 18

1 a) Natural radioactive substances emit three kinds of radiation. Name the three types and state what each of them is.
   b) How would you show the different penetrative powers of the radiations?
   c) Use is made of radiation in hospitals. Describe briefly how it is used and for what purpose.
   d) Special precautions are necessary when radioactive substances are used or stored in a hospital. State two precautions and the reasons for them.
   e) When the Chernobyl disaster took place, certain radioactive isotopes fell on some parts of the United Kingdom. The activity increased above the background count.
   i) What is meant by an *isotope*? ii) What is a *radioactive* isotope?
   iii) What is meant by *background count*? iv) It was said that, after some weeks, what had fallen would no longer be a hazard. How is it possible that it could no longer be dangerous?
   f) What is meant by *uranium fission*? How does this process produce heat in a nuclear reactor?   (MEG Nuffield 88)

Figure 18.23

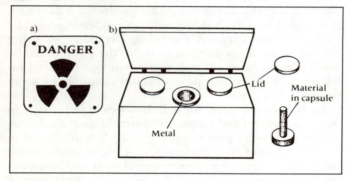

2 The warning symbol in Fig. 18.23a is on a cupboard door.
   a) i) What sort of materials would you expect to find inside the cupboard?
      ii) Why are the materials dangerous?
   b) Fig. 18.23b shows the container in which the material is stored. It is a solid wooden box with holes lined with metal. i) Which metal is used for the lining? ii) Why is this metal necessary?
   c) Write down **two** safety precautions that should be taken when using the material in the capsule.
   d) You would see the danger symbol i) at which type of power station, ii) in which department of a hospital?   (NEA [A] 88)
3 a) A radioactive source is known to emit one type of radiation *only* i.e. α or β or γ. The source was placed in a holder as shown in fig. 18.24, first without the magnet present and then with the magnet present. A suitable detector of radiation was placed at positions 1, 2 and 3 and the count rates recorded. These are shown in the table below.
   i) What is the reason for placing the two metal plates in front of the source?
   ii) Estimate a value for the background count in counts per minute.

	Counts per minute	
	magnet not present	*magnet present*
Detector position 1	26	295
Detector position 2	300	28
Detector position 3	28	26

Figure 18.24

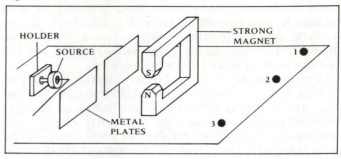

   iii) State the type of the radiation emitted from the source giving a reason for your answer.
   b) One function of the liver in a body is to remove unwanted substances from the blood. If small amounts of radioactive sulphur are injected into a person's blood, the liver will absorb it. After being absorbed by the liver the radioactive sulphur can be detected using a camera which *only* responds to electromagnetic radiation. In the case of a healthy liver a bright image should be observed.
      i) Which type of radiation should the radioactive sulphur emit if it is to be detected by this camera?
   The sulphur is made radioactive by the addition of $^{99}_{43}$Tc, one of the isotopes of technetium, to it. ii) Explain what isotopes are. iii) How many protons and neutrons are in the nucleus of this technetium atom?
   The half-life of $^{99}_{43}$Tc is 6 hours.
      iv) A sample of sulphur contains 5 μg of technetium. What mass in μg of technetium was present 24 hours earlier?
      v) Why is it important that $^{99}_{43}$Tc has a short half-life?
   The pictures of two people's livers built up by the camera are as shown in Fig. 18.25.
      vi) What can you conclude about the two livers?
   c) Describe briefly one industrial use of radioactive substances, apart from nuclear power stations.   (NISEC part 88)

Figure 18.25

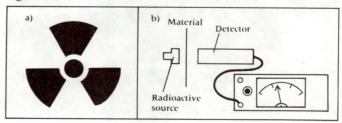

4 a) Fig. 18.26a shows an important safety sign. What would this warn you about?
   b) There are three types of radiation given out by radioactive sources. They are called: alpha particles (α-particles), beta particles (β-particles), gamma rays (γ-rays)

Figure 18.26

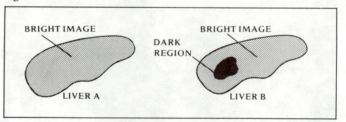

Fig. 18.26b shows a radioactive source which gives out all three types of radiation. A detector is put near the source and shows the radioactivity. Different pieces of material are then placed between the source and the detector.

Copy and complete the following table to show the effect of different materials on the three types of radiation.

Type of material	Radiation(s) stopped by material	Radiation(s) reaching detector
Tissue paper		
Thin sheet of aluminium	alpha beta	gamma
Thick sheet of lead		

c) A source of beta particles (with a very long half-life) and a detector can be used to check the thickness of paper in a paper mill, Fig. 18.27. i) Explain how the device works. ii) Why are beta particles used instead of alpha particles or gamma rays? iii) What is meant by the term *half-life*? iv) Why is it important to use a radioactive source with a long half-life? (SEG 88)

Figure 18.27

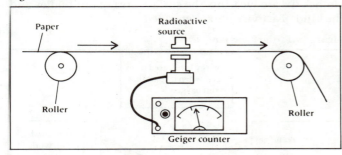

5 In an experiment to measure the half-life of a radioactive element the following results were obtained.

Count rate/counts per minute	1 000	250	125
Time/seconds	0	110	160

a) State clearly what is meant by the half-life of a radioactive element.
b) From the results in the table calculate
  i) **two** different values for the half-life of the element and
  ii) the average half-life of the element. (JMB)

6 The element thorium $^{234}_{90}$Th is radioactive. It decays by emitting beta particles and has a half-life of 24 days.
  i) What is a beta particle?
  ii) Calculate the number of protons and neutrons in the nucleus of an atom of thorium.
  iii) Calculate the number of protons and neutrons in the nucleus formed when a thorium atom emits a beta particle.
  iv) Calculate the time taken for 1 g of thorium to decay leaving $\frac{1}{8}$ g of thorium unchanged. (JMB)

7 A nucleus of the radioactive isotope $^{222}_{86}$Rn emits an α-particle when it decays to a nucleus of the element Po. Complete the equation representing this event:

$$^{222}_{86}\text{Rn} \rightarrow \text{Po} + \text{He}$$

What is the atomic number of the $^{222}_{86}$Rn nucleus? (C)

8 This question is about radioactivity. The graph in Fig. 18.28a shows the variation in activity of a sample of an isotope of radon.
  a) Explain what is meant by the terms: i) activity; ii) isotope.
  b) Using the graph, obtain a value for the half-life of the sample. Explain clearly how you obtained your value.
  Fig. 18.28b shows the tracks produced in a diffusion cloud chamber by a radioactive source.
  c) i) What type of radiation could produce tracks like this? Give two reasons. ii) Describe what might have happened at P and explain your answer.
  d) An element X decays by giving off an alpha particle. Copy and complete the equation below, showing what took place.
  $^{238}_{92}\text{X} = {}_{---}\text{Y} + {}_{---}$ (SEG [ALT] SPEC)

Figure 18.28

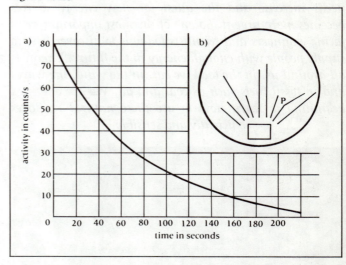

9 A doctor suspects that one of the kidneys of a patient is not working properly.
  The doctor injects him with a radioactive impurity, 1 MBq of iodine-131 (half-life 8 days) which travels to his kidneys. A normal kidney gradually removes all impurity from the blood in about 20 minutes.
  The radioactivity in his kidneys is monitored using gamma ray detectors. The graph in Fig. 18.29 shows the results.
  a) Which kidney do you think is not working properly? Explain your answer.
  b) Why do you think that a gamma ray emitter was used as the injected impurity instead of a beta emitter?
  c) Why would it be less suitable to use as an impurity a material with a very short half-life? (SEB 87)

Figure 18.29

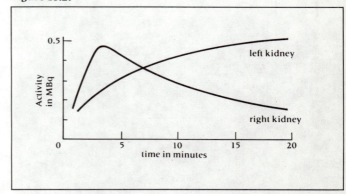

19

Energy

All living things need energy. Energy makes things grow, keeps us warm and sustains life. Without energy all living things would die. We can invent machines to do our work and help us explore the universe but they all need energy to drive them. Without energy supplies we could do very little because all our machines would stop working.

As more and more people live on the earth their energy needs increase and the quest for new energy sources becomes more urgent. So one of the most important tasks facing engineers and scientists today is to discover how to supply people with enough energy in the future. Energy is all around us, in the food we eat, in the wind, the waves and in fossil fuels such as coal and oil. We are only just beginning to learn how to use these natural resources efficiently and sensibly.

19.1
ENERGY AND PEOPLE

Food provides our bodies with the energy we need to do our work and keep warm. But we have found many other ways of using energy to make our lives more comfortable and to help us do our work.

Energy and our bodies

Food is required by our bodies to provide materials for the repair and replacement of cells and for us to grow. But most of our food is needed to provide energy. As we breathe we absorb oxygen through our lungs. When this oxygen combines with glucose molecules in a process called oxidation, energy is released. Energy can be obtained by this process from carbohydrates, fats and proteins. Any surplus food containing energy is converted to extra fat which is stored around our bodies. When extra energy is needed our bodies use some of this stored energy (fat) and we lose 'weight'.

Fig. 19.1 shows how our bodies use their energy supply. The amount of energy we need varies greatly according to what we are doing but at all times we lose energy as heat. Even the loss of heat energy depends on several factors such as the surrounding temperature, the clothes we are wearing, the surface area of our bodies and the kind of activity we are engaged in.

Figure 19.1 *Our bodies need energy*

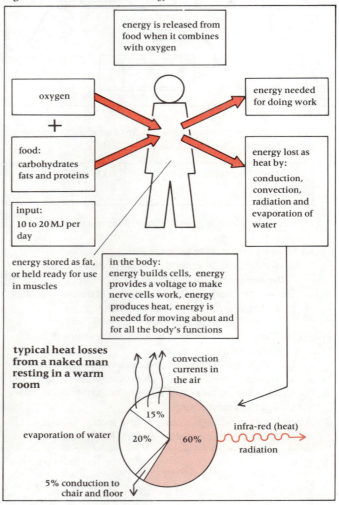

As an example, the figure shows in a pie diagram the percentage of heat lost in each of the four basic ways. The total rate of loss of heat from a naked adult in a warm room is about 100 joules per second or 100 watts, which is equivalent to the power of an electric lamp.

It is interesting to see that while resting a person loses most heat by radiation. This is really infra-red radiation which is electromagnetic radiation of longer wavelength than light (see chapter 23). When people are working very hard or perhaps running in a race, they need to lose much more heat. The greatest increase in heat loss is produced by sweating, which can increase the rate of loss of heat by a factor of 10. As the sweat evaporates, its latent heat of vaporisation is taken from the body surface and helps to keep it cool.

An average rate of loss of heat of 120 watts would produce a total daily loss of about 10 000 000 joules or 10 MJ of energy. (Energy = power × time, power measured in watts and time in seconds.)

We need energy for many purposes apart from keeping warm. The need for energy can be seen and felt when we are working hard, lifting heavy objects, riding a bicycle or climbing a mountain. In addition, our bodies need energy just to keep them going. The building and repairing of new cells, the pumping of blood round our bodies, the electrical energy that is used for signalling by our nerve cells and all body movements and muscle contractions all need a supply of energy.

So we must eat food to provide enough energy to keep us warm and for all these other needs. The amount of energy needed to do physical work varies greatly according to the kind of work and time spent working. Table 19.1 gives a few examples of the rate of working of a 70 kg person for various activities. Averaged over 24 hours this person's total energy need is likely to be somewhere between 10 MJ and 20 MJ per day.

Table 19.1

Activity of a 70 kg person	Rate of working/joules per second or watts
sleeping	80
sitting reading	120
typewriting or piano playing	160
walking slowly	250
running, swimming or hard physical work	500 to 800
walking up stairs	1300

Energy can damage our bodies

Your body can be damaged if it receives too much energy too quickly. For example, if you put your hand into a fire you get burnt. This is because heat energy received too quickly damages the cells. Burns treated very quickly with lots of cold water do less damage because the cold water removes the heat energy from the body tissue.

A bullet is a small metal object which is quite harmless if you hold it in your hand, but when it is fired at high speed by a gun it has kinetic energy which enables it to penetrate your body and do damage. It is the energy of the bullet which is dangerous. Similarly it is the kinetic energy of a fast-moving car and its occupants which causes damage and injury in a collision.

In the previous chapter we learnt that the danger associated with ionising radiations is due to the absorption of energy by the tissues of the body. An alpha particle, like a bullet, is quite harmless when it has given up its kinetic energy. If your body absorbs a lot of energy from radiation, the resulting damage can kill cells.

The kinetic energy of a bullet enables it to pass through an apple so causing much damage.

Energy can do things for us

By eating food we can do work. We can lift things and throw things; we can climb mountains and we can make machines that do even more work for us.

For a long time people used animals to increase the amount of work they could do. They also invented tools and machines such as the lever, the plough, the wheel and the pulley. These machines expanded human activity, giving people more control over the world around them. *Any machine that can do work for us must have a supply of energy,* for work is done only when energy is moved from one place to another or is changed from one form to another.

Important though the inventions of the plough and the wheel were, the work they could do was still limited by the human or animal energy input. Because of energy

People lose a lot of energy as infra-red (heat) radiation. More infra-red radiation is emitted from hot areas of the skin than from cooler areas. In this photograph, taken using an infra-red sensor which scans across the body, the hottest areas from which most energy is radiated appear lightest.

conservation, the energy or work output of a machine can never exceed the energy input.

The use and control of fire has played an important role in our survival and development. Significantly, fire was the first new source of energy that we learnt to use.

Peking man, perhaps as long as 400 000 years ago, is known to have used fire. Fire helped these early people in several ways. It gave heat during the ice ages and cold winters, light in the dark nights and an increased range of food through cooking. But it was to be a very long time before the energy released by fire was used to drive machines.

Only when we learnt to harness the energy of running water to turn a water wheel and the energy of the wind to turn a windmill did we become free of the limitations of using human and animal sources of energy to drive our machines. Water and wind do not get tired, do not need feeding and do not need rests, although they are not always available when or where people want them. Water wheels were used as early as a few hundred years BC, but fire was not used to drive machines until the eighteenth century AD.

In 1698 Thomas Savery invented a machine to pump water from mines. Fire was used to produce steam from a boiler. The steam was cooled in a cylinder and the partial vacuum so created sucked water up from the mine shafts. By 1712 Thomas Newcomen had built a steam engine with a piston in a cylinder, and by the 1760s James Watt had converted the up and down motion of a piston into rotational motion. Now, at last, people had learnt to use fire to drive machines. This was a breakthrough in the use of energy. By burning coal, people could release energy to do work when and where they wanted.

The largest water wheel in Europe is to be found at Laxey, in the Isle of Man.

Coal provided the energy needed and fire the means of converting it to drive the new machines of the industrial revolution in the late 18th century.

Energy at home

Only 200 years after we learnt to use the energy released by fire to drive a steam engine, our homes are full of energy-hungry machines which make our lives more comfortable by keeping us warm and doing many of the daily jobs around the house. For example, we have machines which wash clothes and dishes, mix food, clean floors, polish, drill and sew for us. While some of these machines need our attention, they all take away most of the human effort by using another source of energy to do the work.

How we use energy in the home

One way of looking at the energy we use is to make a study of its use in our own homes.

- *Find out what kinds of energy are used at home.*
- *How is your house heated?*
- *How is the water heated?*
- *Look for energy-using equipment and note the kind of energy supplied in each case.*
- *Find out the annual consumption of each kind of energy by your household.*

You can do this by looking at the bills received over a recent 12 month period.

- *Calculate the total number of units of each kind of energy used.*

As each type of energy is measured in different units it is necessary to convert them all to a common unit for comparison. Use the information in table 19.2 to convert the energy units to joules or megajoules.

Table 19.2

Energy source	Unit of supply	Equivalent
electricity	kilowatt-hour	$3.6 \times 10^6 \text{J} = 3.6 \text{MJ}$
gas	therm	$1.1 \times 10^8 \text{J} = 110 \text{MJ}$
coal	1 tonne = 1000 kg	$2.8 \times 10^{10} \text{J} = 28\,000 \text{MJ}$
oil	1 tonne ≈ 250 gallon	$4.4 \times 10^{10} \text{J} = 44\,000 \text{MJ}$
	1 litre	$3.8 \times 10^7 \text{J} = 38 \text{MJ}$

LEB
London Electricity

CUSTOMER ACCOUNTS OFFICE				DATE OF ACCOUNT		
261 CITY ROAD LONDON EC1V 1LE				18 JAN 84		

METER READING		UNITS USED	UNIT PRICE (pence)	V.A.T. code	AMOUNT £
PRESENT	PREVIOUS				
04892	E 02714	2178	5.470	0	119.13
STANDING CHARGE 18/11/83 TO 31/ 1/84				0	6.07

E=Estimated Reading. Please read the advice given on the back of this bill.
C=Your own reading.

YOUR REFERENCE NUMBER	YOU CAN PHONE US ON	NORMAL READING DATE	AMOUNT TO PAY
045.0616/016.229	01-251 5262	13 JAN 84	£ 125.20

Form No. 1.112/1A 0883

This electricity bill states that 2178 units of electrical energy have been used. These 2178 units are kilowatt–hours, equivalent to $2178 \times 3.6 \text{MJ} = 7840.8 \text{MJ}$.

- *What is the total annual energy consumption at home?*
- *What proportion of the total energy was used for heating?* (This may be a combined figure for heating the house and for hot water.)
- *How could the greatest savings in energy consumption be made?*

The total energy used by a household in the UK can vary over a wide range but a family using central heating can easily use 10^{11} joules per year. Fig. 19.2a gives an example of how the annual use of energy in a home in the UK might divide up, but this varies greatly with the following factors:

a) the size of the house,
b) the location of the house,
c) the amount of insulation and type of construction of the house,
d) the type of heating used and thermostat temperature setting,
e) the number of people living in the house,
f) the life style of the people in the house.

The most striking discovery is that most of our energy needs are connected with keeping warm. This becomes clear when we discover that most of the energy is supplied as gas, oil or coal and this is used mainly for heating the house and heating the water. (The proportion of this energy which is used for cooking is relatively small.)

Electrical energy is used for all other purposes, including lighting and electrical equipment but, unless electricity is used as the main source of heating energy, it will represent only a small proportion of the total energy supplied to a house.

The Department of Energy figures for 1979, fig. 19.2b show that 81% of the energy supplied to homes in the UK was supplied as gas, solid fuels and oil. Apart from cooking, these fuels are used only for heating. Some of the 19% supplied as electricity is also used for heating rooms and water.

Note that the amount of solar energy used in this country is too small to show on the pie diagram of fig. 19.2b. We shall discuss the use of alternative sources of energy later.

Figure 19.2

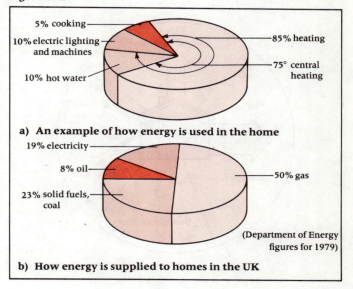

5% cooking
10% electric lighting and machines
10% hot water
85% heating
75° central heating

a) **An example of how energy is used in the home**

19% electricity
8% oil
23% solid fuels, coal
50% gas

(Department of Energy figures for 1979)

b) **How energy is supplied to homes in the UK**

Saving energy at home

As energy becomes more difficult to supply in the future and therefore more expensive to buy, it is in everyone's interest that we save as much energy as we can. It is easy to see that we could save a lot of energy by reducing the amount needed to heat our houses.

To keep a house warm all that is needed is to replace heat energy at the same rate as it is being lost. So that if no heat was being lost to the cold air and ground outside, no heat woud be needed at all. The rate of loss of heat from a house can be reduced by:

a) having a lower temperature inside the house,
b) increasing the insulation around the house.

We have no control over the outside temperature but if we set the room thermostats at a lower temperature and get used to living in a cooler house (perhaps by wearing more clothes) we can save energy by having a smaller temperature difference between the inside and outside of our houses.

Insulation is the name we use for any material which is a bad conductor of heat and can be used to reduce the flow of heat energy from one place to another. Fig. 19.3 shows the proportion of heat that is lost in different ways from a house which has no special insulation against heat losses. These figures are based on a brick-built semi-detached house, and are typical of a 1930s house without any improvements in insulation.

Some reduction in the heat energy lost in each of the five main ways can be made as shown in fig. 19.4. While the greatest saving can be made by insulating the roof and walls of a house, the question of cost should also be taken into account. For example, it costs very little to provide effective draught excluding strips round doors and windows and so the cut in heating bills covers the cost in less than a year. Double glazing will reduce the heat loss by conduction through the windows by about 50%, (only 5% of the total lost from the house). Although valuable, a saving of 5% in the heating costs may mean it takes over 10 years to pay for the double glazing.

Figure 19.3 *The main heat energy losses from a house in the UK (a typical 1930s style semi-detached house without improved insulation)*

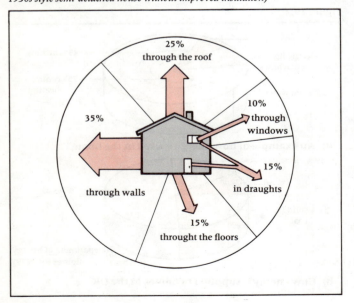

Figure 19.4 *How insulating a house saves energy*

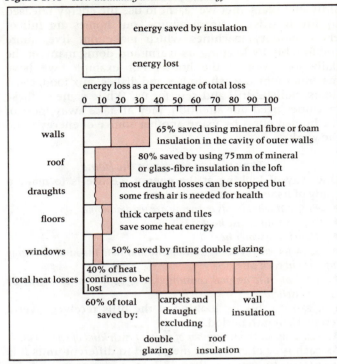

An alternative to double glazing, which is much more effective at reducing heat loss, is to fit insulated window shutters. Folding shutters made of a good insulating material typically 13 mm thick, can be fitted inside windows. When closed at night, insulated shutters can reduce the heat lost through a window to as little as 17% of the loss through a single-glazed window. When the shutters are folded to the sides of the windows in the day they allow heat and light from the sun to enter the room. The heat received during the day can be shut in during the night, and very little extra heating is needed.

Heat losses by conduction through floors can be reduced by including a layer of insulation when a house is being built. Thick fitted carpets or tiles are probably the most practical step that can be taken in an existing house.

These folding shutters fitted inside a window are made of a good insulating material and are much more effective than double glazing at keeping heat in.

Worked Example
Energy consumption

A household uses 5280 'units' (=kilowatt hours) of electrical energy and 1290 therms of gas in a full year. Calculate the total energy used in a year in joules and find the percentage of this which is supplied as gas.

Using the conversion factors given in table 19.2, we have:

$$1290 \text{ therms of gas} = 1290 \times 110 \text{ MJ}$$
$$= 141\,900 \text{ MJ}$$
$$= 1.42 \times 10^{11} \text{ J}$$
$$5280 \text{ units of electricity} = 5280 \times 3.6 \text{ MJ}$$
$$= 19\,008 \text{ MJ}$$
$$= 0.19 \times 10^{11} \text{ J}$$
$$\therefore \text{ total energy used} = (1.42 + 0.19) \times 10^{11} \text{ J}$$
$$= 1.61 \times 10^{11} \text{ J}$$

The percentage of this energy which is supplied as gas is given by:

$$\frac{\text{gas energy used}}{\text{total energy used}} \times 100\% = \frac{1.42 \times 10^{11} \text{ J}}{1.61 \times 10^{11} \text{ J}} \times 100\%$$
$$= 88\%$$

Answer: The total energy used in a year is 1.61×10^{11} joules and 88% of this is supplied as gas.

Other energy saving ideas at home

While most energy can be saved by reducing draughts, fitting insulation and by living at a lower temperature, there are many other ways of making small but valuable reductions in the energy we use at home. Some ideas are given in the following list.

* Lag (wrap with insulating material) hot-water tanks and pipes.
* Turn down the temperature setting on the hot-water thermostat.
* Block off unused fire places to stop heat going up the chimney.
* Fit linings to curtains and always close them at night.
* Do not heat rooms that are not used and keep the doors shut.
* Fit aluminium foil on insulation boards behind radiators on outer walls to reflect radiated heat back into the room.
* Fit separate thermostats to the radiators in each room and turn all except the living room to a lower setting.
* Do not wash up in running water, and stop hot taps dripping.

Make full use of the energy needed to heat an oven by cooking several things at the same time.

* Use full loads of washing in a washing machine and use low-temperature washing powder and programmes.
* Use a washing line and free wind and sunshine instead of a tumble-dryer whenever possible.
* Do not put more water in a kettle than you need for your hot drinks (as long as the element is covered); do not leave the kettle boiling.
* Do not put warm food into a fridge or freezer.
* Switch off all unnecessary lights; use low-power lamps and install fluorescent lighting where it is acceptable.
* Plan cooking to make full use of a heated oven, put lids on pans and turn down gas flames to suit the size of pan.

Calculating heat losses by conduction

The rate of loss of heat by conduction through a wall, roof, floor or window (measured in joules per second or watts) depends on the following factors:

a) the temperature difference between the air inside and the air outside the house (measured in kelvins),
b) the surface area of the structure through which the heat flows (measured in square metres) and
c) the ability of the material to conduct heat, called its **U-value**, measured in $W/(m^2 K)$.

The U-value of a material is the rate of heat flow through it by conduction, per unit surface area and unit temperature difference between the two sides.

$$\text{U-value} = \frac{\text{rate of loss of heat energy}}{\text{surface area} \times \text{temperature difference}}$$

We can calculate the rate of loss, or flow, of heat through a wall using the relation:

$$\text{rate of heat loss} = \text{U-value} \times \frac{\text{surface}}{\text{area}} \times \frac{\text{temperature}}{\text{difference}}$$

So a good conductor of heat will have a high U-value and a good insulator will have a low U-value. The examples of U-values given in table 19.3 show how the heat loss through a particular surface can be reduced. For example, the use of 75 mm of insulation in the loft reduces the U-value from 2.2 to 0.45 $W/(m^2 K)$, which is equivalent to the 80% saving shown in fig. 19.4 (0.45 is about 20% of 2.2, and is the heat which still escapes with the improved insulation).

Table 19.3 Some U-values for materials used in insulating houses; good insulators have low U-values

Material	Insulation thickness	U-value / $\frac{W}{m^2 K}$
tiled roof with no insulation	—	2.2
mineral or glass fibre roof insulation	75 mm	0.45
double brick wall with air cavity between	—	1.7
same wall filled with mineral fibres or foam	75 mm (cavity widths vary)	0.6
single-glazed window (6 mm glass)	—	5.6
double-glazed with air gap	air gap 20 mm	2.9
single-glazed with insulated shutter	air gap 20 mm, shutter insulation 13 mm	0.95

Worked Example
Heat loss

Calculate the rate of heat loss through a window of size 0.80 m × 2.00 m when the temperatures are: outside −5°C, inside 20°C (a) if the window is single-glazed, (b) if the window is double-glazed with a 20 mm air gap and (c) if the window is covered with a shutter made of a 13 mm thick insulating material.

Using the U-values given in table 19.3, compare the saving in heat loss achieved by using the two methods of insulation over a period of 24 hours.

Rate of heat loss $= \text{U-value} \times \dfrac{\text{surface}}{\text{area}} \times \dfrac{\text{temperature}}{\text{difference}}$

a) by single-glazed window $= 5.6\dfrac{W}{m^2\,K} \times 1.6\,m^2 \times 25\,K = 224\,W$

b) by doubled-glazed window $= 2.9\dfrac{W}{m^2\,K} \times 1.6\,m^2 \times 25\,K = 116\,W$

c) by single-glazed window with air gap and insulated shutter $= 0.95\dfrac{W}{m^2\,K} \times 1.6\,m^2 \times 25\,K = 38\,W$

So double-glazing saves $224\,W − 116\,W = 108\,W$

and an insulated shutter saves $224\,W − 38\,W = 186\,W$

Now energy saved = power × time, so the energy saved by the double-glazing in 24 hours is:

$$108\,W \times (24 \times 60 \times 60)s = 9.3 \times 10^6\,J = 9.3\,MJ$$

and the energy saved by the insulated shutter in 24 hours is:

$$186\,W \times (24 \times 60 \times 60)s = 16.1 \times 10^6\,J = 16.1\,MJ$$

Answer: Double-glazing saves 9.3 MJ in 24 hours, and an insulated shutter saves 16.1 MJ in the same time.

Assignments

Find out
a) the kind of insulation used in your own home,
b) the thickness of the loft insulation in your home,
c) the temperatures at which room thermostats are set,
d) whether the hot-water tank is lagged.

Suggest
e) ways of improving the insulation of your own home,
f) ways your family could save energy at home.

Explain why a person needs energy to stay alive.

Try questions 19.1 to 19.8

Roof insulation is important for all buildings because so much heat loss can be stopped this way for very little cost.

19.2
FORMS OF ENERGY

The total amount of energy in existence does not change, because energy cannot be created or destroyed. We may talk about needing energy and using or consuming it, but in fact we can only convert energy from one form to another. When we have converted some energy to a different form we can show that there is always the same amount of energy after the change as before. We call this the **conservation of energy**.

Energy has many forms

Perhaps the forms of energy which are the most obvious are those involved where work is being done and the effects of the energy changes can be seen. But some energy is less obvious because it is stored and is waiting to be released by a conversion process.

Stored energy: chemical energy

Food in all its variety is our store of chemical energy. So plants and animals are a form of energy which can be converted in our bodies by a chemical reaction called oxidation.

All the **fossil fuels** such as coal, oil and gas are stored forms of chemical energy, which are usually converted by burning. (Burning is another chemical reaction involving oxidation.) For example, a car uses stored chemical energy in the form of petrol which it burns to drive its engine and make it move. The car battery also stores energy in a chemical form and uses a chemical reaction to supply electrical energy to various devices in the car. All batteries store chemical energy.

This oil worker is releasing some of the energy stored in his body which he obtained from food.

Stored energy: potential energy

Potential energy is stored by an object when it is in a particular position or condition. Objects which are able to fall down have stored energy caused by gravity. This form of potential energy is therefore called **gravitational potential energy**, and depends on the *raised position* of an object above the ground. As we saw in section 8.4, the gravitational potential energy of an object can be calculated from the formula

$$E_p = mgh$$

So gravitational potential energy is stored in water at the top of a water fall, in a high-level reservoir and in the heavy driving weight of a wound-up grandfather clock. The stored energy in a reservoir can be used to produce electricity, and the stored energy in a grandfather clock is used to drive the clock mechanism.

A cyclist has gravitational potential energy when she reaches the top of a hill. She has used a lot of stored chemical energy in climbing the hill working against gravity. At the top of the hill, the gravitational potential energy she has gained will drive her down the hill with increasing kinetic energy.

Objects which are elastic in some way can store energy when they are stretched, twisted or bent. The energy stored by such an elastic object can be called **elastic potential energy** and depends on the *strained condition* of the object. Some examples include the stretched elastic of a catapult, the wound-up spring of a clockwork motor and the bent condition of a diving board when a diver is about to jump from the end.

The enormous amount of energy released in an atomic explosion comes from the splitting of a vast number of atomic nuclei. This energy has been stored in the atomic nuclei since the atoms were formed.

Stored energy: nuclear energy

Energy is released when radioactive decay occurs. The energy radiated when atoms decay has been stored in their nuclei since they were formed, perhaps since the formation of the solar system.

In nuclear power stations we have found ways of speeding up and controlling the release of the energy stored in the nuclei of atoms. The atomic bomb is the result of an uncontrolled chain reaction of energy released from a very large number of atomic nuclei in a very short space of time.

36 uranium fuel pins are assembled to form each fuel element which will provide the store of energy for an AGR nuclear power station.

Kinetic energy: of moving objects

Any object has energy if it is moving. The energy of motion is known as kinetic energy and can be calculated from the formula

$$E_k = \tfrac{1}{2}mv^2$$

see section 8.4. Energy must be supplied to an object to get it moving, and energy must be taken away from an object to make it stop. The process of changing the kinetic energy of an object involves doing work. For example, a car engine does work to get a car moving and give it kinetic energy, and the brakes do work to stop the car and remove its kinetic energy.

The springboard briefly stores potential energy while it is bent downwards, and then converts this into kinetic energy for the diver by giving him an upwards push.

Kinetic energy: of atoms and molecules, internal energy

All atoms and molecules have some kinetic energy because they all have some kind of motion. The atoms (or molecules) in a gas or liquid move about randomly with various amounts of kinetic energy according to their speeds. The atoms in a solid also have some kinetic energy as they vibrate about their fixed locations in the solid. When an object is heated energy is transferred from an external source to the atoms inside the object. To absorb this extra energy, the atoms of the object must increase their kinetic energy. So heating an object makes the atoms (or molecules) of a gas or liquid move faster and the atoms of a solid vibrate more energetically.

The extra kinetic energy possessed by the atoms inside the heated object is actually the heat energy it has absorbed, which now forms part of the total **internal energy** of the object.

The internal energy is all the energy possessed by the atoms and molecules inside a substance or an object. This includes forms of both kinetic and potential energy.

The absorption of heat energy and the resulting increase in internal energy of an object can be detected as a rise in temperature of the object. The kinetic energy possessed by atoms in their random motion is often called **random thermal energy**.

Sound energy

Unlike the random motion of molecules associated with heat energy, the motion of molecules through which a sound wave passes has an organised and periodic pattern. A longitudinal wave motion passes sound energy from molecule to molecule in the form of a mechanical vibration with a particular frequency and amplitude (chapter 21).

Dolphins, like bats, use sound energy (ultrasound of frequency around 200 kHz) to locate objects in their path. They can identify many objects by the kind of echo received back. A dolphin will also use sound energy to send warning cries and call signs to other dolphins. The sound energy sent through the water makes water molecules vibrate in a longitudinal wave motion.

Electrical energy

Electrical energy is produced by a conversion process from another form of energy. For example, electrical energy from a battery comes from a store of chemical energy. The electrical energy from a power station generator is obtained from the store of chemical energy in a fossil fuel, or perhaps from nuclear energy.

Electrical energy is a very convenient form of energy which can easily be carried to the particular place where energy is needed. It is also a very high-grade form and has many specialised applications for which no other energy form would be suitable. Perhaps the best example is the way computers and communication systems depend on electrical energy. There are also many applications of electric motors which could not be replaced with an alternative such as a steam engine.

The electrical energy is converted to another form of energy when an electric current I flows through a machine or device which has an electrical resistance R. The energy converted can be calculated using the formula: $W = I^2 R t$, see p 251.

Electromagnetic radiation

Energy can also exist as radiation without belonging to matter in any form. Electromagnetic radiation can travel in a vacuum. When this radiation interacts with the atoms and molecules of matter it is absorbed and converted into many other forms of energy.

When **radio wave energy** is absorbed by the metal aerial of a radio or TV receiver it produces a small amount of electrical energy in the form of a small alternating current.

When **infra-red radiation** is absorbed by an object it causes a rise in temperature and a gain of internal energy.

When **light energy** is absorbed by growing plants it is converted into stored chemical energy. Solar cells can convert light energy into electrical energy.

When **ultraviolet light** is absorbed by atoms it is sometimes converted into light energy. This is called fluorescence.

When **X-ray** or **gamma-ray energy** is absorbed by matter it causes ionisation of atoms and molecules.

These metal dishes reflect and focus radio waves to concentrate the radio wave energy onto a small sensitive receiver. Here it is converted into the energy of an electric current.

Energy changes its form

We have already found that energy often changes from one form to another and that machines do work by converting or transferring energy. We can study energy changes by setting up experiments in the laboratory, which are working models of larger scale processes. In all these examples no energy is 'used up' or destroyed; the total amount of energy is conserved.

- *Light a match.*

Before the match burns what kind of energy does it hold?

What else is needed for the match to burn?

What are the final forms of energy?

Is this process reversible?

The forms of energy and changes that occur can be explained by an energy flow diagram. The blocks contain the different forms of energy and the arrows show energy changes or conversion processes. The linked forms of energy make an energy chain. Sometimes the flow of energy along this chain can be reversed, sometimes it cannot. Energy often leaks out of the chain.

- *Connect a battery to a lamp.*

Is this process reversible?

Is any energy lost before it reaches the lamp filament?

Feel the lamp as well as looking at it.

What are the final forms of energy?

- *Knock a large nail into a block of wood with a hammer.*

Is this process reversible?

Feel the hammer and the nail after several blows.

What are the final forms of energy in the hammer and nail? What other forms of energy are produced and what happens to some of the stored chemical energy in your body?

- *Wind up a steel clock spring. Then let it drive a dynamo which is connected to a lamp.*

If we tried to use the electrical energy output from the dynamo to drive an electric motor to wind up the spring again, what do you think would happen?

Where, and in what form has energy been lost along the energy chain?

When the brake or ratchet is released on the spring it always unwinds and its stored energy 'runs down'. To reverse this process some more stored energy is needed to wind it up again.

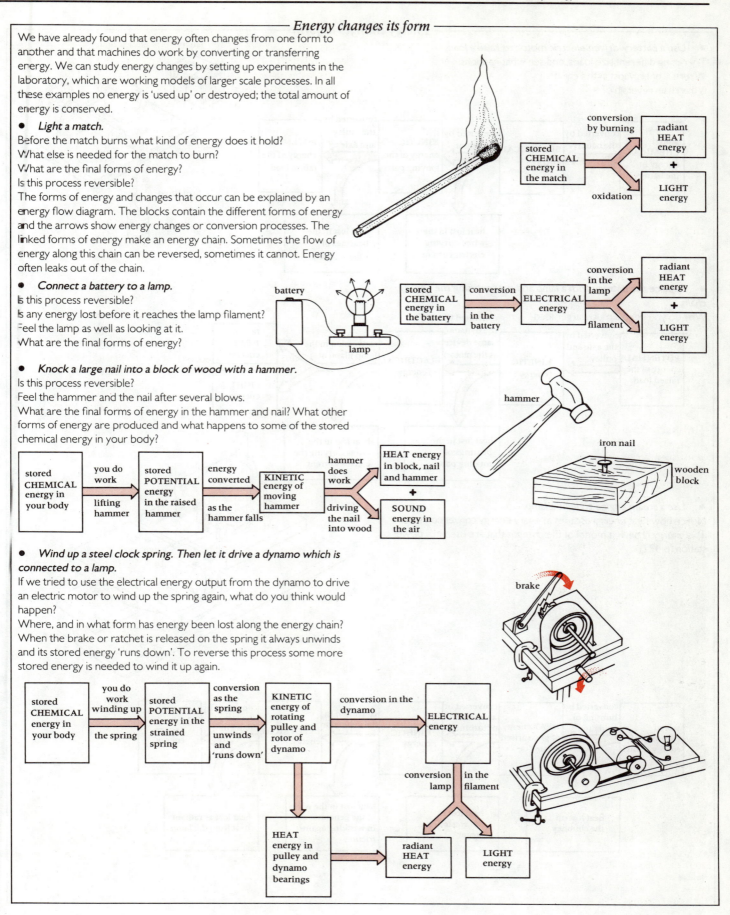

Energy changes its form

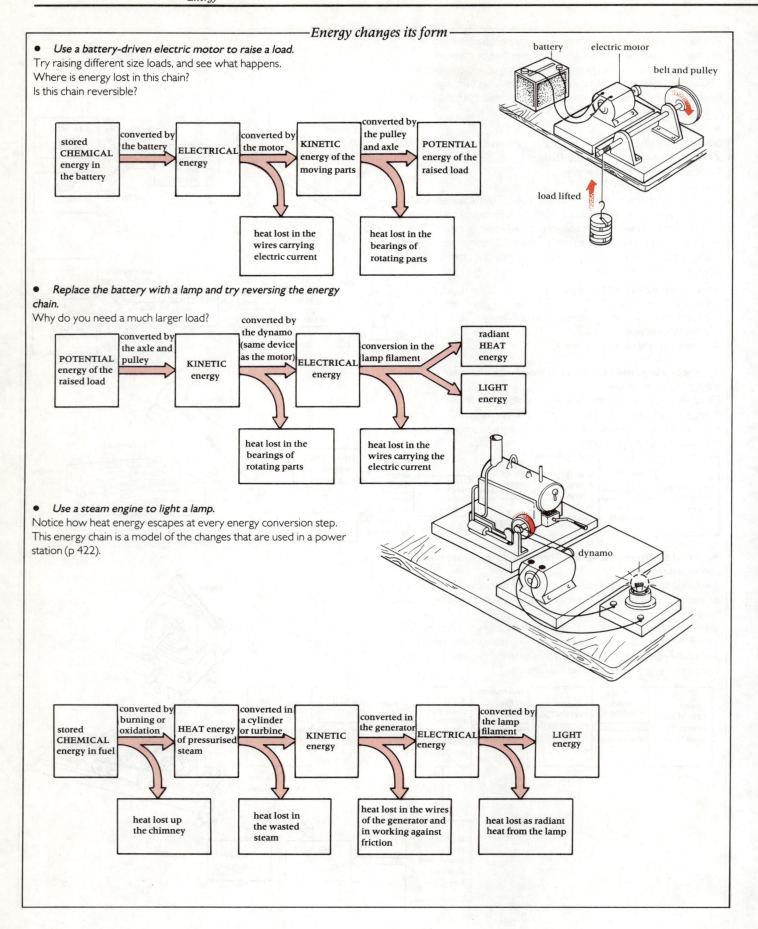

- **Use a battery-driven electric motor to raise a load.**
Try raising different size loads, and see what happens.
Where is energy lost in this chain?
Is this chain reversible?

| stored **CHEMICAL** energy in the battery | → converted by the battery → | **ELECTRICAL** energy | → converted by the motor → | **KINETIC** energy of the moving parts | → converted by the pulley and axle → | **POTENTIAL** energy of the raised load |

heat lost in the wires carrying electric current

heat lost in the bearings of rotating parts

battery   electric motor   belt and pulley

load lifted

- **Replace the battery with a lamp and try reversing the energy chain.**
Why do you need a much larger load?

| **POTENTIAL** energy of the raised load | → converted by the axle and pulley → | **KINETIC** energy | → converted by the dynamo (same device as the motor) → | **ELECTRICAL** energy | → conversion in the lamp filament → | radiant **HEAT** energy / **LIGHT** energy |

heat lost in the bearings of rotating parts

heat lost in the wires carrying the electric current

- **Use a steam engine to light a lamp.**
Notice how heat energy escapes at every energy conversion step.
This energy chain is a model of the changes that are used in a power station (p 422).

dynamo

| stored **CHEMICAL** energy in fuel | → converted by burning or oxidation → | **HEAT** energy of pressurised steam | → converted in a cylinder or turbine → | **KINETIC** energy | → converted in the generator → | **ELECTRICAL** energy | → converted by the lamp filament → | **LIGHT** energy |

heat lost up the chimney

heat lost in the wasted steam

heat lost in the wires of the generator and in working against friction

heat lost as radiant heat from the lamp

Energy changes its form

● *Build a working model of a hydroelectric power station.*

A hydroelectric scheme avoids many of the energy losses found when lighting a lamp using a steam engine, because it uses the running water instead of steam to drive the turbines. Can you explain where some energy is still lost as heat in this scheme?

Can you work out a scheme for reversing this energy chain?

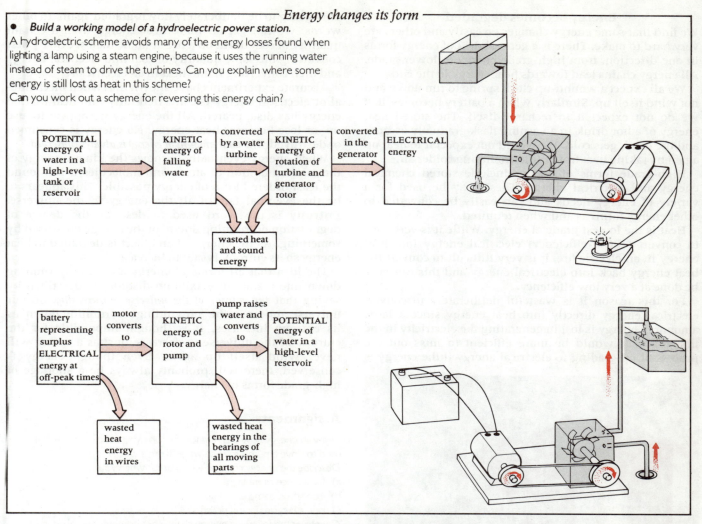

```
POTENTIAL          →    KINETIC       converted    KINETIC       converted    ELECTRICAL
energy of               energy of     by a water    energy of     in the       energy
water in a              falling       turbine       rotation of   generator
high-level              water                       turbine and
tank or                                             generator
reservoir                                           rotor

                        ↓ ↓
                   wasted heat
                   and sound
                   energy
```

```
battery           motor         KINETIC        pump raises     POTENTIAL
representing      converts       energy of      water and       energy of
surplus          to             rotor and      converts        water in a
ELECTRICAL                      pump            to              high-level
energy at                                                       reservoir
off-peak times

     ↓                            ↓
wasted          wasted heat
heat            energy in the
energy          bearings of
in wires        all moving
                parts
```

Energy storage schemes at Ffestiniog and Dinorwig

In 1961 Britain's first pumped storage system began providing electricity at Ffestiniog in North West Wales. The four generators at Ffestiniog can each deliver 90 MW of electrical power in under 1 minute from start-up. The more modern and much larger scheme at Dinorwig in Snowdonia National Park can deliver 1800 MW of electrical power continuously for 5 hours from a full reservoir.

From where do these generators get their energy supply? At times of off-peak demand for electricity the surplus electrical energy is used to pump water from a low-level reservoir to a high-level reservoir. At Ffestiniog and Dinorwig the same machines which work as generators can be also be driven in reverse as pumps. By using it to drive these pumps, the spare electrical energy is stored in the form of gravitational potential energy of the water which it pumps up to the high-level reservoir.

During times of peak demand for electricity, the energy stored in the reservoir is converted back into electrical energy as in a hydroelectric power station. For every 4 units of electrical energy used to pump water, the schemes return about 3 units of electrical energy after storage. How efficient are the schemes? Where has the other unit of energy gone?

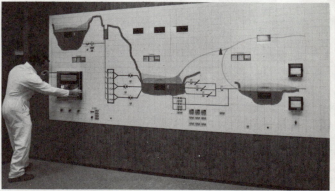

Energy becomes degraded

We find that some energy changes go easily and others are very hard to make. There is a general drift of energy forms in one direction, from high-grade energy to lower grade. All energy chains lead towards heat energy in the end.

We all expect a wound-up clock spring to run down and not wind itself up. Similarly when a battery becomes 'flat' we do not expect it to recharge itself. The stored heat energy of a hot drink in a vacuum flask gradually escapes and the drink gets cold. We would not expect it to heat up again by taking heat from the surrounding colder air.

High-grade forms of energy include stored chemical energy and electrical energy. These may be used for a variety of energy needs and can easily be converted to other energy forms as and when required.

Heat is the lowest grade of energy. While it is very easy to convert (100% efficiency) electrical energy into heat energy in an electric fire, it is very difficult to convert the heat energy back into electrical energy and this can only be done at a very low efficiency.

For this reason it is wasteful deliberately to convert electrical energy directly into heat energy since a large amount of energy is lost in generating the electricity in the first place. It would be more efficient to miss out the process of up-grading to electrical energy if the energy is only going to be deliberately down-graded again. In other words, we can get far more heat energy from a certain quantity of coal burnt on a fire at home than we do by first converting the same quantity of coal into electrical energy and then converting it into heat in an electric fire.

Accurate experiments have shown that when mechanical or electrical energy is converted fully into heat that no energy has disappeared. All the energy we appear to 'use up' or lose becomes heat energy. No energy is consumed or destroyed. *The total amount of energy is always conserved.*

When energy eventually becomes the kinetic energy of the random motion of atoms and molecules it has found the least ordered form of energy possible. This appears to be the eventual fate of all the energy in the universe. **Entropy** is the word used to describe the degree of degradation or 'running down' of the energy possessed by something. As the energy of an object is degraded to heat energy so its entropy is said to increase.

The idea that all forms of energy are steadily running down into a state of maximum disorder is described by saying that *the entropy of the universe is increasing.* So our universe appears to be 'unwinding' or 'running down' as its entropy increases. It is thought impossible for the entropy of the universe to decrease; just as a spring will never wind itself up again. Even though energy is conserved, there will probably always be a shortage of high-grade forms of energy.

Assignments

Draw an energy flow diagram for the following process: a windmill is used to drive a pump to raise water from a well.

Describe the energy changes which occur when you:
a) file a piece of metal,
b) iron some damp washing,
c) ride a bicycle up and down a hill.

Construct the following energy change experiments and produce energy flow diagrams showing where energy is lost or wasted:
d) Compare a battery lighting a lamp with the same battery driving an electric motor, turning a dynamo, lighting the same lamp.
e) Use a battery (or other chemical source of energy) to turn a flywheel. Then use the energy stored in the rotating flywheel to drive a dynamo and then to light a lamp.
f) Wind up a clock spring by hand. Use the energy stored in the clock spring to raise a load. Then use the energy stored in the raised load to drive a dynamo and then to light a lamp. Compare this energy chain with lighting the same lamp by turning a dynamo by hand.

Redraw and complete table 19.4 for the energy conversions in a motor car.

Try question 19.10.

Table 19.4

Machine or process	Original energy form	Final energy form(s)
using the battery		
recharging the battery		
dynamo or alternator		
starting motor		
engine		
brakes		
headlamps		
horn		
windscreen wiper motor		
heated rear window element		

Clocks with springs or weights need winding up. As the clock mechanism turns, the stored energy is gradually converted into heat and the clock 'runs down'. This energy change naturally goes one way only; the clock cannot wind itself up again from the heat in the air.

19.3
ENERGY SOURCES AND USES

What sources of energy are available for us to use? What are the advantages and disadvantages of using each source and what are the best ways of making full use of these finite resources? What alternative sources of energy could be used in the future?

The sources of energy available to us

Solar energy

Life is sustained by the energy from the sun. Without the sun all plants and animals would die. The sun provides 99.98% of the energy which naturally flows through the surface environment of the earth. Put another way, the sun supplies 5000 times more energy than all other sources combined. This emphasises the importance of solar energy to us.

Solar energy arrives at the earth at the rate of 1.7×10^{17} watts, which is 170 thousand million million joules every second! Fig. 19.5 shows what happens to this solar radiation energy.

a) About 30% of the radiation energy is reflected straight back into space. The icecaps and particles in the atmosphere form good reflectors. It is thought that an increase in the area of ice covering the surface of the earth, or an increase in atmospheric dust might reflect even more radiation back into space and so cause the earth to cool towards another ice-age. Most of the short-wavelength electromagnetic radiation, i.e. UV light, is reflected.

b) About 47% of solar radiation is absorbed and con-

verted into heat energy or internal energy. During the day the earth's surface is warmed up, and during the night this energy is radiated back into space as radiant heat energy, i.e. the long-wave electromagnetic radiation called infra-red.

c) About 23% of solar radiation is absorbed causing evaporation of water from the oceans and lakes and from the land and plants. The high latent heat of vaporisation of water means that a lot of absorbed radiant heat energy is used to convert liquid water into water vapour. Convection currents carry the water vapour up into the atmosphere where it forms clouds. Much of this energy then becomes stored as potential energy in glaciers, icecaps, high-level lakes and rivers. As the water runs back to the sea, the potential energy is eventually converted into heat energy.

d) A small amount of energy, about 0.2% of the total, drives the convection currents in the oceans and the atmosphere. Some of this energy appears as wind power and wave power, which are forms of kinetic or mechanical energy.

e) An even smaller amount of energy, only 0.02% of the total, is absorbed by growing plants via the chemical process of photosynthesis. This energy is converted into a store of chemical energy. Some will form food for animals, but it will all eventually decay. Some of the chemical energy is converted into heat energy when humans and the animals do work; some is converted into heat as the plants decay and some is converted and stored as fossil fuels (coal, oil and gas), which may also eventually be burnt and converted finally into heat energy.

Figure 19.5 *What happens to all the energy arriving at the earth from the sun?*

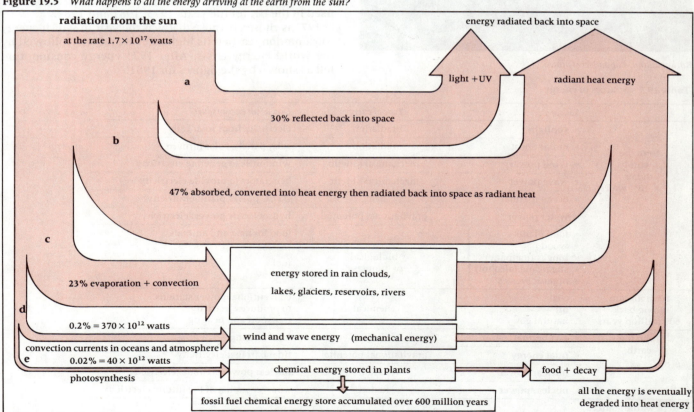

Geothermal energy

The other sources of energy are available from the earth itself. The centre of the earth, is at about 4000°C, much hotter than its surface. Heat energy, known as geothermal energy, flows from the hotter centre towards the cooler surface of the earth. On average the flow of heat from the centre of the earth is at the rate of 0.06 watts per square metre. Over the whole surface of the earth this amounts to about 30×10^{12} watts. Most of this heat energy arrives by conduction through the rocks in the earth's crust, but a small amount arrives by convection in hot water springs and lava in volcanoes.

Karkar Island, Papua New Guinea.

Tidal energy

The tides raise the water in the sea roughly twice a day. If the high-level water at high-tide is trapped then it is possible to use this as a source of potential energy. The energy given to the sea water comes from the gravitational forces between the earth and the moon. So tidal energy comes from the potential and kinetic energy of the motion of the moon round the earth.

Radioactive or nuclear decay

The spontaneous and natural breaking up of atomic nuclei releases energy. This occurs in the earth's crust as the nuclei of naturally occuring unstable or radioactive atoms emit radiation. The energy released is added to the flow of geothermal heat arriving at the surface of the earth.

The energy sources we use

Since we began to use energy to help with our work, what sources have we used? We started by burning wood, then coal and increasingly we have used oil and natural gas. Recently we have been using up the finite chemical store of energy which accumulated over the last 600 million years in the form we call fossil fuels. These reserves of energy would take another 600 million years to replace if plants were allowed to decay at the same rate as long ago. In effect these chemical energy sources are not renewable because we cannot wait that long!

In table 19.5 we can see the full range of energy sources which are available for us to use. Sources of energy which are available from the sun on a daily basis (such as sunlight, wind and waves) and sources which can be replaced fairly quickly (such as wood and food) are called renewable sources. Over the last 200 years we have been using up the non-renewable sources of energy. The box opposite gives details of the primary sources of energy used in the UK for the years 1973, 1981 and 1987.

1973 is chosen because that was the year in which energy consumption reached its highest ever and was followed by the world energy crisis. After 1973 energy consumption fell as shown by the figures for 1981.

Table 19.5   Sources of energy

	Source	Type of energy	Uses as an energy source
from the sun today 1.7×10^{17} watts	* sunlight	radiation	electricity from solar cells
	* radiant heat	radiation	solar panels, solar furnace
	* wind power	mechanical kinetic	windmills to generate electricity
	* wave power	mechanical kinetic	generators to provide electricity
	* ocean thermal	heat	OTEC power plant: electricity
	* water power	gravitational potential	hydroelectric power: electricity
	* food plants * wood * biogas (methane) * sugarcane (alcohol) * organic wastes	biological or 'biomass'	food for man and animals fuel
from the sun going back 600 million years	coal oil natural gas	chemical	conventional power stations to produce electricity, transport and heating
from the earth	* geothermal	heat	local heating
	* tidal energy	gravitational potential	tidal barrage: electricity
	nuclear power	fission	nuclear power stations: electricity
	nuclear power	fusion	? power station of the future: electricity

* Renewable sources.

Energy sources in the UK

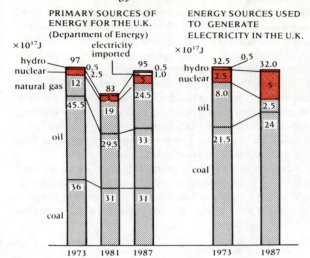

PRIMARY SOURCES OF ENERGY FOR THE U.K. (Department of Energy)

ENERGY SOURCES USED TO GENERATE ELECTRICITY IN THE U.K.

- The primary sources of energy are those forms in which the energy is initially obtained. The red shaded parts of the bar charts show just how little of the primary energy sources used in the UK was not from fossil fuels.
- Notice how the amount of oil used has fallen, being mostly replaced by natural gas from the North Sea.
- In 14 years the contribution from nuclear energy has doubled.
- The amount of electricity being generated in 1987 is almost the same as it was in 1973 but the energy sources used have changed. Oil is now used very little for generating electricity.
- In 1987 the UK imported 11.6 TWh of electricity from Europe. (1 TWh = 1 terawatt hour = 1000 million kWh = 3.6×10^{15} J.)

Comparing forms of transport

Vehicle	Advantages	Disadvantages
Bicycle	No pollution, no fossil fuel, good exercise, cheap travel	Range limited by time and fitness, some accident risk, unpleasant in bad weather
Car	High speed, convenience, comfort	Air and noise pollution, greedy for fuel, some accident risk
Bus	For 20+ people more economical than car, safe travel, comfort	Limited to bus routes, uses fossil fuel, air and noise pollution
Train	For 50+ people most economical, very safe, high speed, comfort	Limited number of stations, diesel trains use fossil fuel and pollute
Electric train	No pollution, near 100% energy conversion to E_k	
Aircraft	Very high speed, comfort, very safe	Limited to airports, very greedy for fuel, use fossil fuel, air and noise pollution

Energy and transport

In 1987 almost 30% of the energy consumed in the UK was used for transport. Transport uses over 50% of the oil. It is therefore important to examine how efficiently transport users consume the energy.

The simplest way is to compare how far different forms of transport go for a unit of energy. The bar chart below compares distances in metres per megajoule of energy. On this chart the car appears to be by far the most economical. However, it is misleading because we should take into account the number of passengers carried. The second bar chart is obtained by multiplying the distances by the number of passengers carried. This shows that a well-loaded train is the most economical form of transport.

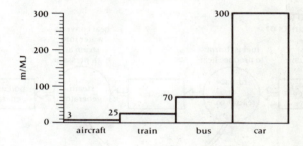

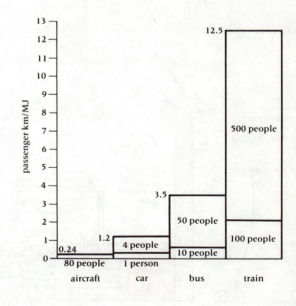

Energy conversions in a power station

1 Heat energy is produced by burning coal or oil in a furnace or from the fission of uranium nuclei in the core of a nuclear reactor.

2 Water absorbs the heat energy in a boiler or heat-exchanger and is turned into steam at a high pressure. So the steam generator stores the heat energy in the steam. There is now extra internal energy in the steam, which in this case we can think of as a form of potential energy. It is similar to the energy stored in an inflated balloon or in a cylinder of pressurised gas. There is potential energy stored in the gas because of its pressurised condition. Similarly, steam under pressure is capable of doing mechanical work so we shall call it potential energy.

3 Converting heat energy into mechanical energy is the most inefficient of the energy conversion processes. This is because we are reversing the natural tendency of mechanical energy to be degraded by friction into heat energy. Steam turbines extract about 50% of the energy stored in the high-pressure steam and convert it into mechanical energy of rotation, i.e. kinetic energy. For maximum efficiency the steam is

expanded to a partial vacuum from its high-pressure state. During this rapid expansion the steam drives the blades of the turbine. Then it is condensed back to water for re-use in the boiler. To condense the steam a large flow of cooling water is needed to absorb its latent heat. The warmed water is exposed to the air in cooling towers where heat is lost to the atmosphere in the form of warm air and evaporated water. This deliberate loss of heat to the atmosphere as the water is cooled alone accounts for the wastage of about 50% of the total energy input to the power station energy chain. Schemes are being planned for making use of this heat energy to provide heating and hot water for domestic and industrial consumers near to power stations.

4 Generators convert the kinetic energy of the rotating turbine into electrical energy by electromagnetic induction. Here friction converts some mechanical energy into heat and the resistance of the wires in the generator, transformer and transmission lines also produce waste heat.

Energy conversions in a power station

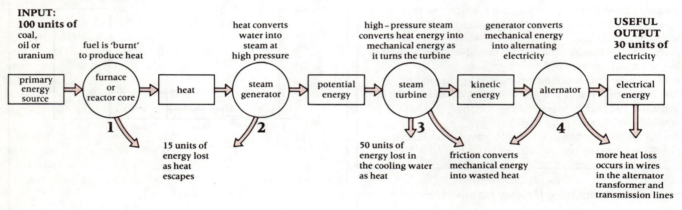

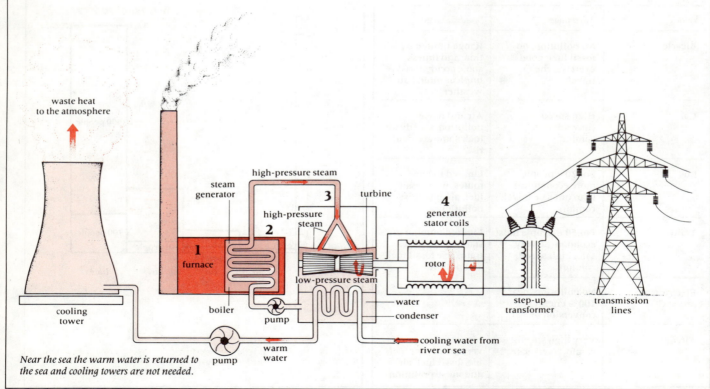

Near the sea the warm water is returned to the sea and cooling towers are not needed.

Generating electricity today

Electrical energy is not a natural or primary source of energy and it cannot be stored very easily. So electrical energy is energy 'on the move' and it has to be consumed as it is produced. During the conversion of primary energy sources into electrical energy about 70% of the energy input is wasted and lost as heat energy.

The electrical energy produced is, however, a very convenient form of high-grade energy with many applications. It also has the advantage that it can be distributed very quickly all over the country wherever wires take it.

In the UK electricity is generated in power stations using coal, oil and nuclear energy as the primary energy sources and a useful addition is made from hydroelectric schemes and pumped water storage schemes like the one at Dinorwig.

The basic energy chain is the same in all power stations. The main forms of energy in the chain and the processes which make the conversions are shown in the box. In addition the lay-out of a coal-fired (or oil-fired) power station is shown schematically.

In this photograph of Didcot power station taken during construction, we can see the blades of the steam turbines before the upper part of their housing was fitted.

The nuclear energy alternative

We have seen that most of our electrical energy is generated from coal and oil. There is no doubt that the world's coal and oil reserves are limited and that the time will come (perhaps within twenty or thirty years) when the demand for oil will exceed the rate at which it can be supplied. Apart from the coming shortage of oil, there are powerful arguments for using oil for other purposes. For example, oil is a basic raw material from which plastics and fertilisers can be made. This suggests that oil is too valuable as a raw material to be wasted by burning to produce heat.

In 1956 the world's first nuclear-powered electricity generating station began working at Calder Hall. The United Kingdom Atomic Energy Authority (UKAEA) now has many years of experience in designing, building and commissioning nuclear power stations. Based on this experience, the building of more nuclear power stations in Britain is one possible alternative to the use of oil and coal for our future supply of electrical energy.

The only major difference between a nuclear power station and a coal- or oil-fired station is that the heat which produces steam for the turbines is provided by nuclear fuel in a reactor, and not by oil or coal burned in a furnace. So how does nuclear fuel produce heat?

Energy from the nucleus by fission

The nucleus of the uranium nuclide $^{235}_{92}U$ (which has 235 nucleons) has the very rare property of splitting into two roughly equal halves when it captures a neutron. The splitting of the nucleus is known as nuclear fission.

The products of the fission of a single uranium-235 nucleus are:

a) Two new elements, known as fission products, two examples of which are given in fig. 19.6. (Most of the fission products are unstable and emit radiation. The disposal of these radioactive fission products as waste material from the used uranium fuel is the cause of a major difficulty with nuclear power stations.)

b) 2 or 3 fast-moving free neutrons with lots of kinetic energy.

c) An average of 3.2×10^{-11} J of energy from each fission of a uranium-235 nucleus. This energy is released as the kinetic energy of the neutrons and as radiation. This kinetic energy is the source of the heat energy obtained from a nuclear reactor.

Where has this energy come from?

The nuclear fission equation shows that the number of nucleons is conserved, fig. 19.6. But an accurate adding up of the total mass on each side of the equation would show that there is a small loss of mass on the right-hand side. This lost mass is converted into energy.

Figure 19.6 *Nuclear fission*

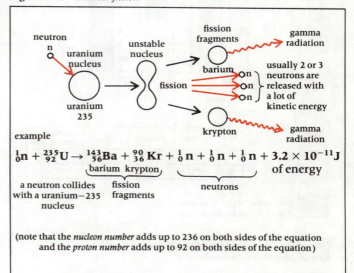

$$^{1}_{0}n + ^{235}_{92}U \rightarrow ^{143}_{56}Ba + ^{90}_{36}Kr + ^{1}_{0}n + ^{1}_{0}n + ^{1}_{0}n + 3.2 \times 10^{-11}J$$

(note that the *nucleon number* adds up to 236 on both sides of the equation and the *proton number* adds up to 92 on both sides of the equation)

In 1905 Albert Einstein worked out the relation between mass and energy in his now famous formula:

$$E = mc^2$$

The energy E released in joules can be found if the mass loss m is given in kilograms and $c = 3 \times 10^8$ m/s, the speed of light.

The *fission* of one uranium-235 nucleus gives off much more energy than would be released by *natural* radioactive decay.

Chain reactions

If the neutrons which shoot out of one dividing ^{235}U nucleus are captured by other nearby ^{235}U nuclei then more fissions will occur, shooting out even more neutrons. In this way it is possible to set up a chain reaction, fig. 19.7.

Figure 19.7 *A chain reaction*

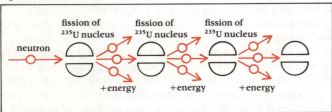

About 85% of the ^{235}U nuclei which capture a neutron quickly undergo fission, shooting out 2 or 3 more neutrons. So we can see that if all these neutrons were captured by ^{235}U nuclei the number of fissions would rapidly increase,

Figure 19.8 *Controlling chain reactions in a reactor core*

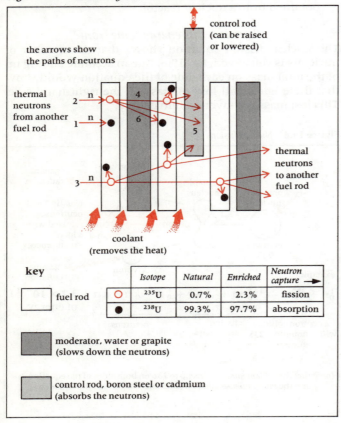

	Isotope	Natural	Enriched	Neutron capture →
○	^{235}U	0.7%	2.3%	fission
●	^{238}U	99.3%	97.7%	absorption

key

☐ fuel rod

▨ moderator, water or graphite (slows down the neutrons)

▨ control rod, boron steel or cadmium (absorbs the neutrons)

leading to a nuclear explosion. But this does not normally happen and in a nuclear reactor we can control the chain reactions. Consider what happens to the neutrons shown entering a fuel rod in fig. 19.8.

Neutron number 1 is captured by the nucleus of a ^{238}U atom in the fuel rod. Natural uranium contains as little as 0.7% of the isotope ^{235}U, so capture by ^{238}U is very likely to happen, as these form 99.3% of the atoms.

However, it has been found that the probability of ^{235}U nuclei capturing neutrons can be increased by slowing the neutrons down. This also reduces the probability of being captured by ^{238}U nuclei. The neutrons, shot out of a ^{235}U nucleus during fission, have a lot of kinetic energy and are known as fast neutrons, because of their speed. These fast neutrons are rarely captured by ^{235}U nuclei.

A material known as a **moderator** (usually water or graphite) is used to slow down the neutrons so increasing their chances of being captured by ^{235}U nuclei and causing fission. *The job of the moderator is to keep the chain reaction going.* Slowed-down neutrons are known as **thermal neutrons** because their slow speed and low kinetic energy is matched to the thermal (heat) energy of the surrounding material. The main types of nuclear reactor are classified by the speed of the neutrons in their cores. Reactors using thermal neutrons to keep the chain reaction going are called **thermal reactors**.

Thermal neutrons numbers 2 and 3 have an improved chance of being captured by a ^{235}U nucleus.

Neutron number 2 causes fission releasing energy and 3 more neutrons, numbers 4, 5 and 6.

Neutron number 4 is slowed down by the moderator and then captured by another ^{235}U nucleus in the next fuel rod. Fission occurs again but all three neutrons are lost. One is captured by a ^{238}U nucleus in the same fuel rod and two are absorbed by the **control rod**.

The control rod, made of boron steel or cadmium, which absorb neutrons, can be raised or lowered into the reactor core to control the rate of the chain reactions. The further into the core a control rod is lowered the more neutrons it will absorb and the more chain reactions it will stop.

When a balance occurs between the neutrons being released in fissions of ^{235}U nuclei and the neutrons being absorbed by ^{238}U nuclei and the control rods, a self-sustaining chain reaction takes place. This is called a **critical reaction**. Heat energy is continually produced at a constant rate in a critical chain reaction from the slowing down and capture of the neutrons in the moderator material, fuel rods and control rods.

Removing heat from the reactor core

To be able to use the heat produced by fission it must be removed from the reactor core. Heat is absorbed by a **coolant** which may be a liquid or a gas. (Carbon dioxide gas, pressurised water and liquid sodium have been used.) The coolant is piped into the core where, as it passes close to the hot fuel rods and moderator, it absorbs much of the heat. The heated coolant then exchanges its heat with the water in a steam generator before it passes back into the reactor core. Fig. 19.9 gives a schematic diagram of the advanced gas-cooled reactor or AGR, which is Britain's most advanced thermal reactor. Table 19.6 compares the features of the main types of thermal reactors.

Figure 19.9 *An AGR thermal reactor*

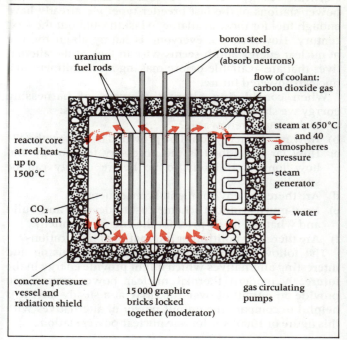

uranium
fuel rods

boron steel
control rods
(absorb neutrons)

flow of coolant:
carbon dioxide gas

steam at 650°C
and 40
atmospheres
pressure

steam
generator

water

reactor core
at red heat
up to
1500°C

CO₂
coolant

concrete pressure
vessel and
radiation shield

15 000 graphite
bricks locked
together (moderator)

gas circulating
pumps

Table 19.6 Some thermal reactors compared

Reactor name	magnox (named after the magnesium alloy which contains the fuel.)	AGR (advanced gas-cooled reactor)	PWR (pressurised water reactor)
Where built	UK, the first thermal reactors	UK, improved thermal reactor	USA (one may be built in the UK)
Fuel	uranium metal (0.7% ²³⁵U)	uranium(IV) oxide (2.3% ²³⁵U: enriched†)	uranium(IV) oxide (3.2% ²³⁵U: enriched†)
Moderator (to slow down neutrons)	graphite	graphite	pressurised ordinary water does both jobs, (high pressure is needed to stop water boiling.)
Coolant (to remove the heat)	carbon dioxide gas (at 400°C and 20 atmospheres)	carbon dioxide gas (at 650°C and 40 atmospheres)	
Control of chain reactions	neutron-absorbing control rods of boron steel	neutron-absorbing control rods of boron steel	neutron-absorbing control rods and emergency boron injection into coolant
Biological shielding and safety	steel pressure vessel and concrete shield	concrete pressure vessel	steel pressure vessel and pressure containment building

† Enrichment means that the proportion of the isotope ²³⁵U which undergoes fission is artificially increased above the naturally occurring proportion of 0.7%.

Inside the sphere of the AGR reactor at Windscale we can see the cap or top of the reactor core in the foreground. The machine used for replacing nuclear fuel elements in the reactor core stands in the background. This reactor was shut down in 1981 after nearly 20 years of operation.

Fast reactors

Fast reactors use fast neutrons for producing energy in their cores. They are also called fast breeder reactors because they can be used to *breed* a new nuclear fuel called plutonium from the otherwise wasted isotope of uranium, ²³⁸U. So fast breeder reactors use fast neutrons, but they do not breed fast.

Each uranium-235 fission releases 2 or 3 neutrons but only one is needed to keep the chain reactions going in the reactor core. Instead of absorbing the spare neutrons in control rods and slowing them down by a moderator, they can be used to breed plutonium fuel.

Fast neutrons absorbed by ²³⁸U nuclei form the new element plutonium-239 ($^{239}_{93}$Pu). ²³⁹Pu, like ²³⁵U, can be used as a reactor fuel, because it also has the valuable property of being split in half when it captures an extra neutron. Plutonium fission also releases a surplus of neutrons with a lot of energy.

The fuel used in a fast reactor is a mixture of plutonium(IV) and uranium(IV) oxides, clad in a stainless steel can. This fuel needs no moderator as the neutrons released by fission are captured as fast neutrons.

The coolant used to carry the heat away is liquid sodium metal. This is a good conductor and is able to remove the heat from the core effectively. As sodium does not boil at the 650°C temperature of the core, a high-pressure containment vessel for the coolant is not needed.

Breeding plutonium

While some plutonium is consumed in the core of a fast reactor (part of the fuel) it is possible to arrange a 'blanket' of ^{238}U around the reactor core. Fast neutrons escaping from the reactor core are captured by the ^{238}U which is converted into new plutonium-239. The balance between consumption of plutonium in the core and production of plutonium in the blanket is delicate, but it is possible to design a fast reactor which will give a net gain of plutonium. Such a reactor could provide energy and also breed more fuel than it uses. This makes the fast (breeder) reactor an attractive alternative for the future. By this means, fast reactors could convert the otherwise unusable isotope ^{238}U into new plutonium-239 fuel which could then be used to keep other reactors going for many years.

Alternative ways of harnessing energy in the future

There are plentiful supplies of coal and there is more renewable energy available than we are ever likely to need, so why is there an energy problem? The main reason is that natural gas and oil, on which we rely heavily at present, are being used up rapidly and will have to be replaced. We have already seen how we could harness nuclear energy as the primary energy source for our future electrical energy needs. We could build new nuclear power stations of the fast breeder type; we already have enough fuel for these available to last throughout the next century. However, not everyone is happy about the use of nuclear energy and it seems wise to look at the alternatives that are available before making a commitment to a nuclear-powered future.

When considering alternative ways of harnessing energy we should ask these important questions:
a) Is the source renewable?
b) How much energy could be provided by this method?
c) How far developed is the necessary technology and how long is it likely to be before energy can be supplied?
d) Are there any environmental consequences?
e) How much would a scheme cost to develop and build and what is a likely price for the energy supplied?
f) Are there any political or international implications?

The following are some of the more promising and interesting alternatives which might provide energy in the future. A modern thermal nuclear power station can provide an output of over 1000 MW at a steady rate. It is helpful to compare the output power of alternatives with this figure of 1000 MW for one nuclear power station.

— *Wind energy* —

Four designs for windmills are shown:
a) Typical traditional British windmill.
b) Large horizontal axis wind turbine (modern design).
c) Large vertical axis wind turbine concept (British design); in the background is a 400 kV electricity pylon.
d) Large vertical-axis wind turbine concept.

The axis of these windmills is the axle about which the blades of the windmill rotate, which is sometimes horizontal and sometimes vertical.

The traditional windmill gave an output of only a few 100 watts but it has been shown that a modern design of windmill with a blade span of 50 metres or more could generate over 1 MW when the wind blows well. So if enough wind blows for half of the year, then about 2000 large windmills could provide the same power as one 1000 MW nuclear power station. One objection to these windmills is that to get the best winds they would have to be sited on the tops of hills and along the coastline in places where they would spoil the scenery.

A 25 m diameter wind turbine built alongside the Carmarthen Bay power station in West Wales was the first windmill to feed electrical energy into the national grid on an experimental basis in 1982. Its power ouput was about 200 kW, but it is expected to provide the equivalent of only about 40 kW of power averaged over a full year.

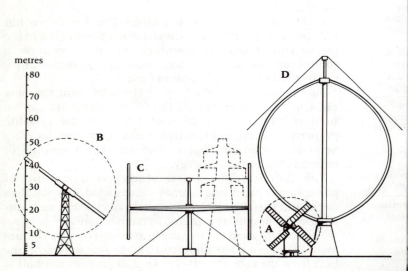

Wave energy

There is little scope for new hydroelectric power schemes in the UK, so experiments on extracting energy from water in another form are being tried. Waves well out to sea in the Atlantic carry about 80 kW of power per metre of wave frontage. The rocking-boom type of wave energy generator shown is capable of extracting and converting about 50% of this energy. So the power available from 25 km of the rocking-boom type of wave generator would be:

$$\tfrac{1}{2} \times 80 \, \frac{kW}{m} \times 25\,000\,m = 1\,000\,000\,kW = 1000\,MW$$

If 25 km of wave generators could provide the same output as one nuclear power station then the total available power from waves using the sites shown on the map could reach 120 000 MW. This is equivalent to 120 nuclear power stations and twice the total present electricity generating capacity in the UK. Peak wave energy would be available in winter coinciding with peak energy demand. The disadvantage of this scheme is the high cost of building and maintaining the wave generators. They would also be very vulnerable and difficult to defend.

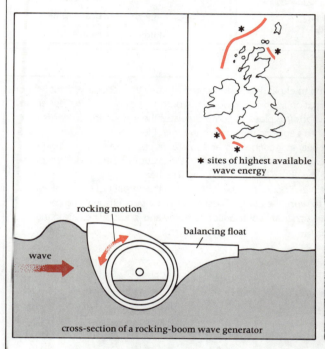

* sites of highest available wave energy

cross-section of a rocking-boom wave generator

(labels: rocking motion, balancing float, wave)

Ocean thermal energy conversion, OTEC

There is up to 20 °C temperature difference between surface water and the water several hundred metres below the surface. This temperature difference can be used to extract heat energy from the oceans.

Warm water taken from the sea near the surface is used to vaporise a volatile liquid such as ammonia or propane. The evaporating liquid takes its latent heat energy from the warm water, which is cooled by 1 to 2 °C. The expanding vapour has energy which is used to drive a turbine, which in turn produces electricity from a generator. The vapour must be condensed again to provide a low pressure region for more vapour to expand into. Cold water is brought up from the deep sea to condense the vapour. The pumps used to circulate the liquid use only a tiny fraction of the heat energy extracted from the sea. The power output of an OTEC plant might be 100 to 160 MW, or about $\frac{1}{10}$ of a nuclear power station. Such a plant would gradually cool down the sea in the surrounding area.

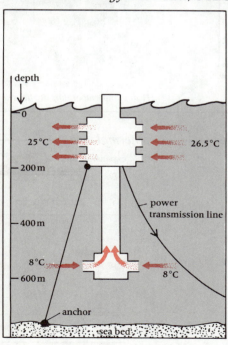

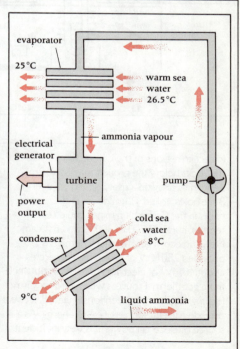

Tidal energy

A particularly promising site for a tidal barrage is the Bristol Channel. It has been estimated that up to 5000 MW of power (equivalent to 5 nuclear power stations) could be extracted from the potential energy stored in the water raised by the tides into the high-level basin. This scheme would also be expensive to build and would have important environmental effects on the whole Bristol Channel area.

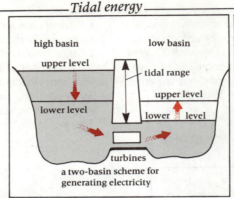

a two-basin scheme for generating electricity

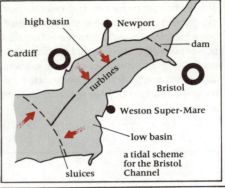

a tidal scheme for the Bristol Channel

Heat pumps

Heat pumps are devices for pumping heat energy from a low-temperature source to a higher temperature source. So a heat pump moves heat energy in the opposite direction to its natural flow from hot to cold. For example, a heat pump can be used to rescue waste heat from bath water or from extracted hot air. Heat pumps are not a source of heat energy in themselves but they transfer heat energy from one place where it is of little use to another place where it is of more use. In this way a heat pump is able to provide us with useful heat energy.

A refrigerator is a good example of a heat pump which pumps heat out of a cabinet to keep it cold. A heat pump working as the reverse of a refrigerator can be used to heat a house or factory. A volatile fluid is allowed to expand to a vapour and remove the necessary latent heat of vaporisation from its surroundings (outside the house). This happens in the *evaporator* pipes. A pump circulates the vapour and compresses it back into a liquid in the *condenser* pipes. Here the vapour discharges its latent heat as it condenses, and so warms up its surroundings inside the house. Heat is absorbed outside in the evaporator and is delivered inside the house by the condenser. The heat energy delivered to the inside of the house may be 3 or 4 times the amount of electrical energy used to work the pump! These devices are very efficient and could make a significant contribution to home and industrial heating in the future.

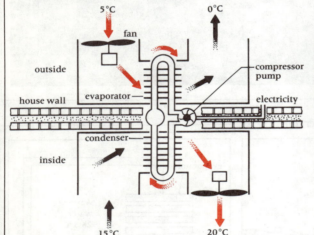

Heat pump evaporators are housed in these roof-top cabinets.

Geothermal energy

In certain regions of the UK the temperature of the rocks below the ground may be warm enough to provide a useful source of energy. Bore holes drilled 12 km down in the granite rocks in the Cornish tin mines could reach temperatures of 200 to 300 °C. In Los Alamos, New Mexico, a geothermal scheme has provided 250 MW of heat which generates about 40 MW of electricity. If the heat obtained from geothermal sources were used for district heating, or perhaps greenhouse heating, it could provide a useful source of extra energy. As a source of electrical power it is very inefficient.

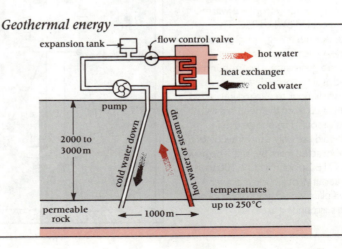

Solar cells

Direct conversion of solar radiant energy into electricity seems a very attractive idea. However, in the UK solar energy arrives at an average of 200 watts per square metre and is converted at only 5% efficiency into electrical energy. So 100 square kilometres of solar cell collectors would provide

$$\frac{5}{100} \times 200 \, \frac{W}{m^2} \times 100 \, Mm^2 = 1000 \, MW.$$

(100 square km = $100 \times 1000^2 \, m^2 = 100 \, Mm^2$). Therefore the equivalent of one nuclear power station would be 100 square kilometres of land surface which is a very large area to find on a small island.

A solar cell powered electric fence in Hungary to fence in cattle. The device which converts sunlight directly into electrical energy can also store enough energy to continue supplying electricity during a period of up to 48 hours of darkness and cloudy weather.

An attractive alternative idea has been suggested for using solar cells to collect energy. A large panel of solar cells perhaps 3 km × 4 km could be assembled in a synchronous orbit around the earth. The panel would orbit hovering over a fixed receiving antenna 7 km across on the earth, and the energy would be converted to microwaves for transmission down to earth. Although solar cells have been used successfully to power satellites and the space station Skylab, no scheme for beaming energy down to the surface of the earth in large quantities has been tried. Such a scheme would be very expensive to build.

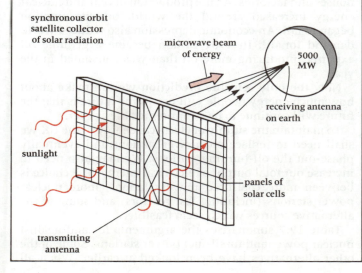

Solar furnace

A concave paraboloidal reflector can be used to focus and concentrate the radiant heat energy from the sun. At the focus the heat energy is as concentrated as that produced by burning fossil fuels. A solar furnace or boiler can be installed at the focus. However in the UK the weather is too often cloudy for such a furnace to provide a reliable or economic energy source.

The French solar furnace in Odielle in the Pyrenees uses over 20 000 electronically controlled mirrors to reflect the sun's radiation energy into a furnace housed in the building at the focus of the mirrors. The temperature in the furnace can quickly reach over 3000°C.

Solar panels

A more successful solar radiant energy absorber is the solar panel used for domestic water heating. It is estimated that about 5% of the total energy used in Britain is for heating hot water in our homes. If each house was fitted with about 4 or 5 square metres of solar panels, about 50% of the energy needed for hot water could be obtained this way. One disadvantage of solar panels is that they provide least heat energy in the winter when the demand for heat is greatest.

The transparent cover traps the solar radiation in the manner of a greenhouse and a black surface improves the rate of absorption of radiation.

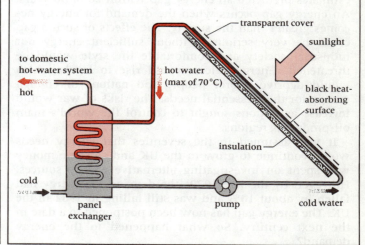

Nuclear fusion

An experiment has been set up at the Culham Laboratory near Oxford to try to harness the energy released by **thermonuclear fusion**. The device, known as JET (joint European torus), is the first to be built anywhere and might achieve controlled nuclear fusion.

Fusion, the 'melting' together of two atomic nuclei, is the main process providing the sun's energy. On earth fusion has been achieved only in the uncontrolled explosion of the hydrogen bomb.

There are several possible fusion reactions and the one thought likely to work requires the following two fuels:

deuterium, $^2_1$H,   a naturally occurring isotope of hydrogen, found in water,

lithium, $^7_3$Li,   plentiful element which can be extracted from ores and from sea water.

A 2000 MW fusion power station would require only 0.5 kg of deuterium and 1.8 kg of lithium per day. This energy output is equivalent to 18 000 tonnes of coal per day.

A possible fusion reaction involves deuterium and tritium nuclei fusing together to form helium:

$$^2_1\text{H} + ^3_1\text{H} \rightarrow ^4_2\text{He} + ^1_0\text{n} + 3 \times 10^{-12}\text{J}$$

deuterium   tritium   helium   neutron   kinetic
fuel   energy

The neutrons released in this reaction are used to provide the tritium from the other fuel, lithium, in the following neutron-absorbing fission process:

$$^7_3\text{Li} + ^0_1\text{n} \rightarrow ^3_1\text{H} + ^4_2\text{He} + ^1_0\text{n}$$

lithium   neutron   tritium for   helium   neutron
fuel   absorbed   fusion

Once the fusion process is started up, the lithium fuel will be converted into tritium by neutron absorption.

To make the deuterium and tritium nuclei fuse together an extremely high temperature of 100 million kelvin is needed. At this temperature the atoms lose all their electrons and the bare nuclei move in what is called a **plasma** at very high speeds. Only at these speeds of about a million metres per second is it possible for two approaching nuclei to overcome the electrostatic repulsion force between their like positive charges and get close enough together to fuse.

Such high-speed, high-temperature charged particles are very difficult to contain in any vessel. The JET device will attempt to do this in a large doughnut-shaped magnetic ring known as a **torus**. This evacuated vessel acts like a 'magnetic bottle'. Electromagnets built around it deflect the charged nuclei into a ring of plasma.

Although the fusion process is the reverse of nuclear fission, in both cases there is a small loss of mass which is released as energy. In the case of fusion, this source of energy is still only a hope for the next century. It could supply a virtually inexhaustible source of energy if we can get it to work.

The JET device, 12 metres high, during the final stages of construction in 1983.

Inside the doughnut-shaped vacuum vessel.

Energy choices

In the early 1970s when the demand for energy (and particularly oil) was growing at an increasing rate, all the estimates predicted an energy gap within 20 or 30 years. An energy gap occurs when the demand for energy becomes greater than the supply. The effects of such a gap could be very serious. Without sufficient energy our industrial society and comfortable life style would be threatened. Energy prices would rise to levels which would deprive poorer people and nations of enough energy for their essential needs. The risks of war would increase as nations sought to control the world's main oil-producing regions.

It was assumed in the seventies that energy needs would continue to grow in the UK and so some money was spent on investigating alternative energy sources. However, by the late 1970s the demand for energy had fallen by about 10% and was still falling in 1982 in the UK. The energy gap has now been postponed to a date in the next century. So what happened to the energy demand?

Higher prices forced everyone to take conservation seriously. For example, many people have insulated their houses and factories. As the production of coal and nuclear energy increased around the world, oil consumption began to fall. An economic depression also reduced world demand for oil. In 1981 Britain became a genuine oil exporter, producing more oil than was consumed in the year.

Now the only sensible prediction we can make about how much energy people will use in the future is that the future will continue to be as unpredictable as ever.

To maintain the supply of electrical energy in the UK we shall need to replace old power stations and eventually phase-out the oil-burning stations. We may also need to increase our total output of electrical energy. The choice is between more coal-fired power stations, more nuclear power stations (including fast reactors) and some of the alternative sources which seem feasible.

Table 19.7 summarises the arguments for and against nuclear power and fossil-fuel power stations. Some of the other alternatives have been looked at earlier. When all

Table 19.7 Comparing fossil fuels and nuclear power

Fossil fuels coal, oil, natural gas	Nuclear power thermal and fast reactors
Obtaining the fuel * extracting fuel is dangerous * coal mining and oil exploration take many lives * coal miners suffer lung diseases	* extracting fuel is dangerous * mining uranium exposes people to dangerous radioactive dust and radon gas causing diseases and cancers
Handling the fuel * impact of fuel transport on roads and local communities * coal dust blows from coal handling and stock piles * fuel oil spillage occurs at sea and in estuaries * visual impact of large coal stores and oil storage tanks	* special safety precautions are necessary for transporting fuel * risk of leakage and theft * fuel processing has special difficulties, and fuel requires reprocessing after use
Using the fuel * produces airborne dust, smoke and gases, including sulphur oxides (responsible for acid rain) and nitrogen oxides causing environmental problems * produces a large bulk of waste ash and dust * power station is safe but produces waste heat in the environment	* great effort must be made to limit the discharge of radioactive gas and particles into the air * limited discharge of radioactive liquids into sea and streams, * waste products require safe storage for a number of years and long-term safe and remote disposal which are hard to achieve * very limited risk of major disaster through failure of pressure vessel and escape of radioactive material into the environment * produces plutonium, which can be used to make nuclear weapons, or as a new fuel for fast reactor nuclear power stations * produces waste heat in the environment

the arguments are considered for and against each energy alternative the decisions are very difficult. A balance is needed between cost, environmental damage, safety and health risks. It is even more difficult estimating our future energy needs, because these are affected by the price of energy, the amount of economic growth and the more efficient use of energy resulting from insulation and other conservation measures.

It is clear that we should:
a) improve energy conservation,
b) thoroughly investigate alternative ways of harnessing energy and keep all options open,
c) take precautions to protect our environment,
d) avoid action or inaction which might lead to an increased risk of war,
e) consider the effects of energy provision, or the lack of it, on employment and the nature of our society.

Assignments

Arrange a visit to your local power station.
Find out
a) What is its primary energy source?
b) What happens to the waste products?
c) What happens to the waste heat?
d) Where is the fuel stored and where does it come from?
e) How many years of energy output from the power station will be needed to pay back the energy used to construct the power station?

Try questions 19.11 to 19.19

Questions 19

1 Describe an experiment you might devise to measure the power of a student while he is riding a bicycle. Explain why you would not expect your results to be very reliable.
2 How much energy would your body need to provide for the following:

a) climbing a hill 500 m high (use your own mass and $g = 10$ N/kg),
b) sweating and evaporation of 1 kg of water from your skin,
c) heat loss by radiation at the rate of 100 watts for 12 hours.
The specific latent heat of vaporisation of water is 2.3 MJ/kg.
3 An electricity account shows that a household has used 1400 units of electrical energy during a three month period.
a) calculate the number of joules of energy used in this time,
b) calculate the average number of joules of energy used per day,
c) calculate the average power consumption in watts,
d) calculate the electricity bill if each unit costs 5.0 p.
4 State the energy changes which occur in the following:
a) gas is used to boil water in a whistling kettle,
b) a child blows up a balloon and then releases it, letting it fly round the room,
c) a guitarist plays an electric guitar over a public address system.
5 A wool blanket is used to cover an electric blanket on a bed. If the electric blanket measures 1.6 m by 1.25 m and the wool blanket has a U-value of 5.0 W/(m² K), calculate the power of the electric blanket needed to keep the temperature under the wool blanket at 25 °C when the air temperature above is 10 °C.
6 This question is about the importance of saving energy in the home and how it can help the country as a whole.
a) A customer of British Gas had double-glazing installed in January 1987. The table below shows the readings of her gas meter for the dates shown.

Date	1986		1987	
	1 January	31 December	1 January	31 December
Readings (units)	6517	7665	7665	8616

If each unit costs 38p, how much money did she appear to have saved in 1987 compared with 1986?
b) The table overleaf shows some information about home insulation.

Type of insulation	Approximate cost (£)	Pay-back time (yr)	% of thermal energy saved
Loft	180	5	10
Double-glazing	5000	70	25
Cavity wall	700	10	20

   i) choose which insulation method would cost the least per pay-back year, (pay-back time is the time it takes for the savings to amount to the original cost of installing the insulation),
   ii) state which method would result in the greatest percentage saving of energy.
c) Besides those listed, there are other smaller-scale methods of insulation against heat loss in the home. Name **two**.
d) i) If the average cost of insulating each uninsulated home in Britain is £1000, and there are 5 million such dwellings, how much would it cost a Government to insulate them all?
   ii) If the saving per year is £500 million, what would be the pay-back time, in years, for the country?
e) There are different sorts of double-glazing, with different air gaps and different thicknesses of glass. Suggest **one** reason in each case for limiting the size of the air gap and the thickness of the glass used. (NEA [B] 88)

Figure 19.10

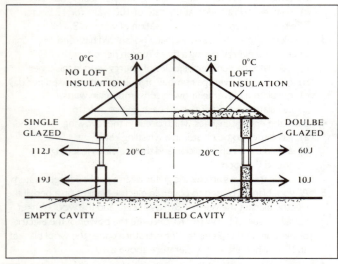

7   Fig. 19.10 shows the heat lost each second from *ONE SQUARE METRE OF SURFACE* for each of the windows, walls and roof in a house with and without insulation. The data are based on an inside temperature of 20°C and an outside temperature of 0°C.
a) Copy and complete the table below where
   'loss'   = heat lost each second from 1 m²
   'saving' = heat 'saved' each second from 1 m² after insulation

	Roof		Wall cavity		Window	
	No insulation	Insulation	Empty	Filled	Single Glazed	Double Glazed
'Loss'	30J	8J				
'Saving'		22J				

b) What other information about the windows, walls and roof would you need in order to decide which form of insulation would 'save' most heat each second?
c) Architects often refer to the 'U' value of a window where:

$$\text{'U' value} = \frac{\text{heat lost each second from 1 m}^2 \text{ of window}}{\text{(inside temperature)} - \text{(outside temperature)}}$$

'U' values are therefore measured in $W/m^2/°C$
   i) Use the information from the diagram to calculate the 'U' values for single and double glazed windows.
   ii) Explain why double glazing a window reduces heat losses.
(SEB 90)

8   Using the U-values given in table 19.3, estimate the rate of loss of heat energy by conduction through the brick walls of a detached house. The house measures 6.0 m by 8.0 m along the ground, and the walls, which are double brick with a cavity between, are 5.0 m high.
a) The temperature outside the house is 2°C and inside it is 22°C and the cavity between the walls is air filled.
b) The temperature outside the house is 2°C and inside it is 17°C and the cavity between the walls is filled with mineral fibres. Comment on the saving of energy and how it depends on each of the two changes made in case (b).

Figure 19.11

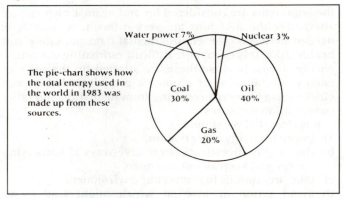

The pie-chart shows how the total energy used in the world in 1983 was made up from these sources.

9   There are five main sources of energy used in the world. These are: coal, nuclear fuel, natural gas, oil and water power.
a) Which **three** energy sources in the above list are known as fossil fuels?
b) i) Which energy source made the biggest contribution according to the pie-chart in Fig. 19.11?
   ii) Only a small fraction of the energy comes from water power. A similar pie-chart **for the United Kingdom** in 1983 would show an even smaller fraction. Why?
c) The table below shows the estimated world reserves of coal, oil and natural gas. The second column shows the quantity which was used in 1983.

	Estimated reserves	Quantity used in 1983
Coal	500 000	1250
Oil	100 000	3000
Natural gas	90 000	1500

(The figures in this table are in 'million tonnes of oil equivalent'.)

   i) If natural gas continues to be used at the 1983 rate, how many years will the reserves last?

ii) If coal continues to be used at the 1983 rate, how many years will reserves last?

iii) It is always difficult to make accurate predictions of how long reserves will last. Why?

d) The pie-chart for the year 2020 is likely to be different from the one shown for 1983. Suggest **two** likely differences and give your reason for each. (SEG 88)

10 Pumped-storage power stations are used to produce electricity during periods of peak demand. Water is stored in one reservoir and allowed to flow through a pipe to another reservoir at a lower level. The rush of water is used to turn turbines which are connected to generators.

a) Why are pumped-storage power stations used to produce electricity in short bursts and not for the continuous generation of electricity?

b) In one such power station 400 kg of water passes through the turbines every second after falling through a vertical height of 500 m. ($g = 10 \text{ m/s}^2$)

Assuming that no energy is wasted, calculate

i) the decrease in gravitational potential energy of 400 kg of water when it falls 500 m;

ii) the power delivered to the turbines by the water;

iii) the power output of the generator;

iv) the current produced by the generator if the output voltage is 20 000 V.

c) At night the water is pumped back to the reservoir to be used again.

i) Why is this done at night? ii) Why is more energy needed to pump the water back than is released when it falls? (SEG 88)

11 The reaction in a nuclear reactor can be represented simply by the equation

$$^{235}_{92}U + ^{1}_{0}n \rightarrow ^{b}_{a}X + ^{d}_{c}Y + 3^{1}_{0}n + \text{Energy}$$

where X and Y are unspecified nuclides.

a) What do the numbers 235 and 92 tell us about the particles in the uranium nucleus?

b) What is the name given to the process described by the equation?

c) Describe the principle governing the energy production.

The energy is produced by converting......................... into...........................

d) By using the reaction equation at the beginning of the question, complete the following equations.

i) a + c = ... ii) b + d = ...

e) What property of the nuclides X and Y makes them a health hazard when the fuel rods are removed from the reactor?

f) The neutron absorbed by the uranium is most likely to be a low energy (thermal) neutron, but the neutrons produced by the reaction are high energy neutrons which must be slowed down.

The neutrons are slowed down by a material called the ....

A suitable material would be ....

g) Explain how energy is produced continuously in the reactor.

h) Explain how the power output of the reactor is controlled. Specific materials should be named.

i) When the fuel rods are removed from the reactor they are contaminated by nuclides with long half-lives. Explain what is meant by the "half-life" of a nuclide.

j) Explain how the contaminated fuel rods may be dealt with safely after removal from the core of the reactor.

k) Some nuclides are deliberately exposed to the radiation in the core of a nuclear reactor so that they may be used later in industry or medicine.

i) Name **one** industrial use and **one** medical use for radioisotopes.

ii) Describe **one** of the processes you have named in (**k**) (i), stating the type of radiation used. (NEA[A]88)

12 a) Fig. 19.12 shows energy conversions in a nuclear power station. The *boxes* represent forms of *energy*.
The *circles* represent *devices* which convert energy into another form. Label the unfilled boxes and circles.

b) State any two problems caused by a nuclear power station and for each problem suggest a suitable precaution. (NEA [A] SPEC part)

13 The sun's source of energy is a nuclear fusion reaction in which mass is converted to energy. The energy E converted by a loss of mass m is given by the Einstein equation

$$E = mc^2$$

where c is the speed of light in a vacuum. The sun is losing mass at a rate of 2×10^9 kilogram per second. (Take the speed of light in a vacuum to be 3×10^8 m/s.)

a) Calculate the power output of the sun.

b) If the total mass of the sun is 2×10^{30} kg and it continues to lose mass at this rate, calculate its remaining life in years. (Take **1** year to be 3×10^7 s.) (O, part)

14 In a nuclear power station, the mass of the fuel decreases as energy is generated. Write down the Einstein mass–energy equation relevant to this process, making clear what each symbol represents.

Calculate the energy output when the mass of fuel decreases by 0.5 kg, assuming an overall efficiency of 40%. (The speed of light in a vacuum is 3.0×10^8 m/s.) (C, part)

15 What potential advantages might there be in developing, as an alternative to thermal nuclear reactors (a) fast (breeder) reactors and (b) nuclear fusion reactors?

16 a) Explain what is meant by a *moderator*.

b) Why is a moderator not needed in a fast (breeder) reactor?

17 a) Explain what is meant by 'an energy gap'.

b) Why is the statement that 'all the world's oil will soon be used up' both inaccurate and misleading?

c) What is meant by a renewable source of energy?

18 Give an account of one alternative renewable source of energy which you think could make a significant contribution to the supply of energy in the UK in the 21st century. Suggest some possible disadvantages of this alternative as well as its obvious advantages.

Figure 19.12

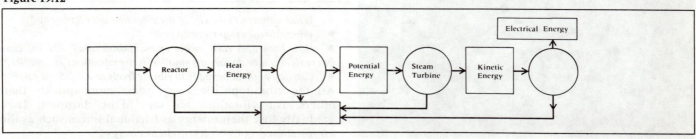

20

Vibrations and waves

*W*e *hear the vibrations of musical instruments and feel the vibrations made by heavy vehicles. We see light and feel the warmth of a fire. Surprisingly, all these sensations have something in common; they all involve the transfer of energy in the form of a wave motion.*

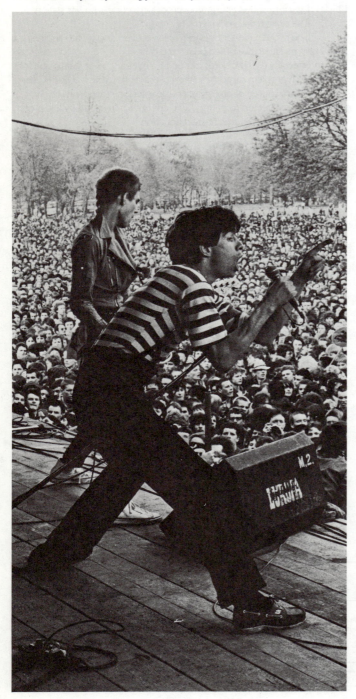

20.1
VIBRATIONS AND OSCILLATIONS

Lots of things which move with a regular to-and-fro motion are said to vibrate or oscillate. Many things vibrate slowly enough for us to study their properties by watching them.

● *Set up some of the examples in fig. 20.1.*

There are others which may be added, like the swinging spot of a galvanometer or the oscillations of a compass needle. Each example is set vibrating by displacing it from its rest position and letting it go; each then vibrates or oscillates naturally in a way which depends on the forces acting on it. But we are not here attempting to explain the different types of vibrations, or the forces acting, only to find their common properties.

Figure 20.1 *Vibrations (the arrows indicate motion, not forces)*

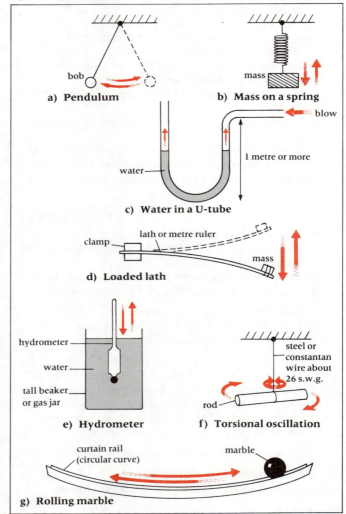

a) **Pendulum**

b) **Mass on a spring**

c) **Water in a U-tube**

d) **Loaded lath**

e) **Hydrometer**

f) **Torsional oscillation**

g) **Rolling marble**

● *What happens to the size of the vibration after letting go?*
● *Where do they stop vibrating?*
● *How long does each swing or oscillation take? Are the time intervals equal or do they get smaller as the vibrations get smaller?*
● *Can any of them be made to vibrate freely at a different rate?*

All the vibrations die away, some more quickly than others. The vibrations are said to be **damped**. They gradually lose their energy as frictional forces, such as the air resistance, convert it into heat energy.

In these examples the oscillator always comes to rest at the same position or level, which is usually the central position in the vibration.

The time taken for an oscillation stays the same even as the movement gets smaller. Not only does the time for each oscillation remain the same from one to the next, but it is also the same every time we start it off. For example, the timing of a pendulum swing is so precisely constant that it is often used to control the speed of clocks.

A complete to-and-fro movement is usually called an **oscillation** or **cycle**. Fig. 20.2 shows that one complete oscillation involves both a forwards and backwards swing of the pendulum or, starting at the mid-point of its swing, an oscillation is completed when the bob passes through the midpoint again moving in the same direction.

Figure 20.2 *An oscillation or cycle*

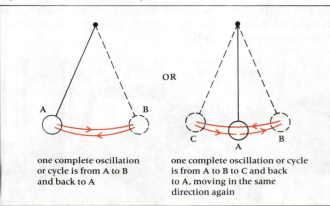

one complete oscillation or cycle is from A to B and back to A

one complete oscillation or cycle is from A to B to C and back to A, moving in the same direction again

Period and frequency

The time taken for an oscillation is called the period or periodic time. More fully:

The period T is the time a vibrating object takes to make one complete oscillation.

The period T, like all time intervals, is measured in seconds.

The frequency f is the number of complete oscillations (or cycles) made in one second.

Frequency is measured in **hertz**, symbol Hz. One hertz, which is defined as one oscillation per second, used to be called one cycle per second, abbreviated to c.p.s.

If in 1 second there are f complete oscillations taking equal times, then each oscillation takes $1/f$ of a second, which is its period T.

$$\text{period} = \frac{1}{\text{frequency}} \quad \text{or} \quad T = \frac{1}{f}$$

Worked Example
Period and frequency

If a vibrating string makes 10 complete oscillations in 1 second it has a frequency f of 10 hertz. What is its period?

The time it takes for 1 complete oscillation is its period T:

$$T = \frac{1}{f} = \frac{1}{10\,\text{Hz}} = 0.1\,\text{s}$$

Answer: the period of the vibrating string is 0.1 seconds.

Displacement and amplitude

*The **displacement** s of a vibrating object is its distance from the rest or central position in either direction.*

*The **amplitude** a is the maximum displacement from the rest or central position in either direction.*

The vibrations we have looked at which have a constant period T, whether the amplitude is large or small, are called **isochronous vibrations** (isochronous means of equal period). This type of vibration is very common and is known as **simple harmonic motion**. However, not all vibrations are isochronous.

● *Try dropping a ball and listen to its bouncing. What do you notice about the time intervals between bounces?*

Producing a graph of the motion of a vibrating object

● *Assemble a pendulum as shown in fig. 20.3.*
● *Dip the paint brush in ink.*
● *Pull the sheet of paper steadily across the floor at right angles to the oscillations so that the paint brush traces the path of its motion on the paper.*

The trace produced shows how the displacement of the pendulum varies with time. Fig. 20.4 shows the important features of this graph:

a) The amplitude a decreases from one oscillation to the next; this is a damped vibration.

b) The period T, shown over several oscillations remains constant.

c) The trace has a wave-like shape known as a sine curve, which is characteristic of all simple harmonic motions.

Figure 20.3 *The 'broomstick' pendulum*

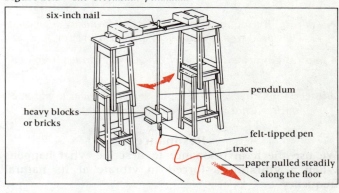

six-inch nail

pendulum

heavy blocks or bricks

felt-tipped pen

trace

paper pulled steadily along the floor

Figure 20.4 *The displacement–time graph for the 'broomstick' pendulum*

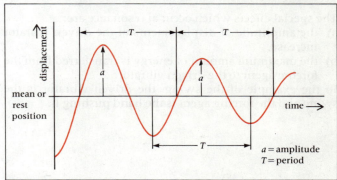

displacement

mean or rest position

time →

a = amplitude
T = period

Natural and forced vibrations

The oscillations of a child on a swing have a certain natural frequency, that is to say, the swinging child left to swing freely will always make the same number of complete oscillations in a certain time. The amplitude of the oscillations decreases unless the lost energy is replaced. To keep the swing moving it must be pushed at exactly the right times, in fact it must be pushed at the same frequency as the swing's own natural frequency. Pushing the swing is forcing it to oscillate. However, a swing can be given regular pushes at frequencies other than its natural frequency. The swinging motion is then said to be **forced**, and the resulting amplitude of the oscillation is small. The best response, or largest amplitude, is always produced when the forcing frequency equals the natural frequency. This effect is called **resonance** and the swing is said to *resonate*. When a vibrating object is forced to vibrate at its natural frequency, the amplitude may increase rather than decrease. This is how the amplitude of the swinging motion is built up; each push at the right moment adds a little extra energy to the vibration.

The vibrating or oscillating object, which may be as large as a bridge or as small as a molecule, is called a **vibrator** or an **oscillator**. Both of these words are used but each is more common in particular cases; for example, the vibrating steel strip used to print dots on ticker tape is called a vibrator but an electrical circuit which generates alternating electricity is called an oscillator. A movement which we can feel or hear is usually called a vibration and the object with such a periodic motion is called a vibrator. In the definitions which follow the word vibrator may be replaced by oscillator when it seems more suitable.

Every vibrator has one or more natural frequencies.

The natural frequency *of a vibrator is that frequency at which it will vibrate freely after a single displacement or push.*

A natural vibration *is one in which an object vibrates freely at its natural frequency.*

A forced vibration *is one in which an object is made to vibrate at the frequency of another oscillator or forcing agent.*

Resonance

We use the term resonance to describe what happens when a vibrator is forced to vibrate at its natural frequency.

Resonance of a vibrator occurs when the forcing frequency equals its natural frequency.

The special effects which occur at resonance are:
a) the amplitude of displacements of the driven vibrator increase,
b) the maximum amount of energy is transferred from the forcing agent to the driven vibrator.

In the example of the swing, the driven vibrator is the swing and the forcing agent is the hand pushing it.

Where does resonance occur?

Resonance in mechanical things: Resonance in mechanical things is very common. A car or a washing machine may vibrate quite violently at particular speeds. In each case, resonance occurs when the frequency of a rotating part (motor, wheel, drum etc.) is equal to a natural frequency of vibration of the body of the machine. In these machines there are usually several natural frequencies at which resonance can build up a vibration to a large amplitude.

The wind, blowing in gusts, once caused a suspension bridge to sway with increasing amplitude until it reached a point where the structure was over-stressed and the bridge collapsed. This happened to the Tacoma Narrows suspension bridge, Washington, USA in 1940. For the same reason, soldiers are instructed to break step when crossing a bridge so that their regular footsteps cannot build up a large-amplitude vibration by resonance with a part of the bridge structure. Factory chimneys and cooling towers have also been set oscillating by the wind, sometimes to the point of collapse.

The Tacoma Narrows suspension bridge caught in the resonant vibration which caused its total collapse in 1940.

Resonance in sound: The story is told of an opera singer who could shatter a glass by singing a note at its natural frequency. A similar example which can easily be tested involves singing into a piano with the damper pedal pressed down so that all the strings are free to vibrate. After singing a steady note, the piano strings of similar natural frequency can be heard to continue the vibration. Without resonance effects we would hardly hear most musical instruments. As we shall learn in the next chapter, the air in a pipe or tube will only vibrate with a large amplitude when it is caused to vibrate at a natural frequency. Similarly the amplitude of vibration of a string or a drum is only large at its natural frequencies.

Resonance in electrical circuits: When we tune a radio or television to a particular station we are selecting a radio transmission at one particular frequency. The detecting circuit is forced to conduct oscillating electric currents at the frequencies of all the radio signals received by the aerial. But the circuit is built to allow a large current to flow at only one frequency, which we can adjust to match the frequency of a particular radio station when we 'tune in'. The forced oscillating currents are all of very small amplitude, except at the selected frequency. At this frequency there is resonance between the forcing radio signal and the oscillating electric current in the tuned circuit.

Demonstrations of resonance

An example of mechanical resonance can be shown using a set of pendulums of different lengths known as Barton's pendulums, fig. 20.5. Several pendulums with a range of lengths from about 20 cm to 80 cm are hung by fine threads from a stronger thread which is not too tightly hung from a rigid support. Each pendulum has a small bob of mass about 10 g. A heavy bob of mass about 100 g attached to another thread forms a driving pendulum D. This pendulum should have the same length as one of the small bob pendulums, R.

● *What happens when the driving pendulum is set swinging.*

The driving pendulum attempts to force each of the other pendulums to oscillate at the same frequency. It sends small amounts of energy via the supporting thread to each of the other pendulums. For these small amounts of energy to add together and build up a large amplitude,

they have to arrive at the same frequency as the natural frequency of the pendulum. Only one pendulum, which has the same natural frequency as the forcing frequency (the one with the same length as the driving pendulum), will show resonance, R in the figure.

It is also worth noting that the pendulums do not all swing in step. Although all the pendulums have the same frequency as the driving pendulum, they do not keep in step with it. We use the word **phase** to describe the way in which these pendulums are in or out of step. The smallest pendulum may be almost in step with the driving pendulum. The two pendulums which are swinging the same way at the same time are said to be *in phase* with each other. The largest pendulum may be swinging the opposite way to the driving pendulum so that they pass each other in the centre of their swings, going in the opposite directions. These two pendulums are said to be *out of phase* by half a cycle or in **antiphase** with each other.

Resonance in sound can be shown using two tuning forks mounted on one box, as shown in fig. 20.6.

● *Set one tuning fork vibrating and then after a moment stop it.* The other tuning fork picks up the vibration by resonance.

● *Load the second tuning fork with a small piece of plasticine on its prongs and repeat the experiment.*

When the second tuning fork is loaded its natural frequency is reduced and the natural frequencies of the two forks are no longer equal. Resonance does not now occur, and the second fork does not pick up enough energy for its vibration to be heard.

Figure 20.6 *Resonating tuning forks*

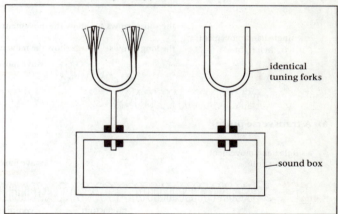

Figure 20.5 *Barton's pendulums – showing resonance*

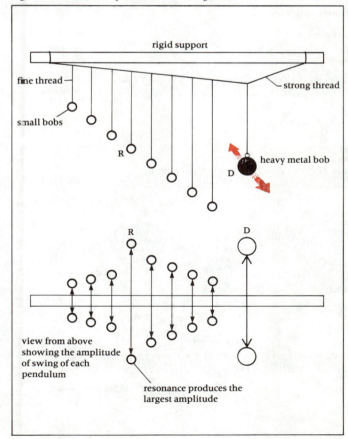

Assignments

Remember

The meanings of the words: period, frequency, amplitude, natural vibration, forced vibration, damped vibration, resonance, in phase and in antiphase.

Investigate

a) Try some examples of resonance yourself. Investigate a swing, a diving board or a piano, for example, and make a set of pendulums as described in the text.

b) Investigate some more machines which rattle or vibrate at particular frequencies like the car or the washing machine. What is the source of the vibration in each case?

Find out how car manufacturers reduce the vibrations and rattles in cars.

Try questions 20.1 to 20.5

20.2
WAVES ON A SPRING

- *Stretch out a slinky spring along a smooth floor or bench top allowing plenty of room on either side for movement of the spring.*
- *Holding one end fixed, generate waves by shaking the other end as shown in fig. 20.7.*

We use the words pulse and wavetrain to distinguish between a single wave motion and a continuous group of waves.

A **pulse** is a short-lived or single wave motion.

A **wavetrain** is a continous group of waves with features which repeat regularly.

Several different types of wave motion are shown in fig. 20.7.

a) *A single transverse pulse* is produced by a quick flick of the hand, a to-and-fro sideways movement at right angles to the slinky spring.

b) *A single longitudinal pulse* is produced by a quick jerk of the hand forwards and backwards along the line of the slinky.

c) *A transverse wavetrain* is produced by swinging the hand to-and-fro sideways (at right angles to the line of the slinky) at a constant frequency of about 4 hertz. If the slinky is long enough the waves can be seen to travel continuously in one direction.

d) *A longitudinal wavetrain* can be seen to travel along the slinky if the hand is oscillated backwards and forwards (in line with the spring) several times at a constant rate.

While watching the waves try to answer these questions.

- *What happens to the amplitude of the pulses and waves as they travel along the slinky and why?*
- *What happens to the speed of the pulses and waves as they travel along the slinky, and is the speed affected by how much the spring is stretched?*
- *For each type of wave motion describe the motion of a single coil in the slinky. (This can be made easier by attaching a light pointer like a drinking straw to one of the coils.)*
- *What happens when pulses reach the fixed end?*
- *What actually travels along the slinky?*

Describing waves

The wave motions seen on a slinky are made up of vibrations of its individual coils; this is a **mechanical wave**. All mechanical waves require a medium or material to travel through and they cause the individual particles of that medium to vibrate or oscillate. We can begin to describe a wave motion by noting what happens to the particles which it causes to vibrate.

Two kinds of waves

There are two main kinds of wave motion which are called transverse and longitudinal:

Transverse waves are ones in which the displacement of the particles is at right angles to the direction of travel of the wave motion.

Transverse waves can be recognised by their crests and troughs.

Figure 20.7   *Wave motions on a slinky*

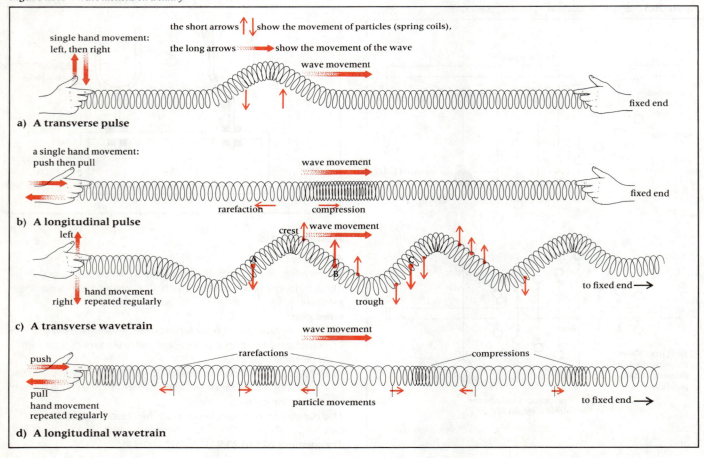

the short arrows ↑↓ show the movement of particles (spring coils),

the long arrows ➡ show the movement of the wave

single hand movement:
left, then right

wave movement

fixed end

a) **A transverse pulse**

a single hand movement:
push then pull

wave movement

rarefaction   compression

fixed end

b) **A longitudinal pulse**

left

crest   wave movement

hand movement
right   repeated regularly

trough

to fixed end ➡

c) **A transverse wavetrain**

wave movement

push

rarefactions   compressions

pull
hand movement
repeated regularly

particle movements

to fixed end ➡

d) **A longitudinal wavetrain**

Longitudinal waves *are ones in which the displacement of the particles is in line with or parallel to the direction of travel of the wave motion.*

Longitudinal waves can be recognised by their compressions and rarefactions. (In compressions the particles are pushed together and in rarefactions they are pulled apart.)

Progressive or travelling waves

In all the examples of wave motions so far described, both transverse and longitudinal, there appears to be something moving or travelling with the waves; but what is actually moving? We can say the following about these waves:

A progressive or travelling wave is the movement of a disturbance which carries energy away from a source.

These waves have the following important features:
a) A progressive or travelling wave carries energy.
b) The medium or material through which a wave travels does not travel with the wave (but there are certain exceptions).
c) The particles of the medium, which are displaced by the wave motion, vibrate about their rest positions, but do not travel with the wave.
d) Each particle in the wave motion vibrates in the same way, but the vibrations have a time lag in the direction of travel of the wave.
e) The shape of the wavetrain or pulse stays the same as it travels through a medium, but its amplitude gets smaller as the energy is lost or the waves spread out.
f) The speed of a wave is not affected by the shape of the waves or their amplitude, but it is affected by the nature of the medium it travels through.

Displacement – position graphs

If we take a photograph of a transverse wavetrain, the instantaneous picture shows the displacement of the particles along the wavetrain at a single moment. The photograph gives a displacement against position graph, fig. 20.8a. (Although this graph looks like a transverse wavetrain it can be used to show the displacement of particles in both transverse and longitudinal waves. The displacement can be the distance the particles are moved from their rest positions either at right angles to, or parallel to, the direction of travel of the wave motion.)

The **wavelength** λ of a transverse wave is most easily understood as the distance between two successive crests or between two successive troughs (successive crests follow next to each other). Similarly the wavelength of a longitudinal wave is seen as the distance between two successive compressions or two successive rarefactions. We usually define wavelength in a way which applies to both transverse and longitudinal waves: (The symbol for wavelength is a Greek letter 'l', written λ and called lambda).

The wavelength λ is the distance between two successive particles which are at exactly the same point in their paths at the same time and are moving in the same direction.

For example: In fig. 20.7c and 20.8a, particles at points A, B and C are at the midpoints in their paths, but only the particles at A and C are moving in the same direction. The wavelength = AC.

The particles in a wavetrain which are a wavelength apart move in step with each other and are said to be in phase with each other.

Figure 20.8 *Graphs of wavetrains*

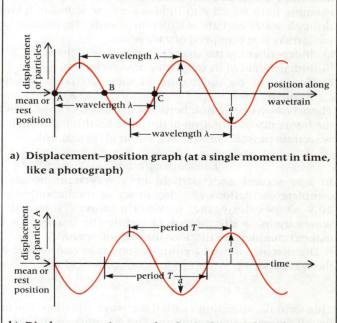

a) **Displacement–position graph (at a single moment in time, like a photograph)**

b) **Displacement–time graph (of a single particle in the wave)**

Displacement – time graphs

Now we look at the motion of a single particle as a wavetrain passes. Fig. 20.8b shows how the displacement of a particle changes as a wavetrain passes, giving a displacement against time graph for the particle. It can be compared with the graph obtained for a single vibrator (the broomstick pendulum of fig. 20.3) and we see that the motion of a particle in a wavetrain is exactly the same as the motion of a single vibrating body. In particular the motion is repeated regularly with a constant period T and has the same characteristic sine wave shape. Now we can see the important connection between vibrations and waves. It follows that the definitions of quantities such as the period, frequency and amplitude of a wavetrain will be the same as for a single vibrator except that we are now describing the motion of a *particle in the wavetrain*.

For a wavetrain:

The period T is the time a particle in the wavetrain takes to make one complete oscillation. (T is measured in seconds.)

The frequency f is the number of complete oscillations made in one second by a particle in the wavetrain. (f is measured in hertz.)

The amplitude a is the maximum displacement of the particles in the wavetrain from their rest positions. (The amplitude of both graphs is the same.)

Wave speed, *c*

Observations of pulses and wavetrains travelling along a slinky suggest that the shape and amplitude of a wave do not affect its speed. When we listen to an orchestra the waves of high-frequency sounds and low-frequency sounds arrive at the listener together, having travelled at the same speed. But waves which travel through different materials are found to travel at different speeds. For example both sound and light waves are found to travel through water and air at different speeds. To summarise we can say that the speed of a wave is:

a) independent of the shape or amplitude of the wave,
b) independent of its frequency or wavelength,
c) dependent on the nature of the material it travels through.

There is, however, a link between the speed of a wave and the frequency of vibration of the particles through which a wavetrain passes. The wave equation gives that link.

The wave equation

In one second each particle in a wavetrain makes *f* complete oscillations (*f* = frequency of oscillation). Fig. 20.9 shows how the wavetrain moves forward *f* wavelengths, a distance of *f*λ, while the particle at A makes *f* complete oscillations during one second.

The distance moved by the wavetrain in one second is the wave speed *c*:

wave speed = frequency × wavelength   or $c = f\lambda$

This formula, sometimes called the wave equation, is true for all types of wave motion. The SI units of the three quantities are: *c* in metres per second (m/s), *f* in hertz (Hz) and λ in metres (m).

Worked Example
Wave speed

A hand displaces a slinky to-and-fro at a frequency of 3 hertz. If the distance between successive crests of the wavetrain is 0.8 metre, calculate the speed of the waves along the slinky.

We have $f = 3\,\text{Hz}$ and $\lambda = 0.8\,\text{m}$, using $c = f\lambda$ gives

$$c = 3\,\text{Hz} \times 0.8\,\text{m} = 2.4\,\text{m Hz} = 2.4\,\text{m/s}$$

Answer: The wave speed *c* is 2.4 metres per second.

Figure 20.9 *The wave equation*

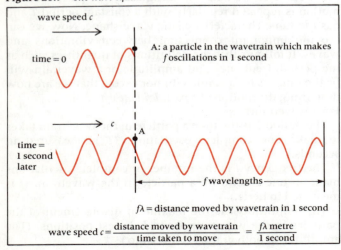

wave speed *c*

time = 0

A: a particle in the wavetrain which makes *f* oscillations in 1 second

c

time = 1 second later

A

f wavelengths

*f*λ = distance moved by wavetrain in 1 second

$$\text{wave speed } c = \frac{\text{distance moved by wavetrain}}{\text{time taken to move}} = \frac{f\lambda \text{ metre}}{1 \text{ second}}$$

Examples of progressive or travelling waves

We shall investigate water waves next because being two-dimensional, (i.e. spread over a surface) they can show more properties of waves than can be seen on a slinky. Water waves can be considered to be transverse waves, because that is what they look like, and floating objects like corks bob up and down on the surface as a wave passes. (In fact drops of water actually roll round in circles under the surface of water waves but this motion is too complicated to be described in this book.)

Light also travels in transverse progressive wavetrains but, as part of the electromagnetic spectrum (chapter 23), it is not a mechanical wave and requires no medium to travel through. Light, like all the other waves in the electromagnetic spectrum, is believed to involve vibrating electric and magnetic fields which travel along at right angles both to each other and to the direction of travel of the waves. Magnetic and electric fields can exist in a vacuum and require no medium. Light waves are discussed in chapter 22.

The most common example of longitudinal waves is provided by sound as it travels through the air producing compressions and rarefactions of the gas molecules. The production and properties of sound waves is the subject of the next chapter.

Assignments

Remember

a) The meanings of the words: pulse, wavetrain, progressive or travelling wave, transverse and longitudinal waves, compression, rarefaction, wavelength.
b) The formula $c = f\lambda$

Test the wave equation

Using a stopwatch and a slinky (or 4 metres or more of flexible rubber or plastic tubing or rope), estimate:

c) the frequency of the oscillations of your hand,
d) the wavelength of the waves generated,
e) the wave speed.
f) Check whether your results agree with the wave equation.

Find out

Name some more examples of wave motions and find out:

g) whether they involve single pulses or wavetrains,
h) whether the displacement is transverse or longitudinal
i) whether it is mechanical (requiring a medium) or electromagnetic (requiring no medium and able to travel through space).

Name a musical instrument and find out:

j) the name of the part of it that vibrates,
k) how it is made to vibrate and
l) how the frequency of the vibration is changed for different notes
m) whether the vibration is transverse or longitudinal.

Try questions 20.6 to 20.8

20.3
WAVES ON WATER

Some of the properties of water waves can be seen on a pond, river, or even in the bath; but accurate observations are more easily made in a ripple tank. What we discover about water waves helps to explain the nature of other types of waves like sound waves and light waves.

Setting up a ripple tank

The tank shown in fig. 20.10 is like a glass table on legs with an edge round the glass top so that it can hold a shallow layer of water. A lamp positioned above the table casts shadows of the water waves through the glass onto a white sheet of paper or board on the floor below. We look at the wave shadows on the floor, rather than the actual water waves in the tank. These shadows show the shapes of the wavefronts.

● *Close the clip on the outlet tube.*
● *Fill the tank with water to a depth of about 5 mm.*
● *Connect the lamp to a suitable supply, usually a 12 V bench supply or transformer.*
● *To level the tank, look at the two reflections of the lamp, one from the water surface and one from the glass surface, and put wedges under the feet of the tank until the two images of the lamp coincide.*
● *Adjust the height and position of the lamp to centre the wave picture on the white sheet on the floor.*

A suitable height is about 50 cm above the tank, but lowering the lamp will broaden the picture if necessary.

● *Fit pieces of metal gauze or sponge around the edges of the tank.*

These act as a 'beach' to scatter or absorb the waves at the edges of the tank, which reduces unwanted reflections from the sides.

● *Now that the tank is ready for use, darken the room.*

Figure 20.10   *A ripple tank*

Investigating single wave pulses

The simplest experiments with water waves can be done by making single circular or straight wave pulses.

● *To make circular wave pulses try dipping a finger or pencil into the water, but avoid touching the glass bottom and shaking the tank. You can also try dropping water drops from a dropping pipette.*
● *To make straight waves, lay a length of thick dowel rod in the water and give it a quick forwards roll.*

It is worth investigating some of the reflection experiments with these single wave pulses before going on to use continuous wavetrains produced by a vibrator.

Generating continous wavetrains

Although continous wavetrains can be produced by repeatedly dipping in a pencil or rocking a dowel rod in a regular way, it is much easier to use a mechanical vibrator to generate continous wavetrains.

● *Hang a small electric motor mounted on a wooden beam by rubber bands from the support above the ripple tank as shown in fig. 20.11a.*

An eccentric (off-centre) metal disc on the axle of the motor causes the beam to vibrate at the same frequency as the revolutions of the motor. The speed of the motor is controlled by a rheostat in series with a low-voltage d.c. supply, or a dry battery (usually 3 or 4.5 V).

● *To generate continous straight waves adjust the height of the vibrating beam so that it just touches the water surface.*

Smoother waves are produced if the surface of the wooden beam is rubbed in the water so as to wet it thoroughly and evenly (fig. 20.11b).

● *To generate continous circular waves fit a small spherical dipper to the beam and raise it so that the dipper just touches the water surface (fig. 20.11c).*

Figure 20.11   *Making continuous wavetrains*

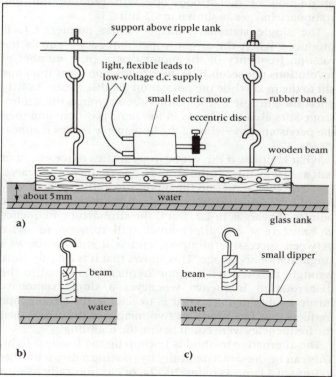

- *Adjust the speed of the motor to vary the frequency and wavelength of the continuous wavetrains.*

The progressive waves move quite quickly across the water surface and any attempt to measure wavelength or to draw wave diagrams from the moving wave shadows on the floor would be quite difficult. A standard method of 'freezing' or slowing down rapid motion is to use a stroboscope.

- *Use a hand-held stroboscope, if available, to view the wave patterns. Alternatively, mount a motor-driven stroboscope just below the lamp so that the ripple tank is illuminated by regular flashes of light instead of the steady light.*

The stroboscope principle

Persistence of vision causes our eyes to merge images, so that we do not notice the gaps between the individual frames of a movie picture or between the scans of the electron beam on a television picture. A stroboscope chops up the picture we see so that our eyes receive separate glimpses of the picture at regular intervals.

When we use a stroboscope to view something which has a regular repetitive motion, if the frequency of the glimpses matches the frequency of the repetitive motion then the motion will appear to stop.

The stroboscope effect can be produced either by chopping up the light illuminating the moving object, fig. 20.12a, or by chopping up the light reflected from it (which amounts to chopping up the eye's view), fig. 20.12b. This figure shows an effective way of demonstrating the stroboscope effect. A black disc with a white arrow glued or painted on it is rotated by an electric motor. Without a stroboscope, the disc appears blurred. If the disc with the arrow is now viewed through a hand-held stroboscope (b), as the speed of rotation of the stroboscope is increased a point is reached where a stationary image of the pointer on the motor disc is seen. Increasing the speed of rotation of the stroboscope will produce multiple stationary images as shown in C2 and C3.

The single stationary image of the pointer, C1, is produced when the stroboscope frequency f equals the rotation frequency of the motor disc n (n = number of revolutions per second). If the stroboscope turns from one slit to the next while the pointer on the disc rotates exactly once, the glimpse of the pointer seen through the stroboscope slits always appears in the same position; and then the persistence of vision of the human eye makes it appear steady.

What happens if the stroboscope rotates twice as fast or half as fast as this? If the stroboscope frequency is *doubled* ($f = 2n$), the pointer on the motor disc only gets half way round between successive glimpses, so two images of the pointer appear, as in C2. But if the stroboscope frequency is *halved* ($f = \frac{1}{2}n$), the pointer will turn round twice between successive glimpses, and will still appear as a single stationary image. This shows that it is easy to think wrongly that the stroboscope frequency is equal to the disc rotation frequency whenever a single stationary image is obtained. The test is to double the stroboscope frequency. If this produces two images of the arrow then the frequencies were equal before the doubling.

The alternative method is to chop up the incident light. This can be done mechanically by rotating a disc with slits in front of a lamp as in fig. 20.12a, or electronically using a

xenon flash tube. In this case, if the light flashes at the same frequency as the pointer rotation frequency, the pointer will always be illuminated in the same position and hence appear stationary.

Stroboscopes are used in several areas of physics, for example to study motion, but here they can be used to 'freeze' water waves and later, sound waves on strings.

Figure 20.12 *Demonstrating the stroboscope principle*

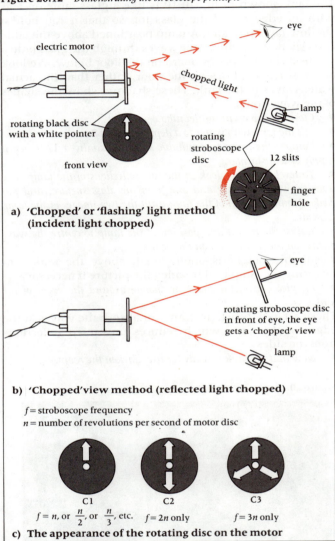

a) **'Chopped' or 'flashing' light method (incident light chopped)**

b) **'Chopped' view method (reflected light chopped)**

f = stroboscope frequency
n = number of revolutions per second of motor disc

C1   C2   C3

$f = n$, or $\frac{n}{2}$, or $\frac{n}{3}$, etc. $f = 2n$ only $f = 3n$ only

c) **The appearance of the rotating disc on the motor**

Wave diagrams and wavefronts

When recording observations from ripple tank experiments, or when answering questions about the properties of waves, it is helpful to know what a wave diagram can and should show and how to construct it correctly. Even sketch diagrams must give the correct information.

Wave diagrams show the positions of wavefronts at a particular instant; they are like a photograph of the wave shadows seen on the screen of the ripple tank. The shadow patterns can represent the positions of wavefronts.

A wavefront is an imaginary line which joins a set of particles which are in phase (in step) in a wave motion.

For example: all the particles along a crest of a wave are in phase and can be considered as a wavefront. In effect, when we draw a wavefront we draw the shape of the wave as seen from above, i.e. a plan view. The shape of the wave seen from the side would be a wave profile.

Points to remember when drawing wave diagrams.

a) In water of a constant depth the waves travel at a constant speed, therefore we draw the wavefronts equally spaced and parallel.

b) The direction of travel of the waves, which should be shown, is always at right angles to the wavefronts.

c) When drawing wave diagrams of reflection or refraction it helps to bear in mind the equivalent light-ray diagrams given in chapters 1 to 3, because, as we shall see, the direction of travel of the waves obeys the same laws as the direction of light rays.

d) Sources of circular waves and their images formed by reflection should be labelled, as should any focus or centre of curvature.

Reflection of straight and circular wavefronts

● *Using first the straight beam to produce parallel straight waves, then a single dipper to produce circular waves, investigate reflection by a straight metal strip acting as a plane reflector. Then repeat your investigations using a curved metal strip as both a concave and a convex reflector.*

● *Note the shape, spacing, speed and direction of travel of the reflected wavefronts and draw wave diagrams as explained above.*

If you are using a stroboscope it should be possible to measure roughly the wavelengths and distances of sources from the reflectors. Figs. 20.13 to 20.17 show the more important results that can be obtained. In all the results the incident and reflected waves can be seen to travel at the same speed and to obey the laws of reflection. The similarities between the reflection of light rays and the reflection of waves is pointed out in the caption to each figure. Bold arrows are used to represent the speed of the waves and all have the same length when the speed does not change.

Figure 20.14 *Circular waves reflected by a plane reflector (some wavefronts are omitted for clarity)*

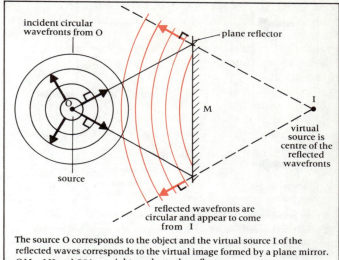

The source O corresponds to the object and the virtual source I of the reflected waves corresponds to the virtual image formed by a plane mirror. OM = MI and OI is at right-angles to the reflector

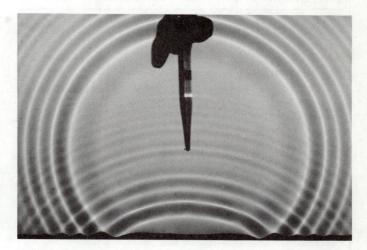

Figure 20.13 *Reflection of straight waves by a plane reflector compared with reflection of a light ray*

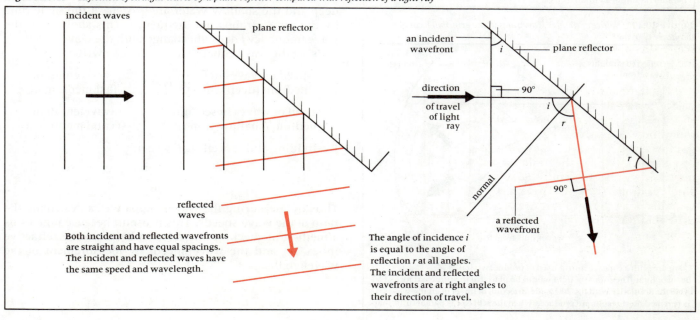

Both incident and reflected wavefronts are straight and have equal spacings. The incident and reflected waves have the same speed and wavelength.

The angle of incidence *i* is equal to the angle of reflection *r* at all angles. The incident and reflected wavefronts are at right angles to their direction of travel.

Figure 20.15 *Straight waves reflected by a concave reflector*

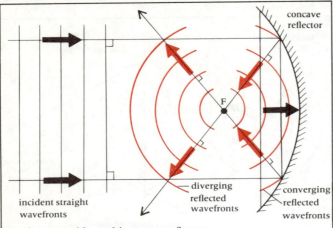

F = the principal focus of the concave reflector.

The reflected wavefronts are circular, converging towards and passing through the principal focus F. Compare with a concave mirror converging parallel rays of light.

Figure 20.16 *Straight waves reflected by a convex reflector*

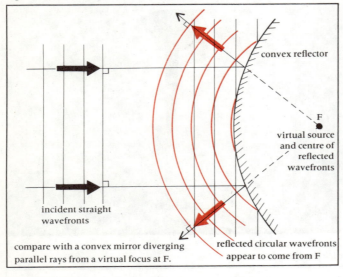

compare with a convex mirror diverging parallel rays from a virtual focus at F.

Figure 20.17 *Source of circular wavefronts at the principal focus of a concave reflector*

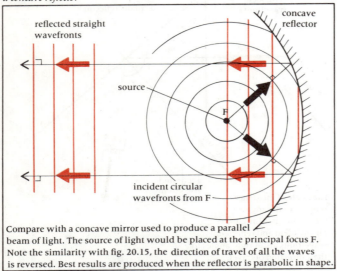

Compare with a concave mirror used to produce a parallel beam of light. The source of light would be placed at the principal focus F. Note the similarity with fig. 20.15, the direction of travel of all the waves is reversed. Best results are produced when the reflector is parabolic in shape.

Refraction of waves by a plane boundary

● *Lay a thick glass plate flat in the ripple tank, separated from the glass base by metal washers. (Without the washers it is difficult to remove the glass plate.)*

● *Add more water to the tank until it just covers the glass plate.*

A drop of detergent added to the water will reduce unwanted surface tension effects. It is important to check that the tank is level. The plate can be arranged so that straight wavefronts arrive parallel to the edge of the glass plate, fig. 20.18, or at an angle of incidence *i*, fig. 20.19. A stroboscope is particularly useful in these experiments because by freezing the waves it is possible to make measurements.

● *Measure the wavelength in shallow water λ_1 and in deep water λ_2 (measure over several wavelengths).*

● *Measure the angles of incidence i and refraction r.*

● *From each of these two measurements calculate the refractive index and compare the values.*

In fig. 20.18 the waves can be seen to travel more slowly in the shallow water. In addition the waves in both the deep and the shallow water are 'frozen' at the same time by the same stroboscope frequency. This shows that the wave frequency *f* does not change with the wave speed.

Using the wave equation $c = f\lambda$ we can write:

speed of waves in first medium (deep water) $c_1 = f \times$ wavelength in first medium λ_1

speed of waves in second medium (shallow water) $c_2 = f \times$ wavelength in second medium λ_2

By dividing and cancelling *f* we get:

$$\frac{c_1}{c_2} = \frac{\lambda_1}{\lambda_2}$$

Thus by measuring the wavelengths we can calculate the ratio of the wave speeds. This is useful because it gives us a method of checking the relation between the refractive index $_1n_2$ and the wave speed, by measurement of the wavelengths.

$$\text{refractive index } _1n_2 = \frac{c_1}{c_2} = \frac{\lambda_1}{\lambda_2}$$

• *No refraction occurs for waves incident normally at a boundary (wavefronts parallel to the boundary), so to investigate the relation between refraction and wave speed, turn the glass plate to give an oblique boundary as shown in fig. 20.19. In fig. 20.19, i is the angle of incidence, and r is angle of refraction.*

When the wavefronts arrive at an angle i to the boundary, the change in wave speed can be seen to cause a change of direction as well.

• *Measure the angles i and r.*
• *Calculate the refractive index $_1n_2$ for the two media using*

$$_1n_2 = \frac{\sin i \text{ (in medium 1)}}{\sin r \text{ (in medium 2)}}$$

The value for the refractive index calculated from the wavelengths agrees with the value obtained from the sines of the angles.

Figure 20.18 *Straight wavefronts passing from deep water to shallow water*

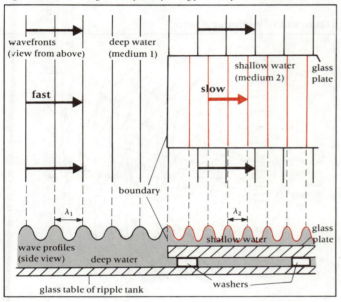

Figure 20.19 *Refraction of straight wavefronts at a plane boundary*

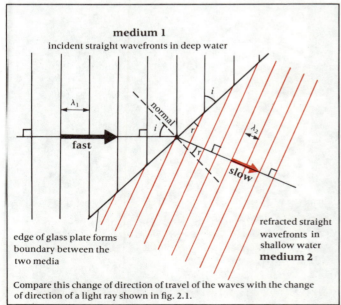

Refraction of waves by a curved boundary

The converging action of a bi-convex lens can be studied by using a bi-convex plate of glass or perspex in the same manner as the straight-edged glass plate.

• *Investigate the converging of straight waves by the lens as shown in fig. 20.20, and the formation of an image of a point source, as shown in fig. 20.21.*

Figure 20.20 *Refraction of straight wavefronts by a converging 'lens'*

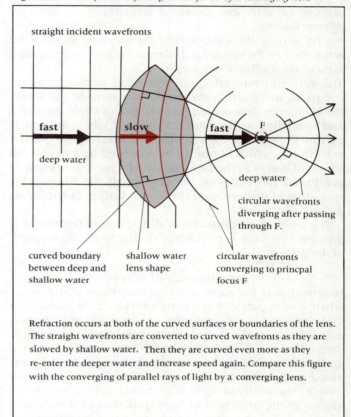

Refraction occurs at both of the curved surfaces or boundaries of the lens. The straight wavefronts are converted to curved wavefronts as they are slowed by shallow water. Then they are curved even more as they re-enter the deeper water and increase speed again. Compare this figure with the converging of parallel rays of light by a converging lens.

Figure 20.21 *A wave model of how a converging lens forms an image of a point object*

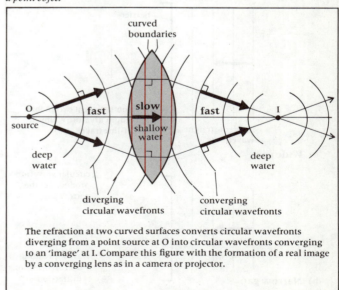

The refraction at two curved surfaces converts circular wavefronts diverging from a point source at O into circular wavefronts converging to an 'image' at I. Compare this figure with the formation of a real image by a converging lens as in a camera or projector.

Diffraction of waves

We shall now investigate some special effects thought to belong only to waves. The effects observed now with water waves will also be demonstrated with sound waves and light waves in the next two chapters.

● *Set up the ripple tank to produce continuous straight waves.*

● *Place two straight metal barriers in the tank parallel to the wavefronts as shown in fig. 20.22.*

● *Try varying the gap or aperture between the two barriers and compare the shape and amount of spreading of the waves for different widths.*

● *Try varying the wavelength of the waves by adjusting the motor speed on the vibrator and compare the amount of spreading of the waves at different wavelengths.*

The very different effects of wide and narrow gaps are compared in fig. 20.22. When the gap is wide the wavefronts emerge almost straight, apart from a slight curvature and spreading at the edges. However, when the gap is narrow the straight wavefronts are converted into circular wavefronts, which appear to be produced by a new point source of waves in the gap. The circular wavefronts spread out round the edges of the gap in all directions. This spreading of waves round corners and edges of barriers is called **diffraction**.

The amount of spreading or diffraction of the waves is greatest when the gap width is similar to the wavelength of the waves.

Interference of waves

Interference is the name given to the effects which occur when two separate wavetrains overlap. It is interesting that waves usually do not seem to bump into one another, rather they pass through each other and merge or combine their effects. For example the different sounds from a group of instruments played together can be heard combined and merged; the various sound waves do not collide. The ability of wave motions to combine together is known as the **principle of superposition of waves**, and what happens when they combine is called **interference**.

To investigate the effects known as interference of waves we superpose (overlap) two separate wavetrains produced by two sources of circular waves.

● *Fit two dippers, about 3 cm apart, to the wooden beam and raise it so that the dippers just touch the surface of the water.*

● *Run the vibrator slowly and observe what happens where the two sets of circular waves overlap.*

● *Increase the speed of the vibrator and note the effect.*

At high speeds the use of a stroboscope is helpful in studying the interference patterns by keeping the wave positions steady.

● *What happens where the crest of one wave overlaps the crest of another wave? (The waves are in phase or in step.)*

● *What happens where the crest of one wave overlaps the trough of another wave? (The waves are in antiphase, or out of step by half an oscillation.)*

The wavefronts drawn in fig. 20.23 show the positions of the wavecrests only.

Figure 20.23 *Interference of waves*

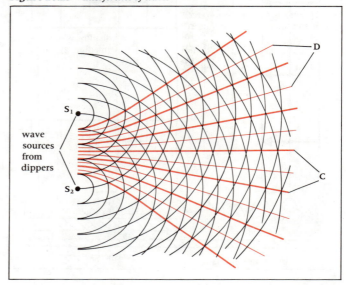

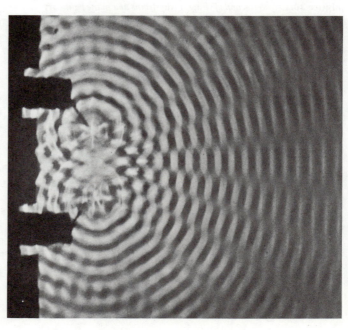

Figure 20.22 *Diffraction of straight wavefronts*

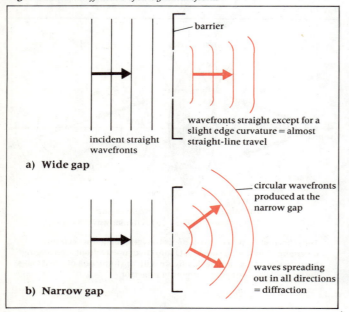

Lines of destructive interference. The water is quite still along several lines, marked D in the figure. These are where the waves are in antiphase. In effect the crest of one wave has filled in the trough of the other to produce no displacement of the water, as shown in fig. 20.24. This effect is called **destructive interference;** the two wave motions have destroyed each other along the lines marked D.

Lines of constructive interference. The amplitude of the disturbance has increased along other lines, marked C, where the waves are in phase. Here the displacement at the crest of one wave has been added to the displacement at the crest of the other wave to produce a larger displacement. This effect, (also shown in fig. 20.24) is called **constructive interference,** the two waves together having constructed a larger amplitude wave. These observations lead us to a statement of the **principle of superposition of waves:**

The displacement of any particle caused by overlapping waves is the sum of the separate displacements caused by each wave at a particular moment.

Figure 20.24   *Superposition of waves*

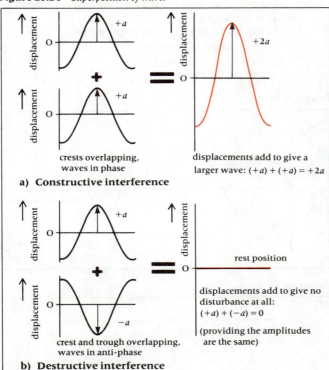

crests overlapping, waves in phase

displacements add to give a larger wave: $(+a) + (+a) = +2a$

a) Constructive interference

crest and trough overlapping, waves in anti-phase

displacements add to give no disturbance at all: $(+a) + (-a) = 0$

(providing the amplitudes are the same)

b) Destructive interference

Double-slit interference

The two dippers may be replaced by two gaps or slits about 1 cm wide and 3 cm apart in a plane barrier. When straight wavefronts reach the two gaps, circular waves emerge on the other side. Where these two sets of waves are superposed the same interference effects are produced as occurred with the two dippers.

With this arrangement it is more difficult to produce good results but it is important because it provides a water wave model of the double-slit experiment in light, known as 'Young's slits' (p 475).

Assignments

Remember
a) The meanings of the words: diffraction, interference, superposition.
b) The conditions under which the following occur: diffraction, constructive interference and destructive interference.

Investigate
Use a stroboscope on some of the following. You will find that different types of stroboscopes and different frequencies are needed for each example. Can you explain why?
c) The back wheel of an inverted bicycle turned at a constant speed.
d) The blades of a fan or the chuck of an electric drill.
e) A record player turntable.
f) A television screen or a film screen from a movie projector.
g) A fluorescent or neon lamp connected to the mains.

Try questions 20.10 to 20.19

Questions 20

1   A student counts his heartbeats and finds that it makes 75 beats in 1 minute. Calculate (a) the frequency of his heartbeat in hertz and (b) the period of his heartbeat.

2   Draw a simple diagram of a child's swing and on it indicate the amplitude of the oscillation.

3   Explain how you would demonstrate the difference between a natural oscillation and a forced oscillation using a swing.

4   Three pendulums P, Q and R hang from the same stretched rubber tube, as shown in fig. 20.25. P and Q are of equal length, but the mass of the bob of P is twice the mass of the bob of Q. R is somewhat shorter and has a bob of much smaller mass.
   a) With reference to a single simple pendulum, explain the meaning of amplitude and frequency and indicate what they depend on.
   b) Pendulum P is set swinging at right angles to the line of the rubber tubing with an amplitude of 110 mm. Describe carefully what happens to P, Q and R as time passes.
   c) After some time, no part of the system is moving. Explain why, in terms of energy and energy conservation.
   d) With reference to the pendulums in the system, explain what is meant by resonance and forced vibration. Comment on the importance of these concepts in the design of a spindrier.   (O)

Figure 20.25

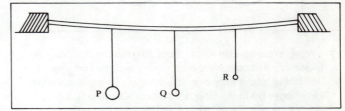

5   This question is about things which oscillate. Fig. 20.26a shows two springs connected together. Hanging from the springs are a number of masses. The masses are set oscillating (bouncing). They oscillate between the points marked A and B.
   a) Which of the following is the correct description of one complete oscillation of the masses? i) From A to the middle. ii) From A to B. iii) From A to B and back to A.
   b) There is no other apparatus for you to use in this experiment but you are asked to speed up the oscillations so that the masses complete one oscillation in less time. Give two ways in which you could alter the equipment in order to do this.

Figure 20.26

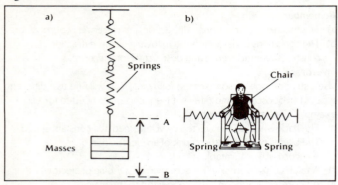

c) A simple description of an oscillation is 'a movement which repeats itself over and over again'. Give two other examples of oscillating systems, not necessarily involving springs. Your examples could include things you have seen inside or outside your school laboratory.

d) Astronauts in space appear to be weightless. However it is important for their health that they keep a daily check on their body mass. Explain fully why a set of bathroom scales would be useless for the job.

e) In fact astronauts use a special chair with springs attached to each side as shown in Fig. 20.26b. They also need a sensitive electronic clock. Describe how you think the astronauts could use this equipment to find out if their body mass changes in the course of the flight. Make sure you say what measurements the astronauts would have to take. (NEA [B] SPEC)

6 Fig. 20.27 shows a wave travelling along a length of rope.
  a) Using the scales on the given axes determine i) the amplitude of the wave, ii) the wavelength of the wave.
  b) If the frequency of the wave is 8 Hz, calculate the speed of the wave in **m/s**. (W 88)

Figure 20.27

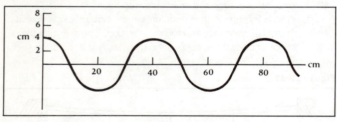

7 Waves enter a harbour at a rate of 30 crests per minute. A girl watches a particular wave crest passing two posts which are 6 metres apart along the direction of travel of the waves; the time from one post to the other is 2 seconds. Calculate (a) the frequency of the wave-motion and (b) the wavelength of the waves. (Nuffield)

8 a) Draw a series of simple diagrams to show how a vibrating tuning fork produces one cycle of its fundamental note.
  b) State the relation between the velocity of a wave, its frequency and its wavelength.
  c) The frequency of the note produced by a particular tuning fork is 320 Hz. Calculate the wavelength of the wave produced in the surrounding air if the speed of sound in air is 340 m/s.
  d) If the sound from the tuning fork in (c) was detected under water, what frequency would be heard? If the wavelength in water is 4.5 m, at what speed does the wave travel through the water? (JMB, part)

9 A vibrator sends 8 ripples per second across a water tank. The ripples are observed to be 4 cm apart. Calculate the velocity of the ripples.

10 A horizontal, white disc has an arrow marked on it, as shown in fig. 20.28. It rotates about a vertical axis through its centre. The frequency of rotation is 100 Hz. Sketch and explain the appearance of the rotating disc when it is illuminated by a stroboscope set at a frequency of 300 Hz. (O & C)

Figure 20.28

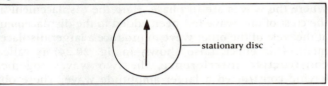

11 Fig. 20.29 illustrates water waves in a ripple tank.
  a) Redraw the figure and on it label clearly
    i) the amplitude of the wave,
    ii) the wavelength of the wave.
  b) State the relationship between the wavelength, frequency and the velocity of the wave.
  c) Calculate the frequency of the wave if the velocity is 30 cm/s and the wavelength is 2 cm.
  d) For a sound wave in air what effect is heard if
    i) the wavelength is decreased,
    ii) the frequency is increased,
    iii) the amplitude is increased? (Joint 16+)

Figure 20.29

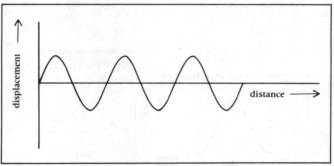

Figure 20.30

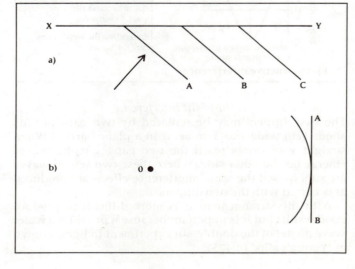

12 a) Fig. 20.30a is an incomplete diagram which shows three
       successive straight waves A, B, and C on water, as they are
       being reflected at a straight barrier XY. Wave-front A is just
       about to be reflected while B and C have already been partly
       reflected.
       Complete Fig. 20.30a showing the positions of the reflected
       parts of the wave-fronts B and C.
    b) Fig. 20.30b is an incomplete diagram which shows a circular
       wave-front originating at O just before reflection at a straight
       barrier AB.
       Complete it by drawing the wave-front as it would be just after
       reflection is complete. Indicate on the diagram where the
       reflected wave-front appears to come from.
    c) If the wavelength of an incident wave is 1.5 cm and the
       frequency of the source at O is 10 Hz, calculate
         i) the wavelength of the reflected wave,
        ii) the speed of the waves over the water.   (NEA [A] SPEC)

Figure 20.31

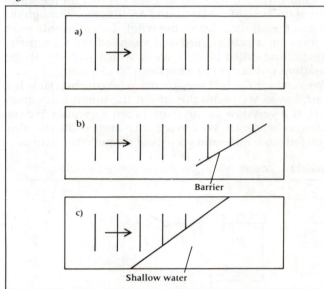

Barrier

Shallow water

13 Fig. 20.31a represents water waves moving from left to right in a
    ripple tank.
    A straight barrier is placed across the ripple tank as shown in
    Fig. 20.31b.
    a) Copy and complete this diagram, showing on it the position of

Figure 20.32

barrier —

i)

barrier —

ii)

the waves after they have met the barrier.
The barrier is removed and a sheet of clear plastic is placed on the
bottom of the ripple tank so as to make a region of shallow water
as shown in fig. 20.31c.
    b) Copy and complete this diagram by including on it the waves in
       the shallow water.
    c) Explain the behaviour of the waves in the shallow water.
                                                         (NEA [B] SPEC)

14 The diagrams in fig. 20.32 show waves in a ripple tank approaching
    (i) a narrow gap in a barrier, and (ii) a wide gap in a barrier. Redraw
    the diagrams and show clearly on them the subsequent
    appearance of the waves.   (L)

15 a) Describe an experiment to show constructive and destructive
       interference of water waves passing through a double slit
       arrangement.
       Your answer should include
         i) a labelled sketch of the apparatus,
        ii) a description of your procedure,
       iii) a plan diagram of the ripple tank showing the wave pattern
            which your experiment would produce, indicating clearly
            where constructive and destructive interference has taken
            place.
    b) With the aid of two clearly labelled diagrams (one for each),
       explain how constructive interference and destructive
       interference was produced in the experiment described in
       part (a).
    c) State the effect on the wave pattern you have drawn in (a)
       (iii) when
         i) the frequency of the source of the waves is increased, the
            slit separation remaining unchanged,
        ii) the separation of the slits is increased, the frequency of the
            source remaining unchanged.   (JMB)

16 Fig. 20.33 represents a plan view of a horizontal ripple tank. Two
    dippers, A and B, vibrate in phase with the same frequency. At a
    point P constructive interference is observed, and at a nearby
    point Q destructive interference is observed.
    a) On the same axes sketch two graphs of displacement against
       time for the vibration of the water at P, one for the waves from
       A, the other for the waves from B. On the same axes sketch a
       third graph showing the displacement of the water at P due to
       both sets of waves arriving at P together.
    b) In exactly the same way, sketch three graphs of displacement
       against time for the water at Q, one for waves from A, another
       for waves from B and the third for both sets of waves arriving
       at Q together.
    c) i) State a relationship between the distances AP and BP.
       ii) State a relationship between the distances AQ and
           BQ.   (JMB)

Figure 20.33

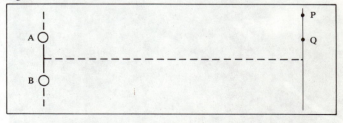

Sound waves

We live in a world filled with a great variety of sounds which affect us in many ways. Many kinds of vibrations send out waves which carry sound energy to our ears. What are these waves like?

21.1
TRAVELLING SOUND WAVES

We know that sound can travel through solids, liquids and gases. There is plenty of evidence of this, for example, sounds from distant sources reach our ears through the air and sounds can be heard through solid walls. Whales communicate by sound over great distances through the sea. So can sound travel through space or through a vacuum the way light does?

Vibrations in a vacuum

We can investigate this by mounting an electric bell or an alarm clock inside a large glass jar, called a bell jar, fig. 21.1. The bell should have a hammer which can be seen vibrating inside the jar, so that we can check whether the bell is working even if we cannot hear it. The bell jar is sealed with grease to a flat metal plate which has an outlet connected to a vacuum pump. We use this to remove the air from inside. If the bell assembly rested on the metal plate at the base of the jar, sound could escape through the plate to the air outside. So the whole bell assembly must be hung on a rubber cord and supplied with electricity through fine coiled flexible wires which will absorb the vibrations rather than passing them on to the air.

We can hear the bell very clearly when the bell jar is full of air, but as we pump the air out the sound fades away until, at a very low air pressure called a vacuum, we can no longer hear it all. We can see the hammer still vibrating but it fails to send sound waves out through the vacuum.

Figure 21.1 *Vibrations in a vacuum can be seen – but not heard*

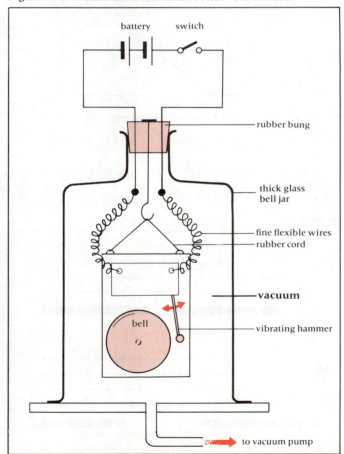

Sound travels as a longitudinal mechanical wave motion

Sound cannot travel through a vacuum. We cannot hear the bell in a vacuum because there is no medium for a sound wave to disturb. Similarly we cannot hear the nuclear explosions on the sun because there is no medium in space. Astronauts cannot speak to each other on the moon without using radio waves, because there is no air on the moon through which sound waves can travel.

When a vibrating object disturbs the surrounding medium the particles of the medium are displaced in the same direction as the resulting sound wave will travel. The particles of the medium are first pushed away by the vibration and then bounce back after collision with more particles further from the source; thus a longitudinal wave motion is established through the medium (p 439).

Sound is a longitudinal wave motion and, being mechanical, requires a medium to travel through.

The medium is simply whatever form of matter the vibrating source of sound is surrounded by or immersed in: it is usually air, but can be any gas, liquid or solid.

Evidently sound waves can travel and are therefore travelling or progressive waves, but, as we shall see in the next section, sound waves can also form another type of wave, called a stationary wave, which appears to stand still and not travel.

Sound waves in air

What happens when a sound wave travels through air? Fig. 21.2 shows how sound waves from a loudspeaker produce compressions and rarefactions of the (invisible) air molecules. When molecules pushed forwards (to the right in the figure) meet molecules bouncing backwards (to the left), after collisions with other molecules in front, a region of compression is produced where the air pressure is higher. In between the compressions are rarefactions where the number of molecules is reduced

and the air pressure is lower. Thus we may describe a progressive sound wave in air as a *travelling pressure wave* in which regions of increased air pressure travel along where the air molecules are compressed together separated by regions of reduced air pressure at the rarefactions. Fig. 21.2b shows how the pressure variations correspond to the compressions and rarefactions shown in (a).

The distance between two successive compressions or two pressure maxima is the wavelength λ of the longitudinal sound wave.

After a sound wave has passed through the air, the oscillating molecules continue to move about randomly, colliding with each other as usual in the same general location as they were before the wave. The travelling sound wave carries energy through the air without carrying the air molecules along with it.

Sound waves displayed on an oscilloscope

By using an oscilloscope to display the electrical oscillations produced by an audio-frequency generator we are able to make comparisons between what our ears hear and the waveforms producing the sound. In fig. 21.3 an audio-frequency generator supplies an oscillating electric voltage to both an oscilloscope and a loudspeaker at a frequency which our ears can hear (audio frequency). While the loudspeaker cone vibrates in and out producing longitudinal sound waves, the electron beam of the oscilloscope oscillates up and down drawing a transverse waveform on the screen at the same frequency.

To spread the waveform out on the screen the oscilloscope time-base must be switched on at a suitable speed to produce a time scale in the *x*-direction in which several up and down oscillations of the electron beam are traced out on the screen. The trace produced is a displacement against time graph of the electrical oscillations. Any change in the vibrations producing the sound can now be seen and heard simultaneously.

Figure 21.2 *Longitudinal sound waves travelling through air*

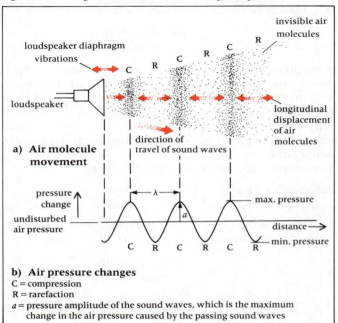

a) **Air molecule movement**

b) **Air pressure changes**
C = compression
R = rarefaction
a = pressure amplitude of the sound waves, which is the maximum change in the air pressure caused by the passing sound waves

Figure 21.3 *Hearing and 'seeing' sound waves*

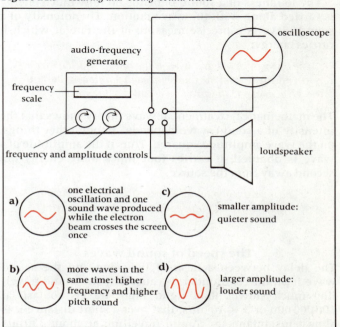

a) one electrical oscillation and one sound wave produced while the electron beam crosses the screen once

b) more waves in the same time: higher frequency and higher pitch sound

c) smaller amplitude: quieter sound

d) larger amplitude: louder sound

Frequency and pitch

The frequency of the electrical oscillations can be read from a scale on the generator. As the frequency is increased the number of waves produced in a certain time increases and we notice two effects: the number of waves displayed on the oscilloscope screen increases as in fig. 21.3b), and the pitch of the sound rises. We find a direct link between the wave frequency and the pitch of the sound: *the pitch of a sound depends on frequency.* (As we shall see later, the relation between the notes on a musical scale depends on the *ratio* of their frequencies rather than their actual frequencies. In this demonstration a doubling of the frequency raises the pitch by an octave, whatever the value of the frequency.)

Amplitude and loudness

The amplitude of the electrical oscillation can be changed and the effect simultaneously seen on the oscilloscope screen and heard from the loudspeaker. A smaller amplitude wave (c) sounds quieter and a larger amplitude wave (d) sounds louder. Thus we observe that: *the loudness of a sound depends on the wave amplitude.*

We also observe that changing the amplitude has no effect on the frequency or pitch of the sound. The relation between loudness and amplitude is, however, not a simple one.

Amplitude and intensity

The loudness of a sound depends on other factors as well as the wave amplitude. The human ear itself is not equally sensitive to all frequencies. So the perceived loudness of a sound (that is, how loud a person thinks a sound is) depends upon the response of a particular ear at that particular pitch. Loudness is therefore partly a subjective judgment (depending on a particular listener) of a particular sound. What one person thinks is too loud, another may find enjoyable.

The loudness of a sound depends on how much energy is carried along with the wave motion. The intensity of a sound wave is a precise measure of the rate at which it carries energy:

The intensity of a sound wave is the rate at which it carries energy away from the source through a unit area at right angles to the direction of travel of the wave.

The mathematical treatment of wave theory shows that the intensity of a sound wave depends, among other things, on the wave amplitude squared. Thus if the amplitude of a wave is doubled, it carries four times more energy per second away from the source.

The intensity of a sound wave is directly proportional to the wave amplitude squared; intensity $I \propto a^2$.

The speed of sound waves

The delay between a flash of lightning and the sound wave it produces, thunder, is caused by the sound travelling much slower than light. Light travels so fast, at $300\,000\,000$ or 3×10^8 m/s, that over a short distance it is almost instantaneous. Sound travelling at about 330 m/s

in air often takes several seconds to reach us from the location of the lightning. For example, a 3 second delay would mean that the sound had travelled 3×330 metres or about 1 kilometre, which is quite close in a thunder storm. Light travels 1 kilometre in about 3.3 microseconds (3.3×10^{-6} s), which is almost a million times quicker than sound.

Measuring the speed of sound by timing echoes

The accuracy with which the speed of sound can be measured depends mainly upon the timing accuracy, since the distance travelled can be measured easily with much greater precision.

For example, a distance of 100 metres probably can be measured by a tape measure to a precision of 1 centimetre, which gives an accuracy of 1 part in 10 000 (100 m = 10 000 cm), whereas the time for sound to travel the same distance (about 0.3 seconds) can at best be measured by stopwatch to a precision of $\frac{1}{100}$ second which, if correct, would only give an accuracy of 1 part in 30 ($\frac{1}{100}$ in $\frac{30}{100}$ of a second). Since stopwatch readings are unreliable due to human reaction times and only become more accurate when measuring relatively long times, attempting to measure 0.3 s by a hand-operated stopwatch would be very inaccurate indeed. How, then, can we improve the accuracy of a direct measurement of the speed of sound using a stopwatch? Essentially we need to measure a much longer time.

Using an echo method doubles the time and better still, using multiples of the echo travel time can greatly increase the total time measured. The following method uses these ideas.
- *Measure a distance of 100 m at right angles to a large wall. (There should be no other large reflecting surfaces nearby.)*
- *Make a sharp clapping sound by banging two blocks of wood together. Repeat the sound at regular time intervals to coincide exactly with the echoes.*
- *Starting at zero as a stopwatch is started, count the number of claps and stop the stopwatch at 50 or a 100 claps.*
- *Calculate the speed of sound using the formula*

$$\text{speed} = \frac{\text{distance travelled}}{\text{time taken}}$$

Remember that the distance is doubled, there and back, and the stopwatch reading must be divided by the number of claps counted. The value obtained by this method is the speed of sound in free air and, as we shall see, it varies with air temperature and is affected by the wind. Accurate values of the speed of sound are quoted at a particular temperature.

Some typical results for this experiment are

distance from wall	$= 100$ m
$\therefore$ distance sound travels	$= 200$ m

time taken for 50 claps	$= 30.3$ s
$\therefore$ time interval between claps	$= 30.3/50 = 0.606$ s

$$\begin{aligned} \text{speed} &= \frac{\text{distance travelled}}{\text{time taken}} \\ &= \frac{200\,\text{m}}{0.606\,\text{s}} \\ &= 330\,\text{m/s} \end{aligned}$$

Notice that by measuring the time for 50 claps to only the nearest 0.1 of a second, we are able to get a result giving three significant figures (appendix C2).

The accuracy of the time measurement can be greatly improved by using electronic timing instead of the hand-operated stopwatch. Electronic timing is used in athletics to give times claimed to be accurate to $\frac{1}{100}$ second. All that is needed are switches which start and stop an accurate clock without delays. Switches are available which respond instantly to sound or light or touch.

It is often easier to calculate the speed of sound from the formula $c = f\lambda$. Several methods of measuring the speed c involve measurement of the wavelength λ of a sound of accurately known frequency.

What affects the speed of sound in a gas?
The speed of sound c depends on two main factors:
a) the nature of the gas,
b) the temperature of the gas ($c \propto \sqrt{T}$, where T is the temperature of the gas in kelvins.)

The effect of the type of gas
If the mouthpiece of a recorder or whistle is connected to a gas tap with a long piece of rubber tubing (several metres long), it will sound when the gas is turned on. The pitch, and therefore the frequency of the note produced, rises suddenly as the air is blown out of the instrument and is replaced by gas. If the tube is now disconnected from the gas tap and blown down by mouth, the frequency of the note falls as air replaces the gas again.

One important difference between air, which is mostly nitrogen and oxygen, and the gas from a gas tap, which is methane, is that air is much denser than methane. Another difference between air and methane gas is the number of atoms in their molecules. Nitrogen and oxygen molecules have two atoms, methane CH_4 has five atoms. It can be shown that both these differences in the nature of the gases affect the speed of sound, but that the difference in density has the greater effect on the speed of sound in this example. *Sound travels more quickly in gases of lower density.*

We can explain the effect as follows:
a) The wavelength λ of the sound produced depends only on the length of the recorder or pipe and does not change.
b) From the equation $c = f\lambda$, we get $f = c/\lambda$, and therefore if λ does not change, the frequency f of the sound is directly proportional to the speed c.
c) Thus an increase in the frequency of the sound when the recorder is gas filled indicates an increase in the speed of sound in the less dense methane gas.

The effect of the pressure of a gas:
While the speed of sound is different in gases of different densities, it does not follow that the speed of sound in a particular gas always changes as its density changes. Newton showed that:

$$\text{the speed of sound in a gas} \propto \sqrt{\left(\frac{\text{pressure}}{\text{density}}\right)}$$

Thus any change in a gas which affects both its pressure and density in the same way will produce no change in the speed of sound through it. For example, if the pressure of a gas is doubled by halving its volume at constant temperature (Boyle's law, p 190), its density will also double and there will be no change in the value of pressure/density and hence no change in the speed of sound. *The speed of sound in a gas is not affected by changes of pressure.*

The effect of the temperature of a gas
A gas expands as its temperature rises; if this expansion occurs at constant pressure (Charles' law, p 188), then the speed of sound c will change. This is because a change of density has occured without a change in pressure. It can be shown that:

$$\text{the speed of sound in a gas} \propto \sqrt{\left(\frac{\text{temperature on}}{\text{the kelvin scale}}\right)}$$
$$c \propto \sqrt{T}$$

This relation gives us a formula for calculating the speed of sound at different temperatures:

$$\frac{c_1}{c_2} = \sqrt{\left(\frac{T_1}{T_2}\right)}$$

where c_1 is the speed of sound at T_1 and c_2 is the speed at T_2, with T_1 and T_2 being temperatures on the kelvin scale.

Worked Example
Temperature and the speed of sound
If the speed of sound in air at 324 K is 340 m/s, calculate its speed at 441 K.

We have
$T_1 = 441\,K$
$T_2 = 324\,K$
$c_2 = 340\,m/s$

Using $\frac{c_1}{c_2} = \sqrt{\left(\frac{T_1}{T_2}\right)}$ we have

$$\frac{c_1}{340\,m/s} = \sqrt{\left(\frac{441\,K}{324\,K}\right)} = \frac{21}{18}$$

$$\therefore \quad c_1 = 340\,m/s \times \frac{21}{18} = 397\,m/s$$

Answer: The speed of sound at 441 K is 397 m/s.

Do other factors affect the speed of sound?
Frequency: When we listen to music some distance away from the instruments it is obvious that sounds of all frequencies produced at the same moment reach the listener together. There is no evidence that high-pitch notes travel at a different speed to low-pitch notes. A church bell provides another illustration of this principle, for a bell produces sounds of many different frequencies but sounds exactly the same a long way off as it does near the church. *The speed of sound does not depend on frequency or pitch.*

Amplitude: Loud sounds do not usually travel faster than quiet sounds. For example, the loud sounds of a drum are heard to keep time exactly with all the other sounds from a band, however far it is from the listener. The shock waves produced by a violent explosion do actually travel faster near the explosion, but this is an exception. *The speed of sound is not usually affected by its wave amplitude or loudness.*

Humidity: Water vapour is less dense than both oxygen and nitrogen and so moist air is slightly less dense than dry air. The speed of sound is thus slightly greater in moist air than in dry air.

Wind: A wind moves the air through which the sound waves travel, so the wind velocity must be added to the sound velocity. In this case we are dealing with vector quantities and the two velocities should be added by using the parallelogram of velocities. If the sound and wind are travelling in the same direction their velocities add together; and if the sound is travelling against the wind it travels more slowly.

The speed of sound in solids and liquids

We have seen that sound requires a medium to travel through and that its longitudinal waves set up vibrations of the particles in the medium. Any medium which has particles that can vibrate will transmit sound, but the nature of the medium will affect the speed at which the vibrations are passed from particle to particle and hence the speed of sound through the medium. Sound travels more quickly through a medium in which the atoms are strongly bound together. If you imagine that in a solid all the atoms are joined together by springs then the stronger the springs the faster the sound travels.

The strong binding between atoms in solids means that sound will travel much more quickly through solids than through gases. A comparison can be made by listening for the sound made by tapping a long steel rail some distance away. The sound which travels through the steel arrives well before the sound which comes through the air. The speed of sound through liquids is also faster than through gases, but the relatively weaker binding between atoms in a liquid results in a lower speed of sound in liquids than in solids.

The speed of sound through a material should not be confused with its ability to transmit sound. Some materials, like those which contain a lot of air trapped inside, are good absorbers of sound, but this is not necessarily because sound travels slowly through these materials. Table 21.1 compares the speed of sound in a solid, liquid and gas. Note that the values are only approximate and are temperature dependent.

Table 21.1

Medium	Speed of sound
steel	5000 m/s
water	1500 m/s
air	330 m/s

Reflection of sound waves

That sound waves can be reflected we have no doubt: we have all heard echoes. Indeed, we have already measured the speed of sound using echoes. But what other applications of the reflection of sound waves are of interest? Do sound waves obey the same laws of reflection as the light rays and water waves we have investigated earlier?

● *Place a ticking clock or stopwatch inside a long cardboard tube at its closed end, fig. 21.4.*

● *Point the open end of the tube towards a wall at an angle of incidence i. (The reflecting surface needs to be hard to reduce absorption of the sound.)*

● *With your ear close to the end of an open cardboard tube, listen to reflections of the sound from the board at different angles of reflection r.*

Figure 21.4 *Sound waves obey the laws of reflection*

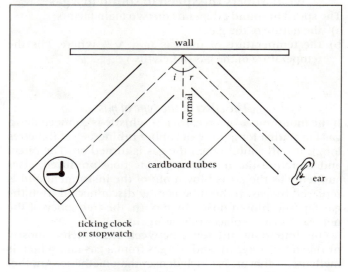

It is found that the reflected sound is loudest when:
a) the angle of reflection *r* is equal to the angle of incidence *i*, and
b) both tubes lie in a plane which is normal to (at right angles to) the reflecting surface.

It is found that sound waves do obey the laws of reflection.

Reverberation

In a cathedral or large hall there are many reflecting walls and surfaces, which form multiple reflections and create the impression that a sound lasts for a long time. A sound produced in a brief moment may linger for several seconds, only gradually fading away. At each reflection some of the sound energy is absorbed and the reflected sound becomes a little quieter. When many echoes merge into one prolonged sound the effect is known as **reverberation**. Too much reverberation causes sounds to become confused and indistinct making it necessary to speak very slowly. Too little reverberation means that the surrounding walls and surfaces are absorbing most of the sound energy very efficiently causing the hall to sound 'dead' and making voices and other sounds appear weak and quiet. A good concert hall has just the right amount of reverberation. The sound reflecting and absorbing properties of a room are called its **acoustics**.

Echoes and reverberation in the Festival Hall, London, are controlled by sound-absorbing materials around the walls, floor and ceiling of the building. Too little absorption and the echoes linger and merge sounds together; too much absorption and the orchestra becomes hard to hear, sounding 'dead' and muffled.

Speaking tubes

Sound waves can be totally internally reflected like light waves (p 27). Speaking tubes are often used for passing messages on ships. A speaking tube is a metal tube with a funnel at each end which will pass sound waves in either direction through the air inside, fig. 21.5. Sound will travel through a bent tube, being totally internally reflected at the inside surfaces of the metal tube, provided the bends are not so tight that the angle of incidence i becomes less than the critical angle from air to metal. Inside the pipe, sound waves travel more slowly in the air than in the surrounding metal walls of the pipe. This is similar to the total internal reflection of light waves when they travel more slowly inside glass than in the surrounding air.

Figure 21.5 *A speaking tube*

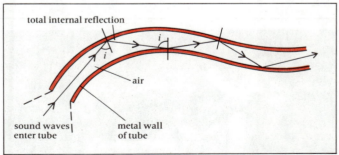

The frequency spectrum of sound waves

If a signal generator is connected to a loudspeaker we can listen to sound waves at any frequency or pitch selected on the generator. At the low-frequency end of the hearing range it is difficult to say when sound becomes a sensation of vibration. Below about 20 hertz the vibrations are felt rather than heard and are called **subsonic** (below sound).

At the high-frequency end of the audible range the limit is also difficult to find exactly because the ear gradually becomes less sensitive as the frequency rises above about 10 kHz. Also, as we get older, the range of high-frequency sounds which we can hear gradually reduces. Very few people can hear a frequency of 20 kHz and this is a convenient upper limit to choose. Thus the full human hearing range is about 20 Hz to 20 kHz as shown in fig. 21.6. Above 20 kHz the waves are known as **ultrasound** (beyound sound). Some animals can hear ultrasonic frequencies and bats use ultrasound in the same way as radar to 'see' with. Many uses have been found for ultrasound where it has several advantages over audible sound.

Figure 21.6 *The frequency spectrum of sound waves*

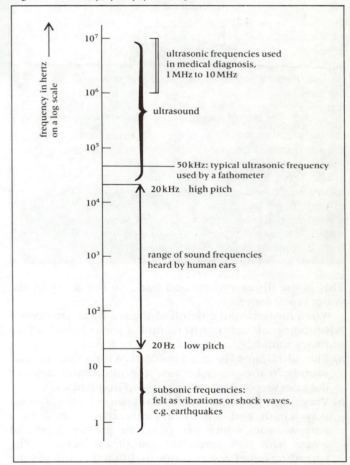

Applications of ultrasound

Ultrasound is usually sent out from the vibrating source or transmitter in brief bursts of energy called **pulses**. When these pulses are reflected by an object an echo is detected a short time later. This is called a **pulse–echo technique**. The time interval between the pulse of ultrasound being sent out and received back can be used to measure the distance to the reflecting surface provided the speed of ultrasound in the medium is known.

$$\text{distance (there and back)} = \text{speed} \times \text{time}$$

This echo method is used by ships in an instrument called a **fathometer** to measure the depth of water below a ship, fig. 21.7. By measuring the time interval between the sending out of a pulse of ultrasound and its echo arriving back from the sea bed, the depth of water can be calculated. For example, if the time interval is 0.8 s and the speed of ultrasound in water is 1500 m/s, the depth of water is calculated as follows:

$$\text{distance travelled by ultrasound} = \text{speed} \times \text{time}$$

$$\therefore \quad \text{distance} = 1500\,\frac{\text{m}}{\text{s}} \times 0.8\,\text{s} = 1200\,\text{m}$$

Figure 21.7 *Measuring the depth of the sea using a fathometer*

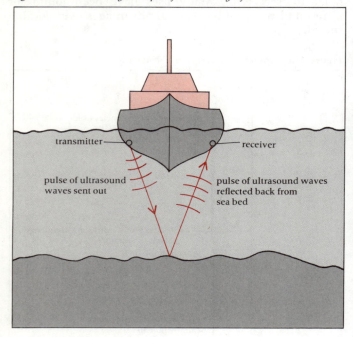

transmitter — receiver

pulse of ultrasound waves sent out

pulse of ultrasound waves reflected back from sea bed

This is the distance there and back, so the depth of the water is 600 metres.

When measuring the depth of the sea there are several advantages of using ultrasound ('sonar') instead of ordinary sound:

a) The ultrasound is not confused with other natural sounds in the water because the ultrasound detector does not respond to ordinary sound frequencies.

b) Very high frequency ultrasound waves have a short wavelength and are more easily concentrated in a narrow beam which can penetrate greater depths of water with less spreading out of the waves. (The spreading out of waves is due to diffraction and has the effect of reducing the wave amplitude so that it cannot travel as far.)

c) Using waves of shorter wavelength means that smaller sized objects or details can be 'seen' or located. Smaller waves can be reflected by smaller objects making them 'visible' in the ultrasound echo. This effect is known as improving the *resolution* of the image.

Some blind people wear spectacles which have an ultrasound transmitter and receiver which warns them of obstacles ahead. Ultrasound pulses are sent out in front and when an echo is received it is converted to an audible sound which tells the person how far away the obstacle is.

Ultrasound in hospital

In hospitals ultrasound is used to obtain images of internal parts of the body. Ultrasound pulses are sent into the body by a transmitter placed in good contact with the skin. Reflections or echoes are received from any surfaces within the body which have either a different density or a different structure or elasticity. The time delay of the echoes gives the depth within the body of the reflecting surfaces, and the reflections in

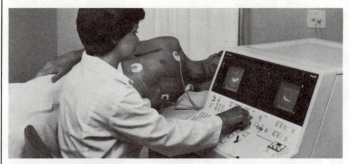

The hand-held ultrasound probe provides an image of the moving heart.

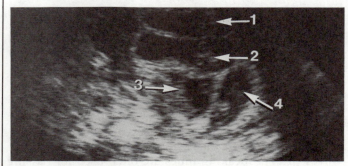

This utrasound image shows four individual embryos in a mother's womb.

different directions can be used to build up an image of something inside. Ultrasound is particularly useful in medicine because:

a) It is thought to be much *safer* than X-rays, which are known to damage cells by ionisation. (Ultrasound is used to get pictures of unborn babies which might be harmed by X-rays.)

b) It can be used *continuously* to watch the movement of an unborn baby or a person's heart, without any injury or risk to the patient.

c) It can measure the *depth* of an object below the body surface from the time delay of the echo, whereas an X-ray picture is flat with no indication of depth.

d) It can detect some *differences* between soft tissues in the body which X-rays cannot. In this way it is sometimes able to find tumours or lumps inside the body.

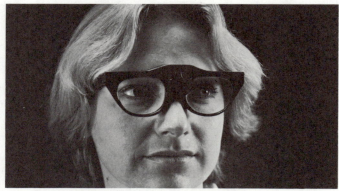

Ultrasound spectacles help a blind person to estimate the distance away of something in front of her.

Refraction of sound waves

Refraction occurs when the speed of the waves changes. The speed of sound waves in air is affected by the air temperature, so if sound waves pass through layers of air at different temperatures they will be refracted, or turned in a different direction. Fig. 21.8 shows how, on a summer's evening when the air near the ground becomes cool, refraction makes it easier to hear distant sounds across the countryside. The sound waves are bent or refracted down towards the ground. This can be explained by showing that the wavefronts are further apart in the warm air (higher up) where the sound travels faster, likewise they are closer together nearer the ground where the sound travels slower in the cooler air. The direction of travel of the sound energy is at right angles to the wavefronts and this can be seen to bend downwards due to the refraction of the sound waves.

On a day when the ground is very hot and the lower layers of air are the hottest, the sound waves are bent upwards away from the ground making it more difficult to hear over any distance. This effect can be compared with the refraction of light which produces a mirage (p 27).

Diffraction of sound waves

As well as reflection and refraction we find that sound waves also show diffraction effects. We notice that sound spreads round corners; sounds can be heard coming round a building from the far side. Sound does not come through an open door in a narrow beam, but fans out so that it can be heard in any direction. These examples are simple evidence of what is called diffraction. As we found with water waves in a ripple tank, waves which are further apart, and have longer wavelengths similar to the size of the gap they are passing through, are diffracted most. For example a doorway may be about 1 metre wide which is very similar to the wavelength of many sounds. Sound waves may have wavelengths as short as 20 cm or as long as 10 m. The short-wavelength, high-pitch sounds tend to be more directional because they are diffracted less than long-wavelength, low-pitch sounds. When we listen to music from a loudspeaker the high-pitch sounds can be heard best in front of the loudspeaker and not so well at the side or behind the speaker; this is what we mean when we say that high-pitch sounds are more directional, they are not spread out as much by diffraction.

Interference of sound waves

As with all wave motions, sound waves can be super-posed, that is they combine together rather than collide when they overlap. As we shall see in the next section, when sound waves travelling in opposite directions are superposed they can produce waves that appear to stand still, called *standing waves*. In general overlapping sound waves produce regions of louder sound by constructive interference and regions of quiet by destructive inter-ference.

When two similar loudspeakers are connected to the *same* audio-frequency generator they will produce sound waves of identical frequency and very similar amplitude. If the coils in the loudspeakers are connected to the generator the same way round, when a current flows through them their sound-producing cones or diaphragms will move forwards together sending out sound waves which are in step, or in phase.

● *Set two speakers, facing the same way, about* 0.5 *to* 1.0 m *apart and select sound frequencies in the range* 500 Hz *to* 2 kHz *on the generator.*

● *Walk slowly across the room parallel to the two loudspeakers and a few metres in front of them, fig.* 21.9.

● *Estimate the distance between places where you hear a loud sound with a quiet place in between.*

The listener hears variations in the loudness of the sound as he moves across the room. A loud sound is produced where the waves from the two speakers arrive in phase and so interfere contructively. A quiet sound is heard where destructive interference occurs, which is where a wave crest from one speaker is cancelled out by a wave trough from the other speaker.

Figure 21.9 *Demonstrating interference with sound waves*

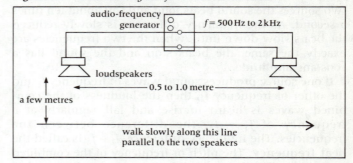

Figure 21.8 *Refraction of sound waves in the air*

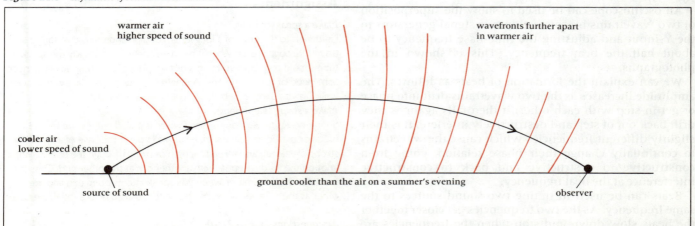

• *You can also investigate the effects of changing the spacing of the speakers, the wavelength of the sound waves or the distance of the listener from the speakers.*

All three of these factors change the spacing of the loud and quiet places in the room. If the speakers are moved closer together the loud places move further apart, that is the distances are inversely related. If the wavelength of the sound waves is increased or the listener moves further away from the speakers, in both cases the loud places are found further apart in direct proportion with the increases. It helps to remember these effects when we investigate the same type of interference with two wavetrains of light.

Note that it is necessary to use two sources of sound which are of identical frequency, similar amplitude and which set off from the loudspeakers together in phase. These conditions are provided by connecting the two speakers to the same generator.

Beats

Another special effect which occurs when sound waves are superposed is called beats. The name 'beats' describes what we hear: the loudness of the sound varies regularly in a beating or throbbing manner. **Beats** are the variations in amplitude of the sound waves formed when two sound waves of slightly different frequencies are superposed.

• *Connect two audio-frequency generators in parallel to one loudspeaker.*
• *Adjust the two amplitudes to be about equal and the two frequencies to be slightly different, say 250 Hz and 260 Hz.*
• *Listen to the combined sound waves from the speaker and slowly increase the 250 Hz frequency and notice what happens to the loudness of the sound.*

With a frequency difference of about 10 Hz between the two sources the sound beats loud and soft about ten times a second. As the frequency difference is slowly reduced the beats slow down until, when the two frequencies are exactly the same, the beats stop and the sound has a constant amplitude.

If one source produces sound waves of frequency f_1 and the other of frequency f_2, then the loudness of the combined waves is heard to rise and fall regularly at a frequency of $f_1 - f_2$, the difference between the frequencies. The frequency difference, $f_1 - f_2$ is called the **beat frequency**. The pitch or frequency of the combined wave is equal to the mean of the two separate frequencies, $\frac{1}{2}(f_1 + f_2)$.

An oscilloscope can be used to show the superposition of two wavetrains by connecting two signal generators to the Y-input and adjusting the time-base frequency to be about half the beat frequency. This is shown in the photographs.

We can explain the formation of beats as follows. The amplitude increases as the two wavetrains drift into phase or get in step with each other; it then falls again as they drift back out of step and destructively interfere. It is their slightly different frequencies which cause the wavetrains to continually change their phase relation, going from constructive to destructive and back to constructive interference at the beat frequency.

Beats can be used in tuning two sound sources to the same frequency. As the two frequencies get closer together the beats slow down and stop when the frequencies are

exactly equal. This is one of the methods used to tune musical instruments so that they make exactly the same frequency sounds as other instruments. A tuning fork of accurate frequency is often used in tuning other instruments.

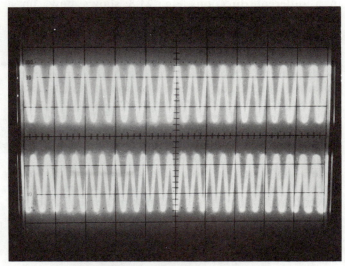

Waves with equal amplitude and of frequencies 250 Hz and 260 Hz before being added together or superposed.

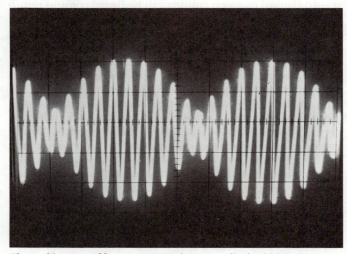

The resulting wave of frequency 255 Hz has an amplitude which varies or 'beats' at a frequency of 10 Hz.

Assignments

Make a speaking tube

Fit funnels to the ends of a long length of pipe and show that sound travels through it even when it has many bends in it. Ideally the speaking tube should connect two rooms or two places far enough apart to need some kind of 'telephone' link. A length of garden hose pipe works quite well. Do you think a metal tube would work any better?

Make a string telephone

Sound waves travel along a string which is pulled taut between two metal or plastic containers. Make a small hole in the base of a tin can or strong plastic tub and thread a length of string through the hole leaving a knot on the inside. Repeat with another can at the other end. One can works as the transmitter when you speak into its open end and the other as the receiver when its open end is placed near another person's ear.

Try questions 21.1 to 21.10.

21.2
STANDING WAVES

Sometimes a wave motion appears to stay in one position or to stand still; when this happens we call it a standing or stationary wave. Standing waves are formed in musical instruments when resonance occurs.

Standing waves on a stretched string

Stationary or standing waves can be produced on a string by using a vibrator to make transverse waves, as shown in fig. 21.10.

• *Take about* 2 m *of fine string and attach one end of it to a fixed stand.*

• *Pass the other end through the hole in the central shaft of the vibrator and attach it to a slotted-mass hanger. The tension should not be too great, or the amplitude of the vibrations will be too small to see.*

• *Connect the vibrator to a signal generator.*

• *Start with a vibration frequency as low as* 1 Hz *and slowly increase it.*

At first the vibrator produces forced vibrations in the string, which have only a small amplitude.

• *Switch off the vibrator and pluck the string in the middle so that it vibrates at its natural frequency f_1. This will be somewhat greater than* 1 *or* 2 Hz.

• *Switch on the vibrator again and slowly increase its frequency until it reaches the natural frequency of the string.*

At this point the vibration of the string builds up to a very large amplitude, and resonance occurs. At this frequency we can also see a *standing wave*.

We can slow down the motion of the string by using a stroboscope in a darkened room. When the stroboscope flash frequency equals the vibrator frequency the string appears to be frozen in a wave shape. When the frequencies are slightly different the wave can be watched in slow motion and we see that the string vibrates transversely (up and down), but the wave form itself does not move along the string. This wave, not travelling along the string but staying in the same place, is an example of a standing wave.

• *Slowly continue to increase the vibrator frequency.*

The string will return to forced vibrations of small amplitude until at a new frequency f_2 resonance occurs again producing another large-amplitude vibration. We find that this frequency is twice the natural frequency, and is given

by $f_2 = 2f_1$. This time, however, the string vibrates with two loops and at its midpoint the string remains still. Under stroboscopic light the two halves of the string can be seen to move in opposite directions and to form the shape of a full wave. Further increases of the vibrator frequency produce more resonances at frequencies f_n, given by

$$f_n = nf_1 \qquad \text{where } n \text{ is any whole number } (n = 1, 2, 3...)$$

Fig. 21.11 shows the simpler ways in which a string can vibrate in resonance. All of these resonant vibrations are called **harmonics**. The harmonics are numbered according to the ratio of their frequencies to the first natural frequency f_1.

The first natural frequency of vibration f_1 is the simplest mode, or way of vibrating, for a string and we call it the **fundamental**. For example, we can say that the fourth harmonic has a frequency f_4 which is four times the fundamental frequency f_1.

When a string is made to vibrate in a musical instrument the sound produced is the combined effect of many harmonics vibrating at the same time on the same string.

Figure 21.11   *Resonant vibrations on a string, standing waves*

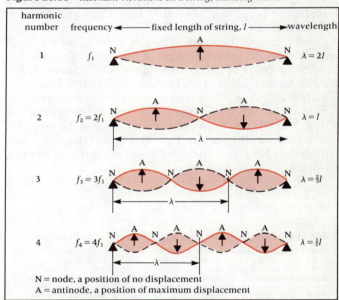

N = node, a position of no displacement
A = antinode, a position of maximum displacement

Figure 21.10   *Demonstrating standing waves on a string*

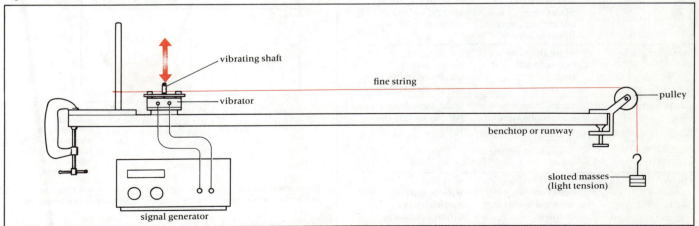

How is a standing wave formed?

The vibrator in the standing wave demonstration sends a travelling wave along the string to the end where the pulley is. Here the travelling wave is reflected back along the string to the vibrator. The two travelling wavetrains, going in opposite directions, are superposed on the string and interfere, so producing constructive and destructive interference at different points along the string. Constructive interference causes a large displacement of the string and destructive interference produces a stationary, undisplaced bit of string. When these stay in the same place on the string a standing wave is set up.

A place where there is no displacement of the string at any time is called a **node**. A place where the string vibrates with maximum amplitude is called an **antinode**.

Obviously the fixed ends of a string must be two nodes and midway between two nodes there will be an antinode. This is shown on the fundamental vibration in fig. 21.11. The harmonics which can be produced occur when the string is divided up into equal sections, which vibrate with large amplitude in between extra nodes. So, for example, the fourth harmonic has three extra nodes spaced evenly along the string dividing it into four equal lengths.

The distance between two adjacent nodes is half a wavelength $\frac{1}{2}\lambda$, as can be seen in figs. 21.11 and 21.12. It follows that the condition necessary for resonance to occur on a vibrating string exists when an exact number of half wavelengths fits into the length of the string. The smallest possible number of half wavelengths, which is one, forms the fundamental resonant vibration. The fourth harmonic, which divides the string into four exact half wavelengths, occurs when the wavelength λ is half the length of the string. The relation between wavelength and the length of the string is shown in fig. 21.11 for the first four harmonics. Note that the fundamental counts as the first harmonic.

Properties of a standing wave

The properties of standing waves are summarised in fig. 21.12.

Standing waves are produced by interference as a result of the superposition of two waves when a travelling wave is reflected back along its incident path.

Standing waves occur by resonance only at the natural frequencies of vibration of a string or medium.

Standing waves have *nodes* where there is no displacement at any time. (Nodes are formed by *destructive interference* as the wave travelling to the right tries to displace the string upwards and the reflected wave travelling to the left tries to displace it downwards.) Only standing waves have nodes.

In between the nodes are positions called *antinodes*, where the displacement has maximum amplitude resulting from *constructive interference* between waves travelling in opposite directions.

The distance between two adjacent nodes is half a wavelength $\frac{1}{2}\lambda$. Similarly, the distance from one antinode to the next antinode is $\frac{1}{2}\lambda$. The distance between a node and the next antinode is $\frac{1}{4}\lambda$. Thus, for a standing wave, the wavelength is *twice* the distance between two adjacent nodes.

The *fundamental* mode of vibration is the natural vibration with the longest wavelength and lowest frequency. It usually has the largest amplitude and provides the dominant pitch of a musical sound, that is the loudest and most obvious pitch.

Table 21.2 compares the main features of standing and progressive waves. It shows the important differences and provides a useful summary of their properties.

Figure 21.12 *Properties of a standing wave*

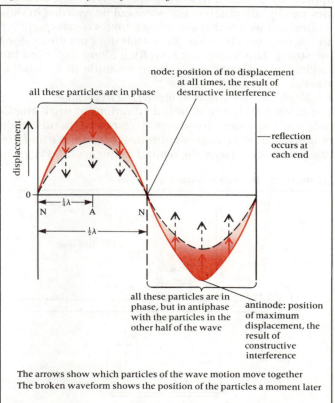

node: position of no displacement at all times, the result of destructive interference

all these particles are in phase

reflection occurs at each end

all these particles are in phase, but in antiphase with the particles in the other half of the wave

antinode: position of maximum displacement, the result of constructive interference

The arrows show which particles of the wave motion move together
The broken waveform shows the position of the particles a moment later

Table 21.2 Comparison of standing waves and travelling waves

	Standing waves	Travelling waves
alternative name	stationary waves	progressive waves
appearance and nature	waveform does not move through medium, but has nodes at fixed places; energy is not carried away from the source	waveform moves through the medium away from its source, carrying energy, but not the medium, with it
wavelength λ	twice the distance between adjacent nodes	the distance between two successive particles which are in phase
amplitude	varies from zero at a node to maximum at an antinode, depends on position along the wave	is the same for all particles along the wave
phase	all the particles between two adjacent nodes are in phase	over one wavelength all particles have different phases

The vibrations of a stretched string

A sonometer is a laboratory instrument used to investigate the vibrations of a stretched string or wire. The wire is fastened to a fixed peg at one end of a wooden box and tensioned by hanging slotted masses on the other end of the wire, which hangs over a pulley, fig. 21.13. The hollow wooden box acts as a sound amplifier, just like those in many musical instruments. Two fixed bridges, F, usually a metre apart, provide a constant length of wire which can vibrate. A third, movable bridge, M is used to vary the length l, by sliding it along the sonometer.

When the wire is plucked in the middle it vibrates at its natural and fundamental frequency f. We are going to investigate how this natural frequency depends on the length l of the wire, on its tension T and on its mass per unit length m (literally the mass, in kg, of 1 metre of wire). We shall adjust the length or tension of the wire to make its frequency the same as that of a particular tuning fork of known frequency.

Musicians can tune the frequency of the wire to be the same as a tuning fork by sounding them both together and listening for beats (p 458). As the two frequencies become equal the beats slow down until a constant amplitude sound is heard. An alternative method of tuning the wire is to place a very small paper rider on the wire at its midpoint, where the antinode should be. Now hold the tuning fork very firmly on a bridge of the sonometer so that its vibrations are picked up by the wire. When the natural frequency of the wire is different from the tuning fork frequency, the forced vibrations will have a very small amplitude and the paper rider will remain quite still. When we adjust the length or tension of the wire so that its natural frequency of vibration becomes the same as the tuning fork, resonance occurs and the amplitude of vibration of the wire quickly builds up and throws the paper rider off the wire.

How does the frequency depend on length?

To investigate how the frequency of a stretched wire depends on its length the tension must be kept constant.

- *Select a tension which makes the fundamental frequency of the string about the same as, or a little lower than, the lowest frequency tuning fork in a set of forks (usually middle C at 256 Hz).*
- *For each of the forks in the set, starting with the lowest frequency, find the length of wire which has the same fundamental frequency.*

We find that as the frequency of the forks increases the position of the movable bridge M has to be adjusted for shorter lengths of the wire to produce resonance.

- *Make a table of results for the values of the tuning fork frequencies f, the wire lengths l and also $f \times l$.*

When the values of f and l are multiplied together a constant result is obtained.

We find that

$$f \times l = \text{constant}$$

or

$$f \propto \frac{1}{l}$$

The frequency of a stretched wire is inversely proportional to its length.

If we plot a graph of f against $1/l$, as shown in fig. 21.14, the straight line confirms the relation $f \propto 1/l$.

Figure 21.14 *Showing how frequency depends on length*

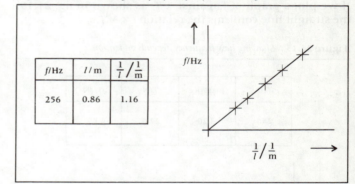

f/Hz	l/m	$\dfrac{1}{l} / \dfrac{1}{\text{m}}$
256	0.86	1.16

Figure 21.13 *The sonometer*

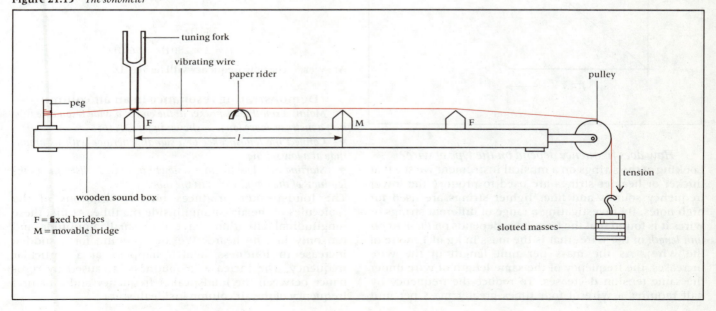

tuning fork

vibrating wire

paper rider

pulley

peg

F   M   F

l

tension

wooden sound box

F = fixed bridges
M = movable bridge

slotted masses

How does frequency depend on tension?

Now, we keep the length of the wire constant while we change the tension to make the frequency of the wire the same as each tuning fork, in turn.

- *Adjust the tension by adding extra slotted masses to the hanger on the end of the wire. When the frequency of the wire goes above the tuning fork frequency, remove the last mass added and try a smaller mass.*
- *When resonance occurs, record the values of the tuning fork frequency f and mass m of the slotted masses.*

The tension T provided by a mass m is given by mg. If m is in kg and $g = 10\,\text{N/kg}$, T will be in newtons.

- *Calculate the tension T and then the square-root of the tension $\sqrt{T}$ for each of the results. Finally, calculate the values of $f/\sqrt{T}$, recording them in your table of results.*

We find that a large increase in tension is needed to produce a small increase in frequency, in fact to *double* the frequency of the wire requires a *fourfold* increase in tension. The results show that

$$f/\sqrt{T} = \text{constant}$$

or

$$f \propto \sqrt{T}$$

The frequency of a stretched wire is directly proportional to the square root of its tension.

If we plot a graph of f against $\sqrt{T}$, as shown in fig. 21.15, the straight line confirms the relation $f \propto \sqrt{T}$.

Figure 21.15 *Showing how frequency depends on tension*

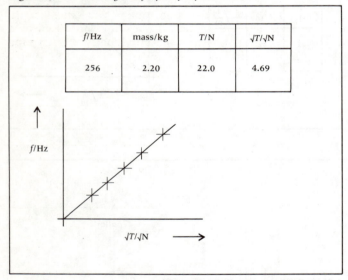

f/Hz	mass/kg	T/N	$\sqrt{T}/\sqrt{N}$
256	2.20	22.0	4.69

How does frequency depend on the type of wire?

Looking at the strings on a musical instrument we see that thicker or heavier strings are used to produce the lower frequency sounds and finer lighter strings are used for high notes. By investigating a range of different strings or wires it is found that the frequency depends on *the mass per unit length* of the wire (that is the mass in kg of 1 metre of the wire). As the mass per unit length of the wire increases, the frequency of the same length of wire under the same tension decreases. To reduce the frequency by half requires a wire of four times greater mass per unit

length. We find the relation between the frequency f of a wire and its mass per unit length m to be:

$$f \propto \frac{1}{\sqrt{m}}$$

The frequency of a stretched wire is inversely proportional to the square root of its mass per unit length.

The fundamental frequency of a stretched wire depends on three things which, combined in a single formula, is found to be

$$f = \frac{1}{2l}\sqrt{\frac{T}{m}}$$

where f = frequency in Hz
l = length in m
T = tension in N
m = mass per unit length in kg/m

Questions which require a knowledge of these relations can often be done without using the full formula and it may be easier to use the following relations:

From $f \times l = \text{constant}$, we get:

$$f_2 l_2 = f_1 l_1 \qquad \text{or} \qquad \frac{f_2}{f_1} = \frac{l_1}{l_2}$$

when T and m are constant.
From $f/\sqrt{T} = \text{constant}$, we get:

$$\frac{f_2}{\sqrt{T_2}} = \frac{f_1}{\sqrt{T_1}} \qquad \text{or} \qquad \frac{f_2}{f_1} = \sqrt{\frac{T_2}{T_1}}$$

when l and m are constant.

Worked Example
The frequency of a stretched string

If a stretched string has a fundamental frequency of 280 Hz, what will its frequency become if its tension is increased to four times its original value and its length remains the same?

We can use the formula

$$\frac{f_2}{f_1} = \sqrt{\frac{T_2}{T_1}}$$

We are told $T_2 = 4T_1$ or $T_2/T_1 = 4$ and $f_1 = 280\,\text{Hz}$. So we have:

$$\frac{f_2}{280\,\text{Hz}} = \sqrt{4},$$

$$\therefore \quad f_2 = 2 \times 280\,\text{Hz} = 560\,\text{Hz}$$

Answer: The new frequency will be 560 Hz.

Demonstrating resonance in an air column

- *Mount a small loudspeaker at the end of a wide resonance tube half a metre or more long, as shown in fig. 21.16.*
- *Connect the loudspeaker to a signal generator with its amplitude at a low setting.*
- *Starting at a low frequency slowly increase it, listening to the loudness of the sound waves in the tube.*

The loudspeaker produces forced vibrations of the molecules of the air column inside the tube. At first these longitudinal vibrations have very small amplitude and can only just be heard. We are listening for a sudden increase in loudness, which happens at a particular frequency. The increase in loudness is caused by resonance between the loudspeaker frequency and a natural frequency of the air column inside the tube.

Figure 21.16 *Demonstrating resonance in an air column*

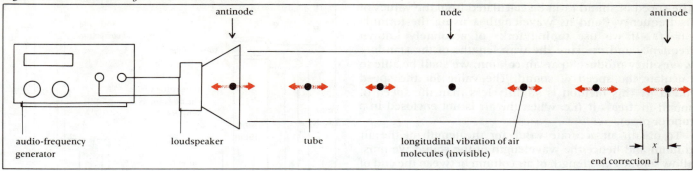

The first resonance occurs at the fundamental or lowest natural frequency of the air column. Resonance also occurs at higher frequencies, which are the harmonics that fit into the same length of pipe or tube. Fig. 21.17 shows how a pipe or tube which is closed at the end has less harmonics than one which is open. The open pipe has all possible harmonics while the closed pipe is limited to the odd number harmonics. A different combination of harmonics is one of the factors which gives a musical instrument a different tone.

It is interesting that a sound wave can be reflected at both a closed end and an open end of a tube. In the case of the *closed end*, air molecules are compressed at the closed end and the *compression* is reflected or 'bounces' back down the tube.

The *open ended* tube, however, allows free movement of air molecules at, and beyond, the end of the tube. This produces a *rarefaction* of molecules in this region. In effect what happens is that a rarefaction is reflected back down the tube so that an antinode is formed just beyond its open end.

Describing standing waves in pipes or tubes

We can summarise the main features of standing waves in pipes as follows:

a) At the end where the vibration of air molecules is produced, and at *open* ends of pipes where the air molecules can vibrate freely, there is always an *antinode*. The position of the antinode is a short distance outside the open end of a pipe.

b) At a *closed* end of a pipe, where longitudinal movement of molecules is prevented, there is always a *node*.

c) The wavelength of the sound wave must fit the length of the pipe as shown in fig. 21.17. The length of the air column which vibrates is slightly longer than the length of the pipe, since the antinodes are just outside the end of the pipe.

d) The fundamental vibration in a closed pipe has a wavelength which is twice as long as the wavelength of the vibration in an open pipe of the same length. This makes the frequency of the sound produced by a closed pipe half that of the same length of open pipe and so its pitch is an octave lower.

Figure 21.17 *Harmonic vibrations in air columns*

harmonic number	frequency	open pipe	wavelength	closed pipe	wavelength
1	f_1 = fundamental	A N A $\longleftarrow l \longrightarrow$	$\lambda = 2l$	A N $\longleftarrow l \longrightarrow$	$\lambda = 4l$
2	$f_2 = 2f_1$	A N A N A $\longleftarrow \lambda \longrightarrow$	$\lambda = l$	no second harmonic	
3	$f_3 = 3f_1$	A N A N A N A $\longleftarrow \lambda \longrightarrow$	$\lambda = \frac{2}{3}l$	A N A N	$\lambda = \frac{4}{3}l$
4	$f_4 = 4f_1$	A N A N A N A $\longleftarrow \lambda \longrightarrow$	$\lambda = \frac{1}{2}l$	no fourth harmonic	
		all harmonics possible, the same as for a stretched string		only odd-number harmonics possible $f = f_1 \times 1,3,5,7$, etc.	

The speed of sound in an air column

The speed of sound c can be calculated from the values of its frequency f and its wavelength λ using the formula $c = f\lambda$. If we use tuning forks of accurately known frequency and measure the wavelengths of the standing waves they produce in an air column, we shall be able to calculate the speed of sound. The value for the speed found by this method is slightly less than the speed of sound in 'free' air (i.e. when the air is not enclosed in a tube or pipe).

To obtain an accurate value for the length of the air column and hence the wavelength of the sound we must allow for the extra length of air column between the end of the tube and the antinode. This extra length is known as the 'end correction', and is shown as a length x in fig. 21.18. It is possible to find two resonance positions for each tuning fork frequency.

The shorter air column (b) which will resonate has a length of a quarter of the wavelength of the sound wave produced by the tuning fork ($\frac{1}{4}\lambda$). At resonance there is a node at the closed end and an antinode at a distance x above the open end of the pipe as shown in fig. 21.18b.

The longer pipe, which resonates at the same frequency, must also have a node at the closed end and an antinode at the open end. We can see in fig. 22.18c that this is possible when the air column is effectively three times longer. The column has an extra node and antinode which allows a sound wave of the same wavelength to fit into it. As both the short and long lengths of pipe include the same end correction x, subtracting the two actual lengths of the tubes will eliminate the end correction and give a value exactly equal to half the wavelength of the sound ($\frac{1}{2}\lambda$).

● *Set a tuning fork vibrating by striking it loosely against a firm rubber surface such as the sole of a shoe or a rubber bung.*
● *Hold it near the open end of the tube so that its vibrating prongs produce a longitudinal vibration of the air column.*
In the apparatus shown in fig. 21.18, the length of the air column depends on the water level in the inner tube. The length is therefore adjusted by sliding the inner tube in and out of the water.
● *Slowly raise the inner tube until the first resonance position is reached, when it will sound the same note as the tuning fork.*
The combined vibrations of the tuning fork and the air column will make a noticeably louder sound.
● *Find the position where the sound is loudest by raising and lowering the tube through the resonance position.*
● *Measure the length of the tube down to the water level (l_1).*
● *Now raise the inner tube to form an air column about three times as long as before. Using the same tuning fork, find the exact length l_2 at which resonance produces the loudest sound again.*
● *Calculate the speed of sound in the column using the following steps.*
The effective length of the shorter air column is

$$l_1 + x = \tfrac{1}{4}\lambda$$

The effective length of the longer air column is

$$l_2 + x = \tfrac{3}{4}\lambda$$

Subtracting these two effective lengths eliminates the unknown end correction x

$$l_2 - l_1 = \tfrac{1}{2}\lambda$$

This gives the wavelength

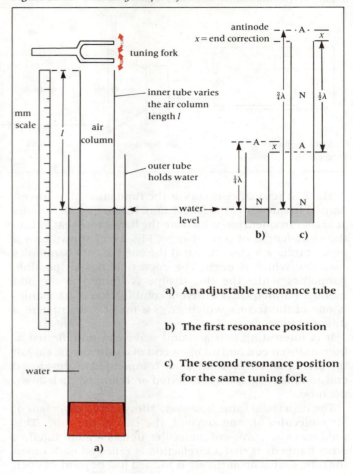

Figure 21.18 *Measuring the speed of sound in an air column*

a) An adjustable resonance tube

b) The first resonance position

c) The second resonance position for the same tuning fork

$$\lambda = 2(l_2 - l_1)$$

Now we can calculate the speed of sound c, from $c = f \times \lambda$:

$$c = f \times 2(l_2 - l_1)$$

If this experiment is repeated for a range of tuning forks of known frequencies, a more accurate value for the speed c can be found. If a graph of f against $1/\lambda$ is plotted, the gradient will give the value of c.

Worked Example
The speed of sound in a tube

A tuning fork of frequency 256 Hz produced resonance in a tube of length 32.5 cm and also in one of length 95.0 cm. Calculate the speed of sound in the air column in the tube.

Using the same symbols as above we have:

$$l_1 = 32.5 \text{ cm} \quad \text{and} \quad l_2 = 95.0 \text{ cm}$$

$$\therefore \quad l_2 - l_1 = 62.5 \text{ cm}$$

and
$$\lambda = 2(l_2 - l_1)$$
$$= 2 \times 62.5 \text{ cm} = 125 \text{ cm} = 1.25 \text{ m}$$

Now $c = f\lambda$

$$\therefore \quad c = 256 \text{ Hz} \times 1.25 \text{ m} = 320 \text{ m/s}$$

Answer: The speed of sound in the tube is 320 m/s.

Assignment

Try questions 21.11 to 21.14

21.3
MUSIC AND NOISE

Some combinations of sounds are called music while others are just a noise. What is the difference between musical sounds and noise? How are these different kinds of sounds produced and how do we detect and measure them? How can we prevent or reduce a noise which is a nuisance?

Musical sounds

What makes a sound musical? People enjoy, and dislike, very different kinds of music, but despite its great variety, all music usually contains recognisable common features. Most music uses sounds of particular and constant frequencies, combined together in various ways. Certain combinations of frequencies produce sounds with interesting musical qualities or character. Music also usually has a pattern or rhythm of sounds which are used according to a theme or plan.

Noise, however, is usually random in frequency, constantly varying without plan or purpose.

Musical scales and intervals

The pitch of a note depends on the frequency of its vibrations. In music the frequency of a note is determined by a **musical scale**. A musical scale is a range of notes of increasing pitch whose frequencies bear particular relationships to each other. It is the *ratio of the frequencies* of two notes which determines their musical relationship and quality rather than their actual frequencies.

The ratio of the frequencies of two notes is called the musical interval between them.

The simplest interval or frequency ratio of 2 to 1 is called an **octave**. For example the note labelled A has a frequency of 220 Hz. The note which has a frequency of twice this A′ = 440 Hz. The note which has a frequency of note A.

Many different ways of dividing up an octave are possible. The Europeans and the Chinese use totally different sets of musical intervals for their music. African music uses much smaller intervals with twice or four times as many notes as in the European scale. Scottish folk songs are based on a five note scale, which matches the black keys on a piano.

European music has evolved using a scale based on four intervals which correspond to simple fractions of the length of a vibrating string. Various slightly different versions of this scale have led to a compromise which best suits keyboard instruments such as the piano and organ. Nowadays these instruments are tuned to an *equal tempered scale* which divides the octave into twelve equal small intervals called **semitones**. Each pair of adjacent notes has the same frequency ratio; thus when an octave interval of ratio 2:1 is divided to give 12 equal tempered intervals these will have the ratio $^{12}\sqrt{2:1}$, which is about 1.06:1.

The notes on this scale are usually labelled with the letters from A to G, which are repeated for each octave. These seven letters fit the white keys on a piano. The other five notes in an octave, provided by the black keys, are labelled according to the white note immediately above or below. For example, the black note above C, called 'C sharp' and written $C^{\#}$, is also the black note below D which is called 'D flat', written D^{b}.

Standard pitch

When the intervals between all the notes in a scale have been fixed, a *standard frequency* must be chosen for one of the notes so that musical instruments may play together using the same frequencies. The agreed international pitch or frequency standard for concert music is based on the frequency of 440 Hz for note A in the musical scale. (A scientific standard pitch is used in physics which is based on the frequency of 256 Hz for note C, often called 'middle C'. The scientific standard of pitch is slightly different from the concert pitch.)

The twelve intervals in an octave, tuned to standard concert pitch, are shown in fig. 21.19 on a keyboard.

The interval from A to A′ has a frequency ratio: $\frac{440}{220} = \frac{2}{1}$ and is called an **octave**.

The interval from A to $A^{\#}$ has a frequency ratio: $\frac{233}{220} = 1.06$ and is called a **semitone**.

On the *equal tempered scale* shown on the keyboard, the interval between all adjacent notes is the same, that is, their frequency ratio is the same: $\frac{233}{220} = \frac{247}{233} = \frac{262}{247} = \cdots = 1.06$.

Figure 21.19 *The scale of notes on a piano*

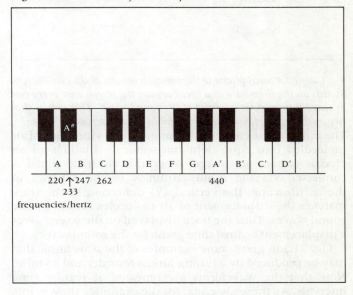

Timbre or quality

The same note of the same pitch or frequency produced by different instruments sounds different. The sounds produced by musical instruments are made up of a fundamental vibration with many harmonics added or superposed. The fundamental vibration, being of lowest frequency and greatest amplitude, gives each sound its basic pitch. The added harmonics of higher frequency (multiples of the fundamental) give the characteristic quality or flavour to the sounds from each particular instrument.

The quality of a sound, or its **timbre**, as it is called, is determined by the following factors:
a) the particular harmonics present in addition to the fundamental vibration,
b) the relative amplitude of each harmonic, i.e. the loudness of each harmonic compared with the fundamental,
c) the transient sounds produced when the vibration is started (transient means lasting a brief time).

Displaying the waveforms of sounds

We can compare the quality of different sounds by displaying them on an oscilloscope screen as shown in fig. 21.20a.

Figure 21.20 *Displaying the waveforms of different sounds*

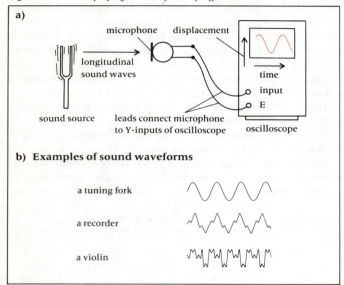

a)

b) **Examples of sound waveforms**

a tuning fork

a recorder

a violin

● *Connect a microphone to the input terminals of the oscilloscope. Switch on the time-base at a speed to scan the screen once every two or three time periods or cycles of the sound wave. This means that two or three complete waveforms should appear across the screen.*
The microphone converts the longitudinal vibrations of air molecules into equivalent transverse oscillations of an electric current or voltage. When this electrical signal is connected via the input terminals to the Y-plates of the oscilloscope, the vertical (Y) deflection of the trace matches the displacement of air molecules caused by the sound waves. Thus the trace displayed on the screen gives a displacement against time graph for the sound waves.

Fig. 21.20b gives some examples of the waveforms that may be produced by a tuning fork, a recorder and a violin. Note that the waveforms are repeated at regular time intervals. As these are equal for the examples shown, this means that they all have the same frequency and pitch. The different shapes of the waveforms, due to different combinations of harmonics, show how varied is the quality of the sounds from different instruments. The tuning fork produces a sine waveform which is a single frequency only. Such a sound, which has no harmonics, is called a 'pure' note and is particularly suitable for use in tuning other instruments to the same frequency.

When the sound of an instrument is imitated by an electronic organ, for the sound to be similar, it must produce the same waveform electrically. Several harmonics are added to the fundamental and their amplitudes are adjusted until, as near as possible, the synthetic waveform matches the natural waveform of the instrument.

Detection and measurement of sound

Most methods of detection of sound involve the conversion of a mechanical vibration into an electrical oscillation of the same frequency and characteristics. Even the human ear produces an electrical signal which is then passed to the brain. The devices which convert mechanical vibrations or sound waves into another form of oscillation are called **transducers**. For example, a microphone is a transducer which converts sound waves in the air into an oscillating electric current or voltage. The recording head of a tape recorder is also a transducer; it converts the oscillating electrical current fed through its coil into a varying magnetic field which then magnetises the metal-coated tape moving past it. Transducers also convert signals back to sound waves in the air. The stylus or pick-up of a record player and the loudspeaker are both transducers which convert the signals from one form to another.

Fig. 21.21 displays the stages of conversion involved in playing a record or in tape recording. Each change is made by a transducer. The sound vibrations themselves exist in various forms, such as the bumps in the grooves of a record or the varying magnetisation of a tape. These alternative forms are called **analogues**. Analogues contain all the sound information, but in a totally different form. The analogue of a sound must contain variations of frequency and amplitude exactly matching the original mechanical vibrations which produced the sound waves. The electric current flowing from a microphone or to a loudspeaker is also an analogue of the sound waves it represents.

Figure 21.21

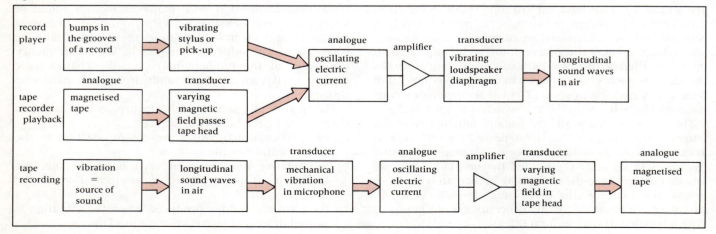

The ear

Our ears have several remarkable abilities, which unless we are deaf, we take for granted every day. The change from the quietest to the loudest sound we can hear may involve an increase in amplitude of over a million times and from the lowest pitch to the highest pitch notes we can hear, the frequency increases about a thousand times. As well as these large ranges, we are able to select what we listen to. We hear what we want to hear, filtering out what we don't. We can pick out a single instrument when a complete orchestra is playing, or concentrate on one person's conversation in a room full of talking people. We can also recognise frequency ratios with great precision and are sensitive to very small differences in frequency. We have a memory for the quality of sounds so that we can recognise a particular musical instrument or a particular person's voice by its characteristic combination of harmonics and transient sounds. The combination of ear and brain has yet to be matched by machine in these skills of recognition and identification of particular sources of sounds, such as voice patterns.

Many of the skills of the ear are provided by the brain in its computer-like selection and analysis of the electrical signals it receives from the ear. The ear converts all the sounds it receives into electrical signals in the cochlea. This is a shell-like structure where different nerve endings are stimulated by vibrations of different frequencies. The amount of stimulation, which determines the strength of the electrical signal, indicates the amplitude of that particular frequency. The diagram below shows how the ear delivers sound waves to the nerve endings in the cochlea.

The ear has three separate compartments called the **outer ear**, **middle ear** and **inner ear**. These are separated by skin membranes or 'windows' which the sound waves are passed through. The outer ear

funnels the longitudinal sound waves down the ear canal to the **ear drum**, the first skin membrane. This is made to vibrate transversely by the impact of the sound waves. Three linked bones, known as the **ossicles**, mechanically transfer the energy of the vibrating ear drum to the much smaller **oval window**. This is the second skin membrane which separates the air-filled middle ear from the fluid-filled inner ear. The ossicles in the middle ear are named after their shapes, looking like a hammer, an anvil and a stirrup, but they work like a set of levers which amplify the movement of the ear drum and produce a larger disturbance at the oval window. The concentration of vibrational energy on to the much smaller area of the oval window also has the effect of increasing the amplitude of vibration.

The inner ear has two separate functions. The **semicircular canals**, not shown in the diagram, help us to keep our balance. The shell-like structure, the **cochlea**, is the other part of the inner ear responsible for converting the mechanical vibrations received by the ear into electrical signals to be passed on to the brain. The cochlea is completely fluid-filled and the sound vibrations travel as longitudinal waves in its spiral coils. On their journey round the coils of the cochlea, sound waves of various frequencies disturb and stimulate particular nerve endings which then send electrical signals to the brain. The brain 'knows' the frequency of the sound by the nerves which send the message. At the other end of the coils of the cochlea there is another flexible membrane or window which allows the fluid to move when the oval window vibrates. This is necessary because the fluid inside the cochlea is incompressible. The block diagram explains the steps by which sound reaches your brain.

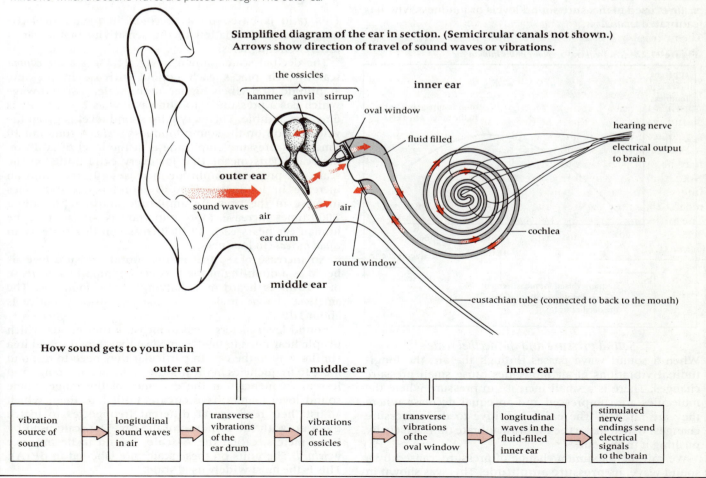

Simplified diagram of the ear in section. (Semicircular canals not shown.) Arrows show direction of travel of sound waves or vibrations.

the ossicles

hammer anvil stirrup

inner ear

oval window

fluid filled

hearing nerve

electrical output to brain

outer ear

sound waves

air air

ear drum

round window

cochlea

middle ear

eustachian tube (connected to back to the mouth)

How sound gets to your brain

outer ear   middle ear   inner ear

| vibration source of sound | → | longitudinal sound waves in air | → | transverse vibrations of the ear drum | → | vibrations of the ossicles | → | transverse vibrations of the oval window | → | longitudinal waves in the fluid-filled inner ear | → | stimulated nerve endings send electrical signals to the brain |

Microphones and sound level indicators

A microphone does the same job as the ear; it is a transducer that converts mechanical sound waves in the air into matching electrical oscillations or signals. The ear sends the electrical signals along nerve conductors to the brain, while a microphone sends them along wires to a recording machine or loudspeaker.

The electrical oscillations produced by a microphone may also be measured by a voltmeter to indicate the loudness or amplitude of the vibrations of the sound wave. Whether a sound is being recorded or measured, the microphone used should respond to the same range of frequencies of sound as the human ear, roughly 20 Hz to 20 kHz. The moving-coil type of microphone has a very even response over most of this frequency range and is widely used for the recording of music. The carbon microphone, however, has a relatively poor frequency response, but is still the most suitable device for use as the mouthpiece of a telephone, for the reasons discussed on p 336.

When we measure the loudness of a sound, the response of the human ear must be taken into account as well as that of the microphone. The human ear does not respond equally well to sounds of different frequencies, as shown in fig. 21.22. This graph shows how sounds in the middle of the frequency range appear loudest to the ear. In addition, the ear compares the loudness of two sounds in terms of the *ratio* of the two amplitudes rather than the difference of their amplitudes. Because of this a special scale is used to measure sound levels or loudness which is calibrated in *decibels*.

Figure 21.22   *The frequency response of the human ear*

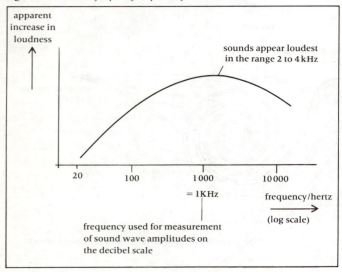

fig. 21.2. The range of pressure amplitudes produced by common sounds is very great; for example, the noise of heavy traffic may produce a pressure change 10 000 times greater than a whisper. The ear does not hear traffic as being 10 000 times louder than a whisper; it compares them in terms of the *ratio* of their pressure amplitudes. The following example illustrates how the ear responds.

Suppose a sound wave produces a pressure change of 1 Pa at the ear drum. If the pressure amplitude was doubled to 2 Pa, we would notice the sound get louder. Another louder sound produces a pressure change of 10 Pa, but if this pressure amplitude was also increased by 1 Pa to 11 Pa, we would not notice any change in loudness. To produce the same apparent increase in loudness it would be necessary to double the pressure amplitude, this time to 20 Pa. In general, adding a little extra amplitude to a quiet sound is very noticeable, but adding the same extra amplitude to a louder sound is not noticeable at all. You would notice a whisper in a library but not at a discotheque.

To represent the relative loudness of different sounds in a way which is similar to the natural response of the human ear we use a special scale which compares the pressure amplitude of a sound wave with a standard value. The scale has units called **decibels**, symbol dB.

The sound loudness scale measures the ratio

$$\frac{\text{pressure amplitude of a sound wave}}{\text{standard pressure amplitude of } 2 \times 10^{-5}\,\text{Pa}}$$

This ratio is based on a specified frequency of 1 kHz because of the variation in the sensitivity of the ear at different frequencies.

The decibel scale, shown in fig. 21.23, is a *logarithmic* scale which places each tenfold increase in pressure amplitude 20 decibels higher on the scale. A sound wave which has a pressure amplitude as low as 2×10^{-5} Pa is only just audible. This very faint sound level is given the value of 0 dB on the sound loudness scale. A sound of 10 times that pressure amplitude is at the level of 20 dB on the scale. This means that for every extra 20 dB on the scale, the pressure amplitude increases 10 times, thus an increase in loudness of 40 dB represents a 100 times increase in the pressure amplitude. 60 dB represents a 1 000 times increase. More important is the fact that the human ear hears each 20 dB increase on this scale as an *equal increase* in loudness.

An increase of +10 dB in the sound pressure level is heard as a doubling of the loudness. Similarly a decrease of −10 dB is heard as a halving of the loudness. The smallest change in loudness that we usually notice is about 3 dB.

Sound level meters used to measure the sounds which people hear need to be designed so that they respond in a similar way to the ear. In particular they need to respond more to frequencies in the middle of the hearing range and less to frequencies at the extremes of the range. Some sound level meters have circuits built into them which adjust their readings at different frequencies to match them up with the response of the ear. The scale used by these meters, called the **A-scale**, is said to be *frequency weighted*. The units on these scales are labelled in **dB(A)**. This is the most widely used scale.

Sound pressure and the decibel scale

When a sound wave passes through the air, the longitudinal vibrations of air molecules cause small pressure changes. There is a small increase in pressure where the molecules are compressed and an equal decrease where they are rarefied. The ear is sensitive to these pressure changes, which make the ear drum vibrate transversely by pushing it in and out.

We call the maximum change of pressure, caused by a sound wave, its **pressure amplitude**. This was shown in

Figure 21.23 *Sound pressure levels*

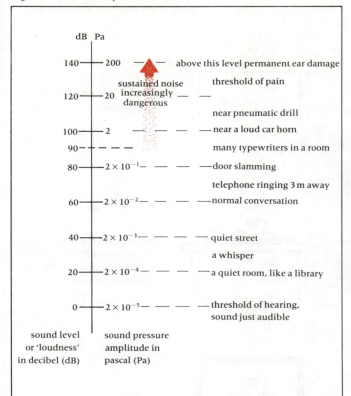

```
       dB  Pa
      140 — 200 ———→   above this level permanent ear damage
                       threshold of pain
      120 — 20 ——— —
                       near pneumatic drill
      100 — 2 ——— ———  near a loud car horn
       90 — ———        many typewriters in a room
       80 — 2 × 10⁻¹ — ——— door slamming
                       telephone ringing 3 m away
       60 — 2 × 10⁻² — ——— normal conversation
       40 — 2 × 10⁻³ — ——— quiet street
                       a whisper
       20 — 2 × 10⁻⁴ — ——— a quiet room, like a library
        0 — 2 × 10⁻⁵ — ——— threshold of hearing,
                            sound just audible
```

sustained noise
increasingly
dangerous

sound level   sound pressure
or 'loudness'   amplitude in
in decibel (dB)   pascal (Pa)

The music played by these young Chinese musicians, when listened to from about 3 metres distance will have a sound level of about 70 dB(A). This level of sound is very comfortable to listen to.

The sound level from this pneumatic drill when heard from close by can be as high as 110 dB(A). This noise level can damage people's hearing and the drill operator must wear ear protection.

The problem of noise

An interesting sound to one person, may be described as a noise by another person. One person's work may disturb another's rest, and one person's leisure music cause others to complain. Noise generally is an unwanted sound which a person would rather not hear. So to what extent are we compelled to live in a noisy world? How much of the noise to which we are exposed is necessary? How is it produced? Can it be prevented or suppressed?

Ear protection must now be worn in many noisy work places.

Noise pollution

Noise is now sometimes described as a form of pollution. By this we mean that other, pleasant or important sounds become spoilt or difficult to hear because of the unwanted noise. The sound of a noisy aircraft flying overhead may make it impossible to hear someone speaking nearby, or spoil the enjoyment of listening to music. Noise can be so loud that it becomes dangerous. The effects of exposure to loud noise get more serious as the level and time of exposure increase. Some people are regularly exposed to sound levels above 90 dB. This may happen in a factory, when driving a large vehicle, or when listening to very loud music at close range. Their hearing may be only temporarily impaired, but it can become permanently damaged with a serious loss of hearing ability. The symptom of 'ringing in the ears' after exposure is a warning sign to avoid frequent repetition of the experience. Temporary blindness and nausea can also be caused by very loud noise. However, even comparatively short exposures to loud noises can have undesirable effects. A person may become less alert and less capable of carrying out a skilled job accurately. Persistent noise causes many people to become irritable, short-tempered, tired and distressed.

So serious is the problem of noise pollution that much is now being done to reduce noise at its source and to protect people who are regularly exposed to it.

Some particular sources of noise make a major contribution to the noise around us. Traffic and aircraft, industrial sources and certain leisure activities are the most common problems. Noise pollution has increased as vehicles and aircraft have got both bigger and more numerous. Only now when the problem has become serious have laws been introduced which limit the noise an aircraft is allowed to make, and which require efficient silencer systems to be fitted to vehicles. Only in recent years have we seen tractor drivers, pneumatic drill operators and machine operators in factories wearing ear protection.

Noise reduction

It is often difficult and expensive to reduce noise at its source. To do this we must have some understanding of how a noise is produced. Machinery often vibrates if it has moving or rotating parts. The vibrations are passed on to surrounding floor, building or vehicle and then on through the air as a sound wave. It follows that if we can reduce the amplitude of the vibration itself or if it can be isolated or insulated from its surroundings, the noise it produces will be reduced.

An example of how a vibration can be reduced is the balancing of a rotating shaft in an engine. A badly balanced shaft makes the whole engine vibrate and also tends to increase wear on bearings. A balanced shaft runs more smoothly with less vibration and less noise.

An example of *insulating* a source of vibrations from its surroundings is the use of a felt mat under a typewriter. The mat insulates the machine from the desk top. Without the mat, the desk top amplifies the sound by picking up the vibrations from the typewriter. Car engines are often supported on mountings formed by bonding rubber blocks to metal brackets. These insulate the engine vibrations from the surrounding car body by *absorbing* the vibrations rather than transmitting them to the rest of the car.

Many simple steps can be taken to reduce noise in a house or office. Double glazing with an air gap of several centimetres and heavy curtains reduce noise from outside. Even a row of trees or an earth bank between a house and a road is effective. Carpets and other floor and wall coverings such as cork and polystyrene products reduce noise by absorbing it and preventing reflections and echoes. In factories particularly noisy machines can be enclosed in a box or separate room. To be effective the source of vibration should not be in direct contact with the surrounding walls and the enclosure should be airtight.

Among the leisure activities that can be considered potentially dangerous are the firing of guns and listening to greatly amplified music, whether live or recorded. Keeping further away from the source of these sounds reduces the sound level considerably, but for performers or operators, who may experience regular exposure to high sound levels, the only thing to do is to wear some form of ear protection.

Some steps to reduce noise.

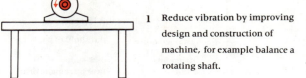

1 Reduce vibration by improving design and construction of machine, for example balance a rotating shaft.

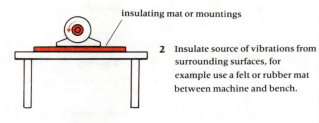

insulating mat or mountings

2 Insulate source of vibrations from surrounding surfaces, for example use a felt or rubber mat between machine and bench.

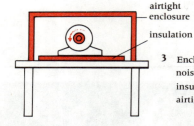

airtight enclosure

insulation

3 Enclose source of vibrations and noise. Enclosure should be insulated from machine and airtight if possible.

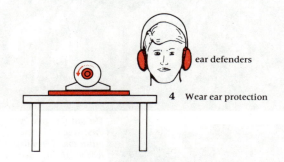

ear defenders

4 Wear ear protection

unbalanced shaft causes vibrations

Very noisy machine on a bench which amplifies the vibrations by resonance

Electronic noise

The telephone or communications engineer and the recording engineer making records and tapes all have a special noise problem. All the transducers which convert sound from one form to another and all the storage systems tend to add unwanted noise in the form of **hiss** or **hum**.

For example a telephone mouthpiece uses a carbon microphone in which the granules of carbon generate a background of hiss or *white noise* caused by the random movement of the granules. (White noise, like white light, contains a broad spectrum of frequencies and has no particular pitch.) A record player adds unwanted noises while converting the mechanical vibrations from the record grooves to electrical signals in the pick-up, and during amplification. A tape recorder adds hiss to the recorded sound because stray magnetism in the recording head and the metal oxide coating on the tape are converted into unwanted noise.

As more and more noise is added to a sound it becomes increasingly difficult to hear clearly. Many improvements in recording and communication electronics are concerned with reducing or filtering out unwanted noise. One of the advantages to be gained from the new generation of **digital** sound recordings is a significant reduction in unwanted noise levels.

Assignments

Test your hearing range

Connect a loudspeaker to the low-resistance or speaker output of an audio signal generator. Find the lowest pitch and highest pitch sounds which you can just hear. Because the ear responds less well to low and very high frequencies, it is necessary to increase the volume or gain to maximum at these frequencies. A good human ear should be able to hear frequencies somewhat lower than 30 Hz and higher than 15 kHz.

Examine the waveforms of the sounds produced by several musical instruments. Use the arrangement shown in fig. 21.20a to display sound waveforms on the screen of an oscilloscope. In a class or in a school there will be several people who can play different instruments. It is helpful to compare sounds of the same frequency from each instrument. Attempt to produce a steady note of constant pitch and amplitude. Make a comparison with a tuning fork, which should produce a pure sine wave.

Measure the noise level (in dB(A)) of a variety of sounds using a noise level meter. It is very interesting to measure the noise level at different places around a school. Readings can also be collected near a road or railway, in a local shop or factory and at home.

Make a list of practical methods used to reduce noise. Investigate at school, at home and in local industry what methods are used to reduce noise pollution. A local radio studio, factory, office, hospital or airport may welcome a visit from a group of students interested in noise reduction.

Try questions 21.15 to 21.19

Questions 21

1 Describe an experiment to show that sound waves cannot pass through a vacuum. (JMB, part)

2 a) Sketch the waveform of a pure note, and label its amplitude a and wavelength λ.
   b) Sketch the waveform of a pure note of twice the frequency and double the intensity of (a) (O, part)

3 In a simple experiment to determine the speed of sound, an observer with a stopwatch stands on a flat stretch of sand and an assistant standing at a measured distance of 800 m fires a pistol. The observer starts his stopwatch when he sees the flash of the pistol and stops it when he hears the sound of the shot. The time intervals obtained for three experiments are: 2.2, 2.3, and 2.1 s. Calculate a value for the speed of sound in air. (C, part)

4 A person claps his hands at approximately $\frac{1}{2}$ second intervals in front of a wall 90 m away. He notices that each echo produced by the wall coincides with the next clap.
   a) Calculate an approximate value for the speed of sound.
   b) If you were using the above as a basis for an experimental method to determine the speed of sound, what procedure would you adopt to obtain high accuracy in the timing part of the experiment? (L, part)

5 A student standing between two walls, as shown in fig. 21.24, shouts once and finds that the time interval between hearing the first and second echoes is 2.0 seconds. What value would the student obtain for the speed of sound in air? (AEB)

Figure 21.24

6 When a sound is made in a large room more than one echo may be heard. Describe and explain this effect.

7 A fishing boat uses ultrasound of frequency 6.0×10^4 Hz to detect fish directly below. Two echoes of the ultrasound are received, one after 0.09 s coming from a shoal of fish and the other after 0.12 s coming from the sea bed. If the sea bed is 84 m below the ultrasound transmitter and receiver, calculate
   a) the speed of the ultrasound in water,
   b) the wavelength of the ultrasound waves in water and
   c) the depth of the shoal of fish below the boat.

8 a) Explain the meaning of the terms *longitudinal* and *transverse* as applied to progressive waves. Give one example of a longitudinal wave and one example of a transverse wave.
   b) Describe in detail a method to find the speed of sound in air. In your account you should indicate what you measure, how you measure it and how you process your measurements to find the speed of sound.
   c) An aircraft A is flying at a constant speed of 270 m/s at a constant height of 8100 m above the surface of the earth. The aircraft directs a radar beam of wavelength 1 cm at a target T on the Earth's surface. After 90 μS an echo from the target is detected on the aircraft.
      If the frequency of the radar waves is 3×10^{10} Hz, find:
      i) the speed of the radar waves, showing how you obtain your answer,
      ii) the distance AT between aircraft and target,
      iii) the time which elapses before A is vertically above T.
      $[1\,\mu s = 1 \times 10^{-6}\,s]$
   d) Explain why *sound waves* cannot normally be used by aircraft in flight to measure their distance from the ground.
   e) Aircraft pilots communicate with air traffic controllers on the ground using electromagnetic waves. A radio transmitter converts electricity to radio waves and a radio receiver performs the reverse function. Using this information describe briefly all the major energy changes which occur to the sound of a pilot's voice before it can be heard by an air traffic controller and in each case give the name of the transducer which causes the change. (NISEC 88)

Figure 21.25

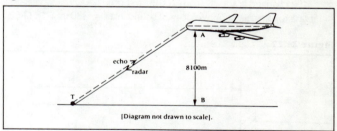

[Diagram not drawn to scale].

9 A wheel has 50 spokes and rotates at 10 revolutions per second. Calculate the frequency of the note obtained by holding a card lightly against the spokes as they rotate. What would you expect to observe if a source of sound of frequency 502 Hz was placed near the card?
   Describe briefly how you would use a stroboscope to check the rate of rotation of the wheel. (S, part)

10 Explain what is heard when two notes of the same intensity and frequencies 256 Hz and 258 Hz are sounded together. (O, part)

11 A small paper rider is placed on a steel wire which is stretched over two fixed bridges which define the length of wire free to vibrate. When a certain vibrating tuning fork A is placed with its stem on one of the bridges, no effect is seen but when A is

replaced by another vibrating tuning fork B, the rider jumps off the wire. Explain this effect.   (JMB, part)

12   a)   In Fig. 21.26, when the wire was plucked and the tuning fork struck, the tuning fork had the higher pitch.
i)   State **two** ways by which the wire could be made to produce the same note as the tuning fork.
ii)   State **one** way in which the vibration of the string and the sound waves in the air are the same, and **one** way in which they are different.
b)   A young scientist said "I think thunder and lightning should really be called lightning and thunder".
Why does this seem a sensible suggestion?   (W 88)

Figure 21.26

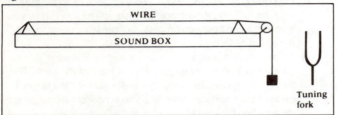

13   a)   Fig. 21.27 shows a piston inside a glass tube. When the air in the tube is made to vibrate, a note is heard.
i)   Express the wavelength of the note, in terms of *l*, when the note is the fundamental.
ii)   Write down an expression for the frequency of the fundamental note.
iii)   What would happen to the frequency of the note if the piston were moved along the tube to the right?
b)   Explain what is meant by *resonance* and name one example in which it occurs in everyday life.
c)   Explain, with the aid of a labelled diagram, how you would demonstrate that sound needs a medium for its transmission whereas light does not.
d)   A source of sound of variable frequency is held in front of the open end of the tube in fig. 21.31, and the distance *l* is 0.33 m. Explain, in detail, what happens as the frequency of the source is changed gradually from 200 Hz to 600 Hz.
Data for this question: Velocity of sound in air = 330 m/s.   (AEB)

Figure 21.27

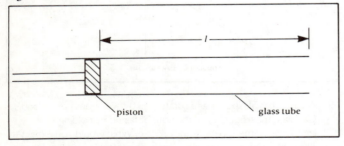

14   The table below shows the results of an experiment in which the length of a stretched wire (a sonometer) was varied until it vibrated in unison with each of several tuning forks taken in turn. The tension of the wire was kept constant throughout the experiment.

Frequency of tuning fork/Hz	256	288	341	384	512
Length of wire/mm	942	837	707	628	471

a)   What do you understand by the frequency of a tuning fork?
b   Plot a graph of the frequency of the tuning fork against the length of the wire.
c)   Use the graph to determine the frequency of the tuning fork which will vibrate in unison with a wire of length 754 mm at the same tension as in the experiment.   (L, part)

15   The chart in Fig. 21.28 shows the number of students in a class who can hear notes of different frequency.
a)   State TWO conclusions, about the hearing of the students, which can be drawn from this chart.
b)   What name is usually given to frequencies above 20 kHz?
c)   Give one medical application of frequencies above 20 kHz.   (SEB 90)

Figure 21.28

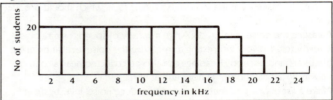

16   A microphone is connected to a cathode ray oscilloscope. A loudspeaker giving a single note is switched on. The trace shown in Fig. 21.29 appears on the screen:
a)   Which of the measurements P, Q, R, S i) is the amplitude, ii) represents the time for one complete oscillation?
b)   Copy Fig. 21.29 and on the same axes sketch the appearance of the screen when i) a louder note is used, ii) a higher frequency note is used, iii) a note of higher pitch is used.   (NEA [A] 88)

Figure 21.29

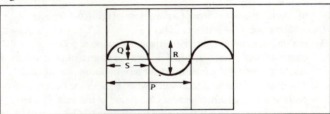

17   a)   Explain what is meant by the musical interval the *octave*.
b)   How many octaves span the interval between the frequencies 32 Hz and 256 Hz?   (O, part)

18   a)   A pupil places a loudspeaker, which is producing a musical note, on a radiator. To his amusement, the sound passes along the pipes to other parts of the school.
i)   Describe the movement of the metal 'particles' of the pipe caused by sound waves passing along the pipe.
ii)   State the name of this type of wave.
b)   The speed of the waves along the pipe is greater than the speed of the waves in air.
i)   How does the wavelength of the waves in the air differ from the wavelength of the waves in the pipe?
ii)   How does the frequency differ in each situation?
c)   Footsteps in an empty house seem loud and often eerie. This is not so in a similar furnished house. Explain why.
d)   If you lived in a street which carried heavy traffic, suggest **three** ways, other than altering the traffic flow, by which you could reduce the noise entering your home.   (Joint 16+)

19   State **two** ways in which some materials help to suppress noise.   (JMB, part)

Light waves

We cannot see waves in a ray of light yet we imagine that it contains millions of very short waves. In this chapter we use the idea of light waves to explain some of the properties of light.

Diffraction of white light by a diffraction grating, below, spreads the light into several rainbow coloured fans of light. The upper portion was taken with film sensitive to UV light.

22.1
THE NATURE OF LIGHT

Scientists have long sought an explanation of the nature of light. The experimental evidence has often appeared to be conflicting and has led to rival theories.

In 1801 Thomas Young was the first person to produce evidence of the wave nature of light when he produced interference fringes in the light which passed through two narrow slits. Young was not the first person to imagine that light might have a wave nature; as early as 1690 Huygens suggested that light could travel as a wave motion.

An alternative theory, proposed by Newton, suggested that light travelled as small particles called *corpuscles*. The particle theory could apparently account for reflection and refraction of light by applying Newton's laws of motion to the small corpuscles. However, the particle theory of light failed to account for the properties of light which we call interference and diffraction. Here the wave theory triumphed!

The wave theory successfully explains the following properties of light:

a) *reflection*,
b) *refraction*, resulting from a change in the speed of light,
c) *diffraction*, the spreading of light waves round corners and through apertures,
d) *interference*, the superposition of light waves resulting in constructive and destructive interference, and
e) *polarisation*, a property of waves we have not seen with water waves and which cannot happen with sound waves.

We shall look at the polarisation of light waves first because it tells us something about the nature of the waves which the other properties cannot.

Polarisation

If we place a single sheet of polaroid film in front of a light source we can see that some light passes through but everything looks much darker. Polaroid film transmits about 30% of the light and its darkening effect is well known through its use in sunglasses. However, it is not this reduction in the amount of light transmitted that makes it special, but the fact that it polarises light, which is not so well appreciated or understood.

When we place a second sheet of polaroid film overlapping the first a surprising thing happens. As we slowly rotate one of the sheets there is one position where they allow no light through at all. We would expect the two sheets together to make things appear even darker, but why should they transmit no light when set in a certain way? To explain this effect we can use a wave model.

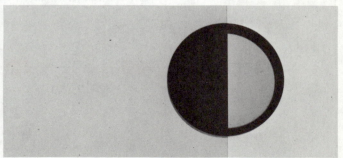

A model of polarisation

- Take two boards A *and* B *with narrow vertical slits and firmly hold them parallel to each other, some distance apart, fig. 22.1.*
- *Pass a length of rope through the two slits, fixing it at one end and holding the other in your hand.*
- *Move your hand so that it makes waves in all directions at right angles to the rope up to slit* A.

The transverse waves arriving at A are due to vibrations in both vertical and horizontal planes (and other planes in between these), but the waves which emerge through slit A and travel on to slit B are caused only by vibrations in the vertical plane. The only waves allowed through are those parallel to the slit. The waves between A and B are called **polarised waves** and the slit A, acting like a sheet of polaroid film, is called a **polariser.**

Polarised waves are waves which lie in one plane only. That plane, containing both the direction of travel of the wave and the direction of the displacement in the wave motion, is called the **plane of polarisation.**

The waves beyond slit A are said to be *vertically polarised*, i.e. they are vibrating up and down in a vertical plane. When these waves arrive at slit B they are polarised in a plane which is parallel to the slit and so they can pass through as shown in fig. 22.1a.

In case (b), however, the vertically polarised waves cannot pass through the horizontal slit B because they are vibrating in a plane which is at right angles to the slit.

We can see that the slits would have no effect on the longitudinal wave motion, such as sound, because there is no vibration at right angles to the direction of travel of the wave. *Longitudinal waves cannot be polarised.*

This mechanical wave model explains what we think happens with the two sheets of polaroid film. When the sheets are arranged so that they both transmit light waves vibrating in the same plane, as in case (a), light can pass through them. However, when one of them is turned through 90° so that its plane of polarisation is at right angles to the other one, no light can pass through, as in case (b). The fact that light can be polarised is very convincing evidence that:

a) light has a wave-like nature, and
b) its waves are transverse rather than longitudinal.

Facts about light waves

a) Light waves are part of the *electromagnetic spectrum of travelling waves*.
b) Light waves are *transverse* and can be *polarised*.
c) Light waves carry energy.
d) Light waves are emitted and absorbed by matter.
e) Light waves can travel through both a vacuum and some matter, but involve no movement of that matter along with the wave motion.
f) Light waves travel at 3.0×10^8 m/s in a vacuum.
g) Light waves can be *reflected*, obeying the laws of reflection.
h) Light waves are *refracted* as their speed changes in different media. The change of speed causes a change of wavelength but no change of frequency.
i) Light waves show *diffraction* and *interference* effects under certain circumstances.
j) Light also shows *particle-like* properties under different circumstances. These particles of energy, called *quanta* or *photons*, are believed to exist at the same time as the wave-like nature (p 344).

Figure 22.1 *Demonstrating polarisation*

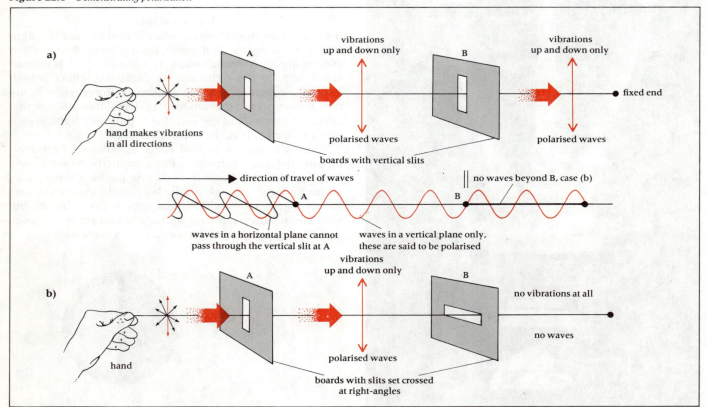

22.2
INTERFERENCE OF LIGHT WAVES

The most striking feature of interference between light waves is that in some places where they overlap we find that light added to light produces darkness. This unexpected result, which seems to conflict with our ordinary experience, can be explained as destructive interference of waves.

Double-slit interference, Young's experiment

When light from the same source passes through two narrow slits which are close together the effect known as interference can be seen. A laser may be used to demonstrate the pattern produced by a double slit but great care must be taken to follow the correct safety rules otherwise eyes may be damaged by the intense light of the laser beam (see appendix D11).

A laser mounted firmly above a bench, pointing away from the viewers towards a screen or wall, will produce a single bright spot of light. When a narrow double slit is placed in front of the laser a pattern of dots can be seen on either side of the original central spot. (See appendix D12 about making double slits.)

● *What do you notice about the spacing of the dots?*
● *What is the effect of using double slits which are closer together or further apart?*

Interference occurs with ordinary light as well as laser light so now try to produce a similar pattern using a small laboratory light source such as a 12 V lamp with a straight or 'line' filament.

● *Set the lamp filament parallel to the double slits as shown in fig. 22.2. Carefully align the apparatus.*

a) The adjustable single slit shown in the figure can be used to adjust the brightness and clarity of the pattern.
b) The interference pattern will be seen more easily if the double slits are surrounded by a black screen so that light reaches the eye only through the slits.
c) The pattern can be viewed on a translucent screen or through an eyepiece with a measuring scale. A suitable translucent screen can be made from tracing or greaseproof paper held flat in a card frame.
d) The experiment requires a good blackout and often it is necessary to cut out stray light reflections, but the most usual cause of not seeing the pattern is misalignment of the apparatus.
e) Adjustments to the alignment can be made by tracing the path of the light along the bench with a sheet of white paper.

● *Can you see vertical bright and dark lines on the screen or through the eyepiece? These are called* interference fringes. *How many bright fringes can you see?*
● *Are the fringes sharp or blurred?*
● *Are the fringes equally spaced and equally bright?*
● *Insert red, green and blue filters in front of the lamp. For each colour answer the previous questions and note any differences.*
● *Try using double slits with a different slit separation and repeat the observations, again noting any differences.*

The fringes should appear as in the photograph in fig. 22.2. For a single colour of light there are equally spaced light and dark bands with blunt edges. The central fringes are brighter. The fringes are closer together for green light than for red, and the blue fringes have about half the spacing of the red ones. The slit separation has an inverse effect, i.e., for slits closer together the fringes are further apart.

Figure 22.2 *A practical arrangement for observing Young's double-slit interference fringes*

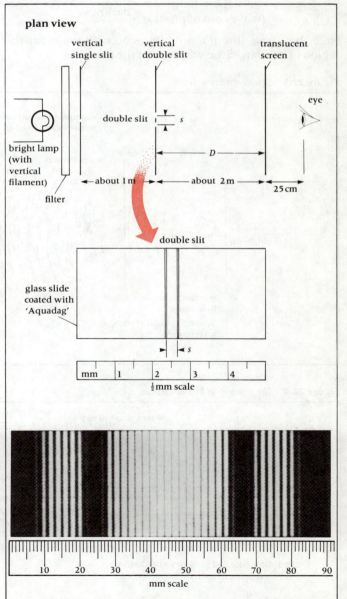

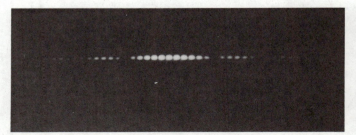

The narrow beam of laser light, after passing through a pair of narrow slits 0.25 mm apart, produces this pattern of bright spots which are equally spaced and symmetrical on either side of a central spot. The central spot has the same size and spacing as its neighbours.

A wave explanation of double-slit interference

Fig. 22.3 shows how wavefronts from the lamp or single slit arrive at the double slits, S_1 and S_2, in phase. We can think of this as a wave crest arriving at both slits at the same moment. Each slit diffracts the light waves, spreading them out, so that in the space between the double slits and the screen there is a region where the two beams of waves overlap. Where waves overlap superposition results in constructive and destructive interference.

As we have already seen in chapter 20, constructive interference occurs when crests overlap, i.e. the waves are in phase (in step). These are the conditions for bright fringes to be formed by *constructive interference*:

$$\frac{\text{light} + \text{light}}{\text{(waves in phase)}} = \text{brighter light}$$

Destructive interference occurs when a crest and a trough overlap, i.e. the waves are out of phase (out of step). Thus dark fringes are produced by *destructive interference*:

$$\frac{\text{light} + \text{light}}{\text{(waves out of phase)}} = \text{darkness}$$

Note that these interference effects occur throughout the region of overlap of the waves from the two slits.

Path difference

Fig. 22.4 shows how constructive interference is produced at a point P on the screen. The distance from slit S_2 to P is greater than from slit S_1, by an amount S_2Q ($S_2Q = S_2P - S_1P$). This is called the *path difference* of the two wavetrains.

When the path difference S_2Q equals a *whole* wavelength, the waves arrive at P in phase and produce a bright fringe. Similarly, when S_2Q is two wavelengths another bright fringe will be produced at a position $2x$ from C and so on. When there is no path difference the waves arrive in phase to form the central bright fringe at C.

At the point midway between the bright fringes at C and P the path difference is *half* a wavelength, the waves arrive out of phase and destructive interference produces a dark fringe. The fringe pattern is symmetrical about C.

In general, if n is a whole number ($n = 0, 1, 2, 3$ etc.) we have:

Bright fringes are produced where the

$$\text{path difference} = n\lambda$$

Dark fringes where the

$$\text{path difference} = (n + \tfrac{1}{2})\lambda$$

Figure 22.3 *Double-slit interference*

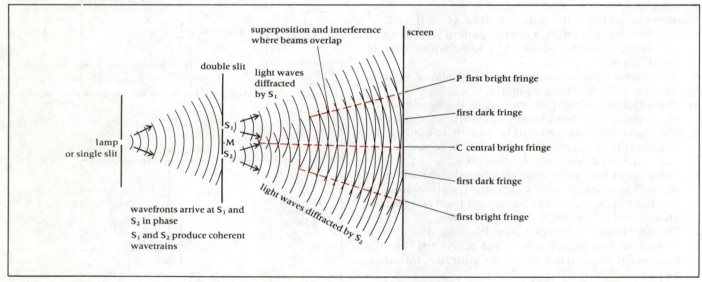

Figure 22.4 *The geometry of Young's fringes*

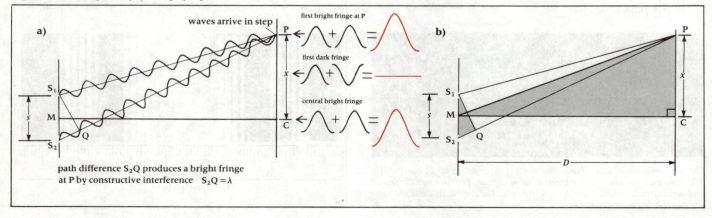

Calculation of the wavelength of light

Referring to fig. 22.4b, we have:

s is the separation of the slits S_1 and S_2,

M is the midpoint of the slits,

D is the distance from the double slits to where the fringes are measured on the screen,

C is the position of the central bright fringe,

P is the position of the first bright fringe,

x is the separation of adjacent fringes, $x = $ PC.

Note that D is more than a thousand times bigger than s, the diagram has been drawn out of proportion to make it clearer.

For the first bright fringe at P

$$\text{path difference } S_2Q = \lambda.$$

The shaded triangles are really very long and thin making MP very nearly parallel to S_2P.

$$\therefore \quad \angle S_1S_2Q = \angle MPC \qquad \text{(alternate angles)}$$

Angle S_1QS_2 is very nearly a right angle

$$\therefore \quad \angle S_1QS_2 = \angle PCM$$

So the shaded triangles PCM and S_1QS_2 have two equal angles and are therefore similar triangles. It follows that

$$\frac{S_2Q}{PC} = \frac{S_1S_2}{MP}$$

and over a long distance D, MP = MC = D. By substituting we get

$$\frac{\lambda}{x} = \frac{s}{D}$$

This gives the **wavelength formula**

$$\lambda = \frac{xs}{D}$$

If we rearrange this formula to give the **fringe separation** x, we get:

$$x = \frac{\lambda D}{s}$$

This formula agrees with the observations made about the interference fringes which can be summarised as follows:

a) $x \propto \lambda$ The fringe separation is directly proportional to the wavelength, i.e. fringes formed by red light waves are proportionally further apart than those formed by blue light waves of shorter wavelength.

b) $x \propto 1/s$ The fringe separation is inversely proportional to the slit separation s, i.e. double slits closer together produce fringes which are proportionally wider apart.

c) $x \propto D$ The fringe separation is directly proportional to the distance of the screen from the slits, i.e. when the screen is further away the fringes are proportionally futher apart.

The conditions for interference fringes

Bright and dark interference fringes can be seen only under certain special conditions. If, for example, we replace the two slits with two separate light sources, fringes cannot be seen. The necessary conditions can be summarised in the following three points.

a) The light waves from the two slits must have exactly the same wavelength and frequency. (With waves of different wavelength the positions where the crests coincided would always be changing.)

b) The two sets of waves must have roughly equal amplitude, otherwise the larger one will swamp the smaller one and fringes will not be seen.

c) The two sets of waves must originate from the same light source. The light waves which pass through the two slits are said to be **coherent** when they come from the same source. Two separate lamps produce waves which are incoherent.

Coherent sources of waves

Two separate lamps do not emit light waves in the same continuous and regular manner as the two dippers mounted on a vibrating beam in a ripple tank. The dippers send out two sets of waves which are closely related to each other because they are linked to the *same source of vibration*, i.e. the beam. The two sets of waves produced by the two dippers are said to be *coherent*. The important thing is that the two sets of waves are really produced by the *same* vibrator. It follows then that any change which occurs in one set of waves also occurs in the other. In particular they will always have the same phase relation at a particular position where they interfere. The result is that the positions of constructive and destructive interference always occur in the same positions (for a particular wavelength). Only when the positions of the interference fringes are fixed can we see them.

When the double slit is illuminated by a single lamp, each wavetrain passes through both slits at the same time. Any change in the phase of the waves, as different wavetrains reach the two slits, happens at *both* slits. The wavetrains which emerge from the two slits bear the same phase relation to each other at all times and are therefore *coherent* waves. The wavetrains from two separate sources of light would each change in phase independently and randomly with each new wavetrain emitted from the sources. No constant phase relation can exist between two sets of waves emitted randomly. Such waves are said to be *incoherent* and do not produce interference fringes in fixed positions.

Fig. 22.5 shows the difference between a pair of incoherent wavetrains and a pair of coherent ones. In both cases the wavetrains begin and end randomly but only in case (b), where the waves are coherent, does the phase relation between the two parallel sets of waves remain constant. As shown in the figure, the waves in (b) are always in phase or in step with each other as they come through the two slits.

Measuring wavelength by double-slit interference

● *Set up the experiment as shown in fig. 22.2 and insert a colour filter (red or blue) in front of the lamp.*

● *Measure the distance D from the double slits to the screen using metre rulers.*

● *Measure the width of about 5 spaces between fringes, (5x). To do this with a screen, make pencil marks on the screen to show the spacing of fringes for both colours. Bright or dark fringes will give the same result, being equally spaced, but be sure you have counted spaces between fringes and not fringes. Measure the distance marked on the screen using a $\frac{1}{2}$ mm scale.*

The slit separation *s*, being a fraction of a millimetre, is most difficult to measure and limits the accuracy of the final result. A satisfactory method is as follows.

• *Hold the double slits against a ½mm scale and view them under a microscope.*

This way an estimate accurate to about 0.1mm is possible. A travelling microscope with a vernier scale would give a more accurate value.

• *Now substitute your results for D, x and s in the formula*

$$\lambda = \frac{xs}{D}$$

and calculate a value for the wavelength of red and blue light.

Worked Example
The wavelength of light

In a double-slit interference experiment with blue light the following measurements were obtained:
 distance of screen from double slit (D) = 1.82 m
 slit separation (s) = 0.30 mm
 distance between 6 fringes = 5 spaces (5x) = 12 mm
From these results calculate the wavelength of blue light.
We have
 D = 1.82 m = 1.8 m (to 2 significant figures)
 s = 0.30 mm = 0.000 30 m = 3.0×10^{-4} m
 $5x$ = 12 mm

$$\therefore \quad x = 2.4\,\text{mm} = 0.0024\,\text{m} = 2.4 \times 10^{-3}\,\text{m}$$

We use

$$\lambda = \frac{xs}{D} = \frac{(2.4 \times 10^{-3}\,\text{m}) \times (3.0 \times 10^{-4}\,\text{m})}{1.8\,\text{m}}$$

$$= \left(\frac{2.4 \times 3.0}{1.8}\right) \times 10^{-7}\,\text{m} = 4.0 \times 10^{-7}\,\text{m}$$

Answer: The wavelength of blue light was found to be 4.0×10^{-7} m or 0.0004 mm.

Assignments

Remember

a) The formula for the wavelength of light, calculated from double-slit interference fringes.:

$$\lambda = \frac{xs}{D}$$

b) The information about light waves given on p 474.
c) The conditions necessary for interference fringes to be seen in light coming from double slits.

Make a corrugated cardboard wave model of double slit interference. Cut two strips of corrugated cardboard, about 25 cm long and ¼ cm wide across the corrugations so that two cardboard 'wavetrains' are produced. Put a sheet of paper on a drawing board and pin each strip to the drawing board through a hole at one end. The strips are turned on their cut sides so that they look like the wavetrains in fig. 22.4a. The fixed ends of the 'wavetrains', representing the slits S_1 and S_2, should be about 5 cm apart. Draw a line parallel to S_1 and S_2 about 20 cm away to represent the screen. Gently pull the cardboard wavetrains tight and overlap them so that they are in phase or out of phase along the line of the screen. Mark the positions where the fringes would occur on the screen.

Try questions 22.1 to 22.9

Figure 22.5 *Incoherent and coherent wavetrains*

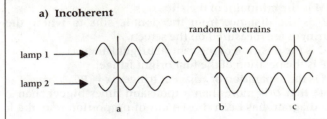

a) Incoherent

lamp 1 →
lamp 2 →

random wavetrains

a   b   c

At a, b and c the phase relation between the wavetrains from lamp 1 and lamp 2 keeps changing randomly. The wavetrains are said to be incoherent.

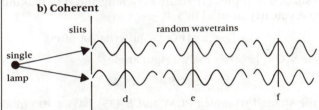

b) Coherent

single lamp

slits

random wavetrains

d   e   f

At d, e and f the phase relation between the two sets of wavetrains is constant even though successive wavetrains are emitted randomly. The parallel sets of wavetrains are said to be coherent.

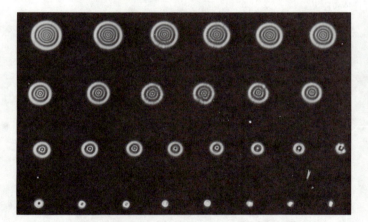

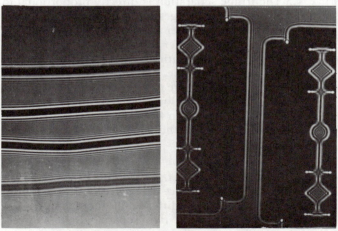

Diffraction fringes formed by light passing through holes of various sizes, and spreading round the edges of four fine hairs and the edges of razor blades.

22.3
DIFFRACTION OF LIGHT WAVES

Now we shall see that light can spread round corners where rays, travelling in straight lines, should not go. This effect, which is due to the diffraction of light waves, gives support to the theory that light has a wave-like nature.

Diffraction by a single slit

Waves are diffracted when they meet an obstruction in their path. For the spreading of the waves to be noticeable, the size of the obstruction or gap through which the waves pass must be similar to the wavelength of the waves. The results from interference experiments suggest that the wavelength of light is in the range 0.4 µm to 0.7 µm. So, if we use a very narrow slit, say about 0.5 µm wide, the light waves will spread out in a fan-shaped beam as if they were produced at the slit from a point source. We saw this effect with water waves in a ripple tank, fig. 20.22b. If however the slit is very much wider, say 1 mm, then the light passing through it will form a shadow with fairly sharp edges and will appear to travel in straight lines. When the slit width is somewhere between these two values the effects are very interesting and provide more evidence of the wave-like nature of light.

Making a single slit

We can make a single slit by the method suggested in appendix D12. A pinpoint, however, will only produce a slit of one particular width and it is interesting to see the effect of varying the slit width. If two razor blades or knife edges are mounted in a frame so that one is fixed and the other can be moved towards it, a narrow and adjustable single slit is formed between the two blade edges.

Diffraction fringes from a single slit

The diffraction pattern produced by a single slit may be demonstrated by mounting the slit in front of a laser and viewing the pattern produced on a screen or wall. (The arrangement is very similar to that used to demonstrate double-slit interference, see p 475 and appendix D12.)

The single slit also produces extra light spots on both sides of the central spot on the laser beam, fig. 22.6a.
- *Look carefully at the brightness and spacing of the light spots. Compare this pattern with that produced by a double slit.*

Figure 22.6

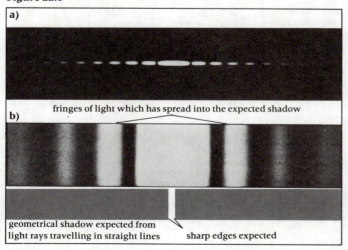

fringes of light which has spread into the expected shadow

geometrical shadow expected from light rays travelling in straight lines

sharp edges expected

- *Produce your own diffraction pattern using a line filament lamp instead of a laser.*
- *Set the single slit parallel to the lamp filament.*
- *From a distance of a few metres, view the lamp through the slit. (Never do this with a laser!)*
- *Slowly reduce the width of the slit until thin light and dark fringes appear on each side of and parallel to the central bright line of the slit.*
- *Note the effect on the fringes of slowly reducing the width of the slit further.*
- *Compare the fringes formed by a single slit with those made by the double slits. Look carefully at the way the brightness and spacing of the fringes changes across the pattern.*

The photograph of diffraction fringes formed by a single slit, fig. 22.6, shows the following features:
a) a bright central band situated where the light is expected to arrive by straight-line travel,
b) narrow fringes of light in the region where there should be totally dark shadow and
c) at the outside edges, fringes gradually getting fainter and closer together.

Diffraction by other objects

Diffraction occurs at all edges where waves can spread round into the shadow region. Thus a narrow object like a fine wire or a human hair can show diffraction fringes at its edges similar to those produced by a single slit. Light and dark fringes are seen at both edges rather than a sharp boundary between light and dark. But why are these fringes not visible all the time? Although diffraction does occur all the time, the fringes are visible only under special conditions. The fringes shown in the photograph opposite are produced by a distant point source of monochromatic light (single colour, single wavelength) placed centrally behind the object in a room with no other light sources. The first bright fringe slightly overlaps the actual edge of the object but the other fringes are outside the edge. Note how the fringes are not equally spaced and how the first bright fringe is actually brighter than the region of uniform illumination away from the edge. The extra brightness accounts for the 'missing' light energy in the dark fringes.

The diffraction grating

We have seen that two narrow slits produce an interference pattern of equally spaced bright and dark fringes. The fringes themselves are not very sharp, fading gradually from brightness to darkness. The effect of bringing the two slits closer together is to increase the spacing of the fringes, but it also makes them appear thicker and even less sharp.

If we increase the number of parallel slits from two to three, to five, to ten and so on, but keep the separation of adjacent slits the same, it has the effect of making the bright fringes sharper without changing their spacing. A **diffraction grating** is a set of multiple parallel slits in which the separation of all the slits is exactly equal. As many as 500 slits or lines can be cut in a space 1 mm wide which means that the separation of adjacent slits is as small as $\frac{1}{500}$ mm or about 2 µm. A slit itself can be as narrow as about 1 µm which is very close to the wavelength of light, 0.4 to 0.7 µm.

To make a diffraction grating as fine as this requires a highly accurate, temperature-compensated machine which cuts the grooves on a sheet of glass. The gaps through which light is transmitted, i.e. the slits, are actually the spaces in between the cut grooves. The grooves themselves scatter light in other directions, acting as the opaque sections in between the slits. Plastic copies can be made from these gratings by using them as a mould. These copies, or replicas, which are much cheaper than a real glass grating, are protected by sandwiching them between two thin glass plates.

A diffraction grating is so called because each of its thousands of very narrow slits diffracts light waves, spreading them out over a very wide angle. The waves from each of the many slits then overlap and interference occurs in the same way as it did with only two slits. The interference fringes produced by the grating with thousands of slits are different in several ways from those formed by only two slits. Fig. 22.7 shows the features of the fringes produced by a diffraction grating from a monochromatic light source, i.e. a source of light of one colour and one wavelength.

a) The increase in the number of slits from 2 to several thousand makes the bright fringes extremely sharp; they are now called 'lines'.

b) The squeezing of the slits closer and closer together opens out the angle between the bright fringes or lines so that they appear as sharp lines with wide dark spaces between them.

c) The slits themselves, being very narrow, cause the light waves to be spread by diffraction over a very wide angle. This makes the bright fringes or lines visible over almost the full 180° angle (90° on either side of the straight-through direction).

The diffraction grating produces a spectrum

● *Place a single bright line-filament lamp with its filament vertical on a bench at one end of a darkened room.*

● *View the lamp filament through a diffraction grating by holding the grating near to one eye and make the following observations.*

● *Does it matter which way round the grating is held?*

● *How many spectra can you see?*

● *What happens in the centre?*

● *Which colour is deviated by the largest angle? Can you explain this, and how does it compare with the deviation produced by a glass prism?*

● *Change the grating for one with a different number of slits per millimetre, i.e., with a different slit separation. What difference does this make?*

The spectrum produced by a diffraction grating from a source of white light is shown in fig. 22.8. We can see several spectra symmetrically arranged on each side of a central white fringe or line. The outer spectra may overlap each other, but otherwise there is darkness between the edges of the spectra. We notice that red light of the longest wavelength is deviated most, which is the opposite result to the deviation produced by a prism. The dispersion (separation of the colours) produced by the diffraction grating is also much greater than by a prism.

Figure 22.7 *Interference fringes produced by a diffraction grating from a monochromatic light source*

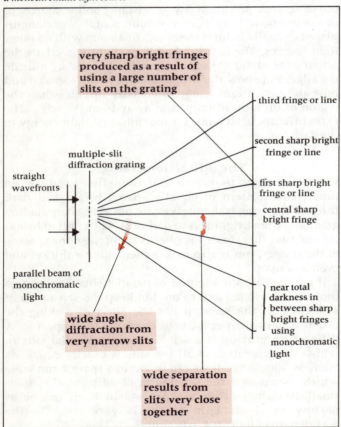

Figure 22.8 *The diffraction grating produces a spectrum*

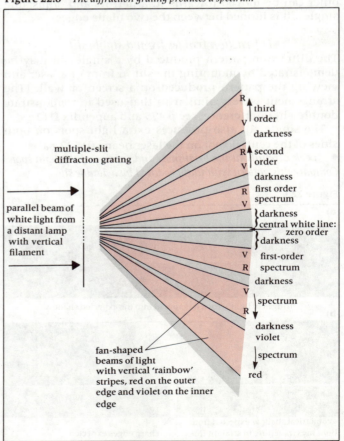

Gratings with a large number of slits per mm, called fine gratings, produce the greatest deviation of the spectra from the central line. This is the result of the slits being closer together. When the deviation is larger the dispersion is also larger, but there are fewer spectra possible in the 90° available. Gratings with fewer slits per mm, called course gratings, produce more spectra but each has less dispersion.

Angle of deviation and wavelength

Fig. 22.9a explains how, in a certain precise direction, the waves of a particular wavelength, coming from all the slits, will be in step and in phase. The circular wavefronts (not shown) emerging from each of the slits combine together to form a new straight wavefront along a line where the waves are all in phase. Then in a particular direction the waves of a particular wavelength reinforce each other by constructive interference and produce a bright line or fringe of light of one particular colour.

The condition for constructive interference to occur between the waves from all the slits is that the path difference between waves from adjacent slits must be a *whole* number of wavelengths. In fig. 22.9a the path difference between adjacent slits is shown as exactly one wavelength and the constructive interference is said to produce the **first-order spectrum**. Monochromatic light (of single wavelength) will produce a single line in a particular direction θ_1 in fig. 22.9b, in the first-order spectrum. White light (of all wavelengths) will produce a spectrum of colours dispersed at different angles according to their wavelengths. We can see that longer wavelengths will increase the deviation angle θ so that it is red light which is deviated most.

Figure 22.9 *The diffraction grating formula*

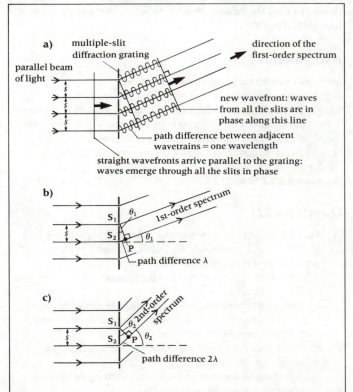

a) multiple-slit diffraction grating

parallel beam of light →

direction of the first-order spectrum

new wavefront: waves from all the slits are in phase along this line

path difference between adjacent wavetrains = one wavelength

straight wavefronts arrive parallel to the grating: waves emerge through all the slits in phase

b)

θ_1 1st-order spectrum

S_1

S_2

θ_1

P

path difference λ

c)

2nd-order spectrum

θ_2

S_1

S_2

P θ_2

path difference 2λ

When the path difference is two whole wavelengths between waves from adjacent slits, fig. 22.9c, the waves again reinforce each other by constructive interference to produce the **second-order spectrum**. Higher order spectra are produced when the path differences between adjacent slits are greater numbers of whole wavelengths. Fig. 22.9c shows how the higher order spectra occur at greater angles of deviation θ_2, etc. The number of orders is limited by the maximum angle of deviation, $\theta = 90°$. The central bright line is called the **zeroth-order spectrum**.

In fig. 22.9b, $\angle S_2S_1P = \theta_1$, the angle of deviation for a particular wavelength in the *first-order* spectrum. In triangle S_2S_1P:

$$\sin \theta_1 = \frac{S_2P}{S_2S_1} = \frac{\text{path difference}}{\text{slit separation}} = \frac{\lambda}{s}$$

In fig. 22.9c, $\angle S_2S_1P = \theta_2$, the angle of deviation for a particular wavelength in the *second-order* spectrum and this time,

$$\sin \theta_2 = \frac{2\lambda}{s}$$

In general we have a formula for any order spectrum, the *n*th order, where *n* is a whole number:

$$\sin \theta = \frac{n\lambda}{s}$$

This is the relation between the angle of deviation θ and the wavelength of light λ for a diffraction grating with a spacing between adjacent slits of s. The two important factors which it confirms are:
a) the deviation θ is dependent upon the wavelength λ, $\sin \theta \propto \lambda$,
b) the deviation θ increases as the slit separation s decreases, $\sin \theta \propto 1/s$,
Note the comparison with the double-slit interference formula: λ and s are in the same positions in both formulas.

The diffraction grating at work

This very important tool is the key to the analysis of light from many different sources. It provides scientists and engineers from many fields with detailed information about the sources from which light comes. Different kinds of spectra are produced when light is emitted by different processes or has passed through different media. For example the astronomer can estimate how fast a star is moving by measuring the wavelengths of light from the star. This is possible because the wavelengths are made slightly longer by the movement of the star away from us. The diffraction grating enables us to measure accurately the small change in wavelength of the light. Astronomers call this increase in wavelength the *red-shift* since an increase in wavelength moves the colours towards the red end of the spectrum.

When a diffraction grating is used for accurate measurements of wavelengths, it is mounted on the turntable of a spectrometer. The optical parts of the spectrometer must be adjusted correctly to produce a *pure spectrum* of the light source. Details of this procedure are found in more advanced books.

Figure 22.10 *Measuring the wavelength of light using a diffraction grating*

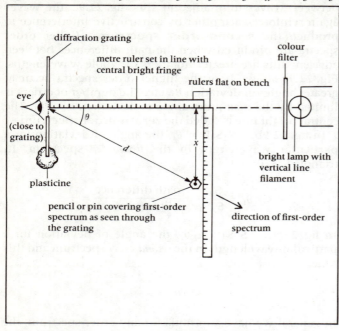

Measuring the wavelength of light using a diffraction grating

● *In a darkened room set up a single bright lamp with its line filament vertical and near bench level.*

● *Select a particular colour by using a filter.*

● *Some distance away, say 2 metres, mount a diffraction grating in a holder at bench level. (Plasticine supports a grating easily, but avoid getting it on the surface of the grating.)*

● *Set a metre ruler in line with the central bright fringe seen through the grating as shown in fig. 22.10. One end of this ruler should be touching the grating.*

● *Set another metre ruler at right angles to the first as shown.*

● *Get someone else to move a pencil or large pin as a pointer along the second ruler until it covers the first-order spectrum seen through the grating, with your eye close up to the grating.*

● *Measure the distances x and d in the figure and find the value of* $\sin \theta$ *from* $\sin \theta = x/d$.

If the grating has, say, 300 slits per mm, the slit separation s is $\frac{1}{300}$ mm.

● *Calculate the wavelength of light using the rearranged formula:*

$$\lambda = \frac{s \sin \theta}{n}$$

$n = 1$ for the first-order spectrum.

Some *typical results* using a green filter are
  $x = 15$ cm,
  grating slit separation $s = \frac{1}{300}$ mm,
  $d = 101$ cm, which, to two significant figures, is 100 cm.

$$\sin \theta = \frac{x}{d} = \frac{15\,\text{cm}}{100\,\text{cm}} = 0.15$$

and using $\lambda = s \sin \theta$ when $n = 1$, we have

$$\lambda = \tfrac{1}{300}\,\text{mm} \times 0.15 = 0.0005\,\text{mm}$$

The wavelength of green light is 0.000 000 5 m or 5.0×10^{-7} m.

The wavelength of laser light can be measured by mounting a fine diffraction grating in front of a laser. (See appendix D11.) The laser beam is diffracted to produce a row of bright spots on the wall in front of the laser. Since the spread of the diffracted beams can be at any angle up to 90° on either side of the central straight-ahead beam, there is a danger of the wide-angle beams being directed towards observers in the laboratory. For this reason the grating rulings should be set horizontally so that the beam is spread out vertically.

The central spot is the zeroth-order maximum and each spot on either side of it is a higher order spectrum; $n = 1$, 2, 3 etc. The wavelength of the laser light can be calculated using the formula

$$\lambda = \frac{s \sin \theta}{n}$$

A mystery

With the evidence provided by interference and diffraction experiments there can be no doubt that under the conditions described light does show a wave-like nature. Similar experiments for other radiations in the electromagnetic spectrum show that all the members of the spectrum possess a *wave-like nature,* in some circumstances.

Yet there also is evidence that all of these forms of radiation travel in small separate bundles of pure energy, called *quanta of energy* or *photons.* In 1905 Einstein was the first person to explain successfully the strange properties of the photoelectric effect (p 343). He developed the idea, suggested by Planck, that light existed in these small individual quanta of energy which had no mass but could 'rain' upon a surface giving up their energy in 'drops'.

Is there a conflict between the two theories? Can light and other radiations be both waves and photons? We have to accept that this *dual nature of radiation* does appear to be a true description.

Assignments

Investigate diffraction by a gramophone record

A gramophone record has very fine grooves cut equally spaced on its surface. Diffraction grating effects can be produced by reflection from these regular grooves. Tilt a record at an angle nearly level with your line of vision, so that the grooves appear much closer together. Now observe the reflection of light from a distant small source of light. Describe what you see.

Try questions 22.10 to 22.13

Questions 22

1 Fig. 22.11 shows the refraction of a beam of red light as it enters glass from air. The lines AA, BB, CC show a series of three adjacent wave crests at a given moment.

   a) Copy the diagram and add to it the position of four wave crests DD, EE, FF, GG.

   b) Explain *in terms of waves* why the beam bends as shown.
       When the colour of the light is changed to blue, the beam bends a little more than the red light did.

   c) What does this tell you about the behaviour of blue light compared with that of red light in glass?
       Refraction does not necessarily prove that light has wave properties, but other experiments do.

d) Draw a labelled diagram to show how you would arrange apparatus to show that light has wave properties.

e) Describe what you would expect to see and explain how the idea that light is a wave motion accounts for it.

f) If you first did the experiment with red light, and then repeated it with blue light, what difference (other than the change of colour or any change of brightness) would occur and how is it explained?

g) Suggest how the apparatus might have to be changed if you wish to try to prove that infra-red radiation (or ultraviolet, if you prefer) has wave properties. (Nuffield)

Figure 22.11

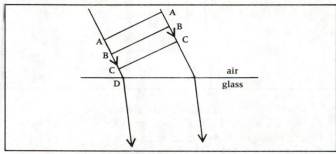

2 Imagine you are wearing a pair of polarising sunglasses and that you hold a second pair of similar glasses in front of them. Describe and explain what you would expect to see as you slowly rotate a lens of the second pair of glasses, keeping it in front of and parallel to a lens of the glasses you are wearing.

3 Interference fringes are formed on a screen when monochromatic light is passed through two narrow slits which are close together. State how, if at all, and explain why the separation of these fringes increases if

a) the screen is moved towards the slits.

b) the slits are made narrower, but the separation is unchanged.

c) a more intense light source is used.

d) light of a longer wavelength is used.

e) the separation of the slits is increased.

4 Fig. 22.12 shows rays of red light passing through two narrow slits and reaching a screen. A series of bright and dark bands are produced on the screen.

a) Explain why the first dark band is formed.

b) Explain why the first dark band is followed by the first bright band.

c) If the red light is now replaced by green light, how will the pattern on the screen change? (Joint 16+)

Figure 22.12

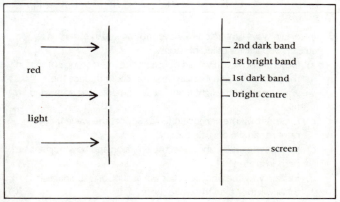

5 A beam of light of a single wavelength passes through parallel slits S_1 and S_2, which are close together. The pattern of bright and dark fringes seen on the screen is shown in fig. 22.13, and some of these fringes are labelled A–E, with D the central bright fringe.

At which one of the points, A–E, has light from S_1 travelled the same distance as light from S_2?

At which one of the points, A–E, has the light from S_2 travelled further than the light from S_1, by a distance of one and half wavelengths of the light used? (Nuffield)

Figure 22.13

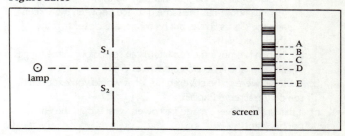

6 When plane wavefronts from a monochromatic light source (a source of light of a single wavelength) fall upon a narrow slit. *diffraction* of the wavefronts occurs.

When there is a second slit close to the first, as in fig. 22.14, interference between the diffracted light from each slit causes a pattern of bright and dark patches ('fringes') on the screen. In certain positions on the screen the *phase difference* between the two beams gives rise to *constructive interference* and in other positions to *destructive interference*.

Explain, with the aid of a diagram in each case, the meanings of the terms in italics.

Describe and explain the effect on the fringe pattern caused by

a) using a source of light of a shorter wavelength,

b) using slits of the same width but having a larger gap between them,

c) obscuring one of the slits. (O & C)

Figure 22.14

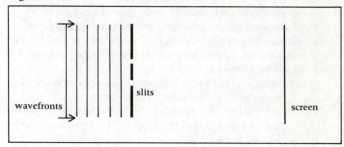

7 In an experiment to measure the wavelength λ of a source of light using Young's slits to form an interference pattern on a screen, the band separation y is given by the formula:

$$y = \frac{\lambda D}{s}$$

where s is the distance between slits and D is the distance from the slits to the screen. Calculate y for yellow light ($\lambda = 5.9 \times 10^{-7}$m) if $s = 0.25$mm and $D = 50$cm.

Name a possible source of yellow light and indicate how it is situated for this experiment. What changes would you expect in the observed pattern if (a) red, (b) white light had been used? Give reasons. (S, part)

8 Groups of students are each using a narrow source of white light, a red filter and double slits to form interference fringes on a screen, as in fig. 22.15.

a) Four comments are made by some of the students. For each comment write a paragraph which will explain to the student the answer to his question. It may help to include diagrams in your explanation.

   i) 'There are only two slits, but there is a series of bright and dark fringes. Why is this?'

   ii) 'When I move the screen closer to the slits the fringes get closer together. Why should they do this?'

   iii) 'My neighbour has the same distance between the screen and the slits as I have, but his fringes are closer together than mine. Why?'

   iv) 'When I remove the red filter I can see colour in the fringes, although the light source is white. Why?'

b) If one slit were to be covered up, what would you expect to see on the screen? Explain.

c) If this single slit were made narrower, what would the effect be? Explain. (Nuffield)

Figure 22.15

9 a) Fig. 22.16 represents the fringe pattern obtained in a double-slit experiment when monochromatic red light was used.

   i) Explain clearly, using the wave theory of light, why dark and red fringes occur.

   ii) State clearly how the pattern would change if monochromatic blue light were used, the rest of the apparatus remaining unchanged.

   iii) What deduction could be made about the difference between red and blue light from the two fringe patterns?

b) In the experiment referred to in (a) the two slits were 0.0002 m apart and the distance from the double-slit to the screen on which the fringes were formed was 4 m.

   i) Sketch and label the arrangement of the apparatus, showing the positions of the source, slits and screen.

   ii) On the screen, the distance between the first red (bright) fringe and the eleventh red fringe was 0.13 m. Calculate the fringe separation.

   iii) Calculate a value for the wavelength of the red light used in the experiment.

c) What is meant by dispersion? Sketch a diagram to show the path taken by a narrow beam of white light when it passes through a glass prism. (JMB)

Figure 22.16

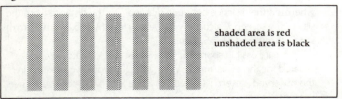

shaded area is red
unshaded area is black

10 Fig. 22.17 shows a series of straight waves approaching a gap.

a) Complete the diagram to show the waves after passing through the gap.

b) Redraw the figure and mark in the wavelength of the waves.

c) What effect would be observed if the wavelength of the waves was reduced?

d) What effect would be observed if the width of the gap was increased? (Joint 16+, part)

Figure 22.17

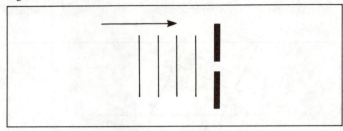

11 Describe the essential features of a diffraction grating.

When a source of light of wavelength λ is suitably set up with such a grating the angle A of deviation for a diffracted image is given by $\sin A = N\lambda$, where N is a constant for the grating. Calculate the angle A for yellow light of wavelength 5.9×10^{-7} m; $N = 10^{6}$ m^{-1}.

If the diffraction pattern for a source shows a wide range of spectral lines, which colours would you expect to be diffracted through the greatest angles? Give a reason. (S, part)

12 In the apparatus represented in fig. 22.18 a beam of white light is incident normally on a diffraction grating and a diffraction pattern is formed on the screen.

a) With the help of a diagram on which is indicated the central position O, describe the pattern appearing on the screen.

b) The grating has 800 rulings per millimetre. What is the grating spacing, and what wavelength will appear at a diffracted angle of 30°?

c) State one significant difference which would result from using a grating with fewer rulings per millimetre.

d) What pattern would be observed on the screen if the white light were replaced by a sodium lamp? (O)

Figure 22.18

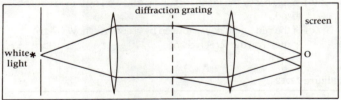

13 In certain ways light acts as a wave motion and in others as a stream of particles called photons.

a) Describe how you would demonstrate interference of light in the laboratory and explain how you could deduce the wavelength of the light from measurements that you could take.

b) Explain how interference effects support the wave theory rather than the particle theory.

c) What is meant by photoelectric emission and how might it be demonstrated?

d) Explain why the photoelectric effect supports the particle theory rather than the wave theory. (O)

23

Electromagnetic radiation

*M*any members of the family called electromagnetic radiation, or electromagnetic waves, have featured in earlier chapters. For example, we have often heard that light waves and gamma rays are part of the electromagnetic spectrum. Now we can see the links and common properties which make many different kinds of radiation part of one continuous spectrum. So which radiations belong to this family? What are the common family features? And what makes each member of the family distinctive and different?

The images of a human hand below are formed by radiation from three different parts of the electromagnetic spectrum: reflected light, emitted infra-red (heat) and transmitted X-rays.

An electromagnetic family

Fig. 23.1 shows the members of the electromagnetic radiation family and how they fit together in a continuous spectrum. The spectrum is arranged with the highest frequency of radiation at the top and the lowest frequency at the bottom. Note that the scale is a logarithmic or constant ratio 'ladder' with each rung of the ladder representing an increase in frequency of ×10. The frequency scale increases upwards but has no definite beginning or end.

Figure 23.1 *Electromagnetic spectrum*

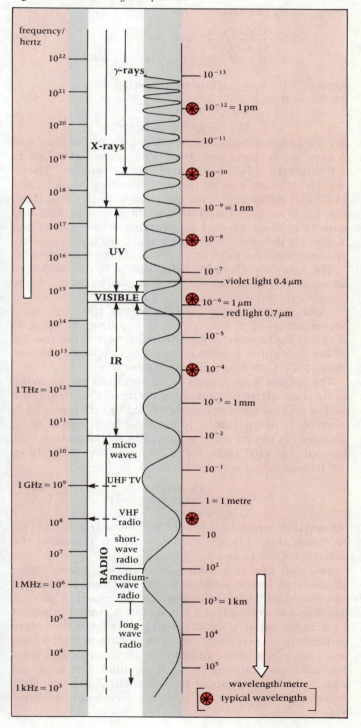

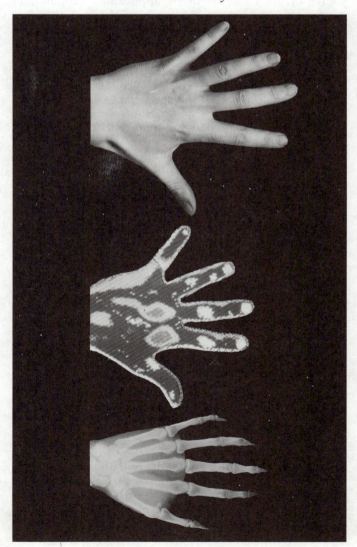

From the wave equation $c = f\lambda$ (p 440), we can calculate the wavelength λ for any frequency f of electromagnetic radiation, because all electromagnetic radiation has the same speed of 3×10^8 m/s in a vacuum. In fig. 23.1 the shortest wavelengths are at the top of the spectrum ladder and the longest wavelengths are at the bottom.

Worked Example
Wavelength and frequency
Calculate the wavelength in a vacuum of green light of frequency 5.0×10^{14} *Hz. The speed of electromagnetic radiation in a vacuum is* 3.0×10^8 *m/s.*

Using $\lambda = c/f$ we have

$$\lambda = \frac{3.0 \times 10^8 \text{m/s}}{5.0 \times 10^{14}/\text{s}}$$
$$= 0.6 \times 10^{8-14} \text{m} = 0.6 \times 10^{-6} \text{m} = 0.6\,\mu\text{m}$$

Answer: The wavelength of green light in a vacuum is 0.6 micrometres.

The range of wavelengths
We find that the shortest wavelengths of gammas and X-rays are in the range from 1 nanometre (10^{-9} m) to 1 picometre (10^{-12} m), and even shorter. These wavelengths are similar to the dimensions of molecules (10^{-9} m), atoms (10^{-10} m) and atomic nuclei (10^{-14} m), see fig. 17.4. It is therefore not surprising to find that gamma rays are emitted by decaying atomic nuclei and X-rays are generated when high-speed electrons penetrate an atom, reaching its inner electrons or even its nucleus.

The X-ray wavelengths, being similar to the spacing of atoms in a crystal structure (10^{-10} m), can be used to study crystal structures.

Light waves, in the middle of the spectrum, have wavelengths of a few tenths of a micrometre (0.4 to $0.7\,\mu$m). These wavelengths, being similar to the spacing of the lines cut in a diffraction grating (a few micrometres apart, p 479), are easily diffracted by such a device.

The longest wavelengths in the spectrum belong to the radio waves. Here we notice that the metal rods used in TV aerials and the telescopic aerials fitted to VHF radios have a length which is similar to the radio wavelengths to be detected. In fact an aerial of length $\frac{1}{4}\lambda$ works well. So, for example, a VHF radio wave of frequency 100 MHz or 10^8 Hz will have a wavelength of

$$\lambda = \frac{c}{f} = \frac{3 \times 10^8 \text{ m/s}}{1 \times 10^8/\text{s}} = 3 \text{ m}$$

An aerial of length $\frac{1}{4}\lambda = \frac{3}{4}$ m or 75 cm would be suitable for receiving this radio frequency, which is typical of the aerial fitted to a radio.

You will find it helpful in recognising where a particular wavelength belongs in the spectrum to think of something with similar dimensions. For example, you can say that a wavelength of a few metres is similar in length to a radio aerial thus suggesting a radio wave. More difficult is a wavelength of perhaps 10 nm. This is too short for a light wave (diffraction grating spacings are a few μm), and too long for an X-ray (molecule sizes and spacing of atoms are about 1 nm or smaller). Between light waves and X-rays we can label the wavelength as ultra-violet.

Examples of typical wavelengths are given in table 23.1 and are indicated in fig. 23.1. Note how these wavelengths can be remembered in a series: gamma rays 10^{-12} m, X-rays 10^{-10} m, UV 10^{-8} m, light is near 10^{-6} m, IR 10^{-4} m and radio waves are 10^{-2} m upwards.

The spectrum is continuous
In the spectrum of white light produced by a hot filament lamp there is a range of colours which gradually change from a deep red through orange to yellow and so on to deep violet. There are no sudden changes of colour and no gaps. We describe such a spectrum as *continuous*.

The electromagnetic spectrum is also continuous. There are no gaps in it and no frequencies anywhere in the range which do not exist. Like the colours of light, the different kinds of radiation gradually change from one to another as their properties gradually change. So there is no sharp boundary between one type of radiation and the next. The dividing lines shown on fig. 23.1 are given only in order to help us give a name to particular wavelengths in the spectrum. There is often a large overlap at the boundaries.

For example, a wavelength of 1 mm can equally well be called either infra-red or microwave radiation (a part of the radio wave family). The name used may depend on the source of the radiation or its application. Gamma and X-rays can have the same wavelengths and therefore occupy the same part of the spectrum. The choice of name given to this radiation depends entirely on how it is generated.

The gradual change of frequency through the spectrum is the same as the gradual change of colour which we can see in the visible part of the spectrum. As the frequency changes through the X-ray part of the spectrum we describe the range as changing from 'soft' X-rays through to 'hard' X-rays as the frequency increases.

In the case of infra-red radiation we now use computers to convert the range of invisible frequencies into false colours of light. Infra-red photographs shown in these false colours allow us to 'see' the different infra-red frequencies and 'see' hot and cold regions with our eyes. (Higher frequency infra-red radiation comes from the hotter regions photographed.)

The common properties
The important properties which are shown by all members of the electromagnetic spectrum are given in the following list

All electromagnetic waves
1. transfer energy from one place to another
2. can be emitted and absorbed by matter
3. do not need a medium to travel through
4. travel at 3.0×10^8 m/s in a vacuum
5. are transverse waves (p 438) and can be polarised (p 473)
6. can be superposed (p 446) and produce interference effects
7. obey the laws of reflection and refraction discovered for light
8. can be diffracted (p 446)
9. carry no charge

It is basically the nature of all those radiations which links them in one spectrum. It is known that all parts of the spectrum transfer energy from a source to a receiver and that the energy travels in the form of an oscillating electric and magnetic field, from which the name electromagnetic waves is obtained.

Table 23.1   Detecting and using the electromagnetic spectrum

Name and typical wavelength	Sources	Detectors	Special properties and uses
γ gamma rays $1\,\text{pm} = 10^{-12}\,\text{m}$	nuclei of radioactive atoms & cosmic rays	photographic film Geiger–Müller tube	very penetrating } very dangerous } high-energy photons used to kill cancerous growths used to find flaws in metals used to sterilise equipment
X-rays $100\,\text{pm} = 10^{-10}\,\text{m}$	X-ray tubes	photographic film fluorescent screen	same radiation as γ-rays, only the source is different also very penetrating and dangerous used to take X-ray pictures: radiography used to treat skin disorders used to study crystal structures: X-ray crystallography
UV ultraviolet light $10\,\text{nm} = 10^{-8}\,\text{m}$	the sun very hot objects arcs and sparks mercury vapour lamps	photographic film photo cells fluorescent chemicals	absorbed by glass causes many chemical reactions damages and kills living cells, causes sun burn UV lamps used in medicine for skin treatment but dangerous to eyes fluorescence used in washing powders and to detect forgeries
visible light $0.6\,\mu\text{m}$ $= 0.6 \times 10^{-6}\,\text{m}$ (green)	the sun hot objects lamps lasers	eye photographic film photocells	refracted by glass and focused by the eye essential for photosynthesis and plant growth used for communication systems: laser and optical fibres used to identify elements in chemistry flame tests
IR infra-red light $100\,\mu\text{m} = 10^{-4}\,\text{m}$	the sun warm and hot objects such as fires and people	special photographic film semiconductor devices such as LDR & photodiode skin	causes heating when absorbed, makes skin feel warm used for heating: radiators and fires emit IR used for photography through haze and fog, IR is not scattered as much as visible light IR photographs taken by satellite provide special information, see text
radio $3\,\text{m}$ (VHF)	microwave ovens TV and radio transmitters using electric circuits and aerials	aerials connected to tuned electric circuits in radio and TV sets	spread round hills and buildings by diffraction microwaves used for cooking used for radio, TV, telephone and satellite communications used for radar detection of ships, aircraft and missiles used in radioastronomy

Sound waves are *not* part of this spectrum because they involve mechanical vibrations of the atoms or molecules of a medium for their existence. Table 23.2 compares sound waves with light waves.

Table 23.2   Comparing sound waves and light waves

Feature	Sound waves	Light waves
transfer energy	yes	yes
can travel through	solids, liquids and gases but need a medium	vacuum, gases, some solids and liquids no medium needed
travel at	$330\,\text{m/s}$ in air $1\,500\,\text{m/s}$ in water $5\,000\,\text{m/s}$ in steel	$3.0 \times 10^8\,\text{m/s}$ in air and vacuum $2.2 \times 10^8\,\text{m/s}$ in water $2.0 \times 10^8\,\text{m/s}$ in glass
type of wave	longitudinal	transverse
can be reflected refracted diffracted polarised	yes yes yes no	yes yes yes yes
typical frequency typical wavelength	$1\,\text{kHz}$ $10\,\text{cm}$ or $1\,\text{m}$	$5 \times 10^{14}\,\text{Hz}$ $500\,\text{nm} = 5 \times 10^{-7}\,\text{m}$
part of the electromagnetic spectrum	no	yes

Above visible light: ultraviolet, X-rays and gamma rays

These radiations are above light in the spectrum in the sense that they have higher frequencies and higher photon energies than light. (Their wavelengths are therefore shorter.)

As its name suggests, *ultraviolet* light is beyond the violet end of the visible light spectrum. Ultraviolet light or UV has a wavelength range of $0.4\,\mu\text{m}$ to about $1\,\text{nm}$. Being next to visible light, UV is also present in solar radiation and the radiation from other very hot objects. A high temperature is needed to produce the high-energy photons of UV radiation. Sparks and electric arcs used in welding metals at very high temperatures emit UV light. Because of this UV light, welders must wear protective goggles which filter out the UV, which would be harmful to the eyes.

The higher frequency and energy of UV light also causes fluorescence in a number of chemicals. When the atoms of a fluorescent material absorb UV light they re-emit the energy as visible light, which makes the material appear to glow. A document may have a signature written in invisible ink containing a fluorescent substance which will show up in UV light. Some soap powders contain a fluorescent chemical which will absorb the UV in sunlight and release it as extra visible light making the washing appear 'whiter than white'.

The spark discharge inside a fluorescent lamp emits UV light. The inside surface of the glass tube is coated with a chemical which absorbs this UV and converts it into extra visible light. The whole lamp is thus made more efficient at producing visible light.

X-rays and *gamma rays* have higher frequencies than UV light, higher photon energies, greater penetration of matter and greater danger to living cells in plants and animals. Table 23.1 lists some of the properties and uses of X-rays and gamma rays. We have already discussed the properties of these radiations and the production of X-rays is described on p 352. The source of gamma rays, their detection and properties are described in chapter 18 about radioactivity. The effects of X-rays and gamma radiation on people are discussed in section 18.3. Table 23.3 gives further information about the use of X-rays in relation to their frequency and energy.

Table 23.3   Some uses of X-rays of different frequencies

Voltage used to accelerate electrons in X-ray tube	Typical X-ray frequency	Application
4 MV	10^{21} Hz	killing deep cancer growths
400 kV	10^{20} Hz	inspection of welds in steel pipes
70 kV	1.7×10^{19} Hz	chest X-ray (diagnostic use)
20 kV	5×10^{18} Hz	treatment of skin diseases

Below visible light: infra-red light

Infra-red radiation has a wavelength range of $0.7\,\mu m$ to about 1 cm. All warm or hot objects lose heat energy by emitting infra-red or IR radiation. When objects absorb IR radiation they gain energy and become hotter. Both emission and absorption of infra-red radiation by an object go on at the same time. If more energy is emitted than is absorbed, then the temperature of the object falls and vice versa.

Infra-red radiation is invisible radiant heat energy. When an object is red-hot it emits red visible light as well as infra-red radiation. At a lower temperature there is less heat energy available in an object and so it emits radiation photons of lower energy and lower frequency. In this sense infra-red radiation is *below* red visible light as its name suggests.

When we warm ourselves near a fire the red glow is attractive to see but it is the invisible IR radiation not the red light which warms our skin. IR lamps are used to provide heat treatment for various illnesses. Experimental methods of detecting IR radiation were described in chapter 10.

Everyday objects and people are sources of infra-red radiation and they emit this radiation at night as well as during the day. Special photographic films which are sensitive to IR can take pictures in the dark without a flashlight. The hottest parts of an object emit more IR radiation of higher frequency and have more effect on the film. So IR photographs can be used to detect temperature differences.

For example, an aerial photograph of a town, taken on IR film, shows as white shapes those houses losing a lot of heat through their roofs. Well-insulated houses appear

This infra-red photograph of an oil storage depot reveals how many tanks are full of oil.

darker. A satellite IR photograph can show, for example, where crops are infected by a disease, or where a missile installation is being built by revealing the changes in surface temperature so produced.

Radio waves

The range of radio wavelengths is very wide, extending from about 1 cm to hundreds of kilometres. There is considerable overlap with the infra-red part of the spectrum in the region also known as *microwaves*. Various parts of the radio wave spectrum have distinctive uses and well-known names, such as microwaves and VHF radio. So within the radio wave part of the electromagnetic spectrum we find another family of waves.

Microwaves

Microwaves have the shortest wavelengths in the radio wave family, being typically a few centimetres. The microwaves used in microwave ovens, of frequency 2450 MHz and wavelength 12 cm, produce a heating effect like IR radiation. (This is not surprising since they overlap at this part of the spectrum.)

Microwaves are used particularly for communication. Radio links between satellites and ground stations usually use microwaves. Some telephone links also use microwaves transmitted between small dish-shaped aerials often mounted on hill-tops and the roofs of high buildings.

Radar systems use microwaves to find the direction and distance of objects which reflect the microwaves back to a large receiving aerial mounted near the transmitter. Good conductors of electricity such as human bodies and metal objects like ships, aircraft and missiles, form suitable reflectors of microwaves.

Ultra-high frequency radio waves, UHF

Radio waves of frequency about 10^9 Hz or 1 GHz are called UHF radio waves. Television pictures are transmitted as modulated radio waves of this frequency. For example, the wavelength of BBC 1 pictures transmitted from the Crystal Palace transmitter at 511 MHz is 0.587 metres. These UHF radio waves are detected by an aerial of length equal to a quarter of this wavelength which is about 15 cm long. This is the length of a typical UHF TV aerial rod.

Very high frequency radio waves, VHF

These radio waves have a typical wavelength of about 3 metres and a typical frequency of about 10^8 Hz or 100 MHz. Radio broadcasts using the FM (frequency modulation) system of carrying sound waves, use this range of radio frequencies. For example, BBC Radio 2, transmitting at about 90 MHz VHF, has a wavelength of 3.3 m and needs an aerial of length ($\frac{1}{4}\lambda$) about 83 cm.

Short wave radio waves occupy the same part of the radio wave spectrum. The different name is usually used to indicate that the radio waves are amplitude modulated, AM.

Medium wave radio uses a frequency typically about 10^6 Hz or 1 MHz. In the UK the band of radio waves known as 'medium waves' covers the frequency range 530 kHz to 1 600 kHz. These radio waves have medium wavelengths of typically a few hundred metres.

While VHF radio waves are received by direct transmission from transmitter to receiver aerial over short distances, medium wave radio is received over much longer distances where direct reception is not possible because of the curvature of the earth. These radio waves are sent up to the earth's ionosphere, which consists of layers of electrically charged or ionised gas molecules about 80 to 400 km above the surface of the earth. This electrically charged blanket around the earth acts as a reflector for medium and long wave radio waves and is used to bounce them back down to distant radio receivers.

Long wave radio is very similar to medium wave radio but uses even longer waves of lower frequencies in the range 150 kHz to 280 kHz. BBC Radio 4 is transmitted at a frequency of 200 kHz on the long wave radio band.

Assignments

Remember

a) the parts of the electromagnetic spectrum and how they fit together in order of increasing frequency or wavelength.

b) a typical wavelength or frequency for each part of the spectrum.

c) the effects of different kinds of radiation on people.

d) one or two interesting uses of each part of the spectrum.

e) how each part of the spectrum is produced.

Name two or three differences between sound waves and electromagnetic waves and explain why sound waves are not part of the electromagnetic spectrum.

Use the formula $c = f\lambda$ to calculate

f) the wavelength of Radio 4 (BBC) long wave transmission at 200 kHz.

g) the wavelength of ITV channel 4 transmitted from Crystal Palace at UHF, 543 MHz.

h) the frequency of radio waves of wavelength 0.8 m received from space by the Jodrell Bank mark 1 A radio telescope.

Explain the meaning of the following words in italics:

continuous spectrum, *transverse* wave, *ultra*violet and *infra*-red.

Try questions 23.1 to 23.8

Questions 23

1 Write out and complete the following paragraph.
   In a vacuum, electromagnetic waves all have the same . . . . Light waves have a greater . . . . . . than radio waves. Another type of electromagnetic radiation is . . . . . . and this can be detected by . . . . An example of a non-electromagnetic wave is . . . . . .
   (Joint 16+, part)

2 How would you show that there is an invisible radiation beyond the red end of the visible spectrum? (JMB, part)

3 Name **one** region of the electromagnetic spectrum which has a wavelength greater than visible light. State briefly (i) how the radiation may be produced, (ii) how the radiation may be detected, (iii) one practical application of the radiation. (JMB, part)

4 Visible light and X-rays are both types of electromagnetic wave. State (i) **four** similarities and (ii) **one** difference between the two types of wave. (JMB, part))

5 Consider the following types of wave: infra-red, ultraviolet, sound, X-rays, radio.
   a) Which of these is not part of the electromagnetic spectrum?
   Of those that are electromagnetic,
   b) which has the shortest wavelengths,
   c) which has the lowest frequencies,
   d) which is unlikely to be detected by a photographic method,
   e) what is significant about their speed? (S, part)

6 Orange light has a wavelength in air of 6.0×10^{-7} m. Calculate the frequency of this light given that the speed of light in air is 3.0×10^8 m/s.
   Which of the following characteristics of light, *speed, wavelength* and *frequency*, does **not** change when light passes from air into water? (C, part)

7 Radio transmission is possible with a certain spectrum of waves that travel through space at 3×10^8 m/s.
   a) Draw a diagram of the complete electromagnetic spectrum, labelling the various types of radiation.
   b) State two differences between radio waves and the other types of radiation in the electromagnetic spectrum.
   c) What is the importance of the ionosphere in radio wave propagation?
   d) What is the frequency of the radio transmission on wavelength 1 500 m? (O)

8 The table below gives some examples of electromagnetic waves (not to scale) and their wavelengths. They **all** have the same speed of 300 000 000 m/s.

Wave	Radio	Micro-waves	Infra-red	Visible light	Ultra-violet
Wavelength	100 m	0.1 m	1 μm	$\frac{3}{10} \rightarrow \frac{7}{10}$ μm	$\frac{1}{100}$ μm

   a) Which of the examples given has the longest wavelength and which the shortest?
   b) What is the frequency of radio waves? ($f = c/\lambda$)
   c) Microwaves are examples of electromagnetic waves and can be used for cooking if they are of the correct frequency (2 500 000 000 hertz). They reflect well off metals and can pass through most non-metals.
      i) Why do you think that the walls of a microwave oven are made of steel?
      ii) Microwave ovens are assumed to cook more efficiently than gas or electric ovens. Give a reason why this might be thought to be a reasonable assumption.
      iii) Water molecules inside food absorb the energy of the microwaves. What is the microwave energy changed to inside the food? (NEA [B] part 88)

APPENDIX A

Quantities, units and symbols, SI

	Symbol for the quantity	Quantity	Unit for the quantity	Symbol for the unit
length mass and time	l	length	metre	m
	A	area	square metre	m²
	V	volume	cubic metre	m³
	m	mass	kilogram	kg
	ρ	density	kilogram per metre cubed	kg/m³
	t	time	second	s
	T	period	second	s
	f	frequency	hertz (= per second)*	Hz
force and pressure	F	force	newton (= kilogram metre per second squared)	N
	W	weight	newton	N
	M	moment of a force	newton metre	N m
	p	pressure	pascal (= newton per square metre)	Pa
mechanics and motion	s	distance, displacement	metre	m
	W	work	joule (= newton metre)	J
	E	energy	joule	J
	P	power	watt (= joule per second)	W
	v, u	speed, velocity	metre per second	m/s
	c	speed of waves	metre per second	m/s
	a	acceleration	metre per second squared	m/s²
	g	acceleration of free fall	metre per second squared	m/s²
	g	gravitational field strength	newton per kilogram	N/kg
	p	momentum	kilogram metre per second	kg m/s
	p	impulse	newton second	N s
heat and temperature	Q	heat energy	joule	J
	θ	temperature	degree Celsius	°C
	T	absolute temperature	kelvin	K
	C	heat capacity	joule per kelvin	J/K
	c	specific heat capacity	joule per kilogram kelvin	J/(kg K)
	L	latent heat	joule	J
	l	specific latent heat	joule per kilogram	J/kg
	α	linear expansivity	per kelvin	per K
electricity	Q	electric charge	coulomb (= ampere second)	C
	I	electric current	ampere	A
	E	electromotive force (e.m.f.)	volt (= joule per coulomb)	V
	V	potential difference (p.d. or voltage)	volt	V
	R	resistance	ohm (= volt per ampere)	Ω
	ρ	resistivity	ohm metre	Ω m
	C	capacitance	farad (= coulomb per volt)	F
	W	electrical energy	joule	J
	B	magnetic flux density	tesla	T
radioactivity	A	activity (of a radioactive source)	becquerel (= per second)	Bq
	D	absorbed dose (of ionising radiation)	gray (= joule per kilogram)	Gy
	H	dose equivalent (of ionising radiation)	sievert (= joule per kilogram)	Sv

* Where a special name exists for the units of a quantity the equivalent units are given in brackets.

APPENDIX B

Prefixes for SI units

		Prefix	Symbol	Example	
multiple	10^{12}	tera	T	terametre	Tm
	10^{9}	giga	G	gigawatt	GW
	10^{6}	mega	M	megajoule	MJ
	10^{3}	kilo	k	kilogram	kg
submultiple	10^{-1}	deci	d	decibel	dB
	10^{-2}	centi	c	centimetre	cm
	10^{-3}	milli	m	milliampere	mA
	10^{-6}	micro	μ	microcoulomb	μC
	10^{-9}	nano	n	nanosecond	ns
	10^{-12}	pico	p	picofarad	pF

APPENDIX C

Numbers and quantities
(Ref. ASE report, 1981, 'SI units, signs, symbols and abbreviations', ISBN 0 902786 75 X.)

1 Writing numbers
When a number has many digits we group them in threes either side of the decimal point. A comma should not be used to separate the groups, e.g. 2390000 and 0.000625.

2 Significant figures
The first significant figure in a number is the first digit from the left other than 0, e.g. in the number 0.06189 the first significant figure is 6.

The number of significant figures is the number of digits counting from the left from the first significant figure, e.g. in the number 0.06189 there are four significant figures. Note that the number 2390000 has only three significant figures. The zeros in front of the decimal point are important to the size of the number but are not counted as significant figures.

When doing a calculation your working and answer should use the same number of significant figures as are given in the numbers in the question. If, for example, a question gives all the numbers with three significant figures and your calculator reads an answer such as:

981.264 51 you should write down only 981,
2.368 4238 you should write down only 2.37,
4231 679.1 you should write down 4230000.

In deciding the third and final significant figure you ask the question: Is the next digit greater or less than 5? In the number 2.368 4238 the fourth digit is 8 which is greater than 5. So the third significant figure is nearer to 7 than to 6 and we write 2.37 rather than 2.36. All the other digits after the third significant figure are written as zero, as in the example 4230000.

3 Standard form or scientific notation
Every number can be written in 'standard form'. Particularly in the cases of very small and very large numbers, standard form is preferred because it is a briefer notation which is less likely to be copied incorrectly.

In standard form the decimal point always appears after the first significant figure. The number of times the number must be multiplied or divided by 10 to have the correct value or size is written as a power of ten, called the *exponent*. So 10^3 means multiply the number by 10 three times and 10^{-4} means divide the number by 10 four times. The 3 and -4 are called exponents or powers of ten. The following examples show how numbers are written in standard form.

Number	Expressed in standard form
300 000 000	3×10^8
330	3.3×10^2
100	1×10^2 or just 10^2
0.1	1×10^{-1} or just 10^{-1}
0.022	2.2×10^{-2}
0.000 012	1.2×10^{-5}
9.81	9.81×10^0 or just 9.81 because $10^0 = 1$

Electronic calculators usually display numbers in standard form which is often called 'scientific notation'. The calculator display shows the power of ten as two separate digits on the right of the display. (They sometimes follow a capital E, for exponent.) To enter the power of ten or exponent, press the key labelled $\boxed{\text{Exp}}$ (exponent) or $\boxed{\text{EE}}$ (enter exponent). To enter a negative exponent press the key labelled $\boxed{+/-}$.

For example, to enter the number 2.3×10^{-4}, press the keys on your calculator as follows:

$$\boxed{2} \; \boxed{.} \; \boxed{3} \; \boxed{\text{EE}} \; \boxed{+/-} \; \boxed{4}$$

the display should now show 2.3^{-04}.

When you write down a number displayed in this form on your calculator you should use standard form. For example, a calculator displaying 1.98^{02} is written as 1.98×10^2.

4 Using quantity algebra
In this book the letters in a formula or equation stand for the physical quantities themselves and include both the magnitude and units of those quantities. This is called *quantity algebra*. For example:
a) v stands for speed which is a physical quantity. If the speed of a car is 25 m/s then v stands for 25 m/s and not just the number 25.
b) using the formula $W = Fs$
   if $F = 2\,\text{N}$ and $s = 3\,\text{m}$
   then we write $W = 2\,\text{N} \times 3\,\text{m}$
   which gives $W = 6\,\text{N}\,\text{m}$ or 6 J
(We can see how the unit of work, the joule, is equivalent to a newton metre.)

5 Labelling axes on graphs and heading columns in tables of readings
The points plotted on a graph are pure numbers and have no units. So we divide the physical quantity plotted on an axis by its units to obtain a pure number.
a) if a current value is given by $I = 5\,\text{A}$ then dividing current by amperes gives a pure number: $I/\text{A} = 5$.
   When an axis is labelled for current values we use the notation: I/A.
b) if a density value is given by $\rho = 0.96\,\text{kg/m}^3$ or $0.96\,\dfrac{\text{kg}}{\text{m}^3}$ then an axis would be labelled as

$$\rho \Big/ \frac{\text{kg}}{\text{m}^3}$$

The same type of labelling is used for columns of readings.

6 Use of Δ to mean 'change of'
In this book the symbol Δ (a Greek capital letter D or delta) is used to indicate a difference or change in a quantity. This is usually used for a small change or difference. For example:

Δt is the time difference between the beginning and end of an event rather than the time of day,
ΔT is the temperature difference or temperature rise between two thermometer readings,
Δl is the change in length or expansion of an object whereas l stands for its full length.

APPENDIX D

Answers to questions

Questions 1

3 0.6 m

8 a) Virtual image same distance behind mirror as object is in front & at 90°
   b) i) IR, light
      ii) concave parabolic
      iii) polished reflector: efficient and regular reflection of IR & light

9 55°

15 a) 30 cm
   b) +30 cm, real, inverted and same size as object
   c) +60 cm, real, inverted and 3X magnified
   d) −30 cm, virtual, erect and 3X magnified, behind mirror

Questions 2

2 a) X = angle of incidence
      Y = angle of refraction
   b) Z = critical angle, at P:
      total internal reflection:
      angle of incidence > critical angle
      used on red rear reflectors
   c) see p. 29

5 a) 48.6°, b) 1.34

6 b) 1.29, c) 51°

7 a) 2.0×10^8 m/s, 41.8°

Questions 3

4 a) × 1/3
   b) i) 1.5 m from lens ii) 375 mm

5 $f = +8$ cm, height = 1.5 cm

6 −20 cm, ×5

9 b) 12.0 cm

11 a) see p. 39, top lens has long focal length; used as objective in a refracting telescope
   b) i) see p.39 ii) power = 1/f(in m)
      iii) see fig. 3.9a
      iv) image formed further away at the student's near point

12 a) long sight (hypermetropia) or lack of accommodation (presbyopia) if he is older than 45
   b) further away
   c) converging (convex meniscus) lens
   d) see fig. 3.17

13 a) 33 cm b) −1.0 m or −100 cm

15 a) 0.05 m or 5.0 cm

16 0.08 m or 8.0 cm tall

18 b) converging (bi-convex) lens
   d) inverted, diminished
   e) shorter exposures possible in poorer light, camera can be hand-held, wide & narrow angles of view possible with different lenses, moving objects can be 'frozen' by short exposures

19 60 mm or 6.0 cm

21 a) to detect radiation of different kinds (wavelengths)
   b) i) Yerkes ii) Jodrell Bank
   c) Mount Palomar, λ/D = 1.2×10^{-7}
   d) collect more radiation & ∴ detect weaker sources

e) Sirius

Questions 4

1 3.0×10^{-4} m, 8.28×10^3 s, 10^{-2} kg, 2.2×10^{-4} kg, 30 g, 30 mm, 0.25 m, 4.0×10^{-3} m², 9.2×10^{-4} m², 10^4 mm²

2 2.15 cm or 21.5 mm

3 23.94 mm

4 a) i) 16 cm³ ii) 41 cm³
   b) i) 25 cm³ ii) ρ = m/V = 3.4 g/cm³

6 a) 0.1 mm b) 4 g c) 3×10^6 mm³ or 3×10^3 cm³ d) 2/3 g/cm³ e) 67 g

7 a) ρ = m/V = 7.0 g/cm³
   b) i) 54 g ii) ρ = 2.7 g/cm³
      iii) steeper gradient

8 i) 1.3×10^3 kg/m³ or 1.3 g/cm³
   ii) 6.4×10^{-5} m³ or 64 cm³

9 5.5×10^3 kg/m³ or 5500 kg/m³ or 5.5 g/cm³

10 a) Time 10 or 20 complete swings
   c) i) gradient = $\dfrac{\Delta y}{\Delta x} = \dfrac{2.8 - 1.6}{0.7 - 0.4} = \dfrac{1.2}{0.3} = 4.0$
      ii) g = 40/4.0 = 10 m/s²

Questions 5

1 a) 400 N b) 10^4 N c) 3.0×10^{-3} N

2 a) 17 N, direction of both forces
   b) 7 N, direction of 12 N force
   c) 13 N, direction between the two forces at 23° to the 12 N force (tan⁻¹ 5/12)

4 a) weight, velocity
   b) Its direction
   c) i) 5 N right, ii) 1 N left, iii) 2 N down, iv) zero

5 b) i) 3464 N ii) 2000 N

6 10 Nm, 8.66 Nm

7 360 cm from the pivot, other side

8 a) i) $W = mg$ = 50 × 10 = 500 N
      ii) moment = Fd = 500 N × 2 m = 1000 Nm
      iii) $W = mg$ = 30 × 10 = 300 N
      iv) moment = Fd = 300 N × 2 m = 600 Nm
      v) extra moment needed = 400 Nm
         $F \times 4$ m = 400 Nm, F = 100 N down
      vi) 500 N + 300 N + 300 N + 100 N = 1200 N
   b) i) vehicle easily turns over onto its side
      ii) blows over
      iii) loading should keep centre of gravity low and central
      iv) & v) see p. 86

9 Support beams across a gap & load their centres. Measure sag as load is increased. Increase load until beam breaks.

10 a) $Fd = Fd$
      10 MN × 4 m = L × 8 m, L = 5 MN
      weight of centre span = 10 MN
   b) downwards
   c) compression along top surface, tension along underside. Steel reinforcement is set in tension inside the concrete

11 0.11 Nm at the 52.5 cm mark

Questions 6

1 a) 4 N/m² or 4 Pa b) 0.5 Pa

c) 1.2×10^5 Pa
d) 4.8×10^6 Pa or 4.8 MPa
2 a) 0.5 N b) 1600 N or 1.6×10^3 N
3 10^5 Pa
4 a) 0.25 m² b) 0.5 m³ c) 1500 kg
d) 15000 N or 15 kN
e) 60 kPa or 6.0×10^4 Pa
5 a) i) $W = mg = 12.5$ N
ii) $A = 20$ cm²
iii) $\max.P = F/\min.A = 0.3125$ N/cm²
b) i) $V = 160$ cm³, $\rho = 7.8$ g/cm³
6 b) i) 8×10^3 kg/m³ or 8000 kg/m³
ii) 0.64 N iii) 3200 Pa or 3.2 kPa
7 a) 460000 Pa or 460 kPa b) 560 kPa
c) 8.0 m d) 8.695 m
8 c) i) & ii) 5000 Pa or 5 kPa,
125000 Pa or 125 kPa
9 a) 0.003 m³ or 3.0×10^{-3} m³
b) 24 N c) 2 kPa
10 c) 1.02×10^5 Pa or 102 kPa
11 a) $P = F/A = 320$ kPa
b) pedal acts as a lever,
force exerted by foot is smaller
c) i) force required from foot becomes
too large or pressure too low
ii) pressure too large
d) total area = 4×0.0075 m² = 0.03 m²
$W = PA = 200$ kPa $\times 0.03$ m² = 6 kN
$m = W/g = 600$ kg
e) extra weight flattens the tyres and increases the
area in contact with the ground
12 b) i) 40 cm³
ii) A, 100 g or 0.1 kg; B, 1.0 N
13 a) 100 g or 0.1 kg
b) 1.0 N in air, 0.36 N in liquid
c) 0.68 N
14 For $M = 90$ g, $h = 15$ cm, 5.0 cm²

Questions 7

1 a) 100 Nm or 100 J
b) 40 kJ c) 200 J
2 a) 500 N b) 150000 J or 1.5×10^5 J
c) 125 J/s or 125 W
3 a) 1.0 kW b) 10 kW
4 i) $W = Fs = 100$ N $\times 2$ m = 200 J
ii) $P = W/t = 22$ J/5 s = 40 W
5 a) $\times 3$ b) changes direction of the force L c) to allow
for expansion & contraction of the wires while
keeping constant tension d) tension = 3L =
6000 N
6 a) 2500 J b) 200 N
8 a) 1400 J b) 1800 J
9 a) i) $W = Fs = 6000$ J
ii) $W = 12000$ J iii) avoid lifting yourself, reduce
work
iv) safety
b) $P = W/t = Fs/t = 200$ W
c) i) 100 N ii) 3/4
11 a) 15 times b) 6 m c) 33.3 N
12 a) 2.0×10^4 Pa or 20 kPa
b) 2.0×10^4 Pa c) 100 N
13 a) i) 300 N
ii) 6.0×10^6 Pa or 6 MPa

b) 600 N c) 10

Questions 8

1 a) 150 km/h b) 41.7 m/s
2 25920 m or 25.92 km
3 200 s
4 2.0 m/s, 6.0 m/s
5 a) 0.3 m/s b) 800 s
6 1000 m/s at 36.9° to flight path of space ship
7 a) 50 s b) 20 m
8 a) constant speed 0 to 30 s, constant deceleration 30
to 60 s
b) 40 s
c) i) $20 - 7.5 = 12.5$ m/s
ii) $a = \Delta v/\Delta t = 1/3$ m/s²
iii) $s = $ (average v)$\Delta t = 300$ m
d) compare areas under graphs
9 b) i) 2.5 m/s² and -1.25 m/s²
ii) 20 m
10 c) 0 to 30 s, constant $a = 0.5$ m/s²,
30 to 50 s, constant $v = 15$ m/s,
50 to 60 s, constant deceleration = $(-)1.5$ m/s²
d) t/s: 10 20 30
s/m: 25 100 225
11 a) $R = 120$ N $- 20$ N $= 100$ N in the direction of
Mary's pull
b) i) 6 m/s ii) 10 m/s
iii) sprint speed = $s/t = 100$ m/10 s = 10 m/s
12 b) 2.0 m/s² c) 50 m d) 15 m/s
13 b) i) 4.0 s ii) 60 m
14 a) weight or gravity, friction or air resistance b)
larger area of parachute increases frictional force,
zero resultant force on man, zero acceleration,
terminal speed
15 30 m/s², 120 m/s
16 a) 10 N b) 0.5 N c) 25 kN
17 0.4 m/s²
18 2 N
19 a) 32 kg m/s b) 1500 kg m/s
c) 120 Ns d) 200 Ns
20 a) i) $P = 10000$ N $- 6000$ N:
vertical forces in equilibrium
ii) constant $v = $ no acceleration = no resultant force
horizontally
b) $m = W/g = 1000$ kg
c) i) Resultant $F = 800$ N $- 400$ N = 400 N to the left
ii) $a = F/m = 0.4$ m/s²
d) i) see p. 141
ii) head restraints, collapsing steering column,
crumple zones
21 a) momentum = $mv = -30000$ kg m/s
b) using conservation of momentum
$0 = (100$ kg $\times v) + (-30000$ kg m/s$)$
$v = 30$ m/s
c) speed limit + 10% = $(26 + 2.6)$ m/s
30 m/s > 28.6 m/s: prosecute
d) i) $a = \Delta v/\Delta t = -30$ m/s²
ii) $F = ma = 60000$ N to right
22 a) $a = \Delta v/\Delta t = 2.5$ m/s²
b) area of triangle = 20 m
area of rectangle = 40 m
total area = 60 m
c) $E_k = \frac{1}{2}mv^2 = 3500$ J

24 a) i) graph curves upwards
   ii) Yes, when $v = 0$ no braking distance is needed
   iii) distance travelled in equal times is directly proportional to speed

b) i) $E_k = \frac{1}{2}mv^2 = 200\,000\,J$
   ii) 30 m
   iii) $W = Fs$, $200\,000\,J = F \times 30\,m$
   $F = 6700\,N$

c) $95\,m/4.2\,m = 23$ car lengths

25 d) 3000 N e) 6000 m/s f) $\times 1/\sqrt{2}$
g) $-12.5\,m/s$

Questions 9

1 $2.0 \times 10^{-6}\,mm$ or $2.0 \times 10^{-9}\,m$ or 2.0 nm

3 i) solid ii) gas iii) liquid

4 a) to make smoke particles visible as they scatter the light

b) large enough to be seen under a microscope but small enough to be visibly moved when air molecules collide with them

c) random, agitated motion of 'dancing' specks of light

d) molecules are in constant motion which is random and varied in both speed and direction

e) no difference in the motion, just harder to see

5 a) 0 to 20 N, linear (straight) part of the graph

b) spring designed to be compressed

c) coils touching each other

d) load on each spring = 10 N
   length of spring = 9.5 cm
   compression = $16 - 9.5 = 6.5\,cm$

6 i) 10 cm ii) 20 N iii) 60 N iv) 21 cm

7 a) see p. 165, fig. 9.19
b) No, graph not straight

8 i) 7.5 N ii) 0.75 kg
iii) $12.5\,g/cm^3$ or $12.5 \times 10^3\,kg/m^3$

Questions 10

1 a) A UV, B light, C IR
b) long wavelength IR
c) condensation of water vapour on cold surface from humid air inside
d) double glazing, drier air, more ventilation

9 a) $1.2 \times 10^{-2}\,m$ or 1.2 cm
b) 53°C or 53 K

10 c) $9 \times 10^{-6}/K$

11 a) 0°C, about 4°C

13 a) i) temperature ii) mercury
   iii) alcohol
b) 100°C, 373 K
c) i) reading stays constant when removed from patient, more accurate scale reading to 0.2°C
   ii) $37 + 273 = 310\,K$
d) pyrometer or thermocouple

15 a) 50 J b) 150 kJ d) 937.5 J/(kgK)

16 i) less than 30°C
ii) more than 30°C

17 a) i) copper is a good conductor of heat
   ii) black surfaces absorb IR radiation best
   iii) to reduce heat loss by conduction & convection
   iv) the process in which shorter wavelength IR

passes through the glass front and is absorbed. Longer wavelength IR emitted by the warm water cannot escape through the glass.

b) i) 20% of 126 MJ = 25.2 MJ
   ii) $\Delta T = 35°C - 5°C = 30°C$
   iii) $mc\Delta T = 25.2\,MJ$
   $m = 200\,kg/day$ or $40\,000\,kg/year$
   iv) 25.2 MJ = 7 kWh
   v) cost = $7 \times 8p = 56p/day$

21 330 kJ/kg

22 a) 672 s
b) 2250 kJ/kg or 2.25 MJ/kg

23 a) 4200 J will raise the temperature of 1 kg of water by 1°C
b) i) Heat = power $\times$ time = $mc\Delta T$
   $c = 4500\,J/(kg°C)$
   ii) Heat loss by conduction to aluminium pot, by air convection above water, by evaporation
c) Heat required for 1 kg of water to be converted into steam without any temperature rise
d) i) $P = IV = 10\,A \times 250\,V = 2500\,W$
   ii) mass vaporised = $2.7 - 1.9 = 0.8\,kg$
   heat = power $\times$ time = ml
   $l = 2250\,000\,J/kg$ iii) 100°C

24 a) missing values: 2083, 3125, 4348
c) p is inversely proportional to V
d) $0.00025\,m^3$

25 d) i) $54.5\,mm^3$ ii) $-266°C$

26 b) 46°C

27 10 m

Questions 11

2 a) i) very small particles, a constituent of matter
   ii) because electrons have a negative charge
   iii) outer regions of atoms
b) i) repulsion – like charges
   ii) repulsion again
c) i) It would attract your hair, small pieces of paper or a stream of water ii) the charges are conducted away through the metal and your arm

3 b) electrons flow from the top of the aircraft to its underside: top +, underside –
c) cancel each other out: remains neutral

4 a) positive b) like charges repel
c) powder loses its charge

Questions 12

1 0.5 A

2 720 C

3 240 s or 4 minutes

4 3 A

5 4 A in direction JX

6 i) true ii) true
iii) false iv) true

7 6 kJ

8 6 V

9 250 s, 12 V

10 $R = V/I = 11.7\,\Omega$

11 $V = IR = 2.5 \times 12.5 = 31.25\,V$

12 $R = V/I = 0.05\,\Omega$

13 a) $I = V/R = 6\,A$ b) 15 A c) 2 A

15 b) X = switch, Y = battery, Z = variable resistor or rheostat
   c) b
   d) connect voltmeter in parallel with Z, i.e. one wire to each end of Z
   e) i) none ii) current at G = 2 × current at F
   f) i) b ii) b
   g) varies current and hence brightness

17 13 A

18 a) $10\,\Omega$ b) $6\,\Omega$ c) $2\,\Omega$ d) $2\,\Omega$

20 $5\,\Omega$, $7.5\,\Omega$

21 a) R = $2\,\Omega$, P = $3\,\Omega$, Q = $4\,\Omega$
    or R = $2\,\Omega$, P = $4\,\Omega$, Q = $3\,\Omega$
   b) R = $4\,\Omega$, P = $3\,\Omega$, Q = $2\,\Omega$
    or R = $4\,\Omega$, P = $2\,\Omega$, Q = $3\,\Omega$

22 0.6 A through all resistors and through the battery

23 $3\,\Omega$

24 A_1 = 3A, A_2 = 2A, A_3 = 1A, A_4 = 3A

25 b) i) 2 A ii) 6 V iii) $1\,\Omega$
   c) i) 2 V ii) 1.5 A

26 b) i) $2\,\Omega$ ii) $9\,\Omega$ iii) 2/3 A iv) 4 2/3 V v) 4/9 A

27 a) i) B $10\,\Omega$ ii) B 2/3 Ω iii) B $3\,\Omega$, C 2/3 A

28 b) ii) $2\,\Omega$ iii) $4\,\Omega$

29 a) 0.075 V b) $0.0251\,\Omega$ c) $995\,\Omega$

30 a) see p. 231
   b) R is increasing as I increases: heating effect causes R to rise
   c) i) ×2 ii) divide by 4
   d) i) increases ii) decreases iii) increases
   e) since heat = I^2R, having a large I is more effective than a large R

31 $1360\,\Omega$

32 $1.5 \times 10^{-7}\,\Omega\text{m}$

33 2.0 m

35 a) 8A b) $31.25\,\Omega$ c) 120 kJ

36 a) i)

meter	terminals	name	quantity
1	X = −	ammeter	current
2	Y = +	voltmeter	voltage or p.d.

    ii) T = mg = 0.1 kg × 10 N/kg
     = 1.0 N
    iii) E_P = mgh = 1.5 J
    iv) time = work/power = 5.0 s
    v) Fv = power, v = 0.3 m/s
     $E_k = \frac{1}{2}mv^2$ = 0.00045 J
    vi) P = IV = 0.4 W
    vii) 0.1 W: heat in wires
    viii) efficiency = 75%
   b) more coils, more turns on the coils, larger current

36 c) 2×10^{-6} F
   d) 4×10^{-4} C or 400 μC

37 b) ii) 7.5 μF

Questions 13

1 a) i) liquid NiCad
    ii) liquid lead acid
   b) store much less charge
   c) dry NiCad: portable (no liquid to spill), long working life, stores more charge, can be recharged

more times than dry lead-acid

2 $8 \times 10^{-3}\,\Omega$, 250 A

3 a) $4\,\Omega$ b) 1.2 V c) 0.3 A
   d) 1.44 V e) 1.8 A

4 d) $5\,\Omega$, 0.2 V, 2/3 Ω

Questions 14

2 See p. 277
   a) N-pole on left

3 a) soft iron
   b) magnetism induced by coil is stronger but temporary
   c) F = mg = 0.15 kg × 10 N/kg = 1.5 N
   d) copper, tin, aluminium, zinc, silver, gold
   e) more turns of wire, coils on both sides, more current

Questions 15

3 a) soft iron
   b) needs to be able to change its magnetism rapidly
   c) X
   d) larger current in the coil
   e) gives stronger magnetic field between poles, i.e. across the gap & through the tape

4 a) to prevent corrosion of contacts
   b) iron reeds have temporary induced magnetism only
   c) can be operated by a current
   d) see p. 298

7 a) cylindrical magnet
   b) magnetism is induced which changes direction as the magnet rotates
   c) between the ends of the coil
   d) faster rotation of magnet or more turns on the coil
   e) alternating: emf changes direction for each half-turn of the magnet
   f) no rotation: no induction

9 a) i) low current and low heat losses in wires
    ii) transformers can be used
   b) i) 20 turns
    ii) I = P/V = 0.1 A
   c) i) 40 turns needed to give 12 V output ii) no change but a higher current will be taken from the secondary & also will flow into the primary

10 a) i) in the live input wire
    ii) between the switch and the primary coil
    iii) earth lead connected to core
   b)

secondary	A or B	V_{out}
2500	A	115 V
5000	B	230 V

11 a) 120 V c) 83.3%

13 a) lines hang on insulators
   b) consider environmental, maintenance and cost issues
   c) i) I = P/V = 500 A
    ii) V_{drop} = IR = 500 × 10 = 5000 W
    iii) P = I^2R = 2.5 MW
    iv) converted into heat
   d) high V allows much lower I & hence lower power losses as heat

14 a) i) $I = P/V = 10$ A
  ii) $E = Pt = 2200$W $\times 5 \times 60 \times 60$s $= 39.6$ MJ
  iii) 11 kWh, cost $= 77$p
  iv) light
 b) 13 A: next value above the normal current

15 a) all three wires go to wrong pins, see p. 317, cable is not gripped correctly, fuse rating should be 5 A
 b) as it enters the socket it opens safety blinds which cover the holes for the other two pins

16 a) shiny surface reflects IR radiation forwards, heavy base makes it more stable
 b) ii) $P = 4 \times 0.75$kW $= 3$kW
  iii) 3kW $\times 5$h $= 15$kWh
    cost $= 15 \times 7$p $=$ £1.05
  iv) $I = P/V = 12.5$ A
  v) 13 A

17 a) A unit of electrical energy, see p. 321 b) 511
 c) i)

Time used	kWh
5	1
0.25	0.5
2	6

  ii) cost $= 7.5 \times 6$p $= 45$p

18 a) 3050 W b) 12.2 A c) 43.92 MJ

Questions 16

4 d) 7.28×10^{-16} J e) 4×10^7 m/s
5 $e = 1.6 \times 10^{-19}$ C
  $m = 9.1 \times 10^{-31}$ kg
6 a) 1 V/cm b) 4 V c) 0.5 ms/cm
 d) 5 ms e) $f = 1/T = 200$ Hz
7 b) ii) about 3.64 V
  iii) 1.63 V, 1.15 V
10 a) see p. 348
 b) for the series resistor $R = V/I$
   $= (5-2)$V$/0.01$ A $= 300$kΩ
11 a) 4.7 kΩ
14 422 Ω for base-emitter bias of 0.7 V
15 a) i) 24 Ω ii) 0.25 A
  iii) $V = IR = 0.25$ A $\times 4$ Ω $= 1.0$ V
 b) i) Y $=$ relay
    Z $=$ npn transistor
  ii) (A) the transistor (B) connect a diode as shown in fig. 16.29
  iii) Z is on, Y is activated and contacts will be open, lamp is off
 c) the lamp in the circuit would be switched on when the LDR is in the dark
17 a) 100 μA b) 90 kΩ
18 a) i) sound to electrical
  ii) electrical to sound
 b) transistor is a current amplifier
19 a) 0.7 V
 b) time taken for capacitor to charge up: the current flows to C through R
 c) make either C or R larger
 d) capacitor is discharged, bell stops ringing
20 a) i) $V_{out} = -20 \times +0.5$V $= -10$V
  ii) $V_{out} = -20 \times -2$V $= +40$V
  but saturates at near $+15$V
 b) near ± 15V
 c) max $V_{in} = \pm 15/-20 = \pm 0.75$V
 d) V_{out} is also alternating but inverted with ampli-

tude ± 15V and, because of saturation, clipped at top and bottom

22 a) ± 4.0 V b) gain of 20 c) 0.75 V

24 A B P Q
  0 0 1 1
  0 1 1 0
  1 0 0 1
  1 1 0 1

25 a) Y Z P Q
  1 1 0 0
  1 0 0 1
  0 1 0 1
  1 1 1 0
 b) i) low ii) 3 V iii) 300 Ω iv) to limit the current through the LED

26 a) OR gate
 b) NOT gate: a logic 1 into the OR gate is needed when the flame goes OUT
 c) AND gate: lamp on when pulse generator gives logic 1 pulses AND alarm is raised:
   output Z is logic 1
 d) A B C Z
  0 0 1 1
  1 0 1 1
  0 1 0 0
  1 1 0 1

27 a) Early Bird
 b) 90 minutes
 c) Early Bird
 d) i) Telstar, orbiting in a low orbit, was in a position over the Atlantic in which it was 'visible' in both America & Britain for only 20 mins in each orbit
  ii) Early Bird 'hovers' in a fixed position over the equator where it can be 'seen' at all times in both America & Britain
 e) To focus and increase the strength of weak signals

28 A microphone
 B amplifier
 E transmitting dish & aerial
 F satellite, relays and amplifies microwave signal
 G receiving dish, focuses microwave signal on to aerial
 H separates pulse code from carrier wave
 I decoder, converts PCM digital signal into electrical analogue signal
 K converts electrical signal into sound

Questions 17

1 a) alpha or proton
 b) neutron or proton
 c) alpha or beta d) neutron
2 i) proton, neutron
  ii) proton, neutron
  iii) proton & electron
  iv) electron & beta particle
4 neon-20 has 10 protons and 10 neutrons
  neon-22 has 12 neutrons
6 1/3675 or 0.0272%
7 2000 layers
  1 in 36 000 000
  5×10^{-14} m

Questions 18

2 a) i) radioactive
   ii) emit radiation which can damage human cells
   b) i) lead
   ii) lead absorbs gamma radiation well
   c) see p. 387
   d) i) nuclear ii) radiography and medical physics

3 a) i) so that only radiation passing between the poles of the magnet can reach the detector in any of its positions
   ii) 27
   iii) beta: direction & extent of deflection
   b) i) gamma
   ii) atoms of the same element which have different numbers of neutrons
   iii) 43, 99–43 = 56
   iv) 4 half-lives:
      $5\mu g \times \frac{1}{2} \times \frac{1}{2} \times \frac{1}{2} \times \frac{1}{2} = 0.3125\,\mu g$
   v) so that radioactive material left in the body decays away quickly
   vi) part of liver B is not functioning
   c) see p. 403

4 a) radiation
   b) tissue paper: only alpha stopped
      lead sheet: all alpha & beta & most of gamma stopped
   c) i) a change in thickness of paper alters the amount of beta radiation reaching the detector
   ii) all alpha would be stopped & the thickness of the paper would not affect the intensity of gamma
   iii) see p. 395
   iv) source with constant activity

5 b) i) 55s, 50s ii) 52.5s

6 ii) 90 protons, 144 neutrons
   iii) 91 protons, 143 neutrons
   iv) 72 days

7 atomic number 86

8 a) i) activity = no of decays per second
   ii) see p. 385
   b) 52s
   c) i) alpha ii) collision
   d) $^{238}_{92}X = ^{234}_{90}Y + ^{4}_{2}\alpha$

9 a) left kidney: fails to remove the impurity in 20 minutes
   b) only gamma radiation can pass out through the body to reach the detector
   c) half-life must be significantly longer than the time taken for the kidney to function

Questions 19

2 b) 2.3 MJ + some extra energy to warm up the water
   c) 4.32 MJ

3 a) 5.04×10^9 J
   b) 5.66×10^7 J or 56.6 MJ
   c) 655 W d) £70.00

5 150 W

6 a) 1986   7665 − 6517 = 1148
      1987   8616 − 7665 =   951
      saving   =   197
      cost = 197 × 38p = £74.86
   b) i) loft ii) double-glazing
   c) see p. 411

d) i) £5000 million or £5 × 10⁹
   ii) 10 years
e) thickness of air gap affects the thickness and cost of window frame; thickness of glass affects weight and cost of glass

7 a)

	wall cavity		window	
loss	19 J	10 J	112 J	60 J
saving	9 J		52 J	

   b) area
   c) i) 112 W/m²/20°C = 5.6 W/(m²°C)
      60 W/m²/20°C = 3 W/(m²°C)
   ii) see p. 410

8 a) 8160 W or 8.16 kW
   b) 2160 W or 2.16 kW

9 a) coal, natural gas, oil
   b) i) oil ii) very few hydroelectric schemes in the UK
   c) i) 60 years ii) 400 years
   iii) reserves are estimates only, demand changes

10 a) see p. 417
   b) i) $E_p = mgh = 2$ MJ
   ii) $P = 2$ MW iii) 2 MW
   iv) $I = P/V = 100$ A
   c) i) demand is lowest at night
   ii) some energy is lost as heat

11 a) 235 nucleons: 92 protons + 143 neutrons
   b) nuclear fission
   c) mass into energy
   d) a + c = 92, b + d = 232
   e) unstable, radioactive
   f) moderator, graphite or water
   g) chain reaction, see p. 424
   h) control rods used, see p. 425
   i) see p. 395
   j) stored for a time under water, reprocessed to recover unspent fuel
   k) see p. 402 & 403

12 a) see p. 422 b) see p. 431

13 a) 1.8×10^{26} W
   b) 3.33×10^{13} years

14 1.8×10^{16} J

Questions 20

1 a) 1.25 Hz b) 0.8 s

5 a) (iii)
   b) use a smaller mass or connect the springs side by side
   c) see p. 434
   d) no contact force between an astronaut & the scales: no stretching or compressing of the spring
   e) measure the time period of oscillations

6 a) i) 4 cm ii) 40 cm
   b) $c = f\lambda = 8$ Hz × 0.4 m = 3.2 m/s

7 a) 0.5 Hz b) 6 m

8 c) 1.0625 m d) 320 Hz, 1440 m/s

9 32 cm/s

11 c) 15 Hz

12 a) see fig. 20.13, p. 443
   b) see fig. 20.14, p. 443
   c) i) $\lambda = 1.5$ cm

    ii) $c = f\lambda = 10\,\text{Hz} \times 1.5\,\text{cm} = 15\,\text{cm/s}$

13 a) see fig. 20.13, p. 443
  b) see fig. 20.19, p. 445
  c) shorter λ, lower speed

Questions 21

3 364 m/s
4 a) 360 m/s
5 336 m/s
7 a) 1400 m/s b) 0.0233 m c) 63 m
8 a) see p. 438
    longitudinal wave: sound in air
    transverse wave: light in air
  b) see p. 452
  c) i) $c = f\lambda = 3 \times 10^8\,\text{m/s}$
    ii) $s = vt = 27000\,\text{m}$
       $AT = 13500\,\text{m}$
    iii) horizontal distance = 10800 m
      (use pythagoras)
      $t = s/v = 40\,\text{s}$
  d) too slow, range too short, sound would disturb people on the ground
9 500 Hz
12 a) i) increase tension, shorten length
    ii) same frequency
  b) we see the lightning first

15 a) highest f heard = 20 kHz
    all students could hear up to 16 kHz
  b) ultra sound
  c) see p. 456
16 a) (i) Q (ii) P
17 b) 3 octaves

Questions 22

5 a) D, A
7 1.18 mm
9 b) ii) 1.3 cm or 0.13 m
    iii) 650 nm or 6.5×10^{-7} m
11 36.2°
12 b) 6.25×10^{-7} m, spacing 1.25 μm

Questions 23

6 5×10^{14} Hz
7 d) 200 kHz
8 a) longest: radio, shortest: UV
  b) f = 3 MHz
  c) i) to keep the microwaves inside the oven by reflection
    ii) shorter time needed, less wasted heat
    iii) heat energy, kinetic energy of vibrating molecules

Index